Single Variable

CALCULUS

with Early Transcendentals

PART 1

Paul Sisson & Tibor Szarvas

Executive Editor: Claudia Vance

Executive Project Manager: Kimberly Cumbie

Vice President, Research and Development: Marcel Prevuznak

Editorial Assistants: Danielle C. Bess, Susan Fuller, Margaret Gibbs, Robin Hendrix, Barbara Miller, Nina Waldron

Review Coordinator: Lisa Young

Senior Design Specialist: Tee Jay Zajac

Layout & Original Graphics: Tee Jay Zajac

Graphics: Robert Alexander, Margaret Gibbs, Jennifer Moran, and Tee Jay Zajac

Quant Systems India: E. Jeevan Kumar, D. Kanthi, U. Nagesh, B. Syam Prasad

Cover Design: Tee Jay Zajac

Cover Sculpture:

Arabesque XXIX

12″ H × 10½″ W × 9½″ D

Bubinga Wood

by Robert Longhurst

www.robertlonghurst.com

HAWKES LEARNING SYSTEMS

A division of Quant Systems, Inc.

546 Long Point Road, Mount Pleasant, SC 29464

Mathematica is a registered trademark of Wolfram Research, Inc.

Library of Congress Control Number: 2014942140

ISBN: 978-1-938891-98-4

TABLE OF CONTENTS

Preface

Chapter 3

Differentiation

Chapter 4

Applications of Differentiation

Chapter 5

Integration

Appendices

Answer Key

Index

FROM THE AUTHORS

This book arises from our deeply held belief that teaching and learning calculus should be a fascinating and rewarding experience for student and professor alike, playing a major role in the student's overall academic growth.

This is true not only because calculus ranks among the monumental achievements of the human intellect, thus providing an excellent introduction to higher-order thinking, but because of its wide-ranging applications in mathematics, the sciences, the business world, and the social sciences.

Our goal was to produce a text that builds on the natural intuition and curiosity of the reader, blending a student-friendly style of exposition with precision and depth. We believe that if done well, the calculus sequence should be a highly enjoyable journey of discovery and growth for the student, reflecting the journeys of discovery experienced by those who originally developed calculus centuries ago.

In other words, we strived to produce a book that is not only instructive, but also enjoyable to read—one that takes its readers from intuitive problem introductions to the rigor of concepts, definitions, and proofs in a natural manner.

Some of the distinctive features of our text include the following:

- A large number of examples and exercises in each section that demonstrate problem-solving techniques, reinforce conceptual understanding, and stimulate interest in the subject

- Carefully selected exercises that gradually increase in level of difficulty, ranging from skill-building "drill-and-practice" problems to less routine, more challenging, and occasionally deep theoretical questions

- Multistep, guided exploratory exercises that allow students to discover certain principles and connections on their own

- A rich variety of application problems from within and outside of mathematics and the sciences

- A constant emphasis on modern technology and its potential to enhance teaching and problem solving, as well as being a tool for investigation, reinforcement, and illustration

In summary, we have aimed for a comprehensive, mathematically rigorous exposition that not only uncovers the inherent beauty and depth of calculus, but also provides insight into the many applications of the subject.

We hope you enjoy the journey to the fullest. Let us know how we did and where we can improve!

Paul Sisson and Tibor Szarvas

ABOUT THE COVER

Arabesque XXIX

T he sculpture on the cover of this text is a piece entitled *Arabesque XXIX* by the American artist Robert Longhurst (b. 1949). Although not a mathematician, Longhurst explains his work in this way: "It just so happens that what I produce sometimes turns out with an orientation toward math. My interest is in creating pieces that have appealing form. Proportion and scale are very important as is the craft aspect of my work." His pieces, which can be found at www.robertlonghurst.com, have captured the attention of many mathematicians over the years, and his work has appeared or been discussed in several math texts.

The object depicted by *Arabesque XXIX* is an example of an **Enneper surface**, a simpler version of which was introduced by the German mathematician Alfred Enneper (1830–1885) in 1864. Enneper, with his contemporary Karl Weierstrass (1815–1897), made great strides in understanding and characterizing *minimal surfaces*. A minimal surface is one that spans a given boundary curve with, locally, the least surface area, meaning that the surface area cannot be decreased by deforming any part of the surface slightly (a soap film spanning a boundary formed by a simple closed loop of wire is an example of a naturally occurring minimal surface). Except for the simplest cases, the area of a surface is a concept that is difficult to even define, let alone calculate, without the benefit of calculus; you will learn how calculus is used to define and determine surface area in Chapter 6.

An alternate, but equivalent, characterization is that a minimal surface is one for which each point has zero *mean curvature*, meaning that the maximum curvature of the surface at the point is equal in magnitude (but opposite in sign) to its minimum curvature. The *curvature* of a given curve at a particular point is a characteristic with precise meaning, which you will learn about in Chapter 12. The surface in Figure 1, a portion of the original Enneper surface, has zero mean curvature at every point; this is illustrated for the central point shown in red, as the curvatures of the two dashed curves add to zero. This particular configuration is also an example of a *saddle point*, which you will study further in Chapter 13.

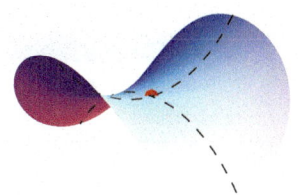

Figure 1

The characterization of minimal surfaces that Enneper and Weierstrass discovered is expressed in terms of integrals of functions of the complex plane, and their work falls into the categories of *differential geometry* and *calculus of variations*, two branches of mathematics that build upon the calculus you'll learn in this text. But the formulas resulting from their integrals are examples of *parametric surfaces*, which you'll study in detail in Chapter 15.

Figure 2

Specifically, the surface of *Arabesque XXIX* is similar to that defined parametrically by

$$x = r\cos\theta - \frac{r^5}{5}\cos 5\theta$$

$$y = -r\sin\theta - \frac{r^5}{5}\sin 5\theta$$

$$z = \frac{2r^3}{3}\cos 3\theta$$

where $0 \le r \le 1.38$ and $0 \le \theta \le 2\pi$. The graph of the surface defined by this parametrization appears in Figure 2.

FEATURES

Chapter Introductions

Each chapter begins with a brief introduction to the historical context of the calculus concepts that follow. Calculus is a human endeavor, and knowledge of how and why a particular idea developed is of great help in understanding it. Too often, calculus is presented in cold, abstract chunks completely divorced from the rest of reality. While a few students may be able to master the material this way, most benefit from an explanation of how calculus originated and how it relates to what people were doing at the time it was created. The introductions connect material learned in previous chapters to upcoming concepts, and illustrate to students the practical and historical importance of the calculus theories they are about to study.

Introduction

Up to this point, we have focused on the concept of the derivative of a function and on developing methods to find and use derivatives. We have seen that a differentiable function f gives rise to a potentially endless list of successive derivatives f', f'', and so on. In particular, we closed Chapter 4 with the question: *If we have perfect knowledge of the derivative of a function, can we use that to completely determine the function itself?*

One way to approach the subject of this chapter is to literally reverse the differentiation process. That is, given a function f, is it possible to find a function F such that $F' = f$? As we learned in Section 4.7, if such a function F exists, we call it an antiderivative of f. But another approach begins with questions that are seemingly unrelated to antidifferentiation, such as *How do we calculate the area bounded between an axis and the graph of a function?* and *If we know an object's instantaneous velocity and its starting point, can we determine its location at any later moment in time?* We call the process of answering such questions *integration*, and the genius of Sir Isaac Newton (1643–1727) and Gottfried Wilhelm Leibniz (1646–1716) lay in their ability to fully grasp the relationship between integration and antidifferentiation.

The *Fundamental Theorem of Calculus* is the name we now give to this relationship. As is often the case with deep fundamental truths, the theorem required a lengthy period of fermentation and the contributions of many others before Newton and Leibniz articulated it in the form we use today. The mathematicians James Gregory (1638–1675) and Newton's teacher Isaac Barrow (1630–1677) were two such forerunners, but the philosophical basis of the Fundamental Theorem of Calculus appeared as early as the 3rd century BC, when Archimedes used the "method of exhaustion" to find the area under a parabola and the volumes of certain objects.

Informally, the underlying philosophy might be stated in this manner:

The sum of all the infinitesimal changes of a function over a region is equal to the net change in the function.

As we will see by the time we finish Chapter 15 of this text, the underlying philosophy takes many specialized forms and is used to solve problems of astonishing diversity, ranging from the areas under parabolas and volumes of objects to problems in electrostatics, gravitational mechanics, and fluid flow. The insights of early Greek mathematicians from millennia ago, expanded and unified by mathematicians of the 17th

Isaac Newton   **Gottfried Leibniz**

Subs

1. The meaning of indefinite integration
2. The Substitution Rule

The Funda between int tool for ca that the no integral sig techniques

TOPICS

1. Each section begins with a list of topics. These concise objectives are a helpful guide for both reference and class preparation. The topics are clearly labeled throughout the section as the relevant material is covered.

Definitions, Theorems, Proofs, Properties, and Procedures

Definitions, theorems, proofs, properties, and procedures are clearly identified and stand out from the surrounding text for easy reference. Step-by-step procedures aid students in learning important problem-solving techniques. The procedures are then clearly illustrated in the examples that follow to help students apply the same steps in exercises at the end of the section. Selected proofs are provided in the text when necessary to establish meaningful connections between mathematical concepts, and then students are asked to prove additional theoretical statements in the exercises.

Theorem

The Substitution Rule for Definite Integrals

If g' is a continuous function on the interval $[a,b]$, and if f is continuous on the range of $u = g(x)$, then

$$\int_a^b f(g(x))g'(x)\,dx = \int_{g(a)}^{g(b)} f(u)\,du.$$

Proof

First, we know that both integrals exist since both integrands are continuous on their respective intervals: $(f \circ g)g'$ is a product of two continuous functions on $[a,b]$, and f is continuous on the interval $[g(a), g(b)]$. Our task is simply to prove that the integrals are equal.

As in the first version of the Substitution Rule, let F be an antiderivative of f.

$$\int_a^b f(g(x))g'(x)\,dx = F(g(x))\Big]_{x=a}^{x=b} \qquad \frac{d}{dx}F(g(x)) = f(g(x))g'(x)$$
$$= F(g(b)) - F(g(a))$$
$$= F(u)\Big]_{u=g(a)}^{u=g(b)}$$
$$= \int_{g(a)}^{g(b)} f(u)\,du$$

Finding Absolute Extrema

Assume that f is continuous on the interval $[a,b]$. To find the absolute extrema of f on $[a,b]$, perform the following steps.

Step 1: Find the critical points of f in (a,b).

Step 2: Evaluate f at each of the critical points in (a,b) and at the two endpoints a and b.

Step 3: Compare the values of the function found in Step 2. The largest is the absolute maximum of f on $[a,b]$ and the smallest is the absolute minimum of f on $[a,b]$.

Definition

Critical Point

Assume that f is defined on an open interval containing c. We say c is a **critical point** of the function f if $f'(c) = 0$ or $f'(c)$ does not exist.

Caution! Notes

Many common errors or difficulties that students encounter when learning calculus are highlighted throughout the text, along with explanations of how to correct them. These warnings help students avoid common mistakes, and aid students in relieving frustration when working through problems.

Caution!

The derivative of a product is *not* the product of the derivatives.

$$\frac{d}{dx}[f(x)g(x)] \neq f'(x)g'(x)$$

Similarly, the derivative of a quotient is *not* the quotient of the derivatives.

$$\frac{d}{dx}\left[\frac{f(x)}{g(x)}\right] \neq \frac{f'(x)}{g'(x)}$$

To see why this is so, it may be helpful to use a geometric interpretation; such an approach is very much in keeping with early Greek mathematics, which almost always viewed a product of two quantities as an area measure of a two-dimensional figure.

Example 3

The function $f(x) = \frac{1}{3}x^3 - 2x^2 + 5x + 1$ is differentiable everywhere, so it satisfies the hypotheses of the Mean Value Theorem over any closed, bounded interval. Find the point(s) satisfying the conclusion of the MVT over the interval $[0, 4]$.

Solution

We begin by calculating the slope of the secant line over the interval.

$$m_{sec} = \frac{f(4) - f(0)}{4 - 0} = \frac{\frac{31}{3} - 1}{4} = \frac{7}{3}$$

The Mean Value Theorem guarantees the existence of at least one point $c \in (0, 4)$ where $f'(c) = 7/3$, but we want to go one step further and actually find the point (or points). Since $f'(x) = x^2 - 4x + 5$, we proceed to solve the equation $x^2 - 4x + 5 = 7/3$.

$$x^2 - 4x + 5 = \frac{7}{3} \qquad \text{Set } f'(x) = m_{sec}.$$

$$x^2 - 4x + \frac{8}{3} = 0$$

$$3x^2 - 12x + 8 = 0$$

$$x = \frac{12 \pm \sqrt{144 - 96}}{6} \qquad \text{Apply the quadratic formula.}$$

$$x = 2 \pm \frac{2\sqrt{3}}{3}$$

These two solutions, approximately equal to 0.85 and 3.15, both satisfy the conclusion of the MVT. (See Figure 8.)

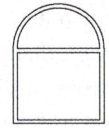

$f(x) = \frac{1}{3}x^3 - 2x^2 + 5x + 1$

Figure 8

Examples

Each section contains numerous examples that illustrate the concepts presented and the skills to be mastered. The examples, which are clearly set off from the surrounding text, demonstrate problem-solving techniques and reinforce conceptual understanding. Many examples contain real-world data to stimulate student interest and illustrate practical applications of calculus.

Applications

The examples and exercises throughout the text contain a rich variety of application problems from within and outside the fields of mathematics and the sciences. These practical applications of the fundamentals of calculus engage students and illustrate the importance of these skills in the real world.

Example 3

Susan is in her car, leading a group of other drivers to a restaura just completed a right turn at an intersection and is accelerating a moment that she is 40 ft from the intersection and heading north the second car in the convoy is still 40 ft away from the interse approaching it at 15 ft/s. How fast is the "as the crow flies" dista Susan and the second car increasing at that particular moment?

Solution

We know how quickly Susan is pulling away from the interse quickly the second car is approaching it, and their respective from the intersection at a given moment in time. We want to dete rate of change of the actual distance between the two cars at tha

A right triangle is again the appropriate diagram, though this tim (as well as the hypotenuse) of the triangle are changing with resp With the labels as shown, we know that $dy/dt = 25$ ft/s when y that $dx/dt = -15$ ft/s when $x = 40$ ft (note that dx/dt is negati decreasing). We want to determine dh/dt at that particular mon

Differentiating both sides of the relationship $x^2 + y^2 = h^2$ with r yields the following:

$$2x\frac{dx}{dt} + 2y\frac{dy}{dt} = 2h\frac{dh}{dt}$$

$$\frac{x\,dx}{dt} + \frac{y\,dy}{dt} = \frac{dh}{dt}$$

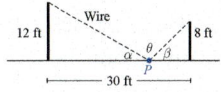

Figure 3

48. The shape of a Norman window can be approximated by a rectangle with a semicircle on top. What dimensions will admit the maximum amount of light if the perimeter of the window is to be P inches?

49. In Exercise 103 of Section 3.6, find the optimum distance s that maximizes the viewing angle.

50. An office building is located right on a riverbank, which is straight. A small power plant is on the opposite bank, 1500 feet downstream from the point directly opposite the office building. The river is 300 feet wide. If we want to connect the power plant and the building by cable, which costs $1700 per foot to lay down underwater and $800 per foot underground, what is the least expensive path for the cable?

51. Two antennas standing 30 feet apart are to be stayed with a single wire. The wire runs from the top of the first antenna, is secured to the ground somewhere between the antennas, and is finally attached to the top of the second antenna. If the height of the first antenna is 12 feet, while that of the second is 8 feet, find the point along the line segment connecting the bases where the wire needs to be staked to the ground if the length of the wire is to be minimal.

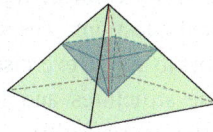

52. If we denote the heights of the antennas in Exercise 51 by h_1 and h_2, respectively, and the distance between them is d, prove that the wire has minimal length if and only if $\alpha = \beta$.

53. In Exercise 51, find the location of P that maximizes the angle θ.

54.* An inverted square pyramid is to be inscribed into a larger square pyramid of volume V, so that the two have a common axis, and the vertex of the inscribed pyramid coincides with the center of the outer pyramid's base. Find the ratio of the pyramids' altitudes so that the volume of the inscribed pyramid is maximal.

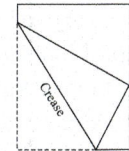

55.* The lower left corner of a letter-sized paper, which is 8.5 in. by 11 in., is folded over to reach the right edge of the paper. Find a way that this can be done so as to produce a crease of minimum length.

56. Find the radius of the base and the height of the right circular cylinder of largest volume that can be inscribed in a sphere of radius R.

57. Repeat Exercise 56, but inscribe a right circular cone instead of a cylinder in the sphere of radius R.

Technology

The text emphasizes the capabilities of modern technology as a tool for investigation, reinforcement, and illustration without distracting from the importance of knowing the fundamental theories of calculus. Technology notes are included throughout to highlight ways in which technology can help to solve problems or explain the concepts taught. Instructors are given the flexibility to include graphing calculator or computer algebra system (CAS) instruction as appropriate. The exercise sets contain separate sections of problems that focus on using technology to perform more advanced computations.

Technology Note

Computer algebra systems such as *Mathematica* provide additional tools for determining limits, but it should always be remembered that software has limitations and can be fooled. *Mathematica* contains the built-in command **Limit** that uses the same mathematical facts we will learn in the next two sections to correctly evaluate many types of limits. Its use is illustrated below. (For additional information on *Mathematica* and the use of the **Limit** command, see Appendix A.)

```
In[1]:= Limit[(Sqrt[x^2+4] - 2) / x^2, x -> 0]

            1
Out[1]=    ―
            4
```

Figure 17

3.3 Technology Exercises

65–70 Use a computer algebra system to find the derivative of $f(x)$. Then graph f along with its derivative on the same screen. By zooming in, if necessary, find at least two x-values where the graph of f has a horizontal tangent line. What can you say about f' at such points? (Answers will vary.)

65. $f(x) = \dfrac{x}{1 + \cos x}$ **66.** $f(x) = \dfrac{1 - \sec x}{1 + \sec x}$

67. $f(x) = \dfrac{\csc x}{x}$ **68.** $f(x) = \dfrac{\sin x}{\cos x + \tan x}$

69. $f(x) = \cos x (\cot x + \tan x)$

70. $f(x) = \dfrac{\cot x}{\sec x + x \cos x}$

71. Find the maximum velocity and acceleration values in Exercise 62 by using a graphing calculator or computer algebra system to graph the velocity and acceleration functions of the oscillating object.

Exercises

Each section concludes with a large selection of exercises designed to allow the student to practice skills and master concepts. The exercises gradually increase in level of difficulty, ranging from skill-building "drill-and-practice" problems to less routine, more challenging, and occasionally deep theoretical questions. The sets are designed to allow instructors to assign exercises based on a wide variety of course structures and to encourage students to master elementary skills or challenge themselves, as appropriate. Throughout the text, multistep, guided exploratory exercises give students the opportunity to discover certain principles and connections on their own. Many of the more difficult problems include helpful hints to point the student in the right direction, without giving away the solution. The most challenging exercises are marked with an asterisk (*). There are also a wide variety of true/false exercises that encourage students to think critically about the concepts learned in the section, requiring them to articulate why a particular statement is invalid.

4.2 Exercises

1–4 Use the graph of the function to visually estimate the value of c in the given interval that satisfies the conclusion of the Mean Value (or Rolle's) Theorem; then check your guess by calculation. If such a c doesn't exist, say why.

1. $f(x) = -x^2 + 6x - 4$ on $[1, 4]$

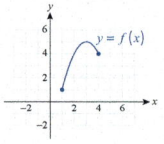

2. $g(x) = -2\sqrt{|x-2|} + 2$ on $[-2, 6]$

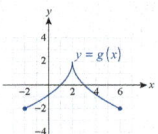

3. $h(x) = \dfrac{x^3}{3} - 2x^2 + \dfrac{11x}{3}$ on $[1, 4]$

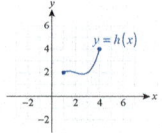

4. $k(x) = -\dfrac{1}{(x-3)^2} + 3$ on $[0, 6]$

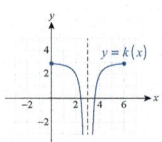

5–8 Prove that the equation has exactly one real solution on the given interval.

5. $x^3 - 3x^2 = 25$ on $[2, 3]$ **6.** $5x^3 + 7x = 9$ on $\mathbb{R}$

7. $\arctan x = 3 - x$ on $[0, 3]$ **8.** $\tan x = \cos x$ on $\left(0, \dfrac{\pi}{2}\right)$

116.* Consider the following function:

$$f(x) = \begin{cases} 2x^2 & \text{if } x \text{ is irrational} \\ 4x^2 & \text{if } x \text{ is rational} \end{cases}$$

Use f to explain why the changing of signs of the derivative is not necessary for a function to have a local extremum.

117. By considering $f(x) = \begin{cases} x & \text{if } x \le 1 \\ 3-x & \text{if } x > 1 \end{cases}$, explain

why the First Derivative Test can't be used for a discontinuous function.

118. Suppose that $f(x)$ and $g(x)$ are at least twice differentiable and that both their first and second derivatives are positive everywhere on an interval I. Which of the following can you prove from these conditions? Prove those statements that are true, and provide counterexamples for the rest.

a. $f(x) + g(x)$ is increasing on I.

b. $f(x) + g(x)$ is concave up on I.

c. $f(x) \cdot g(x)$ is increasing on I.

d. $f(x) \cdot g(x)$ is concave up on I.

e. $f(g(x))$ is increasing on I.

f. $f(g(x))$ is concave up on I.

119.* Use mathematical induction to prove the following generalization of the Second Derivative Test: Suppose that the derivatives of all orders of the function f exist at c, up to $f^{(2k)}(c)$, and that $f'(c) = f''(c) = \cdots = f^{(2k-1)}(c) = 0$, but $f^{(2k)}(c) \ne 0$. Then if $f^{(2k)}(c) < 0$, f has a relative maximum at c; if $f^{(2k)}(c) > 0$, f has a relative minimum at c.

120.* Use mathematical induction to prove that if f is $(2k+1)$-times differentiable at c, $f'(c) = f''(c) = \cdots = f^{(2k)}(c) = 0$, but $f^{(2k+1)}(c) \ne 0$, then f has a point of inflection at c.

121–128 *True or False?* Determine whether the given statement is true or false. In case of a false statement, explain or provide a counterexample.

121. Not all fourth-degree polynomials have inflection points.

122. A function with no inflection points cannot change concavity.

123. If $f'(x)$ is negative on $(-\infty, c)$ and positive on (c, ∞), then f has a minimum at c.

124. If $f(x)$ and $g(x)$ are decreasing, then so is $(f+g)(x)$.

125. If $f(x)$ and $g(x)$ are decreasing, then so is $(f \cdot g)(x)$.

126. A polynomial of degree n cannot have more than $n-1$ extrema on $\mathbb{R}$.

127. If $c \in \mathbb{R}$ is a critical point, then the function has a local minimum or a local maximum at c.

128. If $f''(c) = 0$, then c is an inflection point.

Chapter 1
Review Exercises

1–4 Find the domain and range of the given relation and determine whether the relation is a function.

1. $R = \{(-2,9),(-3,-3),(-2,2),(-2,-9)\}$

2. $3x - 4y = 17$

3. $x = y^2 - 6$

4. $x = \sqrt{y-4}$

5–8 Identify the domain, codomain, and range of the given function.

5. $f : \mathbb{N} \to \mathbb{R}, \quad f(x) = \dfrac{3x}{4}$

6. $g : \mathbb{R} \to \mathbb{R}, \quad g(x) = 5x + 1$

7. $h : \mathbb{R} \to \mathbb{R}, \quad h(x) = \dfrac{1}{x^2 + 1}$

8. $k : \mathbb{N} \to \mathbb{R}, \quad k(x) = 2 + \sqrt{x-1}$

9–12 Find the value of the given function for **a.** $f(x-1)$, **b.** $f(x^2)$, and **c.** $\dfrac{f(x+h)-f(x)}{h}$.

9. $f(x) = (x+5)(2x)$

10. $f(x) = \sqrt[3]{x} + 6(x+4)$

11. $f(x) = \dfrac{3}{x+2}$

12. $f(x) = \sin 2x$

13–14 Find all intervals of monotonicity (intervals where the function is increasing and intervals for the given function.

15–18 Determine if the function is even, odd, or neither and then graph it.

15. $f(x) = \dfrac{1}{3}x^3$

16. $f(x) = \sqrt{x}$

17. $f(x) = -2\sin x$

18. $g(x) = -2\sin^2 x$

19–20 Discuss the symmetry of the given equation and then graph it.

19. $y = |5x|$

20. $x^2 + y^2 = 25$

21–34 Graph the given function. Locate the x- and y-intercepts, if any.

21. $f(x) = 7x - 2$

22. $f(x) = \dfrac{2x-6}{3}$

23. $f(x) = (x-1)^2 - 1$

24. $f(x) = -x^2 - 6x - 11$

25. $f(x) = 4x^3$

26. $f(x) = -\dfrac{2}{x^2}$

27. $f(x) = \dfrac{-x^3 + 7x + 6}{2}$

28. $f(x) = \dfrac{x+1}{x^3 - 4}$

29. $f(x) = \dfrac{\sqrt[4]{x}}{2}$

30. $f(x) = 5|-x|$

31. $f(x) = \left[\!\left[\dfrac{2x}{3}\right]\!\right]$

32. $f(x) = \begin{cases} x^2 & \text{if } x < 1 \\ \dfrac{1}{x} & \text{if } x \geq 1 \end{cases}$

33. $f(x) = -\cot x$

34. $f(x) = \log_{1/2} x$

35–36 Sketch the graph of the given function by first identifying the more basic function that has been shifted, reflected, stretched, or compressed. Then determine the domain and range of the function.

35. $f(x) = (x-1)^3 + 2$

36. $f(x) = 4|x+3|$

37–38 Write an equation for the function described.

Chapter
Review Exercises

Each chapter features a review section that contains a comprehensive overview of exercises from the chapter. These review exercises challenge students to utilize a combination of the skills and concepts learned throughout the preceding sections. Every review section contains true/false exercises to help students master the concepts of that chapter. The medium difficulty level of the review exercises is intended to help students effectively prepare for tests.

Chapter
Projects

Each chapter contains a project that extends beyond the fundamental concepts covered to a more challenging application of calculus. These projects are suitable as individual or group assignments and allow students to extend their knowledge of calculus beyond the typical theories studied. The projects also provide opportunities to explore the aids of technology in understanding and performing more rigorous computations.

Chapter 2
Project

Some years ago, it was common for long-distance phone companies to charge their customers in one-minute increments. In other words, the company charges a flat fee for the first minute of a call and another fee for each additional minute or any fraction thereof (see Exercise 82 in Section 2.5). In this project, we will explore in detail a function that gives the cost of a telephone call under the above conditions.

1. Suppose a long-distance call costs 75 cents for the first minute plus 50 cents for each additional minute or any fraction thereof. In a coordinate system where the horizontal axis represents time t and the vertical axis price p, draw the graph of the function $p = C(t)$ that gives the cost (in dollars) of a telephone call lasting t minutes, $0 < t \leq 5$.

2. Does $\lim\limits_{t \to 1.5} C(t)$ exist? If so, find its value.

3. Does $\lim\limits_{t \to 3} C(t)$ exist? Explain.

4. Write a short paragraph on the continuity of this function. Classify all discontinuities; mention one-sided limits and left or right continuity where applicable.

5. In layman's terms, interpret $\lim\limits_{t \to 2.5} C(t)$.

6. In layman's terms, interpret $\lim\limits_{t \to 3^-} C(t)$.

7. In layman's terms, interpret $\lim\limits_{t \to 3^+} C(t)$.

8. If possible, find $C'(3.5)$.

9. If possible, find $C'(4)$.

10. Find and graph another real-life function whose behavior is similar to that of $C(t)$. Label the axes appropriately and provide a brief description of your function.

For more information please visit:

hawkeslearning.com/calculus

Chapter 1
A Function Primer

Introduction

This chapter begins with the study of *relations* and then moves on to the important class of relations called *functions*. As concepts, relations and functions are powerful and useful generalizations of the equations that one typically studies in introductory math courses. Functions, in particular, lie at the heart of a great deal of the mathematics that you will encounter throughout calculus.

The history of the function concept serves as a good illustration of how mathematics develops. One of the first people to use the idea in a mathematical context was the German mathematician and philosopher Gottfried Wilhelm Leibniz (1646–1716), one of two people usually credited with the development of calculus. Initially, Leibniz and other mathematicians tended to use the term to indicate that one quantity could be defined in terms of another by some sort of algebraic expression, and this (incomplete) definition of function is often encountered even today in elementary mathematics. As the problems that mathematicians were trying to solve increased in complexity, however, it became apparent that functional relations between quantities existed in situations where no such algebraic expression was possible. One example came from the study of heat flow in materials, in which a description of the temperature at a given point at a given time was often described in terms of an infinite sum, not an algebraic expression.

The result of numerous refinements and revisions of the function concept is the definition that you will encounter in this chapter, and is essentially due to the German mathematician Peter Gustav Lejeune Dirichlet (1805–1859). Dirichlet also refined our notion of what is meant by a *variable* and gave us our modern understanding of *dependent* and *independent* variables, all of which you will read about soon.

The proof of the power of functions lies in the multitude and diversity of their applications.

The subtle and easily overlooked advantage of function notation deserves special mention; innovations in notation often go a long way toward solving difficult problems. In fact, if the notation is sufficiently advanced, the mere act of using it to state a problem does much of the work of actually solving it. As you work through Chapter 1, pay special attention to how function notation works—a solid understanding of the notation is a necessary prerequisite to success in calculus.

1.1 Functions and How We Represent Them

Topic One

Relations and Functions

Functions lie at the heart of nearly every aspect of calculus and a solid understanding of them is therefore a necessary step toward success in calculus. This chapter reviews the language and notation of functions, the different ways that functions may be represented, frequently seen functions, and techniques for using and modifying functions.

To begin, it is useful to first define what we mean, mathematically, by the word *relation*; as we will soon see, functions are specialized relations.

Definition

Relation, Domain, and Range

Given two sets A and B, a **relation** R from A to B is a set of ordered pairs of the form (a,b), where $a \in A$ and $b \in B$. (An *ordered pair* is just what it sounds like—a pair of elements written in such a way that it is obvious which is the first element and which is the second.)

Given any particular relation R from A to B, the set of all the first elements (the *first coordinates*) of the ordered pairs is called the **domain** of R, and the set of all second elements (the *second coordinates*) is called the **range** of R. Using set-builder notation, we can write

$$\text{Domain of } R = \left\{ a \mid (a,b) \in R \text{ for some } b \in B \right\}$$

and

$$\text{Range of } R = \left\{ b \mid (a,b) \in R \text{ for some } a \in A \right\}.$$

Although it may not have been stressed at the time, you have undoubtedly had much experience with relations in previous math classes. Indeed, the graph of any equation in x and y, where x and y represent real numbers, is a relation from $\mathbb{R}$ to $\mathbb{R}$, where $\mathbb{R}$ stands for the set of real numbers. More generally, any subset of $\mathbb{R} \times \mathbb{R}$ (the *Cartesian plane*) represents a relation—the subset does not have to arise from an equation or have any obvious mathematical basis.

We will provide four examples of relations, some of which can be defined by a formula or at least a picture, while others will, at first, seem to be "nonmathematical."

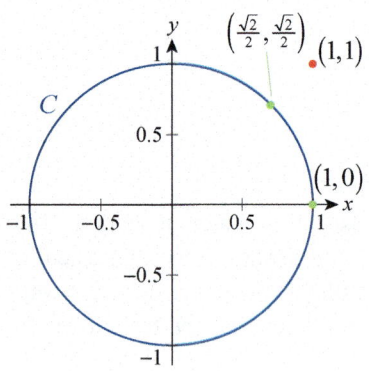

Figure 1

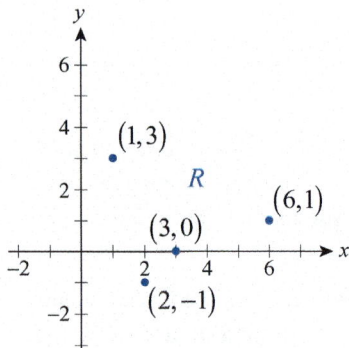

Figure 2

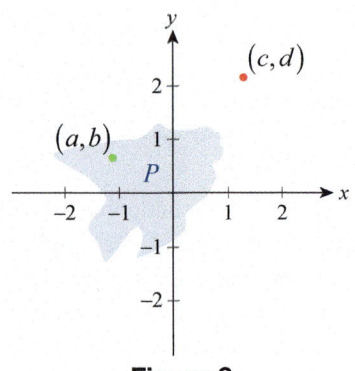

Figure 3

Example 1

a. For real numbers x and y, suppose they are in relation C (i.e., $(x,y) \in C$) if they satisfy the equation $x^2 + y^2 = 1$. For example, $(1,0) \in C$, $(\sqrt{2}/2, \sqrt{2}/2) \in C$, but $(1,1) \notin C$. Figure 1 shows the graph of the equation defining C. (Can you see why we called the relation C?) Note that in this example, the domain of C (the set of x-coordinates) is the set

$$\text{Domain of } C = \{x \in \mathbb{R} \mid -1 \leq x \leq 1\},$$

and similarly,

$$\text{Range of } C = \{y \in \mathbb{R} \mid -1 \leq y \leq 1\}.$$

b. The following set of ordered pairs is a relation from $\mathbb{R}$ to $\mathbb{R}$.

$$R = \{(1,3), (2,-1), (3,0), (6,1)\}$$

Examining the first and second coordinates of the ordered pairs, respectively, we see that

$$\text{Domain of } R = \{1, 2, 3, 6\},$$

while

$$\text{Range of } R = \{-1, 0, 1, 3\}.$$

Notice that just as we did in part a., we can represent this relation in the Cartesian coordinate system (see Figure 2). Also note that relations do not necessarily have any immediate mathematical meaning such as the one in part a. For example, part c. below shows a relation "defined by a picture."

c. Figure 3 can be used to define relation P. Note tha two real numbers x and y are in relation P if the point (x,y) falls in the shaded area or its boundary. For example, as shown by the illustration, $(a,b) \in P$, but $(c,d) \notin P$.

This example is a bit more general than that in part b., but it still satisfies the definition. See part d. for an even more general and nonnumeric example.

d. Suppose that the student population on a particular college campus speaks a total of five languages: English, Chinese, German, Japanese and Spanish. If we represent the above set of languages by $\{E, C, G, J, S\}$, the following relation L from the set of students to the set of languages arises quite naturally.

$$(x,y) \in L \text{ if student } x \text{ speaks language } y$$

Notice that the domain of L is the set of students at the college, while the range is the set of the five given languages. Supposing that Chihiro speaks English and Japanese, but not Spanish, we can express this in symbols by writing

$$(\text{Chihiro}, E) \in L, (\text{Chihiro}, J) \in L, \text{ and } (\text{Chihiro}, S) \notin L.$$

As Example 1 indicates, relations can arise in a multitude of ways. We may encounter relations defined by an equation or a formula, by a picture, by data collected from an experiment, or in words. This characteristic will carry over to our study of functions, which we are now ready to define.

Definition

Functions and Function Notation

A **function** f from a set A to a set B is a relation from A to B with the additional property that <u>each</u> element $a \in A$ is related to <u>exactly one</u> element $b \in B$. Instead of writing $(a, b) \in f$, we typically use **function notation** and write $b = f(a)$. In other words, given $a \in A$, $f(a)$ is the unique element of B associated with a.

The notation $f(a)$ is read "f of a," and when A and B are sets of numbers, $f(a)$ is often referred to as the *value of f at a*. The notation $f : A \to B$ is used to indicate that f is a function (not just a relation) from the set A to the set B.

The underlined words in the above definition are critical. First, note that since *each* element of A is related to an element of B by f, the domain of f is all of A. Second, the requirement that *exactly one* element of B is related to each $a \in A$ means that there is a completely unambiguous dependence in the relationship. This dependence (which has come to be known as *functional dependence*) is the key to the great utility of functions in mathematics.

Caution!

Don't be misled by the parentheses in the function notation to think that $f(a)$ might represent the product of f and a! While parentheses do often indicate multiplication, they are also used in defining functions.

Topic Two
The Language of Functions

A few additional observations and definitions are in order. As noted above, the notation $f : A \to B$ automatically indicates that the domain of f is A. In this context, the set B is called the **codomain** of f. As with relations, the range of f is the set of elements of B that are "used" by f. Using function notation, we can quickly note that the range of f is thus the set $\{ f(a) | a \in A \}$.

The nature of functional dependence is also made clear by two other phrases. If, as above, a is a symbol representing an arbitrary element of the domain of f, then a is called an **independent variable**. And, as above, if b is used to represent an arbitrary element of the range of f, then b is called a **dependent variable**. These phrases often arise when we want to make use of the functional dependence of one variable upon another in a given equation. For example, in the equation $y = 3x^2 - 5$, the value of the variable y is completely determined by whatever value is assigned to the variable x, so y would be labeled the *dependent* variable and x the *independent* variable.

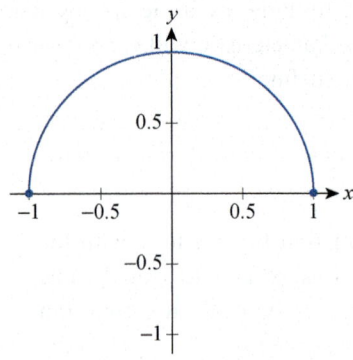

Figure 4　$y = \sqrt{1-x^2}$

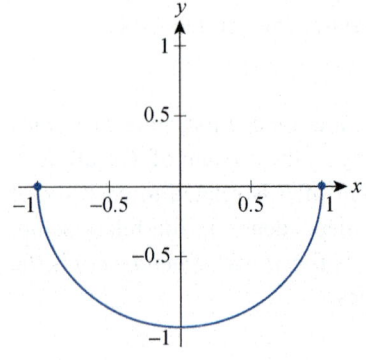

Figure 5　$y = -\sqrt{1-x^2}$

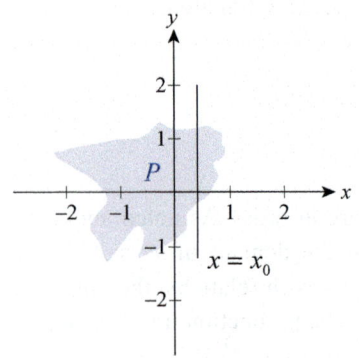

Figure 6

Example 2

Indicate which of the relations from Example 1 are functions, and for those functions make the distinction between the codomain and the range.

Solution

a. The way in which y depends on x in part a. of Example 1 does not demonstrate functional dependence. Consider, for example, the fact that both of the pairs $(0,1)$ and $(0,-1)$ are in the relation C, and thus 0 is related to both 1 and -1. However, if we express y from the equation by considering only the positive square root as follows:

$$y = \sqrt{1-x^2}$$

then the above difficulty is eliminated. Indeed, we obtained a function whose graph is the upper semicircle of radius 1, centered at the origin (see Figure 4). Notice that we can still consider $\{x \in \mathbb{R} \,|\, -1 \le x \le 1\}$ to be the domain of our function; however, the set of elements "used" by f, the range of f, is the following smaller set.

$$\{y \in \mathbb{R} \,|\, 0 \le y \le 1\}$$

Finally, let us also note that in the case of $y = -\sqrt{1-x^2}$ we obtain a different function whose graph is the lower semicircle (see Figure 5).

b. The relation defined in part b. of the previous example is a function. It is easy to check by inspection that each domain element is related to exactly one value.

The codomain of R is $\mathbb{R}$ while the range of R is $\{-1, 0, 1, 3\}$.

c. Examining the graph of the relation given in part c. of Example 1, it is easily concluded that this relation is not a function. Indeed, picking an appropriate x-value $x = x_0$ and drawing a vertical line through it, we see that the y-coordinate of any point falling on the line segment in the interior of the shading is related to x_0 (see Figure 6), thus leaving us with a multitude of choices for y_0.

d. With regards to relation L given in part d. of Example 1, since there are students on campus (such as Chihiro) who speak more than one language, the relation L is not a function either.

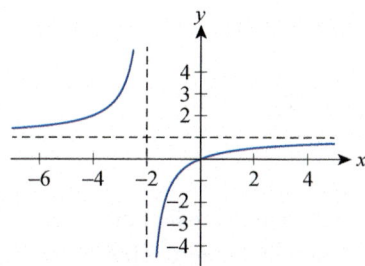

Figure 7 $y = \dfrac{x}{x+2}$

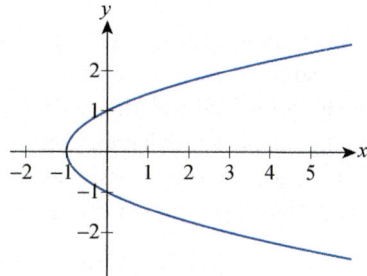

Figure 8 $x = y^2 - 1$

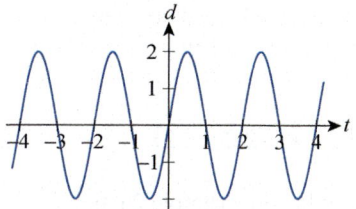

Figure 9 $d = 2\sin \pi t$

Example 3

As we have seen in the preceding examples, we can define a relation or even a function by writing an equation in two variables and considering the set of ordered pairs satisfying our equation. Find the function defined by the given equation, and note the dependent and independent variables.

a. $xy = x - 2y$ **b.** $y^2 - x - 1 = 0$ **c.** $dt - 2t \sin \pi t = 0$

Solution

a. Moving the terms containing y to the left side, we obtain

$$(x+2)y = x, \text{ that is, } y = \frac{x}{x+2}.$$

We see that y is completely and unambiguously determined by x, so this is clearly a functional relationship with y being the dependent and x the independent variable. We often put it this way: y is a function of x.

b. Trying to express y as in part a. leads to the ambiguity

$$y = \pm\sqrt{x+1}.$$

However, it is rather straightforward to see that we can express x as a function of y in this case: $x = y^2 - 1$ defines x as a function of y; that is, y is the independent variable, and x is the dependent variable.

c. We divide both sides of the equation by t for $t \neq 0$ and obtain

$$d = 2\sin \pi t,$$

with d being the dependent, and t the independent variable. (Note that if you ever studied simple harmonic motion, you undoubtedly worked with functions similar to this one.)

In some situations, a function is given to us by means of a formula with no information regarding the domain. In such cases, by convention, the **implied domain** is the largest collection of points for which the formula returns a real number.

Example 4

Find the implied domain for each of the following functions, and also determine the range in each case.

a. $f(x) = \dfrac{x}{x+2}$ **b.** $g(x) = \sqrt{4-x^2}$

Solution

a. The formula for $f(x)$ certainly returns a well-defined real number for all x except for $x = -2$, which makes the denominator zero. Hence the domain of f is the following set.

$$\text{Domain of } f = \left\{ x \in \mathbb{R} \,\middle|\, x \neq -2 \right\}$$

As for the range, it is easy to see that f never assumes $y = 1$, for this would imply the equality of the numerator and the denominator in the formula, which never happens. In fact, being a rational function with the horizontal asymptote of $y = 1$, and having "opposing behavior" on the two sides of its vertical asymptote $x = -2$, we see that f assumes every real value except $y = 1$ (see Figure 10). Thus, the range of f is as follows.

$$\text{Range of } f = \left\{ y \in \mathbb{R} \,\middle|\, y \neq 1 \right\}$$

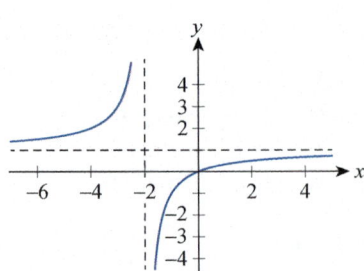

Figure 10 $f(x) = \dfrac{x}{x+2}$

b. Since we cannot have negative values under the square root, the only real numbers for which $g(x)$ makes sense are those satisfying $4 - x^2 \geq 0$.

$$\text{Domain of } g = \left\{ x \in \mathbb{R} \,\middle|\, x^2 \leq 4 \right\} = \left\{ x \in \mathbb{R} \,\middle|\, -2 \leq x \leq 2 \right\}$$

Notice that the graph of g is a semicircle just like in Example 2a, but this time the radius is 2 (see Figure 11). This latter observation implies the following:

$$\text{Range of } g = \left\{ y \in \mathbb{R} \,\middle|\, 0 \leq y \leq 2 \right\}$$

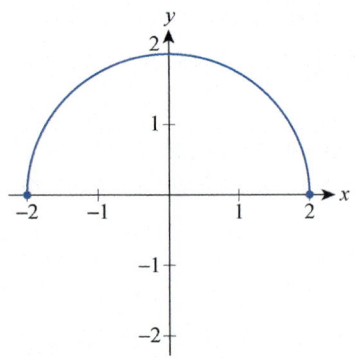

Figure 11 $g(x) = \sqrt{4-x^2}$

Topic Three

Ways of Describing Functions

As we will see, functional relationships arise in many situations that can be described or analyzed with calculus, but the exact nature of the relationships can be portrayed in a variety of ways. The four most common ways are mathematical formulas, tables of data, graphs, and verbal descriptions.

You have undoubtedly seen many examples of mathematical formulas already: such familiar equations as $A = \pi r^2$ and $C = \frac{5}{9}(F - 32)$ serve as illustrations. The first equation relates the area A of a circle to its radius r, and the second equation exactly describes the conversion of Fahrenheit temperature F to Celsius temperature C. Note that the form in which these equations are written leads to a natural interpretation of r and F as independent variables and of A and C as dependent variables.

The functional relationships in these two equations can be made more explicit by writing them in the form $A(r) = \pi r^2$ and $C(F) = \frac{5}{9}(F - 32)$, but it is important to realize that these functions express exactly the same relationships as the equations in the preceding paragraph. Again, context is everything: here, $A(r)$ and $C(F)$ are notational shorthand indicating that A is a function of r and that C is a function of F, not an indication that A is to be multiplied by r or C by F. In a formula such as $A(r) = \pi r^2$, the variable r is sometimes referred to as the **argument** of the function, and it is instructive to realize that the exact symbol used for the argument is irrelevant. That is, $A(s) = \pi s^2$, $A(radius) = \pi (radius)^2$, and $A(\square) = \pi \square^2$ all describe exactly the same function A.

Example 5

Find the following formulas and rewrite them so as to explicitly express the functional relationship between the variables. Identify the argument in each case.

a. The area of an equilateral triangle

b. The stopping distance required by a car traveling at speed v from the moment the brakes are applied

c. The volume of a circular cylinder

Solution

a. The area of an equilateral triangle of side length s is $A = \frac{1}{2}s\frac{s\sqrt{3}}{2}$ (using the well-known fact that the height h of an equilateral triangle is $h = s\frac{\sqrt{3}}{2}$). This formula can be rewritten as

$$A(s) = \frac{\sqrt{3}}{4}s^2,$$

a functional relationship with s being the argument, explicitly showing how the area of an equilateral triangle depends on its side length.

b. The stopping distance d required by a car traveling at speed v when the brakes are applied is estimated by the formula $d = v^2/(2\mu g)$, where μ is the coefficient of friction between the tires and the pavement, and g is the gravity constant. Since μ can also be considered a constant (at least in the short term, when conditions are nearly unchanging) we see that stopping distance depends on initial speed only. This can be expressed by the function

$$d(v) = \frac{v^2}{2\mu g},$$

with v being the argument of function d. (Note the quadratic dependence of d on v! It certainly pays to observe all speed limit signs and warnings.)

c. The volume of a circular cylinder is calculated by $V = \pi r^2 h$, where r is the radius of the base, and h stands for the height of the cylinder. Note that the volume here is a function of two independent quantities. We can express this functional relationship as follows.

$$V(r,h) = \pi r^2 h$$

Both r and h are considered arguments here, and thus $V(r,h)$ is our first example of a *function of two variables*.

Technology Note

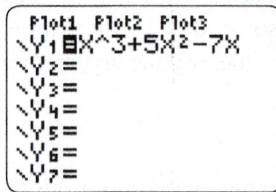

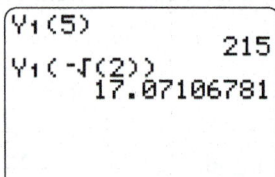

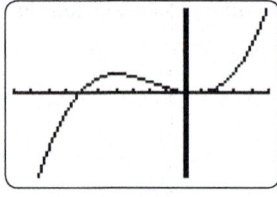

Figure 12

Functions and function notation are immensely important not just as mathematical concepts but also in the way we use technology as a mathematical aid. Graphing calculators and computer algebra systems (CAS) all have methods for defining functions and then making use of them. Figure 12 shows how the function $f(x) = x^3 + 5x^2 - 7x$ is defined using a graphing calculator and how the function is then evaluated at several points and graphed for $x \in [-10, 5]$.

Figure 13 shows how the same function $f(x) = x^3 + 5x^2 - 7x$ is defined in *Mathematica*, evaluated, and then graphed over the specified region. (See Appendix A for a brief introduction to using *Mathematica*.)

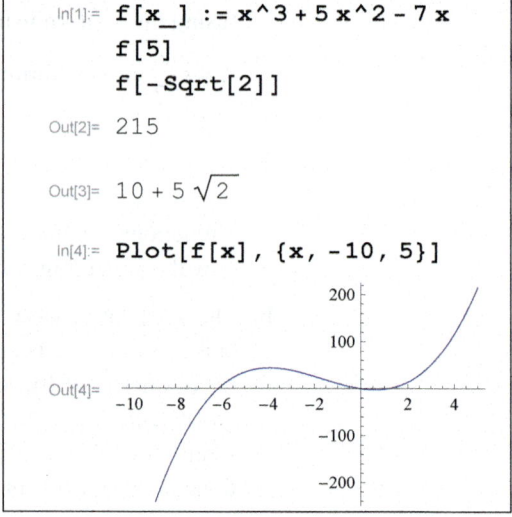

Figure 13

Tables of data, regardless of the setting, almost always implicitly describe a function. For example, any table of population data over two or more points in time (possibly for a city, for a country, or for the number of bacteria in a petri dish) defines population P as a function of time t. Table 1 is a table of the US population in every census year from 1950 to 2010.

Year	Population
1950	151,325,798
1960	179,323,175
1970	203,211,926
1980	226,545,805
1990	248,709,873
2000	281,421,906
2010	308,745,538

Table 1 US Population As Measured by the Census Bureau
Source: www.census.gov

Using function notation (and using the label P as in the preceding paragraph), we would write, for instance, $P(1950) = 151,325,798$ and $P(2010) = 308,745,538$. The first column in Table 1 represents the possible values for the independent variable t, and the second column represents the corresponding values of $P(t)$.

Table 1 only defines seven values for the function $P(t)$, and it is very often the case that we need to be able to extend the knowledge presented in a table to other values of the independent variable. For example, we might want estimates of the US population for every year between 1950 and 2010, or we might desire a projected population for the year 2050. The acts of estimating values of a function based on given data are called **interpolation** and **extrapolation**, and are perhaps best introduced with a picture. Figure 14 contains a plot of the seven data points from Table 1, along with a line that seems to approximate the general behavior of the points well.

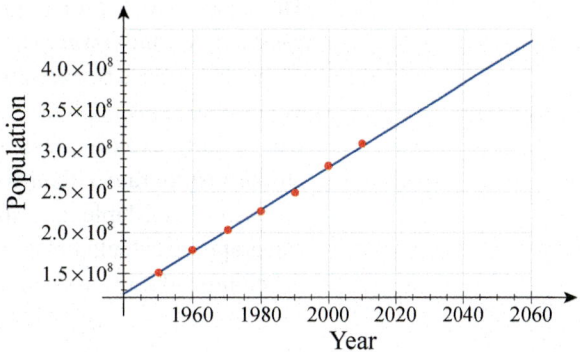

Figure 14 Graph of US Population Data, 1950–2010

To **interpolate** a value from a table or graph means to arrive at an estimate for the value of the implied function *between* two known data points, while to **extrapolate** means to guess at a value *beyond* the given points. In our example, we might interpolate a value for $P(1985)$ to be close to 240 million (2.4×10^8 in scientific notation) and extrapolate a value for $P(2050)$ to be approximately 410 million. However, this census data example is also a good illustration of the limitations of tables and graphs. Figure 15 contains the plot of an expanded version of the US Census table, along with a curve that approximates the behavior.

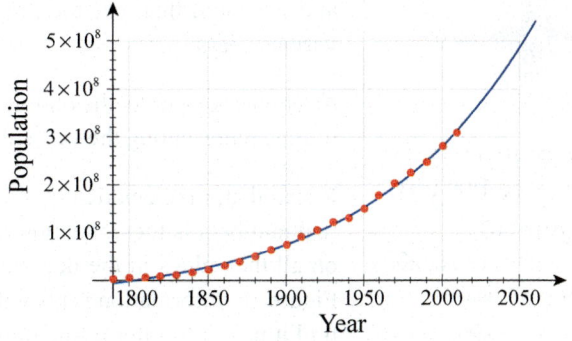

Figure 15 Graph of US Population Data, 1790–2010

If we were asked to extrapolate a value for $P(2050)$ on the basis of this expanded graph, we would probably guess a value of 480 million, significantly different from our first extrapolation of 410 million.

Technology Note

You may be wondering how the blue line and blue curve in Figures 14 and 15 were drawn and whether they can be improved upon. There is no definitive answer to the second question, as *curve fitting* (the act of constructing a curve that closely approximates given data) depends upon making some assumptions about the mathematical nature of the data, and it is often the case that different but equally defensible assumptions can be made. In this case, it was assumed that the seven data points in Figure 14 lay approximately along a straight line (that is, that $P(t) = a + bt$ for some choice of a and b) and that the expanded data set in Figure 15 could be approximated by an exponential curve (specifically, that $P(t) = a + b^t$ for some a and b).

The answer to the first question above (the *how* question) is that the parameters a and b were chosen in each case so that the sum of the squares of the differences between the given data and the fitted curve was as small as possible. This method, called the *least-squares* method of curve fitting, will be fully explored later in this text, but for the moment we will focus on the results rather than the method. Many software packages (such as *Mathematica*, Maple, and spreadsheet programs such as Microsoft Excel) and graphing calculators are capable of calculating a least-squares fit. Using the command **Fit** from *Mathematica*, we arrive at the linear function

$$P(t) = -4.87678 \times 10^9 + 2.57841 \times 10^6 t$$

(whose graph is the blue line in Figure 14)

and the exponential function

$$P(t) = -4.6544 \times 10^7 + 1.00985^t$$

(whose graph is the blue curve in Figure 15).

(See Appendix A for a brief introduction to using *Mathematica*.)

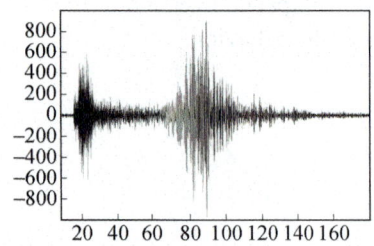

Minutes since 1:00 UTC

Figure 16

Seismogram of December 25, 2004 Sumatra Earthquake

Source: Pennsylvania Geological Survey's website: http://www.dcnr.state.pa.us/topogeo/hazards/earthquakes/sumatra/index.htm (2 Apr. 2014)

The US Census information might have been presented to us visually as nothing more than the exponential graph seen in Figure 15 (without the red data points), and we could then use the graph to estimate values of the implied function for various years.

As another example, the output of a seismograph (such as Figure 16) is a function of time representing the vertical motion of the ground at a given location on Earth.

Recall that, mathematically, the **graph** of a function f that maps real numbers to real numbers is the collection of all points of the form $(x, f(x))$, where x takes on all the values in the domain of f. In some cases, such as the seismogram of Figure 16, information is presented to us first in the form of a picture, and we then find it useful to infer a functional relationship from the picture. In other cases, a function is presented in the form of a table or formula, and we find it useful to construct the graph of the given data and possibly even (as in the US population example) to "fill in" the parts of the implied function that might be missing.

The preceding paragraph gives rise to several questions. The first might be: Is it possible to graph functions other than just real-valued functions of real numbers? And the second might be: Given a picture (such as Figure 16), can we always assume that it does indeed represent the graph of a function?

We will return to the second question when we discuss further the power of graphs. As to the first question, we will see numerous examples throughout this text of ways to graph functions that turn real numbers into ordered pairs of real numbers (that is, functions from $\mathbb{R}$ to $\mathbb{R}^2$) and functions that do the opposite (functions from $\mathbb{R}^2$ to $\mathbb{R}$), along with a few more exotic variations of these ideas. But there is one more visualization tool that is commonly used to give a sense of what a specific function does, sometimes referred to as an *arrow diagram*. We will introduce its use here along with one more example of a function, this time described verbally.

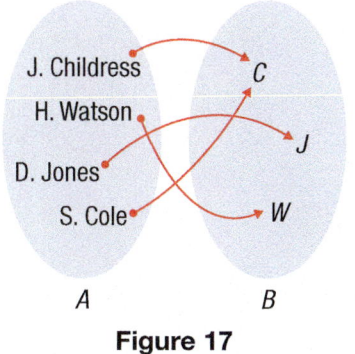

Figure 17

Arrow Diagram of Filing Function F

Consider the act of filing correspondence in a filing cabinet according to the last name of the sender. This act represents a function, which for convenience we might name F for *File*. Thus, for example, a letter from James Childress would be filed in the $F(\text{Childress}) = C$ folder, and one from Deanna Jones would be filed in the $F(\text{Jones}) = J$ folder. Figure 17 is an **arrow diagram** illustrating how F relates the domain set A of correspondents to the codomain set B of file folders.

Note that this act F does indeed satisfy all the requirements to be a function, and that for any specific set of correspondents it would be easy to identify the domain and range of the function. It is also worthwhile to note, however, that the inverse act of relating folders to the correspondents it contains is not likely to be a function, as even the small example depicted in Figure 17 demonstrates. For instance, folder C contains both S. Cole and J. Childress, so it is not the case that <u>each</u> element of set B is related to <u>exactly one</u> element of set A. We will return to this important point in Section 1.4 when we discuss inverse functions.

Example 6

Below, we will give examples of functions that are easily described verbally, the likes of which you meet on a daily basis, perhaps without even thinking of them as functions.

a. In California, license plates of passenger vehicles follow the pattern of a number followed by three letters and then by three digits, such as, "4ZNR915." Since every registered car has a unique license plate with no duplications allowed, we can think of this as a functional relation between the set of registered California passenger cars as the domain and the set of strings following the above pattern as the codomain.

b. When a class of calculus students walks into a classroom and sits down, the students (perhaps unknowingly) define a function with its domain being the students themselves and the codomain being the chairs in the room. Since (at least in the ideal case) each student is able to sit and do so on no more than one chair at a time, we see that this is indeed a functional relationship between the two sets. Note that coming back to class two days later and sitting down in a slightly different order amounts to no more than redefining the function. Also, since every chair may not always be used, there is a clear distinction here between codomain and range.

c. The temperature on a given day in Shreveport, Louisiana is a function of time, with its values being numbers from the Fahrenheit temperature scale. This is a function we can graph on a particular day, as shown below.

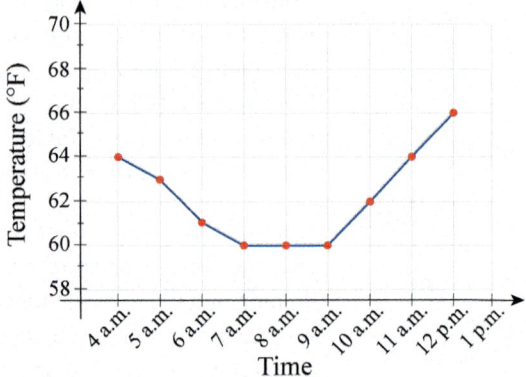

Figure 18

Topic Four

The Power of Graphs

We still have an unanswered question from above: given a graph of a relation, how do we determine whether it represents a functional relationship? For graphs in $\mathbb{R}^2$, there is a simple test to apply.

Definition

The Vertical Line Test

Given a graph of a relation in the Cartesian plane, the graph represents a function if no vertical line passes through the graph more than once. If even *one* vertical line intersects the graph in two or more points, the relation fails to be a function.

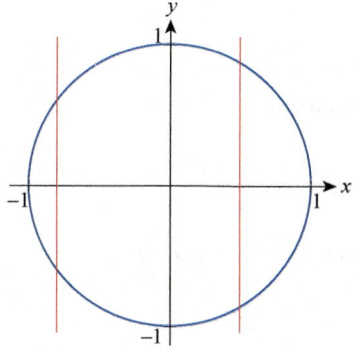

Figure 19

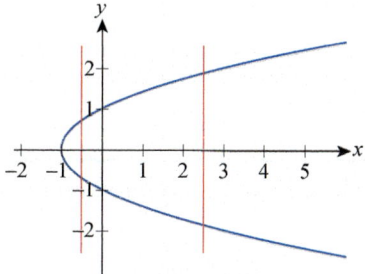

Figure 20

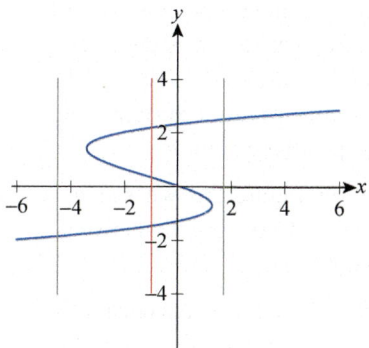

Figure 21

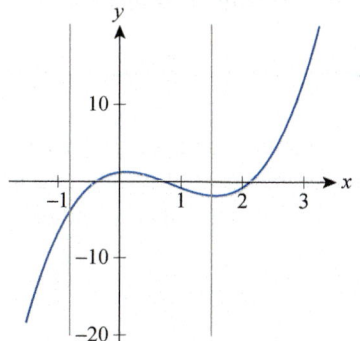

Figure 22

The vertical line test is nothing more than a restatement of the definition of function as it applies to relations from $\mathbb{R}$ to $\mathbb{R}$, but it is worth mentioning because it gives us a quick visual criterion to apply to a large number of graphs. It also immediately answers, in the affirmative, the question of whether graphs generated by mechanical means (such as the seismogram in Figure 16) represent functions. Because the instruments used to generate such graphs will only return one reading for any given point in time, a functional relationship is guaranteed.

Example 7

Below, we provide examples of the use of the vertical line test.

a. The unit circle centered at the origin does not define y as a function of x, since there are plenty of vertical lines intersecting the graph more than once (see Figure 19). In fact, as we have seen in Example 2a, the equation of the unit circle $x^2 + y^2 = 1$ defines both functions $y = \pm\sqrt{1-x^2}$, each having a semicircle for its graph, and thus both satisfying the vertical line test.

b. An easy application of the vertical line test on the graph of $y^2 = x + 1$ shows that y is not a function of x (see Figure 20). We note, however, that again both functions $y = \pm\sqrt{x+1}$ are defined by the original equation. Furthermore, as we noted in Example 3b, our equation does actually define x as a function of y. (Note that because of the change in roles between the variables, from the perspective of the dependent variable x, "vertical lines" are the ones that appear horizontal in the traditional xy-coordinate system. With this interpretation, the "vertical line test" still applies to x as a function of y!)

c. The graph in Figure 21 is not a function either. Note that while a lot of vertical lines intersect the graph only once, it is enough to find one that doesn't.

d. Finally, we mention that the graphs of all functions you worked with in your precalculus course (such as polynomials, rational and trigonometric functions, etc.) do pass the vertical line test, as you can convince yourself easily. As an example, the function $f(x) = 2x^3 - 5x^2 + x + 1$, shown in Figure 22, intersects every vertical line exactly once.

The power of a visual depiction of a relation cannot be overstated (though pictures can be deceptive as well). We will close this section with a brief catalog of function properties that all have strong visual connotations.

Increasing/Decreasing

A real-valued function f is said to be **increasing** on an interval I if whenever x_1 and x_2 are in I,

$$x_1 < x_2 \Rightarrow f(x_1) \leq f(x_2).$$

Similarly, f is said to be **decreasing** on I if whenever x_1 and x_2 are in I,

$$x_1 < x_2 \Rightarrow f(x_1) \geq f(x_2).$$

In either case, f is said to be **monotone** on I, and if the inequality that compares the function values is strict, the function is said to be **strictly increasing** or **decreasing** on I.

Example 8

Identify intervals of *monotonicity* (intervals where the function is increasing or decreasing) for the following functions.

a. $f(x) = x^2$ **b.** $g(x) = x^3 - 3x$ **c.** $h(x) = \cos x$

Solution

a. The graph of $f(x) = x^2$ is a prototypical parabola, opening upward, centered at the origin (see Figure 23). It is clear that f is decreasing on the interval $(-\infty, 0)$ while it is increasing on $(0, \infty)$. Visually, this means that the graph is "falling" as we scan it from left to right along the negative x-axis, while it is "rising" for positive x-values.

b. Examining the graph of $g(x) = x^3 - 3x$ in Figure 24, we see g is increasing on $(-\infty, -1)$ and $(1, \infty)$, while it is decreasing "in between" these half-lines, that is, on the interval $(-1, 1)$.

c. From the graph of $h(x) = \cos x$, we conclude that h is decreasing on the interval $(0, \pi)$ and increasing on $(\pi, 2\pi)$ (see Figure 25). More than that, because of 2π periodicity, $\cos x$ is decreasing on all intervals of the form $(2k\pi, (2k+1)\pi)$ for an arbitrary integer k, while it is increasing on all intervals $((2k-1)\pi, 2k\pi)$.

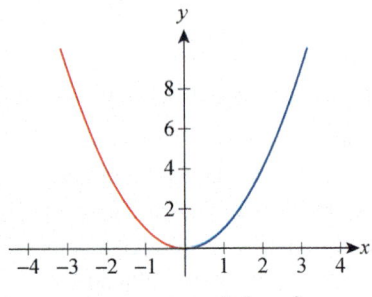

Figure 23 $f(x) = x^2$

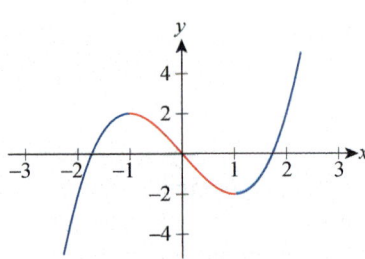

Figure 24 $g(x) = x^3 - 3x$

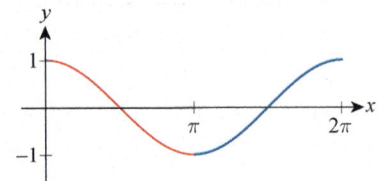

Figure 25 $h(x) = \cos x$

Graphs of functions often display some sort of symmetry, and recognizing the symmetric nature of a given function can provide a great deal of insight into its behavior. The two most elementary kinds of symmetry are termed *even* and *odd*.

Even and Odd Functions

If a real-valued function f has the property that $f(-x) = f(x)$ for all x in its domain, f is said to be an **even function**. Its graph will display symmetry with respect to the y-axis, as shown in Figure 26.

If a real-valued function f has the property that $f(-x) = -f(x)$ for all x in its domain, f is said to be an **odd function**. Its graph will display symmetry with respect to the origin, as shown in Figure 27.

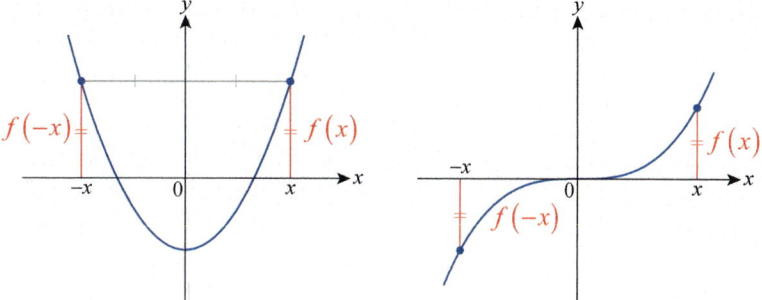

Figure 26 An Even Function   **Figure 27** An Odd Function

More generally, relations between two variables (say, x and y) can also display symmetry. When such relations are graphed in the Cartesian plane, the three most elementary kinds of symmetry are symmetry with respect to the axes and the origin.

Symmetry of Equations

An equation in x and y is **symmetric with respect to**:

1. The **y-axis** if replacing x with $-x$ results in an equivalent equation.

2. The **x-axis** if replacing y with $-y$ results in an equivalent equation.

3. The **origin** if replacing x with $-x$ and y with $-y$ results in an equivalent equation.

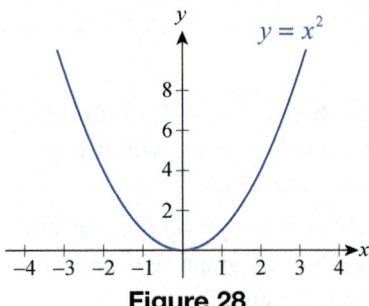

Figure 28

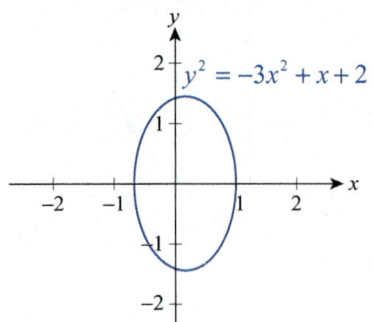

Figure 29

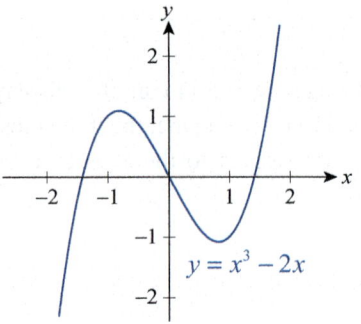

Figure 30

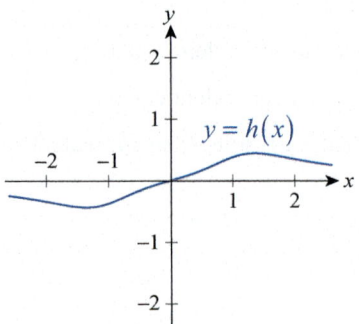

Figure 31

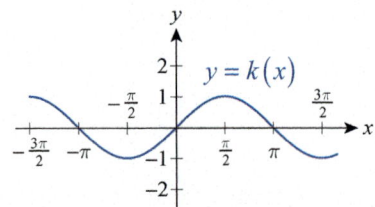

Figure 32

Example 9

Discuss the symmetry of the following functions and equations.

a. $y - x^2 = 0$ **b.** $y^2 + 3x^2 = x + 2$ **c.** $y + 2x = x^3$

d. $h(x) = \dfrac{x^3 + 2x}{2x^4 - x^2 + 7}$ **e.** $k(x) = \sin x$

Solution

a. Since $(-x)^2 = x^2$, we see that replacing x by $-x$ yields an equivalent equation, so the graph must be symmetric with respect to the y-axis. Note that we can rewrite this equation as $y = x^2$, and so its graph is the graph of the function $f(x) = x^2$. This graph, the prototypical parabola, indeed displays symmetry with respect to the y-axis (see Figure 28).

b. If we replace y with $-y$, we obtain an equivalent equation, since $(-y)^2 = y^2$. The same is not true of x, since replacing it by $-x$ turns the right-hand side of the equation into $-x + 2$. Hence we conclude that the graph of this equation is symmetric with respect to the x-axis, but not the y-axis (see Figure 29).

c. If we simultaneously replace x with $-x$ and y with $-y$ in this equation, we obtain

$$-y + 2(-x) = (-x)^3$$
$$-y - 2x = -x^3,$$

and this latter equation is seen to be equivalent to the original, so we conclude that its graph is symmetric with respect to the origin. Finally, note that this graph is the same as that of the function $g(x) = x^3 - 2x$ (see Figure 30).

d. This example is more easily done algebraically instead of graphically. By substituting $-x$ for x we obtain

$$h(-x) = \frac{(-x)^3 + 2(-x)}{2(-x)^4 - (-x)^2 + 7} = \frac{-x^3 - 2x}{2x^4 - x^2 + 7} = -\frac{x^3 + 2x}{2x^4 - x^2 + 7} = -h(x),$$

and conclude that h is an odd function; thus its graph is symmetric with respect to the origin (see Figure 31).

e. The trigonometric function $k(x) = \sin x$ is odd, since $\sin(-x) = -\sin x$, and so its graph displays symmetry with respect to the origin (see Figure 32), a fact you have undoubtedly seen before. Finally, we also note the well-known fact that $g(x) = \cos x$ is even, for we know $\cos(-x) = \cos x$ to be true for all x. These facts are also clearly seen by visually examining these basic trigonometric graphs.

1.1 **Exercises**

1–22 Describe the domain and range of the given relation.

1. $R = \{(-3,1),(-3,5),(-3,-1),(0,0),(1,2)\}$

2. $S = \{(3,-1),(2.6,6),(\pi,0.5),(e,100)\}$

3. $T = \{(4,5.98),(-2,-8),(-2,0),(3,\cos 3)\}$

4. $U = \{(4,4),(4,\pi),(\pi,4),(4,0)\}$

5. $F = \{(\text{Tanisha},\text{swimming}),(\text{Don},\text{biking}),(\text{Peter},\text{skating}),(\text{David},\text{skateboarding})\}$

6. $L = \{(\text{Lin},\text{Chinese}),(\text{Chuck},\text{English}),(\text{Sarah},\text{German}),(\text{Daniel},\text{Hungarian})\}$

7. $A = \{(x,y)\,|\,x \in \mathbb{Z},\ y = 2x+3\}$

8. $B = \left\{(x,y)\,\middle|\,x \in \mathbb{R},\ y = \dfrac{x}{2}\right\}$

9. $C = \{(x,-2x+7)\,|\,x \in \mathbb{Z}\}$

10. $D = \{(2x,5y)\,|\,x \in \mathbb{N},\ y = x+1\}$

11. $3x = y + 5$

12. $\sqrt{2}x - 1.2y = 3$

13. $x = 5$

14. $y = \pi$

15. $x = 3y^2 - 1$

16. $y = |x| - 2$

17.

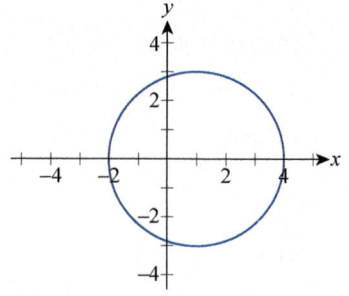

18.

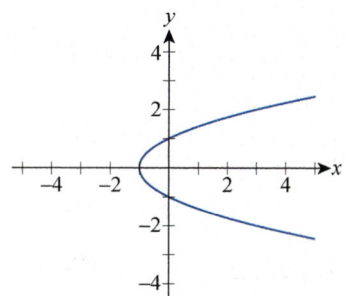

19.

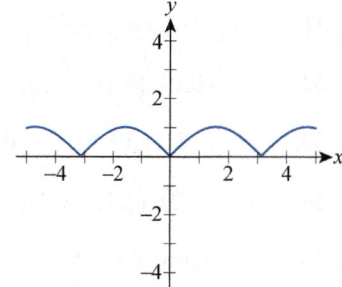

20.

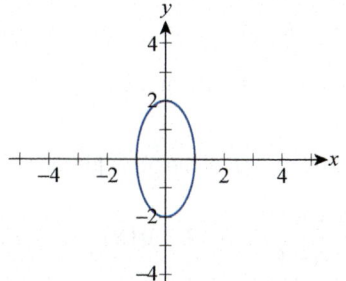

21.

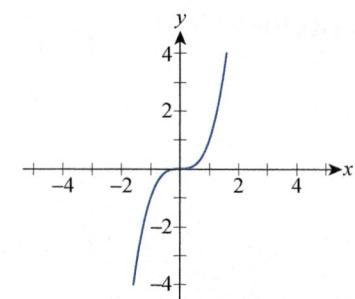

22.

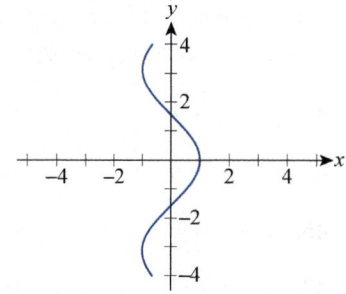

23–26 Describe the domain and range of the given relation. Choose an appropriate domain on which the given relation makes sense. (Answers will vary.)

23. $\{(x,y)|\text{student } x \text{ is registered for course } y\}$

24. $\{(x,n)|x \text{ wears size } n \text{ shoes}\}$

25. $\{(x,y)|y \text{ is the father of } x\}$

26. $\{(P,n)|\text{person } P \text{ weighs } n \text{ pounds}\}$

27–30 List the ordered pairs in the given relation R on the set $A = \{1,2,3,4,5\}$.

27. $(a,b) \in R$ if and only if $a = b$

28. $(a,b) \in R$ if and only if $a < b$

29. $(a,b) \in R$ if and only if $a \mid b$

30. $(a,b) \in R$ if and only if $a + b = 5$

31–34 Determine whether the given relation is a function. If the relation is not a function, explain why.

31. $A = \{(1,3),(-2,4),(0,4)\}$

32. $B = \{(0,0),(0,1),(2,3),(4,5),(6,7)\}$

33. $C = \{(-1,2),(\pi,3),(-1,0),(1,2)\}$

34. $D = \{(1,2),(2,1),(3,4),(4,3)\}$

35–40. Determine whether the relations given in Exercises 17–22 are functions. For those that are not, explain why.

41–52 Determine whether the given equation is a function. If the equation is not a function, explain why.

41. $y = 3x - 4$ **42.** $x = 3y - 4$

43. $x^2 + y^2 = 9$ **44.** $x + y^2 = 9$

45. $x^2 + y = 9$ **46.** $y = \sqrt[3]{x}$

47. $x = x^3 - y$ **48.** $xy = 4$

49. $x = \pi$ **50.** $y = \dfrac{3x}{x^2 + 1}$

51. $F = 5r^2 \pi$ **52.** $V = \dfrac{4}{3}r^3 \pi$

53–58 Express y explicitly as a function of x from the given relation.

53. $\dfrac{x + 3y}{2} = 5$ **54.** $\dfrac{x - 3y}{5} = \dfrac{2y + 7x}{3}$

55. $3x^2 - y = 5 - x + 2y$

56. $x + 7 - 3y = (x - 2)^2 + y$

57. $yx^2 - y = 3x + 1$ **58.** $x + 1 = yx^2$

59–66 Find the value of the given function for **a.** $f(-2)$, **b.** $f(x+1)$, **c.** $f(x+h)$, and **d.** $\dfrac{f(x+h) - f(x)}{h}$.

59. $f(x) = \dfrac{1}{3}x + 2$ **60.** $f(x) = \dfrac{5x - 3}{2}$

61. $f(x) = x^2 - 3$ **62.** $f(x) = 3x^2 - 5x + \dfrac{1}{2}$

63. $f(x) = \sqrt{x}$ **64.** $f(x) = \dfrac{1}{\sqrt{x+2}}$

65. $f(x) = \dfrac{1}{x+1}$ **66.** $f(x) = (x-1)^3 + 5$

67–72 Identify the domain, codomain, and range of the given function.

67. $f: \mathbb{N} \to \mathbb{N}, \quad f(x) = x + 1$

68. $g: \mathbb{N} \to \mathbb{Z}, \quad g(x) = 3x - 2$

69. $h: \mathbb{Z} \to \mathbb{Z}, \quad h(x) = x^2$

70. $F: \mathbb{R} \to \mathbb{R}, \quad F(x) = 2x^4 + 1$

71. $G: [0,\infty) \to \mathbb{R}, \quad G(x) = \sqrt{x}$

72. $H: \mathbb{Q}^+ \to \mathbb{Q}, \quad H(x) = \dfrac{1}{x}$ (Note that $\mathbb{Q}^+$ stands for the set of positive rational numbers.)

73–82 Find the implied domain of the given function.

73. $f(x) = \dfrac{x+1}{x^2 - x - 6}$ **74.** $g(x) = \sqrt{3x + 2}$

75. $h(x) = \dfrac{2}{\sqrt{x^2 - 4x + 3}}$ **76.** $F(t) = \dfrac{1}{\sqrt{4 - t^2}}$

77. $G(s) = \sqrt{2 - s} + \sqrt{s}$ **78.** $D(h) = \dfrac{\dfrac{1}{\sqrt{1+h}} - 1}{h}$

79. $R(x) = \dfrac{1}{|2x+3|}$

80. $H(z) = z^{3/2} - 2$

81. $F(\theta) = \dfrac{2}{1 - \cos\theta}$

82. $\varphi(x) = \dfrac{5}{\sin x - \dfrac{\sqrt{2}}{2}}$

83–88 Turn the formula into a function by finding the argument(s) of the function. Identify any functions of two variables.

83. $C = 2\pi r$

84. $V = \dfrac{4}{3} r^3 \pi$

85. $C = \dfrac{5}{9}(F - 32)$

86. $A = 6a^2$

87. $V = \dfrac{1}{3} b^2 h$

88. $E = \dfrac{1}{2} mv^2$

89–94 Use the vertical line test to decide whether y is a function of x.

89. $y^3 + 1 = x$

90. $2x^2 + 2y^2 = 18$

91. $y^2 + 1 = x$

92. $x = (y - 2)^2$

93. $x = y^3 - 2y$

94. $yx^2 = 1$

95–101 Find all intervals of monotonicity (intervals where the function is increasing or decreasing) for the given function.

95. $f(x) = (x - 1)^2$

96. $g(x) = 4x - x^2$

97. $h(x) = x^3 - 12x$

98. $k(x) = \dfrac{x^2}{x^2 + 1}$

99. $F(x) = |x - 1|$

100. $G(x) = 2x + |3x - 1|$

101. $H(x) = |x + 1| + |x - 2|$

102–110 Discuss the symmetry of the given equation. Give reasons. (**Hint:** See Example 9.)

102. $y = x^2 - 1$

103. $x = y^2 - 1$

104. $x^4 + y^4 = 5$

105. $|x| + |y| = 2$

106. $x - |y| = 2$

107. $xy = 2$

108. $y = \dfrac{2x^3 - x}{x^4 + x^2}$

109. $y^2 + 6x = x^3$

110. $y = (x - 1)^2$

111. Express the perimeter of a square as a function of its area.

112. Express the area of an equilateral triangle as a function of its perimeter.

113. An open-top box is constructed from a 20 in. by 30 in. piece of cardboard by cutting out a square of side length x from each of the four corners and folding up the sides, as shown in the figure below. Express the volume of the box as a function of x.

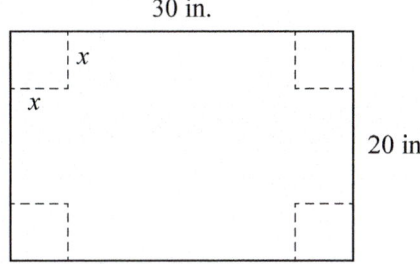

114. Express the surface area A of a cube as a function of its volume V.

115. The height of a circular cone is equal to the diameter of its base. Express its volume V as a function of the radius r of the base.

116. Express the volume of a sphere as a function of its surface area.

117. Knowing that water boils at 212 °F, which corresponds to 100 °C, and the fact that freezing occurs at 32 °F, which is 0 °C, obtain the linear function $C(F)$ that expresses the Celsius temperature C as a function of the Fahrenheit reading F.

118.* The organizers of an educational leadership seminar series have found that the seminar attracts 100 participants when the registration fee is set to $150. They estimate that for each increase of $10 in the registration fee, they will end up with 5 fewer registered participants. Express the revenue R as a function of the registration fee F.

1.2 **A Function Repertory**

The development of calculus can be viewed historically as the creation of an ever-expanding collection of mathematical tools and methods used to model and understand natural phenomena. This use of calculus remains hugely important, and indeed most of this book will focus on just that—the development and application of the tools of calculus. One important subset of these tools consists of the basic functions that, through centuries of experience, have proven to be good models of natural behavior. This section will present a quick overview of the functions we will be using repeatedly. As you will see, familiarity with them and a reasonable level of comfort in their use is very beneficial.

Topic One
Algebraic Functions

The basic functions we use can be split up into two large classes: algebraic functions and transcendental functions. We will begin by providing examples of algebraic functions and work toward a definition of this class.

Definition

Polynomial Functions

A **polynomial function** f of one variable, say the variable x, is any function that can be written in the form $f(x) = a_n x^n + a_{n-1} x^{n-1} + \cdots + a_1 x + a_0$, where each a_i is a constant and n is a nonnegative integer. Each a_i is called the **coefficient** of its associated **term** $a_i x^i$. The highest power of the variable is called the **degree** of the polynomial, and the coefficient of the term containing that highest power is called the **leading coefficient** of the polynomial.

Although polynomial functions with complex number coefficients are useful in some branches of mathematics, every polynomial in this book will be assumed to have real coefficients unless otherwise stated, and hence polynomial functions turn real numbers into real numbers. And since any real number can be raised to any positive integer power, a moment's thought will convince you that the domain (and codomain) of every polynomial function is all of $\mathbb{R}$. The range, however, may not be all of $\mathbb{R}$, and indeed the determination of the precise range for a given polynomial function often lies at the heart of a problem.

Polynomial functions of degree 1 are called **first-degree** polynomials, and their graphs are straight lines. This follows from the fact that such a polynomial has the form $f(x) = a_1 x + a_0$, so the graph is a line with slope a_1 and y-intercept a_0. A polynomial of the form $f(x) = a_0$ is a **constant** polynomial, and its graph is also a line (a horizontal line running through a_0 on the y-axis). The degree of a constant polynomial is 0, unless a_0 happens to be 0 (in which case the polynomial has no nonzero terms and is said to have no degree). Since their graphs are lines, first-degree and constant polynomials are together termed **linear** functions.

Second-degree polynomials have the form $f(x) = a_2x^2 + a_1x + a_0$, and their graphs are parabolas. Such functions also go by the name of **quadratic** functions. Similarly, **third-degree** polynomials are also known as **cubic** functions, and **fourth-degree** polynomials as **quartic** functions. Example 1 illustrates some typical shapes of these frequently occurring functions.

Example 1

Graph the following polynomial functions, and briefly discuss their most notable characteristics.

a. $f(x) = -\dfrac{3}{2}x + 3$ **b.** $g(x) = 2x^2 - 4x - 6$

c. $h(x) = \dfrac{1}{2}(x+2)(x+1)(x-3)$ **d.** $k(x) = -\dfrac{1}{3}x^4 + 2x^3 - \dfrac{1}{3}x^2 - 8x + \dfrac{20}{3}$

Solution

a. The function $f(x)$ is easily recognized as a first-degree function. The slope, which is also the leading coefficient, is negative, meaning that the graph "falls" as we scan it from left to right. The graph confirms the y-intercept being 3, while the x-intercept can quickly be found to be 2 by solving the equation $0 = -\frac{3}{2}x + 3$. Finally, notice that the range of this function is all of $\mathbb{R}$, as is the case for all linear functions, unless they are constant. (See Figure 1.)

b. The function $g(x)$ is a quadratic polynomial with a positive leading coefficient of 2. This means that the graph is an upward-opening parabola. The y-intercept of -6 is easily found by substituting $x = 0$ in the formula. Note that the constant term turns out to be the y-intercept, which is no accident and happens with all polynomials, since substituting $x = 0$ will eliminate all terms but the constant. The x-intercepts, on the other hand, are the two solutions of $2x^2 - 4x - 6 = 0$, namely, -1 and 3. Finally, we note that since the vertex of the parabola is at $(1, -8)$, the range of g is the set $\{y \in \mathbb{R} \mid y \geq -8\}$. (See Figure 2.)

c. The function $h(x)$, being a product of three linear polynomials, is certainly cubic, which is readily seen upon expanding and obtaining the standard form of $h(x) = \frac{1}{2}x^3 - \frac{7}{2}x - 3$. As before, the y-intercept is the constant term, -3, and we see from the factored form of $h(x)$ that the x-intercepts are -2, -1, and 3 (these values being the *zeros* of the polynomial). Finally, notice that, unlike the parabola, this graph "falls" to the left and "rises" to the right, as is the case with all polynomials of odd degree and positive leading coefficient, so the range of $h(x)$ is the entire set of $\mathbb{R}$. (See Figure 3.)

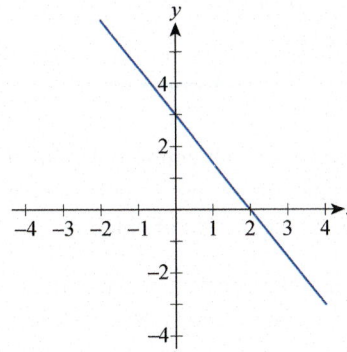

Figure 1 $f(x) = -\dfrac{3}{2}x + 3$

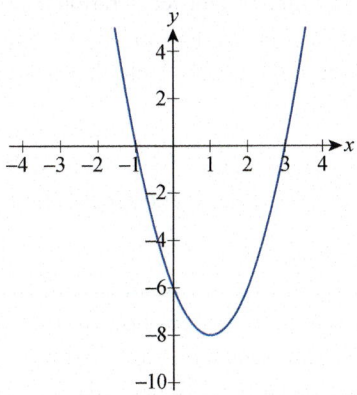

Figure 2 $g(x) = 2x^2 - 4x - 6$

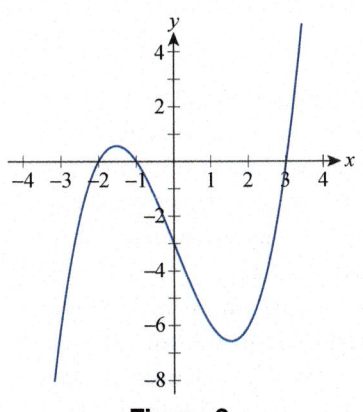

Figure 3

$h(x) = \dfrac{1}{2}(x+2)(x+1)(x-3)$

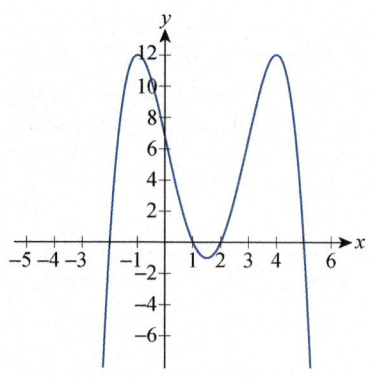

Figure 4

$$k(x) = -\tfrac{1}{3}x^4 + 2x^3 - \tfrac{1}{3}x^2 - 8x + \tfrac{20}{3}$$

d. The function $k(x)$ is an example of a quartic polynomial with a negative leading coefficient, so we expect the graph to be "falling" to both the right and the left, implying that the function's range will be something less than the entire set of $\mathbb{R}$. As in parts b. and c., the y-intercept is the constant term, $\tfrac{20}{3}$. To find x-intercepts, we can use, say, the Rational Zero Theorem along with long or synthetic division to obtain the fully factored form $k(x) = -\tfrac{1}{3}(x-2)(x+2)(x-1)(x-5)$ and conclude that the x-intercepts are ± 2, 1, and 5. From the graph shown in Figure 4, we also see the range to be the set $\{y \in \mathbb{R} \mid y \le 12\}$.

Many familiar natural phenomena can be modeled with relatively simple polynomial functions, and we encounter them frequently. Consider, for example, the following two models from the realm of kinematics (the study of the motion of objects).

Model 1: If an object is moving in a given direction with constant acceleration a, its velocity v (in that direction) at time t is given by the formula $at + v_0$, where v_0 is its initial velocity at time $t = 0$. Using function notation, we would write $v(t) = at + v_0$. Note that v is a linear polynomial.

Model 2: If an object is moving in a given direction with constant acceleration a, its position p (in that direction) at time t is given by the function $p(t) = \tfrac{1}{2}at^2 + v_0 t + p_0$, where p_0 is its position at time $t = 0$ and v_0 is its initial velocity. Note that p is a quadratic polynomial.

As we will see, calculus provides exactly the tools we need to derive these models and fully understand the relationship between position, velocity, and acceleration.

Example 2

A diver jumps from a 24-foot-high diving board so that her position above the water at any given time during the jump is represented by the function $p(t) = -16t^2 + 8t + 24$, where p is measured in feet (ft), and t in seconds (s). Her velocity as a function of time is given by $v(t) = -32t + 8$ ft/s. How much later and with what velocity does the diver hit the water?

Solution

First of all, since the position function $p(t)$ returns the altitude of the diver above the water surface for any specified t-value, we note that even though the diver may not move along a linear path, the position function only applies to one dimension. The same is true of velocity, whereas $v(t)$ returns the instantaneous velocity of the diver at any given time t. Also notice how the constant acceleration caused by gravity, -32 ft/s^2, the initial upward jumping speed of 8 ft/s, and the height of the platform, 24 ft, are "hidden" in the formula for $p(t)$! Indeed, we see from the formula for $v(t)$ that the velocity decreases by 32 ft/s with every passing second; thus there is a negative acceleration of -32 ft/s^2. Substituting $t = 0$ into the formula for $v(t)$ yields the initial velocity $v(0) = 8$ ft/s, while the height of the platform is the initial position $p(0) = 24$ ft.

To find out when the diver hits the water, we turn to the position function and note that at the moment of impact, the position over the water becomes zero, so we solve the quadratic equation $-16t^2 + 8t + 24 = 0$. The solutions are $t = 1.5$ and $t = -1$. We can safely discard the latter, since time cannot be negative, so we conclude that the diver hits the water 1.5 seconds after leaving the diving board.

As for the velocity of impact, we turn to the velocity function $v(t)$. When $t = 1.5$, the velocity of impact is obtained by evaluating $v(1.5) = -40$ ft/s. Note that the negative sign indicates that the direction of motion at the time of impact is downward.

Definition

Rational Functions

A **rational function** f is a function that can be written as a ratio of two polynomials, say $f(x) = p(x)/q(x)$, where p and q are polynomial functions and q is not the zero polynomial. Even though q is not allowed to be identically zero, there may certainly be values of x for which $q(x) = 0$, and at those values f is undefined. Consequently, the domain of f is the set $\{x \in \mathbb{R} \mid q(x) \neq 0\}$.

Chances are, you have spent some time in a previous math class learning techniques for graphing rational functions and identifying characteristics of particular interest. For instance, it is often useful to identify the y-intercept, the x-intercept(s), and all *vertical*, *horizontal*, or *oblique asymptotes* of a rational function. The following example contains illustrations of these objects.

Example 3

Sketch the graphs of the following rational functions, noting all intercepts and asymptotes.

a. $f(x) = \dfrac{x^2 - 1}{x - 1}$

b. $g(x) = \dfrac{x^2 + 1}{x^2 + 2x - 15}$

c. $h(x) = \dfrac{x^2 + 1}{x - 1}$

d. $k(x) = \dfrac{x^3 + x^2 + 2x + 2}{x^2 + 9}$

Solution

a. First of all, note that the domain of f is all real numbers except for 1, and that after canceling the common factor of $x - 1$, the function reduces to $f(x) = x + 1$. This holds whenever $x \neq 1$ and thus the graph of f is simply that of $y = x + 1$ with one point missing from it, since f is not defined at $x = 1$. We denote this with an open circle at the point on the graph corresponding to $x = 1$. Note the x- and y-intercepts of -1 and 1, respectively; there are no asymptotes.

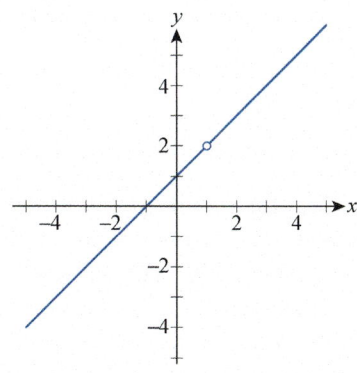

Figure 5 $f(x) = \dfrac{x^2 - 1}{x - 1}$

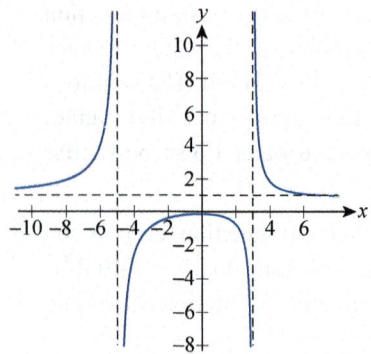

Figure 6 $g(x) = \dfrac{x^2 + 1}{x^2 + 2x - 15}$

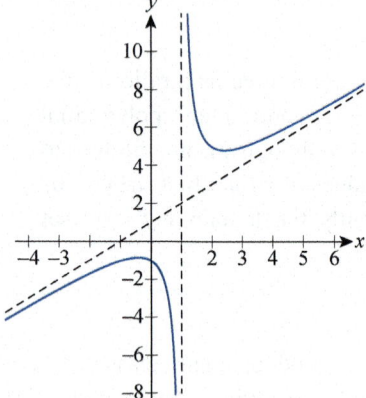

Figure 7 $h(x) = \dfrac{x^2 + 1}{x - 1}$

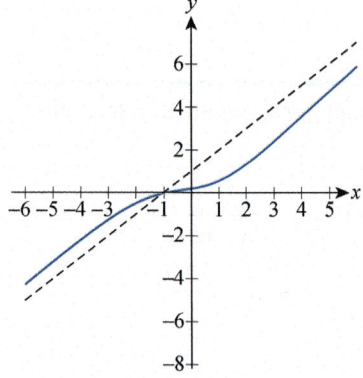

Figure 8

$k(x) = \dfrac{x^3 + x^2 + 2x + 2}{x^2 + 9}$

b. $g(x) = \dfrac{x^2 + 1}{x^2 + 2x - 15} = \dfrac{x^2 + 1}{(x + 5)(x - 3)}$

Since the numerator cannot be factored over the real numbers, we find two vertical asymptotes with equations $x = -5$ and $x = 3$. Also since the numerator and denominator have the same degree and the ratio of the leading coefficients is 1, the line $y = 1$ is the horizontal asymptote. Since the numerator is never zero, there are no x-intercepts, while the y-intercept is the value of the function at 0: $g(0) = -1/15$.

c. Since $x = 1$ makes the denominator zero, and the numerator is not factorable over the real numbers, $h(x)$ has the vertical asymptote $x = 1$ (in stark contrast with $f(x)$ discussed in part a.). Furthermore, since the degree of the denominator is less than that of the numerator, there is no horizontal asymptote. However, since the difference in degrees is just one, we anticipate an oblique (or slant) asymptote, which is revealed by dividing $x - 1$ into $x^2 + 1$ by long or synthetic division.

$$\frac{x^2 + 1}{x - 1} = (x + 1) + \frac{2}{x - 1}$$

Thus we have obtained that the line $y = x + 1$ is a slant asymptote for h. As far as intercepts, there are no x-intercepts since the numerator is never 0, but the function value $h(0) = -1$ yields the y-intercept of -1.

d. Like in part c., the degree of the numerator of $k(x)$ is one more than the degree of the denominator and therefore, k has an oblique asymptote. Polynomial division yields

$$k(x) = (x + 1) - \frac{7x + 7}{x^2 + 9},$$

so we conclude that $y = x + 1$ is the oblique asymptote. Finding $k(0)$ reveals that the function crosses the y-axis at $2/9$. To find x-intercepts, we need the zeros of the numerator polynomial. One of the potential rational zeros is -1, which actually is a zero, and thus the numerator factors as $(x + 1)(x^2 + 2)$, so -1 is the only x-intercept.

Definition

Power Functions

A **power function** in the variable x is a function of the form $f(x) = x^a$, where $a \neq 0$. Note that if a is a positive integer, such a function is an example of a single-term polynomial. If a has the form $1/n$, where n is a positive integer, such a function is a **root function** (and $x^{1/n}$, also written $\sqrt[n]{x}$, is called the nth **root** of x). If $a = -1$, the function is the **reciprocal function** and is an example of a rational function.

While the exponent a in the previous definition can be any nonzero real number, the cases where a is an integer or the reciprocal of an integer are of special interest. Figure 9 contains the graphs of $f(x) = x^a$ for $a = 1$, 3, and 5. Note that the graphs exhibit symmetry with respect to the origin.

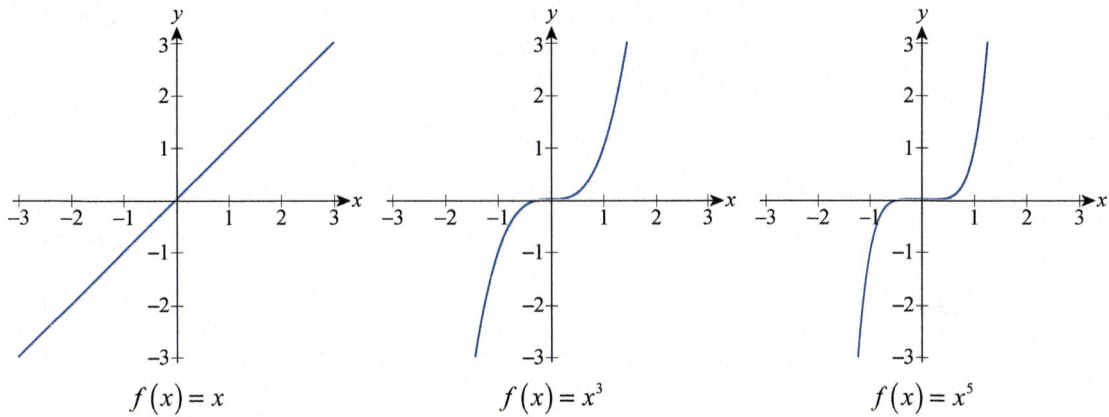

$f(x) = x$ $f(x) = x^3$ $f(x) = x^5$

Figure 9 Odd Exponents

Figure 10 contains the graphs of $f(x) = x^a$ for $a = 2$, 4, and 6. The functions are even and therefore each graph displays symmetry with respect to the y-axis.

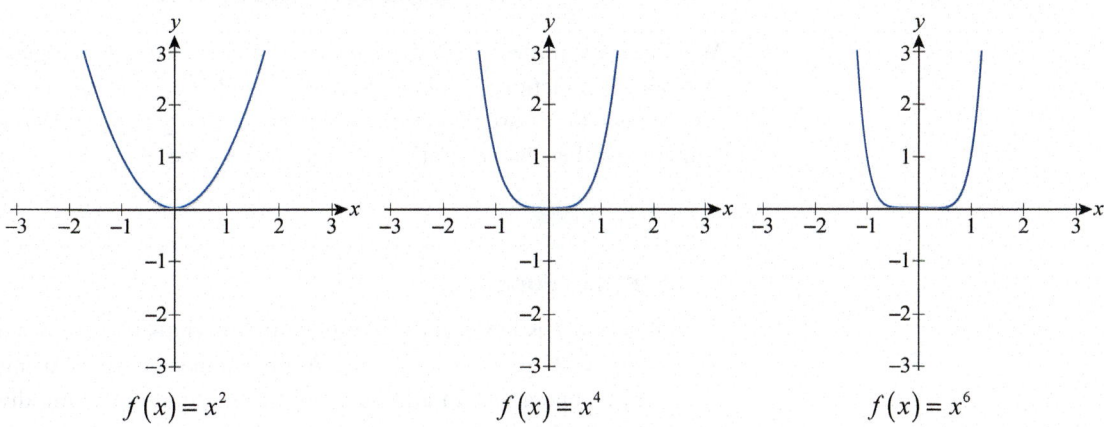

$f(x) = x^2$ $f(x) = x^4$ $f(x) = x^6$

Figure 10 Even Exponents

The graphs of the square root, cube root, fourth root, and fifth root functions are shown in Figure 11. Note that the domain of even root functions is $[0, \infty)$, while the domain of odd root functions is all of $\mathbb{R}$.

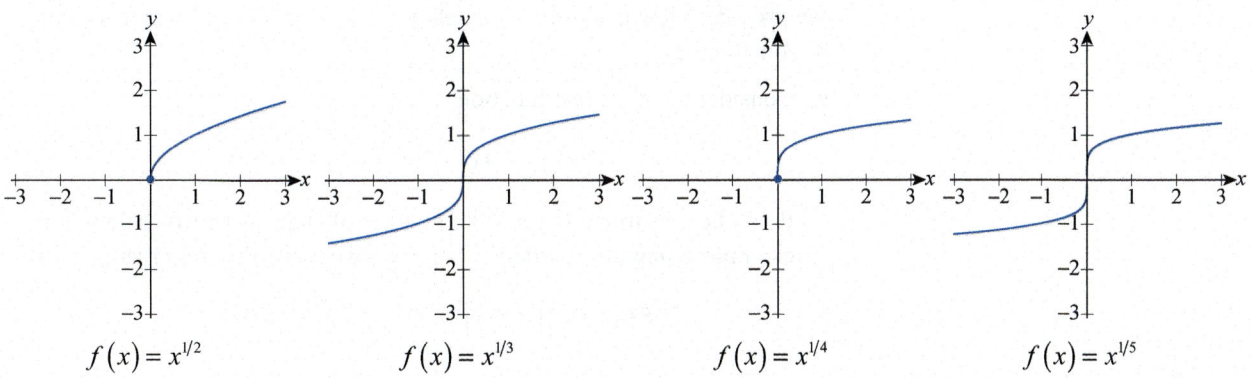

$f(x) = x^{1/2}$ $f(x) = x^{1/3}$ $f(x) = x^{1/4}$ $f(x) = x^{1/5}$

Figure 11 Root Functions

The graph of the reciprocal function $f(x) = x^{-1}$ appears in Figure 12; note that it is another example of an odd function. The function $f(x) = x^{-1}$ also frequently appears in equation form as $y = 1/x$ or $xy = 1$, and the graph of this function (or of the equation) is an example of a hyperbola (whose asymptotes are the coordinate axes in this case). For the sake of comparison, the graphs of $f(x) = x^{-2}$ and $f(x) = x^{-3}$ are also shown.

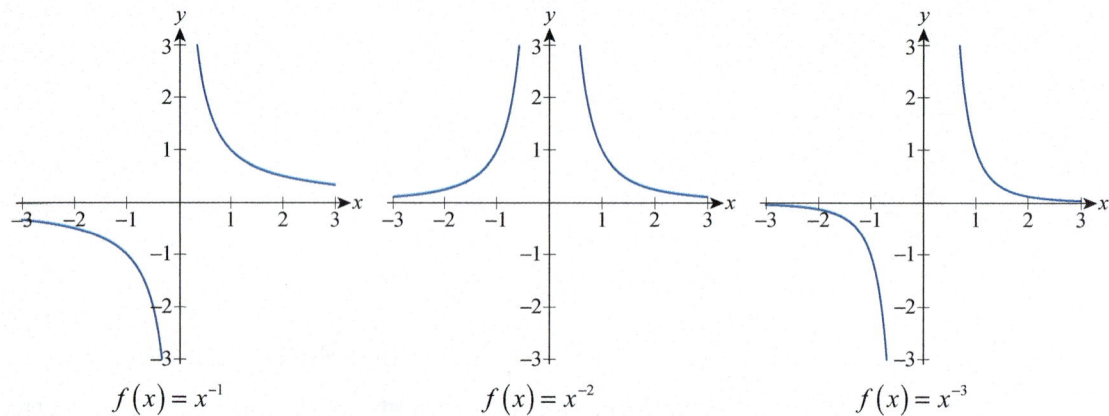

$$f(x) = x^{-1} \qquad f(x) = x^{-2} \qquad f(x) = x^{-3}$$

Figure 12 Negative Exponents

With these examples in mind, we can now define the class of algebraic functions. Polynomial functions, rational functions, and power functions of the form x^r where r is rational are all constructed using only algebraic operations (as defined below), and this characteristic is the basis of our definition.

Definition

Algebraic Functions

A function f is in the class of **algebraic functions** if it can be constructed, starting with polynomials, in a finite number of steps using only the algebraic operations of addition, subtraction, multiplication, division, and the taking of roots.

Example 4

We provide a few examples of algebraic functions and one way each can be constructed.

a. Consider the algebraic function

$$F(x) = -4x^6 + 7x^2 - \sqrt[3]{2x^2 - 3}.$$

It can be constructed in a finite number of steps as shown below. For example, using the notation $f(x) = x^2$, we recognize $F(x)$ as

$$F(x) = -4\left[f(x)\right]^3 + 7f(x) - \sqrt[3]{2f(x) - 3}.$$

Note that this is not the only way to think about constructing F. Can you think of an alternate one?

b. Starting with the "easy" polynomials $g(x) = x$, and $h(x) = 5$, we can construct

$$G(x) = \frac{\sqrt{h(x)g(x) - 2}}{\left[g(x)\right]^3 + 4} + h(x) = \frac{\sqrt{5x - 2}}{x^3 + 4} + 5.$$

c. Starting with the polynomials $p(x) = x + 1$, $q(x) = x - 1$, and $r(x) = x^2 + 1$, we can construct the algebraic function

$$H(x) = \frac{\sqrt[3]{p(x)q(x) + \sqrt{3r(x)}}}{2r(x)q(x)} = \frac{\sqrt[3]{x^2 - 1 + \sqrt{3x^2 + 3}}}{2x^3 - 2x^2 + 2x - 2}.$$

Topic Two
Transcendental Functions

Now that we have defined the class of algebraic functions, the definition of the class of *transcendental* functions is deceptively simple.

Definition

Transcendental Functions

A function f is in the class of **transcendental functions** if it is not algebraic. Such a function cannot be constructed with a finite number of applications of algebraic operations and in this sense *transcends* elementary algebra.

Without previous experience, a student new to mathematics would likely be hard-pressed to come up with an example of a transcendental function. Based only on the definition above, one approach might be to construct a function using an *infinite* number of algebraic operations, and (as we will see) this approach can actually be made to work. But you have doubtless already encountered three important examples of transcendental functions in the past: trigonometric, exponential, and logarithmic functions.

Definition

Trigonometric Functions

The three fundamental trigonometric functions are **sine**, **cosine**, and **tangent** (abbreviated as sin, cos, and tan). Their reciprocals are named, respectively, **cosecant**, **secant**, and **cotangent** (abbreviated as csc, sec, and cot).

Appendix C contains a refresher of the important properties and historical derivation of the trigonometric functions, but a few key facts will be cited here. Recall that, unlike any of the functions seen thus far, trigonometric functions are *periodic*; that is, the complete graph of each one consists of an infinite repetition of one basic shape. Also recall that the argument of a trigonometric function represents an angle measure and that, unless specifically indicated otherwise, angles are assumed to be measured in *radians* (not degrees). Figure 13 illustrates the graphs of the sine, cosine, and tangent functions.

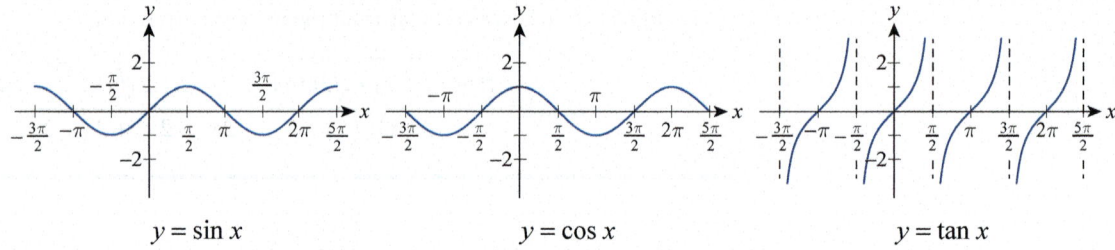

$$y = \sin x \qquad\qquad y = \cos x \qquad\qquad y = \tan x$$

Figure 13 Graphs of Sine, Cosine, and Tangent

Note that the domain for both sine and cosine is $(-\infty, \infty)$, and that they both have a range of $[-1, 1]$. Tangent is defined for all x except $\pm\{\pi/2, 3\pi/2, 5\pi/2, \ldots\}$ (tangent has vertical asymptotes at these values of x), while its range is $(-\infty, \infty)$. The graphs of cosecant, secant, and cotangent can be found in Appendix C.

Example 5

When working with trigonometric functions in calculus, it often surprises students that equivalent answers to particular problems may appear to be completely different. This phenomenon is related to the inherent features of trigonometric functions and to the ways they are related to each other. To illustrate, let us simplify the following trigonometric expressions.

a. $\sin(x - 4\pi)\sec(-x)$ b. $\dfrac{1}{\tan^2(t - \pi) + 1}$ c. $\dfrac{\cos\alpha\cot\alpha}{1 - \cos\left(\dfrac{\pi}{2} - \alpha\right)} - 1$

Solution

a. We start by noting that $\sin x$ and $\cos x$ are 2π-periodic, so $\sin(x - 4\pi)$ simply equals $\sin x$ for all x. Furthermore $\sec x$, being the reciprocal of $\cos x$, is an even function, so $\sec(-x) = \sec x$. Thus we obtain the following.

$$\sin(x - 4\pi)\sec(-x) = \sin x \sec x$$
$$= \sin x \frac{1}{\cos x}$$
$$= \tan x$$

b. As above, we begin by making use of the observation that since $\tan t$ is π-periodic, $\tan(t - \pi) = \tan t$ and so we can write the following.

$$\frac{1}{\tan^2 t + 1} = \frac{1}{\sec^2 t}$$
$$= \cos^2 t$$

Note that here we took advantage of the identity $\tan^2 t + 1 = \sec^2 t$, which can easily be checked.

c. We begin by recalling the following identity.

$$\cos\left(\frac{\pi}{2} - \alpha\right) = \sin\alpha$$

This is easiest to check by noting that since the cosine function is even, we have

$$\cos\left(\frac{\pi}{2} - \alpha\right) = \cos\left(\alpha - \frac{\pi}{2}\right) = \sin\alpha,$$

where the latter equality can easily be checked graphically by noting that if we shift the graph of the cosine function to the right by $\pi/2$ units, we obtain the graph of the sine function (see Figure 13). Using this, and the well-known trigonometric identity of $\sin^2 x + \cos^2 x = 1$, our derivation is as follows.

$$\frac{\cos\alpha\cot\alpha}{1 - \cos\left(\dfrac{\pi}{2} - \alpha\right)} - 1 = \frac{\cos\alpha\,\dfrac{\cos\alpha}{\sin\alpha}}{1 - \sin\alpha} - 1$$

$$= \frac{\cos^2\alpha}{\sin\alpha(1 - \sin\alpha)} - 1$$

$$= \frac{1 - \sin^2\alpha}{\sin\alpha(1 - \sin\alpha)} - 1$$

$$= \frac{(1 + \sin\alpha)(1 - \sin\alpha)}{\sin\alpha(1 - \sin\alpha)} - 1$$

$$= \frac{1 + \sin\alpha}{\sin\alpha} - 1$$

$$= \frac{1}{\sin\alpha} + \frac{\sin\alpha}{\sin\alpha} - 1$$

$$= \frac{1}{\sin\alpha} + 1 - 1$$

$$= \csc\alpha$$

Example 6

The following examples illustrate how trigonometric functions can be used to model well-known real-life phenomena.

a. If an object attached to a spring is pulled down A inches from equilibrium, the resulting oscillating motion is called *simple harmonic motion*. The position function of such a motion can be either $f(t) = A\sin(\omega t)$ or $g(t) = A\cos(\omega t)$, depending on whether the object is in its downward position or at equilibrium when $t = 0$. Here ω is a constant that is related to the characteristics of the spring, and A is the amplitude. Notice how the periodicity of the position functions reflects the oscillating nature of the motion.

b. The change in temperature at a particular location on Earth is referred to by meteorologists as *diurnal temperature variation*. Simply put, the surface receives its *heat influx* from the sun, through what is called *solar radiation*, but the surface also constantly loses heat as a result of its own so-called *terrestrial radiation*. During daytime hours, solar radiation is stronger, causing the Earth to warm up. At night, however, there is a drop in temperature as a result of terrestrial radiation, a process that lasts until about one hour after sunrise, at which time the cycle starts again.

Because of the near-periodic nature of these events, it is reasonable to conjecture that trigonometric functions play a role in the modeling of diurnal temperature variation. British meteorologist Sir David Brunt (1886–1965) gave the following mathematical model for the heat influx F into the ground at the equinoxes.

$$
F = \begin{cases}
F_0\left(\cos(\omega t) - \dfrac{1}{\pi}\right) & \text{if } -\dfrac{\pi}{2} < \omega t < \dfrac{\pi}{2} \\[2ex]
\dfrac{F_0}{\pi} & \text{if } \dfrac{\pi}{2} < \omega t < \dfrac{3\pi}{2}
\end{cases}
$$

where F_0 is a constant, ω is the Earth's angular velocity, and t stands for time, with $t = 0$ at noon. Notice that the first line in the model represents daytime hours, while the second line corresponds to nighttime, characterized by constant heat loss. Also notice the periodicity of the model.

Finally, observe that the above model consists of two separate "rules," each applying to daytime or nighttime hours, respectively. These so-called "piecewise defined" functions will be discussed in more detail at the end of this section.

Definition

Exponential Functions

Given a fixed positive real number a not equal to 1, $f(x) = a^x$ is the **exponential function** with base a.

Note that the distinguishing feature of an exponential function is that the variable occurs in the exponent. The graphs of two particular exponential functions, $f(x) = (1/2)^x$ and $g(x) = 2^x$, are shown in Figure 14.

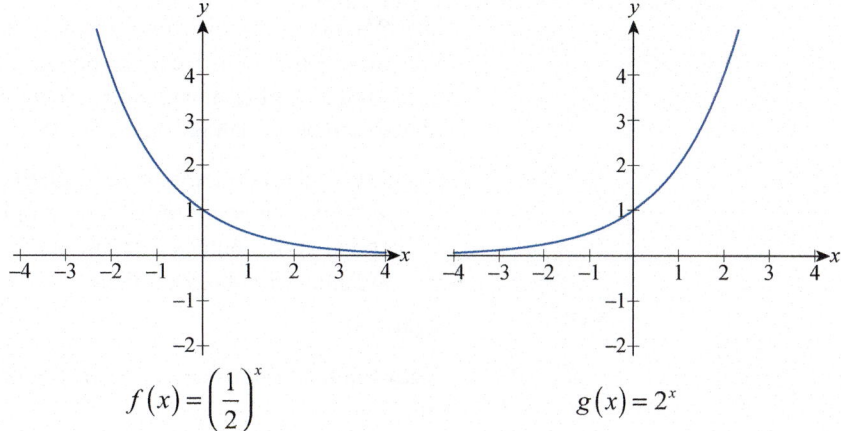

$$f(x) = \left(\frac{1}{2}\right)^x \qquad\qquad g(x) = 2^x$$

Figure 14 Graphs of Exponential Functions

Although exponential functions are not elementary in the sense of, say, polynomial functions, their evaluation is not necessarily difficult. For example, it is easy to compute (and compare) values of the two functions above.

x	$f(x) = \left(\dfrac{1}{2}\right)^x$	$g(x) = 2^x$
-1	$f(-1) = \left(\dfrac{1}{2}\right)^{-1} = 2$	$g(-1) = (2)^{-1} = \dfrac{1}{2}$
0	$f(0) = \left(\dfrac{1}{2}\right)^{0} = 1$	$g(0) = 2^0 = 1$
1	$f(1) = \left(\dfrac{1}{2}\right)^{1} = \dfrac{1}{2}$	$g(1) = 2^1 = 2$
2	$f(2) = \left(\dfrac{1}{2}\right)^{2} = \dfrac{1}{4}$	$g(2) = 2^2 = 4$

One simple but important observation is that, for every base, the domain of an exponential function is $(-\infty, \infty)$ and the range is $(0, \infty)$. Another is that if the base a lies between 0 and 1, the exponential function is strictly decreasing, while an exponential function with base a greater than 1 is strictly increasing. For ease of reference, Appendix B contains a summary of the basic properties of exponential functions.

We study exponential functions because exponential behavior is exhibited in so many natural and man-made phenomena: radioactive decay, rates of temperature change, spread of epidemics, population growth, and compound interest are all good examples. The base e (an irrational number whose first few digits are 2.71828…) is particularly useful in modeling such behavior, a fact that is reflected, as we will soon see, in the nice calculus properties it possesses.

Example 7

Populations in nature, as mentioned previously, are known to grow following an exponential pattern if there are no restrictions such as predators and limits in food supply or space. This means the population size P is an exponential function of time, such as $P(t) = P_0 a^t$, where P_0 is the initial population size corresponding to $t = 0$. This simple model can be used, in the short term and with certain restrictions, to study a wide variety of population growth problems, as illustrated by the following example.

Suppose that 1000 rabbits are released on an uninhabited island that provides practically unlimited food supply, at least for the first few months. If the population doubles every month, how many rabbits are on the island two and a half months after their initial release?

Solution

Knowing the initial size of the population, we start out with the population function

$$P(t) = 1000a^t$$

where a is a yet-unknown base that we will try to determine from the given data.

Note that

$$P(0) = 1000a^0 = 1000,$$

so our function nicely captures the fact that the initial population size is 1000 rabbits. Investigating further, since we know that the population doubles after the first month, we can write

$$2000 = P(1) = 1000a^1 = 1000a,$$

so $a = 2$ and thus

$$P(t) = 1000\left(2^t\right).$$

We can now use a calculator to determine that

$$P(2.5) = 1000\left(2^{2.5}\right) \approx 5657.$$

We conclude that there are approximately 5657 rabbits on the island two and a half months after their initial release.

Definition

Logarithmic Functions

Given a fixed positive real number a not equal to 1, $f(x) = \log_a x$ is the **logarithmic function** with base a. The logarithmic function with base 10 often appears simply as $\log x$, and the logarithmic function with base e usually appears as $\ln x$ (the "ln" stands for "natural logarithm").

Unlike the definitions of polynomial, rational, or exponential functions, the above definition doesn't give us much guidance in how to evaluate logarithms or a sense of their overall nature—it really does nothing more than introduce logarithmic notation and nomenclature. True understanding depends upon knowing that $\log_a x$ is the *inverse function* of a^x, and we will return to the topic of logarithms when we discuss inverses of functions in Section 1.4. Until then, it is enough to note that logarithms are one more important subset of transcendental functions, and that they arise in mathematics for the same reasons exponential functions do. For ease of reference, Appendix B also contains a summary of the basic properties of logarithmic functions.

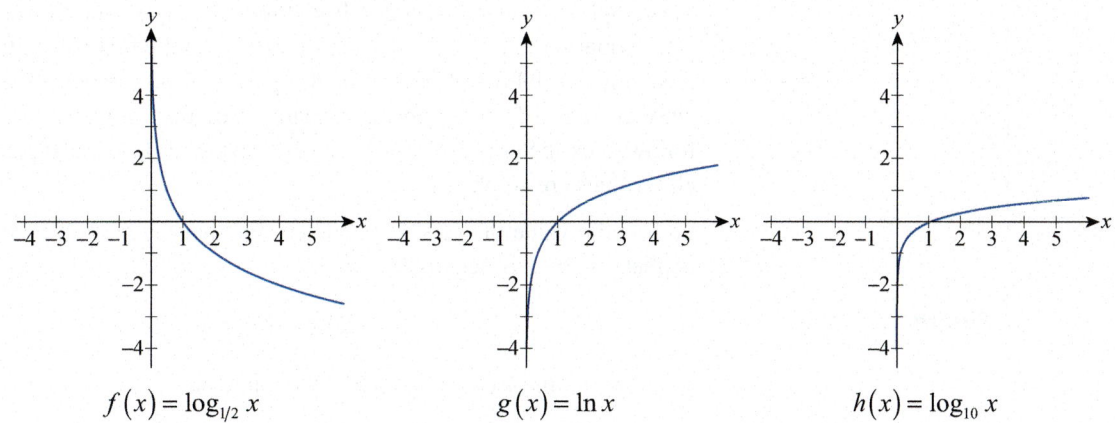

$$f(x) = \log_{1/2} x \qquad\qquad g(x) = \ln x \qquad\qquad h(x) = \log_{10} x$$

Figure 15 Graphs of Logarithmic Functions

Topic Three
Piecewise Defined Functions

We finish this section with a short discussion of functions that are, in a particular sense, pieced together from other functions.

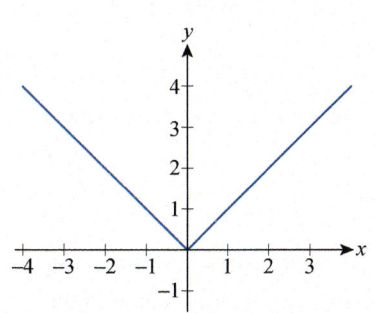

Figure 16
The Absolute Value Function

Consider the function f defined by $f(x) = \begin{cases} -x & \text{if } x < 0 \\ x & \text{if } x \geq 0 \end{cases}$. To evaluate this function for a given x, the first step is to determine which rule to apply, that is, whether x is less than 0 or not. The graph of f simply consists of two pieces, one for $x < 0$ and the other for $x \geq 0$; both pieces are very basic linear functions, and the graph appears in Figure 16.

Although we have used f as a simple example of a function defined in pieces, you have probably seen it before under the name of the **absolute value function**. In this guise, the function can be defined as $f(x) = |x|$; note that the piecewise defined rule is nothing more than the definition of absolute value.

With the absolute value function as a familiar guide, we can easily handle a slightly more complicated example.

Example 8

Sketch the graph of the following piecewise defined function $f(x)$ and find the values $f(-2)$, $f(0)$, $f(1)$, $f(2)$, and $f(3)$.

$$f(x) = \begin{cases} x^2 & \text{if } x < 0 \\ 3x+1 & \text{if } 0 \le x < 2 \\ 7 - \sqrt{x-2} & \text{if } x \ge 2 \end{cases}$$

Solution

This graph is "patched" together from three pieces, as follows. Notice that f is the squaring function for negative x-values, so the first piece of the graph is simply the left-hand branch of the prototypical parabola. For x-values between 0 and 2, the graph is a segment of the straight line $y = 3x + 1$, followed by the graph of $y = 7 - \sqrt{x-2}$ (which is also a half parabola) for all x-values greater than 2.

To find the value of $f(-2)$, we determine that the first of the three rules applies since -2 is negative.

$$f(-2) = (-2)^2 = 4$$

Next, we use the second rule to find the following.

$$f(0) = 3 \cdot 0 + 1 = 1$$

Note that this means that the y-axis as a vertical line meets the graph at $(0,1)$ and misses it at the origin, because $(0,0)$ is not a point on the graph. This is indicated by the empty and full circles in Figure 17.

For $x = 1$ the second rule applies, so finding $f(1)$ is straightforward.

$$f(1) = 3 \cdot 1 + 1 = 4$$

We see that the second rule applies only to x-values strictly less than 2, so we use the third rule to evaluate f at $x = 2$.

$$f(2) = 7 - \sqrt{2-2} = 7$$

Notice, however, that if we had used the second rule with $x = 2$, we would have obtained the same function value! Graphically, this is reflected in the fact that the two "pieces" of the graph meet at the point $(2,7)$, in other words, the graph doesn't "break" like it does at $x = 0$.

To find the last function value, we use the third rule.

$$f(3) = 7 - \sqrt{3-2} = 7 - 1 = 6$$

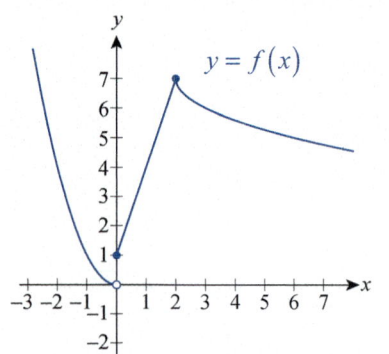

$y = f(x)$

Figure 17

Computer algebra systems such as *Mathematica* usually have built-in commands or syntax for defining and working with piecewise defined functions. In *Mathematica*, that command is (nicely enough) `Piecewise`. Figure 18 shows its use. Note that, in general, graphing technology will not automatically denote "holes" in graphs with empty circles. (See Appendix A for a brief introduction to using *Mathematica*.)

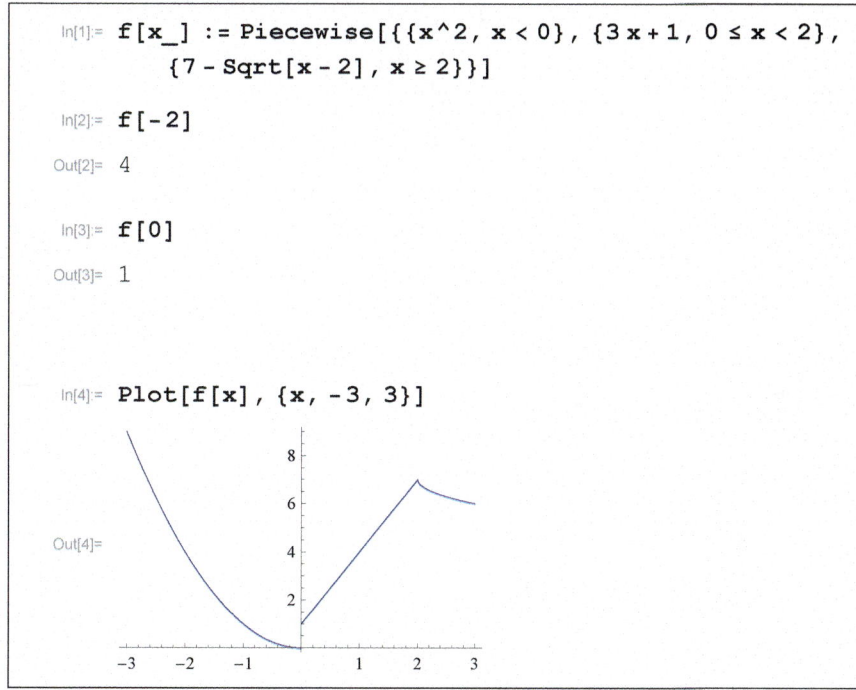

Figure 18

1.2 **Exercises**

1–8 Identify the degree, leading coefficient, intercepts, and range of the given polynomial function, and then graph the function.

1. $f(x) = \dfrac{1}{2}x - \dfrac{3}{2}$

2. $g(x) = -1.2x + 4.8$

3. $h(x) = 2x^2 - 3x - 2$

4. $u(x) = \dfrac{1}{2}x^2 + x - \dfrac{3}{2}$

5. $v(x) = x^3 - 7x + 6$

6. $F(x) = 10 - 8x + \dfrac{x^2}{2} + \dfrac{x^3}{2}$

7. $G(x) = \dfrac{x^4}{4} - 2x^2$

8. $H(x) = 2x^4 + 12x^3 + 2x^2 - 48x - 40$

9–16 Find all asymptotes and intercepts of the given rational function and then sketch the graph of the function.

9. $f(x) = \dfrac{5}{x-1}$

10. $g(x) = \dfrac{x^2-4}{2x-x^2}$

11. $h(x) = \dfrac{x^2+3}{x+3}$

12. $u(x) = \dfrac{x+2}{x^2-9}$

13. $v(x) = \dfrac{x^2-2x-3}{2x^2-5x-3}$

14. $F(x) = \dfrac{3x^2+1}{x-2}$

15. $G(x) = \dfrac{x^2+2x}{x+1}$

16. $H(x) = \dfrac{x^3-27}{x^2+5}$

17–24 Construct the algebraic function in a finite number of steps. (Answers will vary.)

17. $f(x) = \dfrac{\sqrt{x^2-1}}{x+1}$

18. $g(x) = \sqrt[3]{\dfrac{x-1}{-2+x+x^2}}$

19. $h(x) = \sqrt{2x^2+x+1} + 3x(2x+1)$

20. $u(x) = 13x^3(2-x) + 3\sqrt{x} - 5x^2(2x-x^2)$

21. $v(x) = \sqrt{2} + \dfrac{x+3}{\sqrt[5]{2x^2+2x-12}}$

22. $F(x) = \dfrac{\left(x^3-4x^2-7x+10\right)^{2/3}}{\sqrt[5]{x-5}}$

23. $G(x) = \left(x + \left(x + \left(x + (x+1)^3\right)^3\right)^3\right)^3$

24. $H(x) = \sqrt{2x + \sqrt{2x + \sqrt{2x + \sqrt{2x}}}}$

25–34 Simplify the given trigonometric expression.

25. $\dfrac{1 - \cos^2\left(\dfrac{\pi}{2} - x\right)}{\cos x}$

26. $\dfrac{1}{\sec^2 x} + \sin x \cos\left(\dfrac{\pi}{2} - x\right)$

27. $\sin\alpha(\csc\alpha - \sin\alpha)$

28. $\dfrac{1}{1+\cos\alpha} + \dfrac{1}{1-\cos\alpha}$

29. $\cot^2\theta - \cos^2\theta\cot^2\theta$

30. $\cos x(1 + \tan^2 x)$

31. $\dfrac{\sin\beta}{1+\cos\beta} + \cot\beta$

32. $\dfrac{1}{\cos(-t)\csc(-t)}$

33. $\dfrac{1 - \tan^2 x}{\cot^2 x - 1}$

34. $\dfrac{\sin x \tan\left(\dfrac{\pi}{2} - x\right)}{\cos x}$

35–38 Graph the given piecewise defined function. Use empty or full circles as appropriate at the endpoints of the intervals of definition.

35. $F(x) = \begin{cases} -x^2 & \text{if } x \le 0 \\ \dfrac{1}{2}x + 1 & \text{if } x > 0 \end{cases}$

36. $G(x) = \begin{cases} -2x - 4 & \text{if } x \le -2 \\ \dfrac{1}{2}x + \dfrac{3}{2} & \text{if } -2 < x \le 1 \\ \dfrac{1}{x-1} & \text{if } x > 1 \end{cases}$

37. $H(x) = \begin{cases} -x & \text{if } x \le 0 \\ \sin x & \text{if } 0 < x \le \dfrac{\pi}{2} \\ \sqrt[3]{x - \dfrac{\pi}{2}} & \text{if } x > \dfrac{\pi}{2} \end{cases}$

38. $u(x) = \begin{cases} -\sqrt{-x-1} & \text{if } x \le -1 \\ \sqrt{1-x^2} & \text{if } -1 < x \le 0 \\ x^2 & \text{if } x > 0 \end{cases}$

39–42 Create a piecewise defined rule and then graph the function. Use empty or full circles as appropriate at the endpoints of the intervals of definition.

39. $f(x) = |x-1|$

40. $g(x) = \dfrac{x}{|x|}$

41. $h(x) = |\sin x|$

42. $v(x) = |x+2| + |x-3|$

43–45 The greatest integer function is defined as follows: For $x \in \mathbb{R}, [\![x]\!]$ is the greatest integer less than or equal to x. For example, $[\![\pi]\!] = 3$, $[\![1]\!] = 1$, $[\![-1.5]\!] = -2$, and so on.

Use the greatest integer function to sketch the graph of the given function.

43. $f(x) = x - [\![x]\!]$

44. $g(x) = [\![x]\!] - x$

45. $h(x) = [\![\sin x]\!]$

46–48 Simple polynomial functions are used to model real-life phenomena. (**Hint:** See Example 2 for guidance as you work through these problems.)

46. Suppose that while vacationing in Europe, one day you feel a bit dizzy and your host hands you a metric thermometer. Upon checking your temperature, the reading is 39.5 °C. Would you call the doctor? (**Hint:** Recall that the conversion formula between the Fahrenheit and Celsius scales is the linear function $C = \frac{5}{9}(F-32)$. Express F from this formula to answer the question.)

47. Two trains are 630 miles apart, heading directly toward each other. The first train is traveling at 95 mph, and the second train is traveling at 85 mph. How long will it be before the trains pass each other?

48. Jessica started a candle business a few weeks ago, and noticed that the relationship between her total cost in producing the candles and the number of candles produced can be modeled by a linear function. She was able to make 3 candles for a total cost of $29, while 7 candles cost her a total of $41 to produce.

   a. Find a formula for the total investment as a function of the number of candles produced.

   b. Graph the function found in part a. What are the practical meanings of the slope and y-intercept in this particular situation?

   c. How much will be Jessica's total cost in producing 50 candles?

   d. If Jessica plans to invest a total of $320 in the next 3 months, how many candles will she be able to produce?

49–60 Find a formula for the quantity to be optimized, and use the location of the vertex of its graph to solve the problem.

49. A farmer has a total of 200 yards of fencing to enclose a rectangular pen, so that one of the four sides will be the existing wall of a barn. What should the length and width be in order to maximize the enclosed area? (**Hint:** Let x represent the width, and find an expression for the length in terms of x. Then write an expression for the area and analyze the resulting function.)

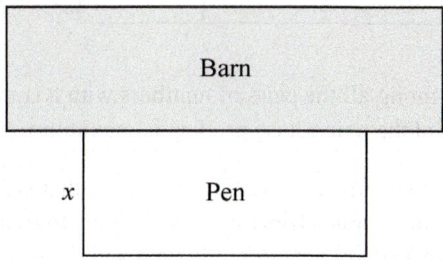

50. A rancher has a rectangular piece of sheet metal that is 20 inches wide by 10 feet long. He plans to fold the metal into a three-sided channel and weld two rectangular pieces of metal to the ends to form a watering trough 10 feet long. How should he fold the metal in order to maximize the volume of the resulting trough?

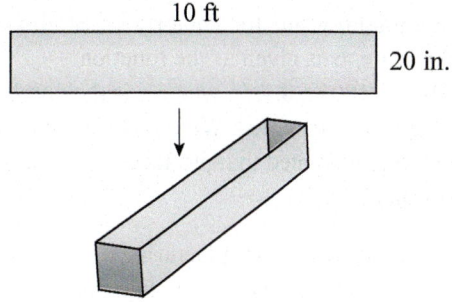

51. Cindy wants to construct three rectangular dog-training arenas side by side using a total of 400 feet of fencing. What should the overall length and width be in order to maximize the area of the three combined arenas? (**Hint:** Let x represent the width, as shown, and find an expression for the overall length in terms of x.)

52. Among all the pairs of numbers with a sum of 10, find the pair whose product is maximum.

53. Find the point on the line $2x + y = 5$ that is closest to the origin. (**Hint:** Instead of trying to minimize the distance between the origin and points on the line, minimize the square of the distance.)

54. Among all the pairs of numbers (x, y) such that $2x + y = 20$, find the pair for which the sum of the squares is minimum.

55. Find a pair of numbers whose product is maximum when two times the first number plus the second number is 48.

56. The total revenue for Thompsons' Studio Apartments is given as the function $R(x) = 100x - 0.1x^2$, where x is the number of apartments rented. What is the number of apartments rented that produces the maximum revenue?

57. The total cost of manufacturing a set of golf clubs is given as the function $C(x) = 800 - 10x + 0.20x^2$, where x is the number of sets of golf clubs produced. How many sets of golf clubs should be manufactured to incur minimum cost?

58. A rock is thrown upward with a velocity of 48 feet per second from the top of a 64-foot-high cliff. What is the maximum height attained by the rock? (**Hint:** Use $h(t) = -16t^2 + 48t + 64$ to describe the height of the rock as a function of time t.)

59. Jason is driving his Mustang GT down a two-lane highway one night, carefully observing the posted speed limit sign of 55 mph. His headlights suddenly illuminate a white-tailed deer, about 120 ft in front of his car, and he immediately hits the brakes. Suppose that the coefficient of friction between his car's tires and the pavement is $\mu = 0.9$. Using the quadratic model from Example 5b in Section 1.1, do you think he will hit the deer? What if he had traveled at 60 mph?

60. A student is throwing a small rubber ball during physical education class at an upward angle so that the horizontal component of the ball's initial velocity is 40 feet per second. If the vertical position function of the ball is given by $h(t) = -16t^2 + 24t + 7$, how far from the student will the ball hit the ground? (**Hint:** First determine how long it will take for the ball to hit the ground. The vertical position h is measured in feet, t in seconds. Ignore air resistance.)

61–72 Trigonometric and exponential functions are used to model real-life situations. (**Hint:** See Examples 6 and 7 for guidance as you work through these problems.)

61. Suppose several potatoes are dumped into the basket of a grocer's scale, which then proceeds to bounce up and down with an amplitude of 4 cm. As discussed in Example 6, a first approximation to this motion can be given by a trigonometric model. Supposing that the constant ω for the above motion is 6π and that $t = 0$ when the potatoes land in the basket, find the position function for this motion. How long does it take for the basket to complete a full period?

62. The size of a local coyote population in a certain California national forest is estimated to cycle annually according to the function $P(t) = 250 + 20\sin(\pi t/6)$, where t is measured in months, starting on March 1st of each year.

a. What is the approximate size of the population on July 1st?

b. When is the population expected to be the smallest, and what is its size then?

63. A certain species of fish is to be introduced into a new man-made lake, and wildlife experts estimate that the population will grow according to $P(t) = (1000)2^{t/3}$, where t represents the number of years from the time of introduction.

    a. What is the doubling time for this population of fish?

    b. How long will it take for the population to reach 8000 fish, according to this model?

64. Assuming a current world population of 6 billion people, and exponential growth at an annual rate of 1.9%, what will the world population be in a. 10 years and b. 50 years?

65. Suppose that a new virus has broken out in isolated parts of Africa, and it is spreading exponentially through tribal villages. The growth of this new virus can be mapped using the following formula where P stands for the number of people in a village, V for the number of infected individuals, and d for the number of days since the virus first appeared:

$$V = P\left(1 - e^{-0.18d}\right)$$

    According to this equation, how many people in a tribe of 300 will be infected after 5 days?

66. The radioactive element polonium-210 decays according to the function $A(t) = A_0 e^{-0.004951t}$, where A_0 is the mass at time $t = 0$, and t is measured in days. The fact that $A(140) = A_0/2$ means that the half-life of polonium-210 is 140 days. What percentage of the original mass of a sample of polonium-210 remains after one year?

67. The half-life of a radioactive material is the time required for an initial quantity to decrease to half its original value. In the case of radium, this is approximately 1600 years.

    a. Determine a so that $A(t) = A_0 a^t$ describes the amount of radium after t years, where A_0 is the initial amount at $t = 0$.

    b. How much of a 1-gram sample of radium would remain after 100 years?

    c. How much of a 1-gram sample of radium would remain after 1000 years?

68. When continuous compounding is used in banking, the balance after t years is described by the formula $A(t) = Pe^{rt}$, where P is the initial amount (or principal) at $t = 0$, and r is the annual interest rate. Suppose Mario made a deposit two years ago, which is compounded continuously at an annual rate of 4.5%. If his current balance is $1094.17, how much was his initial deposit? How much longer would he have to wait until his initial deposit doubles?

69. The function $f(t) = C(1+r)^t$ models the rise in the cost of a product that has a cost of C today, subject to an average yearly inflation rate of r for t years. If the average annual rate of inflation over the next decade is assumed to be 3%, what will the inflation-adjusted cost of a $100,000 house be in 10 years?

70. The concentration of a certain drug in the bloodstream after t minutes is given by the formula $C(t) = 0.05\left(1 - e^{-0.2t}\right)$. What is the concentration after 10 minutes?

71. Carbon-11 has a radioactive half-life of approximately 20 minutes, and is useful as a diagnostic tool in certain medical applications. Because of the relatively short half-life, time is a crucial factor when conducting experiments with this element.

    a. Determine a so that the formula $A(t) = A_0 a^t$ describes the amount of carbon-11 left after t minutes, where A_0 is the amount at time $t = 0$.

    b. How much of a 2-kilogram sample of carbon-11 would be left after 30 minutes?

    c. How much of a 2-kilogram sample of carbon-11 would be left after 6 hours?

72. Charles has recently inherited $8000, which he wants to deposit in a savings account. He has determined that his two best bets are an account that compounds annually at a rate of 3.20% and an account that compounds continuously at an annual rate of 3.15%. Which account would pay Charles more interest?

73–82 *True or False?* Determine whether the given statement is true or false. In case of a false statement, explain or provide a counterexample.

73. The slope of the graph of $y = Ax + B$ is A.

74. The slope of the graph of $y = Ax^2 + Bx + C$ is B.

75. The lines with equations $y = Ax + B$ and $y = -Bx + A$ are perpendicular to each other.

76. A quadratic function can have up to two y-intercepts.

77. If line L_1 has positive slope and L_2 is perpendicular to L_1, then the slope of L_2 is negative.

78. If a polynomial has even degree, then its graph always rises to both the right and the left.

79. All rational functions of the form $p(x)/q(x)$, where $q(x)$ is nonconstant, have at least one asymptote of some kind.

80. Trigonometric functions are transcendental.

81. Logarithmic functions are transcendental.

82. If a population of bacteria grows without restriction from 1000 to 2000 in one hour, then it will grow to 3000 during the next hour.

1.3 **Transforming and Combining Functions**

The repertory of functions in Section 1.2 contains the building blocks for very nearly all the functions we encounter in calculus, but they are just that—building blocks. Typically, we will find it necessary to modify an existing function or combine two or more functions in some way. The most common and useful ways are reviewed in this section.

Topic One
Shifting, Reflecting, and Stretching Functions

Definition

Horizontal Shifting

Let $f(x)$ be a function whose graph is known, and let h be a fixed real number. If we replace x in the definition of f by $x - h$, we obtain a new function $g(x) = f(x-h)$. The graph of g is the same shape as the graph of f, but shifted to the right by h units if $h > 0$ and shifted to the left by h units if $h < 0$.

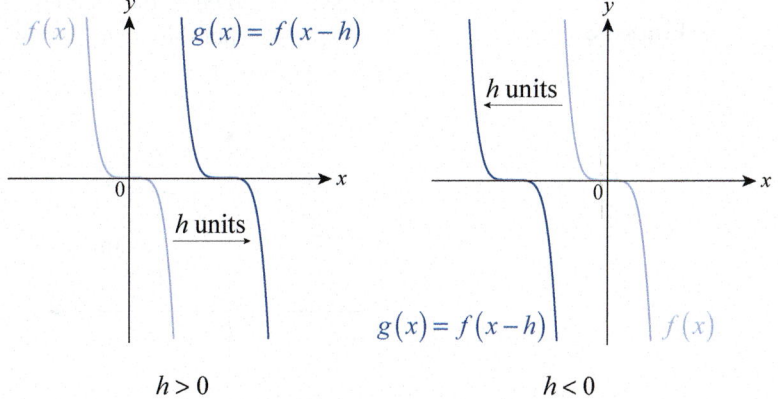

Figure 1 Horizontal Shifting

Caution!

A word of caution is in order here. It is easy to forget that the minus sign in the expression $x - h$ is critical. It may help to remember a few specific examples: replacing x with $x - 5$ shifts the graph 5 units to the *right*, since 5 is positive. Replacing x with $x + 5$ shifts the graph 5 units to the *left*, since we have actually replaced x with $x - (-5)$. With practice, knowing the effect that replacing x with $x - h$ has on the graph of a function will become natural.

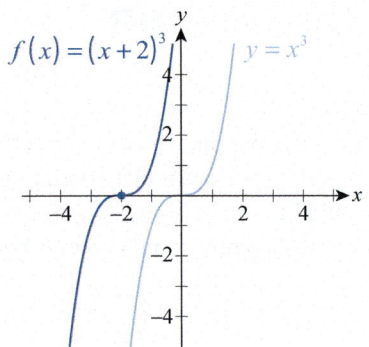

Figure 2

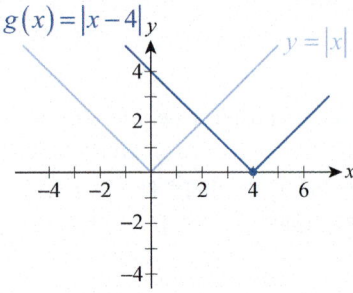

Figure 3

Example 1

Sketch the graphs of the following functions.

    a. $f(x) = (x+2)^3$ **b.** $g(x) = |x-4|$

Solution

a. The shape of $(x+2)^3$ is the same as the shape of x^3, since one expression is obtained from the other by replacing x by $x+2$. We simply draw the basic cubic shape (the shape of $y = x^3$), then shift it to the left by 2 units (see Figure 2). Note, for example, that $(-2, 0)$ is one point on the graph of f.

b. The basic function being shifted is $|x|$. The graph of $g(x) = |x-4|$ has the same shape as the graph of the absolute value function, but shifted to the right by 4 units (see Figure 3). Note, for example, that $(4, 0)$ lies on the graph of g.

Definition

Vertical Shifting

Let $f(x)$ be a function whose graph is known, and let k be a fixed real number. The graph of the function $g(x) = f(x) + k$ is the same shape as the graph of f, but shifted upward by k units if $k > 0$ and shifted downward by k units if $k < 0$.

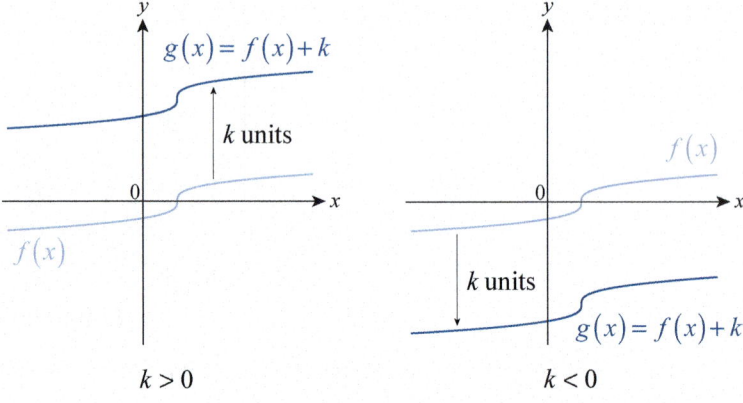

Figure 4 Vertical Shifting

The effect of adding a constant k to a function can be most easily remembered by recalling that every point on the graph of a function f has the form $(x, f(x))$, so adding k shifts every point $(x, f(x))$ to $(x, f(x)+k)$. These new points are above the originals if k is positive and below the originals if k is negative.

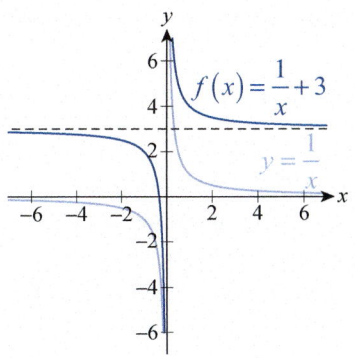

Figure 5

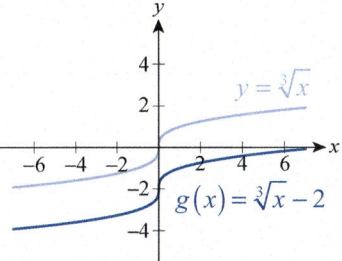

Figure 6

Example 2

Sketch the graphs of the following functions.

a. $f(x) = \dfrac{1}{x} + 3$ **b.** $g(x) = \sqrt[3]{x} - 2$

Solution

a. The graph of $f(x) = (1/x) + 3$ is the graph of $y = 1/x$ shifted up by 3 units (see Figure 5). Note that this doesn't affect the domain: the domain of f is $(-\infty, 0) \cup (0, \infty)$, the same as the domain of $y = 1/x$. However, the range is affected. The range of f is $(-\infty, 3) \cup (3, \infty)$.

b. The basic function being shifted is $\sqrt[3]{x}$. To graph $g(x) = \sqrt[3]{x} - 2$, we shift the graph of $y = \sqrt[3]{x}$ down by 2 units (see Figure 6).

Definition

Reflecting with Respect to the Axes

Let $f(x)$ be a function whose graph is known.

1. The graph of the function $g(x) = -f(x)$ is the reflection of the graph of f with respect to the x-axis.

2. The graph of the function $h(x) = f(-x)$ is the reflection of the graph of f with respect to the y-axis.

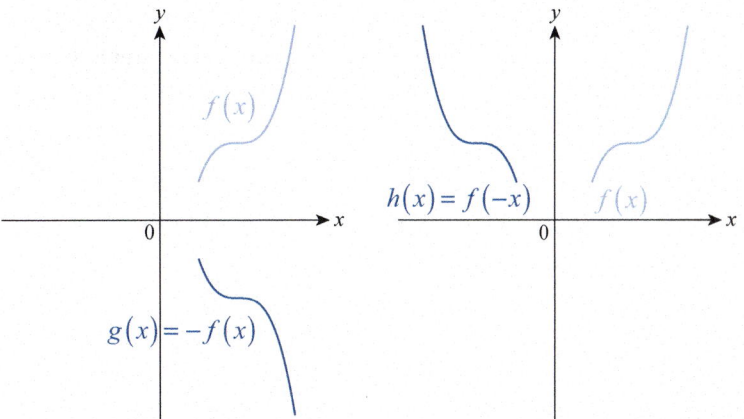

With Respect to the x-Axis   With Respect to the y-Axis

Figure 7 Reflecting with Respect to the Axes

In other words, a function is reflected with respect to the x-axis by multiplying the entire function by -1, and reflected with respect to the y-axis by replacing x with $-x$.

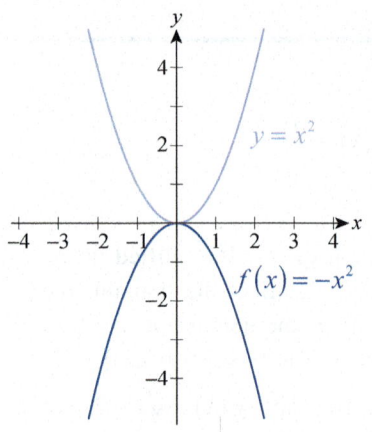

Figure 8

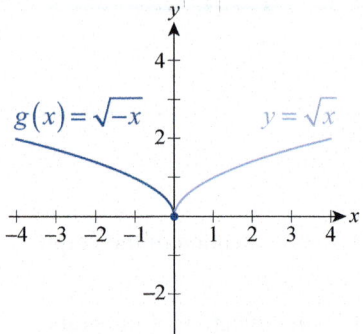

Figure 9

Example 3

Sketch the graphs of the following functions.

a. $f(x) = -x^2$ **b.** $g(x) = \sqrt{-x}$

Solution

a. To graph $f(x) = -x^2$, begin with the graph of the basic parabola $y = x^2$. The entire function is multiplied by -1, so reflect the graph over the x-axis, resulting in the original shape turned upside down (see Figure 8). Note that the domain is still the entire real line, but the range of f is the interval $(-\infty, 0]$.

b. To graph $g(x) = \sqrt{-x}$, we reflect the graph $y = \sqrt{x}$ with respect to the y-axis because x has been replaced by $-x$ (see Figure 9). Note that this changes the domain, but not the range. The domain of g is the interval $(-\infty, 0]$ and the range is $[0, \infty)$.

Definition

Stretching and Compressing

Let $f(x)$ be a function whose graph is known, and let a be a positive real number.

1. The graph of the function $g(x) = af(x)$ is *stretched* vertically compared to the graph of f if $a > 1$.

2. The graph of the function $g(x) = af(x)$ is *compressed* vertically compared to the graph of f if $0 < a < 1$.

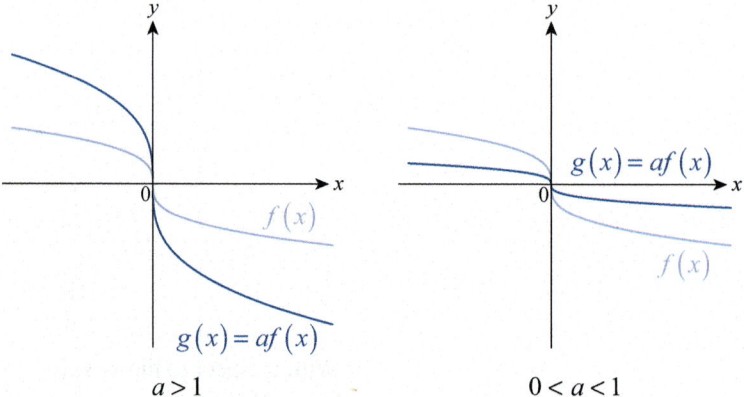

Figure 10 Stretching and Compressing

If the function g is obtained from the function f by multiplying f by a negative real number, think of the number as the product of -1 and a positive real number (namely, its absolute value). The previous definition tells us what multiplication by a positive constant does to a graph, and we already know that multiplying a function by -1 reflects the graph with respect to the x-axis.

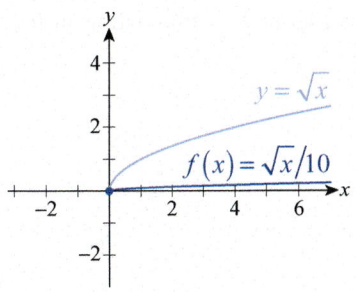

Figure 11

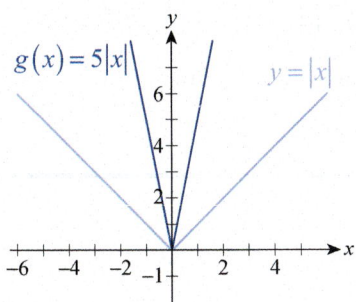

Figure 12

Example 4

Sketch the graphs of the following functions.

a. $f(x) = \dfrac{\sqrt{x}}{10}$ **b.** $g(x) = 5|x|$

Solution

a. Begin with the graph of $\sqrt{x}$. The shape of $f(x)$ is similar to the shape of $y = \sqrt{x}$, but compressed considerably because the second coordinates have been multiplied by the factor of $1/10$, and are consequently smaller (see Figure 11).

b. Begin with the graph of the absolute value function. In contrast to part a., the graph of $g(x) = 5|x|$ is stretched compared to $y = |x|$ (see Figure 12). Every second coordinate has been multiplied by a factor of 5, and is consequently larger.

We can now put all of the above together and consider functions that have been derived through a sequence of transformations from simpler functions.

Order of Transformations

If a function g has been obtained from a simpler function f through a number of transformations, g can be analyzed by looking for the transformations in this order.

Step 1: Horizontal shifting

Step 2: Stretching and compressing

Step 3: Reflecting with respect to the axes

Step 4: Vertical shifting

Consider, for example, the function $g(x) = -2\sqrt{x+1} + 3$. The function g has been "built up" from the basic square root function through a variety of transformations.

Step 1: First, $\sqrt{x}$ has been transformed into $\sqrt{x+1}$ by replacing x with $x + 1$, and we know that this corresponds graphically to a shift leftward by 1 unit.

Step 2: Next, the function $\sqrt{x+1}$ has been multiplied by 2 to get the function $2\sqrt{x+1}$, and we know that this has the effect of stretching the graph of $\sqrt{x+1}$ vertically.

Step 3: The function $2\sqrt{x+1}$ has been multiplied by -1, giving us $-2\sqrt{x+1}$, and the graph of this is the reflection of $2\sqrt{x+1}$ with respect to the x-axis.

Step 4: Finally, the constant 3 has been added to $-2\sqrt{x+1}$, shifting the entire graph upward by 3 units.

These transformations are illustrated in order in Figure 13, culminating in the graph of $g(x) = -2\sqrt{x+1} + 3$.

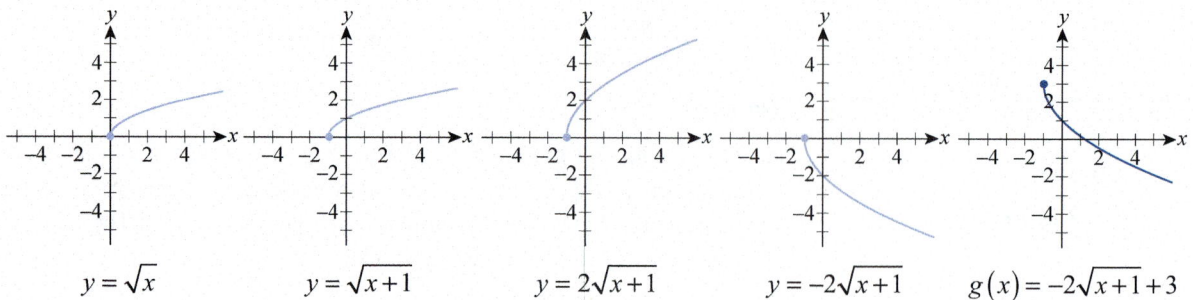

Figure 13 Building the Graph of $g(x) = -2\sqrt{x+1} + 3$

Example 5

Sketch the graph of the function $f(x) = \dfrac{1}{2-x}$.

Solution

The basic function that f is similar to is $1/x$. Now we follow the order of transformations.

Step 1: If we replace x by $x + 2$ (shifting the graph 2 units to the left), we obtain the function $\dfrac{1}{x+2}$, which is closer to what we want.

Step 2: There does not appear to be any stretching or compressing transformation.

Step 3: If we now replace x by $-x$, we have $\dfrac{1}{-x+2}$, which is the same as f. This reflects the graph of $\dfrac{1}{x+2}$ with respect to the y-axis.

Step 4: Since we already found f, we know there is no vertical shift.

The entire sequence of transformations is shown in Figure 14, ending with the graph of f.

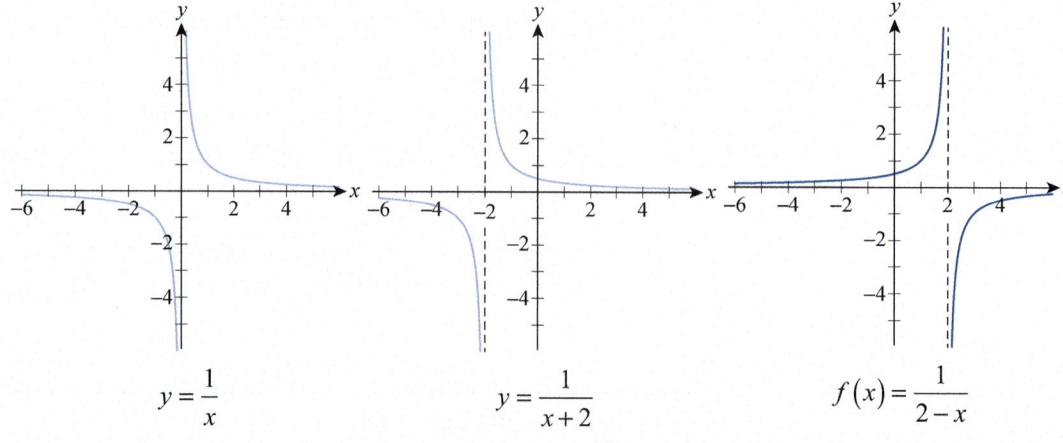

Figure 14

Note: An alternate approach to graphing $f(x) = \dfrac{1}{2-x}$ is to rewrite the function in the form $f(x) = -\dfrac{1}{x-2}$. In this form, the graph of f is the graph of $1/x$ shifted 2 units to the right, and then reflected with respect to the x-axis. The result is the same, as you should verify.

Topic Two
Combining Functions Arithmetically

Shifting, stretching, and reflecting are all modifications applied to a single function. We also often need to combine two or more functions.

We begin with four arithmetic ways of combining two or more functions to obtain new functions. The four arithmetic operations are very familiar to you: addition, subtraction, multiplication, and division. The only possibly new idea is that we are applying these operations to functions, not numbers. But as we will see, the arithmetic combination of functions is based entirely on the arithmetic combination of numbers.

Definition

Adding, Subtracting, Multiplying, and Dividing Functions

Let f and g be two functions. The **sum** $f + g$, **difference** $f - g$, **product** fg, and **quotient** f/g are four new functions defined as follows.

1. $(f+g)(x) = f(x) + g(x)$

2. $(f-g)(x) = f(x) - g(x)$

3. $(fg)(x) = f(x)g(x)$

4. $\left(\dfrac{f}{g}\right)(x) = \dfrac{f(x)}{g(x)}$, provided that $g(x) \neq 0$

The domain of each of these new functions consists of the common elements of the individual domains of f and g, with the added condition that for the quotient function we have to omit those elements for which $g(x) = 0$.

With the above definition, we can determine the sum, difference, product, or quotient of two functions at one particular value for x or find a formula for these new functions based on the formulas for f and g, if they are available.

Example 6

Given that $f(-2) = 5$ and $g(-2) = -3$, find $(f-g)(-2)$ and $(f/g)(-2)$.

Solution

By the definition of the difference and quotient of functions, we obtain the following.

$$(f-g)(-2) = f(-2) - g(-2)$$
$$= 5 - (-3) = 8$$

$$\left(\frac{f}{g}\right)(-2) = \frac{f(-2)}{g(-2)}$$

$$= \frac{5}{-3} = -\frac{5}{3}$$

Example 7

Given the two functions $f(x) = 4x^2 - 1$ and $g(x) = \sqrt{x}$, find $(f+g)(x)$ and $(fg)(x)$.

Solution

By the definition of the sum and product of functions, we obtain the following.

$$(f+g)(x) = f(x) + g(x)$$
$$= 4x^2 - 1 + \sqrt{x}$$

$$(fg)(x) = (4x^2 - 1)(\sqrt{x})$$
$$= 4x^{5/2} - x^{1/2}$$

Note that the domain of f is the entire real line, while the domain of g is $[0, \infty)$.

Since the domain of two functions combined arithmetically is the intersection of the individual domains, $f+g$ and fg both have the domain $[0, \infty)$.

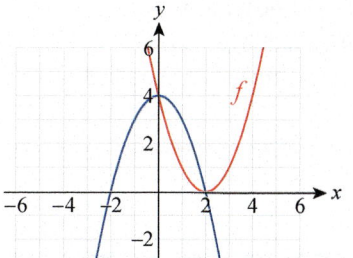

Figure 15

Example 8

Based on the graphs of f and g in Figure 15, determine the domain of f/g and evaluate $(f/g)(1)$.

Solution

From Figure 15, we can see that the domain of both f and g individually is the set of all real numbers, $(-\infty, \infty)$. To find the domain of f/g, we need to check where $g(x) = 0$. Based on the graph, this occurs when $x = -2$ and $x = 2$, so the domain of f/g is as follows.

$$(-\infty, -2) \cup (-2, 2) \cup (2, \infty)$$

Also based upon the graphs, it appears that $f(1) = 1$ and $g(1) = 3$, so $(f/g)(1) = 1/3$.

Topic Three

Composing and Decomposing Functions

A fifth way of combining functions is to form the *composition* of one function with another. Informally speaking, this means to apply one function, say f, to the output of another function, say g. The symbol for composition is an open circle.

Definition

Composing Functions

Let f and g be two functions. The **composition** of f and g, denoted $f \circ g$, is the function defined by $(f \circ g)(x) = f(g(x))$. (The function $f \circ g$ is read "f composed with g" or "f of g," and $f \circ g$ is also referred to as a **composite function**.)

The domain of $f \circ g$ consists of all x in the domain of g for which $g(x)$ is in the domain of f.

The diagram in Figure 16 is a sort of schematic of the composition of two functions. The circles represent sets, with the leftmost circle being the domain of the function g. The arrows indicate the element that x is associated with by the various functions.

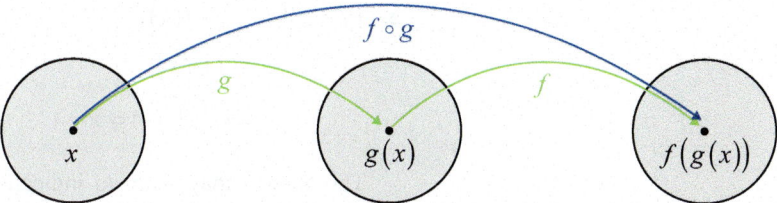

Figure 16 Composition of f and g

As with the four arithmetic ways of combining functions, we can evaluate the composition of two functions at a single point or find a formula for the composition if we have been given formulas for the individual functions.

Caution!

Note that the order of f and g is important. In general, we can expect the function $f \circ g$ to be different from the function $g \circ f$. That is, the composition of two functions, unlike the sum and product of two functions, is not commutative.

Example 9

Given $f(x) = x^2$ and $g(x) = x - 3$, find the following.

a. $(f \circ g)(6)$ **b.** $(f \circ g)(x)$

Solution

a. Since $(f \circ g)(6) = f(g(6))$, the first step is to calculate $g(6)$.

$$g(6) = 6 - 3 = 3$$

Then, apply f to the result: $(f \circ g)(6) = f(g(6)) = f(3) = 3^2 = 9$.

b. To find the formula for $f \circ g$, we simply apply the definition of composition and then simplify.

$$\begin{aligned}
(f \circ g)(x) &= f(g(x)) \\
&= f(x - 3) \\
&= (x - 3)^2 \\
&= x^2 - 6x + 9
\end{aligned}$$

Note that once we have found a formula for $f \circ g$, we have an alternative way of answering the first question: $(f \circ g)(6) = (6)^2 - (6)(6) + 9 = 9$.

Example 10

Let $f(x) = x^2 - 4$ and $g(x) = \sqrt{x}$. Find formulas and state the domains for the following functions.

a. $f \circ g$ **b.** $g \circ f$

Solution

a. $\begin{aligned}
(f \circ g)(x) &= f(g(x)) \\
&= f(\sqrt{x}) \\
&= (\sqrt{x})^2 - 4 = x - 4
\end{aligned}$

The answer may seem to indicate that the domain of $f \circ g$ is all real numbers, but this is incorrect. The domain of $f \circ g$ is actually the interval $[0, \infty)$, because only nonnegative numbers can be plugged into g.

b. $(g \circ f)(x) = g(f(x))$
$$= g(x^2 - 4)$$
$$= \sqrt{x^2 - 4}$$

The domain of $g \circ f$ consists of all x for which $x^2 - 4 \geq 0$, or $x^2 \geq 4$. In interval form, the domain is $(-\infty, -2] \cup [2, \infty)$.

Often, functions can be best understood by recognizing them as a composition of two or more simpler functions. We have already seen one instance of this: shifting, reflecting, stretching, and compressing can all be thought of as a composition of two or more functions. For example, the function $h(x) = (x-2)^3$ can be thought of as the composition of the functions $f(x) = x^3$ and $g(x) = x - 2$.

$$f(g(x)) = f(x-2)$$
$$= (x-2)^3$$
$$= h(x)$$

To "decompose" a function into a composition of simpler functions, it is usually best to identify what the function does to its argument from the inside out. That is, identify the first thing that is done to the argument, then the second, and so on. Each action describes a less complex function, and can be identified as such. The composition of these functions, with the innermost function corresponding to the first action, the next innermost corresponding to the second action, and so on, is then equivalent to the original function.

Decomposition can often be done in several different ways. Consider, for example, the function $f(x) = \sqrt[3]{5x^2 - 1}$. The following illustrates some of the ways f can be written as a composition of functions. Be sure you understand how each of the different compositions is equivalent to f.

1. $g(x) = \sqrt[3]{x}$ $g(h(x)) = g(5x^2 - 1)$
      $h(x) = 5x^2 - 1$ $= \sqrt[3]{5x^2 - 1}$
                                  $= f(x)$

2. $g(x) = \sqrt[3]{x-1}$ $g(h(x)) = g(5x^2)$
      $h(x) = 5x^2$ $= \sqrt[3]{5x^2 - 1}$
                                  $= f(x)$

3. $g(x) = \sqrt[3]{x}$ $g(h(k(x))) = g(h(x^2))$
      $h(x) = 5x - 1$ $= g(5x^2 - 1)$
      $k(x) = x^2$ $= \sqrt[3]{5x^2 - 1}$
                                  $= f(x)$

Example 11

Decompose the function $f(x) = |x^2 - 3| + 2$ into a composition of **a.** two functions and **b.** three functions.

Solution

a. $g(x) = |x| + 2$

$h(x) = x^2 - 3$

$$g(h(x)) = g(x^2 - 3)$$
$$= |x^2 - 3| + 2$$
$$= f(x)$$

b. $g(x) = x + 2$

$h(x) = |x - 3|$

$k(x) = x^2$

$$g(h(k(x))) = g(h(x^2))$$
$$= g(|x^2 - 3|)$$
$$= |x^2 - 3| + 2$$
$$= f(x)$$

Topic Four

Interlude: Recursive Graphics

Recursion, in general, refers to using the output of a function as its input and repeating the process a certain number of times. In other words, recursion refers to the composition of a function with itself, possibly many times. Recursion has many varied uses, one of which is a branch of mathematical art.

Some special nomenclature and notation have evolved to describe recursion. If f is a function, $f^2(x)$ is used in this context to stand for $f(f(x))$, or $(f \circ f)(x)$ (not $[f(x)]^2$!). Similarly, $f^3(x)$ stands for $f(f(f(x)))$, or $(f \circ f \circ f)(x)$ and so on. The functions $f^2, f^3, \ldots$ are called **iterates** of f, with f^n being the n**th iterate** of f.

> *So, Nat'ralists observe, a Flea*
> *Hath smaller Fleas that on him prey,*
> *And these have smaller Fleas to bite 'em,*
> *And so proceed, ad infinitum.*
>
> *- Jonathan Swift*

Some of the most famous recursively generated mathematical art is based on functions whose inputs and outputs are complex numbers. Recall that every complex number can be expressed in the form, $a + bi$ where a and b are real numbers and i is the imaginary unit.

A one-dimensional coordinate system, such as the real number line, is insufficient to graph complex numbers, but complex numbers are easily graphed in a two-dimensional coordinate system.

To graph the number $a + bi$, we treat it as the ordered pair (a, b) and plot the point (a, b) in the Cartesian plane, where the horizontal axis represents pure real numbers and the vertical axis represents pure imaginary numbers.

Benoit Mandelbrot used the function $f(z) = z^2 + c$, where both z and c are variables representing complex numbers, to generate the image known as the Mandelbrot set in 1979. The basic idea is to evaluate the sequence of iterates

$$f(0) = 0^2 + c = c, \quad f^2(0) = f(c) = c^2 + c, \quad f^3(0) = f(c^2 + c) = (c^2 + c)^2 + c, \ldots$$

for various complex numbers c and determine if the sequence of complex numbers stays close to the origin or not. Those complex numbers that result in so-called "bounded" sequences are colored white. We have used similar ideas to generate our own recursive art, as described below.

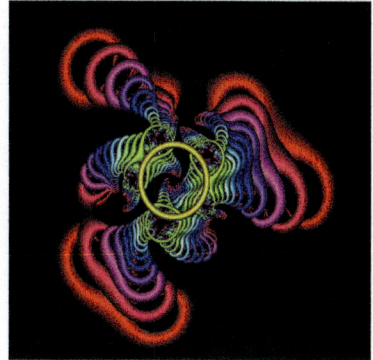

Figure 17   *i* of the Storm

The image "*i* of the storm" reproduced in Figure 17 is based on the function $f(z) = \dfrac{(1-i)z^4 + (7+i)z}{2z^5 + 6}$, where again z is a variable that will be replaced with complex numbers. The image is actually a picture of the complex plane, with the origin in the very center of the golden ring. The golden ring consists of those complex numbers that lie between 0.9 and 1.1 in distance from the origin. The rules for coloring other complex numbers in the plane are as follows: given an initial complex number z not on the gold ring, $f(z)$ is calculated. If the complex number $f(z)$ lies somewhere on the gold ring, the original number z is colored the deepest shade of green. If not, the iterate $f^2(z)$ is calculated. If this result lies in the gold ring, the original z is colored a bluish shade of green. If not, the process continues up to the 12th iterate $f^{12}(z)$, using a different color each time. If $f^{12}(z)$ lies in the gold ring, z is colored red, and if not the process halts and z is colored black.

The idea of recursion can be used to generate any number of similar images, with the end result usually striking and often surprising even to the creator.

1.3 **Exercises**

1–23 Sketch the graph of the given function by first identifying the more basic function that has been shifted, reflected, stretched, or compressed. Then determine the domain and range of the function.

1. $f(x) = (x+2)^3$

2. $G(x) = |x-4|$

3. $p(x) = -(x+1)^2 + 2$

4. $g(x) = \sqrt{x+3} - 1$

5. $q(x) = (1-x)^2$

6. $r(x) = -\sqrt[3]{x}$

7. $s(x) = \sqrt{2-x}$

8. $F(x) = \dfrac{|x+2|}{3} + 3$

9. $w(x) = \dfrac{1}{(x-3)^2}$

10. $v(x) = \dfrac{1}{3x} - 2$

11. $f(x) = \dfrac{1}{2-x}$

12. $k(x) = \sqrt{-x} + 2$

13. $b(x) = [\![x-4]\!] + 4$

14. $R(x) = 4 - |2x|$

15. $S(x) = (3-x)^3$

16. $g(x) = -\dfrac{1}{x+1}$

17. $h(x) = \dfrac{x^2}{2} - 3$

18. $W(x) = 1 - |4-x|$

19. $g(x) = x^2 - 6x + 9$ (**Hint:** Find a better way to write the function.)

20. $h(x) = \dfrac{|x|}{x}$ (**Hint:** Evaluate h at a few points to understand its behavior.)

21. $W(x) = \dfrac{x-1}{|x-1|}$

22. $S(x) = [\![x-2]\!]$

23. $V(x) = -3\sqrt{x-1} + 2$

24–29 Write an equation for the function described.

24. Use the function $f(x) = x^2$. Move the function 4 units to the right and 2 units up.

25. Use the function $f(x) = x^2$. Move the function 6 units up and reflect across the x-axis.

26. Use the function $f(x) = x^3$. Move the function 1 unit to the left and reflect across the y-axis.

27. Use the function $f(x) = \sqrt{x}$. Move the function 5 units to the left and reflect across the x-axis.

28. Use the function $f(x) = \sqrt{x}$. Move the function 3 units down and reflect across the y-axis.

29. Use the function $f(x) = |x|$. Move the function 7 units to the left, reflect across the x-axis, and reflect across the y-axis.

30–33 Use your knowledge about transformations to find a possible formula for the function $f(x)$ given by its graph.

30.

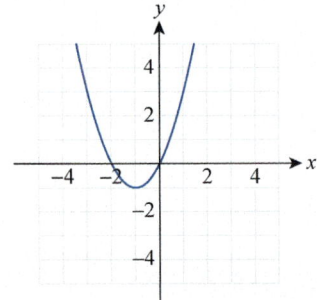

31.

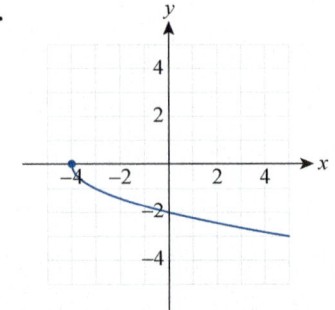

32.

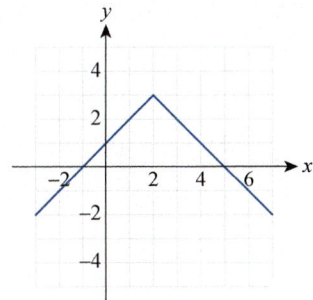

33.

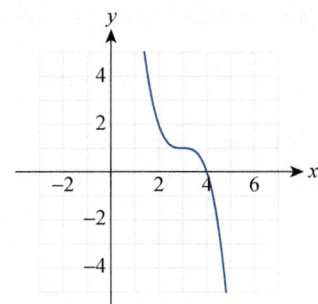

34–45 Use the information given to determine **a.** $(f+g)(-1)$, **b.** $(f-g)(-1)$, **c.** $(fg)(-1)$, and **d.** $(f/g)(-1)$.

34. $f(-1)=-3;\quad g(-1)=5$

35. $f(-1)=0;\quad g(-1)=-1$

36. $f(x)=x^2-3;\quad g(x)=x$

37. $f(x)=\sqrt[3]{x};\quad g(x)=x-1$

38. $f(-1)=15;\quad g(-1)=-3$

39. $f(x)=\dfrac{x+5}{2};\quad g(x)=6x$

40. $f(x)=x^4+1;\quad g(x)=x^{11}+2$

41. $f(x)=\dfrac{6-x}{2};\quad g(x)=\sqrt{\dfrac{x}{-4}}$

42. $f=\{(5,2),(0,-1),(-1,3),(-2,4)\};$
$g=\{(-1,3),(0,5)\}$

43. $f=\{(3,15),(2,-1),(-1,1)\};\quad g(x)=-2$

44.

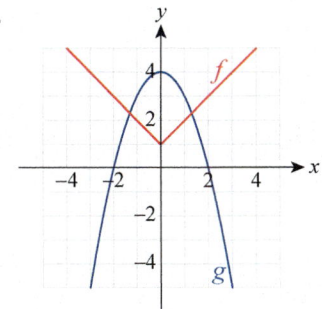

45.

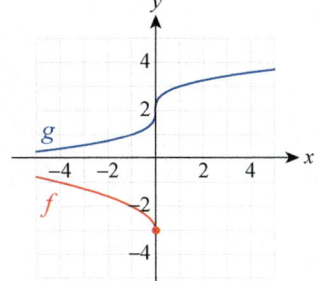

46–53 Find the formula and domain for **a.** $f+g$ and **b.** f/g.

46. $f(x)=|x|;\quad g(x)=\sqrt{x}$

47. $f(x)=x^2-1;\quad g(x)=\sqrt[3]{x}$

48. $f(x)=x-1;\quad g(x)=x^2-1$

49. $f(x)=x^{3/2};\quad g(x)=x-3$

50. $f(x)=3x;\quad g(x)=x^3-8$

51. $f(x)=x^3+4;\quad g(x)=\sqrt{x-2}$

52. $f(x)=-2x^2;\quad g(x)=[\![x+4]\!]$

53. $f(x)=6x-1;\quad g(x)=x^{2/3}$

54–63 Evaluate the expression, if possible, given $f(x)=1/x^2$ and $g(x)=2x+3$.

54. $(f+g)(-7)$ **55.** $(f+g)(-10)$ **56.** $(f-g)(-5)$ **57.** $(f-g)(0)$

58. $(fg)(4)$ **59.** $(fg)(-3)$ **60.** $\left(\dfrac{f}{g}\right)(-2)$ **61.** $\left(\dfrac{f}{g}\right)(0)$

62. $\left(\dfrac{g}{f}\right)(1)$ **63.** $\left(\dfrac{g}{f}\right)(-6)$

64–73 Use the information given to determine $(f\circ g)(3)$.

64. $f(-5)=2;\;\; g(3)=-5$ **65.** $f(\pi)=\pi^2;\;\; g(3)=\pi$

66. $f(x)=x^2-3;\;\; g(x)=\sqrt{x}$ **67.** $f(x)=\sqrt{x^2-9};\;\; g(x)=1-2x$

68. $f(x)=2+\sqrt{x};\;\; g(x)=x^3+x^2$ **69.** $f(x)=x^{3/2}-3;\;\; g(x)=\left\|\dfrac{3x}{2}\right\|$

70. $f(x)=\sqrt{x+6};\;\; g(x)=\sqrt{4x-3}$ **71.** $f(x)=\sqrt{\dfrac{3x}{14}};\;\; g(x)=x^4-x^3-x^2-x$

72.

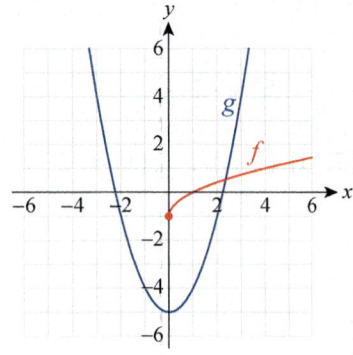

73.

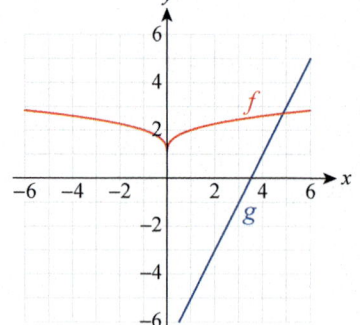

74–87 Find the formula and domain for **a.** $f\circ g$ and **b.** $g\circ f$.

74. $f(x)=\sqrt{x-1};\;\; g(x)=x^2$ **75.** $f(x)=\dfrac{1}{x};\;\; g(x)=x-1$ **76.** $f(x)=\dfrac{4x-2}{3};\;\; g(x)=\dfrac{1}{x}$

77. $f(x)=1-x;\;\; g(x)=\sqrt{x}$ **78.** $f(x)=[\![x-3]\!];\;\; g(x)=x^3+1$

79. $f(x)=x^2+2x;\;\; g(x)=3x^2+5$ **80.** $f(x)=x^2+1;\;\; g(x)=3x^2+5$ **81.** $f(x)=\sqrt{x};\;\; g(x)=2x$

82. $f(x)=\dfrac{1}{x+7};\;\; g(x)=\dfrac{2}{x}$ **83.** $f(x)=\dfrac{1}{x};\;\; g(x)=\dfrac{1}{x}$ **84.** $f(x)=x^2;\;\; g(x)=3x+1$

85. $f(x)=\sqrt[3]{x};\;\; g(x)=x^3$ **86.** $f(x)=\sqrt{x-4};\;\; g(x)=x^2+2$ **87.** $f(x)=\dfrac{3}{1-x};\;\; g(x)=3x^2$

88–93 Write the given function as a composition of two functions. (Answers will vary.)

88. $f(x)=\sqrt[3]{3x^2-1}$ **89.** $f(x)=\dfrac{2}{5x-1}$ **90.** $f(x)=|x-2|+3$

91. $f(x)=x+\sqrt{x+2}-5$ **92.** $f(x)=|x^3-5x|+7$ **93.** $f(x)=\dfrac{\sqrt{x-3}}{x^2-6x+9}$

94. The volume of a right circular cylinder is given by the formula $V = \pi r^2 h$. If the height h is three times the radius r, show the volume V as a function of r.

95. The surface area of a wind sock is defined by the formula $S = \pi r \sqrt{r^2 + h^2}$ where r is the radius of the base of the wind sock and h is the height of the wind sock. As the wind sock is being knitted by an automated knitter, the height h is increasing with time t as defined by the formula $h(t) = \frac{1}{4}t^2, t \geq 0$. Find the surface area S of the wind sock as a function of time t.

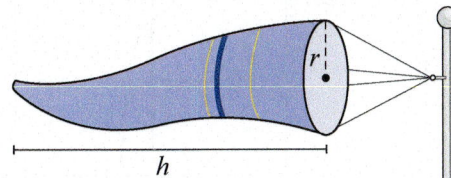

96. The volume of the wind sock described in the previous exercise is given by the formula $V = \frac{1}{3}\pi r^2 h$ where r is the radius of the wind sock and h is the height of the wind sock. If the height h is increasing with time t as defined by the formula $h(t) = \frac{1}{4}t^2$, $t \geq 0$, find the volume V of the wind sock as a function of time t.

97. A widget factory produces n widgets in t hours of a single day. The number of widgets the factory produces is given by the formula $n(t) = 10{,}000t - 25t^2, 0 \leq t \leq 9$. The cost c in dollars of producing n widgets is given by the formula $c(n) = 2040 + 1.74n$. Find the cost c as a function of time t that the factory is producing the widgets.

98. Suppose that $H(x)$ represents the percentage of income spent on a home loan in the year x and $C(x)$ represents the percentage of income spent on a car loan in the year x. If $I(x)$ represents the income in year x, determine the function L that represents the total loan expenses in year x.

99. Given two odd functions f and g, show that $f \circ g$ is also odd. Then verify this fact with the particular functions $f(x) = \sqrt[3]{x}$ and $g(x) = -x^3/(3x^2 - 9)$. (**Hint:** Recall that a function is odd if $f(-x) = -f(x)$ for all x in the domain of f.)

100. Given two even functions f and g, show that the product is also even. Then verify this fact with the particular functions $f(x) = 2x^4 - x^2$ and $g(x) = 1/x^2$. (**Hint:** Recall that a function is even if $f(-x) = f(x)$ for all x in the domain of f.)

101–108 As mentioned in the Interlude, a given complex number c is said to be in the Mandelbrot set if, for the function $f(z) = z^2 + c$, the sequence of iterates $f(0), f^2(0), f^3(0),\ldots$ stays close to the origin (which is the complex number $0 + 0i$). It can be shown that if any single iterate falls more than 2 units in distance (magnitude) from the origin, then the remaining iterates will grow larger and larger in magnitude. In practice, computer programs that generate the Mandelbrot set calculate the iterates up to a predecided point in the sequence, such as $f^{50}(0)$, and if no iterate to this point exceeds 2 in magnitude the number c is admitted to the set. The magnitude of a complex number $a + bi$ is the distance between the point (a, b) and the origin, so the formula for the magnitude of $a + bi$ is $\sqrt{a^2 + b^2}$.

Use the above criterion to determine, without a calculator or computer, if the given complex number is in the Mandelbrot set or not.

101. $c = 0$

102. $c = 1$

103. $c = i$

104. $c = -1$

105. $c = 1 + i$

106. $c = -2$

107. $c = 1 - i$

108. $c = -1 - i$

109–112 *True or False?* Determine whether the given statement is true or false. In case of a false statement, explain or provide a counterexample.

109. The graph of any quadratic polynomial is a transformation of the prototypical parabola.

110. The graphs of $y = f(x)$ and $y = f(-x)$ are reflection images of each other.

111. A cubic function can have up to three x-intercepts.

112. If $f(x)$ is an algebraic function and c is a nonzero constant, then $f(cx) = cf(x)$.

1.3 **Technology Exercises**

113–118 Mentally sketch the graph of the given function by identifying the basic shape that has been shifted, reflected, stretched, or compressed. Then use a graphing calculator or computer algebra system to graph the function and check your reasoning.

113. $f(x) = -2(3-x)^3 + 5$

114. $f(x) = \dfrac{3}{x+5} - 1$

115. $f(x) = \dfrac{-1}{(x-2)^2} - 3$

116. $f(x) = -3|x+2| - 4$

117. $f(x) = -\sqrt{1-x} + 2$

118. $f(x) = \sqrt[3]{2+x} - 1$

119–124 Write a possible equation for the function depicted on the graphing calculator. The function is shown in a $\left[-10, 10\right]$ by $\left[-10, 10\right]$ window.

119.

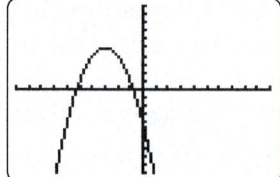

120.

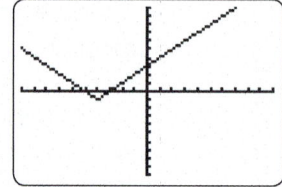

121.

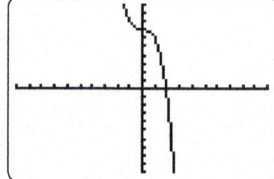

122.

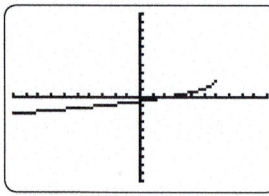

123.

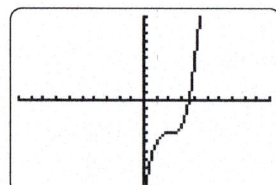

124.

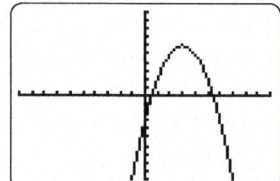

125–127 Use a computer algebra system to determine $(f+g)(x)$, $(fg)(x)$, $(f \circ g)(x)$ and $(g \circ f)(x)$ for the given pair of functions.

125. $f(x) = (3x+2)^2$; $g(x) = \sqrt{x^2 + 5}$

126. $f(x) = \dfrac{1}{3x-5}$; $g(x) = (x+2)^3$

127. $f(x) = \dfrac{x+1}{x-1}$; $g(x) = \dfrac{x-1}{x}$

1.4 **Inverse Functions**

In much of mathematics, the act of "undoing" one or more mathematical operations plays a critical role. For instance, to solve the equation $3x + 2 = 8$, the first step is to "undo" the addition of 2 on the left-hand side (by subtracting 2 from both sides) and the second step is to "undo" the multiplication by 3 (by dividing both sides by 3). In the context of more complex problems, the "undoing" process is often a matter of finding and applying the inverse of a function.

We begin our discussion with the more general idea of the inverse of a relation. Recall that a relation is just a set of ordered pairs; the inverse of a given relation is the set of these ordered pairs with the first and second coordinates exchanged.

Topic One
Inverse Relations and Inverse Functions

Definition

Inverse of a Relation

Let R be a relation. The **inverse of R**, denoted R^{-1}, is the set

$$R^{-1} = \left\{ (b,a) \middle| (a,b) \in R \right\}.$$

Note that this automatically implies the following:

$$\text{Domain of } R^{-1} = \text{Range of } R$$
$$\text{Range of } R^{-1} = \text{Domain of } R$$

Example 1

Determine the inverse of each of the following relations. Then graph each relation and its inverse, and determine the domain and range of both.

a. $R = \left\{ (4,-1), (-3,2), (0,5) \right\}$ **b.** $y = x^2$

Solution

a. $R = \left\{ (4,-1), (-3,2), (0,5) \right\}$ $R^{-1} = \left\{ (-1,4), (2,-3), (5,0) \right\}$

   Domain of $R = \left\{ 4, -3, 0 \right\}$ Domain of $R^{-1} = \left\{ -1, 2, 5 \right\}$

   Range of $R = \left\{ -1, 2, 5 \right\}$ Range of $R^{-1} = \left\{ 4, -3, 0 \right\}$

In Figure 1, R is in blue and its inverse is in red. The relation R consists of three ordered pairs, and its inverse is simply these three ordered pairs with the coordinates exchanged. Note that the domain of R is the range of R^{-1}, and vice versa.

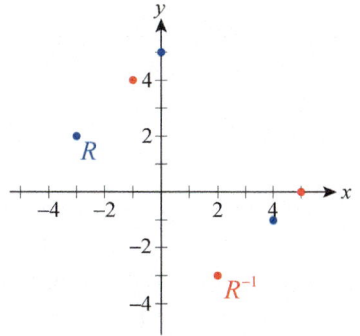

Figure 1

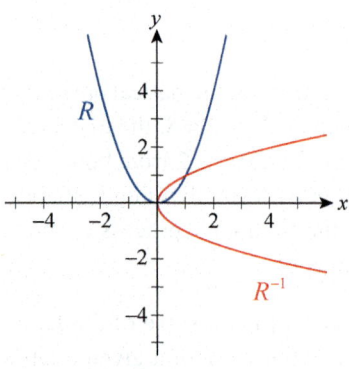

Figure 2

b. $R = \left\{ (x, y) \,\middle|\, y = x^2 \right\}$ $R^{-1} = \left\{ (x, y) \,\middle|\, x = y^2 \right\}$

Domain of $R = \mathbb{R}$ Domain of $R^{-1} = [0, \infty)$

Range of $R = [0, \infty)$ Range of $R^{-1} = \mathbb{R}$

In this problem, R is described by the given equation in x and y. The inverse relation is the set of ordered pairs in R with the coordinates exchanged, so we can describe the inverse relation just by exchanging x and y in the equation. The graphs of R and R^{-1} are shown in Figure 2.

Consider the graphs of the two relations and their respective inverses in Example 1. By definition, an ordered pair (b, a) lies on the graph of the relation R^{-1} if and only if (a, b) lies on the graph of R, so it shouldn't be surprising that the graphs of a relation and its inverse bear some resemblance to one another. Specifically, they are mirror images of one another with respect to the line $y = x$. If you were to fold the Cartesian plane in half along the line $y = x$ in the two examples above, you would see that the points in R and R^{-1} coincide with one another.

The two relations in Example 1 illustrate another important point. Note that in both cases, R is a function, as its graph passes the vertical line test. By the same criterion, R^{-1} in Example 1a is also a function, but R^{-1} in Example 1b is not. The conclusion to be drawn is that even if a relation is a function, its inverse may or may not be a function.

With a bit more thought, we can draw a stronger conclusion from Example 1 about when the inverse of a relation is a function.

In practice, we will only be concerned with the question of whether the inverse of a function f, denoted f^{-1}, is itself a function. Note that f^{-1} has already been defined: f^{-1} stands for the inverse of f, where we are making use of the fact that a function is also a relation.

Caution!

We are faced with another example of reuse of notation. f^{-1} does not stand for $1/f$! We use an exponent of -1 to indicate the reciprocal of a number or an algebraic expression, but when applied to a function or a relation it stands for the inverse relation.

Assume that f is a function. The inverse f^{-1} will only be a function in its own right if its graph passes the vertical line test; that is, only if each element of the domain of f^{-1} is paired with exactly one element of the range of f^{-1}. But this criterion is identical to saying that each element of the range of f is paired with exactly one element of the domain of f. In other words, every *horizontal* line in the Cartesian plane must intersect the graph of f no more than once. We say that functions meeting this condition pass the horizontal line test. (See Figure 3.)

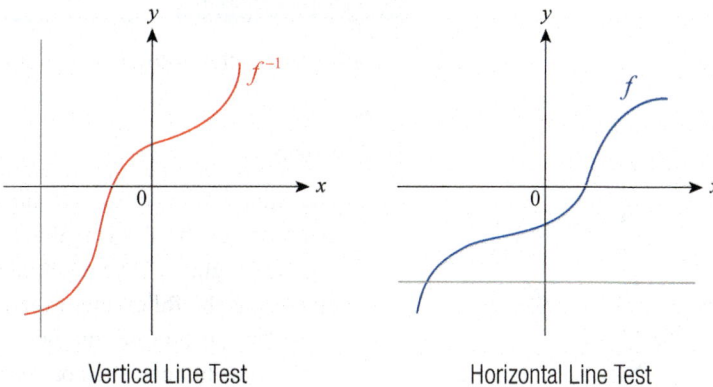

Vertical Line Test   Horizontal Line Test

Figure 3

Definition

The Horizontal Line Test

Let f be a function. We say that the graph of f passes the **horizontal line test** if every horizontal line in the plane intersects the graph no more than once. The inverse of f is a function only if f passes the horizontal line test.

The horizontal line test is only useful if the graph of f is available to study. We can also phrase the above condition in a nongraphical manner as follows. The inverse of f will only be a function if for every pair of distinct elements x_1 and x_2 in the domain of f, we have $f(x_1) \neq f(x_2)$. This criterion is important enough to merit a name.

Definition

One-to-One Functions

A function f is **one-to-one** if for every pair of distinct elements x_1 and x_2 in the domain of f, we have $f(x_1) \neq f(x_2)$. This means that every element of the range of f is paired with exactly one element of the domain of f.

If we now examine Example 1 again, we see that the function R in Example 1a is one-to-one, and so we know that its inverse is also a function. On the other hand, the function R in Example 1b is not one-to-one (plenty of horizontal lines pass through the graph twice), so its inverse is not a function.

Example 2

Determine if the following functions have inverse functions.

a. $f(x) = |x|$

b. $g(x) = (x+2)^3$

Solution

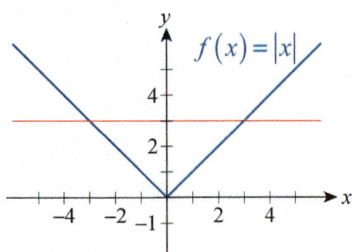

Figure 4

a. The function f does not have an inverse function, a fact easily demonstrated by showing that its graph does not pass the horizontal line test (see Figure 4). An algebraic proof that f does not have an inverse function is the following: even though $-3 \neq 3$, we have $f(-3) = f(3)$. There are an infinite number of pairs of numbers that show f is not one-to-one, but one such pair is all it takes to show that f does not have an inverse function.

b. The graph of g is the standard cubic shape shifted horizontally 2 units to the left. We can see that this graph passes the horizontal line test, so g has an inverse function (see Figure 5). But again, it is good practice to prove this algebraically. Note how each line in the following argument implies the next line.

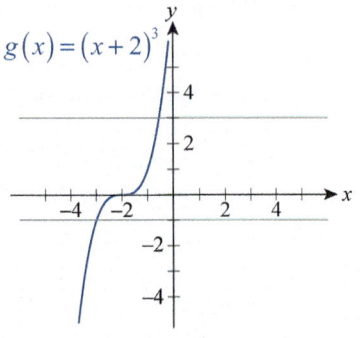

Figure 5

$$x_1 \neq x_2 \Rightarrow x_1 + 2 \neq x_2 + 2$$
$$\Rightarrow (x_1 + 2)^3 \neq (x_2 + 2)^3$$
$$\Rightarrow g(x_1) \neq g(x_2)$$

This argument shows that any two distinct elements of the domain of g lead to different values when plugged into g, so g is one-to-one and hence has an inverse function.

Topic Two

Finding Inverse Functions

In applying the notion of the inverse of a function, we will most often begin with a formula for f and want to find a formula for f^{-1}. This will allow us, for instance, to transform equations of the form

$$f(x) = y$$

into the form

$$x = f^{-1}(y).$$

Before we discuss the general algorithm for finding a formula for f^{-1}, consider the problem with which we began this section. If we define $f(x) = 3x + 2$, the equation $3x + 2 = 8$ can be written as $f(x) = 8$. Note that f is one-to-one, so f^{-1} does exist (as a function). If we can find a formula for f^{-1}, we can transform the equation into $x = f^{-1}(8)$. This is an overly complicated way to solve this equation, but it illustrates the point well.

What should the formula for f^{-1} be? Consider what f does to its argument. The first action is to multiply x by 3, and then the second is to add 2. To "undo" f, we need to negate these two actions in reverse order: subtract 2 and then divide the result by 3.

$$f^{-1}(x) = \frac{x-2}{3}$$

Applying this to the problem at hand, we obtain the following.

$$x = f^{-1}(8) = \frac{8-2}{3} = 2$$

This method of analyzing a function f and then finding a formula for f^{-1} by undoing the actions of f in reverse order is conceptually important, and works for simple functions. For other functions, however, the following algorithm may be necessary.

Finding the Inverse of a Function

Let f be a one-to-one function, and assume that f is defined by a formula. To find a formula for f^{-1}, perform the following steps.

Step 1: Replace $f(x)$ in the definition of f with the variable y. The result is an equation in x and y that is solved for y at this point.

Step 2: Interchange x and y in the equation.

Step 3: Solve the new equation for y.

Step 4: Replace the y in the resulting equation with $f^{-1}(x)$.

Example 3

Find the inverse of each of the following functions.

a. $f(x) = (x-1)^3 + 2$ **b.** $g(x) = \dfrac{x-3}{2x+1}$

Solution

a. We can find the inverse of this function either by the algorithm or by undoing the actions of f in reverse order. The function f subtracts 1 from x, cubes the result, and adds 2; its inverse will first subtract 2, take the cube root of the result, and then add 1.

$$f(x) = (x-1)^3 + 2$$
$$f^{-1}(x) = \sqrt[3]{(x-2)} + 1$$

b. $g(x) = \dfrac{x-3}{2x+1}$ The inverse of the function g is most easily found by the algorithm.

$y = \dfrac{x-3}{2x+1}$ Step 1: Replace $g(x)$ with y.

$x = \dfrac{y-3}{2y+1}$ Step 2: Interchange x and y in the equation.

$x(2y+1) = y-3$ Step 3: Solve the equation for y.

$2xy + x = y - 3$

$2xy - y = -x - 3$

$y(2x-1) = -x-3$

$y = \dfrac{-x-3}{2x-1}$

$g^{-1}(x) = \dfrac{-x-3}{2x-1}$ Step 4: Replace y in the resulting equation with $g^{-1}(x)$.

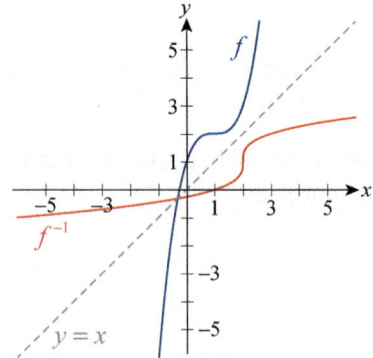

Figure 6
Graph of a Function and Its Inverse

Remember that the graphs of a relation and its inverse are mirror images of one another with respect to the line $y = x$; this is still true if the relations are functions. We can demonstrate this fact by graphing the function and its inverse from Example 3a, as shown in Figure 6.

We can use the functions and their inverses from Example 3 to illustrate one last important point. The key characteristic of the inverse of a function is that it undoes the function. This means that if a function and its inverse are composed together, in either order, the resulting function has no effect on any allowable input.

Properties of Inverse Functions

$f^{-1}(f(x)) = x$ for all x in the domain of f

$f(f^{-1}(x)) = x$ for all x in the domain of f^{-1}

For example, given $g(x) = \dfrac{x-3}{2x+1}$ and $g^{-1}(x) = \dfrac{-x-3}{2x-1}$ (from Example 3b), we have the following:

$$g^{-1}(g(x)) = g^{-1}\left(\frac{x-3}{2x+1}\right) = \frac{-\dfrac{x-3}{2x+1} - 3}{2\left(\dfrac{x-3}{2x+1}\right) - 1}$$

$$= \left(\frac{-\dfrac{x-3}{2x+1} - 3}{2\left(\dfrac{x-3}{2x+1}\right) - 1}\right)\left(\frac{2x+1}{2x+1}\right)$$

$$= \frac{-x+3-6x-3}{2x-6-2x-1} = \frac{-7x}{-7} = x$$

A similar calculation shows that $g(g^{-1}(x)) = x$, as you should verify.

Topic Three
Logarithms as Inverse Functions

One of the best reasons for the existence of logarithmic functions is simply the usefulness of inverse functions. Consider the two simple equations $2^x = 8$ and $2^x = 9$. The first is quickly solved by inspection, and it's easy to guess and verify that $x = 3$ is a solution. With a little more effort and the knowledge that the graph of 2^x is one-to-one (by the horizontal line test), it's even possible to prove that $x = 3$ is in fact the only solution to the equation $2^x = 8$; any other value for x results in 2^x being either less or more than 8. But how do we solve $2^x = 9$?

We can begin to make progress by rephrasing the equation in function form as follows: define the function f by $f(x) = 2^x$ and let y denote an arbitrary real number. Then what we are actually trying to do is solve the equation $f(x) = y$ for the variable x. But that's exactly the process of finding the inverse formula for the function f! If we can determine f^{-1}, we will use the observation that $f(x) = y \Leftrightarrow x = f^{-1}(y)$ to calculate x for a given y. Note that we've already determined that $3 = f^{-1}(8)$; we just need to calculate $f^{-1}(9)$ in order to solve the second equation.

Unfortunately, we've now progressed as far as we can algebraically—it is not possible to express f^{-1} as an algebraic function. We are forced to introduce the class of transcendental functions called logarithms in order to proceed.

Definition

Logarithmic Functions

Let a be a fixed positive real number not equal to 1. The logarithmic function with base a is defined to be the inverse of the exponential function with base a and is denoted $\log_a x$. In symbols,

$$\text{if } f(x) = a^x, \text{ then } f^{-1}(x) = \log_a x.$$

In equation form, the definition of logarithm implies that

$$a^x = y \Leftrightarrow x = \log_a y.$$

Note that a is the base in both equations; it serves as either the base of the exponential function or the base of the logarithmic function.

This logarithmic notation does not do much by itself until we build up a degree of familiarity with the behavior of logarithmic functions and discover their properties. For instance, we can begin by noting that since the range of every exponential function is $(0, \infty)$, this set must be the domain of every logarithmic function. Similarly, the range of any logarithmic function is the domain of its corresponding exponential function, which is $(-\infty, \infty)$. But probably the best way to become familiar with logarithmic functions is to gain experience with their graphs.

Recall that the graphs of a function and its inverse are reflections of one another with respect to the line $y = x$. Since exponential functions come in two forms ($0 < a < 1$ and $a > 1$), logarithmic functions fall into two similar categories. In each of the graphs in Figure 7, the red curve is the graph of an exponential function representative of its class, and the blue curve is the corresponding logarithmic function.

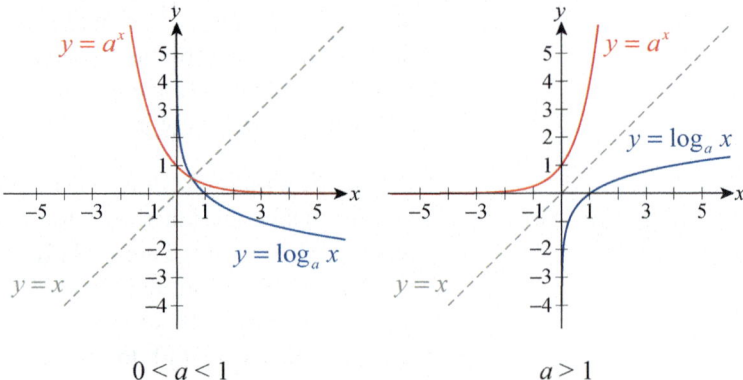

Figure 7 The Two Classes of Logarithmic Functions

Note that the domain of each of the logarithmic functions in Figure 7 is indeed $(0, \infty)$, and that the range in each case is $(-\infty, \infty)$. Also note that the y-axis is a vertical asymptote for both, and that neither has a horizontal asymptote.

We can shift, reflect, stretch, and compress the graphs of logarithms as we can with any other function. The techniques summarized in Section 1.3 are all we need to do the work in Example 4.

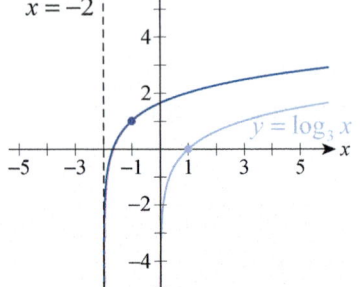

$f(x) = \log_3(x+2) + 1$

Figure 8

$g(x) = \log_2(-x-1)$

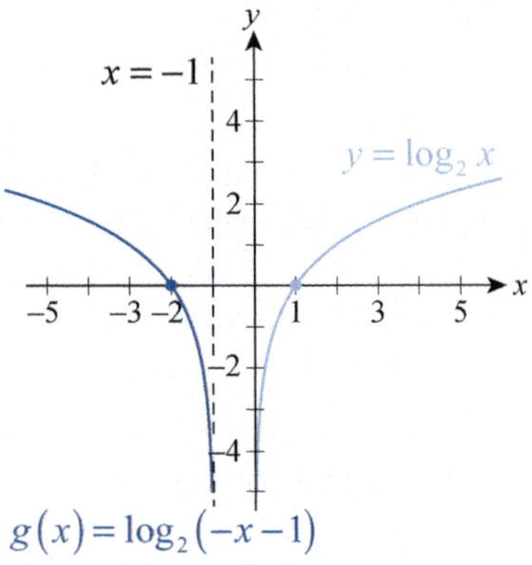

Figure 9

Example 4

Sketch the graphs of the following functions.

a. $f(x) = \log_3(x+2) + 1$ **b.** $g(x) = \log_2(-x-1)$

c. $h(x) = \log_{1/2} x - 2$

Solution

a. Begin by graphing $y = \log_3 x$, which is the basis for the graph of f. We know several points on the graph of $\log_3 x$ exactly; for instance, $(1,0)$ and $(3,1)$ are on the graph of $\log_3 x$.

Since x has been replaced by $x + 2$, we shift this graph 2 units to the left. To find the graph of f, we then shift the result 1 unit up. Note that the vertical asymptote has also shifted to the left (see Figure 8).

b. Again, we start with the basic shape of the graph, and then worry about the transformations. The basic shape of the graph of g is the same as the shape of $y = \log_2 x$. As above, some points on $\log_2 x$ are easily determined, such as $(1,0)$ and $(2,1)$.

To obtain g from $\log_2 x$, the variable x is replaced with $x - 1$, which shifts the graph 1 unit to the right, and then x is replaced by $-x$, which reflects the graph with respect to the y-axis (see Figure 9).

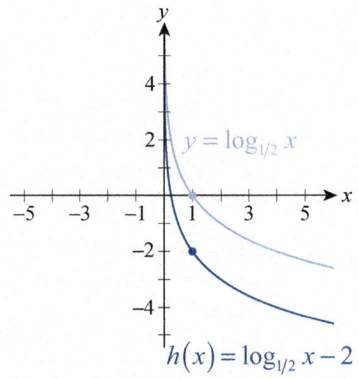

Figure 10

c. We begin with the graph of $y = \log_{1/2} x$. This is a decreasing function, as the base is between 0 and 1. Note that $(1,0)$ and $(1/2,1)$ are two points on the graph of $\log_{1/2} x$.

If we then shift the graph 2 units down, we obtain the graph of the function h (see Figure 10).

Now that we have graphed logarithmic functions, we can augment our understanding of their behavior with a few algebraic observations. First, our work in Example 4 certainly suggests that the point $(1,0)$ is always on the graph of $\log_a x$ for any allowable base a, and this is indeed the case. A similar observation is that $(a,1)$ is always on the graph of $\log_a x$. These two facts are nothing more than restatements of two corresponding facts about exponential functions and are a consequence of the definition of logarithms.

$$\log_a 1 = 0 \text{ because } a^0 = 1$$

$$\log_a a = 1 \text{ because } a^1 = a$$

More generally, we can use the fact that the functions $\log_a x$ and a^x are inverses of one another to write the following.

$$\log_a \left(a^x \right) = x \text{ and } a^{\log_a x} = x$$

We have already mentioned in Section 1.2 the fact that the number e, called the natural base, plays a fundamental role in many important real-world situations, so it is perhaps not surprising that the logarithmic function with base e is worthy of special attention. For historical reasons, the logarithmic function with base 10 is also singled out for further mention.

Definition

Common and Natural Logarithms

- The function $\log_{10} x$ is called the **common logarithm** and is usually written as $\log x$.

- The function $\log_e x$ is called the **natural logarithm** and is usually written as $\ln x$.

The next task to accomplish regarding logarithms is to review some of their algebraic properties. The following table lists three general properties of logarithms alongside the corresponding properties of exponents.

Theorem

Properties of Logarithms and Exponents

In each of the following statements a is assumed to be a positive real number not equal to 1. For the properties of logarithms, x and y represent positive real numbers and r is an arbitrary real number. For the properties of exponents, x and y represent arbitrary real numbers.

	Logarithmic Property	Exponential Property
1.	$\log_a (xy) = \log_a x + \log_a y$	$a^{x+y} = a^x a^y$
2.	$\log_a \dfrac{x}{y} = \log_a x - \log_a y$	$a^{x-y} = \dfrac{a^x}{a^y}$
3.	$\log_a (x^r) = r \log_a x$	$\left(a^x\right)^r = a^{xr}$

We will illustrate the correspondence between the properties of logarithms and the more familiar properties of exponents by proving the first one. The proofs of the second and third will be left as exercises.

Proof

We will start with the right-hand side of the statement and convert the expressions to exponential form. To do this, let $m = \log_a x$ and $n = \log_a y$. The equivalent exponential form of these two equations is:

$$x = a^m \text{ and } y = a^n.$$

Since we are interested in the product xy, note that

$$xy = a^m a^n = a^{m+n}.$$

The statement $xy = a^{m+n}$ can be converted to logarithmic form, giving us

$$\log_a (xy) = m + n.$$

If we now refer back to the original definition of m and n, we have achieved our goal:

$$\log_a (xy) = \log_a x + \log_a y.$$

The properties of logarithms may appear strange at first, but with time they will come to seem as natural as the properties of exponents. In fact, they are the properties of exponents, simply restated in logarithmic form.

In some situations, we will find it useful to use properties of logarithms to decompose a complicated expression into a sum or difference of simpler expressions, while in other situations we will do just the reverse, combining a sum or a difference of logarithms into one logarithm. Examples 5 and 6 illustrate these processes.

Example 5

Use the properties of logarithms to expand the following expressions as much as possible (that is, decompose the expressions into sums or differences of the simplest possible terms).

a. $\log_a \sqrt[3]{\dfrac{xy^2}{z^4}}$

b. $\log \dfrac{2.7 \times 10^4}{x^{-2}}$

Solution

a. $\log_a \sqrt[3]{\dfrac{xy^2}{z^4}} = \log_a \left(\left(\dfrac{xy^2}{z^4} \right)^{1/3} \right)$ Rewrite the radical as an exponent.

$\qquad\qquad = \dfrac{1}{3} \log_a \dfrac{xy^2}{z^4}$ Bring the exponent in front of the logarithm using the third property.

$\qquad\qquad = \dfrac{1}{3} \left(\log_a x + \log_a \left(y^2 \right) - \log_a \left(z^4 \right) \right)$ Expand the expression using the first two properties.

$\qquad\qquad = \dfrac{1}{3} \left(\log_a x + 2 \log_a y - 4 \log_a z \right)$ Apply the third property to the last two terms.

Note that the base is immaterial: a can represent any legitimate logarithmic base.

b. Recall that if the base is not explicitly written, it is assumed to be 10. This is a convenient base when dealing with scientific notation. Note that we can use a calculator to approximate $\log 2.7$.

$$\log \dfrac{2.7 \times 10^4}{x^{-2}} = \log 2.7 + \log \left(10^4 \right) - \log \left(x^{-2} \right)$$

$$= \log 2.7 + 4 + 2 \log x$$

$$\approx 4.43 + 2 \log x$$

Example 6

Use the properties of logarithms to condense the following expressions as much as possible (that is, rewrite the expressions as a sum or difference of as few logarithms as possible).

a. $2 \log_3 \dfrac{x}{3} - \log_3 \dfrac{1}{y}$

b. $\ln \left(x^2 \right) - \dfrac{1}{2} \ln y + \ln 2$

Solution

a. Note that before a sum or difference of log terms can be combined, they have to have the same coefficient. This is usually straightforward to arrange since the coefficient can be moved up into the exponent.

$$2\log_3\frac{x}{3} - \log_3\frac{1}{y} = \log_3\left(\left(\frac{x}{3}\right)^2\right) + \log_3\left(\left(\frac{1}{y}\right)^{-1}\right)$$

Use the third property to make the coefficients appear as exponents.

$$= \log_3\frac{x^2}{9} + \log_3 y$$

Evaluate the exponents.

$$= \log_3\frac{x^2 y}{9}$$

Combine terms using the first property.

Note also that the order in which the properties are applied can vary, though all lead to the same final answer.

b. Again, we begin by rewriting each term so that its coefficient is 1 or −1; the sums or differences that result can then be combined.

$$\ln(x^2) - \frac{1}{2}\ln y + \ln 2 = \ln(x^2) - \ln(y^{1/2}) + \ln 2$$

$$= \ln\frac{x^2}{y^{1/2}} + \ln 2$$

$$= \ln\frac{2x^2}{\sqrt{y}}$$

It should be noted that most calculators, if they are capable of calculating logarithms at all, are only equipped to evaluate common and natural logarithms. Such calculators normally have a button labeled "LOG" for the common logarithm and a button labeled "LN" for the natural logarithm.

It is important to note that we can certainly use calculators to find the decimal form of a logarithmic expression with any base a. To illustrate, we will show the thought process for finding $\log_2 9$.

$$x = \log_2 9$$

Let x stand for the number $\log_2 9$.

$$2^x = 9$$

Convert to exponential form.

$$\ln(2^x) = \ln 9$$

Take the natural logarithm of both sides.

$$x\ln 2 = \ln 9$$

Apply the third property of logarithms.

$$x = \frac{\ln 9}{\ln 2}$$

Solve for x.

$$x \approx 3.17$$

There is nothing special about the natural logarithm, at least as far as this problem is concerned. However, if a calculator is to be used to approximate the number $\log_2 9$, there are (for most calculators) only two good choices: the natural logarithm and the common logarithm. If we had done the work above with the common logarithm, the final answer would have been the same.

$$x = \frac{\log 9}{\log 2} \approx 3.17$$

More generally, a logarithm with base b can be converted to a logarithm with base a through the same reasoning, as summarized below.

Change of Base Formula

Let a and b be positive real numbers, neither of them equal to 1, and let x be a positive real number. Then

$$\log_b x = \frac{\log_a x}{\log_a b}.$$

Topic Four

Inverse Trigonometric Functions

The rationale for inverse trigonometric functions is the rationale for inverses of functions in general. In many situations, we will want to find an angle having a certain specified property, and our method will be to "undo" the action of a given trigonometric function. As a simple example, suppose we need to find an acute angle θ for which $\sin\theta = 1/2$. Chances are you have enough experience with right triangles to recall that $\sin(\pi/6) = 1/2$, so it must be the case that $\theta = \pi/6$. But what if we seek an angle φ for which $\sin\varphi = 0.7$? The problem is similar, but we don't yet have a way to determine φ.

Recall, however, that a function will have an inverse only if it is one-to-one. Recall also that this means the graph of the function must pass the horizontal line test; this is something the trigonometric functions fail to do. Fortunately, there is a way out. By restricting the domain of a trigonometric function wisely, we can make it one-to-one and thus invertible. We will go through the process step by step for the sine function and then briefly show how the other trigonometric functions are dealt with similarly.

There are many ways we could restrict the domain of sine in order to make it one-to-one, but we are guided also by the desire to not lose more than we have to in the restriction. For instance, we could specify that we will only define sine over the interval $[0, \pi/2]$, but by doing so we prevent the newly defined function from ever taking on a negative value (note that $0 \le \sin x \le 1$ for $x \in [0, \pi/2]$). Figure 11 indicates that $[-\pi/2, \pi/2]$ is the largest interval containing $[0, \pi/2]$ that we could choose for the restricted domain; the red portion of the graph is one-to-one and takes on all values between -1 and 1.

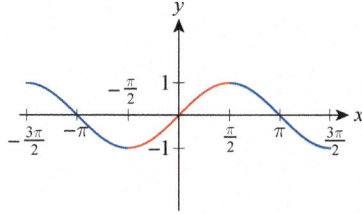

Figure 11
Restricting the Domain of Sine

In practice, context will tell us whether we want to think of sine as being defined on the entire real line or only over the interval $[-\pi/2, \pi/2]$, but the biggest hint will be whether we need to apply the inverse of the sine function. If so, the restricted domain for sine is called for.

Two notations are commonly used for the inverse trigonometric functions. In the case of sine, $\sin x = y$ is equivalent to the equations

$$x = \arcsin y \text{ and } x = \sin^{-1} y.$$

The arcsine notation derives from the fact that $\arcsin y$ is the length of the arc (on the unit circle) corresponding to the angle x. The $\sin^{-1} y$ notation is in keeping with our use of f^{-1} to stand for the inverse of the function f.

Definition

Arcsine

Given $x \in [-\pi/2, \pi/2]$, **arcsine** is defined by either of the following.

$$y = \sin x \Leftrightarrow x = \arcsin y \text{ or } y = \sin x \Leftrightarrow x = \sin^{-1} y$$

In words, $\arcsin y$ is the angle whose sine is y. Since the (restricted) domain of sine is $[-\pi/2, \pi/2]$ and its range is $[-1, 1]$, the domain of arcsine is $[-1, 1]$ and its range is $[-\pi/2, \pi/2]$.

The best way to finish up this introduction to arcsine is with a graph of the function. We already saw that the graphs of a function and its inverse are reflections of one another with respect to the line $y = x$, and this is all we need in order to generate the graph of arcsine. (See Figure 12.)

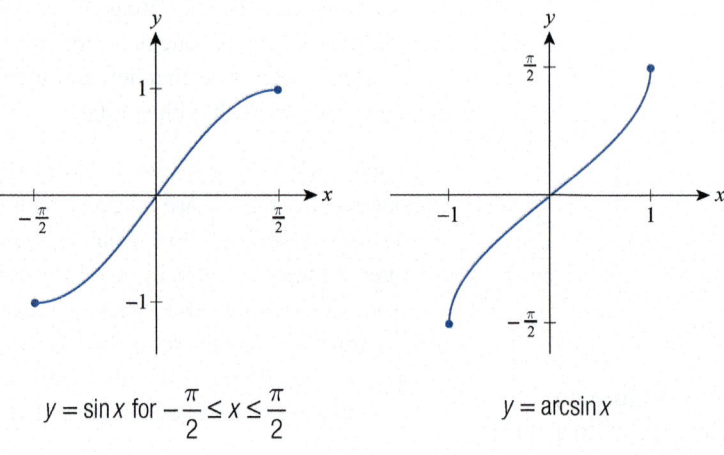

$$y = \sin x \text{ for } -\frac{\pi}{2} \leq x \leq \frac{\pi}{2} \qquad\qquad y = \arcsin x$$

Figure 12

Example 7

With the derivation of arcsine as a guide, construct a definition of arccosine and plot the resulting function.

Solution

As with sine, we first need to restrict the domain of cosine to an interval over which cosine is one-to-one. Picture the graph of cosine in your mind. Most people would probably say that the natural choice for the restricted domain is the interval $[0, \pi]$, and this is indeed the convention. This is all we need in order to make our definition.

Given $x \in [0, \pi]$, **arccosine** (with its two notations) is defined by

$$y = \cos x \Leftrightarrow x = \arccos y \text{ or } y = \cos x \Leftrightarrow x = \cos^{-1} y.$$

To graph the arccosine function, we simply reflect the restricted graph of cosine with respect to the line $y = x$. Since the (restricted) domain of cosine is $[0, \pi]$ and the range is $[-1, 1]$, we know that the domain of arccosine will be $[-1, 1]$ and its range will be $[0, \pi]$. This knowledge serves as a good way to double-check our graph of arccosine.

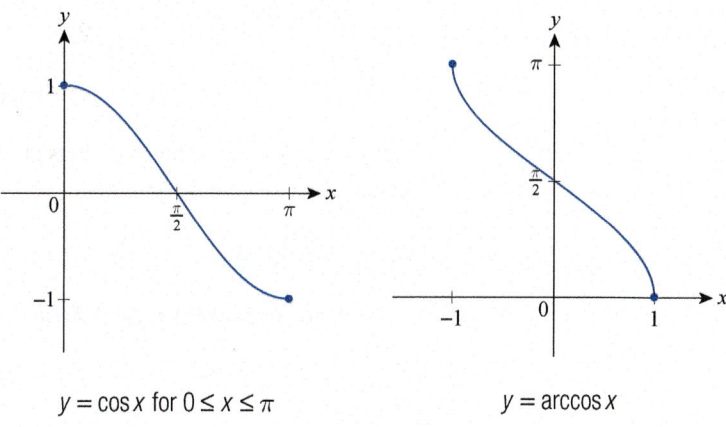

$y = \cos x$ for $0 \le x \le \pi$ $y = \arccos x$

Figure 13

Arctangent is the third commonly encountered inverse trigonometric function, and its definition and graph are arrived at in a similar manner. Figure 14 illustrates the graph of the tangent function restricted to the interval $(-\pi/2, \pi/2)$ along with the graph of arctangent. Note the horizontal asymptotes in the graph of arctangent, corresponding to the vertical asymptotes in the graph of tangent.

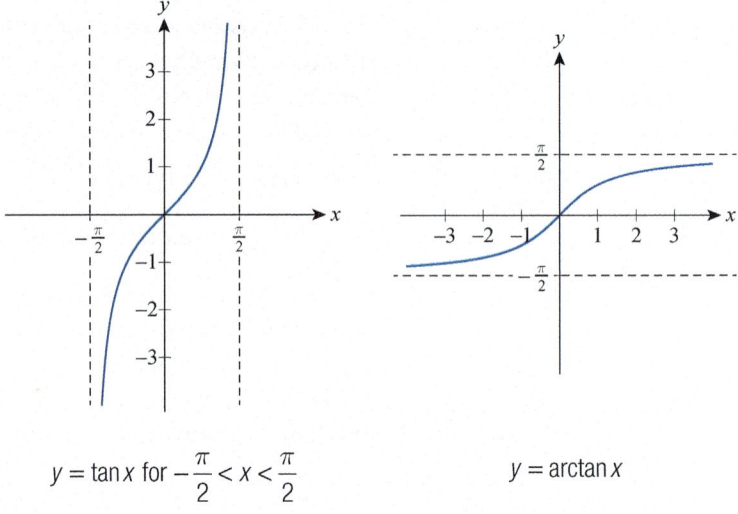

$$y = \tan x \text{ for } -\frac{\pi}{2} < x < \frac{\pi}{2}$$ $$y = \arctan x$$

Figure 14

The following box summarizes facts about the definitions, domains, and ranges of arcsine, arccosine, and arctangent.

Definition

Arcsine, Arccosine, and Arctangent

Function	Notation 1	Notation 2	Domain	Range
Inverse Sine	$\arcsin y = x \Leftrightarrow y = \sin x$	$\sin^{-1} y = x \Leftrightarrow y = \sin x$	$[-1, 1]$	$\left[-\dfrac{\pi}{2}, \dfrac{\pi}{2} \right]$
Inverse Cosine	$\arccos y = x \Leftrightarrow y = \cos x$	$\cos^{-1} y = x \Leftrightarrow y = \cos x$	$[-1, 1]$	$[0, \pi]$
Inverse Tangent	$\arctan y = x \Leftrightarrow y = \tan x$	$\tan^{-1} y = x \Leftrightarrow y = \tan x$	$(-\infty, \infty)$	$\left(-\dfrac{\pi}{2}, \dfrac{\pi}{2} \right)$

The evaluation of inverse trigonometric functions can take several forms, depending on context. One meaning is the actual numerical evaluation of an expression containing an inverse trig function; this may or may not require the use of a calculator. Another meaning is the simplification of expressions containing inverse trig functions, using nothing more than our knowledge of how functions and their inverses behave in relation to one another. We will begin with some numerical examples.

Example 8

Evaluate the following expressions, if possible.

a. $\arcsin\left(\sin\dfrac{3\pi}{4}\right)$

b. $\cos\left(\cos^{-1}(-0.2)\right)$

Solution

a. The potential error in this problem is to assume that $\arcsin\left(\sin(3\pi/4)\right)=3\pi/4$, since arcsin and sin are inverse functions of one another. But $3\pi/4$ lies outside the range of arcsin, which is $\left[-\pi/2,\pi/2\right]$ so we know this can't be the answer. The key is to evaluate the expressions individually:

$$\sin\left(\frac{3\pi}{4}\right)=\frac{1}{\sqrt{2}}$$

$$\arcsin\left(\frac{1}{\sqrt{2}}\right)=\frac{\pi}{4}$$

b. The number -0.2 lies in the domain of arccosine, and all real numbers lie in the domain of cosine, so all the parts of the expression $\cos\left(\cos^{-1}(-0.2)\right)$ make sense and we are safe in stating $\cos\left(\cos^{-1}(-0.2)\right)=-0.2$. If we wanted to explore the expression a bit further, we could note that, from the graph of arccosine, $\cos^{-1}(-0.2)$ is some positive number, and further that $\cos^{-1}(-0.2)$ must be greater than $\pi/2$ since $\cos\left(\cos^{-1}(-0.2)\right)$ is negative. This is indeed the case: a calculator tells us that $\cos^{-1}(-0.2)$ is approximately 1.8.

The last example demonstrated the evaluation of compositions of trig functions with their inverses, but other compositions are also possible. In many cases, a picture aids greatly in the computation.

Example 9

Evaluate the following expressions.

a. $\tan\left(\sin^{-1}\left(-\dfrac{4}{5}\right)\right)$

b. $\cos\left(\arctan 0.4\right)$

Solution

a. Remember that the range of arcsin is $\left[-\pi/2,\pi/2\right]$, and in particular that $\sin^{-1}(-4/5)$ will lie between $-\pi/2$ and 0 (the graph tells us that arcsin of a negative number is negative). If we let $\theta=\sin^{-1}(-4/5)$, then $\sin\theta=-4/5$ and we can sketch the triangle shown in Figure 15 to illustrate the relationship between θ and the given numbers.

The Pythagorean Theorem allows us to calculate x.

$$x=\sqrt{5^2-(-4)^2}=\sqrt{9}=3$$

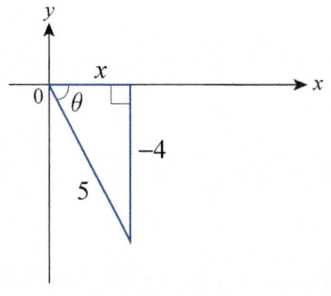

Figure 15

Now we can see that $\tan\theta = -4/3$, so

$$\tan\left(\sin^{-1}\left(-\frac{4}{5}\right)\right) = -\frac{4}{3}.$$

b. We can employ the same method and let $\theta = \arctan 0.4$. This leads to

$$\tan\theta = 0.4 = \frac{4}{10} = \frac{2}{5}$$

and then to the sketch, as seen in Figure 16.

The Pythagorean Theorem gives us $r = \sqrt{5^2 + 2^2} = \sqrt{29}$, and so $\cos(\arctan 0.4) = 5/\sqrt{29}$.

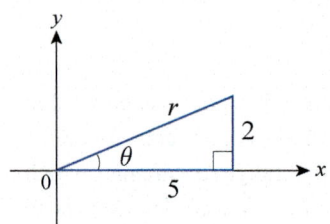

Figure 16

Example 10

Express $\sin\left(\cos^{-1} 2x\right)$ as an algebraic function of x, assuming $-1/2 \le x \le 1/2$.

Solution

Let $\theta = \cos^{-1} 2x$. Then $\cos\theta = 2x$ and we are led to consider a sketch like Figure 17.

In the sketch, we have chosen the simplest lengths for the adjacent side and the hypotenuse that make $\cos\theta = 2x$, though any positive multiples of these lengths would also work. And as always, once the lengths of two sides of the right triangle have been determined, the Pythagorean Theorem provides the length of the third side. Now we can refer to the sketch to see that

$$\sin\left(\cos^{-1} 2x\right) = \sin\theta = \frac{\sqrt{1-(2x)^2}}{1} = \sqrt{1-4x^2}.$$

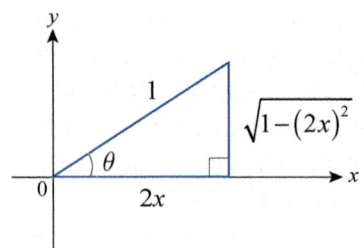

Figure 17

1.4 **Exercises**

1–12 Graph the inverse of the given relation, and state its domain and range.

1. $R = \{(-4,2),(3,2),(0,-1),(3,-2)\}$

2. $S = \{(-3,-3),(-1,-1),(0,1),(4,4)\}$

3. $y = x^3$

4. $y = |x| + 2$

5. $x = |y|$

6. $x = -\sqrt{y}$

7. $y = \dfrac{1}{2}x - 3$

8. $y = -x + 1$

9. $y = [\![x]\!]$

10. $T = \{(4,2),(3,-1),(-2,-1),(2,4)\}$

11. $x = y^2 - 2$

12. $y = 2\sqrt{x}$

13–22 Determine if the given function has an inverse function. If not, suggest a domain to restrict the function so that it would have an inverse function. (Answers will vary.)

13. $f(x) = x^2 + 1$

14. $g(x) = (x-2)^3 - 1$

15. $h(x) = \sqrt{x+3}$

16. $s(x) = \dfrac{1}{x^2}$

17. $G(x) = 3x - 5$

18. $F(x) = -x^2 + 5$

19. $r(x) = -\sqrt{x^3}$

20. $b(x) = [\![x]\!]$

21. $m(x) = \dfrac{13x - 2}{4}$

22. $H(x) = |x - 12|$

23–37 Find the inverse of the given function.

23. $f(x) = x^{1/3} - 2$

24. $g(x) = 4x - 3$

25. $r(x) = \dfrac{x-1}{3x+2}$

26. $s(x) = \dfrac{1-x}{1+x}$

27. $F(x) = (x-5)^3 + 2$

28. $G(x) = \sqrt[3]{3x-1}$

29. $V(x) = \dfrac{x+5}{2}$

30. $W(x) = \dfrac{1}{x}$

31. $h(x) = x^{3/5} - 2$

32. $A(x) = (x^3 + 1)^{1/5}$

33. $J(x) = \dfrac{2}{1-3x}$

34. $k(x) = \dfrac{x+4}{3-x}$

35. $h(x) = x^7 + 6$

36. $F(x) = \dfrac{3-x^5}{-9}$

37. $r(x) = \sqrt[5]{2x}$

38–45 Show that $f^{-1}(f(x)) = x$ and that $f(f^{-1}(x)) = x$.

38. $f(x) = 2x - 3$; $f^{-1}(x) = \dfrac{x+3}{2}$

39. $f(x) = x^2,\ x \geq 0$; $f^{-1}(x) = \sqrt{x}$

40. $f(x) = \dfrac{3x-1}{5}$; $f^{-1}(x) = \dfrac{5x+1}{3}$

41. $f(x) = \dfrac{x-5}{2x+3}$; $f^{-1}(x) = \dfrac{3x+5}{1-2x}$

42. $f(x) = (x-2)^2,\ x \geq 2$; $f^{-1}(x) = \sqrt{x} + 2,\ x \geq 0$

43. $f(x) = \sqrt[3]{x+2} - 1$; $f^{-1}(x) = (x+1)^3 - 2$

44. $f(x) = \dfrac{1}{x}$; $f^{-1}(x) = \dfrac{1}{x}$

45. $f(x) = \dfrac{1}{1+x},\ x \geq 0$; $f^{-1}(x) = \dfrac{1-x}{x},\ 0 < x \leq 1$

46–51 Match the function with the graph of the inverse of the function. The graphs are labeled A through F.

46. $f(x) = x^3$

47. $f(x) = x - 5$

48. $f(x) = \sqrt{x - 4}$

49. $f(x) = x^2$

50. $f(x) = \dfrac{x}{4}$

51. $f(x) = \sqrt[3]{x + 1}$

A.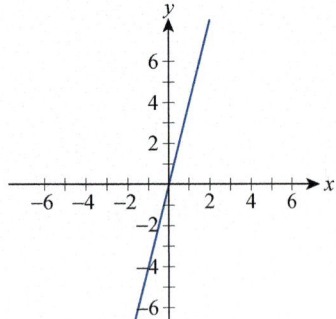

B.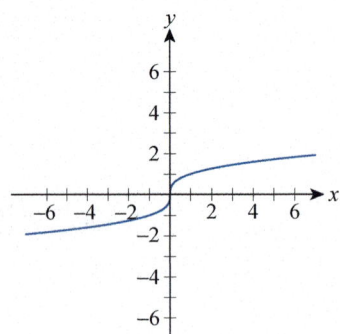

C.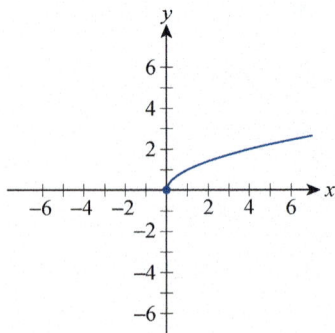

D.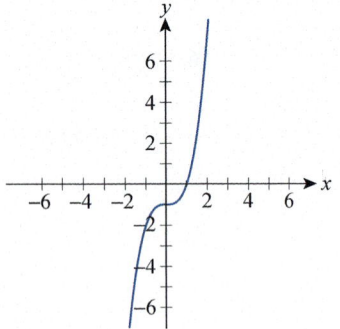

E.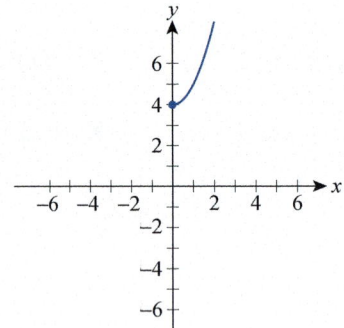

F.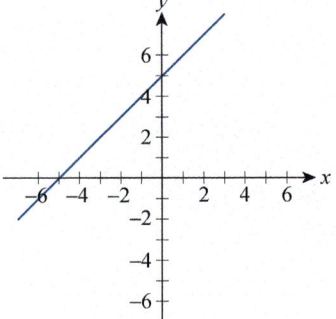

52–60 Match the logarithmic function with its graph. The graphs are labeled A through I.

52. $f(x) = \log_2 x - 1$

53. $f(x) = \log_2(2 - x)$

54. $f(x) = \log_2(-x)$

55. $f(x) = \log_2(x - 3)$

56. $f(x) = 1 - \log_2 x$

57. $f(x) = -\log_2 x$

58. $f(x) = -\log_2(-x)$

59. $f(x) = \log_2 x$

60. $f(x) = \log_2 x + 3$

A.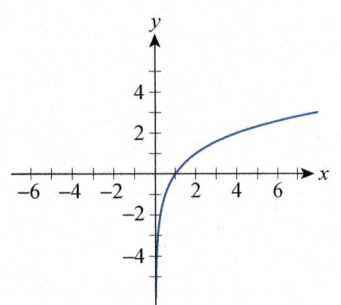

B.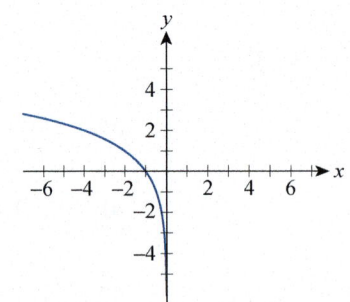

C.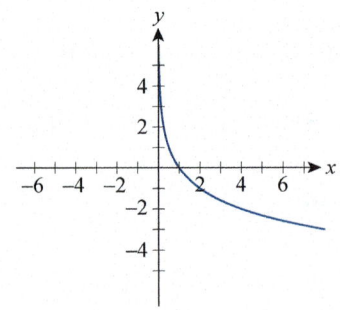

D.
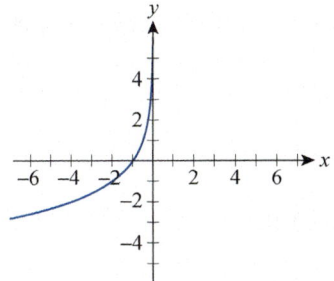

E.
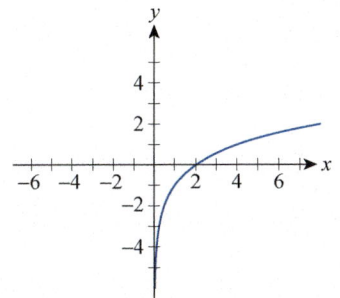

F.
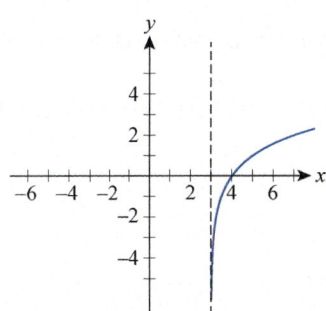

G.
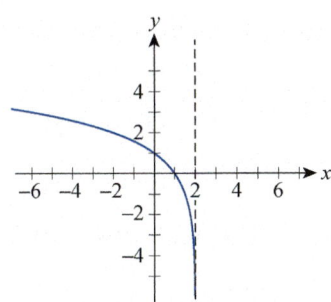

H.
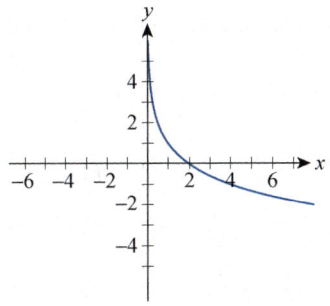

I.
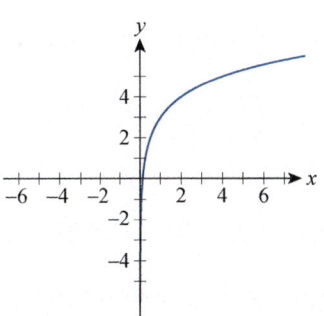

61–72 Sketch the graph of the given function.

61. $f(x) = \log_3(x-1)$

62. $g(x) = \log_5(x+2) - 1$

63. $r(x) = \log_{1/2}(x-3)$

64. $p(x) = 3 - \log_2(x+1)$

65. $q(x) = \log_3(2-x)$

66. $s(x) = \log_{1/3}(5-x)$

67. $h(x) = \log_7(x-3) + 3$

68. $m(x) = \log_{1/2}(1-x)$

69. $f(x) = \log_3(6-x)$

70. $p(x) = 4 - \log_{10}(x+3)$

71. $s(x) = -\log_{1/3}(-x)$

72. $g(x) = \log_5(2x) - 1$

73–78 Evaluate the given expression without using a calculator.

73. $\log_4 16$

74. $\log_5\left(25^3\right)$

75. $\ln\left(e^4\right) + \ln\left(e^3\right)$

76. $\log_4 \dfrac{1}{64}$

77. $\ln\left(e^{1.5}\right) - \log_4 2$

78. $\log_2\left(8^{2\log_2 4 - \log_2 4}\right)$

79–84 Evaluate the given logarithmic expression to two decimal places. (**Hint:** Use the change of base formula.)

79. $\log_6\left(3^4\right)$

80. $\log_7 14.3$

81. $\log_{1/2}\left(\pi^{-2}\right)$

82. $\log_{1/5} 626$

83. $\ln(\log 123)$

84. $\log_{17} 0.041$

85–90 Use the properties of logarithms to rewrite the given expression as a single term that does not contain a logarithm.

85. $5^{2\log_5 x}$

86. $\log_4 16 \cdot \log_x\left(x^2\right)$

87. $e^{2 - \ln x + \ln p}$

88. $e^{5\left(\ln \sqrt[3]{3} + \ln x\right)}$

89. $10^{\log\left(x^3\right) - 4\log y}$

90. $a^{\log_a b + 4\log_a \sqrt{a}}$

91–99 Use the properties of logarithms to expand the given expression as much as possible; that is, decompose the expression into sums or differences of the simplest possible terms. Simplify any numerical expressions that can be evaluated without a calculator.

91. $\ln \dfrac{\sqrt{x^3 pq^5}}{e^7}$

92. $\log_a \sqrt[5]{\dfrac{a^4 b}{c^2}}$

93. $\log\left(\log\left(100x^3\right)\right)$

94. $\log_3\left(9x + 27y\right)$

95. $\log \dfrac{10}{\sqrt{x+y}}$

96. $\ln\left(\ln\left(e^{ex}\right)\right)$

97. $\log_2 \dfrac{y^2 + z}{16x^4}$

98. $\log\left(\log\left(100{,}000^{2x}\right)\right)$

99. $\log_b \sqrt{\dfrac{x^4 y}{z^2}}$

100–105 Use the properties of logarithms to condense the given expression as much as possible, writing the answer as a single term with a coefficient of 1.

100. $\dfrac{1}{5}\left(\log_7\left(x^2\right) - \log_7\left(pq\right)\right)$

101. $\ln 3 + \ln p - 2\ln q$

102. $2\left(\log_5 \sqrt{x} - \log_5 y\right)$

103. $\log\left(x - 10\right) - \log x$

104. $2\log\left(a^2 b\right) - \log\dfrac{1}{b} + \log\dfrac{1}{a}$

105. $3\left(\ln \sqrt[3]{e^2} - \ln\left(xy\right)\right)$

106–111 Evaluate the given expression, if possible.

106. $\cos^{-1}\left(\cos\dfrac{2\pi}{4}\right)$

107. $\sin^{-1}\left(\sin\dfrac{3\pi}{2}\right)$

108. $\tan\left(\tan^{-1} 0.5\right)$

109. $\sin^{-1}\left(\sin\dfrac{7\pi}{6}\right)$

110. $\cos\left(\cos^{-1}\left(-0.8\right)\right)$

111. $\tan^{-1}\left(\tan\dfrac{5\pi}{4}\right)$

112–117 Most calculators are not equipped with arccosecant, arcsecant, and arccotangent buttons, but expressions involving these functions can still be evaluated. For example, to evaluate $\csc^{-1} x$, let $\theta = \csc^{-1} x$.

$$\csc\theta = x$$
$$\dfrac{1}{\sin\theta} = x$$
$$\sin\theta = \dfrac{1}{x}$$
$$\theta = \sin^{-1}\dfrac{1}{x}$$

Use the method described above to evaluate the given expression. (Round your answer to four decimal places.)

112. $\csc^{-1} 5$

113. $\sec^{-1}\left(-0.5\right)$

114. $\cot^{-1} 150$

115. $\cot^{-1}\left(-0.2\right)$

116. $\csc^{-1}\left(-8.9\right)$

117. $\sec^{-1} 2$

118–123 Find the value of the given expression without using a calculator.

118. $\sin\left(\arctan \sqrt{3}\right)$

119. $\cos\left(\sec^{-1}\left(-2\right)\right)$

120. $\tan\left(\operatorname{arccot} 1\right)$

121. $\csc\left(\arccos\left(-\dfrac{\sqrt{3}}{2}\right)\right)$

122. $\tan\left(\sin^{-1}\left(-\dfrac{\sqrt{2}}{2}\right)\right)$

123. $\sec\left(\csc^{-1}\dfrac{2\sqrt{3}}{3}\right)$

124–129 Rewrite the given function as a purely algebraic function.

124. $\tan\left(\cos^{-1} x\right)$

125. $\cot\left(\sin^{-1}\dfrac{2}{x}\right)$

126. $\sec\left(\tan^{-1} 3x\right)$

127. $\tan\left(\sin^{-1}\dfrac{x}{\sqrt{x^2 + 3}}\right)$

128. $\sin\left(\sec^{-1} x\right)$

129. $\cos\left(\tan^{-1}\dfrac{x}{4}\right)$

130–133 Sketch the graph of the given function. Then graph the function using a graphing calculator or computer algebra system to check your answer.

130. $f(x) = \sin^{-1}(x - 3)$ **131.** $f(x) = \sec^{-1} 2x$

132. $f(x) = \arctan \dfrac{x}{2}$ **133.** $f(x) = 2\arccos x$

134–137 An inverse function can be used to encode and decode words and sentences by assigning each letter of the alphabet a numerical value (A = 1, B = 2, C = 3, ..., Z = 26). Example: Use the function $f(x) = x^2$ to encode the word CALCULUS. The encoded message would be 9 1 144 9 441 144 441 361. The word can then be decoded by using the inverse function $f^{-1}(x) = \sqrt{x}$. The inverse values are 3 1 12 3 21 12 21 19, which translates back to the word CALCULUS.

Encode or decode the given message using the numerical values A = 1, B = 2, C = 3, ..., Z = 26.

134. Encode the message SANDY SHOES using the function $f(x) = 4x - 3$.

135. Encode the message WILL IT RAIN TODAY using the function $f(x) = x^2 - 8$.

136. The following message was encoded using the function $f(x) = 8x - 7$. Decode the message.

41 137 65 145 9 33 33 169 113 89 89 33
193 9 1 89 89 1 105 25 57 113 137 145 33
145 57 113 33 145

137. The following message was encoded using the function $f(x) = 5x + 1$. Decode the message.

91 26 66 26 66 11 26 91 126 76 106 91 96
106 71 11 61 76 16 56

138–139 The energy released during earthquakes can vary greatly, but logarithms provide a convenient way to analyze and compare the intensity of earthquakes. Earthquake intensity is measured on the Richter scale (named for the American seismologist Charles F. Richter, 1900–1985). In the formula that follows, I_0 is the intensity of a just-discernible earthquake, I is the intensity of an earthquake being analyzed, and R is its ranking on the Richter scale.

$$R = \log \frac{I}{I_0}$$

By this measure, earthquakes range from a classification of small ($R < 4.5$), to moderate ($4.5 \leq R < 5.5$), to large ($5.5 \leq R < 6.5$), to major ($6.5 \leq R < 7.5$), and finally to greatest ($R \geq 7.5$).

Use this information to solve the problem.

138. The 1994 Northridge, California earthquake measured 6.7 on the Richter scale. What was the intensity, relative to a 0-level earthquake, of this event?

139. The April, 2009 Abruzzo earthquake in Italy was 2,000,000 times as intense as a 0-level earthquake. What was the Richter ranking of this tragic event?

140–141 Sound intensity is another quantity that varies greatly, and the measure of how the human ear perceives intensity, in units called decibels, is very similar to the measure of earthquake intensity. If I_0 is the intensity of a just-discernible sound, I is the intensity of the sound being analyzed, and D is its decibel level, we have the formula $D = 10\log(I/I_0)$. Decibel levels range from 0 for a barely discernible sound, to 40 for the level of normal conversation, to 80 for heavy traffic, to 120 for a loud rock concert, and finally (as far as humans are concerned) to around 160, at which point the eardrum is likely to rupture.

Use the decibel formula given above to answer the question.

140. A construction worker operating a jackhammer would experience noise with an intensity of 20 watts/meter^2 if it weren't for ear protection. Given that $I_0 = 10^{-12}$ watts/meter^2, what is the decibel level for such noise?

141. The intensity of a cat's soft purring is measured to be 2.19×10^{-11} watts/meter^2. Given that $I_0 = 10^{-12}$ watts/meter^2, what is the decibel level of this noise?

142–143 Use inverse trigonometric functions to solve the problem. (Round your answer to four decimal places.)

142. Kim is watching a space shuttle launch from an observation spot 2 miles away from the launchpad. Find the angle of elevation to the shuttle for each of the following heights.

    a. 0.5 miles   **b.** 2 miles   **c.** 2.8 miles

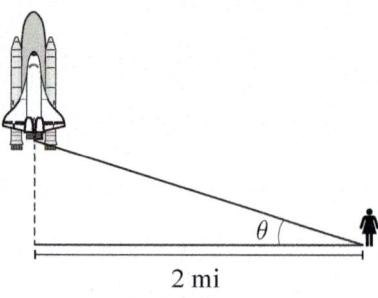

2 mi

143. Jesse is rowing in the men's singles race. The length of the oar from the side of the shell to the water is 7 feet. At what angle is the oar from the side of the boat when the blade is at the following distances from the boat?

    a. 2 feet   **b.** 3 feet   **c.** 5 feet

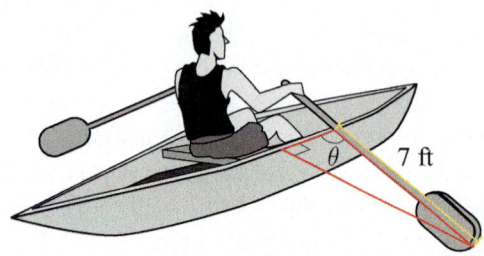

7 ft

144–151 *True or False?* Determine whether the given statement is true or false. In case of a false statement, explain or provide a counterexample.

144. All exponential functions are one-to-one.

145. $\sin(\arcsin x) = x$ for all $x \in [-1,1]$

146. $\arcsin(\sin x) = x$ for all $x \in \mathbb{R}$

147. $\tan(\arctan x) = x$ for all $x \in \mathbb{R}$

148. $\arccos(\cos(3\pi/2)) = 3\pi/2$

149. The domain of $\arcsin x$ is $[-\pi/2, \pi/2]$.

150. The domain of $f(x) = \cot^{-1} x$ is $\mathbb{R}$.

151. The function $f(x) = \sin(\tan^{-1} x)$ can be represented as an algebraic function.

1.5 **Calculus, Calculators, and Computer Algebra Systems**

The technology available at any given time inevitably colors the ways in which mathematics is learned and used. The historical interplay between calculus and technology is extensive, and no discourse on calculus would be complete without a discussion of computational tools. The tools available on graphing calculators and computers are the focus of this section.

Topic One
Graphs via Calculators and Computers

The graphing capabilities of modern calculators and mathematical software are an especially useful technological addition. While pictures can be misleading and must be used with a small amount of skepticism, there is no denying that the ability to quickly sketch curves and surfaces greatly speeds up the process of solving many problems. The details on the use of a particular calculator or software package are best left to the user's manual, but some features are common across all graphing technology.

One of the most basic features is the ability to choose the **display** or **viewing window** when graphing a function. When graphing functions from $\mathbb{R}$ to $\mathbb{R}$, this is simply the choice of the minimum and maximum values for the horizontal and vertical axes. For the purposes of illustration, we will refer to these values as *xMin*, *xMax*, *yMin*, and *yMax* (on some calculators, these are the exact labels for these quantities). The display window is then a rectangle in the plane bounded by these values, $[xMin, xMax]$ by $[yMin, yMax]$, and their choice determines the portion of the function being graphed. It is important to note that this choice effectively gives the user the ability to zoom in (or out) on a particular part of a graph.

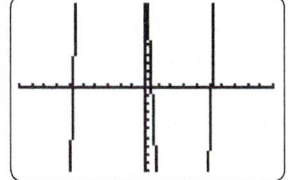

Figure 1

$f(x) = x^3 - 30x + 15$
on $[-10,10]$ by $[-10,10]$

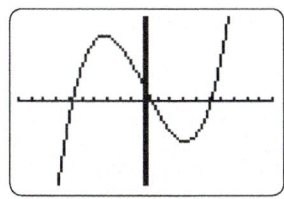

Figure 2

$f(x) = x^3 - 30x + 15$
on $[-10,10]$ by $[-100,100]$

Example 1

Graph the function $f(x) = x^3 - 30x + 15$ in the following viewing windows using a graphing calculator.

a. $[-10,10]$ by $[-10,10]$ **b.** $[-10,10]$ by $[-100,100]$

Solution

a. We set $xMin = -10$, $xMax = 10$, $yMin = -10$, and $yMax = 10$. The resulting graph is displayed in Figure 1. Note that it appears to be cut off at the top and bottom, indicating that the graph continues vertically beyond the viewing window.

b. For this viewing window we just need to change the range of *y*-values to be from $yMin = -100$ to $yMax = 100$. Figure 2 shows the graph in this new viewing window. This is a more complete picture of the graph of $f(x)$, revealing the significant parts of the graph.

The next example illustrates the use of a graphing calculator and a computer algebra system to graph an interesting part of the function $f(x) = x^2 \sin(1/x)$.

Example 2

We will illustrate the graph of the function $f(x) = x^2 \sin(1/x)$, showing the graph at different magnifications or viewing windows.

The first observation we wish to make is that since $\sin(1/x)$ cannot take any values above 1 or below −1, the graph of $f(x)$ cannot go above x^2 or below $−x^2$, as illustrated in Figure 3.

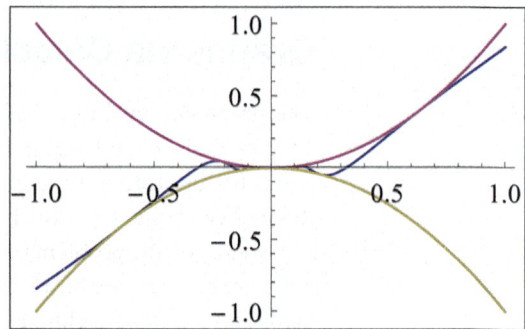

Figure 3b Graphs of x^2, $f(x)$, and $−x^2$

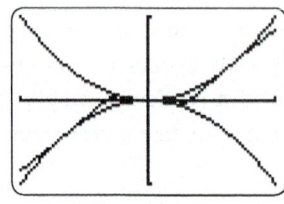

Figure 3a

Graphs of x^2, $f(x)$, and $−x^2$

In order to better understand the graph of $f(x)$, let us take a closer look at $g(x) = \sin(1/x)$. As you might recall from your experience with precalculus, $g(x)$ oscillates between −1 and +1 (like any well-mannered sine function would), but it does so in a surprising way (see Figure 4): it makes infinitely many oscillations near zero on both sides of the y-axis. This is most easily checked by noting that $g(x) = \sin(1/x) = 0$ for $x = 1/(n\pi)$, $n = \pm 1, \pm 2, \dots$. In other words, $g(x) = 0$ for infinitely many values of x that approach 0 as n grows large. In a similar fashion, one can show that $g(x) = 1$ and $g(x) = −1$ infinitely many times on both sides of zero.

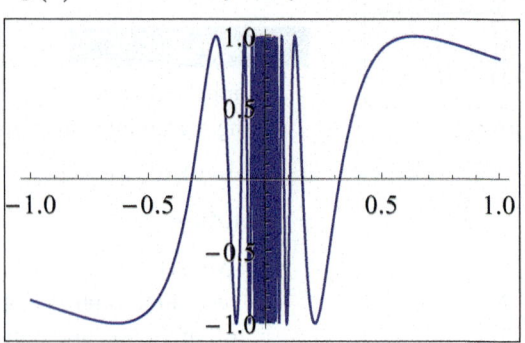

Figure 4b $g(x) = \sin(1/x)$

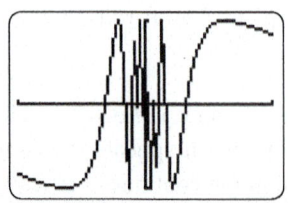

Figure 4a $g(x) = \sin(1/x)$

The effect of multiplying $g(x)$ by x^2 is that $f(x) = x^2 \sin(1/x)$ will still oscillate in a similar fashion, but now between x^2 and $-x^2$ (again, you might recall "damped trigonometric graphs" from your prior studies). The oscillation of $f(x)$ is not very clear in Figure 3 but is well illustrated if we ask our technology to "zoom in" or, equivalently, choose a "smaller" viewing window, as shown in Figure 5.

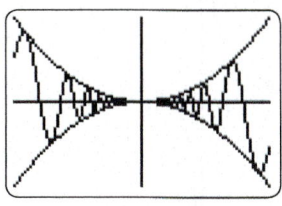

Figure 5a

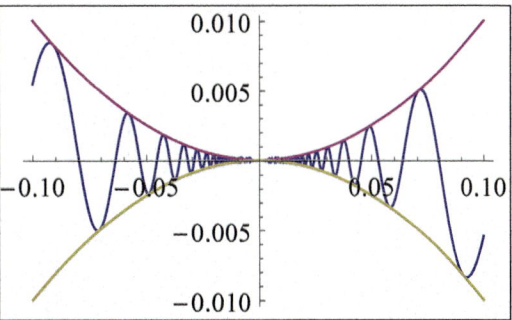

Figure 5b

In order to get a better sense of the infinitude of oscillations, we can zoom in even further, as shown in Figure 6.

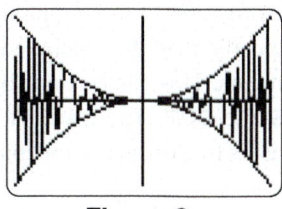

Figure 6a

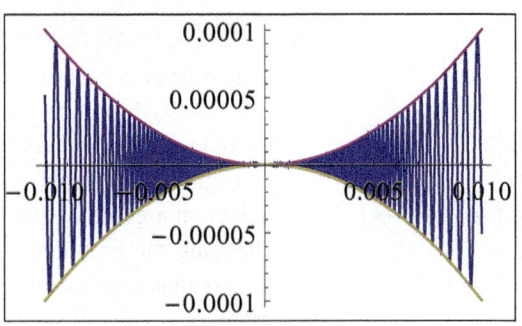

Figure 6b

Finally, we wish to note that $f(x)$ is not defined for $x = 0$; that is, there is no function value for $x = 0$, which is not clear from the graphs we received. Graphing technology, however, is still an extremely useful tool, if we stay aware of its limitations.

All calculators and computers construct graphs by similar means and under similar assumptions, at least until told to do something different. The default behavior, for instance, is to calculate (or *sample*) the value of the function at a certain number of points between *xMin* and *xMax* and then to connect all of the resulting ordered pairs with line segments. This can lead to a very misleading picture in some cases.

As an example, consider the tangent function. Using the identity $\tan x = (\sin x)/(\cos x)$ as a guide, we know that tangent is undefined whenever $\cos x$ is 0, namely $x = \pm\{\pi/2, 3\pi/2, \ldots\}$. But some calculators and software will automatically connect each consecutive pair of sampled points along the graph of tangent and create what appears to be a continuous graph with vertical portions—a result that is not only wrong, but that doesn't even qualify as a function! Figure 7 contains graphs generated by such a calculator and by software without additional guidance.

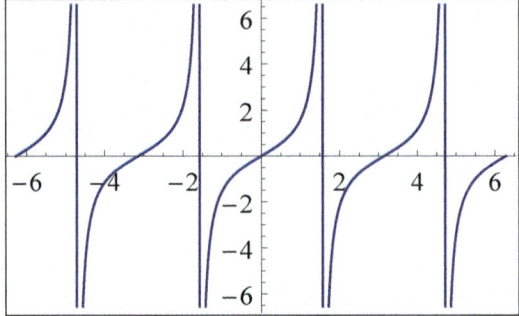

Figure 7b

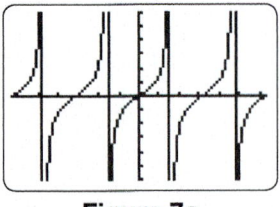

Figure 7a

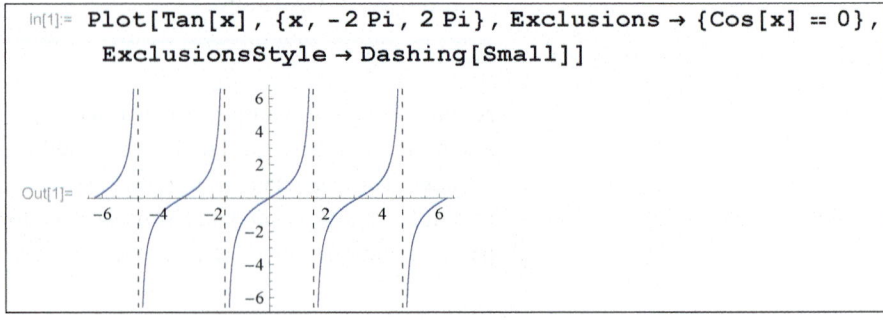

Technology Note

With guidance by the user, the graphs generated by calculators and software can usually be much improved in the cases where the default algorithm leads to misleading (or even wildly inaccurate) pictures. Most graphing calculators, for example, allow the user to specify that the "sampled" points on a graph not be connected by line segments. This can be done by changing the graphing mode from **CONNECTED** to **DOT**. The result is a picture that may not look very smooth, but at least is free of such erroneous artifacts as vertical segments. (See Figure 8.)

Figure 8

Software such as *Mathematica* allows the user much more control. Figure 9 illustrates the use of options in **Plot** that direct *Mathematica* to skip over values of x for which $\cos x$ equals 0 and to indicate those "exclusions" with dashed lines, a notation we commonly use for asymptotes.

```
In[1]:= Plot[Tan[x], {x, -2 Pi, 2 Pi}, Exclusions → {Cos[x] == 0},
        ExclusionsStyle → Dashing[Small]]
```

Figure 9

The process of sampling points of a function in order to sketch its graph can lead to other types of errors, and again it is important to temper the output of a calculator or software package with your mathematical intuition and knowledge. It may seem hard to believe, but the different screenshots in Figure 10 are all depictions of the relatively simple function $f(x) = \sin 25x$.

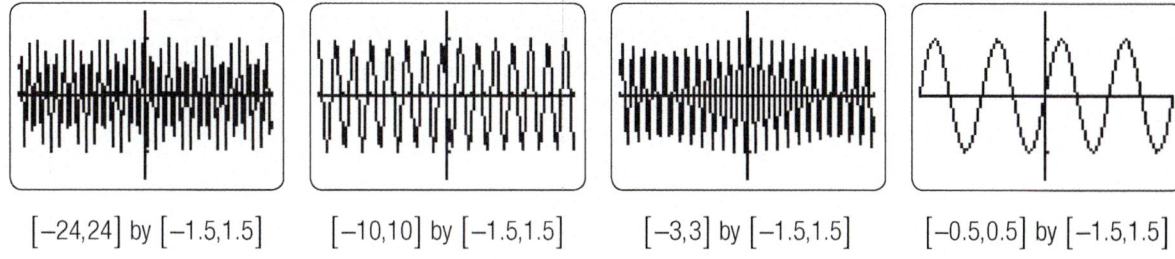

$[-24,24]$ by $[-1.5,1.5]$ $[-10,10]$ by $[-1.5,1.5]$ $[-3,3]$ by $[-1.5,1.5]$ $[-0.5,0.5]$ by $[-1.5,1.5]$

Figure 10 $f(x) = \sin 25x$

What is the explanation for these wildly different graphs? The reason the images in Figure 10 appear to be so different is that the function $f(x) = \sin 25x$ oscillates many more times than can be captured accurately with sampling in a viewing window that is many times the size of the period of the function. Recall that $\sin x$ has a period of 2π, so the function $\sin 25x$ has a period of $2\pi/25 \approx 0.25$. When we use too large a display for this function, the "gaps" between the sampled points hide one or more of the oscillations—most of the behavior of the function is thus missed. If we want to obtain a representation on a calculator that is more likely to be an accurate graph of this function, we should use a viewing window whose horizontal spread is comparable to 0.25, so the last image in Figure 10 is the best choice of the four.

It is easy to conceive of functions that have some attributes best viewed at high magnification, that is, a narrow viewing window and other attributes that show up best from far away (with a wide viewing window). With some technology, the only recourse is to experiment with different magnifications and to examine different regions of the graph until a comprehensive picture has been built up, but computer software today offers us an increasingly large array of tools. We will explore some of these in the next few pages.

Topic Two

Animations and Models

As a simple illustration of a tool not readily available until recently, we will examine the graph of $f(x) = \sin 25x$ once again, this time making use of the **Manipulate** command in *Mathematica*.

Example 3

As we said previously, the graph of $\sin 25x$ oscillates with a high-enough frequency to warrant caution when using a graphing tool. The following screenshots show the use of the **Manipulate** command in *Mathematica*. If you have access to this software package, type in and execute the command as shown and then use the slider to dynamically zoom in and out of the picture. What we are actually doing is dynamically changing the horizontal extent of the viewing window from a minimum of $[-0.1, 0.1]$ to a maximum of $[-10, 10]$. We have also instructed *Mathematica* to increase the number of sampled points (with the **PlotPoints** option) in order to capture the oscillatory behavior more accurately. (See Appendix A for more guidance on using *Mathematica*.)

In[1]:= **Manipulate[Plot[Sin[25 x], {x, -a, a}, PlotPoints → 250], {a, .1, 10}]**

Out[1]=

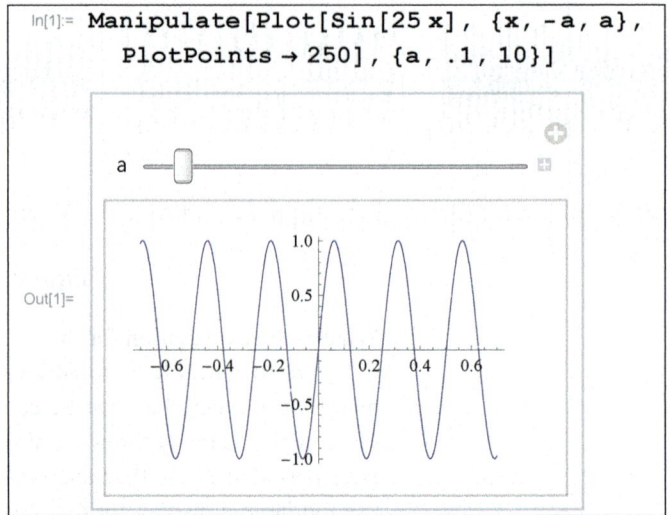

Figure 11a

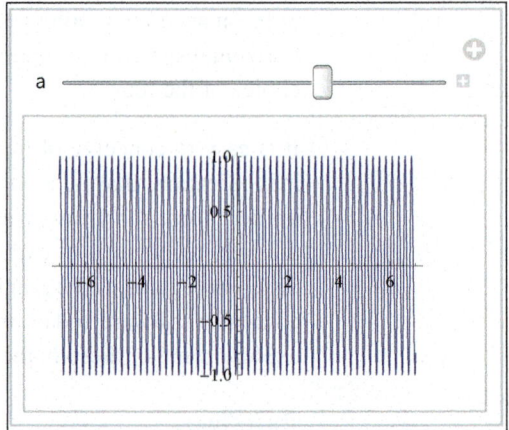

Figure 11b

In Example 3, the variable a is referred to as a *parameter*, a quantity that we want to allow to vary in order to gain insight into some particular behavior. In the usage above, changing a equates to changing the viewing window, but parameters can appear in many different ways. Example 4 illustrates the use of two parameters, one controlling the size of the viewing window and the other appearing in the definition of the function itself.

Example 4

Graph the function $f(x) = \ln x + \dfrac{\sin cx}{c}$.

Solution

The function $f(x) = \ln x + (\sin cx)/c$ exhibits different characteristics depending on both the value of the parameter c and whether we are interested in the behavior close to 0 or on a larger scale. The next screenshots show just a few of the possible choices for c and for the viewing window. Note that the width of the viewing window can vary from $[0, 0.1]$ to $[0, 20]$, and that c can take on values between 1 and 100.

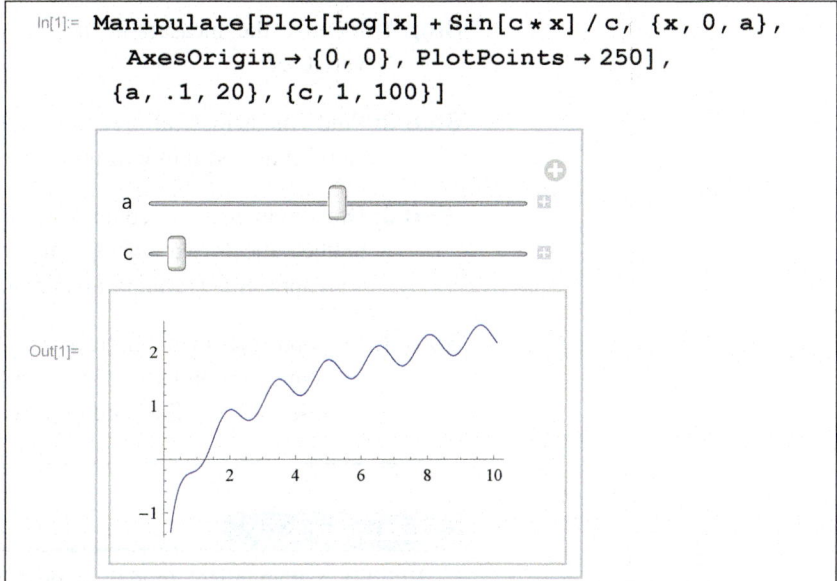

```
In[1]:= Manipulate[Plot[Log[x] + Sin[c * x] / c, {x, 0, a},
         AxesOrigin → {0, 0}, PlotPoints → 250],
       {a, .1, 20}, {c, 1, 100}]
```

Figure 12a

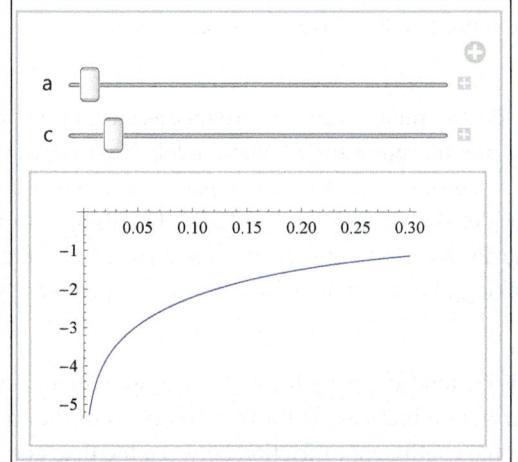

Figure 12b

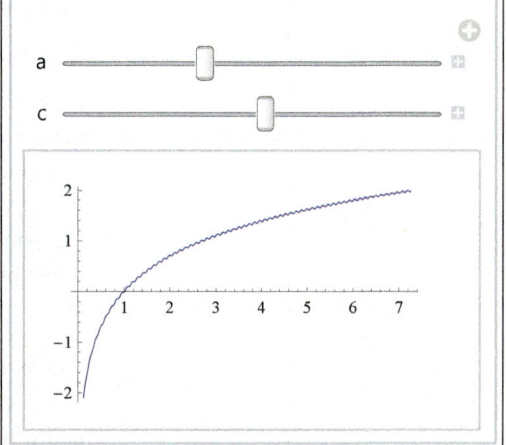

Figure 12c

Throughout this book you will find numerous **mathematical models**, depictions of real-world phenomena using equations, functions, and the tools of algebra and calculus. Many of these models incorporate parameters in a very natural way, and the ability to animate these models, that is, view graphical representations of them for different parameter values, will be especially useful.

Constructing a Mathematical Model

The process of constructing a mathematical model is one you are familiar with—you have worked with many models already, though they may not have been labeled as such. In general, the steps in making a model of a real-world phenomenon are as follows.

Step 1: Identify the measurable or observable quantities and label them as variables.

Step 2: Find mathematical relationships between the variables, making simplifying assumptions about the phenomenon if necessary.

Step 3: Draw mathematical conclusions from the equations, functions, or other mathematical relationships found (this usually amounts to solving equations or performing mathematical operations).

Step 4: Interpret the mathematical conclusions in terms of the real-world phenomenon, and test the descriptive/predictive power of the model against reality. Adjust and improve the model if necessary.

Example 5

Build a mathematical model to predict an automobile's total stopping distance d (i.e., the total distance traveled from the moment the driver decides to stop until the car comes to a complete stop) as a function of its initial speed v. Then construct a graph with animation that is dependent on road conditions and the driver's response time.

Solution

First, we identify time, speed, and distance as the most important measurable quantities for the purposes of our model. Next, observe that the actual stopping distance will consist of two parts: the distance the car travels after the driver decides to stop but before the brakes start to work, that is, the "response distance," and the actual "braking distance" (the distance traveled while braking). Labeling these by d_1 and d_2, respectively, yields

$$d = d_1 + d_2,$$

where d is the total stopping distance. Next, we will let v stand for the speed of the car before braking, M for the total mass of the car and passengers, and μ for the coefficient of friction between the tires and the pavement.

We will make the simplifying assumptions that the brakes come on instantly and operate with constant force. The road surface is assumed to be uniform and level. Finally, we ignore air resistance and heat loss. Note that d_1 can be calculated as

$$d_1 = vt,$$

where t is the time the driver requires to respond. In order to find d_2 note that the car's kinetic energy (the energy stemming from its forward motion) is dissipated during braking, all the way down to zero. Since the braking force F_b arises as a result of friction between the tires and the pavement, we have

$$F_b = \mu M g,$$

where g is the gravity constant. By the law of conservation of energy, the total work done by F_b must equal the car's initial kinetic energy of $\frac{1}{2}Mv^2$, so

$$F_b d_2 = \mu M g d_2 = \frac{1}{2}Mv^2,$$

from which we obtain

$$d_2 = \frac{v^2}{2\mu g}.$$

Thus our mathematical model for the total stopping distance becomes

$$d = d_1 + d_2 = vt + \frac{v^2}{2\mu g}.$$

Since experimental data show the value of t to be approximately 0.75 seconds for the average driver, and $\mu = 0.9$ is realistic on dry pavement, we can complete our model by writing

$$d = 0.75v + \frac{v^2}{1.8g}.$$

In Figure 13, v is assumed to be in meters per second (note that 30 m/s corresponds to 67.1 mph and that $g \approx 9.81 \, \text{m/s}^2$ in the metric system). Observe the quadratic functional dependence, which definitely warns us to observe speed limit signs!

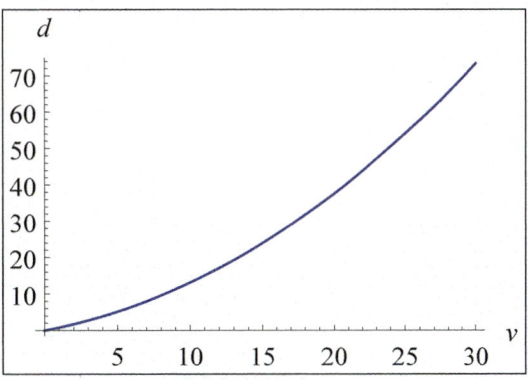

Figure 13

The flexibility of our model is demonstrated by the fact that we can treat the driver's response time t and the friction coefficient μ as parameters, corresponding to the facts that no two drivers are equal and road conditions often change. Exploiting the capabilities of *Mathematica*'s **Manipulate** command, we can easily create animations of the previous graph that are dependent on said parameters.

Notice from the following graphs that while a slower driver response does result in increased stopping distance as predicted by our model, the real drama happens on a slippery road surface. For small μ-values, stopping distance can easily more than double even if the driver has a really short reaction time!

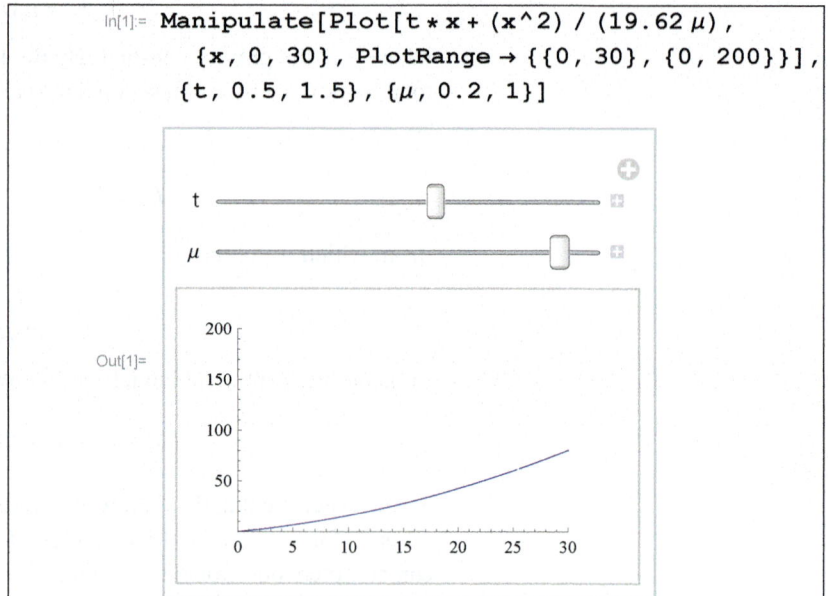

In[1]:= **Manipulate[Plot[t*x + (x^2) / (19.62 μ),**
{x, 0, 30}, PlotRange → {{0, 30}, {0, 200}}],
{t, 0.5, 1.5}, {μ, 0.2, 1}]

Out[1]=

Figure 14a

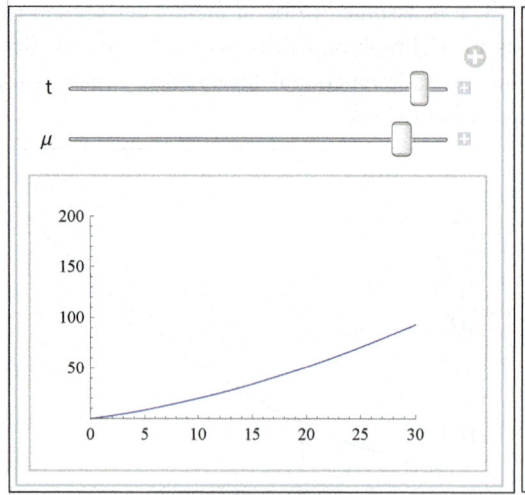

Figure 14b

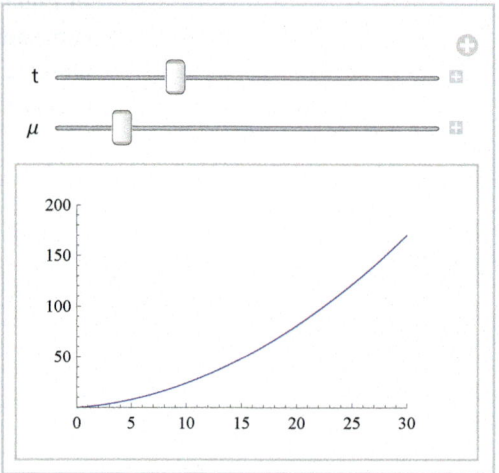

Figure 14c

Topic Three
Least-Squares Curve Fitting

We will close this section with one more example of model building. In some cases, especially where the relevant measurable quantities consist only of collected data, the "simplifying assumptions" in the modeling process may be nothing more complicated than something like "assume the data lie along a straight line" or "assume the data exhibit exponential behavior." In such cases, an **empirical model** may be all that is desired; such a model is based only on the data—no underlying physical principles are used.

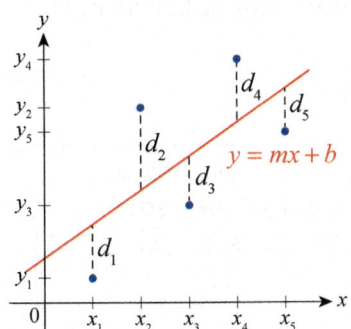

Figure 15

As mentioned briefly in Section 1.1, **least-squares curve fitting** is a useful technique for constructing a function (of a predetermined form) that best approximates a given collection of data points. To illustrate the technique, suppose we wish to fit a line to the n data points $(x_1, y_1), (x_2, y_2), \ldots, (x_n, y_n)$. Every line in the plane has the form $y = mx + b$ for some slope m and y-intercept b, so the task is to determine the best choices for m and b. If we let $d_i = y_i - (mx_i + b)$, then each d_i represents the vertical difference between the ith data point and the line $y = mx + b$. The least-squares method gets its name from the objective of choosing m and b so that $d_1^2 + d_2^2 + \cdots + d_n^2$ is minimized. (See Figure 15.)

As we will see later (Exercise 55 of Section 13.7), calculus allows us to fairly easily determine the following formulas for m and b.

$$m = \frac{\left(\sum_{i=1}^{n} x_i\right)\left(\sum_{i=1}^{n} y_i\right) - n\left(\sum_{i=1}^{n} x_i y_i\right)}{\left(\sum_{i=1}^{n} x_i\right)^2 - n\left(\sum_{i=1}^{n} x_i^2\right)}$$

$$b = \frac{\left(\sum_{i=1}^{n} x_i\right)\left(\sum_{i=1}^{n} x_i y_i\right) - \left(\sum_{i=1}^{n} y_i\right)\left(\sum_{i=1}^{n} x_i^2\right)}{\left(\sum_{i=1}^{n} x_i\right)^2 - n\left(\sum_{i=1}^{n} x_i^2\right)}$$

(Recall that the notation $\sum_{i=1}^{n} x_i$ stands for the sum $x_1 + x_2 + \cdots + x_n$.) These are exactly the formulas that graphing calculators and computer software programs use when determining the line of best fit for a given data set.

v (km/h)	d (m)
40	8.5
48	10
56	11.9
64	13.4
72	15.2
82	16.8
88	18.6
96	20.1
104	21.9
112	23.5
120	25.3

Table 1

```
LinReg
 y=ax+b
 a=.2098812597
 b=.0077026329
```

Figure 17

Example 6

Continuing along the lines of Example 5, suppose that Table 1 contains driver response distances (denoted by d in this example) as a function of initial speed v. Use the least-squares method to find the best-fitting line for the data, and give an interpretation for the slope of the line.

Solution

Note that for the purpose of using our formulas, the numbers in the left column are the values of x_i, while the right column provides the y_i-values. However, using *Mathematica*'s `Fit` command (see Figure 16) or a graphing calculator (see Figure 17), we can obtain the equation of the best-fitting line much quicker:

$$d = 0.21v + 0.008$$

```
In[1]:= Fit[{{40, 8.5}, {48, 10}, {56, 11.9}, {64, 13.4},
          {72, 15.2}, {82, 16.8}, {88, 18.6}, {96, 20.1},
          {104, 21.9}, {112, 23.5}, {120, 25.3}}, {1, x}, x]
Out[1]= 0.00770263 + 0.209881 x
```

Figure 16

Since the constant term is close to zero, ignoring it we obtain

$$d \approx 0.21v,$$

and thus the slope of the best-fitting line is

$$0.21 \approx \frac{d}{v} = t,$$

which is the driver's response time. In order to express t in seconds, we note that since $1\text{ m/s} = 3.6\text{ km/h}$, if $\overline{v}$ is the speed expressed in meters per second, we have $\overline{v} = v/3.6$, and thus t (in seconds) can be obtained as

$$t = \frac{d}{\overline{v}} = \frac{3.6d}{v} \approx 3.6 \cdot 0.21 \approx 0.76.$$

In other words, the slope of the best-fitting line indicates that based upon this particular data set, the average driver response time is approximately 0.76 seconds.

1.5 **Exercises**

1–12 Express the given function (using ( ), ^, ×, ÷, etc.) in a format suitable for entering into a graphing utility.

1. $f(x) = 1 + 3x + \sqrt{x}$

2. $g(x) = 3x - 2 + \sqrt[3]{x}$

3. $h(x) = \dfrac{3x}{\sqrt{x-1}}$

4. $k(x) = \dfrac{1 + \sqrt{x}}{2 - 3x}$

5. $u(x) = \dfrac{(-3x + 16)^4}{3x + 6}$

6. $v(x) = \sqrt{2 + 5x + \sqrt{x}}$

7. $F(x) = \left(2x^{2/3} + 3x^{5/3}\right)^5$

8. $G(x) = \left(9x^{1/5} + 2x^{3/5}\right)^{10}$

9. $H(x) = \ln\left(x^2 + 1\right) + 2^{\sqrt{x}}$

10. $K(x) = \dfrac{e^{\cos x}}{\sqrt{\log\left(x^4 + 2\right)}}$

11. $Q(x) = \dfrac{\arctan x + 1}{\left(\cos\left(\arcsin x\right)\right)^3}$

12. $R(x) = \sqrt{\left(\arccos x\right)^2 + \log_2\left(\tan x\right)}$

13. During the last 5 years, the advertising manager for a corporation has gathered the following data that show the relationship between the advertising budget (in millions of dollars) and the total sales (in thousands of units).

Advertising budget (x) (in millions)	$4.50	$6.50	$3.50	$4.20	$2.60
Sales (y) (in thousands)	37	46	42	32	29

a. Find the least-squares regression line for the data.

b. Estimate the sales if $4 million is budgeted for advertising.

14. Records at a company for the last 5 years show the following relationship between the units sold (in thousands) and the price of a product.

Price (p)	$8.80	$8.00	$7.50	$6.90	$6.20
Units sold (x) (in thousands)	3.8	5.2	7.3	8.0	9.6

a. Find the least-squares regression line for the price in terms of units sold.

b. Estimate the price that should be charged in order to sell 10,000 units.

15. The following data show the amount spent on office-building construction (in thousands) for a particular county during a 6-month period.

Month	Apr	May	Jun	Jul	Aug	Sep
Amount (in thousands)	$24	$24	$30	$49	$68	$69

a. Find the least-squares regression line for the data. (Let $x = 1$ correspond to January, $x = 2$ to February, etc.)

b. Estimate the amount spent on construction in October.

16. The annual revenue (in millions of dollars) for a corporation is given in the following table.

Year	2003	2004	2005	2006	2007	2008
Revenue (in millions)	$66	$82	$127	$201	$310	$315

   a. Find the least-squares regression line for the data. (Let $x = 0$ correspond to the year 2003.)

   b. Estimate the revenue for 2009.

17. The price of livestock futures is the estimated market price of livestock on the delivery date (end of the indicated month). The cattle futures (in cents per pound) for the months February through July are as follows.

Month	Feb	Mar	Apr	May	Jun	Jul
Price (cents per pound)	79.10	76.02	71.80	71.45	71.45	72.50

   a. Find the least-squares regression line for the data. (Let $x = 1$ correspond to January, $x = 2$ to February, etc.)

   b. Estimate the price for August.

18. The total number of foreign tourists visiting the United States between 2000 and 2004, as reported by the US Travel and Tourism Administration, is shown in the following table.

Year	2000	2001	2002	2003	2004
Tourists (in millions)	25.7	26.3	29.7	34.2	38.3

   a. Find the least-squares regression line for the data. (Let x represent the number of years passed since 2000.)

   b. Estimate the number of foreign tourists that visited the United States during 2006.

1.5 Technology Exercises

19–28 Determine whether there are points where we need to be careful in interpreting the result when using graphing technology to graph the given function. Find all those points and explain. Then use a graphing calculator or computer algebra system to sketch the graph, using various viewing windows.

19. $f(x) = x^4 \cos \dfrac{1}{x}$

20. $G(x) = \cos \dfrac{1}{x-2}$

21. $p(x) = \dfrac{1}{2} \tan(3x - 2)$

22. $g(x) = \sec(2x+1)$

23. $q(x) = \dfrac{x^2 - 2x - 1}{x+1}$

24. $r(x) = \dfrac{x^2 + 1}{x^2 - 9}$

25. $h(x) = \dfrac{2x^4 + 1}{2x^4 - 1}$

26. $F(x) = \ln(\cos x)$

27. $s(x) = \cos(\ln x)$

28. $t(x) = \sin(\csc x)$

29–40 Use a graphing calculator or computer algebra system to graph the given function in the window $[-10,10]$ by $[-10,10]$. Explain what appears to be wrong with the picture. Then find a more appropriate window, which reveals the significant parts of the graph, and draw the "improved" graph.

29. $f(x) = \dfrac{3x - 25}{\sqrt{x^2 + 5}}$

30. $g(x) = (40 + 3x)\sqrt{16 - x}$

31. $h(x) = (3x + 4)^2 (5x - 25)^2$

32. $F(x) = (6x + 30)^2 (3x - 15)^2$

33. $G(x) = 35 + 17x - x^2 - x^3$

34. $H(x) = 210 - 80x + x^3$

35. $r(x) = \sqrt[3]{x^3 - x^2 - x - 50}$

36. $u(x) = \sqrt[3]{x^4 - 3x^2 - 3x - 30}$

37. $v(x) = \left(12 - 6x - x^2\right)^{4/3}$

38. $f(x) = \left(x^3 - x - 100\right)^{1/3}$

39. $g(x) = x^2 \sin \dfrac{\pi}{x - 12}$

40. $h(x) = \sec^2 \dfrac{x}{10}$

41–46 Use a graphing calculator or computer algebra system to graph the given function in a suitable window and find the smallest y-value possible. (Use only the given interval, if specified. Round your answer to four decimal places.)

41. $f(x) = x^2 - 104x + 2724$

42. $g(x) = \dfrac{-1 - x^2 - 3x^3}{5^x}$

43. $h(x) = x^3 - 17x + 5; \quad -3 \le x \le 5$

44. $F(x) = \dfrac{\sqrt[3]{x} - 150}{5 + x^2}$

45. $G(x) = x^{1.5} - 8x - 15$

46. $H(x) = x^{1.8} - x - 100$

47–52 Use a graphing calculator or computer algebra system to graph the given function in a suitable window and find the greatest y-value possible. (Use only the given interval, if specified. Round your answer to four decimal places.)

47. $f(x) = 50 - 2^x; \quad -10 \le x \le 10$

48. $g(x) = (x + 1)^5 - 1.5^{x+1}$

49. $h(x) = x^{17} - 17^x; \quad -2 \le x \le 2$

50. $k(x) = x\left(3^{-x}\right)$

51. $F(x) = \dfrac{-2x}{x^2 + 1}$

52. $G(x) = \dfrac{3 - 5x}{\sqrt{3x^2 + 2}}$

53–58 Use a graphing calculator or computer algebra system to graph the given function and describe the characteristics of the graph as c varies. Use different viewing windows.

53. $f(x) = x^2 - cx$

54. $g(x) = \dfrac{1}{2}x^3 - c\left(x^2 + x + 1\right)$

55. $h(x) = e^{cx}$

56. $k(x) = \ln\left(x^2 + cx + 1\right)$

57. $F(x) = \dfrac{x}{c} + \cos \dfrac{c^2 x}{c}$

58. $G(x) = \dfrac{cx^2}{x + cx^3}$

59–64 Use a graphing calculator or computer algebra system to approximate the solution(s) of the given equation, rounded to four decimal places. (**Hint:** Zoom in on the x-intercepts or points of intersection as appropriate for each equation.)

59. $x^3 - 20x - 2 = 0$

60. $2x^3 = 31x + 2$

61. $3\cos x = \sqrt{x}$

62. $\arctan x = \dfrac{1}{100}x^5$

63. $\ln x = x - 2$

64. $x + 5 = e^x$

65–70 Use appropriately large viewing windows on a graphing calculator or computer algebra system to decide which of the given functions eventually "rises faster" toward infinity.

65. $f(x) = \dfrac{1}{2}x^3$; $g(x) = x^2$

66. $f(x) = \sqrt{x}$; $g(x) = x$

67. $f(x) = 5\sqrt{x}$; $g(x) = \dfrac{1}{5}x$

68. $f(x) = \dfrac{1}{2}e^x$; $g(x) = x^2$

69. $f(x) = 5\log x + 5$; $g(x) = \dfrac{1}{2}x^5$

70. $f(x) = 10\arctan x$; $g(x) = 2\sqrt[3]{x}$

71–73 Most graphing calculators and computer algebra systems have regression capabilities to fit curves other than lines to a given data set. Frequently, depending on the tendency of the data, a quadratic, an exponential, or some other type of curve provides for much better approximation. Most often the choice is the modeler's.

Use the regression capabilities of your technology to build a graphical model and then answer the questions.

71. The following table shows daytime temperatures in El Cajon, CA on a particular spring day from 7 a.m. to 12 p.m. Find the best-fitting curve and use it to predict the temperatures at 1 p.m. and 2 p.m.

Time	6 a.m.	7 a.m.	8 a.m.	9 a.m.	10 a.m.	11 a.m.	12 p.m.
Temperature	47 °F	50 °F	55 °F	61 °F	68 °F	73 °F	75 °F

72. The following table shows the winning times of the Olympic men's 100 m dash champions. Find the best-fitting curve and use it to predict the winning times at the next three Olympics.

Year	Men's 100 m dash winning time (s)	Year	Men's 100 m dash winning time (s)	Year	Men's 100 m dash winning time (s)
1896	12.0	1936	10.3	1980	10.25
1900	11.0	1948	10.3	1984	9.99
1904	11.0	1952	10.4	1988	9.92
1908	10.8	1956	10.5	1992	9.96
1912	10.8	1960	10.2	1996	9.84
1920	10.8	1964	10.0	2000	9.87
1924	10.6	1968	9.95	2004	9.85
1928	10.8	1972	10.14	2008	9.69
1932	10.3	1976	10.06	2012	9.63

73. The following table shows acceleration times for the Ferrari Enzo up to 130 mph. Find the best-fitting curve and use it to predict the acceleration times for the Enzo from **a.** 0 to 150 mph and **b.** 0 to 170 mph.

(**Source:** *Car and Driver*)

0–30 mph	0–40 mph	0–50 mph	0–60 mph	0–70 mph	0–80 mph	0–90 mph	0–100 mph	0–110 mph	0–120 mph	0–130 mph
1.5 s	2.0 s	2.7 s	3.3 s	3.8 s	5.0 s	5.8 s	6.6 s	8.0 s	9.2 s	10.3 s

Chapter 1
Review Exercises

1–4 Find the domain and range of the given relation and determine whether the relation is a function.

1. $R = \{(-2,9),(-3,-3),(-2,2),(-2,-9)\}$

2. $3x - 4y = 17$

3. $x = y^2 - 6$

4. $x = \sqrt{y-4}$

5–8 Identify the domain, codomain, and range of the given function.

5. $f : \mathbb{N} \to \mathbb{R}, \quad f(x) = \dfrac{3x}{4}$

6. $g : \mathbb{R} \to \mathbb{R}, \quad g(x) = 5x + 1$

7. $h : \mathbb{R} \to \mathbb{R}, \quad h(x) = \dfrac{1}{x^2 + 1}$

8. $k : \mathbb{N} \to \mathbb{R}, \quad k(x) = 2 + \sqrt{x-1}$

9–12 Find the value of the given function for **a.** $f(x-1)$, **b.** $f(x^2)$, and **c.** $\dfrac{f(x+h)-f(x)}{h}$.

9. $f(x) = (x+5)(2x)$

10. $f(x) = \sqrt[3]{x} + 6(x+4)$

11. $f(x) = \dfrac{3}{x+2}$

12. $f(x) = \sin 2x$

13–14 Find all intervals of monotonicity (intervals where the function is increasing or decreasing) for the given function.

13. $f(x) = (x-2)^4 - 6$

14. $R(x) = \begin{cases} (x+2)^2 & \text{if } x < -1 \\ -x & \text{if } x \geq -1 \end{cases}$

15–18 Determine if the function is even, odd, or neither and then graph it.

15. $f(x) = \dfrac{1}{3}x^3$

16. $f(x) = \sqrt{x}$

17. $f(x) = -2\sin x$

18. $g(x) = -2\sin^2 x$

19–20 Discuss the symmetry of the given equation and then graph it.

19. $y = |5x|$

20. $x^2 + y^2 = 25$

21–34 Graph the given function. Locate the x- and y-intercepts, if any.

21. $f(x) = 7x - 2$

22. $f(x) = \dfrac{2x-6}{3}$

23. $f(x) = (x-1)^2 - 1$

24. $f(x) = -x^2 - 6x - 11$

25. $f(x) = 4x^3$

26. $f(x) = -\dfrac{2}{x^2}$

27. $f(x) = \dfrac{-x^3 + 7x + 6}{2}$

28. $f(x) = \dfrac{x+1}{x^2 - 4}$

29. $f(x) = \dfrac{\sqrt[3]{x}}{2}$

30. $f(x) = 5|-x|$

31. $f(x) = \left\|\dfrac{2x}{3}\right\|$

32. $f(x) = \begin{cases} x^2 & \text{if } x < 1 \\ \dfrac{1}{x} & \text{if } x \geq 1 \end{cases}$

33. $f(x) = -\cot x$

34. $f(x) = \log_{1/2} x$

35–36 Sketch the graph of the given function by first identifying the more basic function that has been shifted, reflected, stretched, or compressed. Then determine the domain and range of the function.

35. $f(x) = (x-1)^3 + 2$

36. $f(x) = 4|x+3|$

37–38 Write an equation for the function described.

37. Use the function $f(x) = 1/x$. Move the function 2 units to the right and 3 units up.

38. Use the function $f(x) = \sqrt[3]{x}$. Move the function 1 unit to the left and reflect across the y-axis.

39–40 Find the formula and domain for **a.** $f + g$ and **b.** f/g.

39. $f(x) = x^2;$ $g(x) = \sqrt{x}$

40. $f(x) = \dfrac{1}{x-2};$ $g(x) = \sqrt[3]{x}$

41–42 Use the information given to determine **a.** $(f \circ g)(x)$, **b.** $(g \circ f)(x)$, and **c.** $(f \circ g)(3)$.

41. $f(x) = -x + 1;$ $g(x) = -x - 1$

42. $f(x) = \dfrac{x^{-1}}{18} - 3;$ $g(x) = \dfrac{x-4}{x^3}$

43–44 Write the given function as a composition of two functions. (Answers will vary.)

43. $f(x) = \dfrac{\sqrt{x+3} + 2}{x^2 + 6x + 9}$

44. $f(x) = |x + 2| + x^2 + 4x + 4$

45–46 Graph the inverse of the given relation, and state its domain and range.

45. $R = \{(-3,5), (2,1), (0,-5), (-1,-2)\}$

46. $y = \dfrac{1}{3}x^2$

47–50 Find the inverse of the given function.

47. $f(x) = \dfrac{2}{7x-1}$ **48.** $f(x) = \dfrac{4x-3}{x}$

49. $f(x) = x^{1/5} - 6$ **50.** $f(x) = \dfrac{6x-7}{2-x}$

51. Show that $f^{-1}(f(x)) = x$ and that $f(f^{-1}(x)) = x$ for the functions given in Exercises 49 and 50.

52–53 Rewrite the given function as a purely algebraic function.

52. $\cos(\sin^{-1} x)$ **53.** $\tan\left(\sec^{-1} \dfrac{x}{2}\right)$

54–55 Use the properties of logarithms to expand the given expression as much as possible; that is, decompose the expression into sums or differences of the simplest possible terms.

54. $\ln \dfrac{x^2 y - xz}{5y^3}$ **55.** $\log \sqrt[3]{\dfrac{yz^2}{x^4}}$

56. If a pebble is shot upward with an initial (vertical) velocity of 56 ft/s, how high does it go? (**Hint:** Use the height function $h(t) = -16t^2 + 56t$.)

57. The half-life of actinium-225 is 10 days. Assuming that A_0 is the mass at time $t = 0$, find the function $A(t)$ that gives the mass remaining after t days. What percentage of the original mass of a sample of Ac-225 remains after 25 days?

58. Prove that the product of an odd function and an even function is odd.

59. Among all the pairs of numbers with a sum of 15, find the pair whose product is maximum.

60. The January 2010 Haiti earthquake was initially reported as a 7.2-magnitude quake on the Richter scale, while the March 2011 earthquake off the eastern shores of Japan was magnitude 9.0. According to these numbers, about how many times more intense was the Japanese earthquake? (**Hint:** See the directions preceding Exercise 138 in Section 1.4.)

61–62 Determine if the given complex number is in the Mandelbrot set. (**Hint:** See the directions preceding Exercises 101–108 in Section 1.3.)

61. $c = -i$ **62.** $c = 2$

63–76 *True or False?* Determine whether the given statement is true or false. In case of a false statement, explain or provide a counterexample.

63. The graph of any function can always be represented by a curve that passes the vertical line test.

64. Any linear function whose graph is a line with negative slope always has an inverse.

65. If f is a function, and $f(a) = f(b)$, then $a = b$.

66. If $(2,5)$ is a point on the graph of an odd function, then $(-2,-5)$ is also on the graph.

67. The graph of a quadratic function is a parabola, so quadratic functions are even functions.

68. In general, $(f \circ g)(x) = (g \circ f)(x)$ holds for all x if both compositions can be formed.

69. If a function can be decomposed into three functions, it can only be done in one way.

70. The graph of $f(x) + a$ is that of $f(x)$ translated vertically by a units.

71. If $c \in \mathbb{R}$ is a constant, and f is any function, then the graphs of $f(cx)$ and $cf(x)$ are identical.

72. When the graph of any function $f(x)$ is reflected about the line $y = x$, the graph of the inverse is obtained.

73. The graph of any function must pass the horizontal line test.

74. If (a,b) is a point on the graph of an invertible function $f(x)$, then (b,a) is on the graph of its inverse.

75. The common logarithm and $y = 10^x$ are inverses.

76. The function $f(x) = \tan^{-1}(x)$ has no asymptotes.

Chapter 1
Technology Exercises

77–79 Mentally sketch the graph of the given function by identifying the basic shape that has been shifted, reflected, stretched, or compressed. Then use a graphing calculator or a computer algebra system to graph the function and check your reasoning.

77. $f(x) = \ln(x+1) + 2$

78. $f(x) = -\dfrac{2}{x-3} + 1$

79. $f(x) = \sin \pi x - 1$

80. The annual expenditures (in millions of dollars) for a corporation are given in the table below.

Year	2006	2007	2008	2009	2010	2011
Expenditures (millions)	$16.2	$17.1	$18.8	$19.6	$21.1	$22.9

   a. Find the least-squares regression line for the data. (Let $x = 0$ correspond to the year 2006.)

   b. Estimate the expenditures for 2012.

81–82 Use a graphing calculator or computer algebra system to approximate the solution(s) of the given equation, rounded to four decimal places. (**Hint:** Zoom in on the x-intercepts or points of intersection as appropriate for each equation.)

81. $x^5 - x^3 - 3 = 0$

82. $x^2 + 6 = 2^{x+1}$

83–84 Use a graphing calculator or computer algebra system to graph the given function and describe the characteristics of the graph as c varies. Use different viewing windows.

83. $u(x) = \dfrac{1 - e^{c/x}}{1 + e^{c/x}}$

84. $v(x) = \dfrac{x}{c^2} \sqrt[4]{c^4 - x^4}$

Chapter 1
Project

As time goes on, there is increasing awareness, controversy, and legislation regarding the ozone layer and other environmental issues. The hole in the ozone layer over the South Pole disappears and reappears in a cyclical manner annually. Suppose that over a particular stretch of time the hole is assumed to be circular with a radius growing at a constant rate of 2.6 kilometers per hour.

Photo Courtesy of NASA

1. Assuming that t is measured in hours, that $t = 0$ corresponds to the start of the annual growth of the hole, and that the radius of the hole is initially 0, write the radius as a function of time, t. Denote this function by $r(t)$.

2. Use function composition to write the area of the hole as a function of time, t. Denote this function by $A(t)$. Sketch the graph of $A(t)$ and label the axes appropriately.

3. After finding $A(1)$, the area of the ozone hole at the end of the first hour, determine the time necessary for this area to double. How much additional time does it take to reach three times the initial area?

4. Are the two time intervals you found in Question 3 equal? If not, which one is greater? Explain your finding. (Use a comparison of some basic functions discussed in Section 1.2 in your explanation.)

5. What are the radius and area after 3 hours? After 5.5 hours?

6. What is the average rate of change of the area from 3 hours to 5.5 hours?

7. What is the average rate of change of the area from 5.5 hours to 8 hours?

8. Is the average rate of change of the area increasing or decreasing as time passes?

9. What flaws do you see with this model? Can you think of a better approach to modeling the growth of the ozone hole?

Chapter 2
Limits and the Derivative

Introduction

This chapter opens with a discussion of two broadly defined problems whose solutions turn out to have much in common. Mathematically, the task of determining the instantaneous velocity of an object and the task of finding the line tangent to the graph of a function both depend on the concept of *limit*. Much of this chapter revolves around developing an intuitive sense as well as a rigorous definition of the concept.

The limit idea inherently involves motion and reflects a major difference between the relatively static world of algebra and the dynamic world of calculus.

As with so many concepts in mathematics, the evolution of the idea spans cultures and ages. Some early thinkers, such as the philosopher Zeno of Elea (ca. 495–ca. 430 BC) and the mathematician Archimedes of Syracuse (ca. 287–ca. 212 BC), developed an appreciation of the power and depth of the concept centuries before later mathematicians overcame the difficulties of rigorously defining and using limits. Zeno, in fact, is chiefly remembered today for the many paradoxes he devised, which illustrate the danger of naïve assumptions about limits and infinity—a variation of one of these paradoxes appears as Example 2 in Section 2.2. Archimedes, whose name usually appears on any list of the greatest mathematicians of all time, used methods we would now classify as belonging to calculus to achieve results that were centuries ahead of their time. Formulas for the area under a parabola and for the volumes and surface areas of certain three-dimensional objects are just a few examples.

Much later, in the seventeenth century AD, the ideas and methods involving limits were brought together and ultimately characterized as "calculus." But even after calculus came to be recognized as an especially rich branch of mathematics, the limit concept retained its somewhat dangerous reputation and continued to trick unwary thinkers. Great discoveries and advances were made using calculus throughout the 1600s and 1700s by a long list of famous mathematicians (many of whom you will read about in the coming chapters), but rigorous definitions of "limit" and related ideas didn't appear until the 1800s. The definition we use today is essentially due to the French mathematician Augustin-Louis Cauchy (1789–1857), who also refined and made rigorous the idea of *continuity* as it applies to functions.

The two problems that open the chapter serve as motivation for the mathematics that follows, but by the end of the chapter it will be apparent that they are only representatives of the many problems that can be solved by means of the *derivative* of a function. The definition of "derivative" and the many methods associated with finding and using derivatives constitute a major portion of calculus and of this text.

2.1 **Rates of Change and Tangents**

The need for calculus can be demonstrated in many ways, ranging from common everyday experiences to problems of a more esoteric nature. We will begin our development of calculus with two easily described problems.

Topic One
The Velocity Problem

The simple relationship *distance = rate × time*, or *d = rt* is often one of the first applications of basic algebra to be studied. This equation can be taken as the definition of average rate *r* of travel for an object moving distance *d* over time *t* (that is, the **average rate** is the *ratio* of *d* over *t*), or it can be used to solve for any one of the variables if the other two quantities are known. It is undeniably useful in situations where an object's average rate is all that is desired (especially so if the object is moving at a constant rate), but it is just as undeniably limited in scope. Most of our everyday experiences involve objects that move at varying speeds, and we don't have to progress very far in mathematics before needing a way to handle more than just *average* rates of travel.

By convention, we often use the word *velocity* or *speed* when discussing a rate of travel that is not necessarily constant (later, we will further refine our usage and differentiate between speed and velocity). To emphasize the distinction with average rate even more, the word *instantaneous* sometimes appears before velocity. One of the best examples of instantaneous velocity is the number you read off your speedometer as you drive—at any given moment, the gauge indicates your velocity *at that instant*.

One form of the "velocity problem" can be put in this context: Suppose you are given the task of calculating a car's instantaneous velocity from perfect knowledge about its distance traveled over various lengths of time but without benefit of the speedometer. How would you do it?

To give us something to visualize and work with, suppose that the car is traveling along a straight road and $d(t)$ represents its location, relative to its starting point, at time *t*; we can set *t* = 0 as the moment when the car starts moving, so $d(0)$ is its initial location. Units of measurement are immaterial at the moment so they will not be specified—it is important only to remember that $d(t)$ represents distance with respect to time *t*. Let's suppose further that we want to determine the car's velocity at time *t* = 10. Knowing nothing more than the relationship *d = rt*, we can easily compute some average rates over time intervals that contain *t* = 10. For example, the average rate of travel from *t* = 9 to *t* = 11 is calculated as follows.

$$\frac{d(11)-d(9)}{11-9}=\frac{d(11)-d(9)}{2}$$

If we suspect that the instantaneous velocity of the car varies quite a bit over the time interval $[9,11]$ and that the average rate is not a good reflection of the velocity at exactly $t = 10$, we can easily compute the average rate over a shorter interval, say $[9.5, 10.5]$.

$$\frac{d(10.5) - d(9.5)}{10.5 - 9.5} = d(10.5) - d(9.5)$$

And there's no need to stop there. Our intuition tells us that as the time intervals get shorter and shorter, the average rate of travel over those time intervals should get closer and closer to the instantaneous velocity at $t = 10$.

With a bit more thought, our intuition may tell us something else. Namely, it really shouldn't matter much if the small intervals we're considering are centered at $t = 10$ or not, as long as the intervals all contain that point in time. After all, the average rate of travel that we are computing is an approximation to the instantaneous velocity at *every* point of the interval, not just the midpoint. In practice, we may find it easier to use intervals that have our point of interest at an endpoint, rather than the midpoint.

At this time, it will be convenient to introduce some more terms and notation. Given a real-valued function f defined at a point x_0, the **difference quotient of** f **at** x_0 **with increment** h is the following ratio.

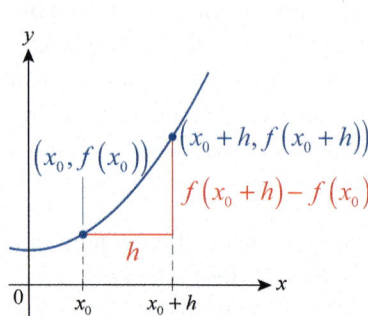

Figure 1

$$\frac{f(x_0 + h) - f(x_0)}{h}$$

Note that if f represents a distance function as described above, this ratio corresponds to the average rate of travel over an interval containing x_0, with x_0 serving as either the left endpoint (if h is positive) or the right endpoint (if h is negative). More generally, if x_0 is contained in the interval $[a, b]$, the ratio

$$\frac{\Delta f}{\Delta x} = \frac{f(b) - f(a)}{b - a}$$

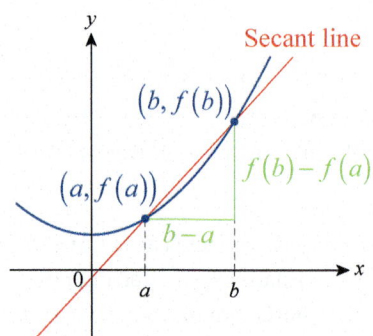

Figure 2

corresponds to the average rate of travel over the interval and also serves as an approximation to the instantaneous velocity at x_0. You have probably seen the symbol Δ before: it is the Greek letter delta and is used to designate a change in the quantity that follows it. The above ratio is read "change in f over change in x" and can be interpreted as the slope of the line (known as the *secant line*) shown in Figure 2.

Example 1

A piece of rock falls into a 256-foot-deep canyon from its rim. If we ignore air resistance, its distance below the rim at time t is

$$d(t) = 16t^2,$$

where d is measured in feet and t is measured in seconds. Estimate **a.** the instantaneous velocity of the rock after 2 seconds and **b.** the velocity of impact.

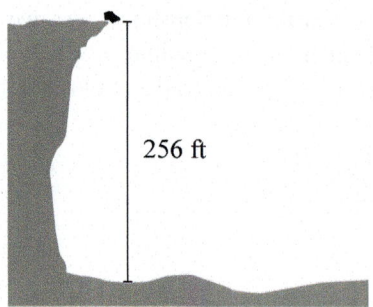

256 ft

Figure 3

Solution

a. We will start by calculating the average velocity of the rock from $t = 2$ to $t = 3$. Note that the elapsed time is $\Delta t = 3 - 2 = 1$, while the distance traveled is $\Delta d = d(3) - d(2)$, so we can set up the following difference quotient to calculate the desired average velocity over the interval $[2,3]$.

$$\frac{\Delta d}{\Delta t} = \frac{d(3) - d(2)}{3 - 2}$$
$$= \frac{16 \cdot 3^2 - 16 \cdot 2^2}{1}$$
$$= 16 \cdot 9 - 16 \cdot 4 = 80 \text{ ft/s}$$

However, this is not expected to be an accurate reflection of the instantaneous velocity at $t = 2$, since the rock is accelerating significantly between $t = 2$ and $t = 3$. We obtain a more accurate guess if we work over a much shorter time interval, say the one from $t = 2$ to $t = 2.1$.

$$\frac{\Delta d}{\Delta t} = \frac{d(2.1) - d(2)}{2.1 - 2}$$
$$= \frac{16 \cdot (2.1)^2 - 16 \cdot 2^2}{0.1}$$
$$= \frac{16 \cdot (4.41) - 16 \cdot 4}{0.1} = 65.6 \text{ ft/s}$$

This is a much better guess, but there is no need to stop here. The following table summarizes the average velocities over time intervals of decreasing lengths, all the way down to a length of 0.01 seconds.

Time interval	$[2, 2.09]$	$[2, 2.08]$	$[2, 2.07]$	$[2, 2.06]$	$[2, 2.05]$	$[2, 2.04]$	$[2, 2.03]$	$[2, 2.02]$	$[2, 2.01]$
Average velocity $\Delta d / \Delta t$	65.44	65.28	65.12	64.96	64.8	64.64	64.48	64.32	64.16

Since the velocity numbers in the table seem to be approaching 64 ft/s as the intervals are becoming shorter, we might guess that number to be the instantaneous velocity of the rock at $t = 2$. Notice also that in the above table, 2 serves as the left endpoint of all of the intervals. Intuitively, we expect the same instantaneous velocity to be obtained if $t = 2$ were the right endpoint, or if it were in the interior of each interval, as long as the interval's length is decreasing and approaching 0. In our next table, we chose $t = 2$ to be the right endpoint of each time interval.

Time interval	$[1.91, 2]$	$[1.92, 2]$	$[1.93, 2]$	$[1.94, 2]$	$[1.95, 2]$	$[1.96, 2]$	$[1.97, 2]$	$[1.98, 2]$	$[1.99, 2]$
Average velocity $\Delta d / \Delta t$	62.56	62.72	62.88	63.04	63.2	63.36	63.52	63.68	63.84

Notice that the average velocities here are all less than in the first table, but are increasing as the time interval is getting shorter, "shrinking" onto the point $t = 2$. This is precisely what should be expected because of the steady acceleration of the rock. In conclusion, both tables support our guess that the rock's instantaneous velocity at $t=2$ is 64 ft/s downward. As a final confirmation, let us calculate the difference quotient over the very short interval $[1.999, 2]$.

$$\frac{\Delta d}{\Delta t} = \frac{d(2) - d(1.999)}{2 - 1.999}$$
$$= \frac{16 \cdot 2^2 - 16 \cdot (1.999)^2}{0.001}$$
$$= 63.984 \text{ ft/s}$$

b. To estimate the velocity of impact, we first need to find when exactly impact occurs. Being in possession of the position function and knowing that at impact the position of the rock is exactly the depth of 256 feet, we can find the time of impact by solving

$$d(t) = 256$$
$$16t^2 = 256$$
$$t^2 = 16,$$

which gives us $t=4$, since we can safely discard the negative root of $t=-4$.

To find the velocity of impact, we use the same strategy as in part a. but this time using $t=4$ as the right endpoint of our intervals for our difference quotients. The table below summarizes our calculations.

Time interval	$[3.91, 4]$	$[3.92, 4]$	$[3.93, 4]$	$[3.94, 4]$	$[3.95, 4]$	$[3.96, 4]$	$[3.97, 4]$	$[3.98, 4]$	$[3.99, 4]$
Average velocity $\Delta d / \Delta t$	126.56	126.72	126.88	127.04	127.2	127.36	127.52	127.68	127.84

As before, we examine the table and conclude that the velocity of impact is approximately 128 ft/s (you may want to further confirm our finding by calculating difference quotients over even shorter time intervals).

Example 2

Daniela lives just a few miles down a straight stretch of highway from her college campus. One morning she left for class, and her position relative to her house was given by the function

$$d(t) = 1600t^2 - 16,000t^3 + 40,000t^4,$$

where d is measured in miles, t in hours, with $t=0$ corresponding to the time when she left her house. Use difference quotients to estimate her instantaneous velocity a. 3 minutes after she left, b. 6 minutes after she left, and c. 7.5 minutes after she left.

From your calculations, can you guess what might have happened?

Solution

a. Just like in Example 1, we will evaluate difference quotients over shorter and shorter time intervals, each using $t = 3$ minutes as the left endpoint. In this example, however, we will denote the length of a time interval by h, and call it an *increment*. With this notation, and by noting that 3 minutes $= 0.05$ hours, a typical difference quotient becomes the following.

$$\frac{d(0.05+h)-d(0.05)}{h}$$

For example, consider the average velocity on the time interval starting at 3 minutes into the drive and lasting for $h = 0.6$ minutes $= 0.01$ hours.

$$v_{ave} = \frac{d(0.05+0.01)-d(0.05)}{0.01}$$
$$= 57.24 \text{ mph}$$

We will get a better approximation of the instantaneous velocity at $t = 3$ minutes if we shorten the time intervals over which we calculate the average velocities. The following table summarizes our calculations over several intervals of decreasing length.

h	0.01	0.005	0.001	0.0005	0.0001
v_{ave}	57.24	58.81	59.79	59.90	59.98

So we conclude that at exactly three minutes into the trip, Daniela was driving at approximately 60 mph.

b. Next, we will estimate the instantaneous velocity at $t = 6$ minutes $= 0.1$ hours. This time, however, we will use negative increments when setting up the difference quotients. For example, with $h = -0.03$, we obtain

$$v_{ave} = \frac{d(0.1-0.03)-d(0.1)}{-0.03}$$
$$= \frac{d(0.1)-d(0.07)}{0.03}$$
$$= 22.92 \text{ mph.}$$

Interestingly, it appears from the above calculation that Daniela is slowing down considerably. In search of the instantaneous velocity, we shorten the time intervals and summarize our calculations in the following table.

h	−0.03	−0.01	−0.001	−0.0005	−0.0001
v_{ave}	22.92	7.96	0.80	0.40	0.08

From the table we must conclude that Daniela slowed down and stopped, for the instantaneous velocity at $t = 6$ minutes appears to be 0. Did she stop due to a traffic jam, or did she turn back? Our next calculation has the answer.

c. If we evaluate several difference quotients with decreasing increments, using $t = 7.5$ minutes $= 0.125$ hours as the left endpoint, using the now-familiar formula

$$v_{ave} = \frac{d(0.125 + h) - d(0.125)}{h},$$

we obtain the following table.

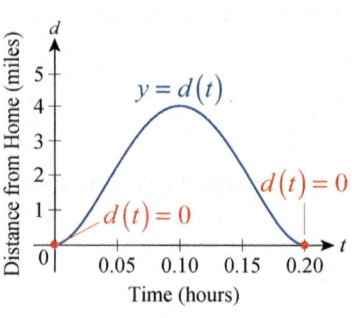

Distance from Home (miles)

$y = d(t)$

$d(t) = 0$

$d(t) = 0$

Time (hours)

Figure 4

h	0.01	0.005	0.001	0.0005	0.0001
v_{ave}	−43.56	−40.65	−38.15	−37.82	−37.56

We estimate the instantaneous velocity at $t = 7.5$ minutes to be −37.5 mph. The negative sign indicates that the direction changed, so Daniela is actually driving back toward her house. (She might have left something at home.) In fact, the graph of the position function is quite revealing, so we show it in Figure 4.

Topic Two

The Tangent Problem

Our second problem will be presented in a purely mathematical context, but, as we will see, it is closely related to the first.

The word "tangent" comes from the Latin word *tangens*, and translates as "touching" (and while there is a connection with the trigonometric function that goes by the name tangent, that connection is not especially instructive at the moment). In mathematics, a line is tangent to a given curve if it "just touches" the curve. In some cases, it is easy to identify such lines; in Figure 5, each of the two red lines is tangent to its associated curve, while the green lines are not. In these two examples, the tangent line touches the curve at exactly one point and doesn't cross the curve.

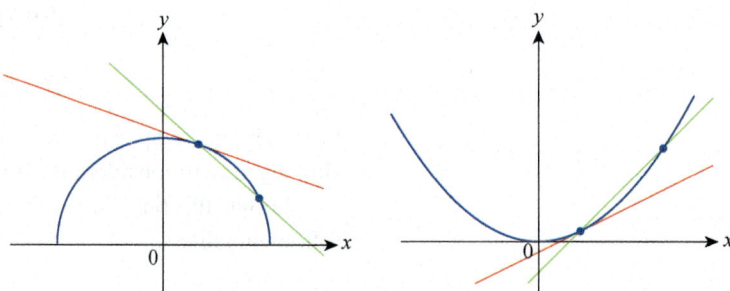

Figure 5 Tangent and Nontangent Lines

In many cases, however, it is not at all clear how to construct the line tangent to a curve at a given point, and, in any event, we will soon need a precise method of construction that results in an equation for a line (as opposed to just a rough sketch). How should we proceed?

Our two simple examples in Figure 5 are perhaps deceptive. If we think a bit harder about what it means for a line to be tangent to a curve, we might be led to the idea that a tangent line should represent the relative rise or fall of the curve at a particular point—in other words, the tangent line should have the same "slope" or trend as the curve at that point. With this interpretation, it's clear that sometimes a tangent line will intersect the curve at more than one point, and that sometimes it is not possible for a unique line to capture the trend at all. Figure 6 illustrates these possibilities.

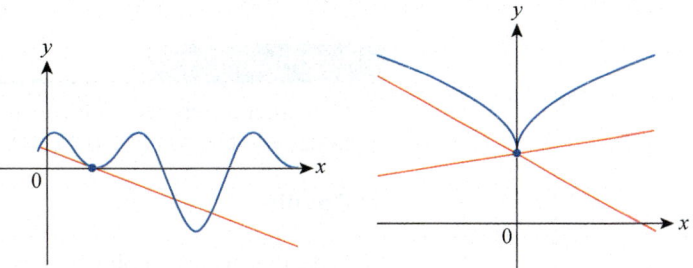

Figure 6 More Tangent and Nontangent Lines

To nail down precisely what we mean by "tangent," and to work toward an equation form for the tangent line, we can follow an approach similar to what we used for the velocity problem. Namely, we will use a process that allows us to construct approximations to the tangent line that are as accurate as we desire.

To set the stage, let's assume that we have a function f for which we want to construct the tangent line at $x = x_0$, that is, at the point $\left(x_0, f\left(x_0\right)\right)$ on the graph of f. Note that all we really need is the slope of the tangent line, since we already know that we want the tangent to pass through the point $\left(x_0, f\left(x_0\right)\right)$. While we don't yet know the slope of the tangent line, we *do* know the slope of any line that passes through $\left(x_0, f\left(x_0\right)\right)$ and a point on the graph of f slightly removed from $\left(x_0, f\left(x_0\right)\right)$; such a line is referred to as a *secant* line (again, not to be confused with the trigonometric function secant). An easy way to denote a "slightly removed" point on the graph is to let $x = x_0 + h$, where h is as close (but not equal) to 0 as we wish it to be. The "slightly removed" point is then $\left(x_0 + h, f\left(x_0 + h\right)\right)$, and the slope of the associated secant line is as follows.

$$\frac{f\left(x_0 + h\right) - f\left(x_0\right)}{\left(x_0 + h\right) - x_0} = \frac{f\left(x_0 + h\right) - f\left(x_0\right)}{h}$$

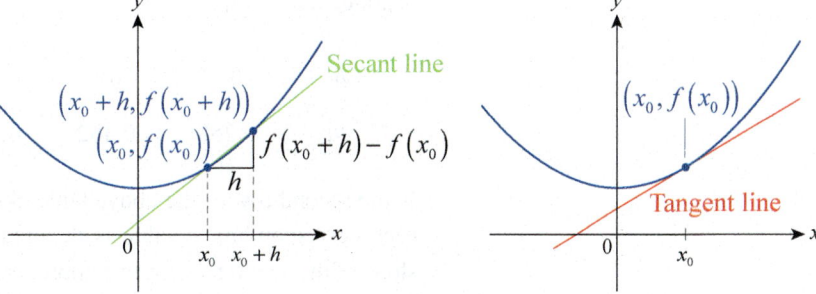

Figure 7

The connection between the tangent problem and the velocity problem is now clear, as we have just constructed a difference quotient again. To make the connection even more apparent, note that in both problems the key concept is "rate of change." In the first, we seek the instantaneous rate of change of the position of an object, and we label the result "velocity." In the second, we seek the rate of change of the graph of a function at a particular point, and we interpret the result as the slope of the line tangent to the graph at that point. The next few examples illustrate how we can actually construct such tangent lines.

Example 3

Construct difference quotients to approximate the slope of the line tangent to the graph of $f(x) = 0.2x^3 - 1.8x^2 + 3.6x$ at the point $(1, 2)$.

Solution

Note that in this problem, $x_0 = 1$. Just as in Examples 1 and 2, we will construct difference quotients using several h-values that are decreasing in magnitude. To start off, letting $h = 1$ we obtain the following.

$$\frac{f(1+1) - f(1)}{1} = f(2) - f(1)$$
$$= 0.2 \cdot 2^3 - 1.8 \cdot 2^2 + 3.6 \cdot 2 - (0.2 \cdot 1^3 - 1.8 \cdot 1^2 + 3.6 \cdot 1)$$
$$= -0.4$$

This means that -0.4 is the slope of the secant line corresponding to $h = 1$. However, given that $x_0 = 1$, our choice of $h = 1$ means that the x-coordinate of the "slightly removed" point in this case is $x = 2$, which is a rather big step away from x_0, so we don't yet expect the above slope to reflect the trend of the graph in any meaningful way. Let us check what happens if $h = 0.5$.

$$\frac{f(1+0.5) - f(1)}{0.5} = 2 \cdot [f(1.5) - f(1)]$$
$$= 2 \cdot [0.2 \cdot (1.5)^3 - 1.8 \cdot (1.5)^2 + 3.6 \cdot 1.5 - (0.2 \cdot 1^3 - 1.8 \cdot 1^2 + 3.6 \cdot 1)]$$
$$= 0.05$$

This shows us that the associated secant line is now nearly horizontal, and its slope went from negative to positive. Decreasing h even further, we evaluate several more difference quotients, and the following table displays our results.

h	0.3	0.1	0.05	0.01	0.001
m_{sec}	0.258	0.482	0.5405	0.58802	0.59880

In the second row of the above table, the slope of the secant associated with each corresponding h-value is denoted by m_{sec}, and we conclude that the slope of the actual tangent line approaches $m = 0.6$. (Later we will learn how to confirm that it actually is equal to 0.6.)

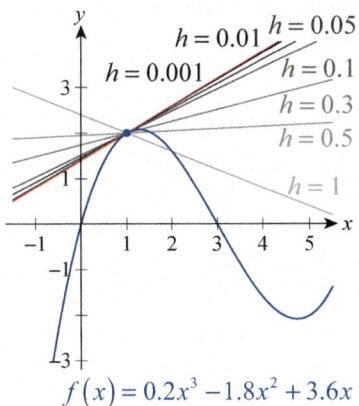

$$f(x) = 0.2x^3 - 1.8x^2 + 3.6x$$

Figure 8

Figure 8 shows the graph of $f(x)$ and the calculated secant lines, as well as the predicted tangent. Note how the secant lines move closer and closer to the tangent as h decreases. Also, the secant line corresponding to $h = 0.001$ is virtually indistinguishable from the tangent line shown in red.

Example 4

Show that the slope of a tangent line of the absolute value function $A(x) = |x|$ at $(0,0)$ can't be inferred from a table of difference quotients.

Solution

In this example, $x_0 = 0$, so using the formula for the difference quotient with, say, $h = 0.5$ we obtain the following.

$$\frac{A(x_0 + h) - A(x_0)}{h} = \frac{|0 + h| - |0|}{h}$$

$$= \frac{|0.5|}{0.5} = 1$$

On the other hand, if h is negative, say $h = -0.5$, the difference quotient is as follows.

$$\frac{A(x_0 + h) - A(x_0)}{h} = \frac{|0 + h| - |0|}{h}$$

$$= \frac{|-0.5|}{-0.5}$$

$$= \frac{0.5}{-0.5} = -1$$

In fact, if we construct a table like before, but this time alternating between positive and negative h-values, something curious happens.

h	0.1	−0.1	0.05	−0.05	0.01	−0.01
m_{sec}	1	−1	1	−1	1	−1

First of all, it seems that ±1 are the only possible values for the slope of the secant, and second, it seems to be +1 when $h > 0$ and −1 when $h < 0$. Notice that this is an observation we can prove fairly easily.

If $h > 0$, since in this case $|h| = h$, the difference quotient becomes

$$\frac{|0 + h|}{h} = \frac{|h|}{h} = \frac{h}{h} = 1,$$

regardless of the value of h. On the other hand, if $h < 0$, then $|h| = -h$, so

$$\frac{|0 + h|}{h} = \frac{|h|}{h} = \frac{-h}{h} = -1,$$

and again, h can be any negative real number.

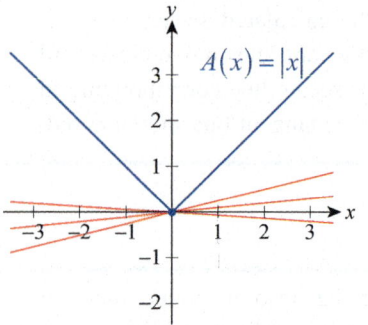

Figure 9

Notice that our results indicate that any secant line corresponding to a positive *h*-value will coincide with the right branch of the graph, while all secant lines obtained from $h < 0$ coincide with the left branch.

What does all this mean? There is no single real number being approached by the difference quotients or slopes as *h* is getting smaller. Graphically, this means that there is no tangent, no single line that would "best capture" the trend of the graph, none that would best align itself to the graph of $A(x)$ at $(0,0)$. The graph has a "sharp turn" there, making it impossible for a tangent line to exist, as shown in Figure 9. We also express this fact by saying that the graph is not "smooth" at $x = 0$.

We might summarize the underlying idea in both the velocity and tangent line problems as follows: *The true value of the quantity we seek can be approximated to any desired degree of accuracy through the use of smaller and smaller increments.* Perhaps it won't come as a surprise that the same technique can be used in a wide variety of settings, not just for velocity and the slope of a tangent. For example, we can approximate the length of a curvilinear segment, such as a portion of the graph of a function, or the area of a plane region bounded by the graph of a function by replacing the graph with series of shorter and shorter line segments. Our last example provides an illustration and serves as a preview of topics we will explore in much greater detail later in the text.

Example 5

Consider the graph of the function $f(x) = x^{3/2}$ on the interval $[0,1]$. By successively dividing $[0,1]$ into smaller and smaller subintervals, estimate

a. the area of the region *R* between the curve and the *x*-axis from $x = 0$ to $x = 1$, and

b. the arc length *s* of the segment of the graph connecting the points $(0,0)$ and $(1,1)$.

Solution

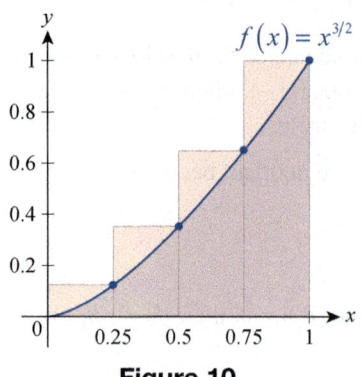

Figure 10

a. First, we divide $[0,1]$ into four subintervals of equal width, by considering $[0,0.25]$, $[0.25,0.5]$, $[0.5,0.75]$, and $[0.75,1]$, respectively. Then we create rectangles of width 0.25 on top of each interval, using the function value at the right endpoint to serve as the height, as shown in Figure 10. Then we simply add up the four areas to serve as our first approximation for the area under the curve.

$$\begin{aligned} S_4 &= A_1 + A_2 + A_3 + A_4 \\ &= 0.25 \cdot f(0.25) + 0.25 \cdot f(0.5) + 0.25 \cdot f(0.75) + 0.25 \cdot f(1) \\ &= 0.25 \cdot (0.25)^{3/2} + 0.25 \cdot (0.5)^{3/2} + 0.25 \cdot (0.75)^{3/2} + 0.25 \cdot (1)^{3/2} \\ &\approx 0.5320 \end{aligned}$$

As is evident from our illustration in Figure 10, this calculation is overestimating the actual area of the region, since the union of the four rectangles completely covers *R* with lots to spare, but we can produce better estimates if we use more and more rectangles by dividing $[0,1]$

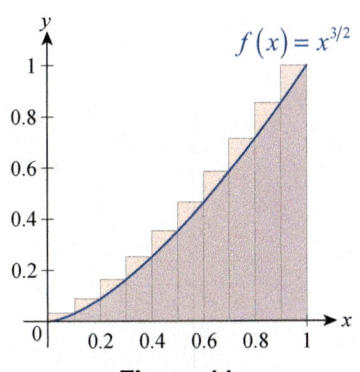

Figure 11

into more and more subintervals. For example, performing a similar calculation with 10 rectangles, each of width 0.1, and again using the function values at the right endpoints for the height of each (see Figure 11), yields the approximation $S_{10} \approx 0.4512$.

We shouldn't be surprised that this number is smaller than our result for S_4, if we compare Figures 10 and 11. S_{10} is still overestimating the true area of R, but as we can see, the error is much less, and we expect it to go down even further if we increase the number of rectangles.

Hand calculations quickly become tedious as n increases, but a computer algebra system or a programmable calculator helps us create a table of values like the one below.

n	50	100	1000	10,000
S_n	0.41005	0.40501	0.40050	0.40005

From the previous table, we conjecture that the true value of the area of R might be $A = 0.4$. Later we will learn how to exactly calculate this type of area.

b. As far as the arc length of the curve connecting $(0,0)$ with $(1,1)$, we will illustrate the calculations by first dividing $[0,1]$ into four subintervals, as we did in part a.

If we label the endpoints as $x_1 = 0$, $x_2 = 0.25$, ..., $x_5 = 1$, and connect the points $(x_i, f(x_i))$ on the graph with $(x_{i+1}, f(x_{i+1}))$ for $i = 1, ..., 4$, a "crude" first approximation for the arc length will simply be the sum of the lengths of the four resulting line segments (see Figure 12). This can be calculated using the Pythagorean Theorem as follows.

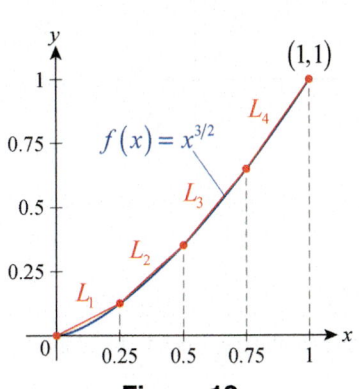

Figure 12

$$s_4 = L_1 + L_2 + L_3 + L_4$$

$$= \sqrt{(0.25)^2 + \left[(0.25)^{3/2}\right]^2} + \sqrt{(0.25)^2 + \left[(0.5)^{3/2} - (0.25)^{3/2}\right]^2}$$

$$+ \sqrt{(0.25)^2 + \left[(0.75)^{3/2} - (0.5)^{3/2}\right]^2} + \sqrt{(0.25)^2 + \left[1 - (0.75)^{3/2}\right]^2}$$

$$\approx 1.4362$$

Upon dividing $[0,1]$ into 10 equal parts, a similar, but a bit longer, calculation yields the much better approximation of

$$s_{10} \approx 1.4389.$$

As before, with the help of a computer or programmable calculator we can generate a table of values such as the following.

n	50	100	1000	10,000
s_n	1.43966	1.43970	1.43971	1.43971

From the table above, we conclude that the true value of s is approximately $s \approx 1.43971$.

2.1 **Exercises**

1–6 Estimate the slope of the tangent line shown in the given graph.

1.
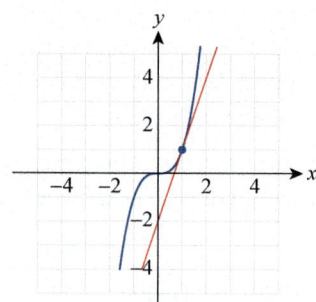

2.
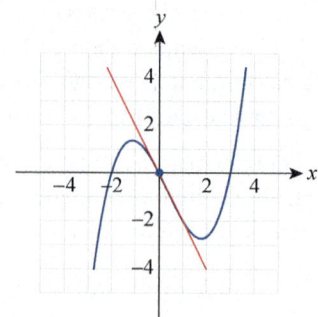

3.
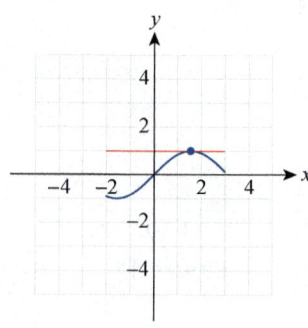

4.
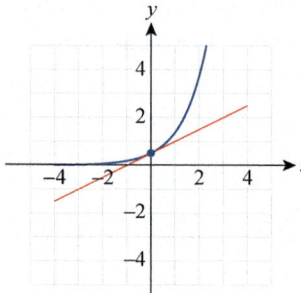

5.
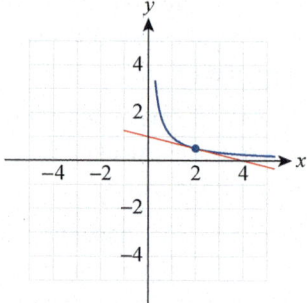

6.
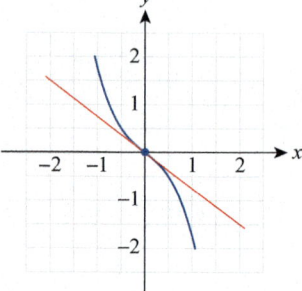

7–18 Use difference quotients to approximate the slope of the tangent to the graph of the function at the given point. Use at least five different h-values that are decreasing in magnitude. (Answers will vary.)

7. $f(x) = 1 - 2x$; $(1, -1)$

8. $g(x) = \dfrac{5}{4}x - 8$; $(8, 2)$

9. $h(x) = \dfrac{1}{3}x^2 - 1$; $(3, 2)$

10. $F(x) = 3 + x - \dfrac{x^2}{2}$; $(4, -1)$

11. $G(x) = \dfrac{1}{4}x^3 - x + 1$; $(-2, 1)$

12. $k(x) = 10 - x^{3/2}$; $(4, 2)$

13. $H(x) = \ln x + 1$; $(e, 2)$

14. $u(x) = \cos x$; $\left(\dfrac{\pi}{2}, 0 \right)$

15. $v(x) = \log 2x - 1$; $(5, 0)$

16. $w(x) = \tan x$; $(0, 0)$

17. $p(x) = -x^4 + 1$; $(1, 0)$

18. $q(x) = x^5 - x + 3$; $(0, 3)$

19. An arrow is shot into the air and its height in feet after t seconds is given by the function $f(t) = -16t^2 + 80t$. The graph of the curve $y = f(t)$ is shown.

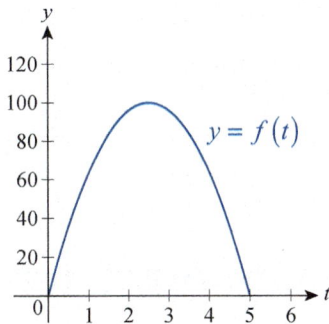

   a. Find the height of the arrow when $t = 2$ seconds.

   b. Find the instantaneous velocity of the arrow when $t = 2$ seconds.

   c. Find the slope of the line tangent to the curve at $t = 2$ seconds.

   d. Find the time it takes the arrow to reach its peak.

20. Suppose that a sailboat is observed, over a period of 5 minutes, to travel a distance from a starting point according to the function $s(t) = t^3 + 60t$, where t is time in minutes and s is the distance traveled in meters.

   a. How far will it travel during the first 6 seconds?

   b. What is the average velocity during the first 6 seconds?

   c. Estimate how fast the boat is moving at the starting point.

   d. Estimate how fast the boat is moving at the end of 3 minutes.

21. A model rocket is fired vertically upward. The height after t seconds is $h(t) = 192t - 16t^2$ feet.

   a. What will be its height at the end of the first second?

   b. What is the average velocity of the rocket during the first second?

   c. Estimate the instantaneous velocity at $t = 0$ seconds.

   d. Estimate the instantaneous velocity at $t = 4$ seconds.

   e. When will the velocity be 0? (**Hint:** You may want to start with the initial velocity you found in part a. and use the fact that under the influence of gravity, when air resistance is ignored, vertical upward velocity decreases by 32 ft/s every second. Once you have a guess, test it by a table of difference quotients.)

22. A particle moving in a straight line is at a distance of $s(t) = 2.5t^2 + 18t$ feet from its starting point after t seconds, where $0 \le t \le 12$. Estimate the instantaneous velocity at **a.** $t = 6$ seconds and **b.** $t = 9$ seconds.

23. The distance, in meters, traveled by a moving particle in t seconds is given by $d(t) = 3t(t+1)$. Estimate the instantaneous velocity at **a.** $t = 0$ seconds, **b.** $t = 2$ seconds, and **c.** at time t_0. (**Hint:** Write the difference quotient corresponding to $t = t_0$, simplify, and try to find the value being approached by the expression as h decreases.)

24. The distance, in meters, traveled by a moving particle in t seconds is given by $d(t) = t^2 - 3t$. Estimate the instantaneous velocity at **a.** $t = 0$ seconds, **b.** $t = 4$ seconds, and **c.*** at time t_0. (See the hint given in part c. of the previous problem.)

25. After start, on a straight stretch of the track, a race car's velocity changes according to the function $v(t) = -1.8t^2 + 18t$, when $0 \le t \le 10$, t is measured in seconds, and $v(t)$ is measured in meters per second.

    a. When does peak velocity occur and what is it? (**Hint:** The graph of $v(t)$ may be helpful.)

    b. When does peak deceleration occur?

    c. Use difference quotients to estimate peak deceleration. Approximately what multiple of $g \approx 9.81 \text{ m/s}^2$ have you obtained?

26. If we ignore air resistance, a falling body will fall $16t^2$ feet in t seconds.

    a. How far will it fall between $t = 2$ and $t = 2.1$?

    b. What is its average velocity between $t = 2$ and $t = 2.1$?

    c. Estimate its instantaneous velocity at $t = 2$.

27. A student dropped a textbook from the top floor of his dorm and it fell according to the formula $S(t) = -16t^2 + 8\sqrt{t}$, where t is the time in seconds and $S(t)$ is the distance in feet from the top of the building.

    a. If the textbook hit the ground in exactly 2.5 seconds, how high is the building?

    b. What was the average speed for the trip?

    c. What was the instantaneous velocity at $t = 1$ second?

    d. What was the velocity of impact?

28–31 Approximate the area of the region between the graph of the function and the x-axis on the given interval. Use **a.** $n = 4$ and **b.** $n = 5$. (Round your answer to four decimal places.)

28. $f(x) = x^2$ on $[0,1]$

29. $g(x) = 16x - x^3$ on $[0,4]$

30. $h(x) = \sin x$ on $[0, \pi]$

31. $F(x) = e^x + 1$ on $[-10, 0]$

32–35 Approximate the arc length of the graph of the function on the given interval. Use **a.** $n = 4$ and **b.** $n = 5$. (Round your answer to four decimal places.)

32. $f(x) = \sqrt{x}$ on $[0,1]$

33. $g(x) = x^3 + x^2$ on $[-1, 0]$

34. $F(x) = \cos x$ on $\left[0, \dfrac{\pi}{2}\right]$

35. $G(x) = \ln x + 1$ on $[1, 2]$

2.1 Technology Exercises

36–39 Use a graphing calculator or computer algebra system to graph $f(x)$ along with three secant lines at the indicated x-value, corresponding to the difference quotients with h-values of 0.2, 0.1, and 0.01, respectively. Can you come up with a possible equation for the tangent? Use technology to test your conjecture.

36. $f(x) = x^2$; $x = 2$

37. $f(x) = -x^3 + x + 1$; $x = \dfrac{\sqrt{3}}{3}$

38. $f(x) = \sin x + \cos x$; $x = 0$

39. $f(x) = 3\sqrt{x}$; $x = 4$

40–43 Use a graphing calculator or computer algebra system to graph the given function $f(x)$ along with

$$D(x) = \frac{f(x + 0.001) - f(x)}{0.001}$$ in the same coordinate system. Explain how the function values of $D(x)$ are reflected on the graph of $f(x)$.

40. $f(x) = x^4$ **41.** $f(x) = x(3 - x)$

42. $f(x) = \sin x$ **43.** $f(x) = \ln x$

44–47 Use a graphing calculator or computer algebra system to find the x-values at which the graph of $f(x)$ does not have a tangent line. Explain.

44. $f(x) = -|x - 1| + 1$ **45.** $f(x) = |x^2 - 4|$

46. $f(x) = |\ln x|$ **47.** $f(x) = (x - 1)^{2/3}$

48–51. Use a computer algebra system to find approximations for the areas in Exercises 28–31 by using **a.** $n = 100$ and **b.** $n = 1000$. (Round your answers to four decimal places.)

52–55. Use a computer algebra system to find approximations for the arc lengths in Exercises 32–35 by using **a.** $n = 100$ and **b.** $n = 1000$. (Round your answers to four decimal places.)

2.2 Limits All Around the Plane

In the previous section we studied two broadly defined problems whose solutions called for the ability to determine the trend of some behavior. Namely, we sought to fully understand the nature of a fraction called the difference quotient as the denominator in the fraction approached 0. This need to exactly determine "limiting" behavior is a characteristic of calculus, and we will devote much of the next four sections to refining our understanding of limits.

Topic One
Limits in Verbal, Numerical, and Visual Forms

We encounter the notion of limiting behavior in many different contexts, mathematical and otherwise. In the velocity problem of Section 2.1, we looked at the behavior of the average rate of travel over shorter and shorter time intervals; that is, we were looking for the value of the rate of travel "in the limit" as the length of the interval shrank to 0. Similarly, in the tangent problem we sought the limiting behavior of the secant lines as the horizontal difference between two points on a graph shrank to 0.

Every limit scenario, when described verbally, contains phrases of a dynamic nature. Examples of such phrases are "as h approaches 0," "as x goes to 3," and "as N grows without bound." In a less mathematical setting, the phrases might be something like "mortgage rates are expected to approach 5 percent" or "assuming the national debt keeps getting bigger and bigger." We might also see tables of data that hint at some limiting behavior.

In order to progress mathematically, we need notation that precisely describes whatever limit situation is under discussion; every such mathematical scenario can be expressed in terms of a function and a variable whose value is approaching a given point. We will begin with an informal introduction of limit notation here and follow up with a formal definition in Section 2.3.

For the purpose of discussion, let us suppose we have a function $f(x)$ defined on an open interval containing c, except possibly at c itself (the reason for this exception will be explained shortly). We say that the limit of $f(x)$ is L as x approaches c, and write

$$\lim_{x \to c} f(x) = L,$$

if the value of $f(x)$ is closer and closer to L as x takes on values closer and closer (but not equal) to c.

Note the small but important caveat: the limit of a function $f(x)$ as x approaches c does not depend *at all* on the value of f at c; in fact, the limit may exist when f is not even defined at c. This technical detail highlights one distinction between the mathematical meaning of "limit" and casual usage of the word—in everyday language, the boundary between limiting behavior and behavior at the limit point is often blurred. The distinction is important mathematically because, as we will see repeatedly, the behavior of a function near the point c, but not at c, is the key.

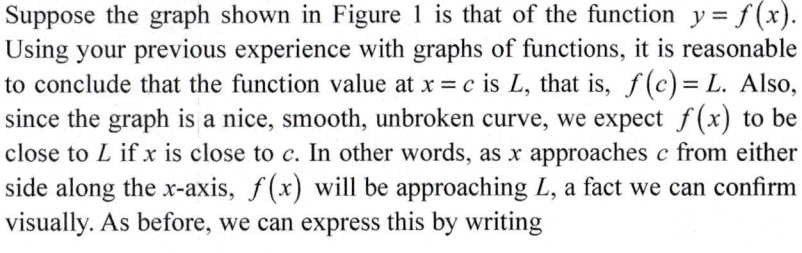

Example 1

Suppose the graph shown in Figure 1 is that of the function $y = f(x)$. Using your previous experience with graphs of functions, it is reasonable to conclude that the function value at $x = c$ is L, that is, $f(c) = L$. Also, since the graph is a nice, smooth, unbroken curve, we expect $f(x)$ to be close to L if x is close to c. In other words, as x approaches c from either side along the x-axis, $f(x)$ will be approaching L, a fact we can confirm visually. As before, we can express this by writing

$$\lim_{x \to c} f(x) = L.$$

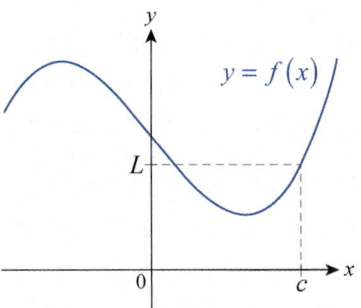

Figure 1

Next, suppose we redefine $f(c)$ to be $L/2$ (or some other number), but do not change anything else about f (let's call the resulting function $\widehat{f}$). Graphically, this means that the point $\left(c, \widehat{f}(c)\right)$ "jumps out" of the graph, and moves down to the $L/2$ level, as shown in Figure 2. In other words, by redefining a single function value, we have relocated *one point* $\left(c, f(c)\right)$ from the graph of f, but we haven't changed a bit the behavior of the function *near* $\left(c, f(c)\right)$. If x takes on values closer and closer to c, but not equal to c, the function value $\widehat{f}(x)$ will still move closer and closer to L, which is the second coordinate of the empty circle (i.e., its altitude over the x-axis). In other words, we can still write

$$\lim_{x \to c} \widehat{f}(x) = L.$$

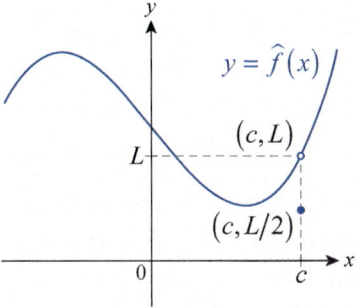

Figure 2

Finally, we will consider one last scenario. Suppose that we delete the function value corresponding to $x = c$ altogether, thus making the function undefined at $x = c$. We will call the resulting function $\tilde{f}$. The empty circle in Figure 3 indicates that one point is "missing" from the graph; that is, the vertical line $x = c$ doesn't meet the graph at all. However, notice that this still doesn't have any influence on the function values *near* $x = c$. As before, when x approaches c, but is not equal to c, the function values $\tilde{f}(x)$ approach L, that is, $\lim_{x \to c} \tilde{f}(x) = L$.

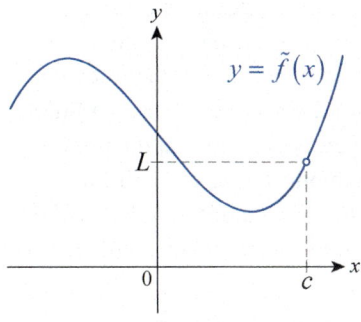

Figure 3

The conclusion to be drawn is that $\lim_{x \to c} f(x)$ may exist without the actual function value $f(c)$ being equal to it or, indeed, even defined.

Example 2

According to a classic paradox, you can never eat an apple for the following reason. A certain amount of time is needed to consume the first half of the apple, and at that point half the apple remains. Additional time is needed to eat half of the remaining half, which constitutes one quarter of the original apple. After that, more time is needed to consume half of the remaining quarter, or one eighth of the apple, and so on. By this reasoning, the process of eating the apple never ends.

Of course, this runs counter to our experience, so where does the problem lie? If you were challenged by an argument such as the one above, could you provide a mathematically correct resolution of the paradox?

Solution

As we shall see, the concept of limits can help resolve the apparent flaw in the argument. Let us assume that a calculus student performs the experiment, and holding a stopwatch in her left hand, eats an apple in exactly 4 minutes. We will make the realistic assumption that the time necessary to eat a certain portion of the apple is directly proportional to its size, and we will monitor the portion of the fruit having been consumed. Proceeding as in the above argument, we first note that half of the apple is gone in the first 2 minutes. At the end of the third minute, another quarter is gone, so we stand at $\frac{1}{2}+\frac{1}{4}=\frac{3}{4}$ of the apple consumed, with one minute left on the clock. At 3 and a half minutes, $\frac{3}{4}+\frac{1}{8}=\frac{7}{8}$ is gone, and at 3 minutes and 45 seconds, $\frac{7}{8}+\frac{1}{16}=\frac{15}{16}$ is gone. With half of the remaining time left, that is, at 3 minutes and 52.5 seconds, $\frac{15}{16}+\frac{1}{32}=\frac{31}{32}$ of the apple is gone, and so on. If we examine the fractions we obtain, it is easy to see that they get closer and closer to 1 as time progresses toward 4 minutes. In fact, a little later, we will have the rigorous mathematical tools to *prove* that the portion of the apple consumed, as time approaches 4 minutes, is as close to 1 as we desire. Mathematically, we say that the limit of these values is 1, that is, the student indeed ate the entire apple in 4 minutes, in accordance with what we would expect from our own experience.

As important as it is to understand what it means when we say that a certain limit exists, it is just as important to understand what it means when a limit fails to exist. We will again defer the technical definition of this notion to Section 2.3, but we can develop our intuitive grasp of the idea and introduce some more notation now.

Assume again that we have a function f defined on an open interval containing the point c, except possibly at c itself. The limit of $f(x)$ as x approaches c then fails to exist if the values of $f(x)$ do not approach any particular real number L. This could occur for various reasons: $f(x)$ might be getting larger and larger (in absolute value), the values of $f(x)$ may "jump" as x passes from one side of c to the other, or the values of $f(x)$ may simply not "settle down" to any particular value no matter how narrowly we focus on the region around c. Figure 4 illustrates these three possibilities.

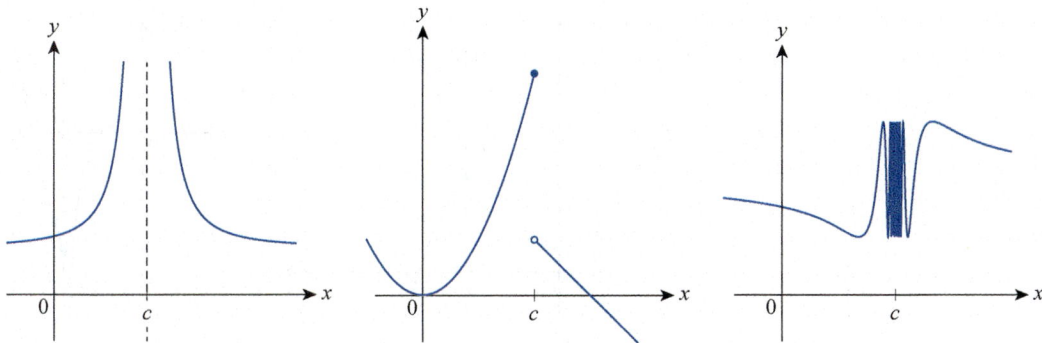

Figure 4 Failure of a Limit to Exist

While potentially confusing, we use the notation

$$\lim_{x \to c} f(x) = \infty \ \text{ or } \ \lim_{x \to c} f(x) = -\infty$$

to indicate the nonexistence of a limit when it fails to exist because the function f either increases without bound (approaches positive infinity) or decreases without bound (approaches negative infinity) as $x \to c$. It should be remembered, however, that in either scenario the limit does not exist since f is not approaching a fixed real number. The notation simply describes the particular *way* in which the limit fails to exist.

Topic Two
Vertical Asymptotes and One-Sided Limits

You have encountered vertical asymptotes before; several examples have already appeared in this text in the review of rational functions and some trigonometric functions such as tangent. The topic under discussion now is not so much the vertical asymptote itself, but the language and notation we use to describe the behavior of a function *near* a vertical asymptote.

A distinguishing feature of a vertical asymptote is that, at least from one side, the function under discussion grows in magnitude without bound. Our first task is to introduce notation that succinctly and accurately describes such behavior. Assume the function f is defined on an open interval with right endpoint c. We write

$$\lim_{x \to c^{-}} f(x) = \infty$$

if the values of $f(x)$ increase without bound as x approaches the point c from the left. Similarly, we write

$$\lim_{x \to c^{-}} f(x) = -\infty$$

if the values of $f(x)$ decrease without bound (approach negative infinity) as x approaches the point c from the left. Figure 5 illustrates these two scenarios.

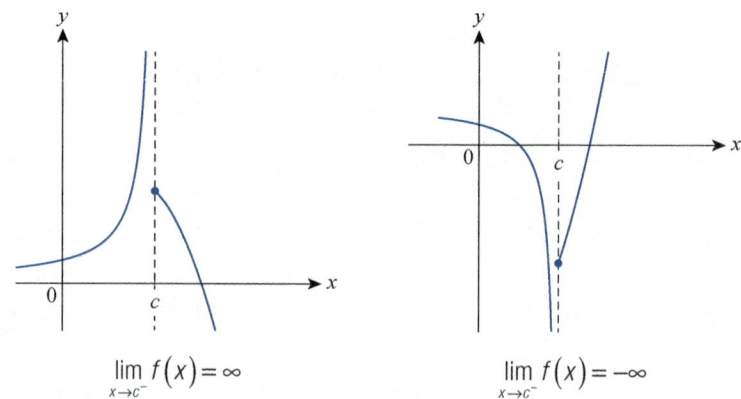

$$\lim_{x \to c^-} f(x) = \infty \qquad\qquad \lim_{x \to c^-} f(x) = -\infty$$

Figure 5 Asymptotic Behavior from the Left

Similarly, we use the notation $\lim_{x \to c^+} f(x)$ in discussing the limit of the function f as x approaches the point c from the right, with the implicit assumption that f is defined on an open interval whose *left* endpoint is c. And as noted before, we need to be careful in interpreting the four statements $\lim_{x \to c^-} f(x) = \infty$, $\lim_{x \to c^-} f(x) = -\infty$, $\lim_{x \to c^+} f(x) = \infty$, and $\lim_{x \to c^+} f(x) = -\infty$; all four indicate that these *one-sided* limits of f at c do not exist, since f is either increasing or decreasing without bound.

You may be wondering why the clause "defined on an open interval" keeps appearing in these limit discussions. The reason is that in order for us to study the limit of a function as x approaches a point c, we need to know that f is defined at all points near c (but not necessarily at c). The exact meaning of *near* will vary from one application to another, but by stipulating that f is defined on an open interval around c, or to one side or the other in the case of one-sided limits, we can be sure that $f(x)$ always represents a real number as x approaches c.

Example 3

Use one-sided limit notation to describe the behavior of the following functions near their vertical asymptote(s).

a. $f(x) = \dfrac{-x}{x-1}$ **b.** $g(x) = \dfrac{x^2+1}{x^2-x-6}$ **c.** $h(x) = \dfrac{1}{(x-1)^2}$

Solution

a. As we can see from the graph in Figure 6, the sole vertical asymptote occurs at $x = 1$, a fact we can deduce from the formula as well. A further examination of the graph reveals that as x approaches 1 from the left, the function values increase without bound, a fact we can express as follows:

$$\lim_{x \to 1^-} f(x) = \infty$$

On the other hand, as x approaches 1 from the right, the function values decrease:

$$\lim_{x \to 1^+} f(x) = -\infty$$

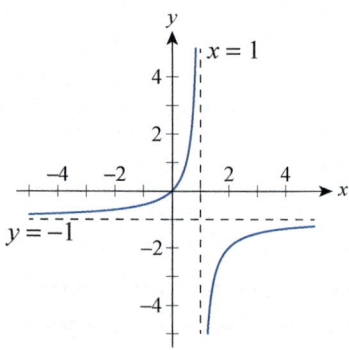

Figure 6 $f(x) = \dfrac{-x}{x-1}$

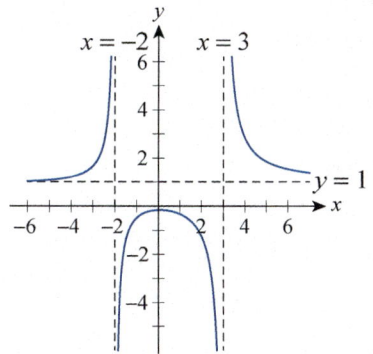

Figure 7 $g(x) = \dfrac{x^2 + 1}{x^2 - x - 6}$

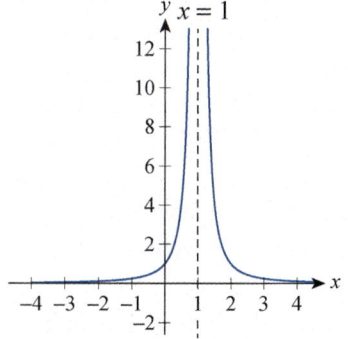

Figure 8 $h(x) = \dfrac{1}{(x-1)^2}$

b. As seen from the graph in Figure 7, $g(x)$ has two vertical asymptotes, at $x = -2$ and $x = 3$, a fact we can confirm algebraically by factoring the denominator of $g(x)$. The function values increase without bound as x approaches -2 from the left and decrease in a similar manner as x approaches -2 from the right. Therefore, we can write $\lim\limits_{x \to -2^-} g(x) = \infty$ and $\lim\limits_{x \to -2^+} g(x) = -\infty$.

Examining the graph on both sides of the vertical asymptote $x = 3$ in a similar manner, we conclude that $\lim\limits_{x \to 3^-} g(x) = -\infty$ and $\lim\limits_{x \to 3^+} g(x) = \infty$.

c. Examining the graph of $h(x)$ in Figure 8, we see that the behavior is the same on both sides of the vertical asymptote $x = 1$; namely, that of the function values increasing without bound as x approaches 1. This makes it unnecessary to distinguish between the two one-sided limits in this case; instead, we can simply write $\lim\limits_{x \to 1} h(x) = \infty$.

Example 4

Deduce the one-sided infinite limits of $f(x) = \dfrac{1}{3 - x}$ at its vertical asymptote without graphing the function.

Solution

We see from the formula that the sole vertical asymptote of $f(x)$ occurs at $x = 3$ (which is where the denominator is 0). We will examine the two one-sided limits separately. Let us assume first that x is approaching 3 from the left. This means that the x-values are moving closer and closer to $x = 3$ through values less than 3 (though approaching it in magnitude). For example, when $x = 2.99$,

$$f(2.99) = \frac{1}{3 - 2.99}$$

$$= \frac{1}{0.01} = 100.$$

Throughout this process of x approaching 3, $f(x)$ stays positive, but its denominator decreases; thus the function values themselves increase in magnitude.

A couple of easy calculations underscore this observation:

$$f(2.999) = \frac{1}{3 - 2.999} = \frac{1}{0.001} = 1000,$$

$$f(2.9999) = \frac{1}{3 - 2.9999} = \frac{1}{0.0001} = 10,000,$$

and so on. The closer x is to 3, the greater the function value $f(x)$ is in magnitude (remember, x never actually reaches the value of 3).

We can now summarize our observations by writing

$$\lim_{x \to 3^-} \frac{1}{3 - x} = \infty.$$

If, on the other hand, x approaches 3 from the right, the denominator and thus the function value becomes negative. As an example,

$$f(3.01) = \frac{1}{3 - 3.01}$$
$$= \frac{1}{-0.01} = -100.$$

It is still true that the closer x is to 3, the greater the function values are in magnitude, but this time, they are all negative. For example, as you can easily check,

$$f(3.001) = -1000,$$
$$f(3.0001) = -10,000,$$

and so on. As before, we can express this by writing

$$\lim_{x \to 3^+} f(x) = -\infty.$$

We can briefly summarize the above as follows: the presence of a vertical asymptote $x = 3$ indicates that the values of $f(x)$ increase or decrease without bound near $x = 3$.

In general, these tendencies may or may not be different on the two sides of the asymptote, that is, for $x < 3$ and $x > 3$, respectively. We were, however, able to find out whether the respective one-sided limits were $+\infty$ or $-\infty$ from the sign of the expression as x approached 3 from a fixed side, even though a graph was not readily available.

With two-sided and one-sided limit notation in hand, we can now provide a more formal definition of vertical asymptote.

Definition

Vertical Asymptote

We say that a function f has a **vertical asymptote** at the point c if at least one of the following statements holds:

$$\lim_{x \to c^-} f(x) = \infty \qquad \lim_{x \to c^+} f(x) = \infty \qquad \lim_{x \to c} f(x) = \infty$$

$$\lim_{x \to c^-} f(x) = -\infty \qquad \lim_{x \to c^+} f(x) = -\infty \qquad \lim_{x \to c} f(x) = -\infty$$

And now that one-sided limit notation has been introduced, it is easy to extend its meaning. We write

$$\lim_{x \to c^-} f(x) = L \ \text{ or } \ \lim_{x \to c^+} f(x) = L$$

if the values of $f(x)$ get closer and closer to L as x approaches c from, respectively, the left or right. The next example illustrates the use of this notation.

Example 5

Use the graph of the function h in Figure 9 to decide whether the given one-sided limits exist. For those that do, find their values.

a. $\lim\limits_{x \to 0^+} h(x)$ **b.** $\lim\limits_{x \to 2^-} h(x)$ **c.** $\lim\limits_{x \to 2^+} h(x)$

d. $\lim\limits_{x \to 2.5^-} h(x)$ **e.** $\lim\limits_{x \to 2.5^+} h(x)$ **f.** $\lim\limits_{x \to 3^-} h(x)$

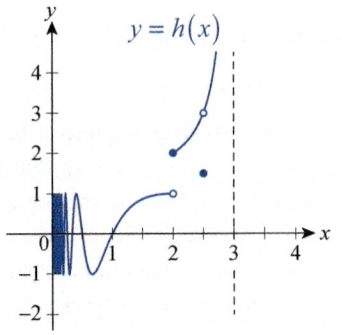

Figure 9

Solution

a. We see from the graph that $h(x)$ oscillates "wildly" between $+1$ and -1 on the right-hand side of $x = 0$. Moreover, the closer x moves to 0, the more oscillations we have; in fact, there are infinitely many oscillations near 0. (Our technology is incapable of graphing infinitely many "waves," which is the reason why we see a "shaded rectangle" immediately to the right of the y-axis.) Therefore, there is no single value being approached by the values of $h(x)$, so we conclude that $\lim\limits_{x \to 0^+} h(x)$ does not exist.

b. As x approaches 2 from the left, the function values are approaching 1, so we conclude

$$\lim_{x \to 2^-} h(x) = 1.$$

Notice, however, that the actual function value is $h(2) = 2$. In general, we cannot expect the limit of a function at c to always equal the function value at c. In other words, as we previously observed in Example 1, what a function does *near* c may not be the same as what it does *at* c.

c. Examining the tendency of the function values as x approaches 2 from the right, we see that

$$\lim_{x \to 2^+} h(x) = h(2) = 2.$$

In other words, the one-sided limit and the function value agree in this case.

d–e. The graph shows that the function values $h(x)$ approach 3 as x approaches 2.5 from either direction, so we can write

$$\lim_{x \to 2.5^-} h(x) = \lim_{x \to 2.5^+} h(x) = 3.$$

Notice that just like in Example 1, since both one-sided limits exist and are equal, we conclude that the *two-sided* limit exists at 2.5, and

$$\lim_{x \to 2.5} h(x) = 3.$$

The last important observation we wish to make here is that the function value at $x = 2.5$ differs from the limit, so once again

$$\lim_{x \to 2.5} h(x) \neq h(2.5).$$

f. Because $x = 3$ is a vertical asymptote, $\lim\limits_{x \to 3^-} h(x)$ does not exist, but we can write

$$\lim_{x \to 3^-} h(x) = \infty.$$

Topic Three

Horizontal Asymptotes and Limits at Infinity

We have just seen examples of when it is appropriate to say that a given limit (two-sided or one-sided) *is* $\pm\infty$; we will now study limits *at* $\pm\infty$.

Just as an unbounded limit at a point c corresponds to a vertical asymptote at that point, there is a connection between horizontal asymptotes and limits at infinity. Recall that a horizontal line passing through L on the y-axis is a horizontal asymptote for the function f if the graph of f approaches the line as x is allowed to either increase or decrease without bound. A given function can have either 0, 1, or 2 horizontal asymptotes, as illustrated in Figure 10; unlike vertical asymptotes, a function can cross a horizontal asymptote any number of times (or not at all).

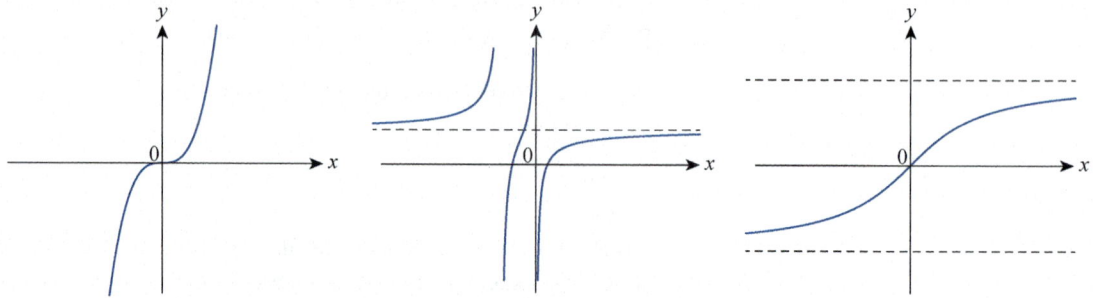

Figure 10 Limits at Infinity

And as with vertical asymptotes, the notion of limit is exactly what is needed to make the definition of horizontal asymptote more precise.

Definition

Horizontal Asymptote

We say that the line $y = L$ is a **horizontal asymptote** for the function f if either of the following statements is true.

$$\lim_{x \to \infty} f(x) = L \text{ or } \lim_{x \to -\infty} f(x) = L$$

Example 6

Identify the horizontal asymptotes of the following functions using their graphs shown in Figures 11 and 12, respectively.

a. $f(x) = \dfrac{4}{\pi}\arctan x$ **b.** $g(x) = e^{-x}\cos 5x$

Solution

a. As shown by the graph in Figure 11, the values of $f(x)$ approach the value of 2 as x increases without bound, and they will be as close to 2 as we wish if we make sure x is large enough. In other words, we can write

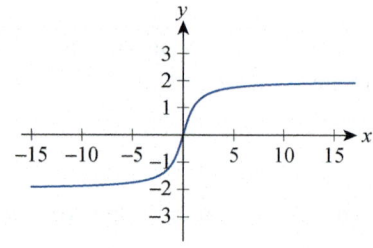

Figure 11 $f(x) = \dfrac{4}{\pi}\arctan x$

$$\lim_{x \to \infty} f(x) = 2$$

and conclude that the line $y = 2$ is a horizontal asymptote for f. Similarly, we conclude that

$$\lim_{x \to -\infty} f(x) = -2,$$

and thus the line $y = -2$ is another horizontal asymptote for f.

b. The graph of g, though oscillating, is approaching the x-axis more and more as x increases (see Figure 12). In other words, as we will soon be able to prove, the function values of $g(x)$ will be as small in absolute value as we wish, if we merely choose x large enough. This is precisely what we need to conclude that the x-axis or $y = 0$ is a horizontal asymptote for g. Using limit notation, this means

$$\lim_{x \to \infty} g(x) = 0.$$

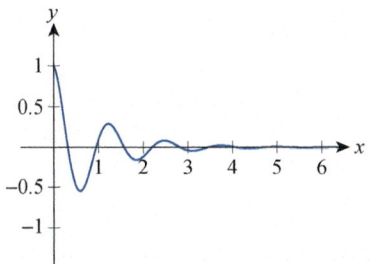

Figure 12 $g(x) = e^{-x} \cos 5x$

As a final remark, since the cosine factor causes $g(x)$ to oscillate, the graph actually crosses its horizontal asymptote infinitely often.

Many functions tend to increase in magnitude without bound as x goes to ∞ or $-\infty$; the first graph in Figure 10 is an example. In such cases, it is natural to use limit notation to provide more information on the behavior of the function. We write

$$\lim_{x \to \infty} f(x) = \infty$$

if the values of f increase without bound as $x \to \infty$, with similar meanings attached to the remaining three possibilities,

$$\lim_{x \to \infty} f(x) = -\infty, \; \lim_{x \to -\infty} f(x) = \infty, \; \text{and} \; \lim_{x \to -\infty} f(x) = -\infty.$$

Example 7

Decide whether the given functions increase or decrease without bound as x goes to ∞ and $-\infty$. Use limit notation to express your conclusion.

a. $f(x) = -x^2$ **b.** $g(x) = x^3 - 7$ **c.** $h(x) = x^5 - 6x^2 + 3$

Solution

a. First, we will examine the function values $f(x)$ as x goes to ∞. Our first observation is that for any positive real number x, if $x > 1$ then $x^2 > x$, so if x tends to ∞, then x^2 will certainly grow in magnitude, without bound. Therefore, $f(x) = -x^2$ will *decrease* without bound as x tends to ∞. Using limit notation,

$$\lim_{x \to \infty} f(x) = -\infty.$$

Next, let us assume x goes to $-\infty$. Like before, x^2 will grow without bound, and since x^2 is positive for any nonzero x, we conclude that $f(x) = -x^2$ again decreases without bound. Thus,

$$\lim_{x \to -\infty} f(x) = -\infty.$$

b. We again let x go to ∞ first. Then by an argument similar to the one given in part a., x^3 grows without bound. This tendency is not influenced by subtracting 7 from x^3, so

$$\lim_{x \to \infty} g(x) = \infty.$$

If on the other hand, x goes to $-\infty$, $x^3 - 7$ will decrease without bound, since the cube of a negative number is negative. Thus we conclude that

$$\lim_{x \to -\infty} g(x) = -\infty.$$

c. As for the function h, we will first argue that its values grow without bound as x tends toward ∞. The reasoning is as follows. It is clear that the first term, x^5 increases without bound as x goes to ∞. And as x grows, the fifth-degree term increases so much faster than the quadratic term, that their difference $x^5 - 6x^2$ still increases. As an illustration, when $x = 3$, $x^5 - 6x^2 = 3^5 - 6 \cdot 9 = 243 - 54 = 189$, but when, say, $x = 10$, $x^5 - 6x^2 = 10^5 - 6 \cdot 10^2 = 100{,}000 - 600 = 99{,}400$. Adding 3 to the previous expression (or adding or subtracting any constant, for that matter) will not change the increasing tendency of the function values, so we conclude that

$$\lim_{x \to \infty} h(x) = \infty.$$

We can argue in a similar fashion in the case of x going to $-\infty$, but since the fifth power of a negative number is negative, x^5 *decreases* without bound in this case. This time, however, subtracting $6x^2$ only helps (but like before, the quadratic term would not be able to change the process, even if we added it to x^5). Again, the constant term is immaterial from the perspective of the tendency of the function values, so we conclude that

$$\lim_{x \to -\infty} h(x) = -\infty.$$

As a final remark, you may recall having discussed the "end behaviors" of polynomial functions in your precalculus course, though at that time most likely without the limit notation.

x	f(x)
0.00001	0.25
0.000001	0.2
0.0000001	0
0.00000001	0

Table 1

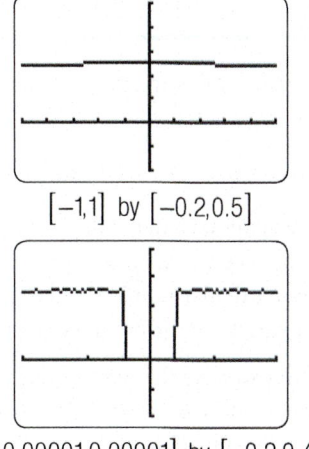

$[-1,1]$ by $[-0.2, 0.5]$

$[-0.00001, 0.00001]$ by $[-0.2, 0.4]$

Figure 13

Exploring $\lim\limits_{x \to 0} f(x)$

Using a Graphing Calculator

Topic Four

Limits and Technology

To this point, we have relied upon graphs, tables, intuition, and elementary reasoning in order to arrive at guesses for limits that we feel confident about. In the next two sections, we will learn a number of mathematical facts (in the forms of theorems and lemmas) that allow us to accurately determine limits of functions without the need for guesswork. But we will occasionally make use of technological aids such as graphing calculators and computer algebra systems in order to guide our explorations. We will close this section with a few examples of the use and misuse of technology in determining limits.

Consider the function $f(x) = \dfrac{\sqrt{x^2 + 4} - 2}{x^2}$.

If we needed to determine the behavior of this function as x goes to 0, without the benefit of the tools we will soon encounter, we might take several different approaches. One approach would be to note that the denominator of the fraction goes to 0 as x goes to 0, so we might guess that the function "blows up", that is, increases in magnitude without bound. That guess would be wrong—our analysis so far misses an important fact. A closer examination of the fraction reveals that the numerator also approaches 0 as x goes to 0, and this common situation can lead to a variety of possible outcomes. A second approach would be to evaluate this function for a number of values of x that are increasingly small; this may or may not lead to a correct guess for the limit, depending on the specific values chosen for x (see Table 1). And a third approach would be to use a graphing calculator (see Figure 13) or computer algebra system (see Figure 14) to graph the function near 0. Again, this may or may not lead to a correct guess, depending on the viewing window specified for the graph.

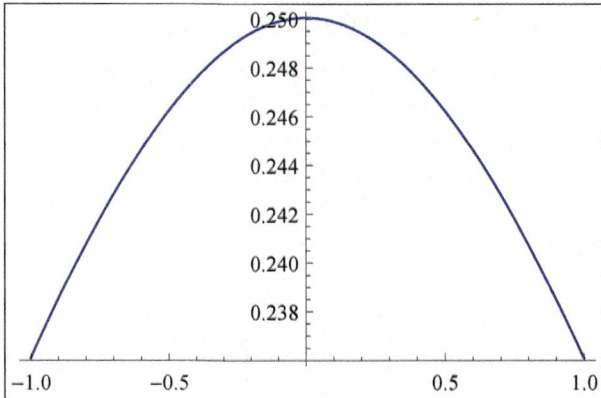

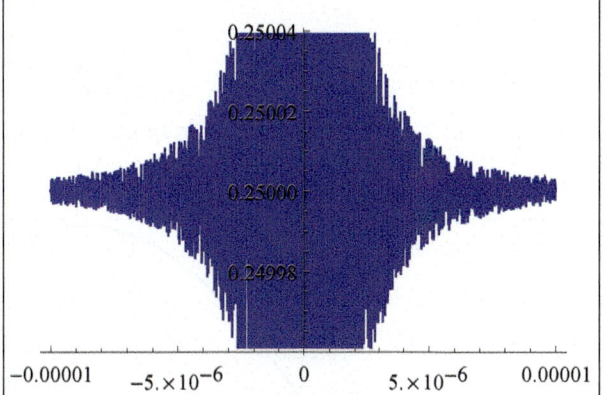

Figure 14 Exploring $\lim\limits_{x \to 0} f(x)$ Using a Computer Algebra System

What is going on, and what results should we trust? The reason we are getting conflicting guidance from our technology is a loss of *significant digits* as the operations of subtraction and division are performed. All numbers used in computation in a calculator or software program are subject to rounding errors, as only a certain number of digits are kept in memory unless specifically directed otherwise. This leads to a loss of precision that is especially problematic in subtracting one number from another that is nearly the same (as in the numerator of $f(x)$) and when calculating the ratio of two small but similar numbers.

A prolonged exploration and a weighing of all the results would probably lead us to eventually guess that the limit of this function as x approaches 0 is $\frac{1}{4}$, but we might be left with a shadow of doubt. Fortunately, the limit theorems of the next two sections will provide us the means to remove all doubt in such situations.

Example 8

Use graphing technology to predict the value of $\lim\limits_{x \to 0} g(x)$, where
$$g(x) = \frac{1 - \cos x}{x}.$$

Solution

Since $\cos x$ approaches 1 as x goes to 0, we once again face the ambiguity of both the numerator and denominator approaching 0 as x goes to 0. Again we don't have enough information to guess at any particular number since, as we said before, a "0/0-type" limit can lead to a variety of different answers.

Using a graphing calculator, we can sketch and examine the graph, which appears to be approaching the origin as x goes to 0 (see Figure 15). This is confirmed by a table of values that we can create with the help of a calculator, using decreasing x-values.

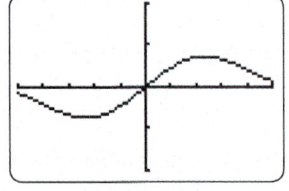

Figure 15
$$g(x) = \frac{1 - \cos x}{x}$$
on $[-5,5]$ by $[-2,2]$

x	$g(x)$
0.1	0.05
0.01	0.005
0.001	0.0005
0.0001	0.00005

From the above observations it is natural to expect that the function values will steadily move toward 0 when x approaches 0, as illustrated by the first graph in Figure 16. However, if we zoom in, again something strange may happen. We used a computer algebra system to illustrate.

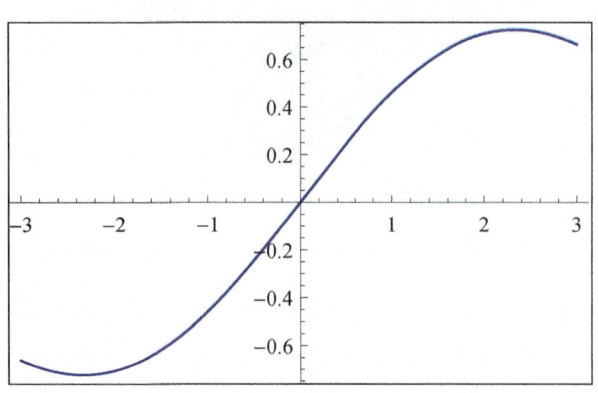

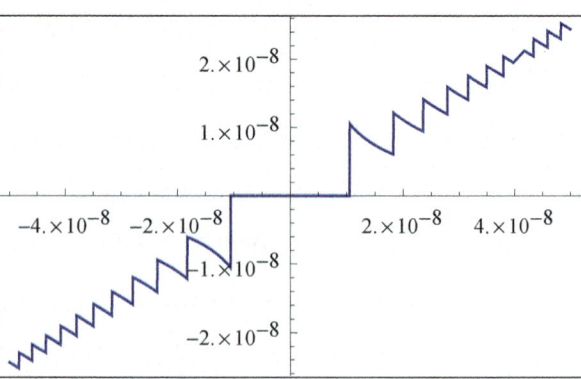

Figure 16 Exploring $\lim\limits_{x \to 0} g(x)$ Using a Computer Algebra System

Again we see that, because of rounding errors and other reasons, our technology can sometimes mislead us by giving seemingly conflicting or inaccurate feedback, and we must be aware of this when using it.

We will stick with our guess that

$$\lim_{x \to 0} g(x) = \lim_{x \to 0} \frac{1 - \cos x}{x} = 0,$$

but realize that we haven't actually proved this at all. We will learn how to do that after discussing limit theorems in upcoming sections.

Technology Note

Computer algebra systems such as *Mathematica* provide additional tools for determining limits, but it should always be remembered that software has limitations and can be fooled. *Mathematica* contains the built-in command **Limit** that uses the same mathematical facts we will learn in the next two sections to correctly evaluate many types of limits. Its use is illustrated below. (For additional information on *Mathematica* and the use of the **Limit** command, see Appendix A.)

```
In[1]:= Limit[(Sqrt[x^2 + 4] - 2) / x^2, x → 0]

Out[1]=  1
        ─
         4
```

Figure 17

2.2 **Exercises**

1–4 Use the graph of the function to find the indicated limit (if it exists).

1. $\displaystyle\lim_{x\to3} f(x)$

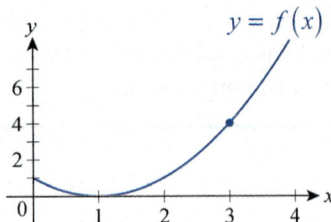

2. $\displaystyle\lim_{x\to1} g(x)$

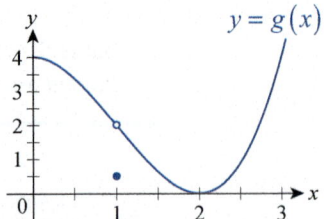

3. $\displaystyle\lim_{x\to0} h(x)$

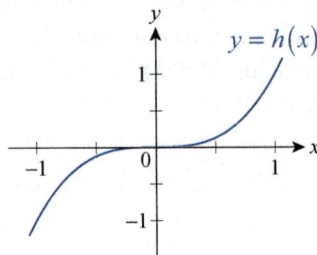

4. $\displaystyle\lim_{x\to1} k(x)$

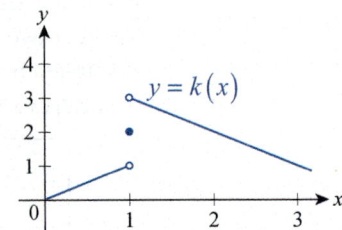

5–12 Create a table of values to estimate the value of the indicated limit without graphing the function. Choose the last x-value so that it is no more than 0.001 units from the given c-value.

5. $\displaystyle\lim_{x\to\sqrt{2}} \frac{x^2-2}{x-\sqrt{2}}$

6. $\displaystyle\lim_{x\to3} \frac{x^3-9x^2+27x-27}{x-3}$

7. $\displaystyle\lim_{x\to1} \frac{x^{10}-1}{x-1}$

8. $\displaystyle\lim_{x\to0} \frac{4\sin x}{3x}$

9. $\displaystyle\lim_{x\to\pi} \frac{2\cos x-1}{1-\sin x}$

10. $\displaystyle\lim_{x\to7^-} \frac{x^2-49}{x-7}$

11. $\displaystyle\lim_{x\to7^+} \frac{x^2+49}{x-7}$

12. $\displaystyle\lim_{x\to0^+} \frac{\sqrt{4+x}}{x}$

13–24 Use one-sided limit notation to describe the behavior of the function near its vertical asymptote(s).

13.

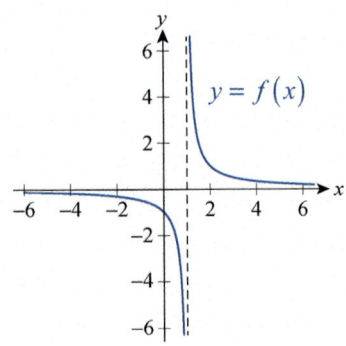

14.

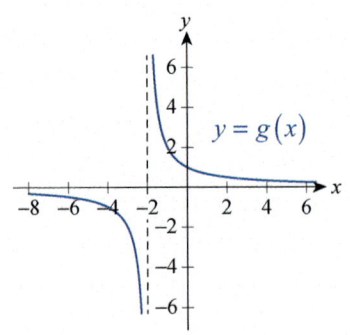

15.

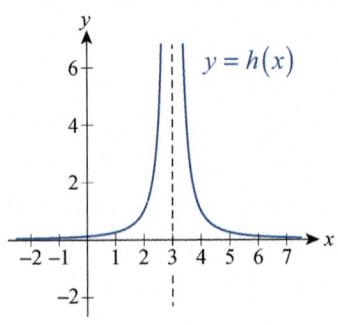

16.

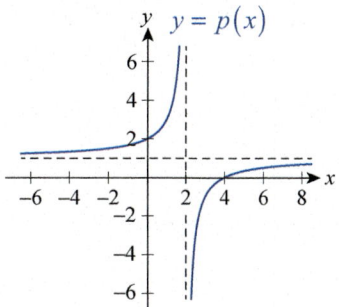

17.

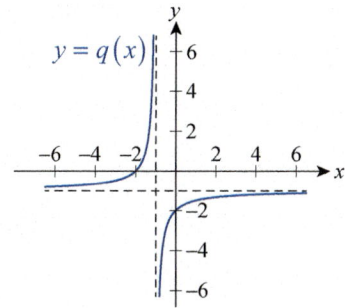

18.

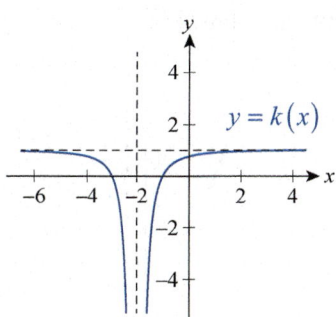

19.

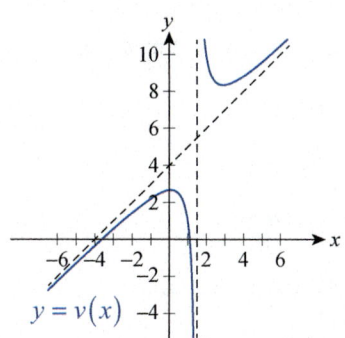

20.

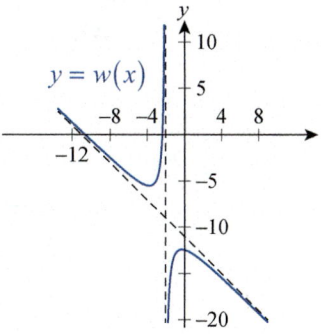

21.

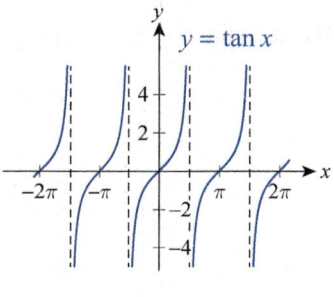

22.

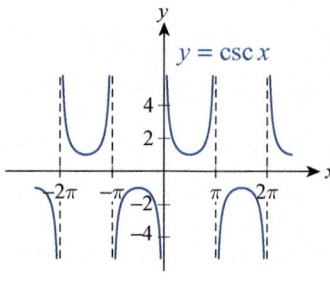

23.

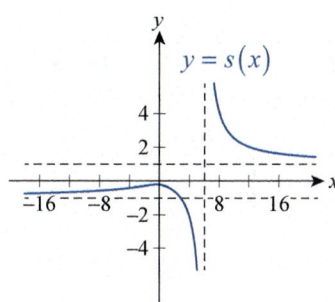

24.

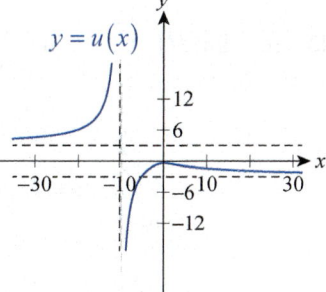

25–36. Consider the functions given in Exercises 13–24. Find their limits at ∞ and $-\infty$ (if they exist). When applicable, use the horizontal asymptote(s) as a guide.

37–46 Use the graph to find the indicated one-sided limits, if they exist.

37. a. $\displaystyle\lim_{x\to 2^-} f(x)$ **b.** $\displaystyle\lim_{x\to 2^+} f(x)$

38. a. $\displaystyle\lim_{x\to 1^-} g(x)$ **b.** $\displaystyle\lim_{x\to 1^+} g(x)$

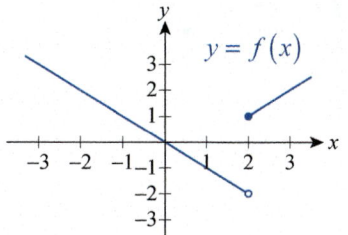

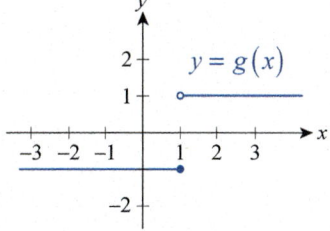

39. **a.** $\displaystyle\lim_{x\to1^-} h(x)$ **b.** $\displaystyle\lim_{x\to1^+} h(x)$

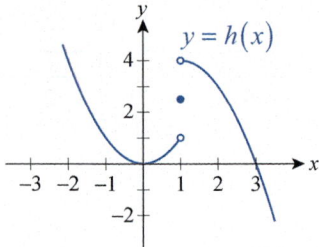

40. **a.** $\displaystyle\lim_{x\to(-\pi/2)^-} F(x)$ **b.** $\displaystyle\lim_{x\to(-\pi/2)^+} F(x)$

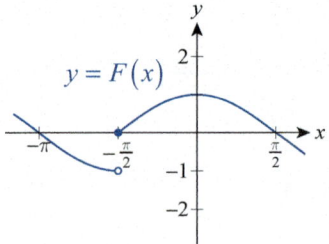

41. **a.** $\displaystyle\lim_{x\to0^-} G(x)$ **b.** $\displaystyle\lim_{x\to0^+} G(x)$

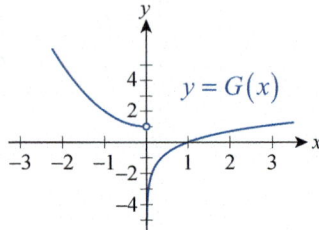

42. **a.** $\displaystyle\lim_{x\to-1^-} k(x)$ **b.** $\displaystyle\lim_{x\to-1^+} k(x)$

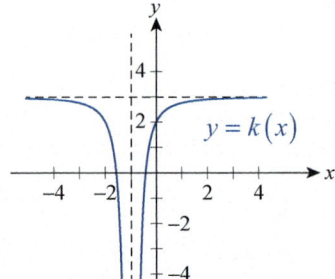

43. **a.** $\displaystyle\lim_{x\to0^-} H(x)$ **b.** $\displaystyle\lim_{x\to0^+} H(x)$

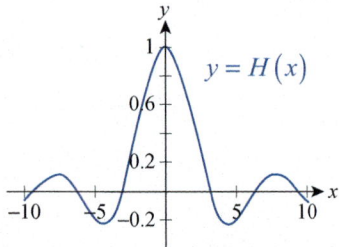

44. **a.** $\displaystyle\lim_{x\to2^-} u(x)$ **b.** $\displaystyle\lim_{x\to2^+} u(x)$

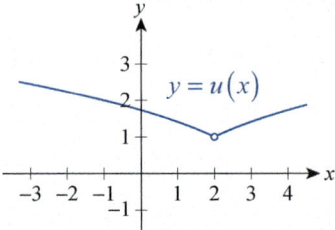

45. **a.** $\displaystyle\lim_{x\to0^-} v(x)$ **b.** $\displaystyle\lim_{x\to0^+} v(x)$

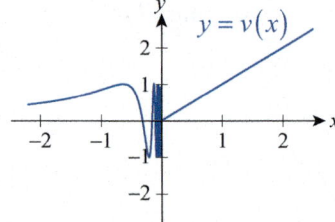

46. **a.** $\displaystyle\lim_{x\to0^-} w(x)$ **b.** $\displaystyle\lim_{x\to0^+} w(x)$

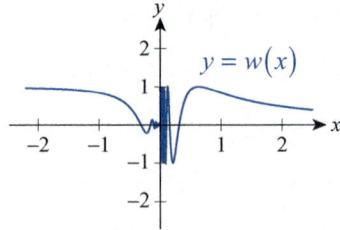

47–58 Use limit notation to describe the unbounded behavior of the given function as x approaches ∞ and/or $-\infty$.

47. $f(x) = x^3$

48. $g(x) = x^2 + 2.1x - 1$

49. $h(x) = -x^4 + 0.2x^3$

50. $k(x) = -0.35x^5 + x + 1.35$

51. $F(x) = \sqrt{x+2}$

52. $G(x) = \sqrt[3]{x+1} - 2.3$

53. $H(x) = |x+2|$

54. $K(x) = -|x+2| - 1$

55. $u(x) = |x-1| + |x+2|$

56. $v(x) = e^{x+2}$

57. $s(x) = -10^{-x} + 1$

58. $t(x) = \ln x - 1$

59–63 *True or False?* Determine whether the given statement is true or false. In case of a false statement, explain or provide a counterexample.

59. If $\lim_{x \to c} f(x)$ does not exist, then $f(x)$ is undefined at $x = c$.

60. If $f(x)$ is undefined at $x = c$, then $\lim_{x \to c} f(x)$ does not exist.

61. If $f(x)$ is defined on $(0, \infty)$ and $y = 0$ is a horizontal asymptote for $f(x)$, then there exists a number M such that if $x > M$ then $f(x) < 1/10^6$.

62. If $f(x)$ has a vertical asymptote at $x = c$, then either $\lim_{x \to c} f(x) = \infty$ or $\lim_{x \to c} f(x) = -\infty$.

63. If $\lim_{x \to c} f(x)$ does not exist, then $\lim_{x \to c^-} f(x) \neq \lim_{x \to c^+} f(x)$ or at least one of $\lim_{x \to c^-} f(x)$ or $\lim_{x \to c^+} f(x)$ does not exist.

2.2 Technology Exercises

64–71 Use a graphing calculator or computer algebra system to decide whether the given limit exists by evaluating the function at several x-values approaching the indicated c-value. Then graph the function to confirm your findings. Do you obtain misleading graphs when choosing small viewing windows?

64. $\lim_{x \to 2} \dfrac{x^2 - 5x + 6}{x - 2}$

65. $\lim_{x \to 3} \dfrac{x - 2}{x^2 - 5x + 6}$

66. $\lim_{x \to -1.5} \dfrac{x^2 - 2.25}{x + 1.5}$

67. $\lim_{x \to 1} \dfrac{x - 1}{\sqrt{x} - 1}$

68. $\lim_{x \to 3} \dfrac{\sqrt{x+1} - 2}{x - 3}$

69. $\lim_{x \to 0} \dfrac{\sin 3x}{2x}$

70. $\lim_{x \to 0^+} \cos \dfrac{1}{x}$

71. $\lim_{x \to 0^+} x \cos \dfrac{1}{x}$

72. Evaluate the function $f(x) = \left(1 + \dfrac{1}{x}\right)^x$ for several consecutive positive integers, and try to observe a tendency. Then use a graphing calculator or computer algebra system to graph $f(x)$ in a large viewing window and try to guess $\lim_{x \to \infty} f(x)$. Have you seen that number before?

73. Write a program in a graphing calculator or computer algebra system to estimate the limit of an input function as x approaches c. (Calculate $f(x)$ successively at x-values increasingly close to c and display the results.) Try your program on Exercises 64–71.

74–81. Use the Limit command specific to your computer algebra system to evaluate the limits in Exercises 64–71. Are your previous results confirmed?

2.3 The Mathematical Definition of Limit

As we have seen, determining the limit of a function at a given point is not necessarily a trivial task. Pictures, tables, or a cursory examination of a formula (if one is available) may not be sufficient to allow us to accurately evaluate a limit, and technological aids can also lead us astray. In this section, we present a rigorous mathematical definition of *limit* that will then let us develop an array of tools that make limit evaluation both precise and tractable.

Topic One

Limits Formally Defined

We still have only an informal notion of what the statement $\lim_{x \to c} f(x) = L$ means. Now that our limit intuition has begun to develop, we can refine our informal notion and make the definition precise and mathematically useful. To be exact, we need to capture the idea that all the values of $f(x)$ become as close to L as we care to specify as x gets close enough to c.

The two parts of that previous sentence that need more work are "as close to L as we care to specify" and "as x gets close enough to c." The most common definition of limit refines those two phrases through the use of two Greek letters, epsilon (ε) and delta (δ), and the following is often referred to as the *epsilon-delta* definition of limit.

> ### Definition
>
> **Formal Definition of Limit**
>
> Let f be a function defined on an open interval containing c, except possibly at c itself. We say that the **limit of $f(x)$ as x approaches c is L**, and write $\lim_{x \to c} f(x) = L$, if for every number $\varepsilon > 0$ there is a number $\delta > 0$ such that $|f(x) - L| < \varepsilon$ whenever x satisfies $0 < |x - c| < \delta$.

The use of the letters ε and δ originated with the French mathematician Augustin-Louis Cauchy (1789–1857). His choice was deliberate, and served to remind the reader of the words *error* and *difference* (the French spellings are similar); you may find this correspondence useful as well. We may now interpret the statement $\lim_{x \to c} f(x) = L$ in this way: if we wish to guarantee that the *error* between $f(x)$ and L is less than the amount ε, it suffices to make sure that the *difference* between x and c is less than the amount δ (but remember that we do not allow x to equal c).

Figure 1 shows the relationship between L, c, ε, and δ for a given function f. The successive images illustrate the way in which δ depends on ε: as we specify smaller values for the error ε, the difference δ correspondingly shrinks. But in each image, any x chosen between $c - \delta$ and $c + \delta$ (except c itself) results in a value for $f(x)$ that is between $L - \varepsilon$ and $L + \varepsilon$.

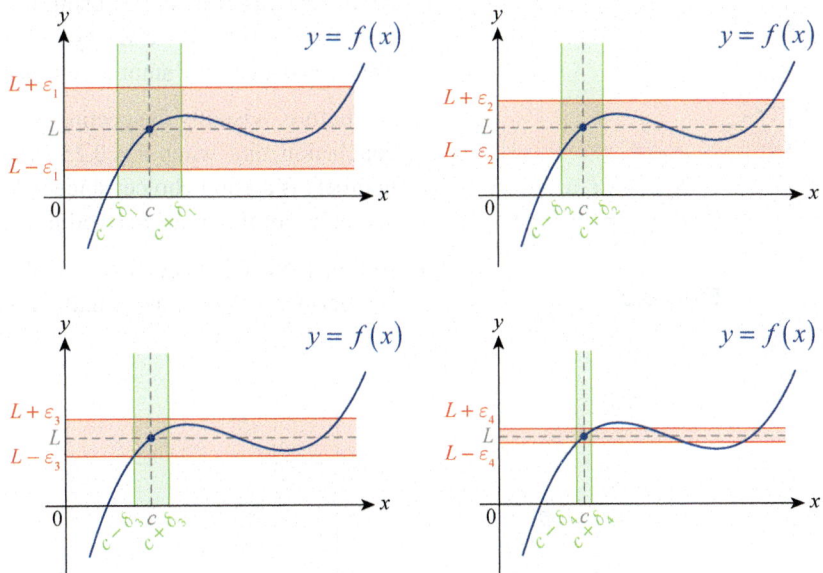

Figure 1 Correspondence between δ and ε

The true value of the formal limit definition is that it allows us to prove limit claims absolutely. In fact, you might consider the correspondence between ε and δ as a series of challenges and responses. If a skeptic wished to challenge the claim that $\lim_{x \to c} f(x) = L$ for the function f shown in Figure 1, each of the four ε's might be presented in turn with the demand "tell me a condition that guarantees $f(x)$ is within ε of L." The response, for each ε, would be the corresponding δ: if $x \neq c$ is chosen between $c - \delta$ and $c + \delta$ then $f(x)$ is guaranteed to fall between $L - \varepsilon$ and $L + \varepsilon$.

Note that the δ in each image is actually the largest possible choice—any positive number smaller than the pictured δ would also serve to guarantee $\left| f(x) - L \right| < \varepsilon$. Also note that, for this function and for the four ε's shown, the largest possible δ is dictated by where the line $y = L - \varepsilon$ intersects the graph of f. In the first and second images, in particular, it is clear that x could be selected from a longer interval to the right of c and still force $\left| f(x) - L \right| < \varepsilon$. Any larger δ than that shown would not suffice, however, for points selected to the left of c.

Example 1

In our first example, we will use the graph of $f(x) = -0.9x^2 + 2$ with $c = 0.5$ to estimate the value of δ corresponding to **a.** $\varepsilon = 1$, **b.** $\varepsilon = 0.5$, **c.** $\varepsilon = 0.25$, and **d.** $\varepsilon = 0.1$.

Solution

a. Like in our discussion before, we see that for $\varepsilon = 1$, the value of δ is dictated by where the line $y = L - \varepsilon$ intersects the graph of f (see Figure 2). If, for example, we choose $\delta = 0.65$, we can ensure that for any $x \neq c$ between $c - \delta$ and $c + \delta$, $f(x)$ will fall between $L - \varepsilon$ and $L + \varepsilon$.

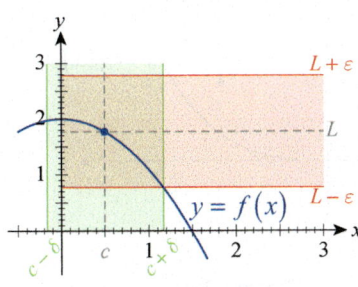

Figure 2

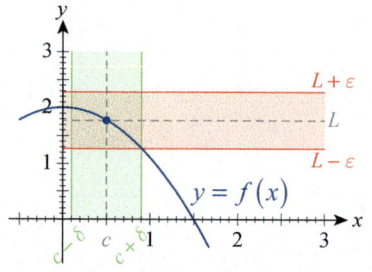

Figure 3

b. As for the case of $\varepsilon = 0.5$, using the illustration in Figure 3, it is clear that if $x \ne c$ falls between, say, 0.1 and 0.9, then $L - \varepsilon < f(x) < L + \varepsilon$. Thus $\delta = 0.4$ (or any smaller number) is a good choice.

c. As before, visually inspecting the graph in Figure 4 leads to the conclusion that when $\varepsilon = 0.25$, the value of $\delta = 0.2$ (or any smaller number) is a good choice, since 0.2 is less than the distance between c and either of the green vertical lines.

d. Again, a visual inspection reveals that $\varepsilon = 0.1$ will require a much smaller δ; $\delta = 0.1$ (or any smaller number) works well (see Figure 5).

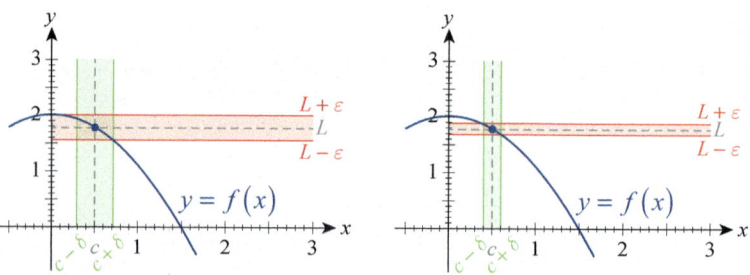

Figure 4   **Figure 5**

Example 2

Consider the following function:

$$g(x) = \begin{cases} x^3 & \text{if } x \ne 0.7 \\ 1 & \text{if } x = 0.7 \end{cases}$$

The empty and full circles in the graph in Figure 6 reflect the fact that x^3 has been redefined at $x = 0.7$: the coordinates of the "hole" are $\left(0.7, (0.7)^3\right) = (0.7, 0.343)$, while $g(0.7) = 1$.

In addition, we can also learn from Figure 6 that there is a $\delta > 0$ corresponding to $\varepsilon = 0.3$ so that if we choose any $x \ne c$ from the interval $(c - \delta, c + \delta)$, the corresponding g-value will fall within ε of $(0.7)^3 = 0.343$.

Figures 7, 8, and 9 are illustrations of the existence of δ-values corresponding to smaller and smaller ε's:

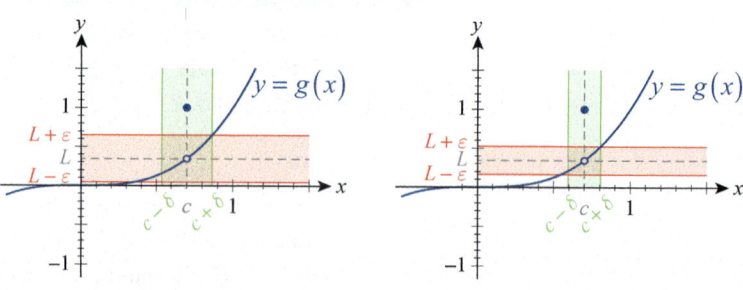

Figure 6   **Figure 7**

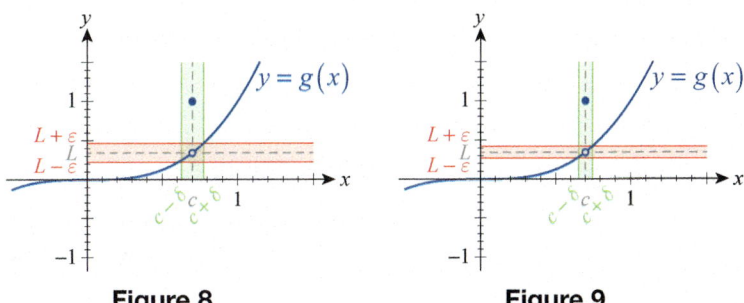

Figure 8   **Figure 9**

From these illustrations, it is at least visually clear that

$$\lim_{x \to 0.7} g(x) = (0.7)^3 = 0.343,$$

even though the function value $g(0.7) = 1$. In other words,

$$\lim_{x \to 0.7} g(x) \neq g(0.7);$$

the limit and the function value at $c = 0.7$ are not equal. More is actually true: we could define $g(0.7)$ to be *any* real number, without affecting the limit. In general, the function value $g(c)$ and the limit $\lim_{x \to c} g(x)$ are entirely independent of each other, that is, the existence and/or value of one does not affect that of the other.

Our formal definitions for the other varieties of limits are similar. Figures 10 and 11 illustrate the construction of the largest δ corresponding to a given ε when one-sided limits are under consideration.

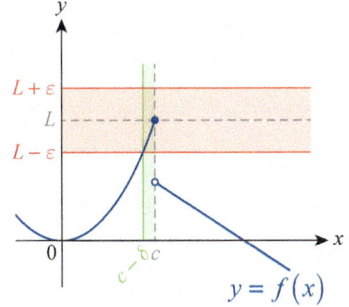

Figure 10 Left-Hand Limit

Definition

Limits from the Left (Left-Hand Limits)

Let f be a function defined on an open interval whose right endpoint is c. We say that the **limit of $f(x)$ as x approaches c from the left is L**, and write $\lim_{x \to c^-} f(x) = L$, if for every number $\varepsilon > 0$ there is a number $\delta > 0$ such that $|f(x) - L| < \varepsilon$ whenever x satisfies $c - \delta < x < c$.

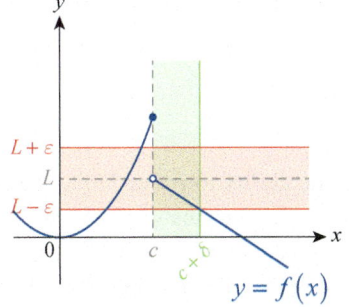

Figure 11 Right-Hand Limit

Definition

Limits from the Right (Right-Hand Limits)

Let f be a function defined on an open interval whose left endpoint is c. We say that the **limit of $f(x)$ as x approaches c from the right is L**, and write $\lim_{x \to c^+} f(x) = L$, if for every number $\varepsilon > 0$ there is a number $\delta > 0$ such that $|f(x) - L| < \varepsilon$ whenever x satisfies $c < x < c + \delta$.

We will wrap up the formal limit definitions with cases in which the symbol ∞ makes an appearance.

Definition

Infinite Limits

Let f be a function defined on an open interval containing c, except possibly at c itself. We say that the **limit of $f(x)$ as x approaches c is positive infinity**, and write $\lim_{x \to c} f(x) = \infty$, if for every positive number M there is a number $\delta > 0$ such that $f(x) > M$ whenever x satisfies $0 < |x - c| < \delta$.

Similarly, we say that the **limit of $f(x)$ as x approaches c is negative infinity**, and write $\lim_{x \to c} f(x) = -\infty$, if for every negative number N there is a number $\delta > 0$ such that $f(x) < N$ whenever x satisfies $0 < |x - c| < \delta$. **Infinite one-sided limits** are defined in a manner analogous to finite one-sided limits.

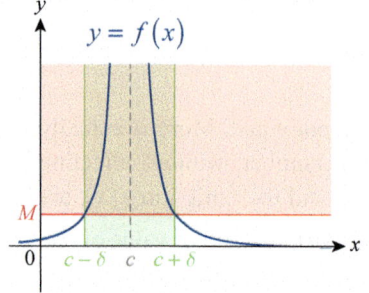

Figure 12

Correspondence between M and δ for an Infinite Limit

If we wish to prove that a given function has an infinite limit at a point c, our task is to prove that given some (presumably large) number M, there is an interval $(c - \delta, c + \delta)$ on which the function is larger than M (except possibly at the point c itself). Figure 12 shows how an appropriate δ can be chosen for the given value of M.

Finally, limits at infinity are defined formally as follows.

Definition

Limits at Infinity

Let f be a function defined on some interval (a, ∞). We say the **limit of f at infinity is L**, and write $\lim_{x \to \infty} f(x) = L$, if for every number $\varepsilon > 0$ there is a number N such that $|f(x) - L| < \varepsilon$ for all $x > N$.

Similarly, for a function defined on some interval $(-\infty, b)$ we say the **limit of f at negative infinity is L**, and write $\lim_{x \to -\infty} f(x) = L$, if for every number $\varepsilon > 0$ there is a number N such that $|f(x) - L| < \varepsilon$ for all $x < N$.

Be sure you understand how these formal definitions of limits at infinity relate to horizontal asymptotes.

Topic Two

Proving a Limit Exists

Although the formal definition of limit gives us the ability to prove limit claims beyond a shadow of a doubt, it does not immediately give us the means by which to determine the value of a given limit in the first place. In other words, we need to know what L is before we can prove that $\lim_{x \to c} f(x) = L$. Fortunately, as we will soon see, our limit definition is the basis for a number of theorems that will make determination of limits much easier.

As a stepping-stone toward those theorems, we will develop our ability to use the limit definition to prove some example claims.

Example 3

Use the ε-δ definition of limits to prove that $\lim\limits_{x\to 3}(2x-1)=5$.

Solution

Suppose an $\varepsilon > 0$ is given. We need to find a $\delta > 0$ such that

$$0 < |x-3| < \delta \Rightarrow |(2x-1)-5| < \varepsilon.$$

Notice that this latter inequality is equivalent to the following:

$$|2x-6| < \varepsilon \Leftrightarrow |2(x-3)| < \varepsilon$$
$$\Leftrightarrow |2|\cdot|x-3| < \varepsilon$$
$$\Leftrightarrow 2|x-3| < \varepsilon$$
$$\Leftrightarrow |x-3| < \frac{\varepsilon}{2}$$

Reading the above chain of equivalent inequalities from the bottom up, this means that $\delta = \varepsilon/2$ (or any smaller number) works. In other words, given $\varepsilon > 0$, if we choose $\delta = \varepsilon/2$, then $0 < |x-3| < \delta$ implies

$$|(2x-1)-5| = |2x-6| = |2(x-3)| = 2|x-3| < 2\delta = 2\cdot\frac{\varepsilon}{2} = \varepsilon,$$

just as we needed to show.

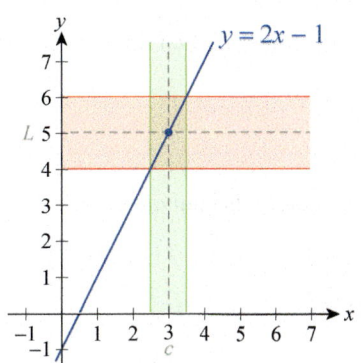

Figure 13

What all this means is that if a skeptic were to challenge you with a small $\varepsilon > 0$ of his or her choosing, you could always respond, for each ε, by picking $\delta = \varepsilon/2$: if $x \ne c$ is chosen between $3 - \delta$ and $3 + \delta$, then $2x - 1$ is guaranteed to fall between $5 - \varepsilon$ and $5 + \varepsilon$. For example, given $\varepsilon = 0.1$, let $\delta = 0.05$. For the reason why x needs to be "twice as close" to 3 as the skeptic's challenge, examine the graph in Figure 13, and note that the slope of the line is 2.

Example 4

Use the ε-δ definition of limits to prove that $\lim\limits_{x\to 4} x^2 = 16$.

Solution

Again, supposing that $\varepsilon > 0$ is given, we need to find a $\delta > 0$ such that

$$0 < |x-4| < \delta \Rightarrow |x^2 - 16| < \varepsilon.$$

Our first observation is that

$$|x^2 - 16| = |(x+4)(x-4)| = |x+4|\cdot|x-4|.$$

Next, we will agree to choose a $\delta > 0$ while also making sure that $\delta < 1$. Note that we can do this, since once a successful $\delta > 0$ is chosen, any smaller number works just fine.

Since $|x-4| < \delta$ translates into $4 - \delta < x < 4 + \delta$, which implies $3 < x < 5$ if $\delta < 1$, we have $|x+4| = x+4 < 5+4 = 9$ and thus

$$|x^2 - 16| = |x+4| \cdot |x-4| < 9|x-4|.$$

So for the given $\varepsilon > 0$, let $\delta > 0$ be chosen so that it is less than the smaller of the numbers 1 and $\varepsilon/9$.

Then if $|x-4| < \delta$,

$$|x^2 - 16| = |x+4| \cdot |x-4| < 9|x-4| < 9\delta < 9 \cdot \frac{\varepsilon}{9} = \varepsilon,$$

this latter inequality finishing our proof.

Example 5

Use the ε-δ definition of limits to prove that $\lim\limits_{x \to \infty} \dfrac{x^2 + 1}{x^2} = 1$.

Solution

As before, let $\varepsilon > 0$ be given. We need to find a number N such that

$$x > N \Rightarrow \left| \frac{x^2 + 1}{x^2} - 1 \right| < \varepsilon.$$

Notice that since $\dfrac{x^2 + 1}{x^2} = 1 + \dfrac{1}{x^2}$, this latter inequality is equivalent to

$$\left| \left(1 + \frac{1}{x^2} \right) - 1 \right| < \varepsilon \Leftrightarrow \left| \frac{1}{x^2} \right| < \varepsilon$$

$$\Leftrightarrow \frac{1}{x^2} < \varepsilon,$$

since x^2 is nonnegative. Also observe that if we choose N large enough so that $1/N^2 < \varepsilon$, then $x > N$ will imply $1/x^2 < 1/N^2 < \varepsilon$. The above choice is always possible, by simply making sure $N > 1/\sqrt{\varepsilon}$.

We can summarize our observations as follows. For a given $\varepsilon > 0$, choose and fix a positive number N satisfying $N > 1/\sqrt{\varepsilon}$. Then if $x > N$,

$$\left| \frac{x^2 + 1}{x^2} - 1 \right| = \left| 1 + \frac{1}{x^2} - 1 \right| = \left| \frac{1}{x^2} \right| = \frac{1}{x^2} < \frac{1}{N^2} < \frac{1}{\left(\dfrac{1}{\sqrt{\varepsilon}} \right)^2} = \frac{1}{\dfrac{1}{\varepsilon}} = \varepsilon,$$

and this latter chain of inequalities finishes our argument.

Example 6

Prove that $\lim\limits_{x\to 2} \dfrac{1}{(x-2)^2} = \infty$.

Solution

For an arbitrary, but fixed, positive number M, we need to exhibit a number $\delta > 0$ such that $1/(x-2)^2 > M$ whenever x satisfies $0 < |x-2| < \delta$. To that end, let us assume that $M > 0$ is fixed, and choose a $\delta > 0$ small enough so that $1/\delta^2 \geq M$. Note that this is possible; for example, $\delta = 1/\sqrt{M}$ (or any smaller number) works: if $0 < |x-2| < \delta$, then

$$\frac{1}{(x-2)^2} > \frac{1}{\delta^2} = \frac{1}{\left(\dfrac{1}{\sqrt{M}}\right)^2} = \frac{1}{\dfrac{1}{M}} = M,$$

which is precisely the inequality we wanted.

As a final remark, we wish to emphasize that $\lim\limits_{x\to 2} \dfrac{1}{(x-2)^2}$ actually does not exist, since ∞ is not a number. What is happening is that the limit fails to exist because the function $f(x) = 1/(x-2)^2$ grows without bound as x approaches 2. We have established this using techniques as in the previous examples, and expressed the same fact using limit notation and the ∞ symbol.

Topic Three

Proving a Limit Does *Not* Exist

We will close this section with an illustration of how we can use our limit definition to prove that a given function does *not* have a limit at a given point. There are occasions when the ability to prove that something does not happen is just as useful as the ability to prove that it does.

Example 7

Prove that for the function $f(x) = \cos(1/x)$, $\lim\limits_{x\to 0} f(x)$ does not exist by showing that the function does not satisfy the ε-δ definition for any possible L at $c = 0$.

Solution

We will first need to think carefully about what it means for the definition to *fail* at $c = 0$. This will happen precisely when no number L works as the limit of the function at 0, in other words, when x can move closer to 0 than *any* given $\delta > 0$ without the corresponding function values $f(x)$ approaching any number. You can think about this, too, as a challenge game. Suppose a skeptic challenges you with a very small number δ, and you are able to show that the function values $f(x)$, corresponding to nonzero x-values with $-\delta < x < \delta$, do not approach any number L. If you are always able to respond to the challenge, no matter how small $\delta > 0$ becomes, you will have proved that $\lim\limits_{x\to 0} f(x)$ does not exist. This is exactly what we endeavor to do in this example.

Suppose that $\delta > 0$ is given. Recall that $\cos 2k\pi = 1$ and $\cos((2k+1)\pi) = -1$ for all $k \in \mathbb{Z}$. Choose and fix a big enough positive integer k such that $1/(2k\pi) < \delta$ (since δ is already fixed, you can achieve this by merely choosing k big enough).

Next, we will use the real numbers $x_1 = \dfrac{1}{2k\pi}$ and $x_2 = \dfrac{1}{(2k+1)\pi}$ as follows. Note that both are less than δ in magnitude, but

$$f(x_1) = \cos\left(\frac{1}{x_1}\right)$$

$$= \cos\left(\frac{1}{\dfrac{1}{2k\pi}}\right)$$

$$= \cos 2k\pi = 1,$$

while

$$f(x_2) = \cos\left(\frac{1}{x_2}\right)$$

$$= \cos\left(\frac{1}{\dfrac{1}{(2k+1)\pi}}\right)$$

$$= \cos((2k+1)\pi) = -1.$$

What this means is that we were able to find x-values less than δ in magnitude, namely x_1 and x_2, for which the corresponding function values $f(x_1)$ and $f(x_2)$ are a full 2 units apart. More importantly, we can do the same, no matter how small a $\delta > 0$ is specified. Thus the values of $f(x)$ cannot be approaching any limit L at all.

To put our argument on a more precise footing, we will show that the definition for the existence of $\lim\limits_{x \to 0} f(x)$ fails, by showing that there is an $\varepsilon > 0$ for which no $\delta > 0$ exists to satisfy the definition.

Let us pick, say, $\varepsilon = 1$, and suppose that there is a $\delta > 0$ such that for some L,

$$|f(x) - L| < 1,$$

whenever $0 < |x - 0| < \delta$. Then proceed to find x_1, x_2 as above. Since both $|x_1| < \delta$ and $|x_2| < \delta$, by assumption we have first of all the inequality

$$|f(x_1) - L| = |1 - L| < 1.$$

What this means is that L is less than 1 unit away from 1; in particular, L is positive.

On the other hand, because of the choice of x_2, we also have the inequality

$$|f(x_2) - L| = |-1 - L| < 1;$$

in other words, L is less than 1 unit from -1 and therefore is negative.

Since such a number does not exist, $\lim\limits_{x \to 0} f(x)$ cannot exist either.

2.3 **Exercises**

1–4 Use the graph to estimate δ corresponding to the given ε satisfying the ε-δ definition of $\lim_{x \to c} f(x) = L$.

1.

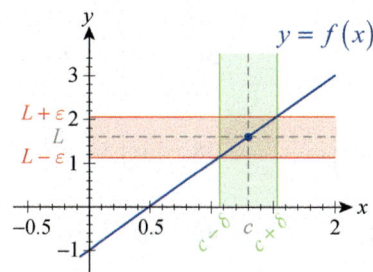

2.

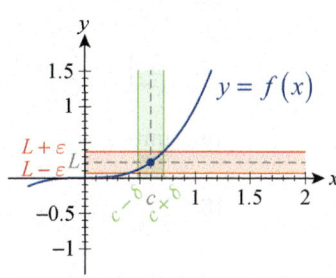

3.

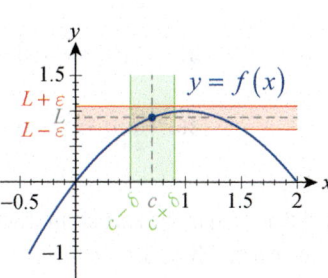

4.

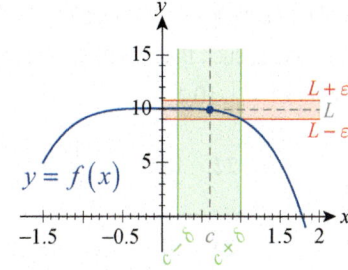

5–10 Calculus students gave the following definitions for the existence of a limit of $f(x)$ at c. Find and correct any errors.

5. "$\lim_{x \to c} f(x)$ exists if for any $\varepsilon > 0$ and real number L there is a $\delta > 0$ such that $0 < |x - L| < \delta$ implies $|f(x) - c| < \varepsilon$."

6. "$\lim_{x \to c} f(x)$ exists and equals L if for any $\varepsilon > 0$ and $\delta > 0$ whenever $0 < |x - c| < \varepsilon$, we have $|f(x) - L| < \delta$."

7. "If there is a real number L such that for an $\varepsilon > 0$ there is a $\delta > 0$ such that whenever $|x - c| < \delta$ and $x \neq c$, we have $|f(x) - L| < \varepsilon$, we say that the limit of the function at c is L."

8. "We say that $\lim_{x \to c} f(x) = L$, if for any $\varepsilon > 0$ there is a $\delta > 0$ such that $|x - c| < \delta \Rightarrow |f(x) - L| < \varepsilon$."

9. "We say that $\lim_{x \to c} f(x) = L$, if for any $\varepsilon > 0$ there is a $\delta > 0$ such that $0 \leq |x - c| \leq \delta \Rightarrow |f(x) - L| \leq \varepsilon$."

10. "If the real number L is such that for any $\varepsilon > 0$ there is a $\delta > 0$ such that $0 < |x - c| < \delta \Rightarrow |f(x) - L| < \varepsilon$, we say that $\lim_{x \to c} f(x) = L$."

11–20 Find a $\delta > 0$ that satisfies the limit claim corresponding to $\varepsilon = 0.1$, that is, such that $0 < |x - c| < \delta$ would imply $|f(x) - L| < 0.1$.

11. $\lim_{x \to 2}(5x - 1) = 9$

12. $\lim_{x \to 1}(3x + 1) = 4$

13. $\lim_{x \to -1}(-x + 2) = 3$

14. $\lim_{x \to 6}\left(4 - \dfrac{x}{2}\right) = 1$

15. $\lim_{x \to 0} x^2 = 0$

16. $\lim_{x \to 8} \sqrt[3]{x} = 2$

17. $\lim_{x \to 1} \dfrac{1}{x} = 1$

18. $\lim_{x \to 0} e^x = 1$

19. $\lim_{x \to 1} \ln x = 0$

20. $\lim_{x \to 0} \cos x = 1$

21–26 Find a number N that satisfies the limit claim corresponding to $\varepsilon = 0.1$, that is, such that $x > N$ (or $x < N$, as appropriate) would imply $|f(x) - L| < 0.1$.

21. $\lim_{x \to \infty} \dfrac{1}{x} = 0$

22. $\lim_{x \to \infty} \dfrac{x^2 - 1}{x^2 + 2x} = 1$

23. $\lim_{x \to -\infty} \dfrac{x + 1}{x} = 1$

24. $\lim_{x \to \infty} \dfrac{2x}{\sqrt{x^2 + x}} = 2$

25. $\lim_{x \to -\infty} e^x = 0$

26. $\lim_{x \to \infty} \arctan x = \dfrac{\pi}{2}$

27–32 For the given function $f(x)$, find a $\delta > 0$ corresponding to $M = 100$, that is, such that $0 < |x - c| < \delta$ would imply $f(x) > 100$ (let $N = -100$ if the limit is $-\infty$, in which case $0 < |x - c| < \delta$ should imply $f(x) < -100$).

27. $\lim_{x \to 0} \dfrac{1}{x^2} = \infty$

28. $\lim_{x \to 0^+} \dfrac{1}{x} = \infty$

29. $\lim_{x \to -1} \dfrac{-1}{(x + 1)^2} = -\infty$

30. $\lim_{x \to 0^+} \ln x = -\infty$

31. $\lim_{x \to (\pi/2)^-} \tan x = \infty$

32. $\lim_{x \to 0^+} \csc x = \infty$

33–46 Use the ε-δ definition to prove the limit claim. (**Hint:** See Examples 3 and 4 for guidance as you work through these exercises.)

33. $\lim_{x \to 1}(2x + 3) = 5$

34. $\lim_{x \to 7} x = 7$

35. $\lim_{x \to c} a = a$

36. $\lim_{x \to 4}\left(\dfrac{1}{4}x + 1\right) = 2$

37. $\lim_{x \to 0}\left(\dfrac{1}{2} - 4x\right) = \dfrac{1}{2}$

38. $\lim_{x \to 9}\left(5 - \dfrac{x}{3}\right) = 2$

39. $\lim_{x \to 1} x^3 = 1$

40. $\lim_{x \to 0} x^2 = 0$

41. $\lim_{x \to 0} \dfrac{1}{2}|x| = 0$

42. $\lim_{x \to -2}|x + 2| = 0$

43. $\lim_{x \to 1^+} \sqrt{x - 1} = 0$

44. $\lim_{x \to 0^-}\left(\sqrt[3]{x} + 1\right) = 1$

45. $\lim_{x \to 1}(x^2 + x) = 2$

46. $\lim_{x \to 3}(3x^2 - 9x + 5) = 5$

47–62 Give the formal definition of the limit claim. Then use the definition to prove the claim. (**Hint:** See Examples 5 and 6 for guidance as you work through these exercises.)

47. $\lim_{x \to \infty} \dfrac{1 + x}{x} = 1$

48. $\lim_{x \to -\infty} \dfrac{2}{x^2} = 0$

49. $\lim_{x \to \infty} \dfrac{1}{\sqrt{x}} = 0$

50. $\lim_{x \to \infty} \dfrac{1 + 3x^3}{x^3} = 3$

51. $\lim_{x \to -\infty} 2^x = 0$

52. $\lim_{x \to \infty}(e^{-x} - 1) = -1$

53. $\lim_{x \to \infty} \dfrac{\sin x}{x} = 0$

54. $\lim_{x \to \infty}(2\arctan x) = \pi$

55. $\lim_{x \to 0^+} \dfrac{1}{x} = \infty$

56. $\lim_{x \to 0} \dfrac{1}{x^4} = \infty$

57. $\lim_{x \to -1} \dfrac{-1}{(x + 1)^2} = -\infty$

58. $\lim_{x \to 0^+} \log x = -\infty$

59. $\lim_{x \to (\pi/2)^+} \tan x = \infty$

60. $\lim_{x \to 0^+} \csc x = \infty$

61. $\lim_{x \to 2} \dfrac{-3}{(x - 2)^2} = -\infty$

62. $\lim_{x \to -2^-} \dfrac{x + 3}{x + 2} = -\infty$

63–67 Decide whether the given limit exists. Prove your conclusion. (**Hint:** See Example 7 for guidance as you work through these exercises.)

63. $\displaystyle\lim_{x\to 0^+} \sin\frac{\pi}{x}$

64. $\displaystyle\lim_{x\to 0^+} x^2 \cos\frac{1}{x}$

65. $\displaystyle\lim_{x\to 0} \frac{|x|}{x}$

66.* $\displaystyle\lim_{x\to 1} f(x)$, where $f(x) = \begin{cases} 1 & \text{if } x \text{ is rational} \\ 0 & \text{if } x \text{ is irrational} \end{cases}$

67.* $\displaystyle\lim_{x\to 0} g(x)$, where $g(x) = \begin{cases} x & \text{if } x \text{ is rational} \\ 0 & \text{if } x \text{ is irrational} \end{cases}$

68. Use ε and δ to state what $\displaystyle\lim_{x\to c} f(x) \neq L$ means.

69. A piston is manufactured to fit into the cylinder of a certain automobile engine. Suppose that the diameter of the cylinder is 82 mm and that the cross-sectional area of the piston is not allowed to be less than 99.89% of that of the cylinder. If both are perfectly round, what does this mean in terms of maximum tolerance for the clearance between the piston and the cylinder wall? (Be sure to identify which function and data take the roles of $f(x)$, c, ε, and δ in this problem.)

70. The tension in a stretched steel wire (in newtons, N) is calculated by the formula
$F = E\dfrac{\Delta L}{L_0} A$, where $E = 2\times 10^{11}$ N/m^2 is the elastic modulus (or Young's modulus) of steel, ΔL is the elongation, L_0 the original length, and A the cross-sectional area (in m^2). Suppose a 1-meter-long steel string of radius 1 millimeter is stretched by 2 millimeters when tuning a musical instrument.

  a. Calculate the tension in the string caused by the above tightening.

  b. If we are not allowed to overload the string by more than 100 N, what is the tolerance in the amount of stretching? (Be sure to identify the function and data taking the roles of c, ε, and δ in this problem.)

71–73 *True or False?* Determine whether the given statement is true or false. In case of a false statement, explain or provide a counterexample.

71. If $f(c) = L$, then as x approaches c,
$\displaystyle\lim_{x\to c} f(x) = L$.

72. If $\displaystyle\lim_{x\to c} f(x)$ exists and equals L, then $f(c) = L$.

73. If $f(x) < g(x)$ for all $x \neq c$, and both $\displaystyle\lim_{x\to c} f(x)$ and $\displaystyle\lim_{x\to c} g(x)$ exist, then $\displaystyle\lim_{x\to c} f(x) < \lim_{x\to c} g(x)$.

2.3 **Technology Exercises**

74–83 Use a graphing calculator or computer algebra system to estimate the given limit. By zooming in appropriately, find δ-values that correspond to $\varepsilon = 0.1$. (Answers will vary.)

74. $\displaystyle\lim_{x\to 5} \frac{x^2 - 5x + 6}{x - 2}$

75. $\displaystyle\lim_{x\to 0} \frac{\sqrt{x+5} - \sqrt{5}}{x}$

76. $\displaystyle\lim_{x\to 3.5} \frac{x^2 - 6.25}{x + 2.5}$

77. $\displaystyle\lim_{x\to 0} \frac{x - 1}{\sqrt{x} - 1}$

78. $\displaystyle\lim_{x\to 0} \frac{\sin 3x}{2x}$

79. $\displaystyle\lim_{x\to -\infty} \frac{2x + 3}{\sqrt{x^2 + 1}}$

80. $\displaystyle\lim_{x\to \infty} \frac{\sqrt{9x^2 + 1}}{x - 2}$

81. $\displaystyle\lim_{x\to -\infty} \frac{2x^2 + 1.5x - 7}{\sqrt{x^4 + 1}}$

82. $\displaystyle\lim_{x\to \infty}\left(\sqrt{x^2 + 3x + 5} - \sqrt{x^2 + 2x + 1}\right)$

83. $\displaystyle\lim_{x\to \infty}\left(1 + \frac{1}{x}\right)^x$

84–89 Use a graphing calculator or computer algebra system to locate a vertical asymptote of the given function. Then for such an asymptote $x = c$ find an appropriate value $\delta > 0$ such that $|x - c| < \delta \Rightarrow |f(x)| > 10$. (Answers will vary.)

84. $f(x) = \dfrac{x^2 - 7}{x^3 + x + 1}$

85. $f(x) = \dfrac{3x + 1}{2x^4 + x - 5}$

86. $f(x) = \ln\dfrac{x^2}{x^2 + 1}$

87. $f(x) = \tan\left(\dfrac{1}{2}x + 3\right)$

88. $f(x) = \csc(2x + 1)$

89. $f(x) = \cot\left(\dfrac{1}{2}\cos x\right)$

2.4 Determining Limits of Functions

We now have a formal definition of *limit* with which to work, and some experience in proving limit claims with epsilon-delta arguments. What we are lacking is a collection of tools allowing us to determine limits in the first place. The theorems in this section will help us do just that and, at the same time, provide the necessary proof that the resulting limits are correct.

Topic One
Limit Laws

We will begin with a table of the basic limit laws and show how they can be used both in immediate applications and in deriving more powerful laws.

Theorem

Basic Limit Laws

Let f and g be two functions such that both $\lim\limits_{x \to c} f(x)$ and $\lim\limits_{x \to c} g(x)$ exist, and let k be a fixed real number. Then the following laws hold.

Sum Law	$\lim\limits_{x \to c}\left[f(x) + g(x) \right] = \lim\limits_{x \to c} f(x) + \lim\limits_{x \to c} g(x)$
Difference Law	$\lim\limits_{x \to c}\left[f(x) - g(x) \right] = \lim\limits_{x \to c} f(x) - \lim\limits_{x \to c} g(x)$
Constant Multiple Law	$\lim\limits_{x \to c}\left[kf(x) \right] = k \lim\limits_{x \to c} f(x)$
Product Law	$\lim\limits_{x \to c}\left[f(x)g(x) \right] = \lim\limits_{x \to c} f(x) \cdot \lim\limits_{x \to c} g(x)$
Quotient Law	$\lim\limits_{x \to c} \dfrac{f(x)}{g(x)} = \dfrac{\lim\limits_{x \to c} f(x)}{\lim\limits_{x \to c} g(x)}$, provided $\lim\limits_{x \to c} g(x) \neq 0$

Proof

We will prove the Sum Law here, and present the proofs of the remaining laws in Appendix E.

For ease of exposition, let $L = \lim\limits_{x \to c} f(x)$ and $M = \lim\limits_{x \to c} g(x)$; we need to show that $\lim\limits_{x \to c}\left[f(x) + g(x) \right] = L + M$ using our ε-δ definition of limit. To this end, assume $\varepsilon > 0$ is given. Then by assumption there exist $\delta_1 > 0$ and $\delta_2 > 0$ such that

$$0 < |x - c| < \delta_1 \Rightarrow |f(x) - L| < \frac{\varepsilon}{2}$$

and

$$0 < |x - c| < \delta_2 \Rightarrow |g(x) - M| < \frac{\varepsilon}{2}.$$

Note that we have found δ_1 and δ_2 so that the differences between the functions and their respective limits are smaller than $\varepsilon/2$. We did so in order to obtain the following consequence for all $x \ne c$ chosen within δ of c, where δ is the smaller of δ_1 and δ_2.

$$|f(x)+g(x)-(L+M)| = |f(x)-L+g(x)-M|$$
$$\le |f(x)-L|+|g(x)-M| \quad \text{Triangle Inequality: } |a+b| \le |a|+|b|$$
$$< \frac{\varepsilon}{2}+\frac{\varepsilon}{2} = \varepsilon \quad\quad\quad \delta \le \delta_1 \text{ and } \delta \le \delta_2$$

Since we have demonstrated that the function $f(x)+g(x)$ is within ε of $L+M$ for all x such that $0 < |x-c| < \delta$, our proof is complete.

Intuitively, the limit laws seem reasonable. For instance, if $\lim_{x \to c} f(x) = L$ and $\lim_{x \to c} g(x) = M$, the Product Law points out (in a precise manner) that for values of x close to c, the function $f \cdot g$ assumes values close to $L \cdot M$. As always, though, it is important to remember that statements about the limit at a point c describe behavior *near* c, not *at* c.

Example 1

For the functions f and g graphed in Figure 1, determine if the following limits exist. If a particular limit exists, evaluate it. If not, give reasons why it fails to exist.

a. $\lim_{x \to 2}[f(x)g(x)]$

b. $\lim_{x \to 2}\left[\frac{3}{7}f(x)+g(x)\right]$

c. $\lim_{x \to 0}[g(x)-f(x)]$

d. $\lim_{x \to 0}[f(x)g(x)]$

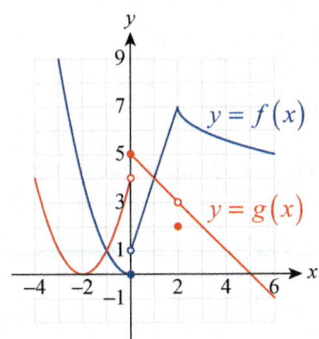

Figure 1

Solution

a. As we can see from the figure, the limits of both f and g exist at $c = 2$. In case of g, we note once again that the actual function value at 2 has no bearing on the value of the limit.

$$\lim_{x \to 2} f(x) = 7 \text{ and } \lim_{x \to 2} g(x) = 3$$

Now we evaluate the limit using the Product Law.

$$\lim_{x \to 2}[f(x)g(x)] = \lim_{x \to 2} f(x) \cdot \lim_{x \to 2} g(x) = 7 \cdot 3 = 21$$

b. Here we are going to use the Sum Law and the Constant Multiple Law.

$$\lim_{x \to 2}\left[\frac{3}{7}f(x)+g(x)\right] = \lim_{x \to 2}\left[\frac{3}{7}f(x)\right] + \lim_{x \to 2} g(x) = \frac{3}{7}\lim_{x \to 2} f(x) + \lim_{x \to 2} g(x)$$

$$= \frac{3}{7} \cdot 7 + 3 = 3 + 3 = 6$$

c. We will first examine the behavior of f near 0. It is clear from the graph that both one-sided limits exist, but they don't agree; thus the two-sided limit of f cannot exist at $x = 0$. In fact, $\lim_{x \to 0^-} f(x) = 0$ while $\lim_{x \to 0^+} f(x) = 1$, and so $\lim_{x \to 0} f(x)$ does not exist.

Similarly for g, $\lim_{x \to 0^-} g(x) = 4$ while $\lim_{x \to 0^+} g(x) = 5$, and thus $\lim_{x \to 0} g(x)$ cannot exist.

In words, since the one-sided limits are unequal, the two-sided limit of g cannot exist at 0. However, the limit laws do apply to one-sided limits, so we can determine those for $\left[g(x) - f(x) \right]$.

$$\lim_{x \to 0^-} \left[g(x) - f(x) \right] = \lim_{x \to 0^-} g(x) - \lim_{x \to 0^-} f(x) = 4 - 0 = 4$$

$$\lim_{x \to 0^+} \left[g(x) - f(x) \right] = \lim_{x \to 0^+} g(x) - \lim_{x \to 0^+} f(x) = 5 - 1 = 4$$

Notice from our findings that something interesting is actually going on here. Even though the two-sided limits of f and g do not exist individually, the one-sided limits of $g - f$ exist and agree at $x = 0$. What this means is that the limit of $g - f$ at 0 actually exists:

$$\lim_{x \to 0} \left[g(x) - f(x) \right] = 4$$

d. Using the Product Law for one-sided limits, we obtain

$$\lim_{x \to 0^-} \left[f(x) g(x) \right] = \lim_{x \to 0^-} f(x) \cdot \lim_{x \to 0^-} g(x) = 0 \cdot 4 = 0, \text{ while}$$

$$\lim_{x \to 0^+} \left[f(x) g(x) \right] = \lim_{x \to 0^+} f(x) \cdot \lim_{x \to 0^+} g(x) = 1 \cdot 5 = 5.$$

Since the one-sided limits of the product function $f \cdot g$ are unequal, we conclude that $\lim_{x \to 0} \left[f(x) g(x) \right]$ does not exist.

Example 2

Let $f(x) = \begin{cases} \dfrac{|x|}{x} & \text{if } x \neq 0 \\ 0 & \text{if } x = 0 \end{cases}$ and $g(x) = \cos x$. Determine whether $\lim_{x \to 0} \left[f(x) g(x) \right]$ exists.

Solution

We will first examine the behavior of $f(x)$ near 0. If x approaches 0 from the left, then since $x < 0$, we have

$$f(x) = \frac{|x|}{x} = \frac{-x}{x} = -1,$$

so f is the constant -1 for all negative x-values.

Therefore,

$$\lim_{x \to 0^-} f(x) = -1.$$

On the other hand, for all positive x-values $|x| = x,$ so

$$f(x) = \frac{|x|}{x} = \frac{x}{x} = 1,$$

thus

$$\lim_{x \to 0^+} f(x) = 1.$$

Since the one-sided limits are unequal, $\lim_{x \to 0} f(x)$ does not exist. Consequently, the Product Law is not applicable, but since it does apply to one-sided limits, we can proceed to examine those separately. Recalling the well-known fact that $\lim_{x \to 0} \cos x = 1,$ we obtain

$$\lim_{x \to 0^-} \left[f(x)g(x) \right] = \lim_{x \to 0^-} f(x) \cdot \lim_{x \to 0^-} g(x) = (-1) \cdot 1 = -1, \text{ while}$$

$$\lim_{x \to 0^+} \left[f(x)g(x) \right] = \lim_{x \to 0^+} f(x) \cdot \lim_{x \to 0^+} g(x) = 1 \cdot 1 = 1.$$

Thus we conclude that the two-sided limit, $\lim_{x \to 0} \left[f(x)g(x) \right],$ does not exist.

The five limit laws that we have listed so far can be quickly extended to a much larger collection when combined with one another and with a few easily proved statements. For instance, if the Product Law is applied to the product of a function with itself, we obtain the statement

$$\lim_{x \to c} \left[f(x)f(x) \right] = \lim_{x \to c} f(x) \cdot \lim_{x \to c} f(x) = \left[\lim_{x \to c} f(x) \right]^2$$

(assuming $\lim_{x \to c} f(x)$ exists). And the same law can now be applied to the product of $f(x)$ and $\left[f(x) \right]^2$ to reach a similar conclusion about the limit of the function $\left[f(x) \right]^3.$ In general, repeated application of the Product Law results in the Power Law, the proof of which can be found in Appendix E.

Theorem

Positive Integer Power Law

Let f be a function for which $\lim_{x \to c} f(x)$ exists, and let m be a fixed positive integer. Then

$$\lim_{x \to c} \left[f(x) \right]^m = \left[\lim_{x \to c} f(x) \right]^m.$$

The two limit statements

$$\lim_{x \to c} 1 = 1 \text{ and } \lim_{x \to c} x = c$$

are certainly reasonable and easily proved, and allow us to extend our list of limit laws to large classes of functions. In particular, we can now state the following two laws, which are examples of what are sometimes referred to as Direct Substitution Laws.

Theorem

Polynomial Substitution Law

Let p be a polynomial function. Then

$$\lim_{x \to c} p(x) = p(c).$$

Theorem

Rational Function Substitution Law

Let p and q be polynomial functions. Then if $q(c) \neq 0$,

$$\lim_{x \to c} \frac{p(x)}{q(x)} = \frac{p(c)}{q(c)}.$$

In Exercises 65–68, you will be guided through the proofs of the above statements.

Example 3

Find the following limits.

a. $\lim_{x \to 2} \left(4x^3 - 5x^2 + 1\right)$

b. $\lim_{x \to -1} \left(2x^4 + x^3 - 3x^2 + 7.5\right)$

c. $\lim_{x \to 3} \dfrac{x^2 + 2}{3x - 1}$

d. $\lim_{x \to 1} \dfrac{2x^3 - 5x + 1}{2x - x^2}$

Solution

a. First using the Sum and Difference Laws, we obtain

$$\lim_{x \to 2}\left(4x^3 - 5x^2 + 1\right) = \lim_{x \to 2}\left(4x^3\right) - \lim_{x \to 2}\left(5x^2\right) + \lim_{x \to 2}(1)$$

$$= 4\lim_{x \to 2} x^3 - 5\lim_{x \to 2} x^2 + 1,$$

where in the last step we used the Constant Multiple Law. Next, the Positive Integer Power Law comes to bear.

$$4\lim_{x \to 2} x^3 - 5\lim_{x \to 2} x^2 + 1 = 4\left(\lim_{x \to 2} x\right)^3 - 5\left(\lim_{x \to 2} x\right)^2 + 1 = 4(2)^3 - 5(2)^2 + 1 = 13$$

Notice that our repeated application of the various rules eventually led us to finding the limit by simply substituting $x = 2$ into the polynomial. This is exactly what the Polynomial Substitution Law allows us to do; so henceforth we don't even have to go through the above, somewhat lengthy, process when finding the limits of polynomials.

b. In this case, we will simply use the Polynomial Substitution Law.

$$\lim_{x \to -1}\left(2x^4 + x^3 - 3x^2 + 7.5\right) = 2(-1)^4 + (-1)^3 - 3(-1)^2 + 7.5 = 5.5$$

c. Notice that our third limit is that of a rational function. Since both the numerator and denominator are polynomials, we can apply the Quotient Law combined with a repeated application of the Polynomial Substitution Law as follows:

$$\lim_{x \to 3}\frac{x^2 + 2}{3x - 1} = \frac{\lim_{x \to 3}\left(x^2 + 2\right)}{\lim_{x \to 3}(3x - 1)} = \frac{3^2 + 2}{3 \cdot 3 - 1} = \frac{9 + 2}{9 - 1} = \frac{11}{8}$$

Notice, however, that what we did was in effect substituting $x = 3$ into the given rational function. This is exactly what the Rational Function Substitution Law says we can always do unless the process results in a 0 denominator.

d. This time, we will simply refer to the Rational Function Substitution Law.

$$\lim_{x \to 1}\frac{2x^3 - 5x + 1}{2x - x^2} = \frac{2(1)^3 - 5 \cdot 1 + 1}{2 \cdot 1 - 1^2} = \frac{-2}{1} = -2$$

In summary, we wish to emphasize that when evaluating limits of a polynomial or rational function, our theorems allow for the calculation to be reduced to a simple matter of evaluating the function at the limit point. The only exception is for those c-values that cause the denominator of a given rational limit to equal 0.

We can gain another significant extension with the addition of the following law, the proof of which is a consequence of the Intermediate Value Theorem of Section 2.5.

Theorem

Positive Integer Root Law

Let f be a function for which $\lim_{x \to c} f(x)$ exists, and let n be a fixed positive integer. Then

$$\lim_{x \to c}\sqrt[n]{f(x)} = \sqrt[n]{\lim_{x \to c} f(x)},$$

with the assumption that $\lim_{x \to c} f(x)$ is nonnegative if n is even.

Exercise 69 shows how the Positive Integer Power Law and the Positive Integer Root Law we have (and, if necessary, the Quotient Law) can be combined to yield the following.

Theorem

Rational Power Law

Let f be a function for which $\lim\limits_{x \to c} f(x)$ exists, and let m and n be fixed nonzero integers with no common factor. Then

$$\lim_{x \to c}\left[f(x) \right]^{m/n} = \left[\lim_{x \to c} f(x) \right]^{m/n},$$

with the assumption that $\lim\limits_{x \to c} f(x)$ is nonnegative if n is even.

Example 4

Use the Rational Power Law to evaluate the following limits.

a. $\lim\limits_{x \to 1}\left(\dfrac{5x+3}{x^2 - 2x + 2} \right)^{7/3}$ **b.** $\lim\limits_{x \to 2}\sqrt{\left(x^4 + 2x^2 + 4\right)^3}$ **c.** $\lim\limits_{x \to 0}\sqrt[3]{4\cos^2 x}$

Solution

a. First of all, we claim that for $f(x) = (5x+3)/(x^2 - 2x + 2)$, $\lim\limits_{x \to 1} f(x)$ exists. In fact, by direct substitution we have the following.

$$\lim_{x \to 1} f(x) = \lim_{x \to 1}\frac{5x+3}{x^2 - 2x + 2} = \frac{5 \cdot 1 + 3}{1^2 - 2 \cdot 1 + 2} = \frac{8}{1 - 2 + 2} = 8$$

It follows that the Rational Power Law applies.

$$\lim_{x \to 1}\left(\frac{5x+3}{x^2 - 2x + 2} \right)^{7/3} = \left(\lim_{x \to 1}\frac{5x+3}{x^2 - 2x + 2} \right)^{7/3} = 8^{7/3} = 2^7 = 128$$

b. To start us off, notice that $\lim\limits_{x \to 2}\left(x^4 + 2x^2 + 4\right) = 28$, a fact easily verified by direct substitution. Also since the above limit is positive and

$$\sqrt{\left(x^4 + 2x^2 + 4\right)^3} = \left(x^4 + 2x^2 + 4\right)^{3/2},$$

the Rational Power Law applies.

$$\lim_{x \to 2}\sqrt{\left(x^4 + 2x^2 + 4\right)^3} = \lim_{x \to 2}\left(x^4 + 2x^2 + 4\right)^{3/2} = \left[\lim_{x \to 2}\left(x^4 + 2x^2 + 4\right) \right]^{3/2} = 28^{3/2}$$

$$= \left(\sqrt{28}\right)^3 = \left(2\sqrt{7}\right)^3 = 2^3\left(\sqrt{7}\right)^3 = 8 \cdot 7 \cdot \sqrt{7} = 56\sqrt{7}$$

c. Since $4\cos^2 x = \left(2\cos x\right)^2$, and since $\lim\limits_{x \to 0}\left(2\cos x\right) = 2$, the Rational Power Law once again applies.

$$\lim_{x \to 0}\sqrt[3]{4\cos^2 x} = \lim_{x \to 0}\left(2\cos x\right)^{2/3} = \left[\lim_{x \to 0}\left(2\cos x\right) \right]^{2/3} = 2^{2/3} = \sqrt[3]{4}$$

Topic Two

Limit Determination Techniques

The limit laws we have stated greatly simplify the determination of many limits and also show that limits of polynomial and rational functions can be found simply by evaluating the function at the limit point (remember that in the case of rational functions the limit point must be in the domain of the function). Functions possessing this property, which we will refer to as the *Direct Substitution Property*, are called *continuous*; much more will be said about these well-behaved functions in the next section. But we will close this section with a discussion of techniques to use when the limit laws and/or direct substitution do not immediately apply.

Many of our techniques make use of the fact that the limit of a function f at a point c is determined entirely by its behavior near, but not at, c. If we can find a function g that is identical to f near c, and if the limit of g at the point c is easy to determine, then we are done. The following examples illustrate the steps typically taken in this process.

Example 5

Find $\displaystyle\lim_{x\to-3}\frac{x^2-9}{x+3}$.

Solution

First of all we note that we cannot use direct substitution to evaluate this limit, for the denominator equals 0 at $x=-3$. In fact, $f(x)=\left(x^2-9\right)/(x+3)$ is not even defined at $x=-3$. However, as far as the limit is concerned, that is not a problem at all (recall that the existence and/or value of $f(c)$ and $\lim_{x\to c} f(x)$ have no bearing on each other). The limit not only exists, but finding it is surprisingly easy, using a bit of algebra:

$$\lim_{x\to-3}\frac{x^2-9}{x+3}=\lim_{x\to-3}\frac{(x+3)(x-3)}{x+3}=\lim_{x\to-3}(x-3)=-3-3=-6$$

A few very important remarks are in order. First of all, canceling the factor of $(x+3)$ is legitimate, since even though x is approaching -3, it never is actually equal to -3 throughout the limit process, so we have not divided by 0. Second, the function $g(x)=x-3$ is actually different from $f(x)$. In fact, they agree everywhere but at $x=-3$; at which point $f(x)$ is undefined, but $g(-3)=-6$ (see Figure 2). However, the behaviors of f and g are identical as x approaches -3, and therefore, so are their limits. (Examining the chain of equalities above, note that we never actually stated the equality of f and g, but merely the fact that their limits at -3 were equal.)

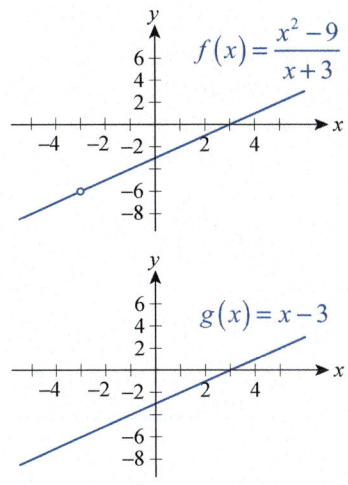

Figure 2

Algebraic techniques similar to the one from Example 5 are common when evaluating limits analytically. We provide further illustrations in the next two examples.

Example 6

Use algebra to evaluate $\lim\limits_{x\to 1}\dfrac{\sqrt{x+3}-2}{x-1}$.

Solution

The trick with a limit such as this is to use the "conjugate product rule" $(a+b)(a-b)=a^2-b^2$. If we multiply both the numerator and denominator by the "conjugate" of the expression containing the radical (in this case, that is $\sqrt{x+3}+2$), we obtain the following.

$$\lim_{x\to 1}\frac{\sqrt{x+3}-2}{x-1}=\lim_{x\to 1}\left(\frac{\sqrt{x+3}-2}{x-1}\cdot\frac{\sqrt{x+3}+2}{\sqrt{x+3}+2}\right)=\lim_{x\to 1}\frac{\left(\sqrt{x+3}\right)^2-2^2}{(x-1)\left(\sqrt{x+3}+2\right)}$$

$$=\lim_{x\to 1}\frac{x+3-4}{(x-1)\left(\sqrt{x+3}+2\right)}=\lim_{x\to 1}\frac{x-1}{(x-1)\left(\sqrt{x+3}+2\right)}$$

$$=\lim_{x\to 1}\frac{1}{\sqrt{x+3}+2}=\frac{1}{\sqrt{1+3}+2}=\frac{1}{2+2}=\frac{1}{4}$$

As in the previous example, the cancellation was legitimate, and even though in the process we did change the function, we did not change the value of the limit.

Example 7

Find the limit: $\lim\limits_{h\to 0}\dfrac{(h+2)^2-4}{h}$

Solution

We will take an algebraic approach very similar to the previous examples, but keeping in mind the fact that in the current problem, the variable is denoted by h, rather than the usual x. This, however, should not cause any difficulties.

$$\lim_{h\to 0}\frac{(h+2)^2-4}{h}=\lim_{h\to 0}\frac{\left(h^2+4h+4\right)-4}{h}=\lim_{h\to 0}\frac{h^2+4h}{h}$$

$$=\lim_{h\to 0}\frac{h(h+4)}{h}=\lim_{h\to 0}(h+4)=4$$

Again, canceling h is legitimate, since h never actually assumes the value 0.

Another technique for determining $\lim\limits_{x\to c}f(x)$ calls for finding two other functions, g and h, such that f is "squeezed between" g and h and for which $\lim\limits_{x\to c}g(x)$ and $\lim\limits_{x\to c}h(x)$ are easier to determine.

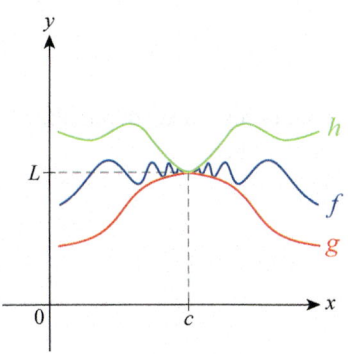

Figure 3
The Squeeze Theorem

Theorem

The Squeeze Theorem

If $g(x) \leq f(x) \leq h(x)$ for all x in some open interval containing c, except possibly at c itself, and if $\lim_{x \to c} g(x) = \lim_{x \to c} h(x) = L$, then $\lim_{x \to c} f(x) = L$ as well.

The statement also holds for limits at infinity, that is, for $c = -\infty$ or $c = \infty$.

The Squeeze Theorem also goes by names such as the Sandwich Theorem and the Pinching Theorem, and a proof is included in Appendix E.

Example 8

Use the Squeeze Theorem to prove $\lim_{x \to 0} x \sin \dfrac{1}{x} = 0$.

Solution

As a first observation, we recall that the sine value of any angle is never greater than 1 or less than -1; that is, we have the well-known inequalities

$$-1 \leq \sin \alpha \leq 1$$

for any α. This certainly means that for any nonzero x, choosing $\alpha = 1/x$,

$$-1 \leq \sin \frac{1}{x} \leq 1,$$

a fact you have seen before. Multiplying all sides of this chain of inequalities by $|x|$, we obtain

$$-|x| \leq x \sin \frac{1}{x} \leq |x|.$$

In words, we can say that while the sine function oscillates between -1 and 1, $x \sin(1/x)$ will oscillate between $-|x|$ and $|x|$. The previous inequality coupled with the fact that $\lim_{x \to 0} (-|x|) = \lim_{x \to 0} |x| = 0$ means that the functions $f(x) = x \sin(1/x)$, $g(x) = -|x|$, and $h(x) = |x|$ satisfy the hypotheses of the Squeeze Theorem. Thus we can simply invoke the theorem, which ensures that the claim

$$\lim_{x \to 0} x \sin \frac{1}{x} = 0$$

is now proven.

Figure 4 shows the Squeeze Theorem at work. Notice how f is "squeezed between" g and h near the origin.

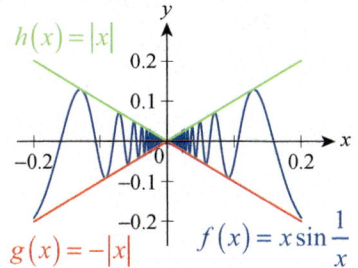

Figure 4

We will end with one last limit theorem that will prove useful in some derivations to follow. (See Appendix E for a proof.)

Upper Bound Theorem

If $f(x) \le g(x)$ for all x in some open interval containing c, except possibly at c itself, and if the limits of f and g both exist at c, then

$$\lim_{x \to c} f(x) \le \lim_{x \to c} g(x).$$

2.4 **Exercises**

1–2 Use the graph to find the given limit.

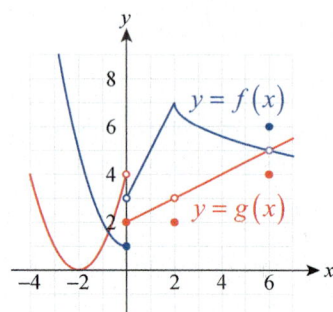

1. a. $\lim_{x \to 0^+} \left[g(x) - 2f(x) \right]$ **b.** $\lim_{x \to 2^+} \left[g(x) f(x) \right]$

2. a. $\lim_{x \to 6} \left[g(x) + f(x) \right]$ **b.** $\lim_{x \to 0^-} \dfrac{f(x)}{2g(x)}$

3–20 Use appropriate limit laws to evaluate the given limit.

3. $\lim_{x \to 4} 5$

4. $\lim_{x \to 4} 5x$

5. $\lim_{x \to 3} (2x + 1)$

6. $\lim_{x \to 1/2} (3 - 4x)$

7. $\lim_{x \to -3} x^2$

8. $\lim_{x \to -2} (-x^5)$

9. $\lim_{x \to 3} (2x^2 - x + 7)$

10. $\lim_{x \to -1} \left(3 + x - \dfrac{5}{2} x^2 \right)$

11. $\lim_{x \to 1/2} (2x^3 - 3x^2 + x - 4)$ **12.** $\lim_{x \to -2} (3x^3 - x^5)$

13. $\lim_{x \to 1} \dfrac{3x - 7}{x + 1}$

14. $\lim_{x \to -1} \dfrac{5x + 3}{x^2 - x}$

15. $\lim_{x \to 3} \left(\dfrac{4x}{11x - x^3} \right)^{1/3}$

16. $\lim_{t \to 1} \left(\dfrac{2t + t^3}{3t^2 + 1} \right)^{3/2}$

17. $\lim_{x \to -2} \sqrt[3]{5x^4 - x^3 + 3x^2 + 2x + 4}$

18. $\lim_{x \to 4} \sqrt{x^4 + 2x^2 + 1}$

19. $\lim_{x \to -3} \left(\dfrac{x^4 - 5x}{x^3 + 2x^2 - 4x} \right)^{4/5}$

20. $\lim_{x \to -5} \sqrt[3]{\left(x^4 + 2x^3 + x^2 \right)^2}$

21–44 Use algebra to evaluate the given limit.

21. $\lim_{x \to 6} \dfrac{x^2 - 36}{x - 6}$

22. $\lim_{x \to -7} \dfrac{x + 7}{x^2 - 49}$

23. $\lim_{x \to 3} \dfrac{3 - 13x + 4x^2}{x - 3}$

24. $\lim_{x \to 4} \dfrac{x^4 - 256}{x^2 - 16}$

25. $\lim_{x \to 5} \dfrac{2x^3 - 7x^2 - 14x - 5}{x^2 - 25}$

26. $\lim_{x \to 2} \dfrac{x^3 - 8}{x^3 - 2x^2 + 2x - 4}$

27. $\lim_{x \to 7} \dfrac{\sqrt{x + 2} - 3}{x - 7}$

28. $\lim_{x \to 9} \dfrac{3 - \sqrt{x}}{x - 9}$

29. $\lim_{x \to 0} \dfrac{\sqrt{x + 5} - \sqrt{5}}{x}$

30. $\lim_{x \to 0} \dfrac{\dfrac{1}{4 + x} - \dfrac{1}{4}}{x}$

31. $\lim_{x \to 2} \dfrac{\dfrac{1}{3} - \dfrac{1}{1 + x}}{x - 2}$

32. If $f(x) = x^2$, find $\lim_{x \to 2} \dfrac{f(x) - f(2)}{x - 2}$.

33. If $g(x) = x^2 - 2$, find $\lim_{h \to 0} \dfrac{g(3 + h) - g(3)}{h}$.

34. If $k(x) = 1 - x + x^2$, find $\lim_{h \to 0} \dfrac{k(2 - h) - k(2)}{h}$.

35. If $p(x) = x^3 + x$, find $\lim\limits_{x \to 1} \dfrac{p(x) - p(1)}{x - 1}$.

36. If $F(x) = \dfrac{1}{x}$, find $\lim\limits_{x \to 1/2} \dfrac{F(x) - F\left(\dfrac{1}{2}\right)}{x - \dfrac{1}{2}}$.

37. $\lim\limits_{h \to 0} \dfrac{\sqrt{x + h} - \sqrt{x}}{h}$

38. $\lim\limits_{x \to 3} \dfrac{2x^2 - 3x - 9}{x^2 - 9}$

39. $\lim\limits_{x \to 2} \dfrac{x^4 - 16}{3x^2 - 5x - 2}$

40. $\lim\limits_{x \to -3} \dfrac{\dfrac{1}{3} + \dfrac{1}{x}}{x^3 + 27}$

41. $\lim\limits_{x \to 2} \dfrac{3 - \sqrt{x + 7}}{x - 2}$

42. $\lim\limits_{x \to 8} \dfrac{8 - x}{\sqrt[3]{x} - 2}$

43. $\lim\limits_{y \to 0} \left(\dfrac{1}{y} + \dfrac{1}{y^2 - y} \right)$

44. $\lim\limits_{x \to 1} \dfrac{\sqrt{x} - 1}{\sqrt[3]{x} - 1}$

45–50 Use $\lim\limits_{x \to c} f(x) = 3$ and $\lim\limits_{x \to c} g(x) = -2$ to find the limit.

45. $\lim\limits_{x \to c} \left[2f(x) - g(x) \right]$

46. $\lim\limits_{x \to c} \dfrac{4f(x) + 3g(x)}{f(x) - \dfrac{1}{2} g(x)}$

47. $\lim\limits_{x \to c} \sqrt{\left[f(x) \right]^4 + 10 \left[g(x) \right]^2}$

48. $\lim\limits_{x \to c} \left(\left[f(x) - 1 \right]^2 \sqrt[3]{g(x)} \right)$

49. $\lim\limits_{x \to c} \left(\left[f(x) \right]^2 + (x - 2)g(x) \right)$

50. $\lim\limits_{x \to c} \left(\dfrac{f(x) + g(x)}{\left[g(x) \right]^2} \right)^{3/2}$

51–58 Use the limit laws to find the one-sided limit.

51. $\lim\limits_{x \to 0^+} \dfrac{x}{|x|}$

52. $\lim\limits_{x \to 0^-} \operatorname{sgn} x \cos x$, where $\operatorname{sgn} x = \begin{cases} 1 & \text{if } x > 0 \\ 0 & \text{if } x = 0 \\ -1 & \text{if } x < 0 \end{cases}$

53. $\lim\limits_{x \to 2^+} \dfrac{\sqrt{x - 2}}{3x + 1}$

54. $\lim\limits_{x \to 1^-} \sqrt{1 - x^2}$

55. $\lim\limits_{x \to (1/3)^-} \dfrac{\sqrt{1 - 3x}}{6x + 5}$

56. $\lim\limits_{x \to 1^+} \left([\![x]\!] - x \right)$ (See the definition of $f(x) = [\![x]\!]$ in Section 1.2, before Exercises 43–45.)

57. $\lim\limits_{x \to 2^+} [\![x]\!] e^x$

58. $\lim\limits_{x \to -1^-} \dfrac{[\![x]\!] (2x^2 + 1)}{x + 3}$

59–64 Use the Squeeze Theorem to prove the limit claim.

59. $\lim\limits_{x \to 0} x^2 \sin \dfrac{1}{x} = 0$

60. $\lim\limits_{x \to 0} |x| \cos x = 0$

61. $\lim\limits_{x \to \infty} \dfrac{\cos x}{x} = 0$

62. $\lim\limits_{x \to -\infty} e^x \sin x = 0$

63. $\lim\limits_{x \to 0^+} x^{3/2} e^{\cos(1/x)} = 0$

64. $\lim\limits_{x \to \infty} \dfrac{\sin^2 x + 1}{2 + x} = 0$

65. Provide a rigorous proof of the limit claim $\lim\limits_{x \to c} 1 = 1$. (**Hint:** Use the fact that for the constant 1 function, $f(x) = 1$ for all x, so in particular, if an $\varepsilon > 0$ is given, $|f(x) - 1| = |1 - 1| = 0$, which makes the choice of δ "easy.")

66. Provide a rigorous proof of the limit claim $\lim\limits_{x \to c} x = c$. (**Hint:** Since $f(x) = x$ in this problem, for a given $\varepsilon > 0$ we need to ensure that $|f(x) - c| = |x - c| < \varepsilon$ as long as $0 < |x - c| < \delta$. This observation makes the choice of δ obvious.)

67. Use Exercise 66 and the basic limit laws to prove the Polynomial Substitution Law. (**Hint:** From Exercise 66 and a repeated application of the Product Law it follows that $\lim\limits_{x \to c} x^k = c^k$. As a next step, from the Constant Multiple Law we can conclude that if $a \in \mathbb{R}$, $\lim\limits_{x \to c} ax^k = ac^k$. From the above claim, a repeated application of the Sum Law will yield the result for a general polynomial.)

68. Use Exercise 67 and the basic limit laws to prove the Rational Function Substitution Law.

69. Combine the Positive Integer Power Law and the Positive Integer Root Law to prove the Rational Power Law. (**Hint:** Assuming first that both m and n are positive, we can write

$$\lim_{x\to c}\left[f(x)\right]^{m/n} = \lim_{x\to c}\left[\left[f(x)\right]^{1/n}\right]^m = \lim_{x\to c}\left[\sqrt[n]{f(x)}\right]^m.$$ Now use the Positive Integer Power Law followed by

the Positive Integer Root Law to obtain that the above limit is equal to $\left[\lim_{x\to c}\sqrt[n]{f(x)}\right]^m = \left[\sqrt[n]{\lim_{x\to c} f(x)}\right]^m$,

from which the result follows. If m is negative, note that $\left[f(x)\right]^{m/n} = 1/\left[f(x)\right]^{-m/n}$, where $-m$ is positive. Thus if we use the Quotient Law along with the previous argument, we obtain

$$\lim_{x\to c}\left[f(x)\right]^{m/n} = \lim_{x\to c}\left[1/\left[f(x)\right]^{-m/n}\right] = 1/\lim_{x\to c}\left[f(x)\right]^{-m/n} = 1/\left[\lim_{x\to c} f(x)\right]^{-m/n},$$ from which the result easily

follows.)

70. Let $D(x) = \begin{cases} 0 & \text{if } x \text{ is rational} \\ 1 & \text{if } x \text{ is irrational} \end{cases}$.

Does $\lim_{x\to 0} D(x)$ exist? Prove your answer.

71. Let $F(x) = \begin{cases} 0 & \text{if } x \text{ is rational} \\ x^2 & \text{if } x \text{ is irrational} \end{cases}$.

Does $\lim_{x\to 0} F(x)$ exist? Prove your answer.

72.* Prove that if $\lim_{x\to c} f(x) = L$ and $\lim_{x\to c} f(x) = K$, then $L = K$. In words, prove that if the limit of f exists at c, then the limit is unique.

73.* Prove that if n and m are positive integers, then

$$\lim_{x\to 1} \frac{x^n - 1}{x^m - 1} = \frac{n}{m}.$$

74.* Prove that if $\lim_{x\to c} f(x) = 0$, then $\lim_{x\to c} \left|f(x)\right| = 0$.

75.* Prove that if $\lim_{x\to c} f(x) = 0$ and $g(x)$ is such that $\left|g(x)\right| \le M$ for some number M (such functions are called *bounded*), then $\lim_{x\to c}\left[f(x)g(x)\right] = 0$.

76.* Prove that in Exercise 75, it is sufficient to require the boundedness of g only on an interval around c (except at c itself).

77.* By finding functions f and g such that $\lim_{x\to c} f(x) = 0$ but $\lim_{x\to c}\left[f(x)g(x)\right] \ne 0$, show that it is necessary to impose a boundedness condition on g in Exercise 75.

78.* Give examples of f and g to show that **a.** the existence of $\lim_{x\to c}\left[f(x) + g(x)\right]$ does not imply the existence of $\lim_{x\to c} f(x)$ and **b.** the existence of $\lim_{x\to c}\left[f(x)g(x)\right]$ does not imply the existence of $\lim_{x\to c} f(x)$.

79. A *concave spherical mirror* is a part of the inside of a sphere, silvered to form a reflective surface. The radius r of the sphere is called the mirror's *radius of curvature*. If the size of such a mirror is small relative to its radius of curvature, light rays parallel to its principal axis are reflected through approximately a single point, called *focus*. In the following illustration, C denotes the center, F_d is the focus, while d is the distance between the incoming ray and the principal axis. Note that according to the Law of Reflection, the incoming and reflected rays make the same size angle α with the radius $\overline{CR}$ (this radius is called *normal* to the mirror surface). One way to determine the *focal length* (the distance between the mirror and the focus) is to find the limiting position of F_d as $d \to 0$. Noting that the triangle $\triangle CRF_d$ is isosceles, by similarity we obtain $\dfrac{CF_d}{(r/2)} = \dfrac{r}{\sqrt{r^2 - d^2}}$. Use this observation to express CF_d and determine the focal length of the spherical mirror by taking the limit as $d \to 0$.

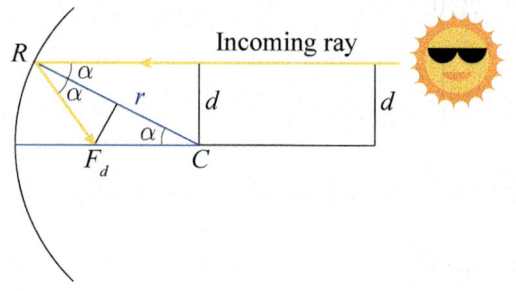

2.5 **Continuity**

Continuity of a function f at a point c was defined informally in Section 2.4 as synonymous with the Direct Substitution Property. That is, if $\lim_{x \to c} f(x) = f(c)$, then we say that f is continuous at c. Continuity, as it turns out, is one of those fundamental concepts that serves as the cornerstone for a great deal of other mathematical ideas. We will devote this section to a more thorough and rigorous study of the notion.

Topic One

Continuity and Discontinuity at a Point

As is often the case with deep, fundamental concepts, a solid understanding of continuity was not easily achieved, and it long evaded mathematicians. A once-common and imprecise definition, still often seen today, is that a continuous function is one whose graph can be drawn without lifting pen or pencil off the paper. This coincides with the actual definition of continuity in the case of sufficiently simple functions but is inadequate for our purposes in calculus.

Definition

Continuity at a Point

Given a function f defined on an open interval containing c, we say f is **continuous at c** if

$$\lim_{x \to c} f(x) = f(c).$$

If the domain of f is an interval containing c either as a left or right endpoint, we also define f to be **right-continuous** or **left-continuous at c** if, respectively,

$$\lim_{x \to c^+} f(x) = f(c) \text{ or } \lim_{x \to c^-} f(x) = f(c).$$

In usage, continuity refers to either the first or the second definition depending on the context.

This definition actually requires f to possess three properties, and it's instructive to break the definition down into these three components. Specifically, in order for f to be continuous at the point c,

1. f must be defined at c;

2. the limit of f at c must exist (one-sided limit when c is an endpoint);

3. the value of the limit must equal $f(c)$.

If any one of the three properties fails to hold, then the function f is **discontinuous** at the point c and we call c a **point of discontinuity** of f. Our first two examples illustrate the identification of points of continuity and discontinuity.

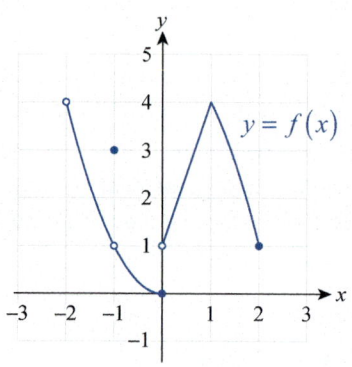

Figure 1

Example 1

Find all points of continuity as well as all points of discontinuity for the function $f(x)$ given by its graph in Figure 1. For any discontinuities, identify those from the three properties on the previous page that fail to hold.

Solution

A careful examination of the graph, paying attention to the empty and full circles, reveals that the domain for this function is $(-2, 2]$. Recalling our studies of limits from the previous two sections, we see that the only point where $\lim_{x \to c} f(x)$ fails to exist is $c = 0$; though there is a discrepancy between the limit and the function value at $c = -1$: $\lim_{x \to -1} f(x) = 1$, but $f(-1) = 3$.

Thus, using the three criteria for continuity, f is continuous at every point of the following intervals: $(-2, -1)$, $(-1, 0)$, and $(0, 2]$. Note that f is defined, its limit exists, and the limit equals the function value at every single point in the above intervals; so they are all points of continuity, including $c = 2$, where we apply the appropriate definition for right endpoints. Thus we can summarize by specifying the set of all points of continuity as follows: $(-2, -1) \cup (-1, 0) \cup (0, 2]$.

To find the points of discontinuity, we first note that $c = -2$ is certainly one; for f is not even defined at $c = -2$ (criterion 1 fails). Therefore, while f is continuous at the right endpoint of its domain, at $c = 2$, it is discontinuous at the left endpoint, namely at $c = -2$. Next, for $c = -1$, as we mentioned above, $\lim_{x \to -1} f(x) \neq f(-1)$, so it is a point of discontinuity (criterion 3 fails). Note, however, that the definition $f(-1) = 1$ will make f continuous; that is, appropriately redefining the function will remove this discontinuity. In contrast, at $c = 0$ the one-sided limits are unequal, so $\lim_{x \to 0} f(x)$ does not even exist. This is not only a point of discontinuity (by virtue of criterion 2 failing), but one that cannot be removed by redefining the function. Such are called *jump discontinuities*; the reason for the name should be clear from the graph.

Lastly, we note that f is certainly discontinuous everywhere outside of $[-2, 2]$, for it is undefined at those points (again, criterion 1 fails). Thus we can summarize the points of discontinuity of f as follows: $(-\infty, -2] \cup \{-1\} \cup \{0\} \cup (2, \infty)$. Be sure you understand the use of different types of parentheses here.

Example 2

Identify and examine the discontinuities of the following functions.

a. $f(x) = \dfrac{x^2 + x}{x}$ **b.** $g(x) = \dfrac{1}{x^2}$ **c.** $h(x) = \sin\dfrac{1}{x}$

Solution

a. The function f is a rational function, and as such it possesses the Direct Substitution Property and thus is continuous at any point where the denominator is nonzero; that is, $f(x)$ is continuous at every $x \neq 0$. Clearly, $x = 0$ is a point of discontinuity since f is not defined there; however, using a bit of algebra we can actually say more. Since

$$\lim_{x \to 0} f(x) = \lim_{x \to 0} \frac{x^2 + x}{x} = \lim_{x \to 0} \frac{x(x+1)}{x} = \lim_{x \to 0}(x+1) = 1,$$

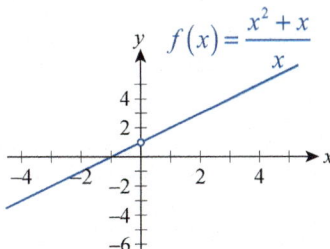

$f(x) = \dfrac{x^2 + x}{x}$

Figure 2

our conclusion is that $\lim_{x \to 0} f(x)$ actually exists, and that the graph of f agrees with that of $y = x + 1$ except for the point corresponding to $x = 0$, where f is undefined. In other words, you can think of the discontinuity of f arising at $x = 0$ as one caused by a single point "missing" from the graph, that is, the graph being the line $y = x + 1$ "punctured" at $(0,1)$, as shown in Figure 2. This is another example of a discontinuity that can be removed; the definition $f(0) = 1$ will make f continuous at $x = 0$.

b. Note that g is undefined and has a vertical asymptote at $x = 0$; moreover, $\lim_{x \to 0} g(x) = \infty$ (see Figure 3). We can argue, as we did in part a., that g is continuous at every $x \neq 0$, everywhere on its domain. The discontinuity at $x = 0$, however, is very different from that of $f(x)$ in part a., since $\lim_{x \to 0} g(x)$ doesn't exist by virtue of the function values approaching infinity. In other words, no definition will make g continuous at 0. This type of discontinuity is called an *infinite discontinuity*.

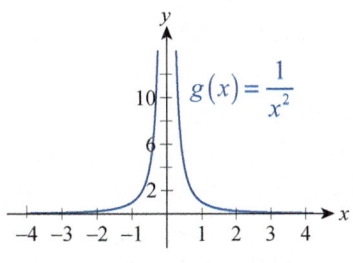

$g(x) = \dfrac{1}{x^2}$

Figure 3

c. By an argument almost exactly like the one we gave in Example 7 of Section 2.3, one can show not only that h is undefined for $x = 0$, but that $\lim_{x \to 0} h(x)$ does not exist either. It follows from that argument that the reason for the nonexistence of the limit is "wild" oscillation: actually, infinitely many oscillations are "squeezed" into arbitrarily small neighborhoods of 0 (see Figure 4). Even good technology cannot do full justice to what is actually going on. It is quite clear, however, that we cannot define the value of $h(0)$ so as to make h continuous. We call this an *oscillating discontinuity*.

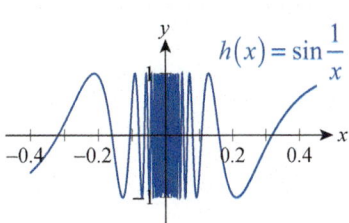

$h(x) = \sin\dfrac{1}{x}$

Figure 4

The points of discontinuity in Examples 1 and 2 illustrate the variety of ways in which one or more of the three properties of continuity can fail. The examples also illustrate that some discontinuities can be remedied by simply redefining the function at the point of discontinuity.

Removable Discontinuity

If a function f has a point of discontinuity at c but $\lim\limits_{x \to c} f(x)$ exists, c is called a **removable discontinuity** of f. The function can be made continuous at c by redefining f at c so that

$$f(c) = \lim_{x \to c} f(x).$$

If c is an endpoint of an interval on which f is defined, replace $\lim\limits_{x \to c} f(x)$ with the appropriate one-sided limit.

If a given point of discontinuity is not removable, it is called **nonremovable**.

Identify the discontinuities in Examples 1 and 2 as removable or nonremovable.

Solution

As we discussed in Example 1, f has a removable discontinuity at $c = -1$, since $\lim\limits_{x \to -1} f(x)$ exists. This not being the case at $c = 0$, the latter is a nonremovable discontinuity. Also, since the right-hand limit of f exists at -2, it has a removable discontinuity there, which can be removed by defining $f(-2) = 4$.

As for the functions in Example 2, we have seen that f is the only one with a removable discontinuity, which occurs at $c = 0$.

The discontinuities of g and h (both at $c = 0$) are nonremovable. The reason in the case of g is that the function values approach infinity near 0, while h has infinitely many oscillations near its point of discontinuity.

If we rephrase the definition of continuity using the epsilon-delta definition of limit, we obtain the following.

Epsilon-Delta Continuity at a Point

Given a function f and a point c in the domain of f, we say f is continuous at c if for every number $\varepsilon > 0$ there is a number $\delta > 0$ such that $|f(x) - f(c)| < \varepsilon$ for all x in the domain of f satisfying $0 < |x - c| < \delta$.

This alternate version of the definition makes it clear that continuity of a function at an endpoint of an interval is really the same idea as continuity elsewhere. The key idea is that f must be defined at c and that all values of $f(x)$ must be close to (meaning within ε of) the value $f(c)$ whenever x lies in the domain of f and is sufficiently close to (within δ of) the point c (see Figure 5).

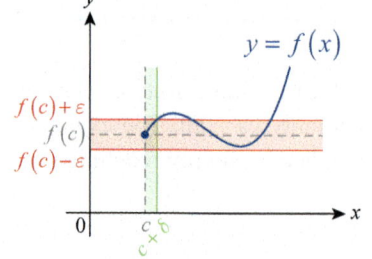

Figure 5

Topic Two
Continuous Functions

Once continuity at a point has been defined, it is natural to extend the meaning of continuity to larger sets. For example, we say that a function is **continuous on an interval** if it is continuous at every point of that interval. For easily graphed functions defined on an interval, this extension agrees well with the intuitive sense that continuity corresponds to the ability to construct a graph without lifting pen from paper.

Our last extension is very similar, but it contains a few subtleties that merit careful consideration.

Definition

Continuity of a Function

A function f is said to be **continuous** (or **continuous on its domain**) if it is continuous at every point of its domain.

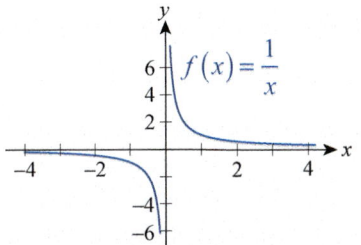

Figure 6

The first subtlety to be aware of is that a continuous function can have points of discontinuity. This seemingly paradoxical statement is entirely consistent with our definitions of continuity at a point and of continuity of a function. Consider, for example, the function $f(x) = 1/x$ (see Figure 6). The domain of this function is $(-\infty, 0) \cup (0, \infty)$, and f is indeed continuous at each point of its domain (you should verify for yourself that $\lim_{x \to c} f(x) = f(c)$ for every point c in the domain). So f is a continuous function, but it is clear that 0 is a point of discontinuity for f since $\lim_{x \to 0} f(x)$ does not exist. Further, it should be noted that f is a continuous function that is not necessarily continuous on every interval of real numbers—for instance, f is not continuous on $[-3, 3]$.

Example 4

Discuss the continuity of $f(x) = \dfrac{x^2 - 1}{x - 1}$. If applicable, identify all points of discontinuity.

Solution

Since f is a rational function, we know that it satisfies the Direct Substitution Property and is thus continuous everywhere except at $c = 1$, where the denominator is 0. $f(1)$ is undefined, making $c = 1$ the only point of discontinuity for f. So once again, our seemingly paradoxical, but correct, conclusion is that f is a continuous function, and its only point of discontinuity is at $c = 1$.

As a final remark, since $f(x) = \dfrac{x^2 - 1}{x - 1} = \dfrac{(x-1)(x+1)}{x - 1} = x + 1$ for all $x \neq 1$, we identify the above discontinuity as removable, since the definition $f(1) = 2$ makes f continuous on the entire real line.

Example 5

Recall $g(x) = 1/x^2$ from Example 2. As we have mentioned, its only point of discontinuity is at $c = 0$, where it is undefined, but g is continuous everywhere on its domain. In this example, we will give a rigorous proof of this latter claim, using the epsilon-delta continuity definition.

Indeed, choose and fix an arbitrary, nonzero $c \in \mathbb{R}$. We will proceed to prove that $g(x) = 1/x^2$ is continuous at c. Note that we may assume $c > 0$ (in the negative case, the argument is similar, or one may take advantage of the fact that g is an even function).

Our epsilon-delta argument will be similar to those given in earlier examples to prove limit claims (this shouldn't come as a surprise, since our epsilon-delta continuity definition is itself closely related to the formal definition of limit). To start us off, suppose that $\varepsilon > 0$ is given. We need to find a $\delta > 0$ such that

$$0 < |x - c| < \delta \Rightarrow \left| \frac{1}{x^2} - \frac{1}{c^2} \right| < \varepsilon.$$

We will first try to bound the quantity $\left| 1/x^2 - \left(1/c^2 \right) \right|$ in order to get an idea of just how small δ needs to be to satisfy the above inequality. Observe that

$$\left| \frac{1}{x^2} - \frac{1}{c^2} \right| = \left| \frac{c^2 - x^2}{x^2 c^2} \right| = \frac{\left| (c - x)(c + x) \right|}{x^2 c^2} = |x - c| \frac{x + c}{x^2 c^2}.$$

Here we are safe to have omitted the absolute value symbols from the last factor, since we can assume that x is close enough to c so that $x + c$ is positive. In fact, we will insist that $\frac{1}{2}c < x < \frac{3}{2}c$ throughout this argument. Note that this is equivalent to requiring that $\delta < \frac{1}{2}c$, which is not a loss of generality since once a successful $\delta > 0$ is chosen any smaller number works just fine.

Next, let's examine the quantity $(x + c)/(x^2 c^2)$. Since x is in the interval specified above, $x + c < \frac{3}{2}c + c = \frac{5}{2}c$, and $x^2 c^2 > \left(\frac{1}{2}c \right)^2 c^2 = \frac{1}{4}c^4$. Thus we have

$$|x - c| \frac{x + c}{x^2 c^2} < |x - c| \frac{\frac{5}{2}c}{\frac{1}{4}c^4} = |x - c| \frac{10}{c^3}.$$

Note that since c is fixed throughout this argument, so is $10/c^3$. Putting the above two chains of equalities and inequalities together, we can summarize what we have obtained so far. If x is sufficiently close to c,

$$\left| \frac{1}{x^2} - \frac{1}{c^2} \right| < |x - c| \frac{10}{c^3}.$$

Now it is time to pick a $\delta > 0$ for the given ε. Let us choose it to be less than the smaller of the numbers $c/2$ and $\left(c^3/10 \right)\varepsilon$. Then if $|x - c| < \delta$,

$$\left| \frac{1}{x^2} - \frac{1}{c^2} \right| < |x - c| \frac{10}{c^3} < \delta \frac{10}{c^3} < \frac{c^3}{10}\varepsilon \frac{10}{c^3} = \varepsilon,$$

and our proof is complete.

Topic Three
Properties of Continuity

Just as the limit laws greatly simplify the task of determining whether a function has a limit at a given point, we have a number of properties of continuity that allow us to quickly answer many questions regarding continuity of functions.

Theorem

Properties of Continuous Functions

Let f and g be two functions both continuous at the point c, and let k be a fixed real number. Then the following combinations of f and g are also continuous at c.

Sums	$f + g$	"A sum of continuous functions is continuous."
Differences	$f - g$	"A difference of continuous functions is continuous."
Constant Multiples	$k \cdot f$	"A multiple of a continuous function is continuous."
Products	$f \cdot g$	"A product of continuous functions is continuous."
Quotients	$\dfrac{f}{g}$, provided $g(c) \neq 0$	"A quotient of continuous function is continuous."

These properties follow immediately from the corresponding limit properties. As an example, we will prove the fourth statement above.

--- **Proof** ---

We assume that f and g are both continuous at c.

$$\lim_{x \to c} (f \cdot g)(x) = \lim_{x \to c} \left[f(x) g(x) \right]$$
$$= \lim_{x \to c} f(x) \cdot \lim_{x \to c} g(x) \quad \text{The limit of a product is the product of the limits.}$$
$$= f(c) \cdot g(c) \quad \text{Both } f \text{ and } g \text{ are continuous at } c.$$
$$= (f \cdot g)(c)$$

As a consequence of these properties, polynomial and rational functions can be immediately classified as continuous functions. That is, every polynomial function and every rational function is continuous at every point of its domain.

Example 6

To illustrate the fact that a rational function is continuous everywhere on its domain, everywhere except for the zeros of the denominator, let us consider

$$r(x) = \frac{x^2 - 2x}{x^2 - 3x + 2}.$$

Since the denominator factors as $x^2 - 3x + 2 = (x-2)(x-1)$, we see that the domain of r is the set $D = \{x \in \mathbb{R} \mid x \neq 1, 2\}$. If $c \in D$, that is, c is an arbitrary real number other than 1 or 2, the Direct Substitution Property applies, and $r(c) = \lim_{x \to c} r(x)$, showing that r is continuous at every point of its domain. Using our limit determination techniques, we can even find out what happens at the two points of discontinuity. Since

$$r(x) = \frac{x^2 - 2x}{x^2 - 3x + 2} = \frac{x(x-2)}{(x-1)(x-2)} = \frac{x}{x-1},$$

it follows that $x = 2$ is a removable discontinuity, and the graph of r agrees with that of $g(x) = x/(x-1)$ for all $x \neq 2$. The discontinuity at $x = 1$, however, is nonremovable. Since the numerator is bounded near 1, while the denominator nears 0, r has a vertical asymptote at $x = 1$. Notice how the graph in Figure 7 supports all of our findings.

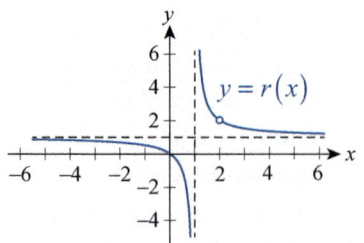

Figure 7

Since sums, differences, products, and quotients of continuous functions are continuous, it is natural to ask if continuity is preserved under other combinations of functions. The following theorem is a bridge toward an important answer to that question.

Theorem

"Limits Pass Through a Continuous Function"

Suppose $\lim_{x \to c} g(x) = a$ and f is continuous at the point a. Then

$$\lim_{x \to c} f(g(x)) = f\left(\lim_{x \to c} g(x)\right) = f(a).$$

In words, we say the limit operation passes inside the continuous function f.

The proof of this theorem can be found in Appendix E—it is a nice example of the use of the epsilon-delta definitions of both limit and continuity. At the moment, though, we are more interested in the following application of the theorem.

Theorem

"A Composition of Continuous Functions Is Continuous"

If g is continuous at the point c, and if f is continuous at $g(c)$, then the composite function $f \circ g$ is continuous at c.

Proof

Because g is continuous at c, we know that $\lim_{x \to c} g(x) = g(c)$, and so by replacing a with $g(c)$ in the previous theorem it follows that

$$\lim_{x \to c} f\big(g(x)\big) = f\Big(\lim_{x \to c} g(x)\Big) = f\big(g(c)\big).$$

Using composition notation, we have

$$\lim_{x \to c}(f \circ g)(x) = (f \circ g)(c).$$

Another useful theorem tells us that inverses of continuous functions are continuous. Its proof can also be found in Appendix E.

Theorem

"The Inverse of a Continuous Function Is Continuous"

If f is one-to-one and continuous on the interval (a,b), then f^{-1} is also a continuous function.

With these theorems in hand, the set of functions that we can classify as continuous instantly expands considerably.

Example 7

Use the above theorems to discuss the continuity of the following functions.

a. $F(x) = \sqrt[3]{\dfrac{2x+1}{x^2 + 5}}$

b. $G(x) = \sqrt{\dfrac{4x^5 - 3x^2 + 7}{9x^3 + 3x}}$

c. $H(x) = \sin \dfrac{2x^3 + x^2 - 3}{x^2 - 2x - 8}$

d. $K(x) = \arcsin \sqrt{1 - x^2}$

Solution

a. Notice that F is a composite function; you can think of it as $F(x) = f\big(g(x)\big)$, where $g(x) = (2x+1)/(x^2 + 5)$, and $f(x) = \sqrt[3]{x}$. Furthermore, g is continuous on all of $\mathbb{R}$, since its denominator is never 0. The same can be said about f, namely that it is continuous on the entire real line. Note that you can see this in at least two ways. Either apply the previous theorem, since f is the inverse of $k(x) = x^3$, which is a continuous function mapping $\mathbb{R}$ onto $\mathbb{R}$; or by referring to the Positive Integer Root Law, you can provide a one-line proof of the continuity of f as follows. If $c \in \mathbb{R}$ is arbitrary,

$$\lim_{x \to c} f(x) = \lim_{x \to c} \sqrt[3]{x} = \sqrt[3]{\lim_{x \to c} x} = \sqrt[3]{c}.$$

Now a simple application of the earlier theorem about continuity of compositions establishes the fact that $F(x) = f\big(g(x)\big)$ is continuous on $\mathbb{R}$.

b. As in part a., we will use the "continuity of compositions" theorem, but while the conditions of the theorem were readily satisfied by both the inner and outer functions on all of $\mathbb{R}$ in the previous problem, we will need to pay a little extra attention here. To start off, we once again use the notation $G(x) = f(g(x))$; however, this time $g(x) = (4x^5 - 3x^2 + 7)/(9x^3 + 3x)$ (see Figure 8), while $f(x) = \sqrt{x}$. According to our theorem, $G = f \circ g$ will be continuous at all points of continuity c of $g(x)$ provided that f is also continuous at $g(c)$. A factoring argument like the one given in part a. shows that g is undefined and has a vertical asymptote at $x = 0$; it is, however, continuous everywhere else. The outer function f, on the other hand, is only defined for $x \geq 0$, but it is continuous on its domain. This latter claim can be established by an argument similar to the one given for the cube root function in part a. In fact, notice that it is easy to generalize that argument and conclude that all root functions are continuous as a result of the previous theorem and the continuity of the functions x^n.

Summarizing our observations up to this point, and using the fact that the composition of continuous functions is continuous, we see that $G(x) = f(g(x))$ will be continuous at all nonzero real numbers c such that $g(c) \geq 0$.

Our next observation is that $g(x) < 0$ on the interval $(-1, 0)$, but is nonnegative everywhere else on its domain, that is, $g(x) \geq 0$ on $(-\infty, -1] \cup (0, \infty)$. (You can algebraically convince yourself of this fact, or simply examine the graph in Figure 8.)

Therefore, we conclude that $G(x) = f(g(x))$ is continuous on the set $(-\infty, -1] \cup (0, \infty)$. The graph of G supports our conclusion. (See Figure 9.)

c. Factoring the denominator easily shows that the inner function of this problem $g(x) = (2x^3 + x^2 - 3)/(x^2 - 2x - 8)$ has two vertical asymptotes, at $x = -2$ and $x = 4$, respectively. However, being a rational function, g is continuous everywhere else on $\mathbb{R}$. As we will show in Example 8, $f(x) = \sin x$ is continuous at every real number, so our conclusion is that $H(x)$ is continuous everywhere except at $x = -2$ and $x = 4$.

d. First of all, we observe that $g(x) = \sqrt{1 - x^2}$ is continuous everywhere on its domain of $[-1, 1]$. Furthermore, the range of g is the interval $[0, 1]$ (note that the graph of g is the upper semicircle of radius 1, centered at the origin). Using the theorem about the continuity of inverse functions, we conclude that $f(x) = \arcsin x$ is continuous at every point of the range of g and hence $K(x) = f(g(x))$ is continuous on its domain of $[-1, 1]$.

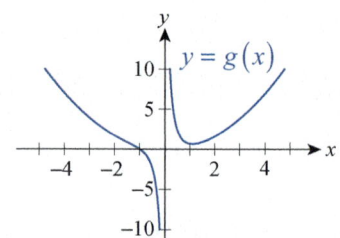

Figure 8

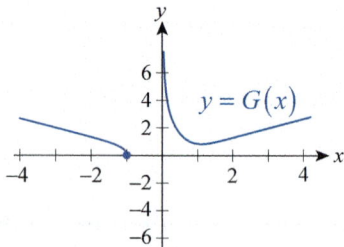

Figure 9

In fact, most familiar functions (those that we bother to give names to) are continuous (though many have points of discontinuity—don't forget the subtle distinction). Trigonometric functions are continuous, a fact that is most easily proven through the use of an alternate formulation of continuity:

f is continuous at the point c if and only if $\lim\limits_{h \to 0} f(c+h) = f(c)$.

Example 8

Use the alternate formulation of continuity to prove that $f(x) = \sin x$ is continuous on $\mathbb{R}$.

Solution

Our goal is to show that for any fixed real number c, $\lim\limits_{h \to 0} \sin(c+h) = \sin c$.

Using the appropriate sum identity known from trigonometry along with the relevant limit laws,

$$\lim_{h \to 0} \sin(c+h) = \lim_{h \to 0} \left(\sin c \cos h + \cos c \sin h\right)$$
$$= \lim_{h \to 0} \left(\sin c \cos h\right) + \lim_{h \to 0} \left(\cos c \sin h\right)$$
$$= \sin c \lim_{h \to 0} \cos h + \cos c \lim_{h \to 0} \sin h,$$

where we also used the fact that during the limit process $\sin c$ and $\cos c$ are constants, so the Constant Multiple Law applies.

Our next step is to recall that as h approaches 0, $\lim\limits_{h \to 0} \cos h = 1$ and $\lim\limits_{h \to 0} \sin h = 0$. (You may recall these facts from the unit-circle definition of sine and cosine for small angles, or simply from the graphs of the sine and cosine functions.) Using these latter two limits we obtain

$$\sin c \lim_{h \to 0} \cos h + \cos c \lim_{h \to 0} \sin h = \sin c \cdot 1 + \cos c \cdot 0$$
$$= \sin c,$$

which is what we needed to prove.

Finally, we note that the continuity of $\cos x$ can be established by a similar argument, or by using the continuity of the sine function along with the identity $\cos x = \sin\left((\pi/2) - x\right)$. Then using the properties of continuous functions, the continuity of the remaining four trigonometric functions $\tan x$, $\cot x$, $\sec x$, and $\csc x$ will follow.

Exponential functions are continuous, a fact largely due to design. For a given base a, the expression $a^{m/n}$ is easily understood for every rational number m/n, and a common way to extend the meaning of a^x to irrational numbers is to define a^x to be the limit of terms of the form a^{r_n}, where $\{r_n\}$ is a sequence of rational numbers approaching x (some work is required in order to show such a definition is unambiguous). And because inverse functions of continuous functions are continuous, the continuity of logarithms and inverse trigonometric functions is assured.

Example 9

Use the above theorems to identify where the following functions are continuous.

a. $f(x) = \dfrac{\arctan(\ln x)}{x^2 + 1}$ **b.** $g(x) = \dfrac{xe^{\cos x}}{\sqrt{x^2 - 1} - 3}$ **c.** $h(x) = \dfrac{\ln x - \sin^{-1} x}{x^2 + x - 2}$

Solution

a. Being the inverses of continuous functions, $\ln x$ is continuous on $(0, \infty)$, and $\arctan x$ is continuous on the entire real line. Thus the composition $\arctan(\ln x)$ is continuous on $(0, \infty)$. Since the denominator of f, $x^2 + 1$ is never 0 on $\mathbb{R}$ (check this), it follows that the quotient $f(x)$ is continuous on $(0, \infty)$.

b. Since all three functions $\cos x$, e^x, and x are continuous on the entire real line, it follows from our theorems that both the composition $e^{\cos x}$ and the product $xe^{\cos x}$ will be continuous on all of $\mathbb{R}$. Next, just as we did in part a., we would like to use the theorem stating the continuity of the quotient of continuous functions, but we need to carefully check the continuity of the denominator of g. First, the square root function requires that the inequality $x^2 - 1 \geq 0$ be true for x. This means, we need $x^2 \geq 1$ to hold, that is, $x \geq 1$ or $x \leq -1$. Since the composition of continuous functions is continuous, we conclude that $\sqrt{x^2 - 1}$ is continuous on $(-\infty, -1] \cup [1, \infty)$. Last, but not least, in order for g to be continuous, its denominator cannot be 0, so we have to exclude the solutions of the equation $\sqrt{x^2 - 1} = 3$. By squaring both sides and adding 1 we obtain $x^2 = 10$, that is, the points $x = \pm\sqrt{10}$ have to be excluded from the domain.

Summarizing our findings, we conclude that $g(x)$ is continuous on the set

$$\left(-\infty, -\sqrt{10}\right) \cup \left(-\sqrt{10}, -1\right] \cup \left[1, \sqrt{10}\right) \cup \left(\sqrt{10}, \infty\right).$$

c. Arguing as we did in part a., $\ln x$ and $\sin^{-1} x$ are continuous, since they are inverses of continuous functions. The domain of $\ln x$ is $(0, \infty)$, while $\sin^{-1} x$ is only defined on $[-1, 1]$. However, notice that $x = 1$ is a zero of the denominator, so by the properties of continuous functions, we conclude that $h(x)$ is continuous on the open interval $(0, 1)$.

Collectively, the theorems in this section allow us to identify a large number of continuous functions very quickly. It might even be possible to do something about the occasional point of discontinuity. Recall that a discontinuity is termed *removable* if the limit at the point exists; in such cases, we can define the **continuous extension** of a function as in the next example.

Example 10

Identify the removable discontinuities and define the continuous extension

of $r(x) = \dfrac{x^2 - 5x}{x^3 - 3x^2 - 10x}$.

Solution

Factoring and canceling yields

$$r(x) = \frac{x^2 - 5x}{x^3 - 3x^2 - 10x} = \frac{x(x-5)}{x(x-5)(x+2)} = \frac{1}{x+2},$$

where the previous equality holds for all x, except for $x = 0$ and $x = 5$. However, as we learned in Section 2.4, both $\lim\limits_{x \to 0} r(x)$ and $\lim\limits_{x \to 5} r(x)$ still exist.

$$\lim_{x \to 0} r(x) = \frac{1}{0+2} = \frac{1}{2}, \text{ while } \lim_{x \to 5} r(x) = \frac{1}{5+2} = \frac{1}{7}$$

Therefore r has two removable discontinuities, at $x = 0$ and $x = 5$. Now we can define $\tilde{r}(x)$, the continuous extension of r, as follows.

$$\tilde{r}(x) = \begin{cases} \dfrac{x^2 - 5x}{x^3 - 3x^2 - 10x} & \text{if } x \neq 0 \text{ or } 5 \\ \dfrac{1}{2} & \text{if } x = 0 \\ \dfrac{1}{7} & \text{if } x = 5 \end{cases}$$

Let us note, however, that although correct and instructive, it is not necessary in this case to use a piecewise defined function to define the continuous extension of r. Alternatively, it is much shorter to define it simply as $\tilde{r}(x) = 1/(x+2)$. Note that this is equivalent to our "piecewise" definition above. Note also that both r and its continuous extension have nonremovable discontinuities at $x = -2$.

We will close this section with one last theorem that makes a connection between our rigorous definition of continuity and the intuitive "graph without interruptions" interpretation of continuity.

Theorem

The Intermediate Value Theorem

If f is a continuous function defined on the closed interval $[a,b]$, then f takes on every value between $f(a)$ and $f(b)$. That is, if L is a real number between $f(a)$ and $f(b)$, then there is a c in the interval $[a,b]$ such that $f(c) = L$.

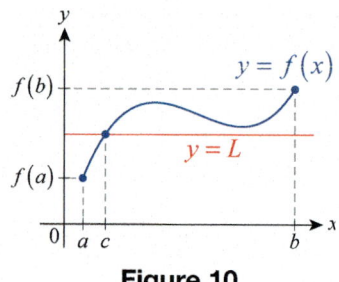

Figure 10

The proof of the Intermediate Value Theorem (IVT) relies upon a property of the real numbers called *completeness*; this concept and the proof of the theorem are presented in courses like Advanced Calculus and Introductory Real Analysis. But the implications of the Intermediate Value Theorem are of interest to us at the moment. Informally, the IVT says that the graph of a continuous function f defined on an interval $[a,b]$ cannot avoid intersecting any horizontal line $y = L$ if L is a value between $f(a)$ and $f(b)$ (see Figure 10). In other words, such a function f has no "breaks" or "jumps" that would allow it to skip over the value L. This property, also referred to as the *Intermediate Value Property*, is often used to prove that an equation of the form $f(x) = L$ must have at least one solution in the interval $[a,b]$. Note that the content of the IVT is precisely the fact that a continuous function f on a closed interval $[a,b]$ possesses the Intermediate Value Property.

Example 11

Use the Intermediate Value Theorem to show that the equation $x^5 + 9x - 4 = 0$ has a solution between 0 and 1.

Solution

Introducing the notation $f(x) = x^5 + 9x - 4$, notice that we are attempting to prove the existence of a solution, or root, of the equation $f(x) = 0$ on the interval $[0,1]$. The existence of such root c would mean that f takes on the value $L = 0$ for some c in $[0,1]$. Does the Intermediate Value Theorem guarantee that? Notice first of all that f is certainly continuous on the closed interval $[0,1]$ (actually, it is continuous on the entire real line, but for the purposes of this problem we can safely ignore what happens outside of $[0,1]$). Next, since 0 and 1 play the roles of a and b in the theorem, respectively, we proceed to examine the values $f(0)$ and $f(1)$.

$$f(0) = 0^5 + 9 \cdot 0 - 4 = -4 < 0$$
$$f(1) = 1^5 + 9 \cdot 1 - 4 = 6 > 0$$

In other words, a change of sign occurs on $[0,1]$; that is, f goes from negative to positive on $[0,1]$. The statement of the Intermediate Value Theorem is precisely the fact that a continuous function cannot do this and avoid intersecting the horizontal line $y = 0$ (the x-axis). More precisely, since $L = 0$ we have $-4 < L < 6$; that is, since 0 is a real number between $f(0) = -4$ and $f(1) = 6$, by the Intermediate Value Theorem there is a c in $[0,1]$ such that $f(c) = L = 0$. In other words, we have

$$c^5 + 9c - 4 = 0.$$

Thus we proved the existence of a root between 0 and 1 for the given equation.

As an important final remark, we note that the Intermediate Value Theorem is an *existence theorem*, in other words, we proved the *existence* of a solution between 0 and 1, without actually specifying what it is. However, in lieu of a formula to solve fifth-degree equations, this is still extremely useful, and there is no need to stop here. Namely, if we know a solution exists, we can "narrow down" the interval around it, by refining our guesses, or by "zeroing in" on the solution. Eventually, we can come up with an approximation of the root with an acceptable level of accuracy, which is the basic idea of so-called *numerical methods* of solving equations. The following Technology Note illustrates this important idea.

Technology Note

Continuing Example 11 by using *Mathematica*, we will illustrate the process of zooming in successively on the root of $x^5 + 9x - 4 = 0$, thereby obtaining an approximation of the solution accurate to two digits after the decimal point. Then we will use the built-in **NSolve** command to obtain *Mathematica*'s numerical approximation of the root.

To start us off, we have graphed $f(x) = x^5 + 9x - 4$ (see Figure 11), and then we zoomed in to obtain that portion of the graph defined on $[0,1]$ (see Figure 12).

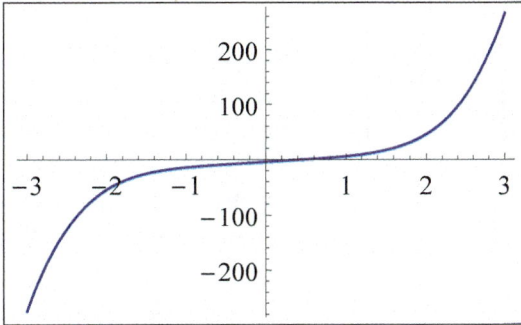

Figure 11

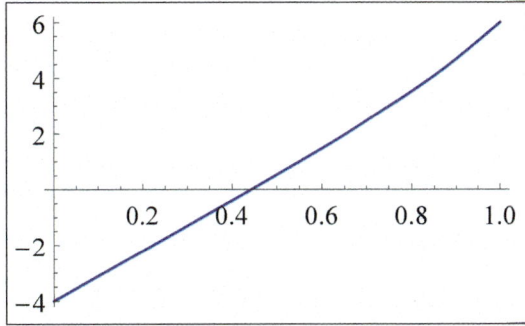

Figure 12

It is clear from both graphs that $f(x) = 0$ has a root between 0 and 1, as we proved in Example 11. Figure 12 suggests that the root is between 0.4 and 0.5, so let's zoom in further to gain more accuracy.

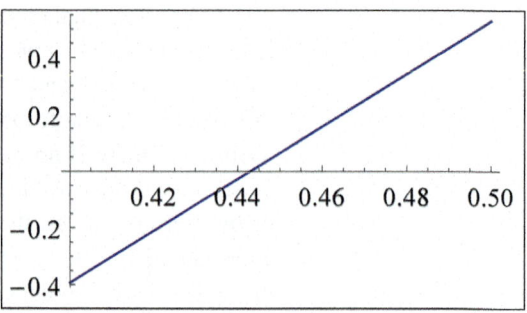

Figure 13

Figure 13 suggests that we should zoom in on that portion of the graph defined on $[0.44, 0.445]$, which we have done (see Figure 14).

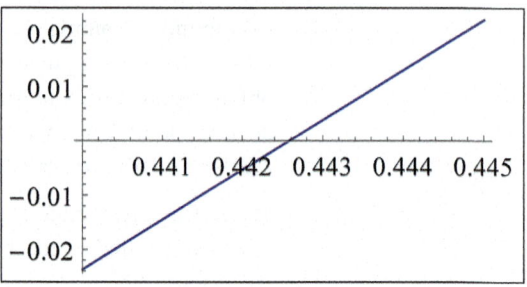

Figure 14

It is clear from Figure 14 that the root is located between 0.442 and 0.443, so we have achieved the desired accuracy; the first two digits after the decimal point are correct. We conclude that the root is $c \approx 0.44$. (Note that the above accuracy could actually have been achieved from carefully eyeballing the graph in Figure 13, while the graph in Figure 14 says even more: we might guess that the root is in fact very close to 0.4426.)

Note that it is also possible to estimate the root by appropriately zooming in on the graph using a graphing calculator (see Figure 15).

Finally, we use the **NSolve** command of *Mathematica* to approximate the solution of $f(x) = 0$. The screenshot in Figure 16 shows the feedback we receive from the software. The appearance of complex roots shows the power of the software, but we can ignore them for now and focus on the sole real root. *Mathematica* approximates it as $c \approx 0.442558$.

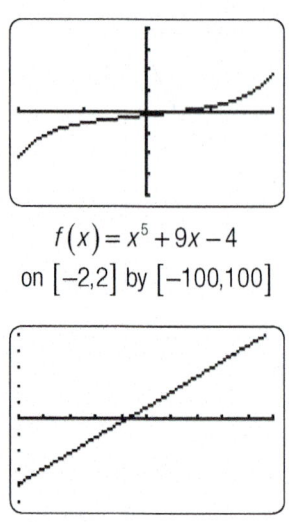

$f(x) = x^5 + 9x - 4$
on $[-2,2]$ by $[-100,100]$

$f(x) = x^5 + 9x - 4$
on $[0.4, 0.5]$ by $[-0.5, 0.5]$

Figure 15 Zooming in on a Root with a Graphing Calculator

```
In[1]:= NSolve[x^5 + 9 x - 4 == 0, x]

Out[1]= {{x → -1.32389 - 1.2337 i},
         {x → -1.32389 + 1.2337 i}, {x → 0.442558},
         {x → 1.10261 - 1.24271 i}, {x → 1.10261 + 1.24271 i}}
```

Figure 16

2.5 **Exercises**

1–2 Find all points of continuity as well as all points of discontinuity for the given function. For any discontinuities, identify those from the three continuity criteria that fail to hold.

1.

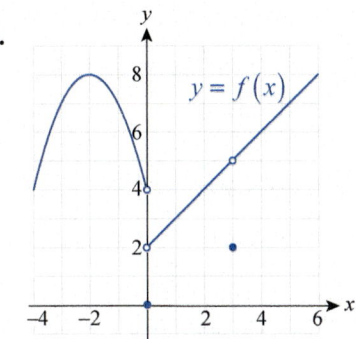

2.

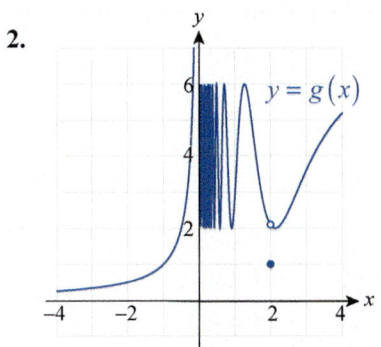

3. Sketch a graph of a function (a formula is not necessary) that has a removable discontinuity at $x = -1$, a jump discontinuity at $x = 2$, but is right-continuous at 2. (Answers will vary.)

4. Sketch a graph of a function that has an infinite discontinuity at $x = 0$ and an oscillating discontinuity at $x = 5$ so that it is still left-continuous at 5. (Answers will vary.)

5–29 Find and classify the discontinuities (if any) of the function as removable or nonremovable.

5. $f(x) = \dfrac{1}{x}$

6. $g(x) = \dfrac{-2}{x-3}$

7. $h(x) = \dfrac{x^2 - 9}{x - 3}$

8. $k(x) = \dfrac{x^2 - 2x}{x^2 + 5x - 14}$

9. $u(x) = \dfrac{x^2 - 9}{x - 2}$

10. $v(x) = \dfrac{x - 1}{x^2 + 2x - 3}$

11. $w(x) = \begin{cases} x + 1 & \text{if } x \le 0 \\ \dfrac{1}{2}x^2 + 1 & \text{if } x > 0 \end{cases}$

12. $f(x) = \begin{cases} \dfrac{1}{2}x - 2 & \text{if } x \le 4 \\ x^3 + 1 & \text{if } x > 4 \end{cases}$

13. $g(x) = \begin{cases} \tan x & \text{if } x \le \dfrac{\pi}{2} \\ \cos x & \text{if } x > \dfrac{\pi}{2} \end{cases}$

14. $h(x) = \begin{cases} \cos x & \text{if } x \le 0 \\ \tan x + 1 & \text{if } x > 0 \end{cases}$

15. $F(u) = \dfrac{u - 4}{\sqrt{u} - 2} \quad (u \ge 0)$

16. $G(s) = \dfrac{s}{\sqrt{s + 4} - 2} \quad (s \ge -4)$

17. $H(t) = \dfrac{t}{\sqrt{t^2 + 2}}$

18. $u(x) = \cos \dfrac{1 - x^2}{1 - x}$

19. $v(t) = |\sin t|$

20. $K(x) = |x + 2| + |x - 1|$

21. $F(t) = \dfrac{t}{t^2 - 1}$

22. $G(t) = \dfrac{t}{t^2 + 1}$

23. $H(x) = |x + 2|$

24. $k(x) = \dfrac{|x - 4|}{x - 4}$

25. $s(x) = [\![x + 2]\!]$

26. $t(x) = 4 - [\![x]\!]$

27. $u(z) = [\![z^2]\!]$

28. $v(x) = x [\![x]\!]$

29. $w(x) = x \left[\!\left[\dfrac{1}{x} \right]\!\right]$

30–33 Use the ε-δ definition to prove that the function is continuous.

30. $f(x) = \dfrac{1}{x}$

31. $g(x) = 3x - 2$

32. $F(x) = x^3$

33. $G(x) = \sqrt{x}$

34–39 Use the theorems of this section to discuss the continuity of the function.

34. $F(x) = \sqrt{\dfrac{x}{x^2 + 7x + 12}}$

35. $G(x) = \sqrt{\dfrac{x^4 - x^3 - 11x^2 + 9x + 18}{2x^3 + x}}$

36. $H(x) = \cos \dfrac{2\ln(x - 3) + 1}{\sqrt[3]{x^2 - 2x - 15}}$

37. $f(x) = \arctan \dfrac{x}{\sqrt{3 - x^2}}$

38. $g(x) = \ln\left(\arcsin(\pi x + 1)\right)$

39. $h(x) = \dfrac{\csc(\pi x + 1)}{\sin\left(\pi e^{x+2}\right)}$

40. Prove the alternate formulation of continuity, that is, the statement that a function f is continuous at the point c if and only if $\lim\limits_{h \to 0} f(c + h) = f(c)$.

41–44 Use the alternate formulation of continuity to prove that the function is continuous.

41. $f(x) = 3x - 5x^2$

42. $g(x) = \cos x$

43. $h(x) = \tan x$

44. $k(x) = e^x$

45–50 Identify the removable discontinuities and define the continuous extension of the function.

45. $f(x) = \dfrac{x^2 + x - 12}{x - 3}$

46. $g(x) = \dfrac{x^3 - 2x^2 - x + 2}{x^2 - 3x + 2}$

47. $h(x) = \dfrac{x - 1}{\sqrt{x} - 1}$

48. $F(x) = \dfrac{\sqrt{x + 1} - 2}{x - 3}$

49. $G(x) = 2^{-1/x^2}$

50. $H(x) = x \cos \dfrac{\pi}{x}$

51–54 Discuss the continuity of the function on the given closed interval.

51. $S(x) = \sqrt{16 - x^2}$ on $[-4, 4]$

52. $T(x) = \left\lVert \dfrac{x}{3} \right\rVert$ on $[0, 3]$

53. $U(x) = \begin{cases} \dfrac{1}{x^2 - 9} & \text{if } |x| < 3 \\ 0 & \text{if } |x| = 3 \end{cases}$ on $[-3, 3]$

54. $V(x) = \begin{cases} x^2 \sin \dfrac{1}{x} & \text{if } x \neq 0 \\ 0 & \text{if } x = 0 \end{cases}$ on $\left[0, \dfrac{1}{\pi}\right]$

55-58 Find the value of a (or the values of a and b, where applicable) such that f is continuous on the entire real line.

55. $f(x) = \begin{cases} 0 & \text{if } x \leq 0 \\ ax & \text{if } 0 < x < 1 \\ 2x + 3 & \text{if } x \geq 1 \end{cases}$

56. $f(x) = \begin{cases} x^3 & \text{if } x \leq 3 \\ ax^2 & \text{if } x > 3 \end{cases}$

57. $f(x) = \begin{cases} -x^2 & \text{if } x < 1 \\ ax + b & \text{if } 1 \leq x \leq 3 \\ (x - 3)^2 + 2 & \text{if } x > 3 \end{cases}$

58. $f(x) = \begin{cases} \cos x & \text{if } x \leq 0 \\ -(x - a)^2 + b & \text{if } 0 < x < 2 \\ \dfrac{1}{2}x - 2 & \text{if } x \geq 2 \end{cases}$

59. Prove that if $f(x)$ is continuous and $f(c) > 0$, then there is a $\delta > 0$ such that $f(x) > 0$ for all $x \neq c$ in the interval $(c - \delta, c + \delta)$.

60. Prove that the Dirichlet function
$$\xi(x) = \begin{cases} 0 & \text{if } x \text{ is rational} \\ 1 & \text{if } x \text{ is irrational} \end{cases}$$
is discontinuous at every real number.

61. Prove that the function
$$f(x) = \begin{cases} 0 & \text{if } x \text{ is rational} \\ x^2 & \text{if } x \text{ is irrational} \end{cases}$$
is continuous only at the single point $c = 0$.

62. Prove that if the functions f and g are both continuous on $\mathbb{R}$ and they agree on the rationals (i.e., $f(x) = g(x)$ for all $x \in \mathbb{Q}$), then $f = g$.

63–68 Decide whether the Intermediate Value Theorem applies to the given function on the indicated interval. If so, find c as guaranteed by the theorem. If not, find the reason.

63. $f(x) = -x^2 + x + 3$ on $[0,3]$; $f(c) = 1$

64. $g(x) = 2x^3 - x^2 - 1$ on $[-1,2]$; $g(c) = 0$

65. $h(x) = \dfrac{x}{x+2}$ on $[0,4]$; $h(c) = \dfrac{1}{2}$

66. $F(x) = \dfrac{2x}{x-1}$ on $[0,2]$; $F(c) = 2$

67. $G(x) = [\![x-2]\!]$ on $[-2,2]$; $G(c) = -\dfrac{1}{2}$

68. $H(x) = \sin\dfrac{3x+2}{2}$ on $\left[-\dfrac{2}{3}, \dfrac{\pi-2}{3}\right]$; $H(c) = \dfrac{1}{2}$

69–74 Use the Intermediate Value Theorem to prove that the given equation has a solution on the indicated interval.

69. $x^3 - 7.5x^2 + 1.2x + 1 = 0$ on $[-1,0]$

70. $2x^3 + x + 10 = 0$ on $[-2,1]$

71. $\cos x = x^2$ on $[0,\pi]$

72. $\ln x - \sqrt{x-2} = 0$ on $[2,5]$

73. $\dfrac{5}{x^2+2} = 1$ on $[-3,-1]$

74. $\cot\dfrac{\pi x}{4} - \dfrac{x}{x+2} = -\dfrac{1}{2}$ on $[1,3]$

75. Suppose that the outside temperature in Columbia, SC on a summer morning at 7:00 a.m. is 74 °F, and it shoots up to 98 °F by 1:00 p.m. Assuming that temperature changes continuously, prove that sometime between 7:00 a.m. and 1:00 p.m. the temperature was exactly 88.35 °F.

76. Prove that if f is continuous and never 0 on the interval $[a,b]$, then either $f(x) > 0$ for every x in $[a,b]$, or $f(x) < 0$ for every x in $[a,b]$.

77.* (Existence of nth roots) Prove that if b is a positive real number and n a positive integer, then there is a positive real number c such that $c^n = b$. (**Hint:** Consider the continuous function $f(x) = x^n$ on the interval $[0, b+1]$.)

78.* Prove that a circle of diameter d has a chord of length c for every number c between 0 and d.

79. Use the function

$$f(x) = \begin{cases} \sin\dfrac{\pi}{x} & \text{if } x \neq 0 \\ 0 & \text{if } x = 0 \end{cases}$$

to prove that the converse of the Intermediate Value Theorem is false; in other words, a function may possess the Intermediate Value Property without being continuous.

80.* (Fixed Point Theorem) Prove that if $f:[a,b] \to [a,b]$ is continuous, then there is a number c in $[a,b]$ with $f(c) = c$ (i.e., c is "fixed," or "not being moved," by f).

81.* A hermit leaves his hut at the foot of a mountain one day at 6:00 a.m. and sets out to climb all the way to the top. He arrives at 6:00 p.m. and realizes that it is too late to go back, so he sets up camp for the night. At 6:00 a.m. the following day, he starts hiking back to his hut, taking the exact same route as the day before. This time, however, it is mostly downhill, so he makes much better time and arrives home at 2:00 p.m. Prove that there is a point along the hermit's route that he passed at exactly the same time on both days. (**Hint:** Apply the Intermediate Value Theorem or the Fixed Point Theorem.)

82. A long-distance phone company charges 31 cents for the first minute and 10 cents for each additional minute or any fraction thereof. Graph the cost as a function of time, find a formula for it, and discuss the significance of its discontinuities. (**Hint:** Use the greatest integer function to construct your answer.)

83.* If Δt denotes the length of the time interval between two events as measured by an observer on a spaceship moving at speed v, and ΔT is the length of the same time interval as measured from Earth, then the formula relating the two quantities is given by

$$\Delta T = \frac{\Delta t}{\sqrt{1 - \dfrac{v^2}{c^2}}},$$

where c is the speed of light. This phenomenon is called *time dilation*, and it follows from the theory of relativity. In essence, it says that a clock moving at speed v relative to an observer is perceived by the same observer to run slower.

a.   Explain why we don't normally notice the time dilation effect in everyday life.

b.   What is the significance of the discontinuity of ΔT (as a function of v)?

84–88 *True or False?* Determine whether the given statement is true or false. In case of a false statement, explain or provide a counterexample.

84. If f is both left- and right-continuous at c, then f is continuous at c.

85. Any function f has an interval (a,b) on which it is continuous.

86. If c is a discontinuity of f, but f does not have a vertical asymptote at c, then c is a removable or jump discontinuity.

87. If $\lim\limits_{x \to c} f(x) = L$, and $f(c) = L$, then f is continuous at c.

88. If c is a discontinuity of f, but $\lim\limits_{x \to c^+} f(x)$ exists, then c is a removable or jump discontinuity.

2.5 Technology Exercises

89–94. Use a graphing calculator or computer algebra system to solve the equations given in Exercises 69–74 to four decimal places.

95–100. Use a graphing calculator or computer algebra system to graph the functions of Exercises 34–39 and explain how the graphs support your discussions of continuity in the aforementioned exercises.

2.6 **Rate of Change Revisited: The Derivative**

We will wrap up this chapter by revisiting the two motivational problems that we began with, making use of the expertise in working with limits that we have since acquired. Our last topic of discussion will be an indication of how the concepts that have been introduced generalize and apply to a wealth of situations.

Topic One

Velocity and Tangent Recap

In Section 2.1, we developed a method for determining instantaneous velocity at a particular point c in time by considering the behavior of the fraction

$$\frac{f(c+h)-f(c)}{h}$$

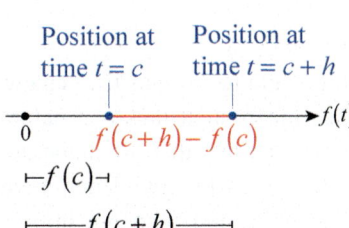

Figure 1

for smaller and smaller values of h. The function $f(t)$ in this setting describes the position of an object at time t moving along a straight line, and the fraction was defined to be the **difference quotient of f at c with increment h**. Although we lacked the terminology and notation at the time, we can now succinctly define the **(instantaneous) velocity of the object at time c** to be

$$\lim_{h \to 0}\frac{f(c+h)-f(c)}{h}.$$

Similarly, we developed a way to arrive at the slope of the line tangent to the function $f(x)$ at the point c by considering the slopes of secant lines that approached the tangent line. Ultimately, we saw that the tangent line slope (if the tangent line exists) is also the limiting behavior of the difference quotient at c, so we can define the **slope of the line tangent to f at c** to also be

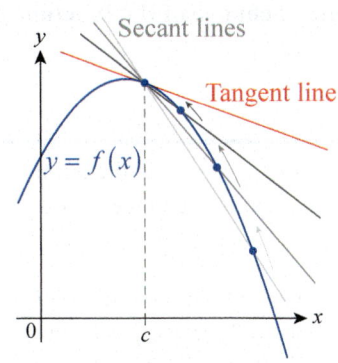

Figure 2

$$\lim_{h \to 0}\frac{f(c+h)-f(c)}{h}.$$

Given this concurrence, it likely comes as no surprise that limits of difference quotients will occupy our attention for some time, and it is appropriate to now introduce some additional notation.

Definition

The Derivative at a Point

The **derivative of the function f at the point c**, denoted $f'(c)$, is

$$f'(c)=\lim_{h \to 0}\frac{f(c+h)-f(c)}{h},$$

provided the limit exists. (Note that $f'(c)$ is read "f prime of c.")

Recall that in Section 2.1 we saw some alternate ways of expressing the same idea. In some contexts, it may be more natural to refer to the point $c + h$ as x, and to express the difference quotient as follows:

$$\frac{f(c+h)-f(c)}{h} = \frac{f(x)-f(c)}{x-c}$$

In this case, the derivative of f at c is obtained by letting x approach c, so an alternate definition of the derivative is

$$f'(c) = \lim_{x \to c} \frac{f(x)-f(c)}{x-c}.$$

One last variation of this idea comes from referring to $f(x)-f(c)$ as the "change in f" or "change in y" (denoted Δf or Δy) and to $x - c$ as the "change in x" (denoted Δx). This is commonly seen when the context makes it natural to refer to the equation $y = f(x)$ and to emphasize the fact that the numerator of the difference quotient represents a vertical difference while the denominator represents a horizontal difference. Using this notation, and with the implicit understanding that c is fixed, the derivative would be denoted as

$$f'(c) = \lim_{\Delta x \to 0} \frac{\Delta f}{\Delta x} = \lim_{\Delta x \to 0} \frac{\Delta y}{\Delta x}.$$

It is important to realize, however, that all these variations represent the same fundamental concept. Regardless of the difference quotient notation used, and regardless of whether we are seeking the instantaneous velocity of an object or the slope of a tangent line, $f'(c)$ indicates the **rate of change of the function f at the point c**.

Example 1

Find the slope of the line tangent to the graph of $f(x) = x^2 - 1$ at the point $(2,3)$. Then use this information to derive the equation of the tangent line.

Solution

As we have discussed above, the slope of the tangent line at the point $(2,3)$ will be the limit of the difference quotient at $c = 2$, or using our new terminology, the derivative of f at the point $c = 2$. Recall that we denote this number by $f'(c)$, and it is calculated as

$$f'(c) = \lim_{h \to 0} \frac{f(c+h)-f(c)}{h}.$$

We now proceed to evaluate the above derivative for the given
$f(x) = x^2 - 1$ at $c = 2$:

$$f'(2) = \lim_{h \to 0} \frac{f(2+h)-f(2)}{h} = \lim_{h \to 0} \frac{\left[(2+h)^2 - 1\right] - (2^2 - 1)}{h}$$

$$= \lim_{h \to 0} \frac{\left[(2^2 + 4h + h^2) - 1\right] - (2^2 - 1)}{h}$$

$$f'(2) = \lim_{h \to 0} \frac{4 + 4h + h^2 - 1 - 4 + 1}{h} = \lim_{h \to 0} \frac{4h + h^2}{h}$$

$$= \lim_{h \to 0} \frac{(4 + h)h}{h} = \lim_{h \to 0}(4 + h) = 4$$

So we conclude that the slope of the requested tangent is $m = f'(2) = 4$.

Notice that we now have all the information necessary to obtain the equation of the tangent line. It is a line with a slope of 4 that passes through the point $(2,3)$. Recall the point-slope form of the equation of a line with slope m that passes through the point (x_0, y_0) is given by

$$y - y_0 = m(x - x_0).$$

Applying this formula to our situation at hand, we obtain the equation

$$y - 3 = 4(x - 2), \text{ or equivalently } y = 4x - 5.$$

The graph of $f(x)$ along with the tangent line are shown in Figure 3.

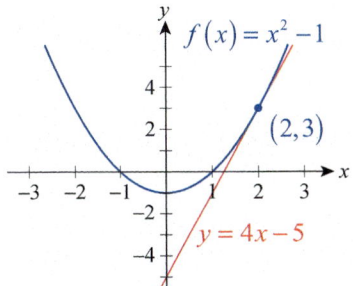

Figure 3

Example 2

A particle is moving along a straight line so that its distance from the start is given by $d(t) = 5t - \frac{1}{2}t^2$, where d is measured in feet and t in seconds. What is the particle's instantaneous velocity at $t = 2$ seconds?

Solution

As we have seen, the instantaneous velocity at time $t = c$ of an object moving along a straight line is the derivative of its position function at $t = c$. Using this observation with our given function d and $t = c = 2$ seconds, we obtain the following:

$$d'(c) = d'(2) = \lim_{h \to 0} \frac{d(2 + h) - d(2)}{h}$$

$$= \lim_{h \to 0} \frac{\left[5(2 + h) - \frac{1}{2}(2 + h)^2\right] - \left[5(2) - \frac{1}{2}(2^2)\right]}{h}$$

$$= \lim_{h \to 0} \frac{\left[10 + 5h - \frac{1}{2}(4 + 4h + h^2)\right] - (10 - 2)}{h}$$

$$= \lim_{h \to 0} \frac{\left(10 + 5h - 2 - 2h - \frac{1}{2}h^2\right) - 8}{h} = \lim_{h \to 0} \frac{3h - \frac{1}{2}h^2}{h}$$

$$= \lim_{h \to 0} \frac{\left(3 - \frac{1}{2}h\right)h}{h} = \lim_{h \to 0}\left(3 - \frac{1}{2}h\right) = 3$$

Thus we conclude that the instantaneous velocity of the particle at time 2 seconds is 3 ft/s.

Example 3

A piece of rock is dropped from a height of 196 feet, and its height above ground level is given by the function $h(t) = 196 - 16t^2$ feet, where t is measured in seconds. Find the instantaneous velocity of the rock at **a.** $t = 1$ second, **b.** $t = 2$ seconds, and **c.** $t = 2.5$ seconds.

What is the velocity of impact?

Solution

a. As before, the instantaneous velocity at time $t = c$ can be found by finding the derivative of the position function at c.

$$h'(c) = \lim_{h \to 0} \frac{\left[196 - 16(c+h)^2\right] - \left(196 - 16c^2\right)}{h}$$

$$= \lim_{h \to 0} \frac{196 - 16\left(c^2 + 2ch + h^2\right) - 196 + 16c^2}{h}$$

$$= \lim_{h \to 0} \frac{-32ch - 16h^2}{h} = \lim_{h \to 0} \frac{h(-32c - 16h)}{h}$$

$$= \lim_{h \to 0} (-32c - 16h) = -32c$$

We now obtain the instantaneous velocity at $t = c = 1$ by a simple substitution.

$$h'(1) = -32(1) = -32 \text{ ft/s}$$

Notice that unlike in our solution of the previous example, rather than substituting $c = 1$ at the outset, we carried c through the computation to obtain a formula for the instantaneous velocity, which we subsequently evaluated at $c = 1$. The advantage of this method lies in the fact that now we can use the formula to easily answer both questions b. and c.

b. $h'(2) = -32(2) = -64 \text{ ft/s}$

c. $h'(2.5) = -32(2.5) = -80 \text{ ft/s}$

Finally, we proceed to find the velocity of impact. The fundamental observation is that impact happens precisely when the position of the rock becomes 0. This helps us determine exactly how many seconds after being dropped the rock hits the ground. Thus we solve the equation $h(t) = 0$.

$$196 - 16t^2 = 0$$
$$16t^2 = 196$$
$$t^2 = 12.25$$
$$t = \pm 3.5$$

Since time is always positive, we will only consider the solution $t = 3.5$ s.

Now we can determine the velocity of impact, which is precisely the instantaneous velocity at the time of impact, by evaluating h' at $c = 3.5$. Notice that we can once again use the formula we derived before:

$$h'(3.5) = -32(3.5) = -112 \text{ ft/s},$$

and we conclude that the rock hits the ground with a velocity of -112 ft/s.

In conclusion, we note that the negative signs in the above answers mean that the direction of velocity is pointing downward throughout the motion. Also, while our answer is close to what we would measure in an actual experiment, it is important to remember that we ignored air resistance in this problem.

Topic Two

The Derivative as a Function

It will frequently be the case that we need to determine the derivative of a function at more than one point (as in our last example) or even at every point of a given interval. In such cases, it is often no more difficult to determine the derivative at a general point than at a specific point. If we are successful in doing so, then what we have achieved is a definition of another function—this new function, having been *derived* from our original function, is referred to as the *derivative* of the function.

Definition

The Derivative of a Function

The **derivative of f**, denoted f', is the function whose value at the point x is

$$f'(x) = \lim_{h \to 0} \frac{f(x+h) - f(x)}{h},$$

provided the limit exists.

This definition is exactly the same as the definition of the derivative at a point— the only difference is that we are extending the definition to all points at which the limit exists.

In usage, names other than f and f' may be more appropriate for the functions under consideration. For instance, in a problem describing the position of an object, labels such as $x(t)$ (if the object is moving along a horizontal line) or $h(t)$ (if the height of a thrown object is being discussed) may be used for the position function. In such cases, it is common to name the derivative of the position function $v(t)$ for velocity. In other words, $v(t)$ represents the instantaneous velocity of the object at time t. It is now appropriate to make a distinction between the ideas of *velocity* and *speed*: if $v(t)$ represents the velocity of an object at time t, $s(t) = |v(t)|$ represents its **speed** at time t.

Example 4

A soccer player kicks a ball vertically upward so that its position relative to ground level is $h(t) = -4.9t^2 + 19.6t + 1$ meters, where t is measured in seconds. Find the velocity and speed of the soccer ball at **a.** $t = 1$ s, **b.** $t = 2.5$ s, and **c.** $t = 3.5$ s.

What is this ball's speed of impact upon its return? (Ignore air resistance.)

Solution

a. Note first of all that the position function realistically reflects the fact that the ball does not start from ground level. Its initial height can be found by substituting $t = 0$ into the position function.

$$h(0) = -4.9(0)^2 + 19.6(0) + 1 = 1 \text{ meter}$$

In order to answer the questions, next we proceed to find $v(t)$, the velocity function of the ball, which is the derivative of $h(t)$. Notice that we are using the alternate notation of Δt instead of h for the increment, so as not to cause confusion with the name of the position function.

$$v(t) = h'(t) = \lim_{\Delta t \to 0} \frac{h(t + \Delta t) - h(t)}{\Delta t}$$

$$= \lim_{\Delta t \to 0} \frac{-4.9(t + \Delta t)^2 + 19.6(t + \Delta t) + 1 - (-4.9t^2 + 19.6t + 1)}{\Delta t}$$

$$= \lim_{\Delta t \to 0} \frac{-4.9t^2 - 9.8t\Delta t - 4.9(\Delta t)^2 + 19.6t + 19.6\Delta t + 1 + 4.9t^2 - 19.6t - 1}{\Delta t}$$

$$= \lim_{\Delta t \to 0} \frac{\Delta t(-9.8t - 4.9\Delta t + 19.6)}{\Delta t}$$

$$= \lim_{\Delta t \to 0} (-9.8t - 4.9\Delta t + 19.6)$$

$$= -9.8t + 19.6$$

Already in possession of the velocity function, it is now easy to find the ball's instantaneous velocity at $t = 1$.

$$v(1) = -9.8(1) + 19.6 = 9.8 \text{ m/s}$$

Note that the speed equals the velocity this time, since the latter is positive.

$$s(1) = |v(1)| = |9.8| = 9.8 \text{ m/s}$$

Notice also that, as one would expect, our function indicates that the ball is slowing down significantly from its initial velocity, which can be found by evaluating

$$v(0) = -9.8(0) + 19.6 = 19.6 \text{ m/s}.$$

b. Substituting again in the velocity function, we find

$$v(2.5) = -9.8(2.5) + 19.6 = -4.9 \text{ m/s}.$$

The negative sign shows that the ball turned back and is now traveling downward. Its speed, however, being independent of direction, is still positive.

$$s(2.5) = |v(2.5)| = |-4.9| = 4.9 \text{ m/s}$$

c. We expect the ball to accelerate as it falls, and that is exactly what we find upon substituting $t = 3.5$ into the velocity function.

$$v(3.5) = -9.8(3.5) + 19.6 = -14.7 \text{ m/s}$$

The negative sign shows that the direction of velocity is still downward, but its absolute value, the ball's speed, has increased.

$$s(3.5) = |v(3.5)| = |-14.7| = 14.7 \text{ m/s}$$

In fact, as it is with any free-falling body when air resistance is negligible, gravity increases the ball's speed by approximately 9.8 m/s every second. (Can you see this from our results?)

To answer the final question regarding the speed of impact, we first need to know when it happens, and then we substitute the appropriate t-value into the velocity function. Since height is 0 upon impact, we can find t by solving $h(t) = 0$, that is,

$$-4.9t^2 + 19.6t + 1 = 0.$$

An application of the quadratic formula yields the positive solution $t \approx 4.05$ seconds. Therefore, the speed of impact is approximately

$$s(4.05) = |v(4.05)| = |-9.8(4.05) + 19.6| = |-20.09| = 20.09 \text{ m/s}.$$

Example 5

Find the derivative of $f(x) = \sqrt{x}$, and use it to determine equations of the tangent lines to the graph of f at the points $(1, 1)$ and $\left(2, \sqrt{2}\right)$.

Solution

We will start with the definition of the derivative, and make use of some of the limit-determination techniques learned in Section 2.4.

$$f'(x) = \lim_{h \to 0} \frac{f(x+h) - f(x)}{h}$$

$$= \lim_{h \to 0} \frac{\sqrt{x+h} - \sqrt{x}}{h}$$

$$= \lim_{h \to 0} \frac{\left(\sqrt{x+h} - \sqrt{x}\right)\left(\sqrt{x+h} + \sqrt{x}\right)}{h\left(\sqrt{x+h} + \sqrt{x}\right)}$$

$$= \lim_{h \to 0} \frac{(x+h) - x}{h\left(\sqrt{x+h} + \sqrt{x}\right)} = \lim_{h \to 0} \frac{h}{h\left(\sqrt{x+h} + \sqrt{x}\right)}$$

$$= \lim_{h \to 0} \frac{1}{\sqrt{x+h} + \sqrt{x}} = \frac{1}{\sqrt{x} + \sqrt{x}} = \frac{1}{2\sqrt{x}}$$

To find the equation of the tangent line at $(1,1)$, we first determine its slope by evaluating the derivative $f'(x)=1/(2\sqrt{x})$ at $x=1$.

$$m = f'(1) = \frac{1}{2\sqrt{1}} = \frac{1}{2}$$

Now the point-slope equation comes to bear, and we obtain

$$y - 1 = \frac{1}{2}(x-1), \text{ or } y = \frac{1}{2}x + \frac{1}{2}.$$

We deal with the equation of the tangent line at $\left(2,\sqrt{2}\right)$ in a similar manner. Since

$$f'(2) = \frac{1}{2\sqrt{2}} = \frac{\sqrt{2}}{4},$$

the requested equation is

$$y - \sqrt{2} = \frac{\sqrt{2}}{4}(x-2), \text{ or } y = \frac{\sqrt{2}}{4}x + \frac{\sqrt{2}}{2}.$$

Notice that we can use f' to find the slope of the tangent at *any* given point $x = c$, if it exists, by simply evaluating $f'(c)$.

In business, $C(x)$ is often used to represent the total cost of producing x items of a certain product, and is referred to as the *cost function*. Similarly, $R(x)$ stands for the total revenue when x items are sold, this latter being called the *revenue function*. Consequently, the *profit function* $P(x)$ satisfies $P(x) = R(x) - C(x)$; or in words, profit equals total revenue minus total cost. The corresponding derivatives, or rates of change, the functions $C'(x)$, $R'(x)$, and $P'(x)$, are called *marginal cost*, *marginal revenue*, and *marginal profit*, respectively. As an application, the following example involves some of the above functions.

Example 6

If a table manufacturer has a cost function of $C(x) = 225 + 2x^2$ along with its revenue function of $R(x) = 117x - \frac{1}{4}x^2$, find **a.** all break-even points and **b.** the marginal profit when $x = 10$, $x = 20$, and $x = 30$.

Solution

a. Break-even points occur when $C(x) = R(x)$, in other words, when revenue levels equal the total cost invested. This leads to the equation

$$225 + 2x^2 = 117x - \frac{1}{4}x^2.$$

After rearranging terms and factoring, this leads to

$$\frac{9}{4}(x-2)(x-50) = 0,$$

thus we conclude that break-even points occur when $x = 2$ and $x = 50$ (see Figure 4).

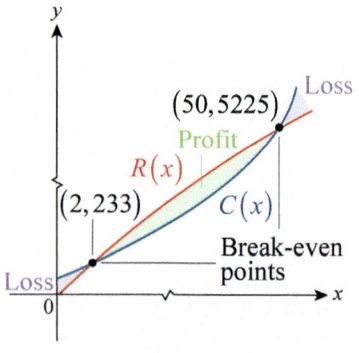

Figure 4

b. Since marginal profit is the rate of change of profit, we must first find the profit function and then calculate its derivative.

$$P(x) = R(x) - C(x)$$
$$= \left(117x - \frac{1}{4}x^2\right) - \left(225 + 2x^2\right)$$
$$= 117x - \frac{1}{4}x^2 - 225 - 2x^2$$
$$= -\frac{9}{4}x^2 + 117x - 225$$

Next, we will use the definition of the derivative to obtain the marginal profit function.

$$P'(x) = \lim_{h \to 0} \frac{P(x+h) - P(x)}{h}$$
$$= \lim_{h \to 0} \frac{-\frac{9}{4}(x+h)^2 + 117(x+h) - 225 - \left(-\frac{9}{4}x^2 + 117x - 225\right)}{h}$$

After expanding, rearranging terms, and canceling, this reduces to

$$P'(x) = \lim_{h \to 0} \frac{h\left(-\frac{9}{2}x + 117 - \frac{9}{4}h\right)}{h}$$
$$= \lim_{h \to 0} \left(-\frac{9}{2}x + 117 - \frac{9}{4}h\right)$$
$$= -\frac{9}{2}x + 117.$$

Now we can calculate the various marginal profits.

$$P'(10) = -\frac{9}{2}(10) + 117 = -45 + 117 = \$72$$

$$P'(20) = -\frac{9}{2}(20) + 117 = -90 + 117 = \$27$$

$$P'(30) = -\frac{9}{2}(30) + 117 = -135 + 117 = -\$18$$

In conclusion, an important remark is in order. Since the number of tables produced is always an integer, the increment h in any difference quotient should be thought of as an integer, so the smallest nonzero h possible is $h = 1$. This explains the following interpretation of marginal profit: $P'(x)$ is an estimate for the increase in $P(x)$ if x is increased by 1. Note also that in this example, $P'(x)$ is actually getting smaller as x gets larger. This shows that the rate of growth of profit per table is actually decreasing at a rate of $18 per table when 30 tables are manufactured and sold. This does not necessarily mean there is a loss, but as production increases, the profit per table is growing less because costs are increasing faster than revenue.

2.6 **Exercises**

1–2 Use the graph to estimate the derivative at the given points.

1. a. $x_1 = -1$ **b.** $x_2 = 1$

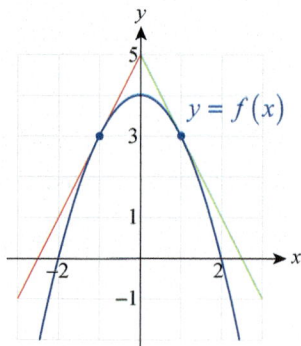

2. a. $x_1 = -2$ **b.** $x_2 = 0$

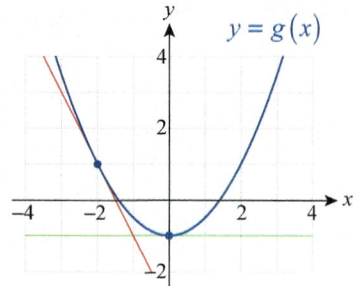

3–14 Find the equation of the tangent line to the graph of $f(x)$ at the given point.

3. $f(x) = x^2 - 2$; $(2,2)$

4. $f(x) = 3x - 2x^2$; $(-1,-5)$

5. $f(x) = \dfrac{1}{2}x + 4$; $(2,5)$

6. $f(x) = 1 - 5x$; $(0,1)$

7. $f(x) = x^3$; $(2,8)$

8. $f(x) = 5x - 2x^3$; $(-1,-3)$

9. $f(x) = \sqrt{x+1}$; $(0,1)$

10. $f(x) = 2\sqrt{1-3x}$; $(-1,4)$

11. $f(x) = \dfrac{1}{x}$; $\left(\dfrac{1}{2}, 2\right)$

12. $f(x) = \dfrac{5}{1-2x}$; $\left(-1, \dfrac{5}{3}\right)$

13. $f(x) = \dfrac{1}{\sqrt{x}}$; $\left(4, \dfrac{1}{2}\right)$

14. $f(x) = \dfrac{2}{\sqrt{x+1}}$; $(3,1)$

15–38 Use the definition (also called the *limit process*) to find the derivative function f' of the given function f. Find all x-values (if any) where the tangent line is horizontal.

15. $f(x) = 2$

16. $f(x) = 2x$

17. $f(x) = 4x + 5$

18. $f(x) = 3 - \dfrac{2}{5}x$

19. $f(x) = 3x^2$

20. $f(x) = 4 - 2x^2$

21. $f(x) = \dfrac{1}{2}x^2 + 5x - 7$

22. $f(x) = x - \dfrac{1}{3}x^2$

23. $f(x) = x^3 + x$

24. $f(x) = 7 + x - 3x^2 + x^3$

25. $f(x) = x^4$

26. $f(x) = \dfrac{1}{2x}$

27. $f(x) = \dfrac{5}{2x - 4}$

28. $f(x) = \dfrac{x - 2}{x + 2}$

29. $f(x) = \dfrac{2x+1}{x-3}$

30. $f(x) = \dfrac{2}{x^2}$

31. $f(x) = \dfrac{1}{x^2+1}$

32. $f(x) = \dfrac{2}{x^2-2x}$

33. $f(x) = \sqrt{5x}$

34. $f(x) = \dfrac{1}{\sqrt{5x}}$

35. $f(x) = \sqrt{2x+1}$

36. $f(x) = \dfrac{1}{\sqrt{x-2}}$

37. $f(x) = \sqrt{x^2+1}$

38. $f(x) = \dfrac{1}{\sqrt{x^2+1}}$

39–44 Find the equation of a tangent line to the graph of the function that is parallel to the given line.

39. $f(x) = x^2 + 3; \quad y - 6x + 1 = 0$

40. $g(x) = 2x - x^2; \quad y - 5 = 4x$

41. $h(x) = \dfrac{1}{2x}; \quad x + 2y = 3$

42. $F(x) = \dfrac{1}{x-3}; \quad y + 4x + 7 = 0$

43. $G(x) = \dfrac{1}{\sqrt{x}}; \quad 54y + x = 1$

44. $H(x) = \dfrac{1}{\sqrt{x^2-7}}; \quad 27y + 4x - 2 = 0$

45–56 Use the alternate form of the definition of the derivative $f'(c) = \lim\limits_{x \to c} \dfrac{f(x) - f(c)}{x - c}$ to evaluate the given slope.

45. $f(x) = 5 - \dfrac{1}{4}x; \quad f'(3.6)$

46. $g(x) = x^2 + 1; \quad g'(-1)$

47. $h(x) = (x+2)^2; \quad h'(3)$

48. $F(t) = \dfrac{1}{t-3}; \quad F'(2)$

49. $G(x) = \dfrac{2}{5-x}; \quad G'(7)$

50. $k(t) = \sqrt{t+5}; \quad k'(11)$

51. $u(x) = 2\sqrt{1-x}; \quad u'(-3)$

52. $v(x) = \dfrac{1}{x^2+1}; \quad v'(0)$

53. $w(s) = \dfrac{1}{\sqrt{s+4}}; \quad w'(5)$

54. $F(t) = t^3 - t; \quad F'(1)$

55. $G(s) = s^4; \quad G'(-2)$

56. $H(x) = \dfrac{2}{\sqrt{x^2+1}}; \quad H'(0)$

57–60 Match the graph of f with the graph of its derivative f' (labeled A–D).

57.

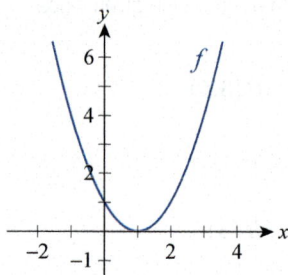

58.

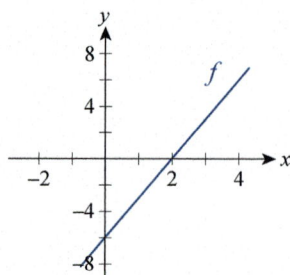

59.

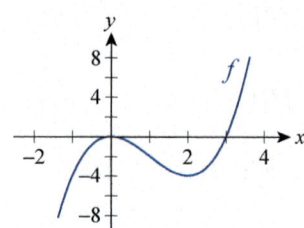

60.

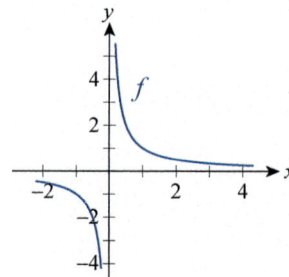

A.

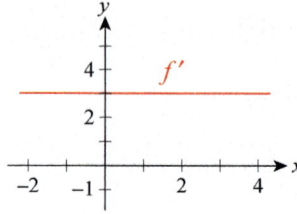

B.

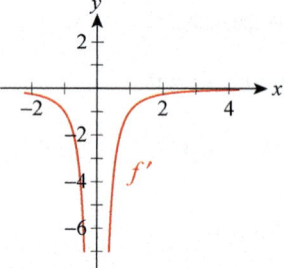

C.

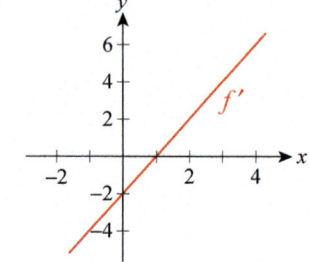

D.

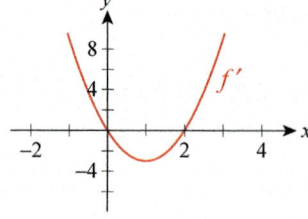

61–65 Sketch the graph of a function f possessing the given characteristics. (A formula is useful, but not necessary.)

61. $f(0)=1$, $f'(0)=0$, $f'(x)<0$ for $x<0$, $f'(x)>0$ for $x>0$

62. $f(1)=0$, $f'(1)=0$, $f'(x)\geq0$ on the entire real line

63. $f(x)>0$ on the entire real line, $f'(x)<0$ on the entire real line

64. $f(1)=1$, $f'(1)=-1$, f' is nonzero on the entire real line

65. $f(1)=5$, $f'(x)=5$ on the entire real line

66. Prove that if $f(x)=c$ (a constant function), then $f'(x)=0$.

67. Use the definition of the derivative to prove that if $f(x)=x$, then $f'(x)=1$.

68. Generalize Exercise 67 to prove that if $f(x)$ is a linear function, then $f'(x)$ is constant.

69.* Use the definition of the derivative to prove that if $f(x)=x^n$ for a positive integer n, then $f'(x)=nx^{n-1}$.

70.* Recall from Section 1.1 that a function f is even if $f(-x)=f(x)$ and odd if $f(-x)=-f(x)$ throughout its domain. Prove that the derivative of an even function is odd and, vice versa, an odd function has an even derivative.

71.* Find the equation of the line tangent to the graph of $f(x)=1/x$ at the point $(c, f(c))$. Prove that the area of the triangle bounded by the tangent line and the coordinate axes is the same for all $c\neq0$.

72. The position function of a moving particle is given by $p(t)=t^2-3t+1$ feet at t seconds. Find all points in time where the particle's speed is 1 ft/s. When does it come to a momentary stop?

73. Repeat Exercise 72 with the position function $p(t)=\frac{1}{9}t^3-t^2+\frac{8}{3}t$.

74. A baseball is hit vertically upward with an initial speed of 80 ft/s. When does it slow down to 32 ft/s? How high does it go and how long is it aloft? (**Hint:** Use the position function $h(t)=-16t^2+80t$. Ignore air resistance.)

75. A rock is thrown upward from the edge of a 150 ft high cliff with an initial velocity of 48 ft/s.

   a. Calculate the velocity and speed of the rock when it is exactly 32 ft above the person's hand.

   b. How high does it go and when does it reach the bottom of the cliff?

   c. What is the velocity of impact?

   (**Hint:** Use $h(t)=-16t^2+48t+150$ as the position function, where h is in feet, t in seconds. Ignore air resistance.)

76. A package is dropped from a small airplane 122.5 meters above the Earth. If we ignore air resistance, how much time does the package need to reach the ground and what is the speed of impact? (**Hint:** The position function is $h(t)=-4.9t^2+122.5$ meters, where t is measured in seconds. Use $g\approx9.81\text{ m/s}^2$.)

77. The following graph is a position function of a student's car relative to her home as she drove to class one morning. From the graph, recreate a possible story of her trip, mentioning distance, velocity, speed, and so forth.

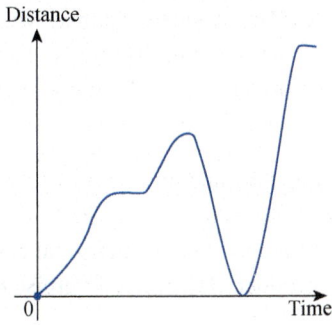

78. A manufacturer has determined that the revenue from the sale of x cordless telephones is given by $R(x) = 94x - 0.03x^2$ dollars. The cost of producing x telephones is $C(x) = 10,800 + 34x$ dollars.

    a. Find the profit function $P(x)$ and any break-even points.

    b. Find $P(200)$, $P(400)$, and $P(600)$.

    c. Find the marginal profit function $P'(x)$.

    d. Find $P'(200)$, $P'(400)$, and $P'(600)$.

79. The owner of a leather retailer has determined that he can sell x attaché cases if the price is $p = D(x) = 46 + 0.25x$ dollars ($D(x)$ is often called the demand function). The total cost for these cases is $C(x) = 0.15x^2 + 6x + 190$ dollars.

    a. Find the profit function $P(x)$. (**Hint:** Find the revenue function $R(x)$ first.)

    b. Find any break-even points.

    c. Find $P(25)$, $P(30)$, and $P(40)$.

    d. Find the marginal profit function $P'(x)$.

    e. Find $P'(25)$, $P'(30)$, and $P'(40)$.

80. The average cost $\overline{C}(x)$ of manufacturing x units of a certain product is $C(x)/x$, where $C(x)$ is the total cost function.

    a. Find the average cost function if $C(x) = 30 + 2x + 0.003x^2$.

    b. What is the rate of change of average cost?

    c. What value of x results in a minimum average cost? (**Hint:** Use the fact that when average cost is a minimum, its rate of change is 0. Alternatively, use technology to graph $C(x)$ for $x \geq 0$ and zoom in on the lowest point.)

81. The average manufacturing cost function of a product is given by $\overline{C}(x) = 20x^{-1} + 3$. Determine the cost function and the marginal cost function for the product. (**Hint:** See Exercise 80.)

2.6 Technology Exercises

82–105. Referring back to the functions given in Exercises 15–38, use a graphing calculator or computer algebra system to sketch the graph of f along with that of f' in the same viewing window. Compare the graphs and describe their relationship.

Chapter 2
Review Exercises

1–4 Use the graph of the function to find the indicated (possibly one-sided) limit, if it exists. Also examine the continuity of the function at the indicated point and classify any discontinuities.

1. $\lim_{x \to 1} f(x)$

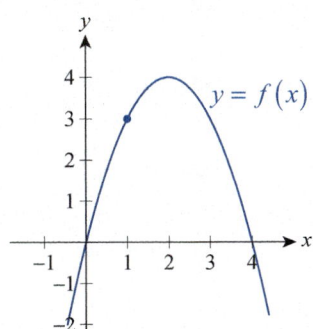

2. $\lim_{x \to 1} g(x)$

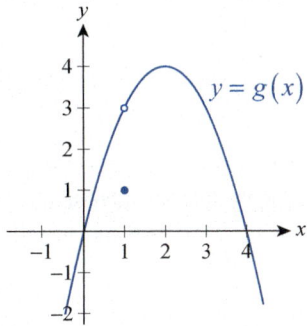

3. $\lim_{x \to 1^-} h(x)$

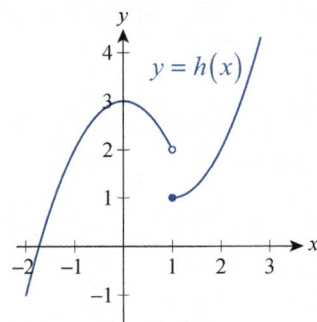

4. $\lim_{x \to -1^+} k(x)$

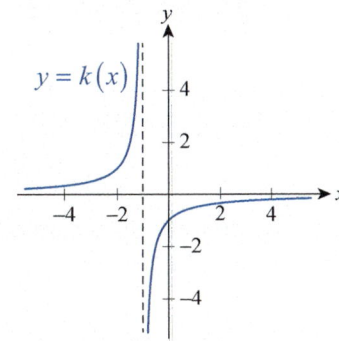

5–8 Use difference quotients to approximate the slope of the tangent to the graph of the function at the given point. Use at least five different h-values that are decreasing in magnitude. (Answers will vary.)

5. $f(x) = 3x + 2;$ $(0, 2)$ **6.** $g(x) = 2 - x^2;$ $(1, 1)$ **7.** $h(x) = \sqrt{x - 1};$ $(2, 1)$ **8.** $k(x) = \sin x;$ $(0, 0)$

9. A pellet is shot vertically upward from an initial height of 6 feet. Its height after t seconds is given by $h(t) = 6 + 608t - 16t^2$ feet. Use difference quotients to answer the questions below.

   a. What will be the pellet's height at the end of the first second?

   b. What is the average velocity of the pellet during the first two seconds?

   c. Estimate the instantaneous velocity at $t = 0$ seconds.

   d. Estimate the instantaneous velocity at $t = 2$ seconds.

   e. When will the velocity be 0?

10–11 Approximate the area of the region between the graph of $f(x)$ and the x-axis on the given interval. Use $n = 4$, as in Example 5a of Section 2.1. (Round your answer to four decimal places.)

10. $f(x) = x^3$ on $[0,1]$ **11.** $f(x) = \ln x$ on $[1,3]$

12–13 Approximate the arc length of the graph of $g(x)$ on the given interval. Use $n = 5$, as in Example 5b of Section 2.1. (Round your answer to four decimal places.)

12. $g(x) = x^{2/3}$ on $[3,8]$ **13.** $g(x) = e^x$ on $[-2,3]$

14–17 Create a table of values to estimate the value of the indicated limit without graphing the function. Choose the last x-value so that it is no more than 0.001 units from the given c-value.

14. $\lim\limits_{x \to 1} \dfrac{x^3 - 1}{x - 1}$ **15.** $\lim\limits_{x \to 0} x^x$

16. $\lim\limits_{x \to 0} \dfrac{\sin 2x}{4x}$ **17.** $\lim\limits_{x \to 0} \left(2x \sin \dfrac{1}{4x}\right)$

18. Use one-sided limit notation to describe the behavior of $f(x) = \dfrac{1}{x-1}$ near $x = 1$.

19–20 Find a $\delta > 0$ that satisfies the limit claim corresponding to $\varepsilon = 0.01$, that is, such that $0 < |x - c| < \delta$ would imply $|f(x) - L| < 0.01$.

19. $\lim\limits_{x \to 0} (3 - 2x) = 3$ **20.** $\lim\limits_{x \to 4} \sqrt{x} = 2$

21–24 Give the precise definition of the limit claim. Then use the definition to prove the claim.

21. $\lim\limits_{x \to 1} (3x + 1) = 4$ **22.** $\lim\limits_{x \to 1} x^2 = 1$

23. $\lim\limits_{x \to 1} \sqrt{x} = 1$ **24.** $\lim\limits_{x \to 2} \dfrac{2}{x} = 1$

25–41 Use algebra and/or appropriate limit laws to evaluate the given limit (one-sided limit where indicated). If the limit is unbounded, use the symbol ∞ or $-\infty$ in your answer.

25. $\lim\limits_{x \to 3} (2x^2 - 3x + 5)$

26. $\lim\limits_{x \to -2} \left(\dfrac{x^3}{4} + 2x^2 - x + 1\right)$

27. $\lim\limits_{x \to 3} \sqrt{x^3 + 2x^2 + 4}$ **28.** $\lim\limits_{x \to -2} \dfrac{2x + 1}{x^2 - x}$

29. $\lim\limits_{t \to 1} \left(\dfrac{3t + 5t^3}{t^2 + 1}\right)^{3/2}$ **30.** $\lim\limits_{x \to 4} \dfrac{x^2 - 16}{x - 4}$

31. $\lim\limits_{x \to -5} \dfrac{x + 5}{x^2 - 25}$ **32.** $\lim\limits_{x \to 5^-} \dfrac{x + 5}{x^2 - 25}$

33. $\lim\limits_{x \to 1^+} \dfrac{x^2 + 1}{x^4 - 1}$ **34.** $\lim\limits_{x \to 1} \dfrac{x^2 - 1}{x^4 - 1}$

35. $\lim\limits_{x \to 3} \dfrac{\sqrt{x+1} - 2}{x - 3}$ **36.** $\lim\limits_{x \to 0} \dfrac{\dfrac{1}{2+x} - \dfrac{1}{2}}{x}$

37. $\lim\limits_{x \to 0^-} \dfrac{2|x|}{x}$ **38.** $\lim\limits_{x \to -2^+} \sqrt{4 - x^2}$

39. $\lim\limits_{x \to 2^-} (\llbracket x \rrbracket + 2x)$ **40.** $\lim\limits_{x \to 1^+} \llbracket x \rrbracket x$

41. If $f(x) = x^2$, find $\lim\limits_{h \to 0} \dfrac{f(x+h) - f(x)}{h}$.

42–43 Use the Squeeze Theorem to prove the limit claim.

42. $\lim\limits_{x \to 0} x \cos \dfrac{1}{x} = 0$ **43.** $\lim\limits_{x \to \infty} \dfrac{\sin x}{\ln x} = 0$

44. Sketch a graph of a function (a formula is not necessary) that is not continuous at $x = 0$ from either direction, but both of its one-sided limits exist at $x = 0$. (Answers will vary.)

45. Sketch a graph of a function that is left-continuous at $x = 0$, but its right-hand limit at $x = 0$ doesn't exist. (Answers will vary.)

46–51 Find and classify the discontinuities (if any) of the given function as removable or nonremovable.

46. $f(x) = \dfrac{x-9}{\sqrt{x}-3}$ $(x \geq 0)$

47. $g(x) = \dfrac{\sqrt{x}+2}{x-4}$ $(x \geq 0)$

48. $h(x) = \dfrac{1}{\sqrt{x^2+1}}$

49. $t(x) = 2 + 2[\![x]\!]$

50. $G(x) = \dfrac{x}{\sqrt{x+1}-1}$ $(x \geq -1)$

51. $k(x) = |x-3| + |x+1|$

52–53 Use the ε-δ definition to prove that the function is continuous.

52. $f(x) = 3x - 1$ **53.** $g(x) = 2x^2$

54. Find the values of a and b such that f is continuous on the entire real line.

$$f(x) = \begin{cases} -1 & \text{if } x \leq -3 \\ ax+b & \text{if } -3 < x < 2 \\ x^2 & \text{if } x \geq 2 \end{cases}$$

55. Use the Intermediate Value Theorem to prove that the equation $2x^5 + x + 1 = 0$ has a solution on the interval $[-1,1]$.

56. Use the Intermediate Value Theorem to show that the graphs of $f(x) = x^3$ and $g(x) = e^{-x}$ intersect.

57–58 Find the equation of the tangent line to the graph of $f(x)$ at the given point.

57. $f(x) = x^2 + x;$ $(1,2)$

58. $f(x) = \sqrt{x};$ $(4,2)$

59–60 Use the definition (also called the limit process) to find the derivative function f' of the given function f. Find all x-values (if any) where the tangent line is horizontal.

59. $f(x) = 2x - x^2$ **60.** $f(x) = \dfrac{3}{x-2}$

61–62 Sketch the graph of a function f possessing the given characteristics. (A formula is useful, but not necessary.)

61. f is continuous at 0, $f(0) = 1$, $f'(x) < 0$ for $x < 0$, $f'(x) > 0$ for $x > 0$, and $f'(0)$ does not exist

62. $g(1) < 0$, $g'(1) > 0$, and $g(2) > 0$, but $g'(2) < 0$

63. Prove that if $f(x)$ is a quadratic function, then $f'(x)$ is linear.

64. A small object is thrown upward with an initial velocity of 12 m/s from the top of a 15 m high building.

   a. How high does it go and when does it reach the ground?

   b. What is the speed of impact?

   (**Hint:** Use $h(t) = -5t^2 + 12t + 15$ as the position function, where h is in meters, t in seconds.)

65. The owner of a small toy manufacturer has determined that he can sell x toys if the price is $p = D(x) = 0.2x + 30$ dollars. The total cost as a function of x is given by $C(x) = 0.1x^2 + 15x + 247.5$ dollars.

   a. Find the profit function $P(x)$.

   b. Find any break-even points.

   c. Find the marginal profit function.

66–73 *True or False?* Determine whether the given statement is true or false. In case of a false statement, explain or provide a counterexample.

66. Instantaneous velocity can be interpreted as the slope of a tangent line.

67. If $\lim_{x \to c} f(x)$ doesn't exist, then $f(x)$ has a vertical asymptote at $x = c$.

68. Any rational function has at least one vertical asymptote.

69. If $\lim_{x \to c} f(x) = A$ and $\lim_{x \to c} g(x) = B$, then $\lim_{x \to c} \dfrac{f(x)}{g(x)} = \dfrac{A}{B}$.

70. If f is defined on $[a,b]$, L is a real number between $f(a)$ and $f(b)$, and $\lim_{x \to c} f(x)$ exists for all $x \in (a,b)$, then there is a c in the interval (a,b) such that $f(c) = L$.

71. If f is continuous at c, then $f(c)$ is equal to both one-sided limits at c.

72. If both one-sided limits of f exist at c, and if f is defined at c, then f is continuous at c.

73. If $g(x) \le f(x) \le h(x)$ for all x in some open interval containing c, and if $\lim_{x \to c} g(x) = \lim_{x \to c} h(x) = L$, then by the Squeeze Theorem $f(c) = L$ as well.

Chapter 2
Technology Exercises

74. Use a computer algebra system to find approximations for the areas in Exercises 10 and 11 by using $n = 100$. (Round your answers to four decimal places.)

75. Use a computer algebra system to find approximations for the arc lengths in Exercises 12 and 13 by using $n = 100$. (Round your answers to four decimal places.)

76. Use a graphing calculator or computer algebra system to verify your answers given for Exercises 14–17.

77. Use a graphing calculator or computer algebra system to approximate the solutions for Exercises 55 and 56. Round your answers to four decimal places.

78–81 Use a graphing calculator or computer algebra system to graph the function, and estimate from the graph the value of the given limit.

78. $\lim_{x \to \infty} x^{1/x}$

79. $\lim_{x \to 0} \dfrac{\arcsin x}{x}$

80. $\lim_{x \to \infty} \left(1 + \dfrac{1}{x}\right)^{2x}$

81. $\lim_{x \to 1} \dfrac{\ln(x^3)}{x - 1}$

Chapter 2
Project

Some years ago, it was common for long-distance phone companies to charge their customers in one-minute increments. In other words, the company charges a flat fee for the first minute of a call and another fee for each additional minute or any fraction thereof (see Exercise 82 in Section 2.5). In this project, we will explore in detail a function that gives the cost of a telephone call under the above conditions.

1. Suppose a long-distance call costs 75 cents for the first minute plus 50 cents for each additional minute or any fraction thereof. In a coordinate system where the horizontal axis represents time t and the vertical axis price p, draw the graph of the function $p = C(t)$ that gives the cost (in dollars) of a telephone call lasting t minutes, $0 < t \leq 5$.

2. Does $\lim_{t \to 1.5} C(t)$ exist? If so, find its value.

3. Does $\lim_{t \to 3} C(t)$ exist? Explain.

4. Write a short paragraph on the continuity of this function. Classify all discontinuities; mention one-sided limits and left or right continuity where applicable.

5. In layman's terms, interpret $\lim_{t \to 2.5} C(t)$.

6. In layman's terms, interpret $\lim_{t \to 3^-} C(t)$.

7. In layman's terms, interpret $\lim_{t \to 3^+} C(t)$.

8. If possible, find $C'(3.5)$.

9. If possible, find $C'(4)$.

10. Find and graph another real-life function whose behavior is similar to that of $C(t)$. Label the axes appropriately and provide a brief description of your function.

Chapter 3
Differentiation

Introduction

Chapter 2 introduced the notion of differentiation by means of several elementary applications.

In this chapter we learn additional notation, develop techniques that allow us to find derivatives of many classes of functions, and explore many more applications.

As we will see, the applications arise from a broad array of disciplines and collectively illustrate the profound power of the derivative.

Each bit of notation found in this chapter reflects the thinking of the mathematician responsible for introducing it, and it is frequently the case that a given mathematical concept can be expressed in a number of equivalent ways. While this may seem needlessly redundant at first, we'll see that the alternatives all offer their own unique advantages, and hence their use persists. Such giants as the Swiss mathematician Leonhard Euler (1707–1783) and the Italian French mathematician Joseph-Louis Lagrange (1736–1813), as well as Sir Isaac Newton (1643–1727) and Gottfried Wilhelm Leibniz (1646–1716), are among those who devised ways of succinctly writing the new ideas of calculus. Each had certain objectives in mind, usually arising from a problem (or set of problems) that they wished to solve, and the notation they created was chosen in pursuit of those objectives.

At the same time, mathematicians were discovering general rules of differentiation that allow derivatives of entire classes of functions to be determined relatively easily. In this chapter, we'll learn how to differentiate polynomial, rational, trigonometric, exponential, and logarithmic functions, among others. We'll also see how arithmetic combinations and/or compositions of functions can be differentiated if we know the derivatives of the component pieces. These techniques are all ultimately based on the difference quotient definition of derivative seen in Chapter 2, but they enable us to differentiate functions without having to laboriously calculate limits of difference quotients.

Finally, we'll learn how derivatives allow us to solve problems from disciplines such as biology, physics, chemistry, and economics. The ability to efficiently and accurately calculate instantaneous rates of change, in all these different settings, allows us to solve problems that would otherwise be intractable.

3.1 **Differentiation Notation and Consequences**

Chapter 2 gave us a few motivational examples for seeking what we now call the derivative of a function, some tools to obtain the derivative, and enough notation to get started. We are ready now to explore the broad ramifications of this new concept, and we begin with a catalog of the terms and notation that will prove useful.

Topic One

Alternative Notations and Higher-Order Derivatives

Recall that the fraction

$$\frac{f(c+h)-f(c)}{h}$$

is called the *difference quotient of f at c with increment h*, and that we let

$$f'(c) = \lim_{h \to 0} \frac{f(c+h)-f(c)}{h},$$

provided this limit exists. This is our definition of the *derivative of f at c*, and we quickly extend it to all points x at which a similar limit exists to define the *derivative of f* as the function whose value at x is given by

$$f'(x) = \lim_{h \to 0} \frac{f(x+h)-f(x)}{h}.$$

If $f'(c)$ exists for a given function f, we say f is **differentiable at c**, and if $f'(x)$ exists for all x in some open interval we say f is differentiable on that interval (requiring the interval to be open is a convenience allowing us to only worry about two-sided limits for the moment). However, as a reflection of the wide breadth of applications and the number of mathematicians who played a role in developing calculus, other notations and terms are frequently used in working with derivatives. If, as is common, the function f appears in the form of the equation $y = f(x)$, any of the following notations may be used to refer to the derivative of the function at x.

$$f'(x) = y' = \frac{dy}{dx} = \frac{df}{dx} = \frac{d}{dx}f(x) = Df(x) = D_x f(x)$$

In words, these notations are informally read, respectively, as "f prime of x," "y prime," "dee y, dee x," "dee f, dee x," "dee, dee x of f," "dee f," and "dee sub x of f" (or close variations of such phrases). Although the wealth of choices can be confusing at first, they all refer to the instantaneous rate of change of the function f or of the dependent variable y with respect to the independent variable x. The symbols d/dx, D, and D_x are called **differentiation operators** (with respect to x, at the moment), and the formal meaning of, for instance, dy/dx is "the derivative of y with respect to x."

If any one of the above notations served well in all applications, the alternative notations would likely fall into disuse. But as we will see, they each have their strengths and weaknesses. The *prime notation* (historically due to Joseph-Louis Lagrange) emphasizes the functional aspect of the derivative, while the notations dy/dx and df/dx (due to Gottfried Wilhelm Leibniz) are especially useful when working with equations. A symbol such as "dy" denotes an object called a *differential* and the notation dy/dx does indeed indicate a ratio of differentials (and the fact that $dy/dx = \lim\limits_{\Delta x \to 0}(\Delta y/\Delta x)$ is no coincidence). However, we are not yet ready to delve into differentials, and for the time being dy/dx should simply be regarded as a synonym for y'.

Example 1

Find df/dx for the function $f(x) = (1+x)/(1-x)$.

Solution

Since df/dx is just alternative notation for $f'(x)$, we have the following.

$$\frac{df}{dx} = f'(x)$$

$$= \lim_{h \to 0} \frac{f(x+h) - f(x)}{h}$$

$$= \lim_{h \to 0} \frac{\dfrac{1+(x+h)}{1-(x+h)} - \dfrac{1+x}{1-x}}{h}$$

$$= \lim_{h \to 0} \frac{\dfrac{\left[1+(x+h)\right](1-x) - (1+x)\left[1-(x+h)\right]}{\left[1-(x+h)\right](1-x)}}{h}$$

$$= \lim_{h \to 0} \frac{\left(1+x+h-x-x^2-xh\right) - \left(1-x-h+x-x^2-xh\right)}{h\left[1-(x+h)\right](1-x)}$$

After combining like terms in the numerator, this expression becomes

$$\lim_{h \to 0} \frac{2h}{h\left[1-(x+h)\right](1-x)} = \lim_{h \to 0} \frac{2}{\left[1-(x+h)\right](1-x)}$$

$$= \frac{2}{\left(1-x\right)^2}.$$

Thus we conclude that for $f(x) = (1+x)/(1-x)$,

$$\frac{df}{dx} = \frac{2}{\left(1-x\right)^2}.$$

Given that it emphasizes the function nature of the derivative, it isn't surprising that the prime notation allows us to easily denote the evaluation of a derivative at a point. In order to indicate the same concept with the differential notation, we need one additional symbol, as follows:

$$\text{given } y = f(x), \ \left.\frac{dy}{dx}\right|_{x=c} \text{ is synonymous with } f'(c).$$

The notation $\left.\dfrac{dy}{dx}\right|_{x=c}$ is read "the derivative of y with respect to x evaluated at $x = c$."

Example 2

Let $f(x) = x^2 - x$. For $y = f(x)$, use both the prime and the differential notations to evaluate $f'(x)$. Use the derivative to find $f'(2)$, and express your answer in both notations.

Solution

As shown in Figure 1, Δx is just an alternate notation for the increment h, while $\Delta y = f(x+h) - f(x)$. Thus the corresponding difference quotient can be rewritten as

$$\frac{f(x+h) - f(x)}{h} = \frac{\Delta y}{\Delta x},$$

and hence

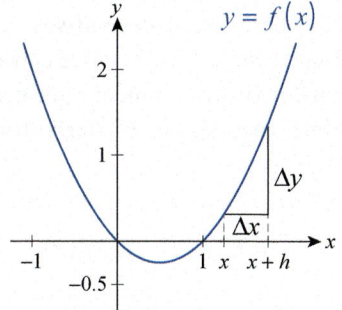

Figure 1

$$
\begin{aligned}
f'(x) &= \lim_{h \to 0} \frac{f(x+h) - f(x)}{h} \\
&= \lim_{\Delta x \to 0} \frac{\Delta y}{\Delta x} \\
&= \frac{dy}{dx} \\
&= \lim_{h \to 0} \frac{\left[(x+h)^2 - (x+h)\right] - (x^2 - x)}{h} \\
&= \lim_{h \to 0} \frac{(x^2 + 2xh + h^2 - x - h) - (x^2 - x)}{h} \\
&= \lim_{h \to 0} \frac{2xh + h^2 - h}{h} = \lim_{h \to 0}(2x + h - 1) \\
&= 2x - 1.
\end{aligned}
$$

Finally, we will use the alternate notations to express the derivative at $x = 2$.

$$f'(2) = \left.\frac{dy}{dx}\right|_{x=2} = \left.\frac{df}{dx}\right|_{x=2} = 2(2) - 1 = 3$$

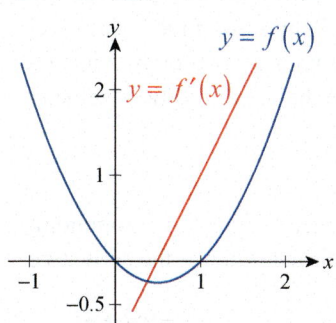

Figure 2

Figure 2 shows the graphs of $y = f(x)$ and that of $dy/dx = f'(x)$ on the same set of axes. The reader is invited to examine the relationship between the rate of change, or "slope" of the graph of f and the function values $f'(x)$ as x ranges over the real numbers. We will examine this relationship in great depth over the coming sections.

Before we leave the topics of notation and terminology, we need to make one last observation. As we have seen, the derivative of a function f is a function in its own right (assuming f is differentiable). Given that, it is reasonable to ask whether the derivative of f' makes sense, and what meaning $(f')'$ might possess.

Such *higher-order derivatives* do indeed make sense, and often possess meaning corresponding to familiar real-world experiences. First, some notation: given a function f and the equation $y = f(x)$, the following all denote the **second derivative of f (or y) with respect to x.**

$$f''(x) = y'' = \frac{d^2 y}{dx^2} = \frac{d^2 f}{dx^2} = \frac{d}{dx}\left(\frac{dy}{dx}\right) = \frac{d}{dx}\left(\frac{df}{dx}\right) = D^2 f(x) = D_x^{\,2} f(x) = D_{xx} f(x)$$

There are, in fact, even more ways to indicate the same idea (such as the $(f')'$ that appeared above), but they all represent the notion of differentiating, with respect to x, a function that is itself a derivative with respect to x. Such symbols are read as "f double prime," "y double prime," "dee squared y, dee x squared," and so on. And the patterns continue with yet higher-order derivatives; although we will do little more than introduce the notation here, **third derivatives** are indicated with symbols such as f''' ("f triple prime") and $d^3 y/dx^3$ ("dee cubed y, dee x cubed"). Beyond this level, the order of the derivative is often indicated with a number inside parentheses (instead of "prime" marks). The nth **derivative of f (or y)** can be written as

$$f^{(n)}(x) = y^{(n)} = \frac{d^n y}{dx^n} = \frac{d^n f}{dx^n} = \frac{d}{dx} y^{(n-1)} = \frac{d}{dx} f^{(n-1)} = D^n f(x) = D_x^{\,n} f(x).$$

What meaning can we attach to such mathematical constructs as the second derivative? Consider again an example in which $x(t)$ denotes the position, at time t, of an object moving along a straight path. As we now know, $x'(t)$ represents the object's instantaneous rate of change of position (at time t), better known as velocity. So $x''(t)$ represents the object's instantaneous rate of change of velocity; that is, $x''(t)$ tells us how quickly the object's velocity is increasing or decreasing at time t. This "rate of change of velocity" goes by the common name **acceleration**.

Example 3

A tennis ball is hit vertically upward so that its position function is given by $h(t) = -16t^2 + 96t + 4$ feet above the ground at t seconds. Find the velocity and acceleration functions. What are the initial position, velocity, and acceleration values?

Solution

As we have seen, velocity is the first derivative of position, commonly labeled as $v(t)$, which we shall obtain using the definition of the derivative.

$$v(t) = h'(t) = \lim_{h \to 0} \frac{\left[-16(t+h)^2 + 96(t+h) + 4\right] - \left(-16t^2 + 96t + 4\right)}{h}$$

After expanding, combining, and canceling like terms in the numerator we obtain

$$v(t) = \lim_{h \to 0} \frac{-32th - 16h^2 + 96h}{h}$$
$$= \lim_{h \to 0} (-32t - 16h + 96)$$
$$= -32t + 96.$$

We see from the position and velocity functions that the initial position and initial velocity, respectively, are

$$h(0) = -16(0)^2 + 96(0) + 4 = 4 \text{ ft}$$

and

$$v(0) = h'(0) = -32(0) + 96 = 96 \text{ ft/s}.$$

Next, as we have discussed above, the acceleration function of the tennis ball is the second derivative of position, in other words, the first derivative of velocity. Using the common notation of $a(t)$ for the acceleration function, once again with the help of the definition of the derivative we obtain

$$a(t) = v'(t)$$
$$= \frac{d}{dt}(-32t + 96)$$
$$= \lim_{h \to 0} \frac{\left[-32(t+h) + 96\right] - (-32t + 96)}{h}$$
$$= \lim_{h \to 0} \frac{-32h}{h}$$
$$= -32 \text{ ft/s}^2.$$

What this means is that the tennis ball's acceleration is constant; that is,

$$a(t) = -32 \text{ ft/s}^2.$$

The reason for the negative sign is our convention of treating "upward" velocities as positive, but in many contexts, the negative sign can be omitted. In fact, any falling object that is under the influence of gravity only (i.e., all other forces including air resistance are negligible) moves at the same constant acceleration. This law of nature was discovered by Galileo Galilei (1564–1642) who was the first scientist to claim that in the absence of air resistance, *all* objects fall with the *same* constant acceleration, *anywhere* on the Earth. This constant acceleration caused by gravity is denoted by g, and if we exit the specific context of our problem for a moment and omit the negative sign, we have

$$g = 32 \text{ ft/s}^2,$$

or using metric units,

$$g = 9.8 \text{ m/s}^2.$$

One way to remember the constant accelerating action of gravity is to note that its effect is to "add" to the velocity of a falling object an "extra" downward velocity of $32\,\text{ft/s}$ every second (or an "extra" downward $9.8\,\text{m/s}$ every second if we are working with metric units).

In conclusion, to answer our last question of the initial acceleration of the tennis ball, we note that since acceleration is constant and pointing downward,

$$a(0) = -32 \text{ ft/s}^2,$$

and our solution is complete.

Although it is seen less often, the third derivative of a position function $x(t)$ also has a common name. Since a relatively large value (in magnitude) for $x'''(t)$ at a particular time t would mean a rapid change in the acceleration of the object, which corresponds to a very sudden change in the object's position, the third derivative is sometimes called the **jerk**.

Example 4

Find the first three derivatives of $f(x) = \frac{2}{3}x^3 - 3x + \frac{1}{3}$.

Solution

We start by calculating f' using the definition.

$$f'(x) = \lim_{h \to 0} \frac{\left[\frac{2}{3}(x+h)^3 - 3(x+h) + \frac{1}{3}\right] - \left[\frac{2}{3}x^3 - 3x + \frac{1}{3}\right]}{h}$$

$$= \lim_{h \to 0} \frac{\frac{2}{3}(x+h)^3 - 3(x+h) - \frac{2}{3}x^3 + 3x}{h}$$

$$= \lim_{h \to 0} \frac{\frac{2}{3}\left[(x+h)^3 - x^3\right] - 3h}{h} \qquad \text{Apply the expansion formula for the cube of a binomial.}$$

$$= \lim_{h \to 0} \frac{\frac{2}{3}\left(x^3 + 3x^2h + 3xh^2 + h^3 - x^3\right) - 3h}{h} \qquad \text{Combine like terms.}$$

$$= \lim_{h \to 0} \frac{2x^2h + 2xh^2 + \frac{2}{3}h^3 - 3h}{h}$$

$$= \lim_{h \to 0} \left(2x^2 + 2xh + \frac{2}{3}h^2 - 3\right) \qquad \text{For a fixed } x, \text{ if } h \to 0, \text{ both } 2xh \to 0 \text{ and } \frac{2}{3}h^2 \to 0.$$

$$= 2x^2 - 3$$

Notice that the derivative of the given cubic polynomial is a quadratic function. Next, knowing that

$$f''(x) = \frac{d}{dx} f'(x),$$

we obtain the second derivative of f as follows.

$$f''(x) = \frac{d}{dx}\left(2x^2 - 3\right)$$

$$= \lim_{h \to 0} \frac{\left[2(x+h)^2 - 3\right] - \left(2x^2 - 3\right)}{h}$$

$$= \lim_{h \to 0} \frac{2(x+h)^2 - 2x^2}{h}$$

$$= \lim_{h \to 0} \frac{2\left(x^2 + 2xh + h^2\right) - 2x^2}{h} \qquad \text{Combine like terms and factor out } h.$$

$$= \lim_{h \to 0} \frac{h(4x + 2h)}{h}$$

$$= \lim_{h \to 0}\left(4x + 2h\right)$$

$$= 4x$$

Perhaps it doesn't come as a surprise that the derivative of the above quadratic function is linear. To obtain the third derivative, we will differentiate yet again. As you might have observed already, the derivative of a linear function is a constant, so that is what we expect f''' to be (see Example 3 above or Exercise 68 of Section 2.6).

$$f'''(x) = \frac{d}{dx}\left(f''(x)\right)$$

$$= \lim_{h \to 0} \frac{4(x+h) - 4x}{h}$$

$$= \lim_{h \to 0} \frac{4x + 4h - 4x}{h}$$

$$= \lim_{h \to 0} \frac{4h}{h}$$

$$= \lim_{h \to 0} 4$$

$$= 4$$

Exercise 66 of Section 2.6 asks the reader to prove that the derivative of any constant function is identically 0, so we would like to note as we conclude this example that the fourth, fifth, and any subsequent higher-order derivatives of f are identically 0.

Figure 3 shows the graph of f along with those of its first three derivatives on the same axes. The reader is invited to compare the graphs and examine how the rate of change of each at every x is reflected by the corresponding value of its derivative.

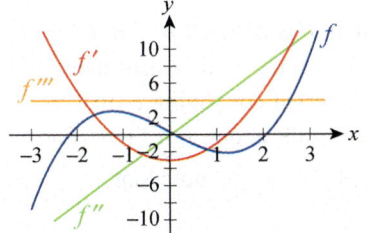

Figure 3

Topic Two

Consequences of Differentiability

Two primary questions will serve well to guide us:

What does the differentiability of a function f at a point c tell us about f?

and

What are the characteristics of a function f that fails to be differentiable at a point c?

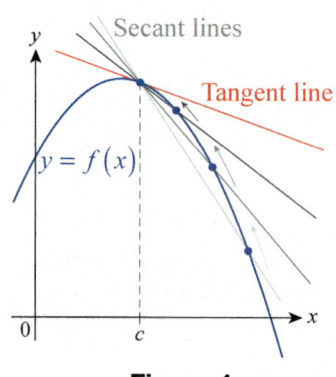

Figure 4

We began to answer these questions with our informal exploration of tangent lines in Chapter 2, and we know that the graph of a function f can only have a line tangent to it at the point c if secant lines that approximate the tangent line behave in a certain way (see Figure 4). Namely, the "limiting behavior" of the secant lines must exist—if so, the result is what we identify as the tangent line. This requirement imposes some restrictions on functions that are differentiable, the first of which follows.

Theorem

Differentiability Implies Continuity

If f is differentiable at c, then f is continuous at c.

Proof

Since we seek to prove that f is continuous at the point c, we need to show $\lim_{x \to c} f(x) = f(c)$. The one fact we have to work with is that $f'(c)$ exists.

By our definition of $f'(c)$, this means that

$$\lim_{x \to c} \frac{f(x) - f(c)}{x - c}$$

exists, and so implicitly we are also given that f is defined at c; in other words, $f(c)$ exists. Further, the fact that the above limit exists means that $f(x)$ is defined for all x in some open interval containing c (that is, "sufficiently close to c"), so it makes sense to talk about $\lim_{x \to c} f(x)$. For $x \neq c$, we can multiply the difference quotient by $x - c$ and note that

$$f(x) = \frac{f(x) - f(c)}{x - c} \cdot (x - c) + f(c).$$

We can now apply the limit laws of Chapter 2 to obtain the following:

$$\lim_{x \to c} f(x) = \lim_{x \to c} \left[\frac{f(x) - f(c)}{x - c} \cdot (x - c) + f(c) \right]$$

$$= \lim_{x \to c} \left[\frac{f(x) - f(c)}{x - c} \cdot (x - c) \right] + \lim_{x \to c} f(c) \qquad \text{The limit of a sum is the sum of the limits.}$$

$$= \lim_{x \to c} \left[\frac{f(x) - f(c)}{x - c} \right] \cdot \lim_{x \to c} (x - c) + \lim_{x \to c} f(c) \qquad \text{The limit of a product is the product of the limits.}$$

$$= f'(c) \cdot 0 + f(c) \qquad \qquad f(c) \text{ exists and } \lim_{x \to c} f(c) = f(c).$$

$$= f(c)$$

We can gain additional insight into the restrictions that differentiability imposes on the behavior of a function by considering the second primary question. If f is *not* differentiable at c, the limit of the difference quotient must not exist. We know from the preceding theorem that a discontinuity at c would cause this condition, but there are other ways for the limit to not exist. The behavior of the function $f(x) = |x|$ at the point $x = 0$ gives us one example. For $x > 0$ the difference quotient is

$$\frac{f(x) - f(0)}{x - 0} = \frac{|x| - 0}{x - 0} = \frac{x}{x} = 1,$$

and for $x < 0$ the difference quotient is

$$\frac{f(x) - f(0)}{x - 0} = \frac{|x| - 0}{x - 0} = \frac{-x}{x} = -1.$$

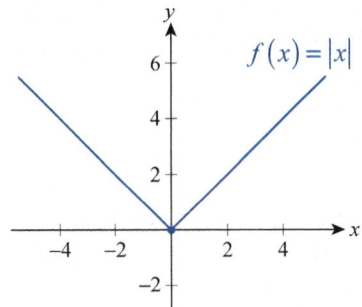

Figure 5 A Corner

The fact that the two one-sided limits do not agree at 0 means that $f(x) = |x|$ is not differentiable at 0; such a point on the graph of a function, where the one-sided difference quotient limits exist but are unequal, is called a **corner** (see Figure 5).

What if one or both of the one-sided difference quotients fails to exist at a point c where f is continuous? One possibility is if one of the one-sided difference quotients tends to $-\infty$ and the other to $+\infty$; such a point on the graph of f is called a **cusp**. In the graph of $f(x) = \sqrt{|x|}$ from Figure 6, note how the secant lines to the right of the cusp all have positive slopes (and increase without bound as x approaches 0), while the secant lines on the left all have negative slopes.

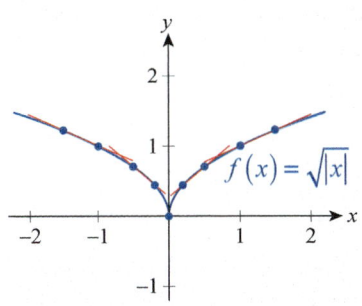

Figure 6 A Cusp

If both one-sided difference quotients have a limit of $+\infty$, or if both have a limit of $-\infty$, the graph of f at c actually has a tangent line but its slope is undefined; we (naturally enough) refer to such a situation as a **vertical tangent line**, and the graphs of $\sqrt[3]{x}$ and $-\sqrt[3]{x}$ both possess this property at 0 (see Figure 7). Note that the secant lines approximating the vertical tangent all have positive slope on the graph of $\sqrt[3]{x}$ and all have negative slope on the graph of $-\sqrt[3]{x}$.

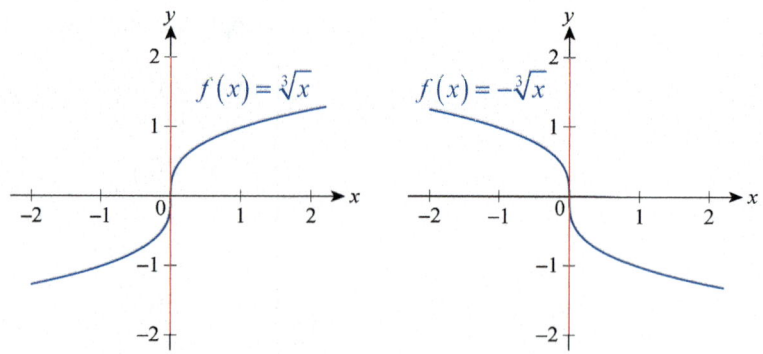

Figure 7 Vertical Tangent Lines

It is also possible for one or both of the one-sided difference quotient limits at a point c to not exist and for the difference quotients to not tend to $+\infty$ or $-\infty$ either. Such a function, which again may even be continuous at c, simply oscillates too much on every interval containing c. Our next example illustrates this phenomenon.

Example 5

Prove that the function $f(x) = \begin{cases} x\sin(1/x) & \text{if } x \neq 0 \\ 0 & \text{if } x = 0 \end{cases}$ is continuous, but not differentiable, at 0.

Solution

To see the continuity of $f(x)$ at 0, observe that since $-1 \leq \sin(1/x) \leq 1$ for all nonzero x, it follows that

$$-|x| \leq x\sin\frac{1}{x} \leq |x|,$$

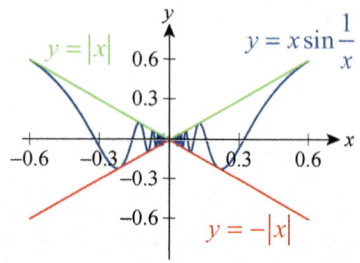

Figure 8

if $x \neq 0$. (See Figure 8.) We can now apply the Squeeze Theorem, since $\lim\limits_{x\to 0}|x| = \lim\limits_{x\to 0}(-|x|) = 0$. Thus, $\lim\limits_{x\to 0} f(x) = 0 = f(0)$, so $f(x)$ is continuous at 0.

To prove the nondifferentiability of f, we will show that the limit of its difference quotients does not exist at 0. This will be due to extensive (in fact, infinitely many) oscillations of f near the origin, as we shall see shortly. Let us first assume that $x > 0$, and examine the difference quotient $[f(x) - f(0)]/(x - 0)$:

$$\frac{f(x) - f(0)}{x - 0} = \frac{x\sin\dfrac{1}{x} - 0}{x - 0} = \frac{x\sin\dfrac{1}{x}}{x} = \sin\frac{1}{x}$$

In other words, proving that the one-sided difference quotient limit does not exist for $x > 0$ is now equivalent to proving that $\lim\limits_{x\to 0^+} \sin(1/x)$ does not exist. To that end, we will present an argument very similar to that given in Example 7 of Section 2.3, albeit in a somewhat abbreviated fashion. (Note that this also means that not only f oscillates "too much" near 0, but so do its difference quotients.)

Suppose that the above limit exists, and $\lim\limits_{x\to 0^+} \sin(1/x) = L$. Then for $\varepsilon = 1$, there is a $\delta > 0$ such that

$$\left| \sin\frac{1}{x} - L \right| < 1$$

whenever $0 < |x| < \delta$.

Recall from your studies of the sine function that for any integer k, $\sin(4k+1)\dfrac{\pi}{2} = 1$, and $\sin(4k+3)\dfrac{\pi}{2} = -1$. In addition, we can choose and fix a big enough positive integer k such that

$$\frac{1}{(4k+1)\dfrac{\pi}{2}} < \delta.$$

Using the notation

$$x_1 = \frac{1}{(4k+1)\dfrac{\pi}{2}} \quad \text{and} \quad x_2 = \frac{1}{(4k+3)\dfrac{\pi}{2}},$$

note that $x_2 < x_1 < \delta$, while $\sin(1/x_1) = 1$ and $\sin(1/x_2) = -1$. By assumption,

$$\left| \sin\frac{1}{x_1} - L \right| = |1 - L| < 1,$$

which implies that $L > 0$, but also

$$\left| \sin\frac{1}{x_2} - L \right| = |-1 - L| < 1,$$

implying in turn that $L < 0$, an obvious impossibility.

Thus $\sin(1/x)$ cannot be approaching any limit L at all as x approaches 0 from the right; that is, $\lim\limits_{x\to 0^+} \sin(1/x) = \lim\limits_{x\to 0^+} (f(x) - f(0))/(x - 0)$ does not exist.

The case of $x < 0$ is handled in an analogous fashion, and it will follow that neither one-sided difference quotient limit exists, and thus $f(x)$ is not differentiable at 0, just as we needed to show.

In summary, and informally, the differentiability of a function f at a point c implies that f is not only continuous at c but also fairly "smooth" and well-behaved—no corners, cusps, or excessive oscillation can occur at c. Keep in mind, though, that it is possible for the graph of f to possess a vertical tangent line at c, and that this may be overlooked because f' will not exist at such a point. However, such points can be identified by comparing the behavior of the one-sided difference quotients.

As we have seen, one-sided limits of difference quotients can be useful when determining the behavior of a function. We will now give these limits a name.

Definition

One-Sided Derivatives

If a function f is defined on an interval of the form $[a, p)$ (that is, at a and on some half-open interval to the right of a), the **right-hand derivative of f at a** is defined to be

$$f'_+(a) = \lim_{h \to 0^+} \frac{f(a+h) - f(a)}{h},$$

provided the limit exists.

Similarly, if f is defined on an interval of the form $(p, b]$, the **left-hand derivative of f at b** is defined to be

$$f'_-(b) = \lim_{h \to 0^-} \frac{f(b+h) - f(b)}{h},$$

provided the limit exists.

We can now extend our definition of differentiability to intervals that are not open. For example, we say f *is differentiable on the interval* $[a,b]$ if f is differentiable on (a,b) and $f'_+(a)$ and $f'_-(b)$ exist.

Example 6

Consider the following piecewise defined function:

$$f(x) = \begin{cases} 2\sqrt{x} & \text{if } 0 \le x < 1 \\ \dfrac{x^2 + 3}{2} & \text{if } 1 \le x \le 2 \end{cases}$$

a. Find the one-sided derivatives of f at 1 and use them to decide whether f is differentiable at $c = 1$.

b. Find the appropriate one-sided derivatives at the endpoints of the domain of f.

Solution

a. First of all, since $\lim\limits_{x \to 1^-} 2\sqrt{x} = \lim\limits_{x \to 1^+} (x^2 + 3)/2 = 2$, f is certainly continuous at 1. But is it differentiable? The one-sided derivatives will have the answer.

$$f'_-(1) = \lim_{h \to 0^-} \frac{f(1+h) - f(1)}{h}$$

$$= \lim_{h \to 0^-} \frac{2\sqrt{1+h} - \frac{1^2+3}{2}}{h}$$

$$= \lim_{h \to 0^-} \frac{\left(2\sqrt{1+h} - 2\right)\left(2\sqrt{1+h} + 2\right)}{h\left(2\sqrt{1+h} + 2\right)}$$

$$= \lim_{h \to 0^-} \frac{4(1+h) - 4}{h\left(2\sqrt{1+h} + 2\right)}$$

$$= \lim_{h \to 0^-} \frac{4}{\left(2\sqrt{1+h} + 2\right)} = 1$$

Thus we see that the left-hand derivative of f at 1 equals 1. The right-hand derivative at the same point is determined as follows.

$$f'_+(1) = \lim_{h \to 0^+} \frac{f(1+h) - f(1)}{h}$$

$$= \lim_{h \to 0^+} \frac{\frac{(1+h)^2+3}{2} - \frac{1^2+3}{2}}{h} \qquad \text{Combine the fractions in the numerator.}$$

$$= \lim_{h \to 0^+} \frac{\frac{2h+h^2}{2}}{h} \qquad \text{Cancel } h.$$

$$= \lim_{h \to 0^+} \frac{2+h}{2} = 1$$

Since the one-sided derivatives agree, we conclude that f is differentiable at $c = 1$. Graphically this means that even though the graph is "patched together" at $c = 1$ from two "pieces," this is done in a "smooth" way, in other words, without the formation of any corner or cusp.

b. Since $a = 0$ is the left endpoint of the domain of f, only the right-hand derivative is defined.

$$f'_+(0) = \lim_{h \to 0^+} \frac{f(0+h) - f(0)}{h}$$

$$= \lim_{h \to 0^+} \frac{2\sqrt{0+h} - 2\sqrt{0}}{h}$$

$$= \lim_{h \to 0^+} \frac{2\sqrt{h}}{h}$$

$$= \lim_{h \to 0^+} \frac{2}{\sqrt{h}} = \infty$$

So the right-hand derivative of f at 0 does not exist.

Finally, at the right endpoint $b = 2$,

$$f'_-(2) = \lim_{h \to 0^-} \frac{f(2+h) - f(2)}{h}$$

$$= \lim_{h \to 0^-} \frac{\dfrac{(2+h)^2 + 3}{2} - \dfrac{2^2 + 3}{2}}{h} \qquad \text{Combine the fractions in the numerator.}$$

$$= \lim_{h \to 0^-} \frac{\dfrac{4h + h^2}{2}}{h} \qquad \text{Cancel } h.$$

$$= \lim_{h \to 0^-} \frac{4 + h}{2} = 2.$$

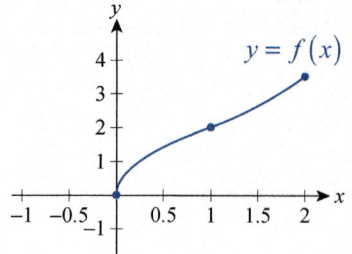

Figure 9

To summarize our findings, the graph of f is joined together "smoothly" from two pieces at $c = 1$, it goes "vertical" at the left endpoint of its domain, and has a "slope" of 2 at the right endpoint, since the right-hand derivative exists there. Figure 9 graphically reflects these observations.

We will close this section with one more theorem that relates the ideas of continuity and differentiability.

Theorem

Darboux's Theorem

If f is differentiable on $[a,b]$, then f' takes on every value between $f'_+(a)$ and $f'_-(b)$. That is, f' has the Intermediate Value Property (IVP) on the interval $[a,b]$: if L is a value between $f'_+(a)$ and $f'_-(b)$, there is a point c in $[a,b]$ such that $f'(c) = L$.

In Section 2.5, we saw that continuous functions on an interval $[a,b]$ have the Intermediate Value Property, so we already know that the function f' has the IVP if it is continuous. As it turns out, derivatives of functions are not necessarily continuous, but the French mathematician Jean-Gaston Darboux proved in 1875 that a function can only *be* the derivative of some other function if it possesses the IVP. In Chapter 5 we will prove a partial converse of this when we see that if a function is continuous, then it is the derivative of some other function.

3.1 **Exercises**

1–12 Find the derivative of the given function at the specified point and express your answer using the differential notation due to Leibniz.

1. $f(x) = 7;\quad x = 1$

2. $g(x) = \dfrac{1}{2}x - 5;\quad x = -1$

3. $h(x) = \dfrac{1}{2}x^2 - 5;\quad x = 0$

4. $F(t) = \dfrac{1}{5}t + 2t^2;\quad t = \dfrac{1}{5}$

5. $G(s) = \dfrac{1}{3}s^3 - s;\quad s = -3$

6. $H(t) = \dfrac{1}{2}t^4 + t^2;\quad t = -2$

7. $K(z) = \dfrac{5}{3z + 1};\quad z = 0$

8. $T(t) = \dfrac{2t - 3}{t + 1};\quad t = \sqrt{5} - 1$

9. $w(z) = \dfrac{1}{z^2 + 2};\quad z = \sqrt{2}$

10. $A(t) = \sqrt{2t};\quad t = 0$

11. $Q(y) = \dfrac{1}{\sqrt{3y}};\quad y = 1$

12. $B(u) = \sqrt{u^2 + 1};\quad u = -2\sqrt{2}$

13–24 Find the derivative of the function and use the differentiation operator D_x to express your answer.

13. $f(x) = \pi^2$

14. $g(x) = 1 - \dfrac{2}{3}x$

15. $h(t) = \dfrac{3}{2}t^2 + 10t - 1$

16. $F(s) = 4 - s + \dfrac{1}{2}s^2 + s^3$

17. $G(y) = \dfrac{1}{3y}$

18. $H(t) = \dfrac{t - 3}{t + 3}$

19. $S(z) = \dfrac{-3}{z^2}$

20. $T(u) = \dfrac{4}{u^2 - 3u}$

21. $R(v) = \sqrt{2v - 3}$

22. $f(y) = \dfrac{2}{\sqrt{y + 2}}$

23. $X(y) = 2y^4$

24. $u(x) = \dfrac{2}{\sqrt{2x^2 + 1}}$

25–33 Find the first, second, and third derivatives of the function. Then graph the function along with its derivatives in the same coordinate system and compare the graphs. (**Hint:** See Example 4.)

25. $f(x) = \dfrac{5}{2}x - 1$

26. $g(x) = x^2 + 5$

27. $h(x) = -\dfrac{1}{2}x^2 + x - \dfrac{3}{2}$

28. $U(x) = -x^3$

29. $V(x) = \dfrac{1}{3}(x - 2)^3$

30. $F(t) = t^4 - 1$

31. $G(x) = 2(x - 1)^4$

32. $H(x) = \dfrac{1}{x}$

33. $K(s) = \dfrac{-2}{s - 1}$

34–37 The graphs of the position, velocity, acceleration, and jerk of a moving particle are given. Decide which one is which, label them accordingly, and explain.

34.

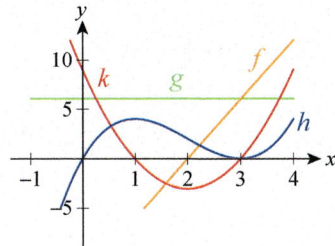

35.

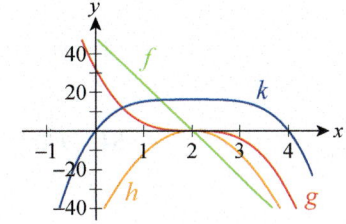

36.

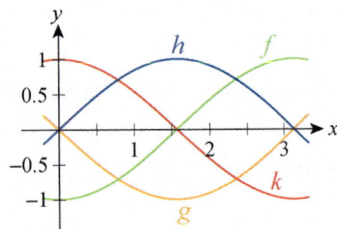

37.

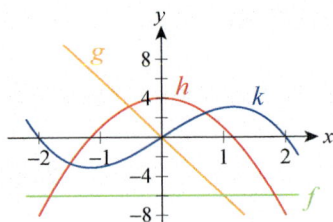

38–43 Use the given graph of the function to find all *x*-values where the function is differentiable.

38.

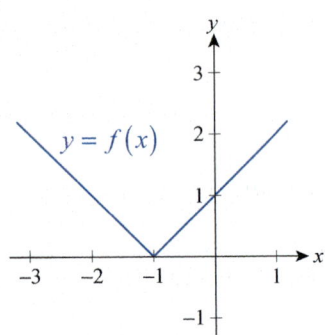

39.

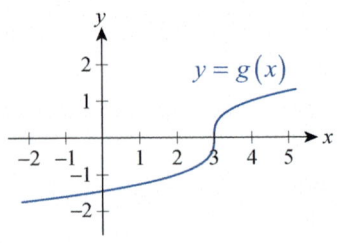

40.

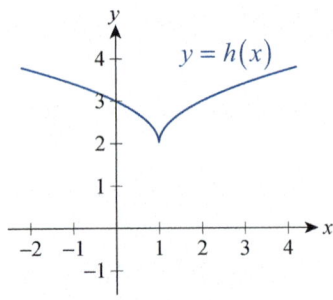

41.

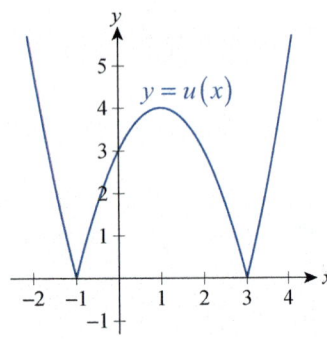

42.

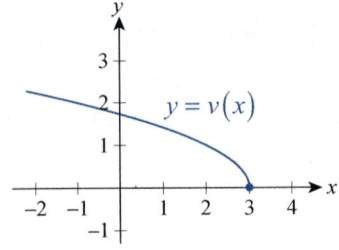

43.

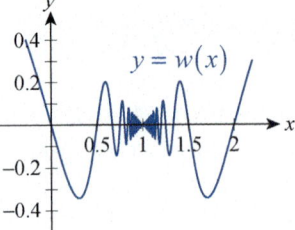

44–58 Find all points where the function is not differentiable. For each of those points, find the one-sided derivatives (if they exist).

44. $f(x) = |x + 5|$

45. $g(x) = |x + 2| - |x - 4|$

46. $h(x) = (x - 1)^{2/3}$

47. $F(x) = \sqrt[3]{x - 1.5} + 2$

48. $H(x) = \sqrt{1.8 - x}$

49. $k(x) = \sqrt{3 - x^2}$

50. $G(x) = \dfrac{x^2}{x^2 - 9}$

51. $m(x) = |x^2 - 6x + 5|$

52. $A(t) = [\![t - 4]\!]$

53. $B(x) = x - [\![x]\!]$

54. $F(t) = \begin{cases} \dfrac{1}{2} t \cos \dfrac{1}{t} & \text{if } t \neq 0 \\ 0 & \text{if } t = 0 \end{cases}$

55. $H(z) = \begin{cases} \sqrt{z} \sin \dfrac{\pi}{z} & \text{if } z > 0 \\ 0 & \text{if } z = 0 \end{cases}$

56. $P(x) = \begin{cases} \sqrt[3]{x - 1} & \text{if } x < 1 \\ (x - 1)^2 & \text{if } x \geq 1 \end{cases}$

57. $G(x) = \begin{cases} 2x + 2 & \text{if } x \leq -2 \\ -\dfrac{1}{2} x^2 & \text{if } x > -2 \end{cases}$

58. $S(t) = \begin{cases} \dfrac{1}{t} & \text{if } t \leq 1 \\ t & \text{if } t > 1 \end{cases}$

59. Prove that the function

$$f(x) = \begin{cases} x^2 \sin \dfrac{1}{x} & \text{if } x \neq 0 \\ 0 & \text{if } x = 0 \end{cases}$$

is differentiable at 0. Contrast this result with Example 5.

60. The position function of a car crashing head-on at 62 mph during a crash test is $x(t) = -196t^2 + 27.78t$, where x is measured in meters and t in seconds. Find the deceleration of the dummy inside the car. What multiple of g is this (where g is the gravity constant)?

61. The position from its starting point of a small plane preparing for takeoff is given by $x(t) = 1.1t^2$ meters (t is measured in seconds).

   a. What is the acceleration of the plane?

   b. How long does it take for the plane to reach the minimum takeoff speed of 33 m/s?

   c. What is the minimum required runway length for this type of plane?

62. The position function of a theme park thrill ride moving along a straight line is $x(t) = \frac{14}{3}t^3 + 10t$ ft ($0 \leq t \leq 3$, t is measured in seconds). Find the velocity, acceleration, and jerk. How far from starting position are the cars at the end of the 3-second time interval?

63. The **symmetric derivative** of a function f at a point c is defined as

$$f'_{sym}(c) = \lim_{h \to 0} \frac{f(c+h) - f(c-h)}{2h}.$$

   a. Prove that if f is differentiable at c, then its symmetric derivative exists and $f'_{sym}(c) = f'(c)$.

   b. Give an example of a function g and a point c such that $g'_{sym}(c)$ exists, but g is not differentiable at c.

64–67 *True or False?* Determine whether the given statement is true or false. In case of a false statement, explain or provide a counterexample.

64. If f is continuous at c, then f is differentiable at c.

65. If f is differentiable at c, then f is continuous at c.

66. If f is differentiable at c, then

$$f'(c) = \lim_{\Delta x \to 0} \frac{f(c - \Delta x) - f(c)}{-\Delta x}.$$

67. If both one-sided derivatives of f at c exist, then f is differentiable at c.

68. Sketch the graph of $f(x) = -x^2 + 2x$ and its derivative on the interval $[0,2]$ in the same coordinate system. Where (on which interval) is f' positive? Where is f' negative? Identify those intervals where f is increasing versus decreasing. Do you see a connection? Can you give an intuitive reason for your findings?

69. Repeat Exercise 68 for the function $g(x) = 1/x^2$. Sketch both g and g' on their entire domains and summarize your observations. Can you formulate a general conjecture?

3.1 **Technology Exercises**

70–75 Use a computer algebra system to graph the given function along with its derivative in the same viewing window and answer the questions of Exercise 68. (**Hint:** Use the "Derivative" feature of your technology to find f' first.)

70. $f(x) = \dfrac{1}{3}x^3 - 5x^2 - 1$ **71.** $f(x) = \dfrac{x^2}{x^2 - 4}$

72. $f(x) = \dfrac{x}{x^2 + 1}$ **73.** $f(x) = -\cos x$

74. $f(x) = \sin^2 x$ **75.** $f(x) = e^{1/(x^2 + 1)}$

76–79 Use a computer algebra system to find the first four derivatives of f; then graph them along with f in the same viewing window and compare the graphs.

76. $f(x) = \dfrac{1}{2}x^5 - 2x^4 - 5x^3 + 7$

77. $f(x) = \arctan x$

78. $f(x) = \dfrac{2x}{x-3}$

79. $f(x) = x\sin\left(\dfrac{1}{10}x - 1\right)$

80–83 Use a graphing calculator or computer algebra system to graph the function and identify all points where the function is not differentiable. Explain.

80. $G(x) = \left(x^2 - 4\right)^{2/5}$

81. $H(t) = |2t - 1|^{2/3}$

82. $P(x) = \begin{cases} \arctan x & \text{if } x < 0 \\ x^{3/2} & \text{if } x \geq 0 \end{cases}$

83. $L(t) = \sqrt{|t|}\,\sin\dfrac{1}{|t|}$

3.2 Derivatives of Polynomials, Exponentials, Products, and Quotients

Just as the limit laws allow us to quickly evaluate many limits without resorting to epsilon-delta arguments, differentiation rules allow us to determine most common derivatives without having to consider limits of difference quotients. Much of the next five sections will be devoted to developing and using these differentiation rules.

Topic One
Elementary Differentiation Rules

We begin with a rule for arguably the simplest sorts of functions—those that have a constant value, say k. Since the graph of $f(x) = k$ is the horizontal line $y = k$, the tangent line to the graph at any point must have slope 0. This is the geometric (and informal) justification for the rule that if $f(x) = k$ then $f'(x) = 0$ for every x. The formal proof of this statement is just as easy: given any x,

$$f'(x) = \lim_{h \to 0} \frac{f(x+h) - f(x)}{h} = \lim_{h \to 0} \frac{k - k}{h} = \lim_{h \to 0} 0 = 0.$$

Theorem

The Constant Rule

If f is a constant function, say $f(x) = k$, then $f'(x) = 0$. In other notation, we write

$$\frac{df}{dx} = \frac{d}{dx}(k) = 0.$$

In the last two sections, we worked through the difference quotient limit process to determine the derivatives of $f(x) = x$ and $f(x) = x^2$. These limits are not terribly difficult: briefly,

$$\frac{d}{dx}(x) = \lim_{h \to 0} \frac{(x+h) - x}{h} = \lim_{h \to 0} \frac{h}{h} = \lim_{h \to 0} 1 = 1$$

and

$$\frac{d}{dx}(x^2) = \lim_{h \to 0} \frac{(x+h)^2 - x^2}{h} = \lim_{h \to 0} \frac{x^2 + 2xh + h^2 - x^2}{h} = \lim_{h \to 0} \frac{h(2x+h)}{h} = \lim_{h \to 0} (2x + h) = 2x.$$

But we don't want to have to work through a similar computation every time we encounter a function of the form $f(x) = x^n$ if, instead, a general rule can be found. Fortunately, this is indeed the case.

Theorem

The Positive Integer Power Rule

If $f(x) = x^n$, where n is a positive integer, then $f'(x) = nx^{n-1}$. In other notation, we write

$$\frac{df}{dx} = \frac{d}{dx}(x^n) = nx^{n-1}.$$

Proof

There are actually several ways to prove this first Power Rule, and because they introduce useful techniques we will provide two common proofs here.

Proof 1

The identity

$$x^n - c^n = (x - c)(x^{n-1} + x^{n-2}c + \cdots + xc^{n-2} + c^{n-1})$$

can be verified by multiplying out the right-hand side and canceling the common terms. With this identity in hand, we can use one of the alternative forms of the difference quotient to obtain the following.

$$
\begin{aligned}
f'(c) &= \lim_{x \to c} \frac{f(x) - f(c)}{x - c} = \lim_{x \to c} \frac{x^n - c^n}{x - c} \\
&= \lim_{x \to c} \frac{(x - c)(x^{n-1} + x^{n-2}c + \cdots + xc^{n-2} + c^{n-1})}{x - c} \\
&= \lim_{x \to c} (x^{n-1} + x^{n-2}c + \cdots + xc^{n-2} + c^{n-1}) \\
&= c^{n-1} + c^{n-2}c + \cdots + cc^{n-2} + c^{n-1} \qquad \text{\textcolor{red}{Note that there are n terms.}} \\
&= nc^{n-1}
\end{aligned}
$$

Proof 2

Alternatively, we can use the Binomial Theorem to expand the term $(x + h)^n$ as follows.

$$
\begin{aligned}
f'(x) &= \lim_{h \to 0} \frac{f(x + h) - f(x)}{h} = \lim_{h \to 0} \frac{(x + h)^n - x^n}{h} \\
&= \lim_{h \to 0} \frac{\left(x^n + nx^{n-1}h + \dfrac{n(n-1)}{2}x^{n-2}h^2 + \cdots + nxh^{n-1} + h^n\right) - x^n}{h} \\
&= \lim_{h \to 0} \frac{h\left(nx^{n-1} + \dfrac{n(n-1)}{2}x^{n-2}h + \cdots + nxh^{n-2} + h^{n-1}\right)}{h} \\
&= \lim_{h \to 0} \left(nx^{n-1} + \dfrac{n(n-1)}{2}x^{n-2}h + \cdots + nxh^{n-2} + h^{n-1}\right) \\
&= nx^{n-1}
\end{aligned}
$$

As we will see, the Power Rule is actually true for all real number exponents, a fact we will prove in stages as we acquire the necessary tools.

Example 1

Use the Constant Rule and Positive Integer Power Rule to determine the derivative of the given function.

a. $f(x) = -5$ **b.** $g(x) = 0$ **c.** $h(x) = \pi^3$

d. $k(x) = x^4$ **e.** $s(t) = t^{15}$ **f.** $w(z) = \dfrac{1}{z}$

Solution

a. Since f is a constant function, the Constant Rule applies. Using the various notations, we obtain

$$f'(x) = \frac{df}{dx} = \frac{d}{dx}(-5) = 0.$$

Recall that there are even more ways to express our answer. For example, $D_x\, 5 = 0$, or if f appears in the form of $y = 5$, we might write $dy/dx = 0$, etc.

b. Again, since g is constant, its derivative will be identically 0, just like that of any other constant function.

$$g'(x) = \frac{dg}{dx} = \frac{d}{dx}(0) = 0$$

c. At first, it might be tempting to try to use the Positive Integer Power Rule on h, but we must be careful to avoid that mistake. Notice that π^3 is a constant, so $h(x)$ is a constant function and when differentiated it behaves like any other constant function.

$$h'(x) = \frac{dh}{dx} = \frac{d}{dx}(\pi^3) = 0$$

d. Since $k(x)$ is a positive integer power of the independent variable x, the Positive Integer Power Rule applies with $n = 4$.

$$k'(x) = \frac{dk}{dx} = \frac{d}{dx}(x^4) = 4x^3$$

e. Notice that the independent variable of s is denoted by t, but the Positive Integer Power Rule applies, just like it would with the variable x.

$$s'(t) = \frac{ds}{dt} = \frac{d}{dt}(t^{15}) = 15t^{14}$$

f. Since $1/z = z^{-1}$ is not a positive integer power of z, we cannot apply the Positive Integer Power Rule for w, but will proceed using the definition of the derivative instead. Our conclusion will be pleasantly consistent.

$$\frac{dw}{dz} = \lim_{h \to 0} \frac{\frac{1}{z+h} - \frac{1}{z}}{h} = \lim_{h \to 0} \frac{\frac{z-(z+h)}{(z+h)z}}{h} = \lim_{h \to 0} \frac{-h}{h(z+h)z} = \lim_{h \to 0} \frac{-1}{(z+h)z} = -\frac{1}{z^2}$$

Notice that if we had applied the Positive Integer Power Rule, we would have obtained

$$\frac{dw}{dz} = \frac{d}{dz}\left(z^{-1}\right) = (-1)z^{-1-1} = -z^{-2} = -\frac{1}{z^2},$$

which is precisely the correct answer. In other words, the pattern of the rule still applies to the exponent of −1. As mentioned immediately preceding this example, the Power Rule actually holds true for any real exponent, a fact we will prove later.

The next rule points out that the derivative of a constant multiple of a given function is no more difficult to determine than the derivative of the function itself.

Theorem

The Constant Multiple Rule

Given a constant k and a differentiable function f, the derivative of $kf(x)$ is $kf'(x)$. In other notation,

$$\frac{d}{dx}\left[kf(x)\right] = k\frac{d}{dx}f(x) = kf'(x).$$

In words, the Constant Multiple Rule is often summed up with a phrase like "constants pass through the differentiation operator," and such a phrase may bring to mind similar behavior we have already seen. Recall that one of the properties of limits is

$$\lim_{x\to c}\left[kf(x)\right] = k\lim_{x\to c}f(x),$$

and in fact it is this property that is used in the proof of the Constant Multiple Rule—you will work through this proof in Exercise 108.

Theorem

The Sum and Difference Rules

If f and g are both differentiable at x, the sum $f+g$ is also differentiable at x and $(f+g)'(x) = f'(x)+g'(x)$. In other notation,

$$\frac{d}{dx}\left[f(x)+g(x)\right] = \frac{d}{dx}f(x)+\frac{d}{dx}g(x).$$

Similarly, $f-g$ is differentiable and $(f-g)'(x) = f'(x)-g'(x)$.

--- **Proof** ---------------

$$(f+g)'(x) = \lim_{h\to 0} \frac{(f+g)(x+h)-(f+g)(x)}{h}$$

$$= \lim_{h\to 0} \frac{[f(x+h)+g(x+h)]-[f(x)+g(x)]}{h}$$

$$= \lim_{h\to 0} \left[\frac{f(x+h)-f(x)}{h} + \frac{g(x+h)-g(x)}{h} \right]$$

$$= \lim_{h\to 0} \frac{f(x+h)-f(x)}{h} + \lim_{h\to 0} \frac{g(x+h)-g(x)}{h}$$

$$= f'(x)+g'(x)$$

To prove the corresponding Difference Rule, we can make use of the Sum Rule and the Constant Multiple Rule.

$$(f-g)'(x) = (f+(-1)g)'(x) = f'(x)+(-1)g'(x) = f'(x)-g'(x)$$

With the Constant Rule, Positive Integer Power Rule, Constant Multiple Rule, and Sum and Difference Rules, we can now quickly differentiate any polynomial. Further, it should be noted that these rules imply that the derivative of a polynomial is always another polynomial.

Example 2

Use the above rules to find the derivative of the given function.

a. $p(x) = 2x^2 + 3x$ **b.** $q(x) = x^3 + \dfrac{3}{2}x^2 - 5x + 7$

c. $f(x) = 3\sqrt{x} - \dfrac{2}{x}$

Solution

a. Since p is a polynomial of two terms, the Sum Rule immediately ensures that we can differentiate it *termwise*; that is, we can differentiate the terms separately and add up the results.

$$p'(x) = \frac{d}{dx} p(x) = \frac{d}{dx}\left(2x^2 + 3x\right) = \frac{d}{dx}\left(2x^2\right) + \frac{d}{dx}(3x)$$

Next, the Constant Multiple Rule and the Power Rule come to bear, and we obtain

$$\frac{d}{dx}\left(2x^2\right) + \frac{d}{dx}(3x) = 2\frac{d}{dx}\left(x^2\right) + 3\frac{d}{dx}(x) = 2(2x) + 3(1) = 4x + 3.$$

Thus we conclude that $p'(x) = 4x + 3$. Notice what we found here is that the derivative of a quadratic polynomial is a linear polynomial. This is indeed the case in general; in fact it is not hard to see that if the degree of p is n, then its derivative has degree $n-1$.

b. Even though $q(x)$ has more than two terms, a repeated use of the Sum Rule still enables us to differentiate termwise, as shown.

$$q'(x) = \frac{d}{dx} q(x)$$

$$= \frac{d}{dx}\left(x^3 + \frac{3}{2}x^2 - 5x + 7\right)$$

$$= \frac{d}{dx}(x^3) + \frac{d}{dx}\left(\frac{3}{2}x^2 - 5x + 7\right) \qquad \text{Apply the Sum Rule for the second time.}$$

$$= \frac{d}{dx}(x^3) + \frac{d}{dx}\left(\frac{3}{2}x^2\right) + \frac{d}{dx}(-5x + 7) \qquad \text{Apply the Sum Rule for the third time.}$$

$$= \frac{d}{dx}(x^3) + \frac{d}{dx}\left(\frac{3}{2}x^2\right) + \frac{d}{dx}(-5x) + \frac{d}{dx}(7)$$

This illustrates the fact that the Sum Rule makes it possible for us to differentiate any polynomial term by term, a fact you will be asked to prove in Exercise 113. This observation along with the first two rules of this section will then imply that the derivative of a polynomial is always another polynomial (Exercise 114). We now complete the differentiation of q.

$$\frac{d}{dx} q(x) = \frac{d}{dx}(x^3) + \frac{d}{dx}\left(\frac{3}{2}x^2\right) + \frac{d}{dx}(-5x) + \frac{d}{dx}(7)$$

$$= \frac{d}{dx}(x^3) + \frac{3}{2}\frac{d}{dx}(x^2) + (-5)\frac{d}{dx}(x) + \frac{d}{dx}(7)$$

$$= 3x^2 + \frac{3}{2}(2x) + (-5)(1) + 0$$

$$= 3x^2 + 3x - 5$$

c. With regards to $f(x)$, recall that from Example 1 we know the derivative of $1/x$ to be $-1/x^2$; and also $\frac{d}{dx}(\sqrt{x}) = \frac{1}{2\sqrt{x}}$. We calculated this latter derivative in Example 5 of Section 2.6, or you might arrive at it by using the yet-unproven Power Rule for a fractional exponent.

$$\frac{d}{dx}(\sqrt{x}) = \frac{d}{dx}(x^{1/2}) = \frac{1}{2}x^{-1/2} = \frac{1}{2}\cdot\frac{1}{x^{1/2}} = \frac{1}{2\sqrt{x}}$$

Armed with these facts, we use a combination of the Difference Rule and Constant Multiple Rule to complete our solution, as follows.

$$f'(x) = \frac{d}{dx} f(x)$$

$$= \frac{d}{dx}\left(3\sqrt{x} - \frac{2}{x}\right)$$

$$= \frac{d}{dx}(3\sqrt{x}) - \frac{d}{dx}\left(\frac{2}{x}\right) \qquad \text{Difference Rule}$$

$$= 3\frac{d}{dx}(\sqrt{x}) - 2\frac{d}{dx}\left(\frac{1}{x}\right) \qquad \text{Constant Multiple Rule}$$

$$= 3\left(\frac{1}{2\sqrt{x}}\right) - 2\left(-\frac{1}{x^2}\right)$$

$$= \frac{3}{2\sqrt{x}} + \frac{2}{x^2} = \frac{3}{2}x^{-1/2} + 2x^{-2}$$

Example 3

Find the equation of the line tangent to the graph of

$$p(x) = 0.3x^5 + 1.1x^4 - x^3 + 0.6x^2 + 2.9x - 0.9$$

at the point $(1,3)$.

Solution

Recall that the slope of the tangent is $p'(1)$, the derivative of p at $x = 1$.

Termwise differentiation yields

$$p'(x) = \frac{d}{dx}(0.3x^5) + \frac{d}{dx}(1.1x^4) - \frac{d}{dx}(x^3) + \frac{d}{dx}(0.6x^2) + \frac{d}{dx}(2.9x) - \frac{d}{dx}(0.9)$$

$$= 0.3\frac{d}{dx}(x^5) + 1.1\frac{d}{dx}(x^4) - \frac{d}{dx}(x^3) + 0.6\frac{d}{dx}(x^2) + 2.9\frac{d}{dx}(x) - 0$$

$$= 0.3(5x^4) + 1.1(4x^3) - 3x^2 + 0.6(2x) + 2.9$$

$$= 1.5x^4 + 4.4x^3 - 3x^2 + 1.2x + 2.9,$$

and thus the slope we are seeking is

$$p'(1) = 1.5(1^4) + 4.4(1^3) - 3(1^2) + 1.2(1) + 2.9 = 7.$$

Now the point-slope form of the equation of a line yields

$$y - 3 = 7(x - 1),$$

or equivalently, the requested equation of the tangent is

$$y = 7x - 4.$$

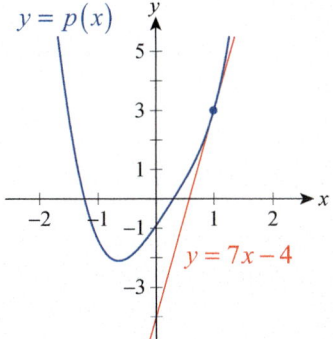

$y = p(x)$

$y = 7x - 4$

Figure 1

The curve and its tangent line are graphed in Figure 1.

Example 4

The distance from the origin of a particle moving along the x-axis is described by the function $x(t) = 2t^3 - 1.5t^2 + 4t$ inches, where t is measured in seconds. Find the instantaneous velocity and acceleration of the particle at $t = 2$ seconds.

Solution

As we have seen before, velocity is the derivative of the position function. Note, however, that in this example x is actually the dependent variable, and differentiation takes place with respect to the independent variable t.

$$v(t) = \frac{d}{dt}x(t) = \frac{d}{dt}(2t^3 - 1.5t^2 + 4t)$$

$$= \frac{d}{dt}(2t^3) - \frac{d}{dt}(1.5t^2) + \frac{d}{dt}(4t)$$

$$= 2\frac{d}{dt}(t^3) - 1.5\frac{d}{dt}(t^2) + 4\frac{d}{dt}(t)$$

$$= 2(3t^2) - 1.5(2t) + 4(1)$$

$$= 6t^2 - 3t + 4$$

Thus the instantaneous velocity of the particle at $t = 2$ seconds is

$$v(2) = 6(2)^2 - 3(2) + 4 = 22 \text{ in./s}.$$

To find the acceleration, we differentiate the velocity function.

$$\begin{aligned}
a(t) = \frac{d}{dt}v(t) &= \frac{d}{dt}\left(6t^2 - 3t + 4\right) \\
&= \frac{d}{dt}\left(6t^2\right) - \frac{d}{dt}\left(3t\right) + \frac{d}{dt}\left(4\right) \\
&= 6\frac{d}{dt}\left(t^2\right) - 3\frac{d}{dt}\left(t\right) + 0 \\
&= 6(2t) - 3(1) \\
&= 12t - 3
\end{aligned}$$

Substituting $t = 2$ into the acceleration function gives the instantaneous acceleration of the particle at $t = 2$ seconds.

$$a(2) = 12(2) - 3 = 21 \text{ in./s}^2$$

Topic Two

Differentiation of Exponential Functions

We now have rules allowing us to easily differentiate polynomial functions; the overall goal is to develop similar rules for many classes of functions, and we consider exponential functions next.

Recall that an exponential function has the form $f(x) = a^x$, where a is a positive real number not equal to 1. Since none of the rules we have developed so far help us in determining f', we must begin with the definition of derivative—the limit of a difference quotient:

$$\begin{aligned}
f'(x) &= \lim_{h \to 0} \frac{f(x+h) - f(x)}{h} \\
&= \lim_{h \to 0} \frac{a^{x+h} - a^x}{h} \\
&= \lim_{h \to 0} \frac{a^x\left(a^h - 1\right)}{h} \qquad \text{Factor out } a^x. \\
&= a^x \lim_{h \to 0} \frac{a^h - 1}{h} \qquad \text{The factor } a^x \text{ does not depend on } h \text{ and passes through the limit.}
\end{aligned}$$

It is worth noting, in particular, that

$$f'(0) = a^0 \lim_{h \to 0} \frac{a^h - 1}{h} = \lim_{h \to 0} \frac{a^h - 1}{h},$$

so in general we can write

$$f'(x) = a^x \lim_{h \to 0} \frac{a^h - 1}{h} = a^x f'(0).$$

In other words, the derivative of $f(x) = a^x$ is a^x multiplied by a constant, namely $f'(0)$. All that remains is to evaluate $\lim_{h \to 0} (a^h - 1)/h$.

Although we can easily construct a table of values of $(a^h - 1)/h$ for fixed values of a and for h as close to 0 as we desire, the proof that the limit above actually exists must be deferred until we have the machinery of Chapter 5. The table below indicates that $f'(0) \approx -0.69$ when $a = \frac{1}{2}$, $f'(0) \approx 0.69$ when $a = 2$, and $f'(0) \approx 1.10$ when $a = 3$. And since $f'(0)$ represents the slope of the line tangent to $f(x) = a^x$ at $x = 0$, these approximate values for $f'(0)$ correspond to the relative rise or fall of the three tangent lines depicted in Figure 2.

h	$\dfrac{\left(\dfrac{1}{2}\right)^h - 1}{h}$	$\dfrac{2^h - 1}{h}$	$\dfrac{3^h - 1}{h}$
0.1	−0.6697	0.7177	1.1612
0.01	−0.6908	0.6956	1.1047
0.001	−0.6929	0.6934	1.0992
0.0001	−0.6931	0.6932	1.0987

Based on Figure 2, it is certainly reasonable to expect that there is a value for a between 2 and 3 such that

$$\lim_{h \to 0} \frac{a^h - 1}{h} = 1,$$

since the value of this limit is approximately 0.69 when $a = 2$ and approximately 1.10 when $a = 3$. This is indeed true, and we give this particular value for a the label e; that is, we define e to be the real number such that

$$\lim_{h \to 0} \frac{e^h - 1}{h} = 1.$$

But again, the rigorous proof that such an e exists awaits us in Section 5.3, Exercise 100. In the meantime, it suffices to know that e, an irrational number approximately equal to 2.7183, is the unique real number with the above property. This leads to the nice fact that

$$\frac{d}{dx}(e^x) = e^x \left(\lim_{h \to 0} \frac{e^h - 1}{h} \right) = e^x;$$

in other words, the derivative of e^x is itself. We will return to the derivative of $f(x) = a^x$ for other values of a in Section 3.4.

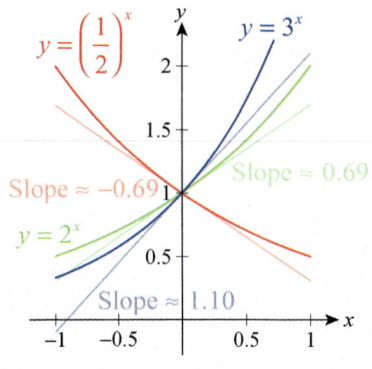

Figure 2
Graphs of a^x and Associated Tangent Lines at $x = 0$

Theorem

Derivative of e^x

$$\frac{d}{dx}\left(e^x\right) = e^x$$

Example 5

Find the first three derivatives of the following functions.

a. $f(x) = -2e^x$ **b.** $g(s) = 3s^2 + e^s$ **c.** $h(x) = ae^x, a \in \mathbb{R}$

Solution

a. Using the Constant Multiple Rule along with the fact that e^x is its own derivative we have:

$$f'(x) = \frac{d}{dx}\left(-2e^x\right) = (-2)\frac{d}{dx}\left(e^x\right) = -2e^x$$

In other words, we find that $f(x)$ is also the derivative of itself. Repeating the above calculation two more times, we conclude that the second and third derivatives of f are also equal to f, that is,

$$f''(x) = f'''(x) = -2e^x.$$

In fact, it is easy to generalize and observe that all derivatives of $f(x)$ are equal to itself. We will have to say more about this observation in part c. of this example.

b. In addition to the rules used in part a., to find the derivative of $g(s)$ we will also need the Sum Rule:

$$\begin{aligned}
g'(s) &= \frac{d}{ds}\left(3s^2 + e^s\right) \\
&= \frac{d}{ds}\left(3s^2\right) + \frac{d}{ds}\left(e^s\right) \\
&= 3(2s) + e^s = 6s + e^s
\end{aligned}$$

A similar calculation yields the second derivative.

$$g''(s) = \frac{d}{ds}g'(s) = \frac{d}{ds}\left(6s + e^s\right) = \frac{d}{ds}(6s) + \frac{d}{ds}\left(e^s\right) = 6 + e^s$$

You might already have figured out the third derivative by observing that differentiation only affects the first term. Here are the details.

$$g'''(s) = \frac{d}{ds}g''(s) = \frac{d}{ds}\left(6 + e^s\right) = \frac{d}{ds}(6) + \frac{d}{ds}\left(e^s\right) = 0 + e^s = e^s$$

Finally, we note in passing that any higher-order derivative of g will now be e^s, since e^s is the derivative of itself.

c. Let us now investigate the derivatives of ae^x, where a is a real constant. We now use the prime notation along with the Constant Multiple Rule:

$$h'(x) = \left(ae^x\right)' = a\left(e^x\right)' = ae^x$$

In other words, the derivative of any constant multiple of e^x is itself. Repeating the above procedure any number of times allows us to conclude that derivatives of all orders of the function $h(x) = ae^x$ are equal to h itself for all $a \in \mathbb{R}$. This is a generalization of the observation we made in part a. of this example.

Topic Three

The Product and Quotient Rules

The Sum and Difference Rules for differentiation are easy to remember ("the derivative of a sum or difference is the sum or difference of the derivatives"), and their proofs are straightforward—for instance, the Sum Rule follows almost immediately from writing down and then rearranging the appropriate difference quotient. It is tempting to assume that the Product and Quotient Rules follow a similar pattern, but that is not the case—the truth is a bit more interesting.

Caution!

The derivative of a product is *not* the product of the derivatives.

$$\frac{d}{dx}\left[f(x)g(x)\right] \neq f'(x)g'(x)$$

Similarly, the derivative of a quotient is *not* the quotient of the derivatives.

$$\frac{d}{dx}\left[\frac{f(x)}{g(x)}\right] \neq \frac{f'(x)}{g'(x)}$$

To see why this is so, it may be helpful to use a geometric interpretation; such an approach is very much in keeping with early Greek mathematics, which almost always viewed a product of two quantities as an area measure of a two-dimensional figure.

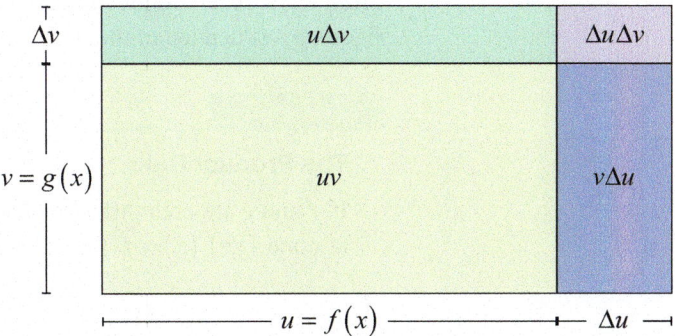

Figure 3 Geometric Interpretation of $(fg)'$

To make such a geometric connection, suppose f and g are two differentiable functions taking on only positive values (so we can interpret those values as being measures of length). Then the product $f(x)g(x)$ represents the area of a rectangle with, say, width $f(x)$ and height $g(x)$. (See Figure 3.) Assume x is fixed and, for ease of exposition, let $u = f(x)$, $v = g(x)$, $\Delta u = f(x+\Delta x) - f(x)$, and $\Delta v = g(x+\Delta x) - g(x)$. Geometrically, $(uv)'$ represents the instantaneous rate of change in the rectangular area uv, so we are seeking a formula for

$$\lim_{\Delta x \to 0} \frac{\Delta(uv)}{\Delta x}.$$

Referring again to Figure 3, the small change in area $\Delta(uv)$ with respect to a small change Δx can be expressed as

$$\Delta(uv) = (u+\Delta u)(v+\Delta v) - uv = u\Delta v + v\Delta u + \Delta u \Delta v,$$

so

$$\frac{\Delta(uv)}{\Delta x} = u\frac{\Delta v}{\Delta x} + v\frac{\Delta u}{\Delta x} + \Delta u\frac{\Delta v}{\Delta x}.$$

Since f is differentiable at x, it is continuous at x and so $\Delta u = f(x+\Delta x) - f(x) \to 0$ as $\Delta x \to 0$. Hence, using several limit laws,

$$\begin{aligned}
(uv)' &= \lim_{\Delta x \to 0} \frac{\Delta(uv)}{\Delta x} \\
&= \lim_{\Delta x \to 0}\left(u\frac{\Delta v}{\Delta x} + v\frac{\Delta u}{\Delta x} + \Delta u\frac{\Delta v}{\Delta x} \right) \\
&= u\lim_{\Delta x \to 0}\frac{\Delta v}{\Delta x} + v\lim_{\Delta x \to 0}\frac{\Delta u}{\Delta x} + \lim_{\Delta x \to 0}\Delta u\lim_{\Delta x \to 0}\frac{\Delta v}{\Delta x} \\
&= uv' + vu' + 0\cdot v' \\
&= uv' + vu'.
\end{aligned}$$

A geometric argument such as the above is valuable in adding to our understanding of the meaning of differentiation and in building intuition, but it must be augmented by a formal proof in order to be complete (for instance, the geometric argument doesn't make sense if f or g takes on negative values). Such a proof is given immediately following the formal statement of the Product Rule, offering a rigorous justification of the rule.

Theorem

The Product Rule

If f and g are both differentiable at x, the product fg is also differentiable at x and $(fg)'(x) = f'(x)g(x) + f(x)g'(x)$. In other notation,

$$\frac{d}{dx}[f(x)g(x)] = \left[\frac{d}{dx}f(x)\right]g(x) + f(x)\left[\frac{d}{dx}g(x)\right].$$

─ **Proof** ─────────────────────────────

$$(fg)'(x) = \lim_{h \to 0} \frac{f(x+h)g(x+h) - f(x)g(x)}{h}$$

$$= \lim_{h \to 0} \frac{f(x+h)g(x+h) - f(x)g(x+h) + f(x)g(x+h) - f(x)g(x)}{h}$$

$$= \lim_{h \to 0} \frac{f(x+h)g(x+h) - f(x)g(x+h)}{h} + \lim_{h \to 0} \frac{f(x)g(x+h) - f(x)g(x)}{h}$$

We begin above, unsurprisingly, with the limit of a difference quotient. In order to make the limit tractable, however, we need to rewrite the difference quotient in such a way that difference quotients for f and g make an appearance, and the easiest way to do this is to add and subtract the expression $f(x)g(x+h)$ in the numerator, as shown in the second step. The third step then splits the difference quotient into two fractions, and we proceed by applying additional limit properties.

$$(fg)'(x) = \lim_{h \to 0} \frac{f(x+h)g(x+h) - f(x)g(x+h)}{h} + \lim_{h \to 0} \frac{f(x)g(x+h) - f(x)g(x)}{h}$$

$$= \lim_{h \to 0} \left[\frac{f(x+h) - f(x)}{h} g(x+h) \right] + \lim_{h \to 0} \left[f(x) \frac{g(x+h) - g(x)}{h} \right]$$

$$= \left[\lim_{h \to 0} \frac{f(x+h) - f(x)}{h} \right] \left[\lim_{h \to 0} g(x+h) \right] + f(x) \left[\lim_{h \to 0} \frac{g(x+h) - g(x)}{h} \right]$$

$$= f'(x)g(x) + f(x)g'(x)$$

Note that we have used the fact that differentiability implies continuity in order to replace $\lim_{h \to 0} g(x+h)$ with $g(x)$ in the last step.

──

Example 6

Use the Product Rule to find the derivative of each function.

a. $F(x) = 5x$ b. $G(x) = x^2$

c. $H(x) = (2x^2 + 1)(x^3 - 4)$ d. $K(x) = x^2 e^x$

Solution

a. In order to differentiate F, we could certainly use the approach of Example 2 and, with the help of the Constant Multiple Rule, arrive at the answer of $F'(x) = 5$. However, our aim here is to demonstrate the truth of the Product Rule, so let's think of F as the product of the constant function $f(x) = 5$ and the function $g(x) = x$ and apply the Product Rule:

$$F'(x) = \frac{d}{dx}[f(x)g(x)]$$

$$= \left[\frac{d}{dx} f(x) \right] g(x) + f(x) \left[\frac{d}{dx} g(x) \right]$$

$$= \left[\frac{d}{dx}(5) \right] x + 5 \left[\frac{d}{dx}(x) \right]$$

$$= (0)x + 5(1)$$

$$= 5$$

b. Again, the Positive Integer Power Rule is perhaps the easiest way to determine G', but let us see how the Product Rule gives us the same answer. Since $x^2 = x \cdot x$, we have

$$G'(x) = \frac{d}{dx}(x \cdot x) = \frac{d}{dx}(x) \cdot x + x \cdot \frac{d}{dx}(x) = 1 \cdot x + x \cdot 1 = 2x,$$

which is exactly what the Positive Integer Power Rule would tell us.

c. With the assignments $f(x) = 2x^2 + 1$ and $g(x) = x^3 - 4$, we see that $H(x) = f(x)g(x)$, so applying the Product Rule gives us

$$H'(x) = \frac{d}{dx}\left[f(x)g(x) \right]$$

$$= \left[\frac{d}{dx} f(x) \right] g(x) + f(x) \left[\frac{d}{dx} g(x) \right]$$

$$= \left[\frac{d}{dx}(2x^2 + 1) \right](x^3 - 4) + (2x^2 + 1)\left[\frac{d}{dx}(x^3 - 4) \right]$$

$$= (4x)(x^3 - 4) + (2x^2 + 1)(3x^2)$$

$$= (4x^4 - 16x) + (6x^4 + 3x^2)$$

$$= 10x^4 + 3x^2 - 16x.$$

Notice that we can check our answer by multiplying the two factors of H together and differentiating the resulting polynomial the usual way:

$$H(x) = (2x^2 + 1)(x^3 - 4) = 2x^5 + x^3 - 8x^2 - 4,$$

so applying our rules,

$$H'(x) = 2\left(x^5\right)' + \left(x^3\right)' - 8\left(x^2\right)' - 0$$

$$= 2(5x^4) + (3x^2) - 8(2x)$$

$$= 10x^4 + 3x^2 - 16x,$$

as we expected. The beauty of mathematics stems partly from the fact that usually there are several possible approaches to the solution of any given problem.

d. While all the previous derivatives could be handled in alternative ways, in the case of the function K there is no apparent way to avoid the Product Rule. Using the prime notation this time, we can write

$$K'(x) = \left(x^2 e^x \right)'$$

$$= \left(x^2 \right)' e^x + x^2 \left(e^x \right)'$$

$$= (2x)e^x + x^2 \left(e^x \right)$$

$$= \left(x^2 + 2x \right)e^x,$$

where we again used the fact that e^x is the derivative of itself.

Example 7

Find $F(1)$ if $F(x) = \sqrt{x} \cdot g(x)$, and $F'(1) = 2$ and $g'(1) = 3$.

Solution

First of all, since we know $F(1) = \sqrt{1} \cdot g(1) = g(1)$, finding $F(1)$ or $g(1)$ are equivalent problems. Keeping this in mind, it is not hard to see that the Product Rule provides us with an equation to determine $g(1)$, as follows.

$$F(x) = \sqrt{x} \cdot g(x)$$

$$F'(x) = \left(\sqrt{x}\right)' \cdot g(x) + \sqrt{x} \cdot g'(x) \qquad \text{We have seen before that } \left(\sqrt{x}\right)' = \frac{1}{2\sqrt{x}}.$$

$$= \frac{1}{2\sqrt{x}} \cdot g(x) + \sqrt{x} \cdot g'(x)$$

Substituting $x = 1$, we have

$$F'(1) = \frac{1}{2\sqrt{1}} \cdot g(1) + \sqrt{1} \cdot g'(1).$$

(Note that in general, if $x = c$, it follows from the Product Rule that

$$(fg)'(c) = f'(c)g(c) + f(c)g'(c).$$

We used this fact with $c = 1$ in our equation above.)

Next, substituting the known quantities of $F'(1) = 2$ and $g'(1) = 3$, our equation becomes

$$2 = \frac{1}{2}g(1) + 3.$$

Solving the above for $g(1)$, we obtain

$$g(1) = -2.$$

Finally, recalling that $F(1) = g(1)$, we conclude that $F(1) = -2$.

As an intermediary step in developing the Quotient Rule, we will first find a rule for the derivative of the reciprocal of a differentiable function, a rule that is useful in its own right.

Theorem

The Reciprocal Rule

If g is differentiable at x and $g(x) \neq 0$, $1/g$ is also differentiable at x and

$$\frac{d}{dx}\left[\frac{1}{g(x)}\right] = \frac{-g'(x)}{\left[g(x)\right]^2}.$$

— Proof ——————————————————————————

$$\frac{d}{dx}\left[\frac{1}{g(x)}\right] = \lim_{h \to 0} \frac{\frac{1}{g(x+h)} - \frac{1}{g(x)}}{h}$$

$$= \lim_{h \to 0}\left[\frac{1}{h}\left(\frac{1}{g(x+h)} - \frac{1}{g(x)}\right)\right]$$

$$= \lim_{h \to 0}\left[\frac{1}{h}\left(\frac{g(x) - g(x+h)}{g(x+h)g(x)}\right)\right]$$

We are hoping to relate the derivative of $1/g$ to the derivative of g if possible, so we are again seeking a way to rewrite the difference quotient so that g' will make an appearance. Using the fact that $1/g(x)$ is independent of h and can be pulled out of the above limit, and using the continuity of g at x, we can rearrange the quotient to obtain the following.

$$\frac{d}{dx}\left[\frac{1}{g(x)}\right] = \left[\frac{1}{g(x)}\right]\lim_{h \to 0}\left[\left(\frac{1}{g(x+h)}\right)\left(\frac{g(x) - g(x+h)}{h}\right)\right]$$

$$= \left[\frac{1}{g(x)}\right]\left[\lim_{h \to 0}\frac{-1}{g(x+h)}\right]\left[\lim_{h \to 0}\frac{g(x+h) - g(x)}{h}\right]$$

$$= \left[\frac{1}{g(x)}\right]\left[\frac{-1}{g(x)}\right]g'(x)$$

$$= \frac{-g'(x)}{\left[g(x)\right]^2}$$

——————————————————————————

One immediate application of the Reciprocal Rule is the extension of the Power Rule to negative integers. Given a positive integer n, note the following.

$$\left(x^{-n}\right)' = \left(\frac{1}{x^n}\right)'$$

$$= \frac{-\left(x^n\right)'}{\left(x^n\right)^2} \qquad \text{Reciprocal Rule}$$

$$= \frac{-nx^{n-1}}{x^{2n}} \qquad \text{Positive Integer Power Rule}$$

$$= -nx^{-n-1}$$

Since the Power Rule is trivially true for x^0, we have now proved the rule for all integers.

Example 8

Use the Reciprocal Rule to find the derivative of $f(x) = 1/\sqrt{x}$.

Solution

Recall that $\left(\sqrt{x}\right)' = 1/\left(2\sqrt{x}\right)$. Using this fact along with the Reciprocal Rule, we can write

$$\frac{d}{dx}\left(\frac{1}{\sqrt{x}}\right) = \frac{-\left(\sqrt{x}\right)'}{\left(\sqrt{x}\right)^2}$$

$$= \frac{-\dfrac{1}{2\sqrt{x}}}{\left(\sqrt{x}\right)^2}$$

$$= \frac{-1}{2x\sqrt{x}} = -\frac{1}{2}x^{-3/2}.$$

Notice that our result is yet another illustration of the validity of the Power Rule for fractional exponents $(n = -1/2$ in this case). We will prove the Power Rule for all real exponents later as we acquire the necessary tools.

Theorem

The Quotient Rule

If f and g are both differentiable at x and $g(x) \neq 0$, f/g is also differentiable at x and

$$\frac{d}{dx}\left[\frac{f(x)}{g(x)}\right] = \frac{f'(x)g(x) - f(x)g'(x)}{\left[g(x)\right]^2}.$$

Proof

The Product and Reciprocal Rules together allow us to quickly arrive at the Quotient Rule, as follows.

$$\frac{d}{dx}\left[\frac{f(x)}{g(x)}\right] = \frac{d}{dx}\left[f(x)\frac{1}{g(x)}\right]$$

$$= f'(x)\frac{1}{g(x)} + f(x)\left(\frac{-g'(x)}{\left[g(x)\right]^2}\right) \qquad \text{Product and Reciprocal Rules}$$

$$= \frac{f'(x)g(x)}{\left[g(x)\right]^2} - \frac{f(x)g'(x)}{\left[g(x)\right]^2}$$

$$= \frac{f'(x)g(x) - f(x)g'(x)}{\left[g(x)\right]^2}$$

Example 9

Use the Quotient Rule to find the following derivatives.

a. $\dfrac{d}{dx}\left(\dfrac{5x^2 + x}{x^2 + 2}\right)$ **b.** $\dfrac{d}{dx}\left(\dfrac{3/x + x}{x - 1/x}\right)$ **c.** $\dfrac{d}{dx}\left(\dfrac{\sqrt{x} + 2/\sqrt{x}}{\sqrt{x}}\right)$

Solution

a. Using the Quotient Rule with $f(x) = 5x^2 + x$ and $g(x) = x^2 + 2$, we obtain the following.

$$\frac{d}{dx}\left[\frac{f(x)}{g(x)}\right] = \frac{d}{dx}\left[\frac{5x^2 + x}{x^2 + 2}\right]$$

$$= \frac{\left(5x^2 + x\right)'\left(x^2 + 2\right) - \left(5x^2 + x\right)\left(x^2 + 2\right)'}{\left(x^2 + 2\right)^2}$$

$$= \frac{(10x + 1)\left(x^2 + 2\right) - \left(5x^2 + x\right)(2x)}{\left(x^2 + 2\right)^2}$$

$$= \frac{\left(10x^3 + x^2 + 20x + 2\right) - \left(10x^3 + 2x^2\right)}{\left(x^2 + 2\right)^2}$$

$$= \frac{-x^2 + 20x + 2}{\left(x^2 + 2\right)^2}$$

b. While we could certainly proceed with the Quotient Rule for this function, this is a situation when it pays to rewrite our quotient before differentiating. Notice that by multiplying both the numerator and denominator by x, we obtain an expression that is free of reciprocals and therefore is more convenient to handle.

$$\frac{3/x + x}{x - 1/x} = \frac{3 + x^2}{x^2 - 1}$$

$$\frac{d}{dx}\left(\frac{3/x + x}{x - 1/x}\right) = \frac{d}{dx}\left(\frac{3 + x^2}{x^2 - 1}\right)$$

$$= \frac{\left(3 + x^2\right)'\left(x^2 - 1\right) - \left(x^2 - 1\right)'\left(3 + x^2\right)}{\left(x^2 - 1\right)^2}$$

$$= \frac{(2x)\left(x^2 - 1\right) - (2x)\left(3 + x^2\right)}{\left(x^2 - 1\right)^2}$$

$$= \frac{(2x)\left(x^2 - 1 - 3 - x^2\right)}{\left(x^2 - 1\right)^2}$$

$$= \frac{-8x}{\left(x^2 - 1\right)^2}$$

c. Again, we can make the expression look simpler if we rewrite before differentiating. We will start by multiplying both the numerator and denominator by $\sqrt{x}$.

$$\frac{\sqrt{x}+2/\sqrt{x}}{\sqrt{x}} = \frac{\sqrt{x}+2/\sqrt{x}}{\sqrt{x}}\cdot\frac{\sqrt{x}}{\sqrt{x}} = \frac{x+2}{x} = \frac{x}{x}+\frac{2}{x} = 1+\frac{2}{x}$$

$$\frac{d}{dx}\left(\frac{\sqrt{x}+2/\sqrt{x}}{\sqrt{x}}\right) = \frac{d}{dx}\left(1+\frac{2}{x}\right)$$

$$= \frac{d}{dx}(1)+2\frac{d}{dx}\left(\frac{1}{x}\right) \qquad \text{Recall that } \left(\frac{1}{x}\right)' = -\frac{1}{x^2}.$$

$$= 0+2\left(-\frac{1}{x^2}\right)$$

$$= -\frac{2}{x^2}$$

What you should remember from the last two differentiations is that it might not always be the best idea to proceed right away with the Quotient Rule whenever you see a quotient. While you can certainly obtain the correct answer that way, often you can save a lot of time just by rewriting the expression before differentiating.

3.2 **Exercises**

1–12 Use the appropriate rules from this section to find the derivative of the given function.

1. $f(x) = 5 - 2x$

2. $g(x) = \frac{4}{5}x + 2$

3. $h(x) = \frac{1}{2} + 2x - 3x^2$

4. $F(x) = x^3 - 2x^2 + \frac{1}{2}x - 77$

5. $G(x) = \frac{1}{2}x^4 + 2x^3 - x^2 + 3.2x + \sqrt{2}$

6. $k(x) = x^{11} - 0.2x^{10} + \frac{\pi}{3}x^3 + \pi$

7. $H(x) = x^8 + \sqrt{2}x^5 - 2x^4$

8. $R(t) = t^{100} - 2t^{59} + \pi t^{38} + et$

9. $S(z) = 4z^3 - 3\sqrt{z} + 11.2$

10. $Q(s) = \frac{1}{3s} - \sqrt{2s} + \sqrt{2s}$

11. $T(r) = \pi r^2 + 2e^r + \pi^2$

12. $N(t) = e^{2+t} + \frac{t+1}{t} + pt$

13–20 Use the Product Rule to find the indicated derivative. Then find the answer without the use of the Product Rule, by multiplying first, and compare your answers.

13. $\frac{d}{dx}\big[(x+2)(3x+5)\big]$

14. $\frac{d}{dx}\big[(3x+7)(x^2+2x)\big]$

15. $\frac{d}{dx}\big[(x^2-6)(2x^2+5x)\big]$

16. $\frac{d}{dx}\big[(2x^3+3x^2)(4x^2-2x+5)\big]$

17. $\frac{d}{dx}\left[\left(\frac{1}{3}x^3+\frac{7}{5}x^5\right)\left(\frac{2}{x}-4x^2\right)\right]$

18. $\frac{d}{dx}\big[(e^x+3)(e^2-5)\big]$

19. $\frac{d}{dt}\big[(3+2\sqrt{t})(4\sqrt{t}-5)\big]$

20. $\frac{d}{ds}\left[s\left(-3-\frac{1}{3}s^3\right)(s^4+2s)\right]$

21–32 Use the Reciprocal Rule or Quotient Rule to determine the derivative of the function.

21. $f(x) = \dfrac{1}{1-2x}$

22. $g(x) = \dfrac{1}{4x - 2x^2}$

23. $h(x) = \dfrac{2}{2x^3 - 5x^2 + 3x + 1}$

24. $F(x) = \dfrac{e}{e^x - \sqrt{x}}$

25. $G(x) = \dfrac{2x+1}{x-5}$ **26.** $k(x) = \dfrac{3x-4}{2x^2+5}$

27. $H(x) = \dfrac{x^3 - 3x^2}{2x^3 + 5x^2 + 1}$ **28.** $A(x) = \dfrac{6\sqrt{x}}{3x-4}$

29. $B(u) = \dfrac{u^2}{\sqrt{u}+1}$ **30.** $f(t) = \dfrac{4-\sqrt{t}}{t^2+3}$

31. $g(t) = \dfrac{3-t}{4 - 5\sqrt{t}}$ **32.** $w(s) = \dfrac{1+2e^s}{3e^s+5}$

33–38 Differentiate the quotient by simplifying it algebraically first.

33. $f(x) = \dfrac{30x^2 - 10x^6}{5x}$

34. $g(x) = \dfrac{1 + 5x + x^2}{x}$

35. $h(x) = \dfrac{3x^{1/2} - 5x^{3/2} + 7x^{5/2} - 9x^{7/2}}{x^{1/2}}$

36. $F(x) = \dfrac{2(\sqrt{x})^3 + 3\sqrt{x}}{\sqrt{x} + (\sqrt{x})^3}$

37. $G(x) = \dfrac{\dfrac{6}{x^2} - \dfrac{5}{x} + 1}{\dfrac{1}{x} - \dfrac{3}{x^2}}$

38. $H(x) = \dfrac{2 - \dfrac{1}{e^x}}{2e^{-x}}$

39–61 Using the rules of this section, differentiate the given function.

39. $f(t) = t^{1/2}(4t + 3)$

40. $g(x) = x^2\left(\sqrt{x} + \dfrac{1}{\sqrt{x}}\right)$

41. $h(s) = \left(5 + \dfrac{1}{s}\right)\left(s^2 + \dfrac{1}{5}\right)$

42. $F(x) = \dfrac{1}{x} + \dfrac{2}{x^2}$

43. $G(x) = 3x^{-5} + 2x^{-3}$

44. $k(s) = s^2\left(\dfrac{3}{s} + \dfrac{1}{s-1}\right)$

45. $H(t) = \sqrt{t}(9 - t^2)$

46. $K(x) = \dfrac{\dfrac{1}{x^2} - 3}{x + 2}$

47. $w(z) = z\left(2 + \dfrac{4}{4 - \sqrt{z}}\right)$

48. $L(T) = T^{-3}(2 - 4T^{-2})$

49. $r(x) = \dfrac{x - a^2}{x + a^2}$ **50.** $Q(t) = \dfrac{at + b}{ct + d}$

51. $F(x) = e^x(2 + \sqrt{x})$

52. $E(s) = \dfrac{2 + se^s}{e^s - s}$

53. $C(x) = \dfrac{a}{a + \dfrac{a}{x}}$ **54.** $D(x) = \dfrac{x}{x + \dfrac{x}{a}}$

55. $G(s) = \dfrac{3s^2}{2e^s + s}$

56. $S(t) = (4 - \sqrt{t})(2 - e^t)$

57. $L(y) = (y^4 - 3y^3)(y^2 - 2y^5)$

58. $h(z) = \dfrac{1}{ae^z + z}$ **59.** $H(s) = \dfrac{a}{b + ce^s}$

60. $T(x) = (x + 2)(2x^2 - x)(x^3 + 5)$

61. $L(t) = t(2e^t + \sqrt{t})\left(\dfrac{1}{t} - 1\right)$

62–67 Find the first, second, and third derivatives of the function.

62. $f(x) = 2x + 5$

63. $g(x) = \dfrac{x}{x+1}$

64. $h(x) = 3\sqrt{x}$

65. $F(x) = 2 - x + 5x^2 - \pi x^3$

66. $V(z) = 2z^2 + \dfrac{2}{z^2}$

67. $W(t) = 3t^2 + 3e^t$

68–71 Find a function f that satisfies the given conditions. (**Hint:** A polynomial is the most natural choice. Answers will vary.)

68. $f(0) = 2$, $f'(0) = 1$, and $f''(0) = -1$.

69. $f(0) = 0$ and f has horizontal tangent lines at $x = 2$ and $x = -2$.

70. $f(0) = 1$, $y = x + 1.5$ is tangent to the graph at $x = 1$, and $y = 5.5 - x$ is tangent to the graph at $x = 3$.

71. $f(1) = 5$, $f'(1) = 8$, f has a horizontal tangent line at $x = -1$, and $f'''(x) = 6$.

72–75 Find a formula for the kth derivative of the function. (**Hint:** Calculate the first few derivatives and try to recognize an emerging pattern.)

72. $f(x) = x^n$

73. $g(x) = \dfrac{1}{x}$

74. $h(x) = xe^x$

75.* $q(x) = x^n e^x$

76. If $f(1) = 2$, $f'(1) = 1$, $g(1) = -1$, and $g'(1) = 3$, find the following function values.

a. $(f - g)'(1)$

b. $(fg)'(1)$

c. $\left(\dfrac{f}{g}\right)'(1)$

77. If $f(3) = -1$, $f'(3) = 5$, $g(3) = 1/2$, and $g'(3) = -2$, find the following function values.

a. $(2f + 5g)'(3)$

b. $(4fg)'(3)$

c. $\left(\dfrac{f}{2g}\right)'(3)$

78–82 Find the equation of the line tangent to the graph of the function at the given point.

78. $f(x) = \dfrac{x^2 + 1}{x}$; $(1, 2)$

79. $w(x) = \dfrac{8}{x^2 + 4}$; $(2, 1)$

(This curve is called the **witch of Agnesi**.)

80. $g(x) = \dfrac{2}{\sqrt{x} + 1}$; $(1, 1)$

81. $h(x) = \dfrac{2e^x}{x^2}$; $(1, 2e)$

82. $k(x) = \dfrac{2x}{2 + x^2}$; $\left(1, \dfrac{2}{3}\right)$

(This curve is called a **serpentine**.)

83. Find the equation of the **normal line** to the graph of $s(x) = 9x/(x^2 + 9)$ at the point $(0, 0)$. (We call a line **normal** to the graph at a point if it is perpendicular to the line tangent to the graph at the same point.)

84. Repeat Exercise 83 for the graph of $h(x) = e^x/(x^4 + 2)$ at $(0, 1/2)$.

85–96 Find all x-values where the graph of the function has a horizontal tangent line, or prove that the graph has no horizontal tangent line.

85. $f(x) = x^2 - 2x$

86. $g(x) = 2x^3 - 3x^2 - 12x + 1$

87. $h(x) = \dfrac{2}{x^2}$

88. $F(x) = \dfrac{2}{x^2 + 1}$

89. $G(x) = \dfrac{1}{2}e^x - 2$

90. $k(x) = x^2 - a$

91. $H(x) = \dfrac{1}{x^2 - a}$

92. $f(x) = \sqrt{x + 5}$

93. $g(x) = \dfrac{x^2}{x^2 + 5}$

94. $F(x) = \dfrac{2x - 1}{x^2}$

95. $P(x) = 2x^3 + 3x^2 + 3x - 5$

96. $Q(x) = e^x - x$

97. The line $y = 8x + b$ is tangent to the graph of $f(x) = ax^2$ at $x = 2$. Find the values of a and b.

98. Repeat Exercise 97 with the line $y = -2x + b$ that is tangent to the graph of $g(x) = -x^2 + ax$ at $x = 2$.

99. Show that the graphs of $y = \frac{1}{2}e^x$ and $y = 1/(x+1)^2$ intersect at $x = 0$ in such a way that their respective tangent lines are perpendicular at their point of intersection.

100–103 Find the equation(s) of the line(s) tangent to the graph of f through the indicated point, which does not lie on the graph of f. (**Hint:** If the point of tangency is $(x, f(x))$, then the slope of the tangent line going through (a, b) is $f'(x) = (f(x) - b)/(x - a)$.)

100. $f(x) = x^2$; $(0, -1)$ **101.** $f(x) = e^x$; $(0, 0)$

102. $f(x) = \sqrt{x}$; $(-2, 0)$ **103.** $f(x) = \dfrac{1}{x}$; $(-1, 0)$

104–107 Assuming that f and g are differentiable functions, differentiate the given expression.

104. $\dfrac{f(x)}{x}$ **105.** $\dfrac{xf(x)}{g(x)}$

106. $e^x g(x)$ **107.** $\dfrac{f(x)e^x}{g(x) + 2}$

108. Use the definition of the derivative to prove the Constant Multiple Rule. (**Hint:** For a given constant k, start out by using the definition
$$\frac{d}{dx}[kf(x)] = \lim_{h \to 0} \frac{kf(x+h) - kf(x)}{h},$$ and let k "pass through" the limit sign.)

109. Use the Product Rule and mathematical induction to provide a third proof of the Positive Integer Power Rule. (**Hint:** The base case of $n = 1$ should be obvious. After setting up the induction hypothesis of $(x^n)' = nx^{n-1}$, find the derivative of x^{n+1} by treating it as $x^{n+1} = x \cdot x^n$ and use the Product Rule along with the induction hypothesis.)

110. Use the Product Rule to arrive at a rule for
$$\frac{d}{dx}[f(x)g(x)h(x)].$$

111. Use the Product Rule to arrive at a formula for $\dfrac{d}{dx}[f(x)]^2$, and then use Exercise 110 to find the formula for $\dfrac{d}{dx}[f(x)]^3$. Do you recognize a pattern?

112. Use mathematical induction to prove the **Generalized Positive Integer Power Rule**:
$$\frac{d}{dx}[f(x)]^n = n[f(x)]^{n-1} f'(x)$$
(The hint provided in Exercise 109 might prove helpful, but you will need to appropriately modify your induction hypothesis.)

113. Use the Sum Rule and mathematical induction to prove that any finite sum of functions $f_1(x) + f_2(x) + \cdots + f_n(x)$ can be differentiated termwise, that is,
$$\left[f_1(x) + f_2(x) + \cdots + f_n(x) \right]' = f_1'(x) + f_2'(x) + \cdots + f_n'(x).$$
In particular, polynomials can be differentiated termwise. (Look at the hint provided in Exercise 109.)

114. Use Exercise 113 and the results of this section to prove that the derivative of a polynomial is always another polynomial.

115. A Formula One race car was moving in a parabolic curve with equation $y = \sqrt{x}$ when it hit an oil patch and the driver lost control at the point $(4, 2)$. The car left the track along the tangent line at the same point. Where did he hit the tire wall if the equation of the tire wall was $y = 4$? (Fortunately, there were no injuries.)

116. The position function of a golf ball rolling on an incline is given by $d(t) = 2t^2 + 3t$, where d is measured in meters, t in seconds. Find the ball's velocity and acceleration at $t = 4$ seconds.

117. The velocity function of a moving particle is given by $v(t) = 50t/(t + 10)$ ft/s. Find its acceleration at **a.** $t = 2$ seconds and **b.** $t = 10$ seconds.

118. The position function of a moving object is given by $p(t) = 2t^3 - 6t$ ft. Find its position and acceleration at the instant when its velocity changes directions.

119. The position function of an object dropped by an astronaut on the Moon is $h(t) = -0.81t^2 + 1.5$, where h is measured in meters, t in seconds. What is the acceleration due to gravity on the Moon? How long does it take for the above object to reach the ground and what is the speed of impact?

120. The radius of a spherical balloon being inflated increases according to the function $r(t) = 2 + 5\sqrt[3]{t}$, where r is measured in centimeters and t in seconds. Find the rate of change of the balloon's volume and surface area with respect to time at $t = 8$ seconds.

121–129 *True or False?* Determine whether the given statement is true or false. In case of a false statement, explain or provide a counterexample.

121. If $y = \pi x^n$, then $y' = n\pi x^{n-1}$.

122. If $y = \pi^n$, then $y' = n\pi^{n-1}$.

123. If $y = \pi/x^n$, then $y' = \pi/(nx^{n-1})$.

124. If $y = e^x$, then $y' = xe^{x-1}$.

125. If $y = \pi e^x$, then $y' = \pi e^x$.

126. If p is a fifth-degree polynomial, then its sixth derivative is 0.

127. If $F(x) = \dfrac{f(x)}{g(x)}$, then $\dfrac{d}{dx}F(x) = \dfrac{\dfrac{d}{dx}f(x)}{\dfrac{d}{dx}g(x)}$.

128. The jerk of a free-falling object is 0.

129. If a polynomial $p(x)$ has degree n, then its derivative has degree $n - 1$.

3.2 **Technology Exercises**

130–135. Use the differentiation capabilities of a computer algebra system to check your answers for Exercises 62–67, and then graph each function along with its derivatives in the same viewing window.

136–140. Referring back to Exercises 78–82, verify your answers by using a graphing calculator or computer algebra system to graph each function and its indicated tangent line in the same viewing window.

3.3 Derivatives of Trigonometric Functions

In this section, we extend those functions for which we have differentiation rules to the class of trigonometric functions. The development of the rules proceeds quickly once we establish two important facts.

Topic One
Two Useful Limits

The derivatives of the six basic trigonometric functions, as with the exponential functions, cannot be determined by applying any of the rules we have developed thus far. So we will begin by working with limits of difference quotients and, as we will soon see, the key to our work will be the evaluation of the following limits.

> ### Lemma
>
> Measuring θ in radians (as is customary),
>
> $$\lim_{\theta \to 0} \frac{\sin \theta}{\theta} = 1 \text{ and } \lim_{\theta \to 0} \frac{\cos \theta - 1}{\theta} = 0.$$

— Proof —

Neither limit is trivial, as the numerator and denominator approach 0 in both cases—the key to the evaluation is determining the relative *rate* at which this happens.

We can construct some bounds on the relative rate for the first limit by referring to Figure 1. For any angle θ between 0 and $\pi/2$, the area of triangle OBD is less than the area of the circle sector OBD, which in turn is less than the area of triangle OBC. Note also that since the radius of the circle is 1, the lengths of the two legs of triangle OAD are $\cos \theta$ and $\sin \theta$, while the length of the leg opposite θ in triangle OBC is $\tan \theta$. Using the area formulas for triangles and circular sectors, we have the following.

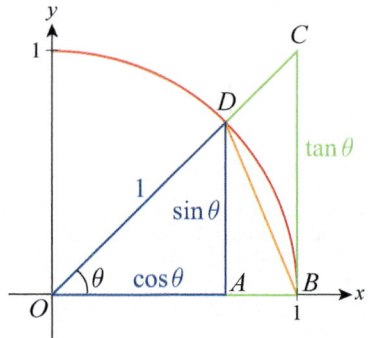

Figure 1 Understanding $\frac{\sin \theta}{\theta}$

$$\text{Area of } \triangle OBD = \frac{1}{2}(\text{base})(\text{height}) = \frac{1}{2}(1)(\sin \theta) = \frac{1}{2}\sin \theta$$

$$\text{Area of sector } OBD = \frac{1}{2}r^2 \theta = \frac{1}{2}(1)^2 \theta = \frac{1}{2}\theta$$

$$\text{Area of } \triangle OBC = \frac{1}{2}(\text{base})(\text{height}) = \frac{1}{2}(1)(\tan \theta) = \frac{1}{2}\tan \theta = \frac{1}{2} \cdot \frac{\sin \theta}{\cos \theta}$$

So the above statements imply

$$\frac{1}{2}\sin \theta < \frac{1}{2}\theta < \frac{1}{2} \cdot \frac{\sin \theta}{\cos \theta}.$$

Since $\sin\theta$ is positive for all θ in this range, dividing by $\sin\theta$ and multiplying by 2 leaves each inequality unchanged.

$$1 < \frac{\theta}{\sin\theta} < \frac{1}{\cos\theta}$$

Taking the reciprocal of each fraction *does* reverse each inequality, so the above is equivalent to

$$\cos\theta < \frac{\sin\theta}{\theta} < 1.$$

This is the sort of upper and lower bound on $(\sin\theta)/\theta$ that we need. Since $\cos\theta$ has a limit of 1 as $\theta \to 0$, the Squeeze Theorem tells us that

$$\lim_{\theta\to 0^+} \frac{\sin\theta}{\theta} = 1.$$

Since $(\sin\theta)/\theta$ is an even function,

$$\frac{\sin(-\theta)}{-\theta} = \frac{-\sin\theta}{-\theta} = \frac{\sin\theta}{\theta},$$

and we also have

$$\lim_{\theta\to 0^-} \frac{\sin\theta}{\theta} = 1,$$

so the first limit is proved.

We can now use this result to prove the second limit, making use of the trigonometric identity $\cos 2x = 1 - 2\sin^2 x$. Replacing $2x$ with θ and subtracting 1 from both sides of this identity, we have

$$\cos\theta - 1 = -2\sin^2\left(\frac{\theta}{2}\right),$$

and dividing through by θ results in

$$\frac{\cos\theta - 1}{\theta} = -\frac{2\sin^2\left(\frac{\theta}{2}\right)}{\theta} = -\frac{\sin^2\left(\frac{\theta}{2}\right)}{\frac{\theta}{2}}.$$

So we calculate the limit as follows.

$$\lim_{\theta\to 0} \frac{\cos\theta - 1}{\theta} = \lim_{\theta\to 0}\left[-\frac{\sin^2\left(\frac{\theta}{2}\right)}{\frac{\theta}{2}}\right]$$

$$= -\lim_{\theta\to 0}\left[\frac{\sin\left(\frac{\theta}{2}\right)}{\frac{\theta}{2}} \cdot \sin\left(\frac{\theta}{2}\right)\right]$$

$$= -\lim_{\alpha\to 0}\left(\frac{\sin\alpha}{\alpha} \cdot \sin\alpha\right) \qquad \text{Replace } \theta/2 \text{ with } \alpha.$$
$$\text{If } \theta \to 0 \text{ then } \alpha = \theta/2 \to 0.$$

$$= -(1)(0) = 0$$

Caution!

The assumption that the angles in the preceding lemma are measured in radians is critical, as our formula for the area of a circular sector is based on radian measure. In Section 3.4, we will learn how to find derivatives of trigonometric functions when using other angle measurements.

Example 1

Use the lemma to find the following limits.

a. $\displaystyle\lim_{x\to 0}\frac{\sin 3x}{7x}$ **b.** $\displaystyle\lim_{x\to 0}\frac{\cos x - 1}{\tan x}$

Solution

a. In order for us to be able to use the lemma, we first multiply and divide by 3:

$$\lim_{x\to 0}\frac{\sin 3x}{7x} = \lim_{x\to 0}\left(\frac{3}{3}\cdot\frac{\sin 3x}{7x}\right) = \lim_{x\to 0}\left(\frac{3}{7}\cdot\frac{\sin 3x}{3x}\right) = \frac{3}{7}\lim_{3x\to 0}\frac{\sin 3x}{3x},$$

where in the last step we used the fact that when x approaches 0, so does $3x$ and vice versa. Next, if we rename $3x$ by calling it θ, we can rewrite our limit as

$$\frac{3}{7}\lim_{3x\to 0}\frac{\sin 3x}{3x} = \frac{3}{7}\lim_{\theta\to 0}\frac{\sin\theta}{\theta}.$$

Now by a direct application of the lemma,

$$\frac{3}{7}\lim_{\theta\to 0}\frac{\sin\theta}{\theta} = \frac{3}{7}\cdot 1 = \frac{3}{7}.$$

b. This limit does not directly resemble either of those in the lemma, but we can rewrite it as follows.

$$\lim_{x\to 0}\frac{\cos x - 1}{\tan x} = \lim_{x\to 0}\frac{\cos x - 1}{\dfrac{\sin x}{\cos x}} = \lim_{x\to 0}\frac{(\cos x - 1)\cos x}{\sin x} = \lim_{x\to 0}\frac{\dfrac{\cos x - 1}{x}\cdot\cos x}{\dfrac{\sin x}{x}}$$

Applying the limit laws and both parts of the lemma, along with the fact that $\lim_{x\to 0}\cos x = 1$, we now obtain

$$\lim_{x\to 0}\frac{\dfrac{\cos x - 1}{x}\cdot\cos x}{\dfrac{\sin x}{x}} = \frac{0\cdot 1}{1} = 0.$$

Topic Two
Derivatives of Sine and Cosine

The derivatives of sine and cosine can now be found quickly, using one more trigonometric identity and the two limits we just studied.

Theorem

Derivative of Sine

$$\frac{d}{dx}(\sin x) = \cos x$$

— Proof —

$$\frac{d}{dx}(\sin x) = \lim_{h \to 0} \frac{\sin(x+h) - \sin x}{h}$$

$$= \lim_{h \to 0} \frac{\sin x \cos h + \cos x \sin h - \sin x}{h} \qquad \sin(u+v) = \sin u \cos v + \cos u \sin v$$

$$= \lim_{h \to 0} \left(\sin x \cdot \frac{\cos h - 1}{h} + \cos x \cdot \frac{\sin h}{h} \right)$$

$$= (\sin x)\left(\lim_{h \to 0} \frac{\cos h - 1}{h} \right) + (\cos x)\left(\lim_{h \to 0} \frac{\sin h}{h} \right) \qquad \begin{array}{l} \text{Both } \sin x \text{ and } \cos x \text{ do not} \\ \text{depend on } h \text{ and can be pulled} \\ \text{out of the limit.} \end{array}$$

$$= (\sin x)(0) + (\cos x)(1) = \cos x$$

Note that a pleasing pattern is beginning to emerge regarding differentiation within classes of functions: the derivative of any polynomial is another polynomial, the derivative of e^x is e^x, and our first derivative of a trigonometric function turns out to be another trigonometric function. While this pattern cannot be relied upon too heavily, we will see it continue to appear from time to time. In particular, we will see that the derivatives of the six basic trigonometric functions remain in the class of trigonometric functions.

Example 2

If we follow the slopes of tangent lines to the graph of $y = \sin x$ starting, say, at $x = 0$, it is not difficult to discover an oscillating "sinusoidal" pattern.

As shown in Figure 2, when $x = 0$ the corresponding tangent is a line through the origin, which appears to have a slope of 1. As x increases, the slopes of the tangent lines decrease, reaching 0 when $x = \pi/2$. The slopes then become negative as x continues to increase from $\pi/2$. When x reaches the value of π, the slope of the corresponding tangent of the sine function is -1. The slopes then start to increase, reaching the value of 0 again at $x = 3\pi/2$, as a close examination of the sine graph reveals. Notice that this is exactly the behavior of the function values of $y = \cos x$, that is, the slope of the tangent to the sine graph at any given x-value appears to equal the corresponding function value of the cosine function, which is exactly what we proved in the previous theorem.

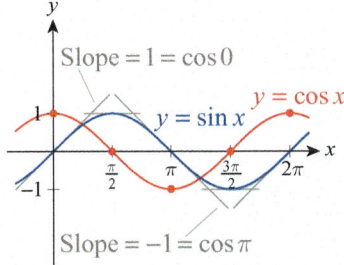

Figure 2

Example 3

Use differentiation rules to find the derivative of $f(x) = (x^2 - 3x)\sin x$.

Solution

The Product Rule along with the previous theorem yields

$$\frac{df}{dx} = \frac{d}{dx}(x^2 - 3x) \cdot \sin x + (x^2 - 3x)\frac{d}{dx}(\sin x)$$
$$= (2x - 3)\sin x + (x^2 - 3x)\cos x.$$

Theorem

Derivative of Cosine

$$\frac{d}{dx}(\cos x) = -\sin x$$

Proof

This can be proven using the methods used to show that the derivative of the sine function is cosine. (See Exercise 54.)

Example 4

Use differentiation rules to find the following derivatives.

a. $\dfrac{d}{dx}(7e^x + 3\cos x)$ **b.** $\dfrac{d}{dx}\left(\dfrac{1 + \sin x}{2 - \cos x}\right)$

Solution

We will need the differentiation rules of the previous section along with the derivatives of sine and cosine, as follows.

a. $\dfrac{d}{dx}(7e^x + 3\cos x) = 7\dfrac{d}{dx}(e^x) + 3\dfrac{d}{dx}(\cos x)$
$$= 7e^x + 3(-\sin x)$$
$$= 7e^x - 3\sin x$$

b. Note that we need to start with the Quotient Rule:

$$\frac{d}{dx}\left(\frac{1 + \sin x}{2 - \cos x}\right) = \frac{(1 + \sin x)'(2 - \cos x) - (1 + \sin x)(2 - \cos x)'}{(2 - \cos x)^2}$$
$$= \frac{(\cos x)(2 - \cos x) - (1 + \sin x)(\sin x)}{(2 - \cos x)^2}$$
$$= \frac{2\cos x - \cos^2 x - \sin x - \sin^2 x}{(2 - \cos x)^2} \qquad \sin^2 x + \cos^2 x = 1$$
$$= \frac{2\cos x - \sin x - 1}{(2 - \cos x)^2}$$

Example 5

If an object is oscillating so that the so-called *restoring force* acting on the object is directly proportional to the negative of its displacement at any time, the resulting motion is called *simple harmonic motion*. This happens, for example, when an object is attached to a spring, pulled or pushed, and then let go. (We are ignoring air resistance, friction, and all other forces here.) Recall from Example 6 of Section 1.2 that the equation of this motion, or the position function of the object, can be either $f(t) = A\sin(\omega t)$ or $g(t) = A\cos(\omega t)$, depending on initial conditions (ω is a constant related to the spring, A is the amplitude).

Suppose an object of mass m is hanging on a spring and is at equilibrium. If we pull it downward by 50 cm and release it, the equation of motion is given by

$$y(t) = -0.5\cos t,$$

where y is the vertical distance from equilibrium, measured in meters with upward displacement considered positive, and t is measured in seconds.

a. Find the position at $t = 10$ seconds.

b. What is the maximum displacement of the object and when does it occur?

c. Find the maximum velocity and the position where it occurs.

d. Find the maximum acceleration value. When does it occur?

Solution

a. The position at $t = 10$ seconds is quickly found by evaluating the position function.

$$y(10) = -0.5\cos 10 \approx 0.4195 \text{ m} = 41.95 \text{ cm}$$

The positive sign means that the object is found approximately 41.95 cm above equilibrium at $t = 10$ seconds.

b. Since displacement is "signed distance" from equilibrium, maximum displacement will occur precisely when the position function assumes its maximum value (positive or negative). Stated more precisely, this happens when the absolute value of the position is maximum. Since the cosine function oscillates between +1 and −1 reaching these maximum absolute values at $k\pi$, $k \in \mathbb{Z}$, we conclude that the absolute value of the position function given in this example is greatest when $t = k\pi$ seconds, for $k \in \mathbb{Z}$, $k \geq 0$. The greatest displacement value is

$$\left|y(k\pi)\right| = \left|-0.5\cos(k\pi)\right| = \left|-0.5\right| \cdot \left|\cos(k\pi)\right| = 0.5(1) = 0.5 \text{ m} = 50 \text{ cm}.$$

Note that this is consistent with our everyday experience. Since there was no initial velocity (no initial "push" or "shove" at the start of motion), we expect the initial displacement of 50 cm to remain maximal throughout the motion. (As we have stated already, we are ignoring all other forces and therefore the fact that the motion actually "dies down" over the long term.) Also, recalling that $\cos(k\pi) = 1$ if k is even and $\cos(k\pi) = -1$ for an odd k, in light of the position function we conclude that the object reaches its maximum displacement of 50 cm *below* equilibrium when $t = k\pi$ and k is a nonnegative even integer. If k is odd, on the other hand, the object is 50 cm *above* equilibrium at $t = k\pi$ seconds.

c. The velocity function is the derivative of the position function.

$$v(t) = y'(t) = -0.5(\cos t)' = -0.5(-\sin t) = 0.5\sin t$$

By an argument similar to the one given in part b., and keeping the graph of the sine function in mind, we see that velocity is greatest when the absolute value of the sine function is 1, that is, when $t = (2k+1)(\pi/2)$, $k \in \mathbb{Z}$. Moreover, since $\sin((2k+1)(\pi/2)) = 1$ for an even k and $\sin((2k+1)(\pi/2)) = -1$ if k is odd, we conclude that velocity is maximal and pointing upward when $t = (2k+1)(\pi/2)$ seconds for a nonnegative even k.

$$v\left((2k+1)\frac{\pi}{2}\right) = 0.5\sin\left((2k+1)\frac{\pi}{2}\right) = 0.5(1) = 0.5 \text{ m/s}$$

On the other hand, if k is odd and positive,

$$v\left((2k+1)\frac{\pi}{2}\right) = 0.5\sin\left((2k+1)\frac{\pi}{2}\right) = 0.5(-1) = -0.5 \text{ m/s};$$

that is, velocity is maximal and is pointing downward in this case. Note that for the above t-values the cosine function is 0, so maximum velocity (or speed) occurs each time the object is passing through the equilibrium position.

d. Since acceleration is the derivative of velocity, we will obtain the acceleration function by differentiating.

$$a(t) = v'(t) = 0.5(\sin t)' = 0.5\cos t$$

Since this is precisely the negative of the position function, we can use our observations made in part b. and conclude that the acceleration values are greatest when the object is in its extreme positions, but acceleration is *positive* when the object's position is negative, that is, the object is accelerating downward when above the equilibrium position and vice versa. The maximum acceleration values are

$$a(k\pi) = 0.5\cos(k\pi) = 0.5(1) = 0.5 \text{ m/s}^2$$

when k is an even nonnegative integer and

$$a(k\pi) = 0.5\cos(k\pi) = 0.5(-1) = -0.5 \text{ m/s}^2$$

when k is odd.

Topic Three
Derivatives of Other Trigonometric Functions

Knowing the derivatives of sine and cosine, we can find the derivatives of the remaining trigonometric functions without having to resort to limits of difference quotients.

Theorem

Derivatives of Tangent, Cotangent, Secant, and Cosecant

$$\frac{d}{dx}(\tan x) = \sec^2 x \qquad\qquad \frac{d}{dx}(\cot x) = -\csc^2 x$$

$$\frac{d}{dx}(\sec x) = \sec x \tan x \qquad\qquad \frac{d}{dx}(\csc x) = -\csc x \cot x$$

Proof

We will prove two of the above statements, and leave the remaining two as exercises (see Exercise 59).

$$\frac{d}{dx}(\tan x) = \frac{d}{dx}\left(\frac{\sin x}{\cos x}\right)$$

$$= \frac{(\sin x)' \cos x - \sin x (\cos x)'}{\cos^2 x} \qquad \text{\textcolor{blue}{Quotient Rule}}$$

$$= \frac{\cos x \cos x - \sin x(-\sin x)}{\cos^2 x}$$

$$= \frac{\cos^2 x + \sin^2 x}{\cos^2 x}$$

$$= \frac{1}{\cos^2 x} = \sec^2 x \qquad \text{\textcolor{blue}{$\sin^2 x + \cos^2 x = 1$}}$$

For the derivative of cosecant, we could either use the Quotient Rule or, more quickly, its specialized form as shown.

$$\frac{d}{dx}(\csc x) = \frac{d}{dx}\left(\frac{1}{\sin x}\right)$$

$$= \frac{-(\sin x)'}{\sin^2 x} \qquad \text{\textcolor{blue}{Reciprocal Rule}}$$

$$= -\frac{\cos x}{\sin^2 x}$$

$$= -\frac{1}{\sin x} \cdot \frac{\cos x}{\sin x} = -\csc x \cot x$$

Example 6

Use the above theorem to find the derivatives of the following functions.

a. $f(x) = \tan x \csc x$

b. $g(x) = \dfrac{1 - \sin x}{\cot x}$

Solution

a. Using the Product Rule, we calculate the derivative as follows.

$$f'(x) = (\tan x)' \csc x + \tan x (\csc x)'$$

$$= \sec^2 x \csc x + \tan x (-\csc x \cot x)$$

$$= \frac{1}{\cos^2 x} \cdot \frac{1}{\sin x} - \frac{\sin x}{\cos x} \cdot \frac{1}{\sin x} \cdot \frac{\cos x}{\sin x}$$

$$= \frac{1}{\cos^2 x} \cdot \frac{1}{\sin x} - \frac{1}{\sin x}$$

$$= \sec^2 x \csc x - \csc x = \left(\sec^2 x - 1\right)\csc x$$

Notice, however, that differentiation will be much easier if we rewrite f before differentiating. Since

$$f(x) = \tan x \csc x = \frac{\sin x}{\cos x} \cdot \frac{1}{\sin x} = \frac{1}{\cos x} = \sec x,$$

our theorem tell us that

$$f'(x) = (\sec x)' = \sec x \tan x.$$

However, is this answer equivalent to the one we obtained above? The answer, reassuringly, is yes, which we can show by using the identity $\tan^2 x + 1 = \sec^2 x$.

$$\left(\sec^2 x - 1\right)\csc x = \tan^2 x \csc x = \tan x \tan x \csc x$$

$$= \tan x \cdot \frac{\sin x}{\cos x} \cdot \frac{1}{\sin x} = \tan x \cdot \frac{1}{\cos x} = \tan x \sec x$$

We have just witnessed an illustration of the fact that when working with trigonometric expressions, sometimes very different-looking answers may be equivalent.

b. Using the Quotient Rule this time, we obtain the following.

$$g'(x) = \frac{(1-\sin x)' \cot x - (1-\sin x)(\cot x)'}{\cot^2 x}$$

$$= \frac{-\cos x \cot x - (1-\sin x)(-\csc^2 x)}{\cot^2 x}$$

$$= \frac{-\cos x \cot x + \csc^2 x - \csc^2 x \sin x}{\cot^2 x}$$

$$= \frac{-\cos x \cdot \dfrac{\cos x}{\sin x} + \dfrac{1}{\sin^2 x} - \dfrac{1}{\sin^2 x} \sin x}{\dfrac{\cos^2 x}{\sin^2 x}}$$

$$= \frac{-\cos^2 x \sin x + 1 - \sin x}{\cos^2 x} \qquad \text{Multiply both numerator and denominator by } \sin^2 x.$$

$$= \frac{1 - \sin x (\cos^2 x + 1)}{\cos^2 x}$$

Example 7

Determine the second derivative of $f(x) = \cot x$.

Solution

Since $\dfrac{d}{dx}(\cot x) = -\csc^2 x$, we will obtain the second derivative of f by differentiating $f'(x) = -\csc^2 x$.

$$f''(x) = \frac{d^2}{dx^2}(\cot x) = \frac{d}{dx}(-\csc^2 x)$$

We can differentiate $-\csc^2 x$ by using the Generalized Positive Integer Power Rule (see Exercise 112 of Section 3.2) or simply by an application of the Product Rule.

$$\frac{d}{dx}(-\csc^2 x) = -\frac{d}{dx}\left[(\csc x)(\csc x)\right]$$

$$= -(\csc x)' \csc x - \csc x (\csc x)'$$

$$= -2\csc x (\csc x)'$$

$$= -2\csc x (-\csc x \cot x)$$

$$= 2\csc^2 x \cot x$$

$$= 2 \cdot \frac{1}{\sin^2 x} \cdot \frac{\cos x}{\sin x} = \frac{2\cos x}{\sin^3 x}$$

Example 8

Find all points on the graph of $g(t) = 3\sin 2t$ where the slope of its tangent line is -3.

Solution

Our plan is to find $g'(t)$ first, and then by solving the equation $g'(t) = -3$, identify all values $\bar{t}$ such that $g'(\bar{t}) = -3$.

In order to differentiate g, we will use the identity $\sin 2t = 2\sin t \cos t$ followed by the Product Rule.

$$
\begin{aligned}
g'(t) &= \frac{d}{dt}(3\sin 2t) \\
&= 3\frac{d}{dt}(\sin 2t) \\
&= 3\frac{d}{dt}(2\sin t \cos t) \\
&= 6\frac{d}{dt}(\sin t \cos t) \\
&= 6\big[\cos t \cos t + \sin t(-\sin t)\big] \\
&= 6\big(\cos^2 t - \sin^2 t\big) \\
&= 6\cos 2t \qquad\qquad \cos 2t = \cos^2 t - \sin^2 t
\end{aligned}
$$

Next, we solve the trigonometric equation $6\cos 2t = -3$.

$$
6\cos 2t = -3
$$

$$
\cos 2t = -\frac{1}{2}
$$

$$
2t = \frac{2\pi}{3} + 2k\pi \quad \text{or} \quad 2t = \frac{4\pi}{3} + 2k\pi, \; k \in \mathbb{Z}
$$

Equivalently, the solutions are

$$
\bar{t} = \frac{\pi}{3} + k\pi \;\text{ or }\; \bar{t} = \frac{2\pi}{3} + k\pi, \; k \in \mathbb{Z}.
$$

The corresponding points on the graph of g where the tangents have a slope of -3 are as follows.

$$
\left\{\left(\frac{\pi}{3} + k\pi, \frac{3\sqrt{3}}{2}\right), \left(\frac{2\pi}{3} + k\pi, -\frac{3\sqrt{3}}{2}\right)\right\}, \; k \in \mathbb{Z}
$$

3.3 **Exercises**

1–12 Use the results of this section to find the indicated limit.

1. $\displaystyle\lim_{x\to 0}\frac{\sin 3x}{2x}$

2. $\displaystyle\lim_{x\to 0}\frac{-\sin\frac{x}{2}}{5x}$

3. $\displaystyle\lim_{x\to 0}\frac{\sin \pi x}{x}$

4. $\displaystyle\lim_{x\to 0}\frac{\tan 4x}{5x}$

5. $\displaystyle\lim_{x\to 0}\frac{\cos 5x-1}{2x}$

6. $\displaystyle\lim_{x\to 0}\frac{\cos x-1}{\sin x}$

7. $\displaystyle\lim_{x\to 0}\frac{x+\tan x}{\sin x}$

8. $\displaystyle\lim_{\beta\to 0}\frac{\csc\beta-\cot\beta}{\beta\csc\beta}$

9. $\displaystyle\lim_{x\to 0}\frac{\sin(\sin x)}{\sin x}$

10. $\displaystyle\lim_{\alpha\to 0}\frac{\tan(\alpha^2)}{\alpha}$

11. $\displaystyle\lim_{t\to 0}\frac{2t+3\tan t}{\sin t}$

12. $\displaystyle\lim_{\theta\to 0}\frac{\cos\theta\sin\theta-\sin\theta}{\theta^2}$

13–30 Differentiate the given function.

13. $f(x)=2\sin x-5\cos x$ **14.** $g(x)=3x^2+2\tan x$

15. $h(x)=x\cos x$ **16.** $F(x)=2.5x(1-\cot x)$

17. $G(x)=2\sqrt{x}\sec x$ **18.** $k(x)=\pi x\sin x+\pi x$

19. $L(x)=-3e^x(\csc x+\cot x)$

20. $f(x)=2\cos 2x-2\cos x$

21. $g(x)=\cot^2 x$ **22.** $h(x)=\dfrac{\tan x}{x}$

23. $F(t)=\dfrac{1-\cos t}{t^2}$ **24.** $W(x)=\dfrac{1+\cos x}{1+\sin x}$

25. $R(z)=\dfrac{e^z+\sin z}{z}$ **26.** $N(w)=\dfrac{2\sqrt{w}-\sec w}{\sqrt{w}}$

27. $B(x)=\dfrac{\frac{1}{\sin x}-\sin x}{\cos x}$ **28.** $G(y)=y\cot y\csc y$

29. $T(s)=s^2 e^s\cot s$ **30.** $r(t)=\dfrac{1}{t\sin t\cos t}$

31–36 Find all points where the function has a horizontal tangent line.

31. $f(x)=\dfrac{1}{2}x+\sin x$ **32.** $g(x)=x+\sin 2x$

33. $h(x)=\sec^2 x$ **34.** $T(s)=\tan s-s$

35. $K(u)=\tan u+\cot u$ **36.** $F(t)=\dfrac{1-\sin t}{1-\cos t}$

37–40 Find all x-values where the tangent line to the graph of the function is parallel to the given line.

37. $f(x)=\sin x+\dfrac{3}{2};\quad y-x=\dfrac{3}{2}$

38. $g(x)=\cot x;\quad y+2x=\pi$

39. $G(x)=\dfrac{x}{3}-\tan x;\quad x+y=5$

40. $F(x)=\sin x\cos x;\quad 2x+2y=7$

41–44 Find the equation of the tangent line to the graph of the given function at the indicated point.

41. $f(x)=2x\cos x;\quad (0,0)$

42. $g(x)=\tan x-\sec x;\quad (0,-1)$

43. $h(x)=2\csc x-\sin x;\quad \left(\dfrac{\pi}{2},1\right)$

44. $k(x)=\dfrac{\cot x}{x};\quad \left(\dfrac{\pi}{4},\dfrac{4}{\pi}\right)$

45. Let us assume that for some function f,
$f(0)=1$ and $f'(0)=2$. Let $F(x)=f(x)\tan x$,
$G(x)=f(x)/\cos x$, and $H(x)=f(x)\sin x\cos x$.
Find $F'(0)$, $G'(0)$, and $H'(0)$.

46–48 Verify the trigonometric identity by differentiating both sides of the equation. (**Hint:** If $f'(x)=g'(x)$, it doesn't necessarily follow that $f(x)=g(x)$. In general, we can only conclude that $f(x)=g(x)+c$ for some constant c.)

46. $\tan x\cot x=1$

47. $(1-\cos\theta)(1+\cos\theta)=\sin^2\theta$

48. $\dfrac{1}{1-\cos x}+\dfrac{1}{1+\cos x}=2\csc^2 x$

49–52 Find $f'(x)$, $f''(x)$, and $f'''(x)$. Observing a pattern, find a formula for $f^{(n)}(x)$.

49. $f(x)=\sin x$ **50.** $f(x)=\cos x$

51. $f(x)=e^x\sin x$ **52.** $f(x)=e^x\cos x$

53. Provide a second proof of the limit statement $\lim\limits_{\theta\to 0}(\cos\theta - 1)/\theta = 0$ by multiplying both the numerator and denominator by $\cos\theta + 1$ to obtain

$$\lim_{\theta\to 0}\frac{\cos\theta - 1}{\theta} = \lim_{\theta\to 0}\frac{\cos^2\theta - 1}{\theta(\cos\theta + 1)}.$$

Then by the Pythagorean identity $\sin^2 x + \cos^2 x = 1$, you obtain

$$\lim_{\theta\to 0}\frac{\cos^2\theta - 1}{\theta(\cos\theta + 1)} = \lim_{\theta\to 0}\frac{-\sin^2\theta}{\theta(\cos\theta + 1)} = \lim_{\theta\to 0}\frac{\sin\theta(-\sin\theta)}{\theta(\cos\theta + 1)} = \lim_{\theta\to 0}\frac{\sin\theta}{\theta}\lim_{\theta\to 0}\frac{-\sin\theta}{\cos\theta + 1}.$$

Conclude the argument by using the first limit statement in the lemma at the beginning of this section.

54. Prove that $\dfrac{d}{dx}(\cos x) = -\sin x$ by mimicking the proof of the theorem "Derivative of Sine." (**Hint:** Recall the angle sum identity you will need $\cos(u + v) = \cos u \cos v - \sin u \sin v$.)

55. Provide an alternative proof of the fact that $\dfrac{d}{dx}(\sin x) = \cos x$ by using the identity
$$\sin x - \sin c = 2\sin\frac{x - c}{2}\cos\frac{x + c}{2}.$$
(**Hint:** Rewrite the difference quotient $\dfrac{\sin x - \sin c}{x - c}$ as $\dfrac{2\sin\dfrac{x - c}{2}\cos\dfrac{x + c}{2}}{x - c}$. Let $c \to x$, and use the lemma from the beginning of this section.)

56. Use the definition of the derivative and the lemma from the beginning of this section to show that $(\sin 3x)' = 3\cos 3x$. Generalize to obtain that if $k \in \mathbb{R}$, $(\sin(kx))' = k\cos(kx)$.

57. Repeat Exercise 56 with $f(x) = \cos(kx)$.

58. Find a constant a such that the graphs of $f(x) = a\sin x$ and $g(x) = a\cos x$ intersect at right angles, that is, their respective tangent lines are perpendicular at their point(s) of intersection.

59. Prove the remaining two cases of the theorem "Derivatives of Tangent, Cotangent, Secant, and Cosecant," namely, the statements
$$\frac{d}{dx}(\cot x) = -\csc^2 x \text{ and } \frac{d}{dx}(\sec x) = \sec x \tan x.$$
(**Hint:** Mimic the proof presented in the text, using the derivatives of sine and cosine along with appropriate differentiation rules.)

60. The cross-section of an ice cream cone is an isosceles triangle, with the angular opening at the bottom being $2t$ (radians). Assuming that the ice cream sits on top of the cone in the shape of a perfect hemisphere, let V_i = volume of the ice cream, V_c = volume of the cone. Express both of these volumes in terms of t, and then compute $\lim\limits_{t\to 0^+}\dfrac{V_i}{V_c}$.

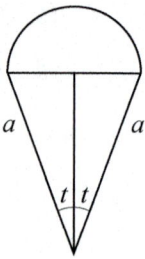

61. An object is tied to the top of an inclined surface of variable angle of elevation so that the rope is parallel to the surface. The tension in the rope is given by $F = mg(\sin x - \mu\cos x)$, where m is the mass of the object, g is the gravity constant, and μ is the coefficient of friction (assume all units are metric units).

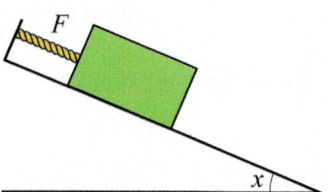

    a. What is the rate of change of F with respect to x?

    b. For what x-value (if any) is this rate of change equal to 0?

62. Suppose an object oscillating in fluid obeys the position function $y = 10e^{-0.2t}\cos(2\pi t)$, where y is the distance from equilibrium, measured in centimeters with upward displacement considered positive, and t is measured in seconds. Such motion is called *damped harmonic motion*. Can you see why?

   a. Find the position, velocity, and acceleration at $t = 3.5$ seconds.

   b. What is the maximum displacement of the object and when does it occur?

   (**Hint:** Use the definition of the derivative to find the derivatives of $e^{-0.2t}$ and $\cos(2\pi t)$. You may also want to review Exercise 57 for the latter.)

63. A 15 ft ladder is leaning against a wall, making an angle of β with the horizontal, when it starts sliding. If x denotes the distance of the bottom of the ladder from the wall, find the rate of change of x with respect to β when $\beta = \pi/6$ (or 30°). Interpret the result.

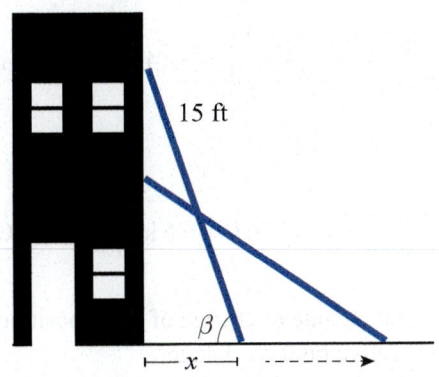

15 ft

β

x

64. A man is pulling his child on a sled at a constant rate, via a rope that makes an angle of α with the horizontal. Since there is no acceleration, the pulling force satisfies the equation $F\cos\alpha = \mu(mg - F\sin\alpha)$, where μ is the coefficient of friction, m is the total mass of the sled and child, and g is the gravity constant.

   a. Express F as a function of α.

   b. Find the rate of change of F with respect to α.

   c. What is the above rate when $\alpha = 60°$?

   d. When (if ever) is this rate of change 0?

3.3 Technology Exercises

65–70 Use a computer algebra system to find the derivative of $f(x)$. Then graph f along with its derivative on the same screen. By zooming in, if necessary, find at least two x-values where the graph of f has a horizontal tangent line. What can you say about f' at such points? (Answers will vary.)

65. $f(x) = \dfrac{x}{1 + \cos x}$

66. $f(x) = \dfrac{1 - \sec x}{1 + \sec x}$

67. $f(x) = \dfrac{\csc x}{x}$

68. $f(x) = \dfrac{\sin x}{\cos x + \tan x}$

69. $f(x) = \cos x(\cot x + \tan x)$

70. $f(x) = \dfrac{\cot x}{\sec x + x\cos x}$

71. Find the maximum velocity and acceleration values in Exercise 62 by using a graphing calculator or computer algebra system to graph the velocity and acceleration functions of the oscillating object.

3.4 **The Chain Rule**

In this section, we introduce a powerful rule of differentiation that dramatically expands the sorts of functions we can differentiate.

Topic One

Proof of the Chain Rule

The Chain Rule gives us a formula for differentiating a composition of two functions, and its nature is pleasantly consistent with our intuition. As an illustration of its meaning, consider the following two questions.

> 1. *Christi typically walks at a pace of 4 mph and rides her bike, on average, at a rate 3 times as fast as she walks. What is her typical biking speed in miles per hour?*
>
> 2. *Tom is climbing a mountain at a rate of 0.5 vertical kilometers per hour, and knows from experience to expect a drop in temperature of 7 °C for every kilometer gained in altitude. What rate of temperature drop is Tom experiencing per hour?*

The answers to both questions involve a product of rates or ratios. In the first question,

$$\frac{\text{miles biked}}{\text{hour}} = \left(\frac{\text{miles biked}}{\text{miles walked}} \right) \left(\frac{\text{miles walked}}{\text{hour}} \right) = (3)(4 \text{ mph}) = 12 \text{ mph}$$

and in the second question,

$$\frac{\text{temp. drop}}{\text{hour}} = \left(\frac{\text{temp. drop}}{\text{vert. km}} \right) \left(\frac{\text{vert. km}}{\text{hour}} \right) = (7 \text{ °C/h})(0.5 \text{ km/h}) = 3.5 \text{ °C/h}.$$

Informally, the Chain Rule tells us that the rate of change of a composition of two functions is the product of their respective rates of change.

Theorem

The Chain Rule

If g is a function differentiable at the point c, and if f is a function differentiable at $g(c)$, then $f \circ g$ is differentiable at c and

$$(f \circ g)'(c) = f'(g(c)) \cdot g'(c).$$

In Leibniz notation, if we let $y = f(u)$ and $u = g(x)$, then

$$\left. \frac{dy}{dx} \right|_{x=c} = \left(\left. \frac{dy}{du} \right|_{u=g(c)} \right) \left(\left. \frac{du}{dx} \right|_{x=c} \right).$$

Before embarking on a rigorous proof of the Chain Rule, it may be instructive to attempt a proof motivated by the example problems above. If, as in the statement of the theorem, we let $y = f(u)$ and $u = g(x)$, then it is tempting to write

$$\frac{dy}{dx} = \lim_{\Delta x \to 0} \frac{\Delta y}{\Delta x}$$

$$= \lim_{\Delta x \to 0} \frac{\Delta y}{\Delta u} \cdot \frac{\Delta u}{\Delta x}$$

$$= \lim_{\Delta x \to 0} \frac{\Delta y}{\Delta u} \cdot \lim_{\Delta x \to 0} \frac{\Delta u}{\Delta x}$$

$$= \lim_{\Delta u \to 0} \frac{\Delta y}{\Delta u} \cdot \lim_{\Delta x \to 0} \frac{\Delta u}{\Delta x} \qquad \text{\scriptsize $\Delta u \to 0$ as $\Delta x \to 0$ since g is continuous.}$$

$$= \frac{dy}{du} \cdot \frac{du}{dx}.$$

Indeed, this argument is perfectly adequate for the two simple problems with which we began, as all the rates under discussion were nonzero constants. The difficulty lies in the fact that, in general, we do not know that $\Delta u \neq 0$ as $\Delta x \to 0$, and we cannot divide by 0. To make the argument rigorous, we need to introduce some techniques that we will explore further in Section 3.9 when we study the linear approximation of functions.

— Proof

Our overall goal is to evaluate

$$\lim_{h \to 0} \frac{f\big(g(c+h)\big) - f\big(g(c)\big)}{h},$$

and to do so we have to find a convenient way to rewrite the term $f\big(g(c+h)\big)$.

Given the function g and the fixed point c, define a new function

$$v(h) = \frac{g(c+h) - g(c)}{h} - g'(c).$$

Note that the differentiability of g at c means $\lim_{h \to 0} v(h) = 0$. Similarly, let $\tilde{c} = g(c)$ and define a function

$$w(k) = \frac{f(\tilde{c}+k) - f(\tilde{c})}{k} - f'(\tilde{c}),$$

and note that $\lim_{k \to 0} w(k) = 0$. We can now rearrange these two equations to obtain

$$g(c+h) = g(c) + \big[g'(c) + v(h)\big]h \qquad (1)$$

and

$$f(\tilde{c}+k) = f(\tilde{c}) + \big[f'(\tilde{c}) + w(k)\big]k. \qquad (2)$$

(While equations (1) and (2) may be somewhat intimidating in appearance, don't be deterred—just keep in mind that both $v(h)$ and $w(k)$ go to 0 as their arguments go to 0.)

We can now rewrite $f\big(g(c+h)\big)$ as follows.

$$f\big(g(c+h)\big) = f\big(g(c)+\big[g'(c)+v(h)\big]h\big) \qquad \text{By (1)}$$
$$= f\big(\tilde{c}+\big[g'(c)+v(h)\big]h\big)$$

Denoting $\big[g'(c)+v(h)\big]h$ by k, we can use (2) to continue:

$$f\big(g(c+h)\big) = f\big(\tilde{c}+\big[g'(c)+v(h)\big]h\big)$$
$$= f(\tilde{c})+\big[f'(\tilde{c})+w(k)\big]k \qquad \text{By (2)}$$
$$= f\big(g(c)\big)+\big[f'(\tilde{c})+w(k)\big]k$$

We now subtract $f\big(g(c)\big)$ from both sides to obtain

$$f\big(g(c+h)\big)-f\big(g(c)\big) = \big[f'(\tilde{c})+w(k)\big]k$$
$$= \big[f'(\tilde{c})+w(k)\big]\big[g'(c)+v(h)\big]h. \qquad k = \big[g'(c)+v(h)\big]h$$

If we now divide our last equation by h, we have

$$\frac{f\big(g(c+h)\big)-f\big(g(c)\big)}{h} = \big[f'(\tilde{c})+w(k)\big]\big[g'(c)+v(h)\big]$$
$$= \big[f'\big(g(c)\big)+w(k)\big]\big[g'(c)+v(h)\big].$$

Note that our definition of k implies $k \to 0$ as $h \to 0$, so taking the limit gives us the desired result.

$$\lim_{h\to 0}\frac{f\big(g(c+h)\big)-f\big(g(c)\big)}{h} = \lim_{h\to 0}\big[\big(f'\big(g(c)\big)+w(k)\big)\big(g'(c)+v(h)\big)\big]$$
$$= \lim_{k\to 0}\big[f'\big(g(c)\big)+w(k)\big]\cdot\lim_{h\to 0}\big[g'(c)+v(h)\big] \qquad k \to 0 \text{ as } h \to$$
$$= f'\big(g(c)\big)\cdot g'(c)$$

Example 1

Differentiate the following functions.

a. $F(x)=\big(x^5+3\big)^2$ 　　　　　　b. $G(x)=\big(x^5+3\big)^{1000}$

c. $H(x)=\sqrt{7x^3-4x+5}$ 　　　　d. $K(x)=e^{\cos x}$

e. $T(x)=\tan\left(\dfrac{3x+1}{2x-5}\right)$

Solution

a. We recognize F as a composite function; with the notation $g(x)=x^5+3$, and $f(x)=x^2$, F can be written as $F=f\circ g$. Since both f and g are certainly differentiable everywhere, the Chain Rule applies, and we obtain the following.

$$F'(x)=(f\circ g)'(x)=f'\big(g(x)\big)\cdot g'(x)=2\big(x^5+3\big)\cdot\big(5x^4+0\big)=10x^4\big(x^5+3\big)$$

b. While one could argue that in part a. we could simply have expanded $F(x)$ and differentiated the resulting polynomial, therefore rendering the use of the Chain Rule unnecessary, this is clearly not the case in part b. While an expansion is theoretically possible, it is certainly not feasible. In fact, $G(x)$ is our first illustration of how much we have gained with the Chain Rule; our task turns out to be just as easy as it was in part a.

$$G'(x) = 1000(x^5 + 3)^{999} \cdot (5x^4) = 5000x^4(x^5 + 3)^{999}$$

c. Here once again the Chain Rule is the only practical way to proceed. We first identify the outer function to be $f(x) = \sqrt{x}$, while the inner function is $g(x) = 7x^3 - 4x + 5$. The Chain Rule then implies

$$H'(x) = (f \circ g)'(x) = f'(g(x)) \cdot g'(x)$$

$$= \frac{1}{2\sqrt{7x^3 - 4x + 5}} \cdot (21x^2 - 4) = \frac{10.5x^2 - 2}{\sqrt{7x^3 - 4x + 5}}.$$

d. We will illustrate Leibniz notation when differentiating $K(x)$. Since the inner function is $u = \cos x$, while the outer function is $y = f(u) = e^u$, we obtain the following.

$$\frac{dy}{dx} = \frac{dy}{du} \cdot \frac{du}{dx} = e^u \cdot (-\sin x) = -e^{\cos x} \sin x \qquad \begin{array}{l} (e^u)' = e^u \\ (\cos x)' = -\sin x \end{array}$$

e. To differentiate $T(x)$, we will need to apply the Quotient Rule in combination with the Chain Rule, since the inner function is a rational function. Notice, however, that the order is important. We start with the Chain Rule, since T is composite, and then use the Quotient Rule to differentiate the inner function.

$$T'(x) = \sec^2\left(\frac{3x+1}{2x-5}\right) \cdot \left(\frac{3x+1}{2x-5}\right)'$$

$$= \sec^2\left(\frac{3x+1}{2x-5}\right) \cdot \frac{(3)(2x-5) - (3x+1)(2)}{(2x-5)^2}$$

$$= -\frac{17}{(2x-5)^2} \sec^2\left(\frac{3x+1}{2x-5}\right)$$

Example 2

Find the first and second derivatives of $f(\alpha) = \tan \alpha$, where α is measured in degrees.

Solution

We know that $f'(x) = \sec^2 x$ if x is measured in radians. In light of the conversion formula between degrees and radians, $x = x(\alpha) = (\pi/180)\alpha$, you can actually think of f as a composite function.

$$f(\alpha) = f(x(\alpha)) = \tan(x(\alpha))$$

Our first application of the Chain Rule yields the first derivative of f.

$$\frac{df}{d\alpha} = \frac{df}{dx} \cdot \frac{dx}{d\alpha} = \sec^2(x(\alpha)) \cdot \frac{\pi}{180} = \frac{\pi}{180} \sec^2\left(\frac{\pi}{180}\alpha\right)$$

In order to obtain the second derivative of f, we will differentiate with respect to α again, using the Power Rule in combination with the Chain Rule.

$$\frac{d^2 f}{d\alpha^2} = \frac{d}{d\alpha}\left[\frac{\pi}{180} \sec^2(x(\alpha))\right]$$
$$= \frac{\pi}{180} \frac{d}{dx}(\sec^2 x) \cdot \frac{dx}{d\alpha}$$
$$= \frac{\pi}{180}(2\sec x)(\sec x \tan x) \cdot \frac{\pi}{180}$$
$$= \frac{2\pi^2}{180^2} \sec^2\left(\frac{\pi}{180}\alpha\right) \tan\left(\frac{\pi}{180}\alpha\right)$$

Example 3

Find the derivatives of the following functions by repeatedly applying the Chain Rule.

a. $F(x) = \sin\sqrt{x^2 + 1}$ **b.** $G(x) = \left(1 + \left(2 + 3x^4\right)^5\right)^6$

Solution

a. What sets F apart from the functions previously considered is the fact that it is a composite function with an inner function that is itself composite. Using the usual function notation,

$$F(x) = (f \circ g \circ h)(x) = f(g(h(x))),$$

where

$$f(x) = \sin x, \; g(x) = \sqrt{x}, \text{ and } h(x) = x^2 + 1.$$

Keeping this in mind, when we come to the point in our first application of the Chain Rule where the inner function is being differentiated, we will need to use the Chain Rule a second time. We work through the following steps.

$$F'(x) = f'\big(g(h(x))\big)\cdot\big[g(h(x))\big]'$$

$$= \left(\cos\sqrt{x^2+1}\right)\left(\sqrt{x^2+1}\right)'$$

$$= \left(\cos\sqrt{x^2+1}\right)\frac{1}{2\sqrt{x^2+1}}\left(x^2+1\right)'$$

$$= \left(\cos\sqrt{x^2+1}\right)\frac{1}{2\sqrt{x^2+1}}(2x)$$

$$= \frac{x}{\sqrt{x^2+1}}\cos\sqrt{x^2+1}$$

b. $G(x)$ is another example of a "compound-composite" function. Using the notation of part a., if we write $G(x)$ as $G(x)=(f\circ g\circ h)(x)$, a close examination of the formula reveals that

$$f(x) = x^6, \quad g(x) = 1+x^5, \quad \text{and} \quad h(x) = 2+3x^4.$$

Proceeding like we did in part a., we obtain the following.

$$G'(x) = (6)\left(1+\left(2+3x^4\right)^5\right)^5\left(1+\left(2+3x^4\right)^5\right)'$$

$$= (6)\left(1+\left(2+3x^4\right)^5\right)^5\left(0+5\left(2+3x^4\right)^4\left(2+3x^4\right)'\right)$$

$$= (6)\left(1+\left(2+3x^4\right)^5\right)^5(5)\left(2+3x^4\right)^4\left(2+3x^4\right)'$$

$$= (30)\left(1+\left(2+3x^4\right)^5\right)^5\left(2+3x^4\right)^4\left(12x^3\right)$$

$$= 360x^3\left(2+3x^4\right)^4\left(1+\left(2+3x^4\right)^5\right)^5$$

Topic Two
Consequences of the Chain Rule

The Chain Rule, in combination with the Power Rule, tells us that if u is a function of the variable x, then

$$\frac{d}{dx}\left(u^n\right) = nu^{n-1}\frac{du}{dx}.$$

This is worth noting because the need to differentiate functions of the form u^n arises often, and the fact that it is true for all real numbers n makes the observation all the more useful. Recall, though, that to this point we have actually only proved the Power Rule for integer exponents. The Chain Rule allows us to now extend the proof to all rational number exponents.

─── **Proof** ─────────────────────────────────────

The Power Rule for Rational Exponents

We will first prove that for every nonzero integer n, $\left(x^{1/n}\right)' = (1/n)x^{(1/n)-1}$.

Given such an n, let $g(x) = x^{1/n}$ and $f(x) = x^n$. Note then that $(f \circ g)(x) = f(g(x)) = \left(x^{1/n}\right)^n = x$, so we already know that $(f \circ g)'(x) = 1$. But by the Chain Rule,

$$(f \circ g)'(x) = f'(g(x))g'(x)$$
$$= n(g(x))^{n-1} g'(x) \qquad\qquad f'(x) = nx^{n-1}$$
$$= n\left(x^{(n-1)/n}\right)g'(x). \qquad\qquad g(x) = x^{1/n}$$

So it must be the case that $n\left(x^{(n-1)/n}\right)g'(x) = 1$. Solving this equation for $g'(x)$, we obtain the desired result.

$$g'(x) = \frac{1}{n\left(x^{(n-1)/n}\right)} = \left(\frac{1}{n}\right)\left(\frac{1}{x^{1-(1/n)}}\right) = \frac{1}{n}x^{(1/n)-1}$$

Now let m be an integer as well; we will use the above result to prove the Power Rule for the rational exponent m/n.

$$\left(x^{m/n}\right)' = \left[\left(x^{1/n}\right)^m\right]'$$
$$= m\left(x^{1/n}\right)^{m-1}\left(x^{1/n}\right)' \qquad\qquad \text{Chain Rule and Power Rule for exponent } m$$
$$= m\left(x^{(m-1)/n}\right)\left(\frac{1}{n}x^{(1/n)-1}\right) \qquad\qquad \text{Power Rule for exponent } 1/n$$
$$= \frac{m}{n}x^{(m/n)-(1/n)+(1/n)-1}$$
$$= \frac{m}{n}x^{(m/n)-1}$$

───

Example 4

Combine the Chain Rule with the Power Rule to differentiate the following functions.

a. $F(x) = \left(3\sin x - x^5\right)^{3/2}$ **b.** $G(x) = \dfrac{1}{\sqrt[3]{3x^2 + 2x - 5}}$

Solution

a. We identify $f(x) = x^{3/2}$ as the outer function, while $g(x) = 3\sin x - x^5$ is the inner function. Now the Chain Rule can be applied. We also need to use the Power Rule to differentiate f:

$$F'(x) = \frac{3}{2}\left(3\sin x - x^5\right)^{(3/2)-1}\left(3\sin x - x^5\right)' = \frac{3}{2}\sqrt{3\sin x - x^5}\left(3\cos x - 5x^4\right)$$

b. Notice that we can rewrite G as

$$G(x) = \left(3x^2 + 2x - 5\right)^{-1/3}.$$

Now we proceed in a manner similar to part a.

$$G'(x) = -\frac{1}{3}\left(3x^2 + 2x - 5\right)^{(-1/3)-1}\left(3x^2 + 2x - 5\right)'$$

$$= -\frac{1}{3}\left(3x^2 + 2x - 5\right)^{-4/3}\left(6x + 2\right)$$

$$= \left(-2x - \frac{2}{3}\right)\left(3x^2 + 2x - 5\right)^{-4/3}$$

$$= -\frac{2x + \dfrac{2}{3}}{\left(\sqrt[3]{3x^2 + 2x - 5}\right)^4}$$

Finally, notice that we could have started out with the Reciprocal Rule and arrived at the same result.

$$G'(x) = \frac{-\left(\sqrt[3]{3x^2 + 2x - 5}\right)'}{\left(\sqrt[3]{3x^2 + 2x - 5}\right)^2}$$

$$= \frac{-\left(\left(3x^2 + 2x - 5\right)^{1/3}\right)'}{\left(\sqrt[3]{3x^2 + 2x - 5}\right)^2}$$

$$= \frac{-\dfrac{1}{3}\left(3x^2 + 2x - 5\right)^{-2/3}\left(6x + 2\right)}{\left(\sqrt[3]{3x^2 + 2x - 5}\right)^2}$$

$$= \frac{-\dfrac{1}{3}\left(6x + 2\right)\left(\sqrt[3]{3x^2 + 2x - 5}\right)^{-2}}{\left(\sqrt[3]{3x^2 + 2x - 5}\right)^2} = -\frac{2x + \dfrac{2}{3}}{\left(\sqrt[3]{3x^2 + 2x - 5}\right)^4}$$

Example 5

Find an equation for the tangent line to the graph of $f(x) = e^{(\sin x + \cos x)}$ at the point $(\pi/2, e)$.

Solution

We shall start by determining $f'(x)$ and evaluating it at $x = \pi/2$. Using the Chain Rule and the fact that e^x is the derivative of itself,

$$f'(x) = e^{(\sin x + \cos x)}\left(\cos x - \sin x\right).$$

Substituting $x = \pi/2$ yields

$$f'\left(\frac{\pi}{2}\right) = e^{(1+0)}\left(0 - 1\right) = -e,$$

which is the slope of the requested tangent. Thus, using the point-slope form of the equation of a line, we obtain

$$y - e = -e\left(x - \frac{\pi}{2}\right),$$

or equivalently,

$$y = -ex + \frac{e(\pi + 2)}{2}.$$

Figure 1 confirms our findings. Note that the function and its tangent line were graphed in a viewing window of $[-1, 4]$ by $[-1, 8]$.

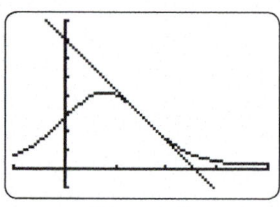

Figure 1

Example 6

Use the Chain Rule to find the derivatives of the following trigonometric functions.

a. $f(x) = \cos 2x$ **b.** $g(x) = \cos^2 x$ **c.** $h(x) = \cos(x^2)$

Solution

All three functions are composite, and therefore we will use the Chain Rule, but careful attention is needed in examining the structure of each function, that is, in identifying the inner and outer components.

a. In the case of f, cosine is the outer function, while $2x$ serves as the inner function. (Using parentheses, we could actually rewrite f as $f(x) = \cos(2x)$). Thus the derivative is determined as follows.

$$f'(x) = (\cos 2x)' = -\sin 2x \cdot (2x)' = -2\sin 2x$$

b. Since $g(x) = \cos^2 x = (\cos x)^2$, we see that the cosine function is the inner function here, and the outer function is $y = x^2$. Thus the Chain Rule in this case yields

$$g'(x) = (\cos^2 x)' = 2\cos x \cdot (\cos x)'$$
$$= 2\cos x \cdot (-\sin x) = -2\cos x \sin x = -\sin 2x.$$

c. Unlike in part b., the "squaring function" $y = x^2$ is the inner function this time, and cosine is the outer. Using the Chain Rule accordingly, we obtain

$$h'(x) = \left[\cos(x^2)\right]' = \left[-\sin(x^2)\right](x^2)' = \left[-\sin(x^2)\right](2x) = -2x\sin(x^2).$$

The Chain Rule, along with the fact that $\left(e^x\right)' = e^x$, tells us that for any function u of x,

$$\frac{d}{dx}\left(e^u\right) = e^u\frac{du}{dx}.$$

We can use this observation to find the derivative of the exponential function a^x for any base $a > 0$. We begin by noting that

$$a^x = \left(e^{\ln a}\right)^x = e^{(\ln a)x},$$

and conclude that

$$\frac{d}{dx}\left(a^x\right) = \frac{d}{dx}e^{(\ln a)x} = e^{(\ln a)x}\frac{d}{dx}\left[(\ln a)x\right]$$
$$= e^{(\ln a)x}(\ln a) = a^x \ln a.$$

We record this result here as our latest differentiation rule. (Note that this rule is consistent with the fact that $\left(e^x\right)' = e^x$, as $\ln e = 1$).

Theorem

Derivatives of a^x and a^u

Given a fixed real number $a > 0$,

$$\frac{d}{dx}\left(a^x\right) = a^x \ln a.$$

More generally, if u is a differentiable function of x, then

$$\frac{d}{dx}\left(a^u\right) = \left(a^u \ln a\right)\frac{du}{dx}.$$

Example 7

Find an equation for the tangent line to the graph of $F(x) = 2^{\tan\left(\sin(x/2)\right)}$ at $(0,1)$. Then find all x-values between -2π and 2π where the tangent line to the graph is horizontal.

Solution

As we have seen in Example 3, F is yet another example of a composite function with a composite inner function, so a repeated application of the Chain Rule will be needed to determine its derivative (be sure you understand each of the following steps).

$$F'(x) = 2^{\tan(\sin(x/2))} (\ln 2)\left(\tan\left(\sin\frac{x}{2} \right)\right)'$$

$$= (\ln 2) 2^{\tan(\sin(x/2))} \sec^2\left(\sin\frac{x}{2} \right)\cdot\left(\sin\frac{x}{2} \right)'$$

$$= (\ln 2) 2^{\tan(\sin(x/2))} \sec^2\left(\sin\frac{x}{2} \right)\cdot\left(\cos\frac{x}{2} \right)\cdot\left(\frac{1}{2} \right)$$

$$= \left(\frac{\ln 2}{2} \right) 2^{\tan(\sin(x/2))} \cos\frac{x}{2}\sec^2\left(\sin\frac{x}{2} \right)$$

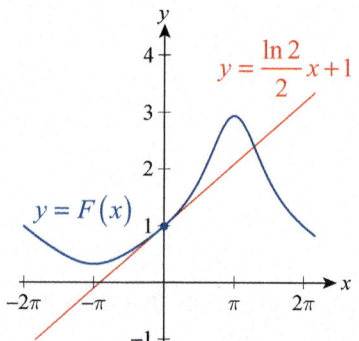

$y = \dfrac{\ln 2}{2}x+1$

$y = F(x)$

Figure 2

At $x = 0$, the derivative is

$$F'(0) = \left(\frac{\ln 2}{2} \right) 2^{\tan(\sin 0)} \cos 0\sec^2\left(\sin 0 \right) = \left(\frac{\ln 2}{2} \right) 2^0 \cos 0\sec^2 0 = \frac{\ln 2}{2},$$

so keeping in mind that the y-intercept of the requested tangent is 1, its equation is

$$y = \frac{\ln 2}{2}x+1.$$

The function and its tangent line are graphed in Figure 2.

Next, we will determine the x-values between -2π and 2π at which $F'(x) = 0$. In other words, we need to solve the equation

$$\left(\frac{\ln 2}{2} \right) 2^{\tan(\sin(x/2))} \cos\frac{x}{2}\sec^2\left(\sin\frac{x}{2} \right) = 0.$$

Since the exponential factor is never 0, after dividing and converting the secant to cosine the equation reduces to

$$\frac{\cos\dfrac{x}{2}}{\cos^2\left(\sin\dfrac{x}{2} \right)} = 0.$$

This latter equation is satisfied if and only if

$$\cos\frac{x}{2} = 0,$$

which is true within the specified interval when $x = \pi$ or $x = -\pi$.

The function and the two horizontal tangent lines are shown in Figure 3.

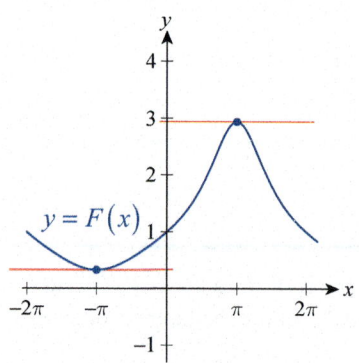

$y = F(x)$

Figure 3

3.4 **Exercises**

1–9 Identify $f(x)$ and $u = g(x)$ such that $F(x) = f(u) = f(g(x))$. Also find $h(x)$ wherever $F(x) = f(g(h(x)))$. (Answers will vary.)

1. $F(x) = (3x - 2.5)^6$

2. $F(x) = 2(x^3 - 5x^2 + \pi)^{-4}$

3. $F(x) = 2\sqrt[3]{x^2 - 9}$

4. $F(x) = \dfrac{-3}{5 + \sqrt{x^3 + x}}$

5. $F(x) = \sin\dfrac{1}{x^2 + 1}$

6. $F(x) = 3\cos\left(\dfrac{\tan x}{2}\right)$

7. $F(x) = \csc(3e^x)$

8. $F(x) = \sec(e^{2+\sqrt{x}})$

9. $F(x) = \dfrac{3}{\sqrt{\ln(x^2 + 1)}}$

10–60 Find the derivative of the given function.

10. $f(x) = (2x^2 + x)^7$

11. $g(x) = 3(x^5 - \pi x^2 + 7.5)^{11}$

12. $h(x) = \dfrac{1}{2}(x^8 + 5x^3 - ex)^{100}$

13. $F(x) = -3(5 + 2\sqrt{x})^{-5}$

14. $G(x) = (2x^2 - 3x + 1)^{2/3}$

15. $k(x) = -5(x^5 - 2x^3 + 10.5x)^{-2/5}$

16. $f(x) = \sqrt{2 - 4x}$

17. $g(x) = \sqrt{x^2 - 5x + 2}$

18. $h(x) = (4x + 5)^{21}(3x - 7)^{13}$

19. $q(x) = 2(x^3 - 5x)^{2/3}(x + 3)^{5/4}$

20. $r(t) = \dfrac{1}{3t + 1}$

21. $k(z) = \dfrac{1}{1 + 5z - 2z^2}$

22. $F(x) = \left(\dfrac{2x - 3}{1 - 7x}\right)^{10}$

23. $S(v) = \left(\dfrac{2v + 1}{v^2 - 5}\right)^{-3}$

24. $G(y) = \left(\dfrac{3y^2 - 1}{2 + 4y}\right)^{7/5}$

25. $T(s) = \left(\dfrac{s^2 - 1}{s^2 + 1}\right)^{-2/3}$

26. $G(x) = \dfrac{(5 - \pi x^2)^2}{(1 + 2x)^3}$

27. $H(x) = \dfrac{\sqrt{x^2 - 2}}{(x^2 + 2)^2}$

28. $R(x) = \sqrt{\dfrac{1}{x^2 - 1}}$

29. $B(t) = \sqrt[3]{\dfrac{t}{2t^2 + 1}}$

30. $K(s) = \sqrt{\dfrac{2s - 5}{3s + 1}}$

31. $t(x) = \sin(\cos x)$

32. $Q(x) = 2\tan(\sin x)$

33. $P(x) = x\tan^2 x$

34. $w(x) = \cot(x^2)$

35. $U(z) = 5\sec^2 z$

36. $R(x) = x\sqrt{\sin x}$

37. $C(x) = \sin^2(\tan x)$

38. $U(v) = \csc\left(\dfrac{v}{\cos v}\right)$

39. $V(x) = e^{\cos x}$

40. $R(\theta) = e^{\theta \tan \theta}$

41. $w(x) = \sin\sqrt{2x + 1} + e^{\tan\sqrt{2x+1}}$

42. $t(x) = 10^{\sqrt{x}}$

43. $f(x) = \pi 2^{\sin(\pi x)}$

44. $u(x) = 2^{x^2} - 4^{\sqrt{x}}$

45. $t(s) = \tan(2^s)$

46. $u(x) = \cot^2(2^{\sin x})$

47. $E(x) = 5^{5^x}$

48. $K(x) = \sqrt[3]{3^x} + 3^{\sqrt[3]{x}}$

49. $N(x) = \cos^2\left(e^{\cos(x^2)}\right)$

50. $u(t) = \tan^3(t^3 + 3^t)$

51. $C(x) = \cos^2(x^2)$

52. $F(x) = 5^{x^5}$

53. $t(s) = \sqrt{\cos(10^s)}$

54. $G(t) = \sec^{-3}(5^t)$

55. $H(s) = \sin(2^s)\tan(2^s)$

56. $w(s) = \sin(\tan(2^s))$

57. $T(z) = \sin(e^z) + e^{\sin z}$

58.* $q(x) = \sin(\cos(\tan(\cot x)))$

59.* $U(\theta) = \theta + \tan(\theta + \tan(\theta + \tan\theta))$

60.* $v(x) = \left(1 + \left(2 + (3 + 4x)^5\right)^6\right)^7$

61–68 Find an equation for the tangent line to the graph of the given function at the specified point.

61. $f(x) = \sqrt{2x^2 + 1};\quad (2,3)$

62. $g(x) = \left(x^2 + 3x + 4\right)^{2/3};\quad (1,4)$

63. $q(x) = \cos(\tan x);\quad (0,1)$

64. $S(x) = \sin\left(x^2\right) + \sin^2 x;\quad (0,0)$

65. $M(x) = \dfrac{e^{\cos x}}{x};\quad \left(\pi, \dfrac{1}{e\pi}\right)$

66. $a(x) = 10^{\sqrt{x}};\quad (1,10)$

67. $h(x) = \dfrac{3x+1}{\sqrt{x^2+3}};\quad (1,2)$

68. $u(x) = \pi^{\pi^{\sin x}};\quad (0,\pi)$

69–76 Find all x-values where the line tangent to the given curve is horizontal.

69. $f(x) = \left(x^2 - 8x + 15\right)^{100}$

70. $g(x) = \dfrac{2x+3}{x^2 - 2}$

71. $h(x) = \sqrt{x^2 + 1}$ **72.** $T(x) = \tan^{10} x$

73. $w(x) = \sec\left(x^2 + 2\right)$ **74.** $t(x) = \cos(\cos x)$

75. $k(x) = e^{x/(x^2+1)}$ **76.** $q(x) = \pi^{\cos^2 x}$

77–84 Determine the second derivative of the function.

77. $p(x) = \left(x^2 + 5\right)^{20}$ **78.** $r(t) = \sqrt{t^2 + 5}$

79. $g(x) = 5\cos^2 x$ **80.** $c(x) = e^{\tan x}$

81. $F(t) = t\sin\left(t^2\right)$ **82.** $d(x) = 5^{5^x}$

83. $G(x) = \sin^2 x + \cos^2 x$

84. $U(s) = \sec\sqrt{s}$

85. Suppose that $f(1) = 1$, $f'(1) = -2$, $g(1) = 1$, and $g'(1) = 5$. If $F(x) = (f \circ g)(x)$ and $G(x) = (g \circ f)(x)$, find $F'(1) + G'(1)$.

86. Let $P(x) = x(x+1)(x+2)\cdots(x+10)$. If $F(x) = (P \circ P)(x)$, find the value of $F'(0)$.

87. Find a formula for the nth derivative of $f(x) = \cos(kx)$, $k \in \mathbb{R}$. (**Hint:** Use the Chain Rule and recognize a pattern.)

88. Repeat Exercise 87 for the function $g(x) = 2^{kx}$.

89. Use the Chain Rule to prove that the function $f(x) = \sin\left(1/x^2\right)$ is differentiable for $x \neq 0$.

90. Use the Chain Rule to construct a second proof of the Quotient Rule. (**Hint:** Rewrite $f(x)/g(x)$ as $f(x) \cdot \left[g(x)\right]^{-1}$.)

91. Use the Chain Rule to prove that the derivative of an even function is odd and vice versa.

92. Find all points where the line tangent to the graph of $y = \sqrt[3]{\cos x}$ is horizontal, as well as those where it is vertical.

93.* A spherical balloon is being inflated so that its radius is increasing at a rate of $dr/dt = 0.1$ in./s. Find the rate at which the volume of the balloon is increasing when its radius is $r = 4$ in. (**Hint:** Notice that $V(t) = V(r(t))$ and use the Chain Rule.)

94.* Pouring sand is forming a conical shape so that the radius of the bottom of the cone is always twice its height throughout the process. If the height of the cone is increasing at a rate of $dh/dt = 0.5$ mm/s, find the rate at which the volume of the cone is increasing when its height is $h = 50$ mm. (See the hint given in Exercise 93.)

95. The position function of a vibrating loudspeaker cone is given by $x(t) = 10^{-3}\cos 1500t$, where distance is measured in meters, time in seconds. As indicated by the position function, the cone is at one of its extreme positions at $t = 0$. Use the above information to find **a.** the maximum velocity of the cone and **b.** the maximum acceleration of the cone.

96. The position function for damped harmonic motion of an object of mass m is $x(t) = Ae^{-\frac{k}{2m}t}\cos(\omega t)$, where A is the amplitude and k and ω are constants specific to the motion. Find the velocity and acceleration functions for this motion.

97. Unless conditions are "extreme," most gases obey the so-called *Ideal Gas Law*, which says $PV = nRT$, where P stands for pressure measured in pascals (Pa), V for volume, n for the number of moles (mol) of gas in the container, T denotes temperature measured in kelvins (K), and R is the *universal gas constant*, which is the same for all gases. Suppose 5 moles of gas are being slowly compressed by a piston in a container so that $dV/dt = -2 \cdot 10^{-8}$ m^3/s. Assuming that temperature is being kept constant at $T = 293$ K throughout the process, find the rate of change of pressure with respect to time when $V = 10^{-3}$ m^3. (Use $R \approx 8.315\,\text{J}/(\text{mol} \cdot \text{K})$.)

3.4 **Technology Exercises**

98–99 The Maclaurin polynomial of order 2 of the function $f(x)$ is used to approximate $f(x)$ near $x = 0$. It is defined as

$$P_2(x) = f(0) + f'(0)x + \frac{1}{2}f''(0)x^2.$$

Find the Maclaurin polynomial of order 2 for $f(x)$. Then use a graphing calculator or computer algebra system to graph f along with its Maclaurin polynomial. (We will learn more about Maclaurin polynomials in Section 10.8.)

98. $f(x) = \cos(\sin x)$ **99.** $f(x) = \dfrac{1}{x^2 + 1}$

3.5 Implicit Differentiation

In this section, we learn how to find the rate of change of one variable with respect to another when the functional relationship between the two is presented implicitly.

Topic One

Derivatives of Implicitly Defined Functions

The functions we have worked with so far have been written (or at least could be written) in explicit form. That is, the dependent variable, say y, has been defined explicitly as a function of the independent variable, say x. And we have now gained experience in working with the resulting equation $y = f(x)$ to determine the derivative y', assuming the function f is differentiable.

But there are many instances where the functional dependence of one variable on another is defined only implicitly by an equation and it is inconvenient or in fact impossible to solve the equation for the dependent variable. Despite this, the rate of change of one variable with respect to the other may still be central to the problem we are attempting to solve. For instance, given an equation in x and y, the notion of dy/dx may make perfect sense and be what we seek—we just need a method to determine dy/dx without first expressing y as a function of x. This method is called *implicit differentiation*.

We will begin with a comparison of implicit differentiation and what we might now call explicit differentiation, using an example where both methods can be applied.

Example 1

Given the equation $y^2 = x$, determine dy/dx.

Solution

We can actually solve this equation for y, but we can express the result in functional form only if we distinguish between the two solutions of the equation. Accordingly, let

$$y_1 = \sqrt{x} \quad \text{and} \quad y_2 = -\sqrt{x}.$$

Note that y_1 corresponds to the upper half of the horizontally oriented parabola shown in Figure 1, while y_2 corresponds to the lower. Using the Power Rule of differentiation,

$$y_1' = \frac{1}{2}x^{-1/2} = \frac{1}{2\sqrt{x}} \quad \text{and} \quad y_2' = -\frac{1}{2}x^{-1/2} = -\frac{1}{2\sqrt{x}}.$$

If we were asked to find the rate of change of y with respect to x at, say, $P_1 = (1,1)$ and $P_2 = \left(0.5, -\sqrt{0.5}\right)$, we would calculate

$$\left.\frac{dy_1}{dx}\right|_{x=1} = \frac{1}{2\sqrt{1}} = \frac{1}{2} \quad \text{and} \quad \left.\frac{dy_2}{dx}\right|_{x=0.5} = -\frac{1}{2\sqrt{0.5}} = -\frac{1}{\sqrt{2}}.$$

These two values represent the slopes of the two tangent lines L_1 and L_2 that are shown in Figure 1.

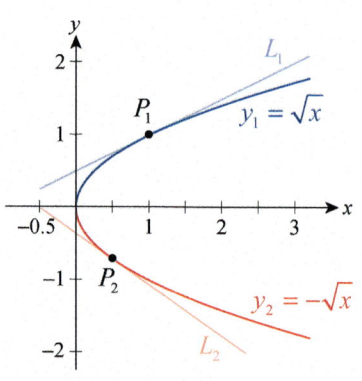

Figure 1

Alternatively, using the method of implicit differentiation, we leave the equation in its original form and differentiate both sides with respect to x. To differentiate the left-hand side, we *implicitly assume* that y is a differentiable function of x and apply the Chain Rule.

$$\frac{d}{dx}(y^2) = \frac{d}{dx}(x)$$

$$2y\frac{dy}{dx} = 1 \qquad \text{Chain Rule}$$

$$\frac{dy}{dx} = \frac{1}{2y}$$

This one formula contains the same information as the two formulas for y_1' and y_2', and using it to calculate the slopes of the lines L_1 and L_2 we obtain the same results:

$$\left.\frac{dy}{dx}\right|_{P_1} = \left.\frac{dy}{dx}\right|_{y=1} = \frac{1}{2(1)} = \frac{1}{2} \quad \text{and} \quad \left.\frac{dy}{dx}\right|_{P_2} = \left.\frac{dy}{dx}\right|_{y=-\sqrt{0.5}} = \frac{1}{2\left(-\sqrt{0.5}\right)} = -\frac{1}{\sqrt{2}}$$

The implicit assumption that y is a differentiable function of x is subtle but important. Fortunately, in our usage the assumption is largely self-regulating—if y is *not* a differentiable function of x at a particular point, that fact usually becomes evident as a result of the process. For instance, in Example 1 we know that dy/dx does not exist at the origin, as the line tangent to the curve $y^2 = x$ is vertical at that point; this is reflected in the fact that the formula $1/(2y)$ is undefined when $y = 0$.

In summary, the method of implicit differentiation is as follows.

Method of Implicit Differentiation

Step 1: Given an equation in x and y, differentiate both sides with respect to x under the assumption that y is a differentiable function of x.

Step 2: Solve the resulting equation for dy/dx.

Example 2

Determine dy/dx for the graph of the equation $x^3 + y^3 = 7xy$ (this famous curve is called the *folium of Descartes*). Use your result to find the coordinates of the rightmost point on the loop. (See Figure 2.)

Solution

We will start by differentiating both sides of the equation with respect to x, treating the variable y as a differentiable function of x (you can actually think of y as $y = y(x)$).

$$\frac{d}{dx}(x^3 + y^3) = \frac{d}{dx}(7xy)$$

$$3x^2 + 3y^2\frac{dy}{dx} = 7\left(y + x\frac{dy}{dx}\right) \qquad \text{Power Rule and Product Rule}$$

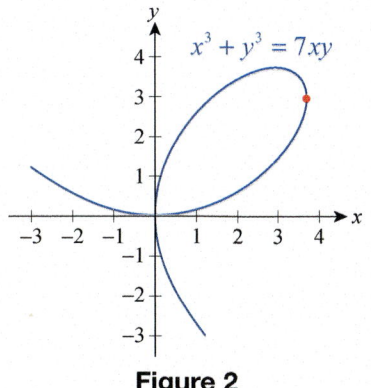

$x^3 + y^3 = 7xy$

Figure 2

Next, we rearrange terms and solve the equation for dy/dx.

$$3y^2 \frac{dy}{dx} - 7x\frac{dy}{dx} = 7y - 3x^2$$

$$\left(3y^2 - 7x\right)\frac{dy}{dx} = 7y - 3x^2$$

$$\frac{dy}{dx} = \frac{7y - 3x^2}{3y^2 - 7x}$$

To find the coordinates of the rightmost point on the loop, note that it is a point where the corresponding tangent line is vertical, in other words, one whose coordinates make dy/dx undefined. This will happen when $3y^2 - 7x = 0$, that is, when $x = \frac{3}{7}y^2$. Substituting this expression for x into the equation of the folium, we obtain

$$\left(\frac{3}{7}y^2\right)^3 + y^3 = 7\left(\frac{3}{7}y^2\right)y,$$

which reduces to

$$\left(\frac{3}{7}\right)^3 y^6 + y^3 = 3y^3 \qquad\qquad \text{Divide by } y^3.$$

$$\left(\frac{3}{7}\right)^3 y^3 = 2 \qquad\qquad \text{Solve for } y.$$

$$y = \frac{7}{3}\sqrt[3]{2}.$$

Note that dividing both sides of the equation by y^3 was justified, for $y \neq 0$ in the problem at hand. Next, the x-coordinate of the point is obtained.

$$x = \frac{3}{7}y^2 = \frac{3}{7}\left(\frac{7}{3}\sqrt[3]{2}\right)^2 = \frac{7}{3}\left(\sqrt[3]{2}\right)^2$$

We conclude that the coordinates of the point in the extreme-right position on the loop are as follows.

$$\left(\frac{7}{3}\left(\sqrt[3]{2}\right)^2, \frac{7}{3}\sqrt[3]{2}\right)$$

Technology Note

We remark that while determining the x- and y-coordinates from Example 2 can be done by hand, a computer algebra system is very useful in shortening the calculations. *Mathematica*'s output is shown in Figure 3. Notice that multiple solutions are given, but we are interested in the one with the largest x-value. (For additional information on *Mathematica* and the use of the `Solve` command, see Appendix A.)

In[1]:= `Solve[{x == (3 / 7) * y^2, x^3 + y^3 == 7 * x * y}, {x, y}]`

Out[1]= $\left\{ \{x \to 0, \ y \to 0\}, \ \left\{x \to \dfrac{7}{3}\,(-2)^{2/3}, \ y \to -\dfrac{7}{3}\,(-2)^{1/3}\right\}, \right.$

$\left\{x \to \dfrac{7 \times 2^{2/3}}{3}, \ y \to \dfrac{7 \times 2^{1/3}}{3}\right\},$

$\left. \left\{x \to -\dfrac{7}{3}\,(-1)^{1/3}\,2^{2/3}, \ y \to \dfrac{7}{3}\,(-1)^{2/3}\,2^{1/3}\right\}\right\}$

Figure 3

Example 3

Determine dy/dx for the curve $3 - y^3 + y^2 + 5y - x^2 = 0,$ and use it to find the equation of the line that is perpendicular to the tangent at $(3, 2)$ (such a line is called *normal* to the curve). At what points does the curve have horizontal tangents?

Solution

Implicit differentiation of the equation defining the curve yields

$$\frac{d}{dx}\left(3 - y^3 + y^2 + 5y - x^2\right) = \frac{d}{dx}(0)$$

$$0 - 3y^2 \frac{dy}{dx} + 2y\frac{dy}{dx} + 5\frac{dy}{dx} - 2x = 0,$$

an equation that we now solve for dy/dx.

$$\left(-3y^2 + 2y + 5\right)\frac{dy}{dx} = 2x$$

$$\frac{dy}{dx} = \frac{2x}{-3y^2 + 2y + 5}$$

At the point $(3, 2),$ the slope of the tangent line is calculated by substituting $x = 3$ and $y = 2$ in the formula for dy/dx.

$$\left.\frac{dy}{dx}\right|_{(3,2)} = \frac{2(3)}{-3(2)^2 + 2(2) + 5} = \frac{6}{-3} = -2$$

Since the normal line is perpendicular to the tangent, its slope is

$$m_n = -\frac{1}{-2} = \frac{1}{2}. \qquad \text{The slopes of perpendicular lines are negative reciprocals.}$$

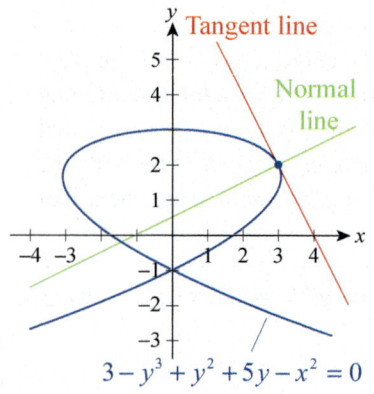

$$3 - y^3 + y^2 + 5y - x^2 = 0$$

Figure 4

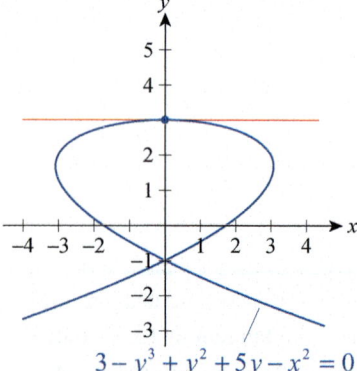

$$3 - y^3 + y^2 + 5y - x^2 = 0$$

Figure 5

Figure 4 confirms that the values we found for the slopes of the tangent and normal lines at $(3, 2)$ are reasonable. So using the point-slope form, the equation of the normal line at $(3, 2)$ can be written as

$$y - 2 = \frac{1}{2}(x - 3),$$

or equivalently,

$$y = \frac{1}{2}x + \frac{1}{2}.$$

Finally, we wish to determine all points at which the tangent to the curve is horizontal. Since the formula we derived for dy/dx is a rational expression, and such expressions are equal to 0 if and only if their numerator is 0 (provided that the denominator does not vanish at the same point), we conclude that the only horizontal tangent line occurs at the point $(0, 3)$; while $x = 0$ also at the point $(0, -1)$, the derivative dy/dx is undefined there.

The curve and its horizontal tangent line are graphed in Figure 5.

Example 4

Given the equation $2y^5 + 4xy^2 - 5x^3 y - x^5 + 10 = 0$, find dy/dx by implicit differentiation.

Solution

As in the previous examples, treating y as a function of x, we differentiate both sides with respect to x.

$$\frac{d}{dx}\left(2y^5 + 4xy^2 - 5x^3 y - x^5 + 10\right) = \frac{d}{dx}(0)$$

$$10y^4 \frac{dy}{dx} + 4\left(y^2 + 2xy\frac{dy}{dx}\right) - 5\left(3x^2 y + x^3 \frac{dy}{dx}\right) - 5x^4 + 0 = 0$$

Next, we rearrange terms and solve for dy/dx.

$$10y^4 \frac{dy}{dx} + 4y^2 + 8xy\frac{dy}{dx} - 15x^2 y - 5x^3 \frac{dy}{dx} - 5x^4 = 0$$

$$\left(10y^4 + 8xy - 5x^3\right)\frac{dy}{dx} = 5x^4 + 15x^2 y - 4y^2$$

$$\frac{dy}{dx} = \frac{5x^4 + 15x^2 y - 4y^2}{10y^4 + 8xy - 5x^3}$$

In conclusion, two important remarks are in order. First, note that the variables x and y play fairly symmetrical roles in the given equation. Therefore, we could have treated x as a differentiable function of the independent variable y. This is not the most common approach, but it is correct and sometimes warranted. Differentiating implicitly, we can then determine dx/dy. We show the differentiation step below (remember that y is the variable and $x = x(y)$ is the function this time).

$$10y^4 + 4\left(\frac{dx}{dy}y^2 + 2xy\right) - 5\left(3x^2 y\frac{dx}{dy} + x^3\right) - 5x^4 \frac{dx}{dy} = 0$$

Solving for dx/dy (the details are left to the reader) yields the final answer, which turns out, not surprisingly, to be the reciprocal of our solution above.

$$\frac{dx}{dy} = \frac{1}{\dfrac{dy}{dx}} = \frac{10y^4 + 8xy - 5x^3}{5x^4 + 15x^2 y - 4y^2}$$

Finally, if a function is defined implicitly in terms of a polynomial equation of degree five or higher, in general we have no hope of expressing the function explicitly with a formula. This follows from the brilliant insights of Niels Henrik Abel (1802–1829) and Évariste Galois (1811–1832), who showed that in general it is impossible to solve a polynomial equation of degree five or higher with a formula (like we can, for example, solve quadratic equations with the quadratic formula).

As another example of its use, we will use implicit differentiation in an alternative proof of the Power Rule for rational exponents.

── **Proof** ─────────────────────────────────

Second Proof of the Power Rule for Rational Exponents

Let m and n be nonzero integers, and let $y = x^{m/n}$; we seek an alternate proof of the formula $y' = (m/n)x^{(m/n)-1}$.

First, raise both sides to the nth power to obtain the equation $y^n = x^m$, and assume that y is a differentiable function of x. Implicit differentiation with respect to x then yields the following.

$$\frac{d}{dx}\left(y^n\right) = \frac{d}{dx}\left(x^m\right)$$

$$ny^{n-1}\frac{dy}{dx} = mx^{m-1} \qquad \text{Chain Rule and Integer Power Rule}$$

$$\frac{dy}{dx} = \frac{m}{n} \cdot \frac{x^{m-1}}{y^{n-1}} \qquad \text{Solve for } dy/dx.$$

$$\frac{dy}{dx} = \frac{m}{n} \cdot \frac{x^{m-1}}{\left(x^{m/n}\right)^{n-1}} \qquad y = x^{m/n}$$

$$\frac{dy}{dx} = \frac{m}{n} \cdot \frac{x^{m-1}}{x^{m-(m/n)}}$$

$$\frac{dy}{dx} = \frac{m}{n} x^{m-1-m+(m/n)} = \frac{m}{n} x^{(m/n)-1}$$

Topic Two

Higher-Order Derivatives

Implicit differentiation can be used repeatedly to obtain higher-order derivatives, as illustrated in the following examples.

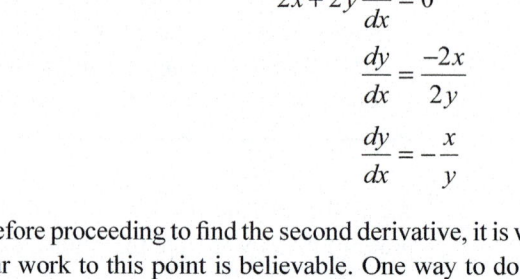

Example 5

Find d^2y/dx^2 for the circle $x^2 + y^2 = 1$.

Solution

Differentiating both sides with respect to x and solving for dy/dx, we obtain the following.

$$\frac{d}{dx}(x^2 + y^2) = \frac{d}{dx}(1)$$

$$2x + 2y\frac{dy}{dx} = 0$$

$$\frac{dy}{dx} = \frac{-2x}{2y}$$

$$\frac{dy}{dx} = -\frac{x}{y}$$

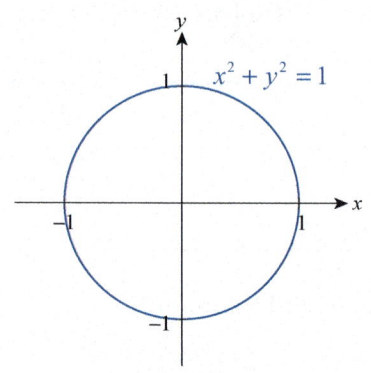

$x^2 + y^2 = 1$

Figure 6

Before proceeding to find the second derivative, it is worthwhile to check that our work to this point is believable. One way to do this is to ask questions such as the following: *where is the first derivative 0, where is it undefined, and what sign does it have in each of the four quadrants?* From Figure 6, we expect dy/dx to be 0 at the two points $(0,1)$ and $(0,-1)$, and that is certainly confirmed by our formula. We also expect an undefined first derivative at the points $(-1,0)$ and $(1,0)$ (i.e., when $y = 0$), and that also agrees with our result for dy/dx. We leave it to the reader to verify that the sign of dy/dx in each of the quadrants is what we would expect (see Exercise 53).

We can now take our result for dy/dx and differentiate with respect to x once more.

$$\frac{d^2y}{dx^2} = \frac{d}{dx}\left(-\frac{x}{y}\right)$$

$$= \frac{y\frac{d}{dx}(-x) - (-x)\frac{d}{dx}(y)}{y^2} \qquad \text{Quotient Rule}$$

$$= \frac{-y + xy'}{y^2}$$

$$= \frac{-y + x\left(-\frac{x}{y}\right)}{y^2} \qquad y' = -\frac{x}{y}$$

$$= \frac{-y^2 - x^2}{y^3} \qquad \text{Multiply the numerator and denominator by } y.$$

$$= -\frac{1}{y^3} \qquad x^2 + y^2 = 1$$

Note that the second derivative is a formula depending only on y. It is again instructive to verify that this formula makes sense for points along the upper half of the circle and for points along the lower half (see Exercise 54).

Example 6

Find d^2y/dx^2 given that $xy + x - 2 = y$.

Solution

As in the previous example, we proceed to find dy/dx first.

$$\frac{d}{dx}(xy + x - 2) = \frac{d}{dx}(y)$$

$$y + x\frac{dy}{dx} + 1 - 0 = \frac{dy}{dx} \qquad \text{Product Rule}$$

$$\frac{dy}{dx}(x - 1) = -y - 1$$

$$\frac{dy}{dx} = \frac{y+1}{1-x}$$

Next, differentiating with respect to x once more we obtain the second-order derivative.

$$\frac{d^2y}{dx^2} = \frac{d}{dx}\left(\frac{y+1}{1-x}\right)$$

$$= \frac{y'(1-x) - (y+1)(-1)}{(1-x)^2} \qquad \text{Quotient Rule}$$

$$= \frac{\dfrac{y+1}{1-x}(1-x) + (y+1)}{(1-x)^2} \qquad y' = \frac{y+1}{1-x}$$

$$= \frac{(y+1) + (y+1)}{(1-x)^2}$$

$$= \frac{2y+2}{(1-x)^2}$$

3.5 **Exercises**

1–12 Use implicit differentiation to determine dy/dx for the given equation. Then check your answer by expressing y explicitly and using differentiation rules.

1. $x + y^2 = 2$

2. $xy = 3$

3. $x^2 - y^2 = 1$

4. $4x^2 + 25y^2 = 100$

5. $3xy^2 = x - 5$

6. $y^2\sqrt{x} = 2x^2 + 1$

7. $y\sqrt{x+2} = xy - 2$

8. $2x^2y - 3y - x - 1 = 0$

9. $2y\cos x - xy = x + 3$

10. $ye^x + 2y - 1 = 0$

11. $\dfrac{2}{x} - \dfrac{3}{y} = 4$

12. $x^2\sqrt{y} - x^2 - 1 = e^2$

13–20 Find dx/dy by implicit differentiation. Then check your answer by expressing x explicitly in terms of y and differentiating with respect to y using differentiation rules.

13. $x - y^2 = 0$

14. $xy - y^3 = 3y$

15. $x^3 + y^3 = 1$

16. $-5xy^2 + 4xy - 3y^2 - y - 2 = 0$

17. $xy + 3\sin y = e^y$

18. $xy = \sqrt{y^2 + 1} - 5x$

19. $\sqrt[3]{8x^3 - 5y^4} = 3$

20. $(y+2)\sqrt{x+3} = \sqrt{y}$

21–28 Use implicit differentiation to find the equations of the tangent and normal lines at point P for the well-known curve.

21. $x^2 + y^2 - \left(x^2 + y^2 - y\right)^2 = 0$; $P(1,0)$

Cardioid

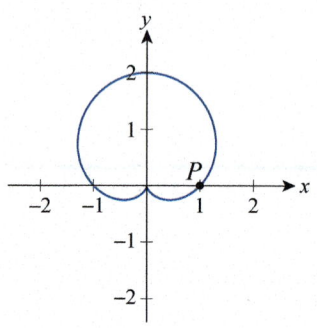

22. $\left(x^2 + y^2\right)^2 - 8y^2x = 0$; $P(2,2)$

Bifolium

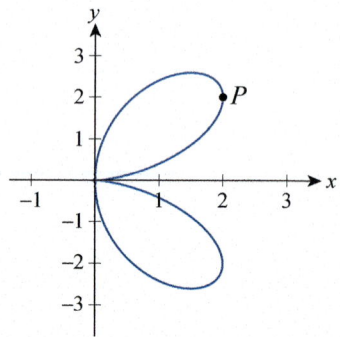

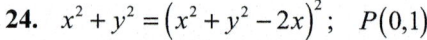

23. $\left(x^2 + 4\right)y = 8$; $P(2,1)$

Witch of Agnesi

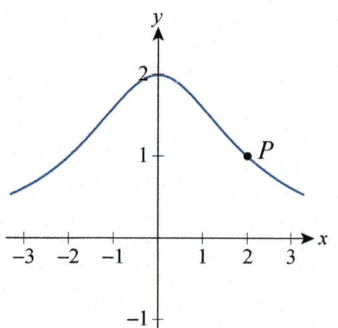

24. $x^2 + y^2 = \left(x^2 + y^2 - 2x\right)^2$; $P(0,1)$

Limaçon

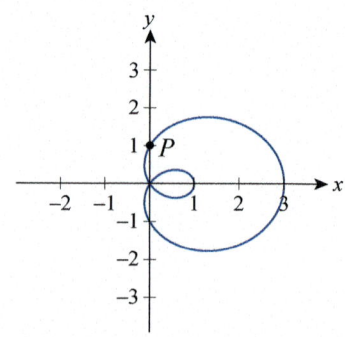

25. $9\left(x^2+y^2\right)=\left(x^2+y^2+2y\right)^2;\quad P(3,0)$

Dimpled Limaçon

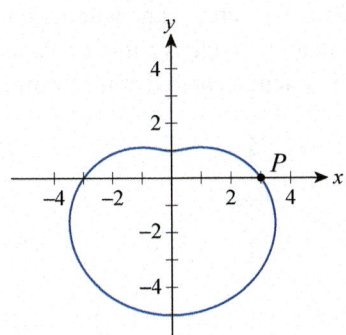

26. $\left(x^2+y^2\right)^2=16xy;\quad P(2,2)$

Lemniscate

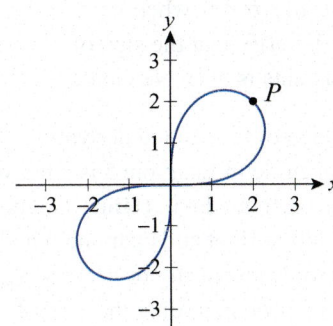

27. $(6-x)y^2=2x^3;\quad P(2,2)$

Cissoid

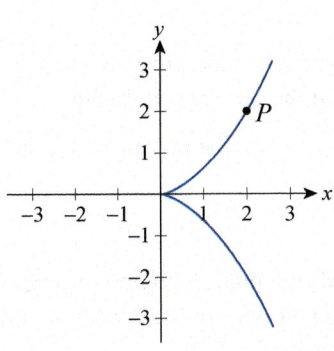

28. $x^{2/3}+y^{2/3}=10;\quad P(-1,27)$

Astroid

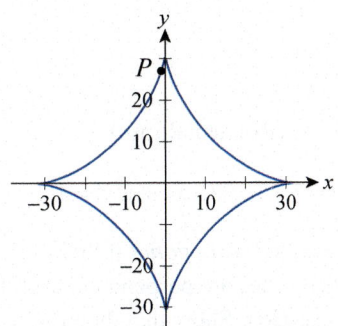

29–44 Find dy/dx by implicit differentiation.

29. $x^4+y^4=1$

30. $\sqrt{x}+\sqrt{y}=4$

31. $x^3y^4-x^4y^3=1$

32. $y=\cos(x-2y)$

33. $(x+y)^3+3=x+y$

34. $e^{xy}=e^x+e^y$

35. $\sin^2 x+\cos^2 y=\tan\left(x^2+y^2\right)$

36. $\sqrt{x^2+y^2}=2x$

37. $\dfrac{x+3y^2}{y-x^2}=2x+1$

38. $\dfrac{y}{x^3}-\dfrac{x}{y^3}=x^3y^3$

39. $\sqrt{2xy}=3y-5x$

40. $y-x=x^4y^4$

41. $\tan x=\sin y-2xy$

42. $e^x\tan x=y+\cos y$

43. $\sqrt{\sin x+\cos x}=\sec(x+y)$

44. $\left(\tan x+\cot y\right)^2=1+x$

45. Find ds/dt by implicit differentiation: $s^2t^3-2t=\sqrt{s}$.

46. Find dt/ds by implicit differentiation: $s\sin t=t\cos s$.

47–52 Find d^2y/dx^2 by implicit differentiation.

47. $4y^2-x^2=4$

48. $y-x=xy-2$

49. $xy^2+5=x$

50. $y^3=xy+1$

51. $x^3+y^3=3$

52. $\sqrt{x}+\sqrt{y}=2$

53. Notice that for a circle centered at the origin, any line tangent to the curve in the first quadrant has negative slope; this is consistent with our observation that $dy/dx < 0$ when $x > 0$ and $y > 0$ (see Example 5). Verify that the sign of dy/dx in each of the quadrants is what we would expect.

54. Verify that the sign of the second derivative $d^2 y/dx^2$ of the circle in Example 5 is what we would expect in each quadrant. (**Hint:** Traverse the circle from left to right and examine whether the first derivative is increasing or decreasing; then draw a conclusion regarding the sign of the second derivative.)

55–58 Find all points on the given curve where it has horizontal or vertical tangent lines.

55. $xy^2 - x^2 y = \dfrac{1}{4}$

56. $x^2 - xy + y^2 = \dfrac{1}{4}$ (Rotated ellipse)

57. $\left(x^2 - 2x + 5\right)y = 5$ **58.** $xy + y^2 x^2 = 1$

59. Two graphs are called *orthogonal* if their respective tangent lines are perpendicular at their point(s) of intersection. Show that the graphs of $x^2 - y^2 = 5$ and $xy = 6$ are orthogonal.

60. Generalizing Exercise 59, show that the families of curves $x^2 - y^2 = a$ and $xy = b$ are orthogonal for $a, b \in \mathbb{R}$. (Such families of curves are called *orthogonal trajectories*.)

61. Repeat Exercise 60 for the families $x^2 + y^2 = a$ and $y - bx = 0$.

62. Repeat Exercise 60 for the families $x^2 + y^2 = ax$ and $x^2 + y^2 = by$.

63. Use implicit differentiation to prove that a tangent line to a circle is always perpendicular to the radius connecting the center and the point of tangency. (**Hint:** We can assume without loss of generality that the circle is a unit circle in the xy-coordinate system, centered at the origin.)

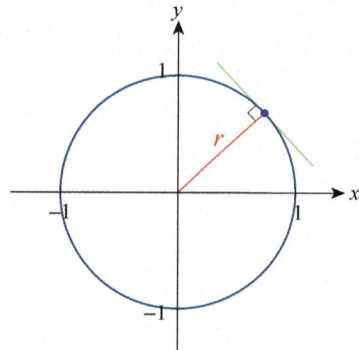

64. Use implicit differentiation to prove that the equation of the line tangent to the ellipse $\dfrac{x^2}{a^2} + \dfrac{y^2}{b^2} = 1$ at the point $\left(x_0, y_0\right)$ is $\dfrac{x_0}{a^2}x + \dfrac{y_0}{b^2}y = 1$.

65. Use implicit differentiation to find the equations of the two tangent lines to the ellipse $2x^2 + y^2 = 2$ through the point $\left(0, 2\right)$.

66. An object of mass m is attached to a spring and is moving along the x-axis so that its position and velocity satisfy the equation $m\left(v - v_0\right)^2 = -kx^2$. Use implicit differentiation to verify Hooke's Law; that is, prove that the restoring force exerted by the spring satisfies $F = -kx$. (**Hint:** Differentiate and use Newton's Second Law of Motion, which states that $ma = F$.)

3.5 **Technology Exercises**

67–70. Use the implicit graphing capabilities of a computer algebra system to graph the curves along with the tangent lines you found in Exercises 55–58 and visually verify that your answers are correct.

71–73. Use a computer algebra system to graph the families of curves in Exercises 60–62 for several different values of the parameters a and b. Visually verify that they are orthogonal.

74. Beautiful, "irregular" curves can be created by using a computer algebra system to plot graphs of equations such as the following:

$$(x^2 - 1)(x - 2)(x - 3) = (y - 1)(y - 2)(y - 3)$$

Graph the above equation and explain why this graph cannot be that of a function. Then try experimenting by slightly modifying the above equation and thus creating your own curves. (Answers will vary.)

75. Repeat Exercise 74 starting with the equation $x^5 - 3x^3 - x^2 = -y^5 + 3y^3 - y^2$.

3.6 Derivatives of Inverse Functions

In Section 1.4, we reviewed the algebraic relationship between a given function and (if it exists) its inverse. As we will see, the relationship extends into the realm of calculus when the function under consideration is also differentiable.

Topic One

The Derivative Rule for Inverse Functions

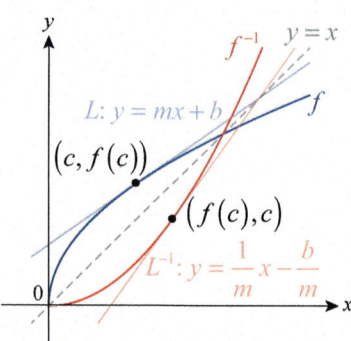

Figure 1

To gain some insight into the formula we will soon develop, suppose L is the line tangent to a function f at the point $(c, f(c))$, and suppose further that L is neither horizontal nor vertical (see Figure 1). Then L can be written in the form $y = mx + b$, with $m \neq 0$. Although we will not provide a formal proof in this text, the fact that f is sufficiently well-behaved to have such a tangent at $(c, f(c))$ means that f^{-1} also possesses a tangent at the point $(f(c), c)$—a fact that is certainly plausible, as the graphs of a function and its inverse are reflections of one another. And since the graph of f^{-1} is the reflection of the graph of f with respect to the line $y = x$, the line tangent to f^{-1} at the point $(f(c), c)$, denoted L^{-1} in Figure 1, is similarly the reflection of the line L. As the reader can verify (by inverting the function $y = mx + b$), the line L^{-1} is described by the equation $y = (1/m)x - (b/m)$; in particular, the slope of L^{-1} is the reciprocal of the slope of L.

This reciprocal relationship between the derivative of f at the point c and the derivative of f^{-1} at the point $f(c)$ is, in fact, seen at every point where f has a nonzero derivative. The formula that precisely states the relationship can take either of two identical forms, both stemming from an application of the Chain Rule. Assuming again the differentiability of f^{-1}, and using the fact that the composition of a function and its inverse (in either order) is the identity function, note that

$$f^{-1}(f(x)) = x$$

$$\frac{d}{dx}\left[f^{-1}(f(x))\right] = \frac{d}{dx}(x) \qquad \text{Differentiate with respect to } x.$$

$$\left(f^{-1}\right)'(f(x)) \cdot f'(x) = 1 \qquad \text{Chain Rule}$$

$$\left(f^{-1}\right)'(f(x)) = \frac{1}{f'(x)}$$

and, using the alternate order,

$$f(f^{-1}(x)) = x$$

$$\frac{d}{dx}\left[f(f^{-1}(x))\right] = \frac{d}{dx}(x) \qquad \text{Differentiate with respect to } x.$$

$$f'(f^{-1}(x)) \cdot \left(f^{-1}\right)'(x) = 1 \qquad \text{Chain Rule}$$

$$\left(f^{-1}\right)'(x) = \frac{1}{f'(f^{-1}(x))}.$$

The choice of which form of the rule to use depends on the context—in practice, one form is often more easily applied than the other. For reference, we repeat the two formulas here as a theorem.

Theorem

The Derivative Rule for Inverse Functions

If a function f is differentiable on an interval (a,b), and if $f'(x) \neq 0$ for all $x \in (a,b)$, then f^{-1} both exists and is differentiable on the image of the interval (a,b) under f, denoted as $f((a,b))$ in the formula below. Further,

$$\text{if } x \in (a,b), \text{ then } \left(f^{-1}\right)'\left(f(x)\right) = \frac{1}{f'(x)},$$

and

$$\text{if } x \in f((a,b)), \text{ then } \left(f^{-1}\right)'(x) = \frac{1}{f'\left(f^{-1}(x)\right)}.$$

Example 1

a. Given $f(x) = x^2$, use the Derivative Rule for Inverse Functions to determine $\left(f^{-1}\right)'(4)$.

b. Given $g(x) = x^5 - x^2 + 5x - 7$, use the Derivative Rule for Inverse Functions to determine $\left(g^{-1}\right)'$ at the point $g(2)$.

Solution

a. Keeping in mind that $f^{-1}(x) = \sqrt{x}$, and using the second form of the rule, we obtain the following.

$$\left(f^{-1}\right)'(4) = \frac{1}{f'\left(f^{-1}(4)\right)}$$

$$= \frac{1}{f'\left(\sqrt{4}\right)}$$

$$= \frac{1}{f'(2)}$$

$$= \frac{1}{2 \cdot 2} = \frac{1}{4} \qquad f'(x) = 2x$$

b. Notice that although we cannot find a formula for $g^{-1}(x)$, the good news is that we don't have to if we use the first form of the Derivative Rule for Inverse Functions.

$$\left(g^{-1}\right)'\left(g(2)\right) = \frac{1}{g'(2)}$$

$$= \frac{1}{5 \cdot 2^4 - 2 \cdot 2 + 5} \qquad g'(x) = 5x^4 - 2x + 5$$

$$= \frac{1}{81}$$

Topic Two
Derivatives of Logarithms

As an immediate application, we will use the second formulation of the Derivative Rule for Inverse Functions to determine the derivative of the natural logarithm.

Theorem

Derivative of ln x

$$\text{For all } x > 0, \ \frac{d}{dx}(\ln x) = \frac{1}{x}.$$

Proof

We begin by defining $f(x) = e^x$, since we already know that $f'(x) = e^x$ and we seek a formula for the derivative of $f^{-1}(x) = \ln x$.

$$\frac{d}{dx}(\ln x) = \left(f^{-1}\right)'(x) \qquad\qquad f^{-1}(x) = \ln x$$

$$= \frac{1}{f'\left(f^{-1}(x)\right)} \qquad\qquad \text{Derivative Rule for Inverse Functions}$$

$$= \frac{1}{e^{f^{-1}(x)}} \qquad\qquad f'(x) = e^x$$

$$= \frac{1}{e^{\ln x}} = \frac{1}{x}$$

More generally, if $u(x)$ is a differentiable function and if $u(x) > 0$, the Chain Rule tells us that

$$\frac{d}{dx}(\ln u) = \frac{1}{u}\frac{du}{dx} \quad \text{or} \quad \frac{d}{dx}\big[\ln u(x)\big] = \frac{u'(x)}{u(x)}.$$

One consequence of this is that for $x < 0$,

$$\frac{d}{dx}\big[\ln(-x)\big] = \frac{1}{-x}(-1) = \frac{1}{x},$$

leading to the following fact that we will use soon.

$$\frac{d}{dx}\big(\ln|x|\big) = \frac{1}{x}$$

Example 2

Determine dy/dx for the following functions.

a. $y = \ln(kx)$, where k is a positive real number

b. $y = \ln(x^3 + 2x)$ **c.** $y = \ln\sqrt[4]{\dfrac{2x+1}{x^3}}$

Solution

a. Defining $u(x) = kx$ and applying the formula we obtained,

$$\frac{d}{dx}\big[\ln(kx)\big] = \frac{(kx)'}{kx} = \frac{k}{kx} = \frac{1}{x}.$$

Notice that what we have found is the fact that $\ln x$ and $\ln(kx)$ have the same derivative. This shouldn't come as a surprise if we recall that by the properties of logarithms,

$$\ln(kx) = \ln k + \ln x;$$

therefore,

$$\frac{d}{dx}\big[\ln(kx)\big] = \frac{d}{dx}(\ln k) + \frac{d}{dx}(\ln x) = 0 + \frac{1}{x} = \frac{1}{x}. \qquad \text{\small \color{blue}$\ln k$ is a constant}$$

Recall that what this means geometrically is the fact that the graph of $\ln(kx)$ can be obtained from that of $\ln x$ by a vertical shift of $\ln k$ units. However, vertical shifting doesn't change the slope of a tangent at any point, which visually explains why the derivatives are equal.

b. By the same formula that we put to good use in part a.,

$$\frac{d}{dx}\big[\ln(x^3 + 2x)\big] = \frac{(x^3 + 2x)'}{x^3 + 2x} = \frac{3x^2 + 2}{x^3 + 2x}.$$

c. We could define $u(x) = \sqrt[4]{(2x+1)/x^3}$ and proceed like above; however, finding $u'(x)$ seems a bit tedious at best. Instead, it is to our advantage to use the properties of logarithms before differentiating. Since

$$\ln\sqrt[4]{\frac{2x+1}{x^3}} = \frac{1}{4}\Big[\ln(2x+1) - \ln(x^3)\Big] = \frac{1}{4}\ln(2x+1) - \frac{3}{4}\ln x,$$

differentiating becomes much more straightforward.

$$\frac{d}{dx}\left(\ln\sqrt[4]{\frac{2x+1}{x^3}}\right) = \frac{1}{4} \cdot \frac{d}{dx}\big[\ln(2x+1)\big] - \frac{3}{4} \cdot \frac{d}{dx}(\ln x)$$

$$= \frac{1}{4} \cdot \frac{2}{2x+1} - \frac{3}{4} \cdot \frac{1}{x}$$

$$= \frac{1}{4x+2} - \frac{3}{4x} = \frac{-4x - 3}{4x(2x+1)}$$

Example 3

Find the equation of the line tangent to the graph of f at the point $(0,1)$.

$$f(x) = \ln\big((e-1)\cos x + e^x\big)$$

Solution

After defining $u(x) = (e-1)\cos x + e^x$, we can proceed by differentiating f to obtain the slope.

$$f'(x) = \frac{d}{dx}\Big[\ln\big((e-1)\cos x + e^x\big)\Big] = \frac{(1-e)\sin x + e^x}{(e-1)\cos x + e^x} \qquad \frac{d}{dx}\big[\ln u(x)\big] = \frac{u'(x)}{u(x)}$$

Therefore, the slope of the desired tangent is

$$m = f'(0) = \frac{(1-e)\sin 0 + e^0}{(e-1)\cos 0 + e^0} = \frac{1}{(e-1)+1} = \frac{1}{e}.$$

Finally, using the point-slope form, the equation of the tangent line is obtained:

$$y - 1 = \frac{1}{e}(x - 0),$$

or equivalently,

$$y = \frac{1}{e}x + 1.$$

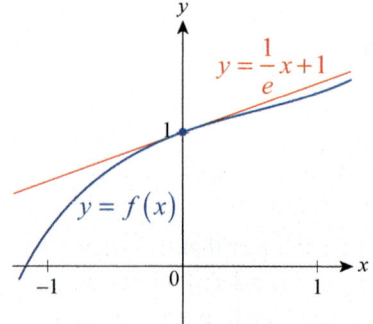

Figure 2

The function and its tangent are graphed in Figure 2.

Using the change of base formula and the formula for the derivative of the natural logarithm, we can arrive at a formula for the derivative of a logarithm of any base.

Theorem

Derivative of $\log_a x$

Given a positive base a, $a \neq 1$, and for all $x > 0$,

$$\frac{d}{dx}\left(\log_a x\right) = \frac{1}{\ln a}\cdot\frac{1}{x}.$$

Proof

Since we now know the derivative of $\ln x$, all we need is the logarithmic change of base formula.

$$\frac{d}{dx}\left(\log_a x\right) = \frac{d}{dx}\left(\frac{\ln x}{\ln a}\right) = \frac{1}{\ln a}\cdot\frac{d}{dx}\left(\ln x\right) = \frac{1}{\ln a}\cdot\frac{1}{x}$$

We can again apply the Chain Rule to obtain the following fact for positive differentiable functions $u(x)$.

$$\frac{d}{dx}(\log_a u) = \frac{1}{u \ln a}\frac{du}{dx} \quad \text{or} \quad \frac{d}{dx}\big[\log_a u(x)\big] = \frac{1}{\ln a}\frac{u'(x)}{u(x)}$$

Example 4

Determine the following derivatives.

a. $\dfrac{d}{dx}\big[\log_5(2x+3)\big]$ **b.** $\dfrac{d}{dx}\left[\log\dfrac{\tan x}{\sqrt{x^2+1}}\right]$

Solution

a. Note that if $u(x) = 2x+3$, $u'(x) = 2$, so by the above formula we obtain the following.

$$\frac{d}{dx}\big[\log_5(2x+3)\big] = \frac{1}{\ln 5}\cdot\frac{2}{2x+3}$$

b. First of all, recall that "log" stands for the common logarithm, whose base is 10. Second, just like in Example 2, we will be better off if we rewrite the logarithmic expression before differentiating.

$$\frac{d}{dx}\left[\log\frac{\tan x}{\sqrt{x^2+1}}\right] = \frac{d}{dx}\left[\log(\tan x) - \frac{1}{2}\log(x^2+1)\right]$$

$$= \frac{1}{\ln 10}\cdot\frac{(\tan x)'}{\tan x} - \frac{1}{2\ln 10}\cdot\frac{(x^2+1)'}{x^2+1}$$

$$= \frac{1}{\ln 10}\cdot\frac{\sec^2 x}{\tan x} - \frac{1}{2\ln 10}\cdot\frac{2x}{x^2+1}$$

$$= \frac{1}{\ln 10}\left(\frac{1}{\sin x\cos x} - \frac{x}{x^2+1}\right)$$

Topic Three

Logarithmic Differentiation

In algebra, logarithms are often applied to both sides of an equation in order to bring exponents down as coefficients, thereby making any variables that were in the exponents much more accessible (recall that $\log(x^r) = r \log x$). This property and other properties of logarithms are often just as useful when we need to differentiate complicated functions. The practice is termed *logarithmic differentiation*, and the procedure to find the derivative of such a function f is as follows.

Logarithmic Differentiation

Step 1: Apply the natural logarithm to both sides of the equation $y = f(x)$, and use the properties of logarithms as appropriate.

Step 2: Under the assumption that y is differentiable, differentiate both sides of the resulting equation implicitly with respect to x.

Step 3: Solve the resulting equation for y'.

Example 5

Use logarithmic differentiation to differentiate the following functions.

a. $f(x) = \dfrac{\sqrt{1+x^2}}{(2x-5)^3}$ **b.** $g(x) = \dfrac{(x^3-3x)^{3/2}(2x-1)^{1/2}}{(5x+2)^4}$ **c.** $h(x) = x^{2x}$

Solution

a. We start by writing

$$y = \frac{\sqrt{1+x^2}}{(2x-5)^3},$$

then take the natural logarithm of both sides and use properties of logarithms.

$$\ln y = \ln \frac{\sqrt{1+x^2}}{(2x-5)^3}$$

$$\ln y = \frac{1}{2}\ln(1+x^2) - 3\ln(2x-5) \qquad \text{Step 1}$$

Next, we differentiate both sides implicitly with respect to x, and solve for y'.

$$\frac{y'}{y} = \frac{1}{2}\cdot\frac{2x}{1+x^2} - 3\cdot\frac{2}{2x-5} \qquad \text{Step 2}$$

$$y' = y\left(\frac{x}{1+x^2} - \frac{6}{2x-5}\right) \qquad \text{Step 3}$$

$$y' = \frac{\sqrt{1+x^2}}{(2x-5)^3}\left(\frac{x}{1+x^2} - \frac{6}{2x-5}\right) \qquad y = \frac{\sqrt{1+x^2}}{(2x-5)^3}$$

b. Proceed in a manner similar to part a.

$$y = \frac{\left(x^3 - 3x\right)^{3/2}\left(2x - 1\right)^{1/2}}{\left(5x + 2\right)^4}$$

$$\ln y = \frac{3}{2}\ln\left(x^3 - 3x\right) + \frac{1}{2}\ln\left(2x - 1\right) - 4\ln\left(5x + 2\right)$$

Implicit differentiation now yields the following.

$$\frac{y'}{y} = \frac{3}{2} \cdot \frac{3x^2 - 3}{x^3 - 3x} + \frac{1}{2} \cdot \frac{2}{2x - 1} - 4 \cdot \frac{5}{5x + 2}$$

$$y' = y\left(\frac{9x^2 - 9}{2x^3 - 6x} + \frac{1}{2x - 1} - \frac{20}{5x + 2}\right)$$

$$y' = \frac{\left(x^3 - 3x\right)^{3/2}\left(2x - 1\right)^{1/2}}{\left(5x + 2\right)^4}\left(\frac{9x^2 - 9}{2x^3 - 6x} + \frac{1}{2x - 1} - \frac{20}{5x + 2}\right)$$

c. Notice first of all that since both the base and exponent are variable expressions, h is neither an exponential function nor a power function, so none of the differentiation formulas we learned up to this point are going to help. The fact that we can still differentiate h is perhaps an illustration of logarithmic differentiation at its best. Simply we follow the three-step process.

$$y = x^{2x}$$
$$\ln y = \ln\left(x^{2x}\right)$$
$$\ln y = 2x\ln x \qquad\qquad \ln\left(x^r\right) = r\ln x$$
$$\frac{y'}{y} = 2\ln x + 2x\frac{1}{x} \qquad\qquad \text{Product Rule on the right-hand side}$$
$$y' = 2y\left(\ln x + 1\right)$$
$$y' = 2x^{2x}\left(\ln x + 1\right)$$

We can use logarithmic differentiation to revisit the Power Rule one last time, and on this occasion we can finally justify the rule for all real exponents.

Theorem

The Power Rule

Under the assumption that x^r is a real number, where r is a fixed real number constant, the derivative of x^r is rx^{r-1}.

─ **Proof** ─────────────────────────────────

We begin with the equation $y = x^r$, but in order to apply the natural logarithm to both sides, we insert an intermediate step to obtain $|y| = |x^r|$. Logarithmic differentiation then gives us the following.

$$|y| = |x|^r \qquad\qquad |x^r| = |x|^r$$

$$\ln|y| = \ln\left(|x|^r\right) \qquad\qquad \text{Assume } x \neq 0.$$

$$\frac{d}{dx}\left(\ln|y|\right) = \frac{d}{dx}\left(r\ln|x|\right) \qquad \text{Differentiate and apply } \ln\left(|x|^r\right) = r\ln|x|.$$

$$\frac{y'}{y} = r \cdot \frac{1}{x}$$

$$y' = r \cdot \frac{y}{x}$$

$$y' = r \cdot \frac{x^r}{x} = rx^{r-1}$$

Note that we are safe in assuming $x \neq 0$ above, as we can easily determine the (possibly one-sided) derivative of x^r at $x = 0$ from a limit of difference quotients for $r \geq 1$, and for $r < 1$ the derivative is undefined at $x = 0$.

─────────────────────────────────

Example 6

Find all points of differentiability for the following functions and determine the derivatives.

a. $f(x) = x^{5/3}$ **b.** $g(x) = x^{1/3}$ **c.** $h(x) = x^{3/2}$

Solution

a. For $x \neq 0$, the Power Rule applies, and we have

$$f'(x) = \frac{5}{3}x^{(5/3)-1} = \frac{5}{3}x^{2/3}.$$

If $x = 0$, we use the definition of the derivative:

$$f'(0) = \lim_{h\to 0}\frac{(0+h)^{5/3} - 0^{5/3}}{h} = \lim_{h\to 0}\frac{h^{5/3}}{h} = \lim_{h\to 0}h^{2/3} = 0.$$

Thus we conclude that $f'(0) = 0$, so f is in fact differentiable everywhere on $\mathbb{R}$, and the formula $f'(x) = \frac{5}{3}x^{2/3}$ does hold for all $x \in \mathbb{R}$.

b. Again, if $x \neq 0$,

$$g'(x) = \frac{1}{3}x^{(1/3)-1} = \frac{1}{3}x^{-2/3} = \frac{1}{3x^{2/3}}.$$

For $x = 0$,

$$g'(0) = \lim_{h\to 0}\frac{(0+h)^{1/3} - 0^{1/3}}{h} = \lim_{h\to 0}\frac{h^{1/3}}{h} = \lim_{h\to 0}\frac{1}{h^{2/3}} = \infty,$$

indicating that $g'(0)$ is undefined (in fact, g has a vertical tangent) at $x = 0$. Thus we conclude that g is differentiable only on $(-\infty, 0) \cup (0, \infty)$.

c. The first important observation about h is that since $x^{3/2} = \sqrt{x^3}$, h is only defined for nonnegative x-values. If $x > 0$, by the Power Rule

$$h'(x) = \frac{3}{2}x^{1/2} = \frac{3}{2}\sqrt{x}.$$

For $x = 0$, the derivative cannot exist, since h is undefined on the left-hand side of 0; however, we may attempt to evaluate the right-hand derivative, using the definition:

$$h'(0) = \lim_{h \to 0^+} \frac{(0+h)^{3/2} - 0^{3/2}}{h} = \lim_{h \to 0^+} \frac{h^{3/2}}{h} = \lim_{h \to 0^+} h^{1/2} = \lim_{h \to 0^+} \sqrt{h} = 0$$

Thus we see that h is undefined and hence cannot be differentiable for $x < 0$, but is differentiable on $(0, \infty)$, with its one-sided derivative also existing at $x = 0$.

Example 7

Use the Power Rule to determine the derivative of $f(x) = \left(\sqrt{\cos x + 1}\right)^{\pi}$.

Solution

First of all, notice that we can rewrite f as

$$f(x) = (\cos x + 1)^{\pi/2}.$$

The Power Rule assures us that the derivative of the power function $y = x^{\pi/2}$ is $y' = \frac{\pi}{2}x^{(\pi/2)-1}$, but in the problem at hand, we need to combine this fact with the Chain Rule.

$$f'(x) = \frac{\pi}{2}(\cos x + 1)^{(\pi/2)-1}(\cos x + 1)'$$

$$= \frac{\pi}{2}(\cos x + 1)^{(\pi/2)-1}(-\sin x)$$

$$= -\frac{\pi}{2}\sin x(\cos x + 1)^{(\pi/2)-1}$$

In conclusion we note that, slightly generalizing the above solution, we can derive what we can call the Generalized Power Rule.

Theorem

The Generalized Power Rule

If $u = u(x)$ is differentiable and x^r is differentiable at $u(x)$, then

$$\frac{d}{dx}(u^r) = ru^{r-1}\frac{du}{dx}.$$

Topic Four

Derivatives of Inverse Trigonometric Functions

Our last application in this section of the Derivative Rule for Inverse Functions will be in determining the derivatives of the inverse trigonometric functions. Recall, from Section 1.4, that an inverse for each of the six common trigonometric functions can be defined, but the precise definition depends on how the domain of each function is restricted (such a restriction is necessary in order to make each function one-to-one). Since there are several possible choices for each restricted domain, there is no one "correct" definition for the six inverse functions—in practice, the restricted domain is often determined on the basis of whatever is most convenient for the particular problem at hand. Fortunately, the formulas for the derivatives of the inverse functions are very similar regardless of the exact definition, and the process outlined below can always be followed to determine the formulas.

With that in mind, the definitions for the inverses that we will use for now are as follows. The graphs of the inverse functions, along with their domains and ranges, are shown in Figure 3.

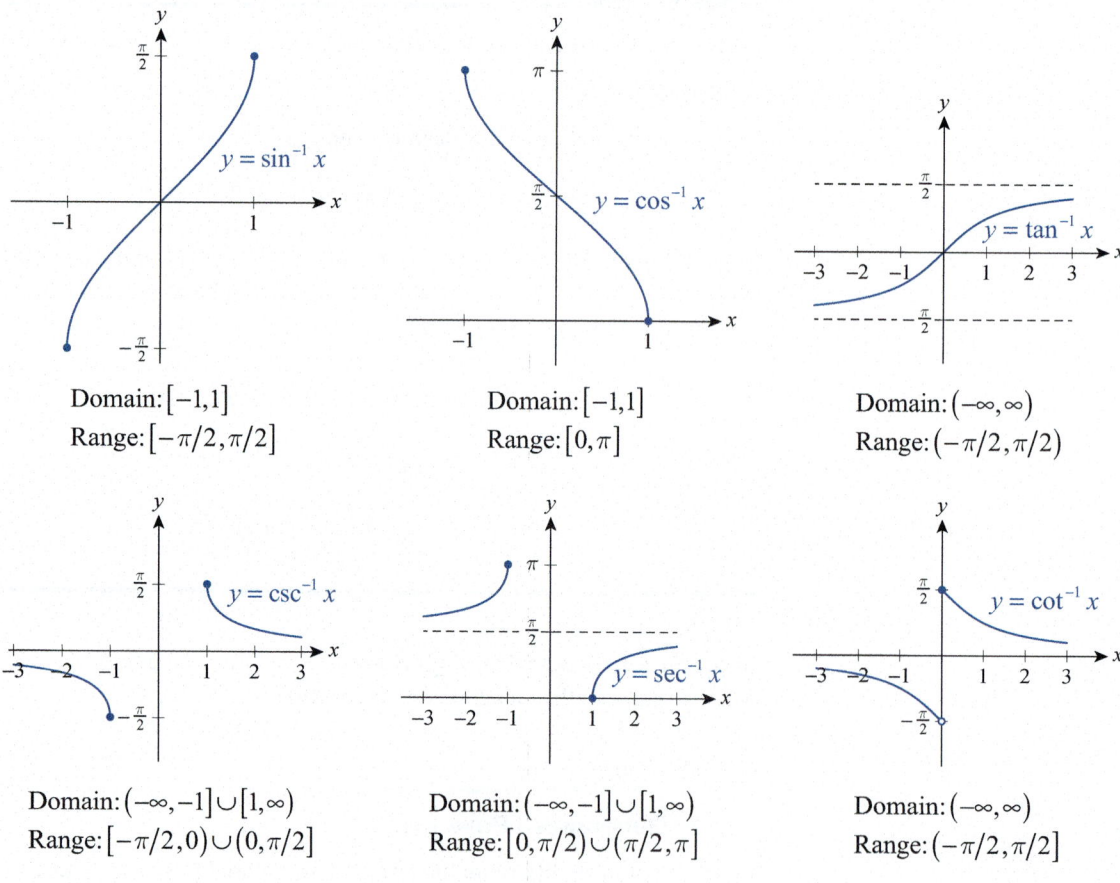

Figure 3 Inverse Trigonometric Functions

One advantage of this particular choice of definitions is that the following identities hold:

$$\csc^{-1} x = \sin^{-1}\left(\frac{1}{x}\right)$$

$$\sec^{-1} x = \cos^{-1}\left(\frac{1}{x}\right)$$

$$\cot^{-1} x = \tan^{-1}\left(\frac{1}{x}\right), \text{ with } \cot^{-1} 0 = \frac{\pi}{2}$$

As a consequence, we will be able to use the Chain Rule to find the derivatives of $\csc^{-1}$, $\sec^{-1}$, and $\cot^{-1}$, once we have formulas for the derivatives of the other three functions. (Different but equally useful identities apply with alternative definitions of the six inverse functions, as illustrated in Exercise 105.)

We are now ready to determine some derivatives, and we will begin with the derivative of arcsine. The process makes use of the techniques outlined in Section 1.4.

$$\left(\sin^{-1}\right)'(x) = \frac{1}{\cos\left(\sin^{-1} x\right)} \qquad\qquad \left(f^{-1}\right)'(x) = \frac{1}{f'\left(f^{-1}(x)\right)} \text{ and } \sin' = \cos$$

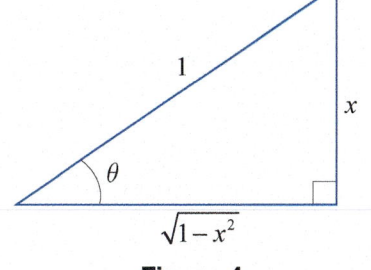

Figure 4

We evaluate expressions such as $\cos\left(\sin^{-1} x\right)$ by constructing a diagram like the one in Figure 4. Start by letting $\theta = \sin^{-1} x$, rewriting this in the equivalent form $\sin\theta = x$, and then labeling the angle, hypotenuse, and opposite side of a right triangle as shown. Such a diagram is the pictorial representation of the statement $\sin\theta = x$, and the Pythagorean Theorem tells us that the adjacent side must consequently have a length of $\sqrt{1-x^2}$. The final steps are then as follows.

$$\left(\sin^{-1}\right)'(x) = \frac{1}{\cos\left(\sin^{-1} x\right)}$$

$$= \frac{1}{\cos\theta} \qquad\qquad \sin^{-1} x = \theta$$

$$= \frac{1}{\sqrt{1-x^2}}. \qquad\qquad \cos\theta = \frac{\sqrt{1-x^2}}{1} \text{ (See Figure 4.)}$$

Example 8

Use the above technique to determine the derivative of $\tan^{-1} x$.

Solution

Again, using the second form of the Derivative Rule for Inverse Functions, we obtain

$$\left(\tan^{-1}\right)'(x) = \frac{1}{\sec^2\left(\tan^{-1} x\right)}.$$

$(f^{-1})'(x) = \dfrac{1}{f'(f^{-1}(x))}$ and $\tan' = \sec^2$

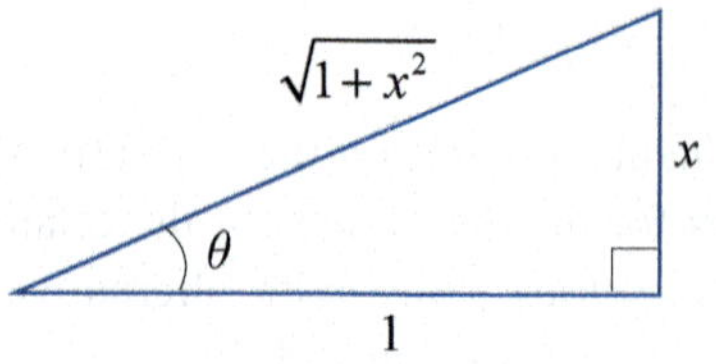

Notice how the labeling in Figure 5 helps determine $\sec^2\left(\tan^{-1} x\right)$. Since

$$\tan\theta = \frac{x}{1} = x$$
$$\theta = \tan^{-1} x,$$

Figure 5

we have

$$\sec^2\left(\tan^{-1} x\right) = \sec^2\theta = 1 + x^2.$$

$\sec\theta = \dfrac{\sqrt{1+x^2}}{1}$

Therefore,

$$\left(\tan^{-1}\right)'(x) = \frac{1}{\sec^2\left(\tan^{-1} x\right)} = \frac{1}{1+x^2}.$$

The derivatives of $\cos^{-1}$, $\sec^{-1}$, and $\cot^{-1}$ are left as exercises (see Exercises 67–69), but we will find the derivative of $\csc^{-1}$ here.

$$\frac{d}{dx}\left(\csc^{-1} x\right) = \frac{d}{dx}\left[\sin^{-1}\left(\frac{1}{x}\right)\right]$$

Use the identity found for $(\sin^{-1})'$.

$$= \frac{1}{\sqrt{1-(1/x)^2}}\frac{d}{dx}\left(\frac{1}{x}\right)$$

Chain Rule

$$= \frac{1}{\sqrt{1-(1/x)^2}}\left(-\frac{1}{x^2}\right)$$

$$= \frac{\sqrt{x^2}}{\sqrt{x^2}\sqrt{1-(1/x)^2}}\left(-\frac{1}{x^2}\right)$$

Multiply by $\dfrac{\sqrt{x^2}}{\sqrt{x^2}}$.

$$= \frac{|x|}{\sqrt{x^2-1}}\left(-\frac{1}{x^2}\right)$$

$\sqrt{x^2} = |x|$

$$= -\frac{1}{|x|\sqrt{x^2-1}}$$

$\dfrac{|x|}{x^2} = \dfrac{1}{|x|}$

For convenience, the derivative rules for the six inverse trigonometric functions are collected as follows, in the more general form that assumes u is a differentiable function of x.

Theorem

Derivatives of Inverse Trigonometric Functions

$$\frac{d}{dx}\left(\sin^{-1} u\right) = \frac{1}{\sqrt{1-u^2}}\frac{du}{dx} \qquad\qquad \frac{d}{dx}\left(\cos^{-1} u\right) = -\frac{1}{\sqrt{1-u^2}}\frac{du}{dx}$$

$$\frac{d}{dx}\left(\tan^{-1} u\right) = \frac{1}{1+u^2}\frac{du}{dx} \qquad\qquad \frac{d}{dx}\left(\cot^{-1} u\right) = -\frac{1}{1+u^2}\frac{du}{dx}$$

$$\frac{d}{dx}\left(\sec^{-1} u\right) = \frac{1}{|u|\sqrt{u^2-1}}\frac{du}{dx} \qquad\qquad \frac{d}{dx}\left(\csc^{-1} u\right) = -\frac{1}{|u|\sqrt{u^2-1}}\frac{du}{dx}$$

Example 9

Use the appropriate rules to find the following derivatives.

a. $\dfrac{d}{dx}\left[\cot^{-1}\left(3x^2\right)\right]$ 　　　　　　　　　**b.** $\dfrac{d}{dx}\left[\sec^{-1}\left(10^{2x}\right)\right]$

Solution

a. $\dfrac{d}{dx}\left[\cot^{-1}\left(3x^2\right)\right] = -\dfrac{\left(3x^2\right)'}{1+\left(3x^2\right)^2}$　　　　　　　$u(x) = 3x^2$

$$= -\frac{6x}{1+9x^4}$$

b. $\dfrac{d}{dx}\left[\sec^{-1}\left(10^{2x}\right)\right] = \dfrac{\left(10^{2x}\right)'}{\left|10^{2x}\right|\sqrt{\left(10^{2x}\right)^2-1}}$　　　　　$u(x) = 10^{2x}$

$$= \frac{10^{2x}\left(\ln 10\right)\cdot\left(2x\right)'}{10^{2x}\sqrt{10^{4x}-1}} \qquad \left|10^{2x}\right| = 10^{2x} \text{ since } 10^{2x} > 0$$

$$= \frac{2\ln 10}{\sqrt{10^{4x}-1}}$$

3.6 **Exercises**

1–15 Use the Derivative Rule for Inverse Functions to determine $\left(f^{-1}\right)'(a)$ for the indicated value of a. (In these and subsequent exercises, the domain of f is assumed to have been restricted so that the inverse exists and is differentiable, whenever appropriate.)

1. $f(x) = x^3$; $a = 8$

2. $f(x) = 2x - 1$; $a = 5$

3. $f(x) = \sqrt{x}$; $a = 3$

4. $f(x) = \sqrt[3]{x + 2}$; $a = -1$

5. $f(x) = x^2 + 5$; $a = 9$

6. $f(x) = x^{3/2}$; $a = 27$

7. $f(x) = \dfrac{2x}{x - 1}$; $a = 4$

8. $f(x) = \dfrac{5}{(x - 1)^3}$; $a = 5$

9. $f(x) = \dfrac{3}{x^2 + 2}$; $a = 1$

10. $f(x) = e^{2x}$; $a = -5$

11. $f(x) = 10^x$; $a = 10$

12. $f(x) = 2^{\sqrt{x}}$; $a = 8$

13. $f(x) = \sin x$; $a = \dfrac{\sqrt{3}}{2}$

14. $f(x) = 2\tan^{-1} x$; $a = \dfrac{\pi}{2}$

15. $f(x) = \sin(x^2)$; $a = \sin 0.01$

16–30 Determine the value of $\left(g^{-1}\right)'(b)$ at the given point (assume that the domain of g is appropriately restricted so that g^{-1} exists). (**Note:** Do *not* attempt to find a formula for g^{-1}.)

16. $g(x) = x^5 + 2x + 1$; $b = g(1)$

17. $g(x) = x^6 - 11x^4 + x$; $b = g(-1)$

18. $g(x) = x^{100} + x^{50} + 1$; $b = g(-1)$

19. $g(x) = \sqrt{x^4 + x^2}$; $b = g(-2)$

20. $g(x) = \left(3x^8 + x^3 + 1\right)^{3/2}$; $b = g(1)$

21. $g(x) = \left(2x^9 - 3\sqrt{x}\right)^{2/5}$; $b = g(1)$

22. $g(x) = x^5 + x + 2$; $b = 2$

23. $g(x) = x^{17} + 2x^{11} - 2x + 3$; $b = 4$

24. $g(x) = \dfrac{x^3 + 8}{\sqrt{x + 1}}$; $b = g(2)$

25. $g(x) = \dfrac{x + 1}{x^3 + 7}$; $b = \dfrac{1}{7}$

26. $g(x) = e^{x^4 - x + 2}$; $b = g(-2)$

27. $g(x) = x\sin x$; $b = \dfrac{\pi}{2}$

28. $g(x) = 10^{\cos(x^3 + x)}$; $b = g(1)$

29. $g(x) = \tan\sqrt{x}$; $b = g(1)$

30. $g(x) = x^3 e^{x^2 + 1}$; $b = e^2$

31–48 Determine the derivative of the given function.

31. $f(x) = \ln(x^3)$

32. $g(x) = (\ln x)^3$

33. $h(x) = \ln(x^2 + 3)$

34. $F(x) = \ln\left(x\sqrt{x^2 + 4}\right)$

35. $G(x) = x\ln\sqrt{x^2 + 4}$

36. $k(x) = \ln\dfrac{2x}{x^2 + 1}$

37. $L(x) = \dfrac{\ln 2x}{x^2 + 1}$

38. $f(x) = \ln\sqrt{\dfrac{x + 3}{2x + 5}}$

39. $g(x) = \ln\sqrt[3]{\dfrac{x + 3}{x - 3}}$

40. $H(x) = \ln(\ln x)$

41. $F(t) = \ln\left(\sqrt{t^2 + 4} + 2t\right)$

42. $L(s) = \ln\dfrac{\sqrt{s^2 + 2}}{s^4 + s^2 + 1}$

43. $T(x) = \ln|\cos x|$

44. $C(t) = \ln(\sin^2 t + 1)$

45. $v(x) = \cos 2x(\ln(\cos 2x))$

46. $F(t) = \dfrac{\log_5 t}{t^2}$

47. $w(x) = x\log x$

48. $t(x) = \log_{3/2}\left(\left(5x^2 + 4\right)^{3/2}\right)$

49–66 Use logarithmic differentiation to find y'.

49. $y = (x+1)(x+2)(x+3)(x+4)$

50. $y = \dfrac{(x+1)(x+2)}{(x+3)(x+4)}$

51. $y = \sqrt[3]{(2x-1)(x-5)(3x+1)}$

52. $y = \dfrac{(x^2-1)^{2/3}(5x^3+3)}{(x^2+x+2)(x^4-10)^{3/4}}$

53. $y = \dfrac{\sqrt[3]{x^3-5x^2+7}(x+2)}{x^{2/3}\sqrt{3x^2+4}}$

54. $y = \sqrt[3]{\dfrac{x^3-2x^2+1}{(x^2-1)(x^3+5)}}$

55. $y = \dfrac{x^2\sqrt[5]{x^3+3}}{\sqrt[4]{x^4+4}}$

56. $y = x^{x^2}$

57. $y = (\sin x)^{1/x}$

58. $y = (2x^2+1)^{\tan x}$

59. $y = (\cos x)^{\sqrt{x}}$

60. $y = (\sqrt[3]{x})^{\sqrt[3]{x}}$

61. $y = (\ln x)^x$

62. $y = \dfrac{(\ln x)^x(x^3-1)}{e^x+2}$

63. $y = x^{x^x}$

64. $y = (\sin x)^{\cos x}$

65. $y = (\ln x)^{\sin x}$

66. $y = (e^x)^x$

67. Mimic the procedure seen in the text to find a formula for the derivative of $y = \cos^{-1}x$.

68. Find a formula for the derivative of $y = \sec^{-1}x$ (see Exercise 67).

69. Find a formula for the derivative of $y = \cot^{-1}x$ (see Exercise 67).

70–93 Determine dy/dx. (Recall that arcsin x is just a different notation for $\sin^{-1}x$, and the same holds for the other inverse trigonometric functions.)

70. $y = \cos^{-1}(x^2)$

71. $y = \tan^{-1}(2x+1)$

72. $y = x\arcsin x$

73. $y = \ln(\arctan x)$

74. $y = (\text{arccot } x)^2$

75. $y = \arccos\sqrt{x}$

76. $y = \tan^{-1}x + \dfrac{x}{1+x^2}$

77. $y = \arccos x - x\sqrt{1-x^2}$

78. $y = \dfrac{\text{arccot }x}{x}$

79. $y = \arctan(e^x)$

80. $y = \text{arccot}(\ln 3x)$

81. $y = \dfrac{1-\arctan x}{1+\arctan x}$

82. $y = \arccos x \cdot \text{arccot } x$

83. $y = \left(\arcsin(x^3)\right)^2$

84. $y = \sec^{-1}(e^{x^2})$

85. $y = \sec^{-1}(x^2+1)$

86. $y = \csc^{-1}(e^{-x})$

87. $y = \sec^{-1}\sqrt{x^2+1}$

88. $y = \sin(\arccos 3x)$

89. $y = (\arctan x)^x$

90. $y = (\arcsin x)^{\ln x}$

91. $y = \cos\left(\text{arccsc}(x^2+1)\right)$

92. $y = \tan\left(\text{arcsec}\sqrt{1+e^{2x}}\right)$

93. $y = \cos\left(\text{arccot}\dfrac{x-1}{\sqrt{2x-1}}\right)$

94–99 Find the equation of the line tangent to the graph of $y = f(x)$ at the indicated x-value. (If needed, round your answer to three decimal places.)

94. $f(x) = \log_2(x^2+1);\quad x = 1$

95. $f(x) = \dfrac{(2+x)2^{\ln x}}{x^2 e^x};\quad x = 1$

96. $f(x) = \arcsin(\ln 3x);\quad x = \dfrac{1}{3}$

97. $f(x) = x\arccos\dfrac{x}{4} - \ln\dfrac{1}{x^2+1};\quad x = 2$

98. $f(x) = (\sin x)^x;\quad x = \dfrac{\pi}{2}$

99. $f(x) = x^{\ln(\arctan x)};\quad x = 1$

100. Differentiate $f(x) = \arcsin x + \arccos x$. What information about f can you glean from your answer?

101. Repeat Exercise 100 for the function $f(x) = \arcsin(1/x) - \text{arccsc } x$.

102. A father is videotaping his child releasing a helium-filled balloon. Assuming that the balloon rises vertically, let the distance between father and child be denoted by s and the height of the balloon, measured from the child, be denoted by h. Find a formula for the angle of elevation α of the camera as it is following the rise of the balloon. Then differentiate with respect to time to find $d\alpha/dt$.

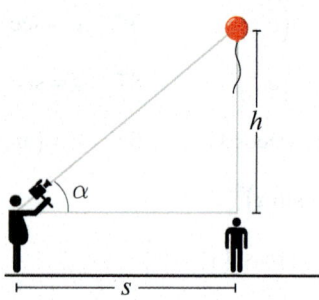

103. The height of the screen of a drive-in movie theater is 20 ft and it is mounted 8 ft above ground level. Find a formula for the angle θ at which the screen is viewed by a driver parked s ft from the screen. Then differentiate to find the rate of change of the viewing angle as a function of the distance s.

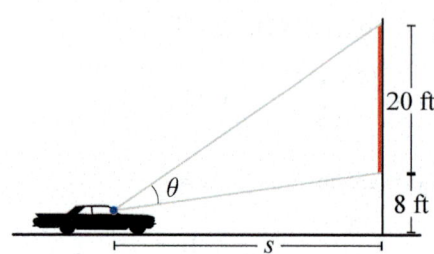

104. An air traffic controller observes a small plane flying horizontally toward the tower and determines from the instrument readings that the distance between the tower and the plane is 10,560 ft, the flying altitude is 5340 ft, and the speed of the plane is 120 mph.

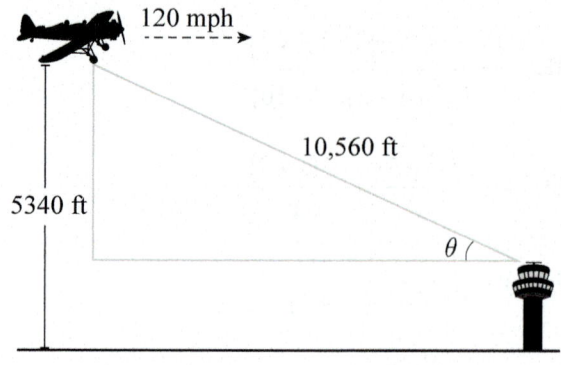

a. Find the angle of elevation θ at which the controller first sees the plane, if the tower is 60 ft high.

b. Find the angular rate of change $d\theta/dt$ when the plane is 1.25 miles from the controller.

105. Give an alternative definition to $\cot^{-1} x$ so as to make the function continuous and satisfy the identity $\cot^{-1} x = (\pi/2) - \tan^{-1} x$. Graph the function. (**Hint:** Appropriately restrict the domain of $\cot x$. You might also think about the relationship between the graph of the function to be defined and that of $\tan^{-1} x$.)

106–109 *True or False?* Determine whether the given statement is true or false. In case of a false statement, explain or provide a counterexample.

106. The tangent lines to the graphs of $\ln x$ and $\ln 3x$ have the same slope for all x.

107. If $y = \log \pi$, then $y' = \dfrac{1}{\ln 10} \cdot \dfrac{1}{\pi}$.

108. The derivative of $\csc^{-1} x$ is negative everywhere.

109. The functions $f(x) = \ln x$ and $g(x) = \log_c x$ are constant multiples, hence so are their derivatives.

3.7 **Rates of Change in Use**

We have already seen how derivatives relate to the physical notions of velocity, acceleration, and jerk, and we now have many examples of the use of differentiation in mathematical applications such as the construction of tangent lines. In this section, we will introduce additional applications in which the rate of change of one quantity with respect to another plays a key role and show how differentiation is used to determine that rate of change.

Topic One

Biology Applications

The ability to mathematically model the population growth (or decline) of a species over time is of great utility, whether the study is relatively small and contained (such as the growth of bacteria in a petri dish over the course of a few hours) or large (such as the number of grey wolves in the western United States over the course of several years). In general, such models can be quite complex and include such factors as food availability, the effects of disease, space constraints, and interactions with other species. But at their most basic, many population models start with an assumption of exponential growth, at least over short periods of time.

The reason for this is that the growth of a population usually depends to a large extent on the number of members capable of producing more members. For instance, in the simple case of bacteria in a petri dish, the more bacteria there are the faster the population will grow, as there are more bacteria to reproduce themselves. If we define $P(t)$ to be the population of a given species at time t, the mathematical statement of the above observation is this: $P'(t) = kP(t)$, where k is a fixed positive constant. That is, the rate of change in the population at time t is proportional to the population at time t.

In future chapters, we will see how to begin with an equation like $P'(t) = kP(t)$ and solve it for $P(t)$. But we can draw upon one of the results of this chapter to make an educated guess now as to the form of $P(t)$. Recall that the number e is the unique exponential base for which $\left(e^x\right)' = e^x$; that is, the exponential function with base e and its derivative are equal. A quick application of the Chain Rule then tells us that for any constant k, $\left(e^{kx}\right)' = ke^{kx}$, an equation very much like $P'(t) = kP(t)$. More generally, note that if we let $P(t) = ce^{kt}$, where c and k are both constants, then $P'(t) = cke^{kt} = kP(t)$. If we specify that the population at time $t = 0$ is P_0, then $P_0 = P(0) = ce^{k(0)} = c$, giving us the following basic population model.

Definition

Exponential Growth Model

$$P(t) = P_0 e^{kt}$$

Example 1

As an example of the use of the exponential growth model, suppose a strain of bacteria being cultured in a petri dish is observed to double in count every hour. That information alone is sufficient to determine the *growth constant k*, under the assumption that t is measured in hours, as follows.

$$P(1) = 2P(0) \qquad \text{Population at 1 hour is twice the initial population.}$$
$$P_0 e^{k(1)} = 2P_0 \qquad \text{Use } P(t) = P_0 e^{kt}.$$
$$e^k = 2 \qquad \text{Cancel } P_0.$$
$$k = \ln 2 \qquad \text{Solve for } k.$$

If the culture began with, say, $P_0 = 15$ bacteria, the population of bacteria (rounded to the nearest integer) and the rate of population growth (rounded to one decimal place) at the 0, 1/2, 1, and 2 hour marks are as follows.

Population $P(t) = 15e^{(\ln 2)t}$	Rate of population growth $P'(t) = (\ln 2)P(t)$
$P(0) = 15e^0 = 15$	$P'(0) = (\ln 2)P(0) \approx 10.4$ bacteria/hour
$P(1/2) = 15e^{(1/2)\ln 2} = 15e^{\ln(2^{1/2})} = 15(2^{1/2}) \approx 21$	$P'(1/2) = (\ln 2)P(1/2) \approx 14.7$ bacteria/hour
$P(1) = 15e^{\ln 2} = 15(2) = 30$	$P'(1) = (\ln 2)P(1) \approx 20.8$ bacteria/hour
$P(2) = 15e^{2\ln 2} = 15e^{\ln(2^2)} = 15(2^2) = 60$	$P'(2) = (\ln 2)P(2) \approx 41.6$ bacteria/hour

Topic 2

Physics Applications

Newton's Second Law of Motion often appears as $F = ma$, where F represents force, m mass, and a acceleration. This is a simplified version of Newton's actual observation, however, which is that the net force on an object is equal to the rate of change (with respect to time) of its momentum: $F = dP/dt$, where momentum P is the product of the object's mass m and velocity v. The Product Rule allows us to rewrite the second law as follows.

Theorem

Newton's Second Law of Motion

$$F = \frac{d}{dt}(mv) = \frac{dm}{dt}v + m\frac{dv}{dt}$$

The fact that mass is constant in many elementary applications accounts for the frequent reduction to the familiar $F = m(dv/dt)$, or $F = ma$.

Example 2

A freight train is slowly moving forward at a constant velocity v under a hopper that is dropping grain at a constant rate into the train's open-topped freight cars. In order to maintain its constant velocity, the train's engine must exert a force (beyond that necessary to counter friction) equal to the rate of change of its momentum. What is that force?

Solution

In this setting, v is constant and hence the dv/dt (or acceleration) factor is 0. But the mass is changing, and so Newton's Second Law tells us that the force necessary to overcome the increasing load is

$$F = \frac{dm}{dt} v + m \cdot 0 = \frac{dm}{dt} v.$$

Using units of kilograms (kg) for mass, meters (m) for distance, seconds (s) for time, and newtons (N) for force, if the train has a constant forward velocity of $\frac{1}{2}$ m/s and is being loaded at a rate of 300 kg/s, the force the engine must exert is

$$F = \frac{dm}{dt} v = \left(300 \text{ kg/s}\right)\left(\frac{1}{2} \text{ m/s}\right) = 150 \left(\text{kg} \cdot \text{m}\right)/\text{s}^2 = 150 \text{ N}.$$

Figure 1

Newton's laws of motion are remarkable for their brevity and versatility—their principles accurately describe the behavior of seemingly unrelated physical situations. As another example of the use of his second law, consider the following experiment.

Example 3

A rope of length L and mass M is held vertically over a scale so that its lower end just touches the scale, and is then allowed to drop onto the scale. What force does the scale register as the rope drops onto it?

Solution

The answer to this question *after* the rope has fully dropped onto the scale and is at rest is easily found with an elementary application of Newton's Second Law: the force registering at that point is the mass of the rope M times the acceleration due to gravity g. The product Mg is what we normally call *weight*, and Mg is indeed what would show on the scale's dial. But the more interesting question is what registers on the scale during intermediate stages when, say, a segment of length x has landed on the scale and the segment of length $L - x$ has yet to hit.

For convenience, let m denote the mass of the segment of length x that has dropped and is lying at rest on the scale. At any moment in time we know that $m = M\left(x/L\right)$ (the total mass times the fraction of the rope on the scale), but it is important to keep in mind that m is continuously changing over time from the moment the rope is dropped to the time when $x = L$. The number registering on the scale at any given moment reflects both the force

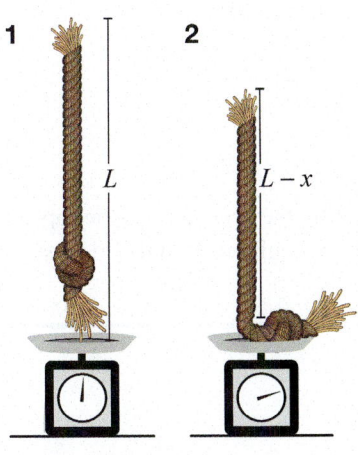

Figure 2

necessary to counter the weight of the mass m and the force necessary to stop the segment of the rope just hitting the scale with velocity v. We will find the net force by determining the rate of change of momentum P (recall that $P = mv$).

Newton's Second Law tells us that

$$F = \frac{dm}{dt}v + m\frac{dv}{dt},$$

and the determination of dm/dt affords a good opportunity to apply the Chain Rule. We already know how m depends on x, but it's also true that m is a function of time. Thus,

$$\frac{dm}{dt} = \frac{dm}{dx} \cdot \frac{dx}{dt} = \left(\frac{M}{L}\right)v,$$

and so

$$F = \frac{dm}{dt}v + m\frac{dv}{dt} = \left(\frac{M}{L}\right)v \cdot v + m\frac{dv}{dt} = \left(\frac{M}{L}\right)v^2 + mg,$$

where we have also replaced the rate of change of velocity dv/dt with the acceleration due to gravity g.

At the moment in time when a segment of length x is lying on the scale, the portion of the rope just hitting the scale has fallen from a height of x. In Exercises 17 and 18 you will study the velocity v of an object with initial velocity 0 that has fallen a distance x, and you will show that $v^2 = 2xg$. This gives us

$$
\begin{aligned}
F &= \left(\frac{M}{L}\right)v^2 + mg \\
&= \left(\frac{M}{L}\right)(2xg) + mg \\
&= \frac{2Mx}{L}g + \frac{Mx}{L}g \qquad\qquad m = \frac{Mx}{L} \\
&= \frac{3Mx}{L}g.
\end{aligned}
$$

Note that this tells us that the force registered by the scale just as the top of the rope hits ($x = L$) is $3Mg$, and then the scale equilibrates and registers Mg soon after.

Topic Three

Chemistry Applications

Chemical reactions produce one or more *product* substances from one or more *reactant* substances. As an illustration, a process in which a molecule of each of two reactants A and B produces a molecule of substance C is represented by

$$A + B \rightarrow C,$$

and chemists often need to understand the rates of change of the concentrations of the individual substances in such a reaction. The notation $[X]$ is used to denote the concentration of a substance X (typically measured in *moles per liter*, where 1 mole $= 6.022 \times 10^{23}$ molecules), so the rates of change of the concentrations of the three substances in the reaction $A + B \rightarrow C$ correspond to the derivatives

$$\frac{d[A]}{dt}, \frac{d[B]}{dt}, \text{ and } \frac{d[C]}{dt}.$$

The **reaction rate** of a chemical process is defined as the rate of change of concentration of one of the products or, equivalently, the negative of the rate of change of concentration of one of the reactants. The negative sign is attached to the derivatives of the concentrations of the reactants because those substances are decreasing with respect to time, while the rate of change of concentration of a product is positive. In the reaction $A + B \rightarrow C$, the reaction rate r is thus defined to be

$$r = -\frac{d[A]}{dt} = -\frac{d[B]}{dt} = \frac{d[C]}{dt}.$$

More generally, the stoichiometric relation between substances in a reaction can be used to determine the reaction's rate. If a molecules of substance A and b molecules of substance B react to produce c molecules of the product C, denoted symbolically as

$$aA + bB \rightarrow cC,$$

then the reaction rate r is

$$r = -\frac{1}{a}\frac{d[A]}{dt} = -\frac{1}{b}\frac{d[B]}{dt} = \frac{1}{c}\frac{d[C]}{dt}.$$

Example 4

Nitrogen gas N_2 and hydrogen gas H_2 react to produce ammonia NH_3 according to the process

$$N_2 + 3H_2 \rightarrow 2NH_3,$$

so the reaction rate r for the process and the rates of change of concentrations of the substances are related by

$$r = -\frac{d[N_2]}{dt} = -\frac{1}{3}\frac{d[H_2]}{dt} = \frac{1}{2}\frac{d[NH_3]}{dt}.$$

If we measure the production of ammonia as $1.2 \text{ mol}/(\text{L}\cdot\text{h})$ (i.e., 1.2 moles of ammonia per liter per hour), then the rate of change of concentration of hydrogen gas is

$$\frac{d[H_2]}{dt} = -\frac{3}{2}\frac{d[NH_3]}{dt} = -\frac{3}{2}\cdot 1.2 \text{ mol}/(\text{L}\cdot\text{h}) = -1.8 \text{ mol}/(\text{L}\cdot\text{h})$$

and the rate of change of concentration of nitrogen gas is

$$\frac{d[N_2]}{dt} = -\frac{1}{2}\frac{d[NH_3]}{dt} = -\frac{1}{2}\cdot 1.2 \text{ mol}/(\text{L}\cdot\text{h}) = -0.6 \text{ mol}/(\text{L}\cdot\text{h}).$$

Topic Four

Business and Economics Applications

Business models often begin with basic assumptions about the cost $C(x)$ of producing x units of a particular product, the revenue $R(x)$ that can be generated by selling x units, and the resultant profit function $P(x) = R(x) - C(x)$. Once these functions are determined, economists and business strategists are then interested in the *marginal cost*, the *marginal revenue*, and the *marginal profit*. These three functions approximate, respectively, the cost, revenue, and profit associated with increasing production from x units to $x + 1$ units.

Although the variable x in this setting can only realistically take on positive integer values, it is convenient to assume the functions C, R, and P have been extended to smoothly changing (in fact, differentiable) functions defined for all positive real x, at least over some interval. Such an assumption is very common in applications, as it then allows us to use the tools of calculus. For example, the cost of increasing production from x units to $x + 1$ units is $C(x+1) - C(x)$, which is also the average rate of change in the cost over the interval $[x, x+1]$.

$$C(x+1) - C(x) = \frac{C(x+1) - C(x)}{(x+1) - x} = \frac{\Delta C}{\Delta x}$$

Under the assumption mentioned above, $\Delta C/\Delta x \to C'(x)$ as $\Delta x \to 0$, so the derivative of the cost function at x is a convenient approximation of the change in cost as production increases from x to $x + 1$ units. This sort of approximation is an example of *linearization*, which, as we will see in Section 3.9, has many important and varied uses. For the moment, we use this observation as the basis for three formal definitions.

Definition

Marginal Cost, Marginal Revenue, and Marginal Profit

$$\text{Marginal Cost} = C'(x)$$
$$\text{Marginal Revenue} = R'(x)$$
$$\text{Marginal Profit} = P'(x) = R'(x) - C'(x)$$

In practice, it is likely that the variable x, representing the number of units produced, varies over time; changes in production rate can be a result of seasonal supply and demand, labor situations, and the availability of raw materials. Calculus allows us to easily incorporate such knowledge into our model and determine, for instance, the rate of change of profit with respect to time:

$$\frac{d}{dt}P(x(t)) = P'(x(t)) \cdot x'(t) = \left[R'(x(t)) - C'(x(t)) \right] \cdot x'(t)$$

Economists often build models on the basis of such quantities as *the number of workers in the labor force* and *the average productivity per worker*. From their understanding of these numbers, they can construct and analyze further quantities, such as gross productivity.

Example 5

Let $w(t)$ represent the number of workers in a given industry at time t, and let $p(t)$ represent each worker's average productivity. Then the gross productivity at time t is given by $g(t) = w(t)p(t)$. Suppose that it is known that at a specific time t the labor force is decreasing at a rate, *relative to the size of the labor force*, of 1 percent. This translates into the statement $w'(t) = -0.01w(t)$ (such "relative" data is often more commonly cited than "absolute" data). Suppose that at the same time it is also known that the average *relative* productivity per worker is increasing at a rate of 3 percent (note again the "relative" modifier). This second fact tells us that $p'(t) = 0.03p(t)$, and we can use these two pieces of information to determine the relative rate of change of gross productivity as follows.

$$
\begin{aligned}
g'(t) &= w'(t)p(t) + w(t)p'(t) && \text{Product Rule} \\
&= (-0.01w)(p) + (w)(0.03p) && \text{The } t\text{'s will be omitted from this point on.} \\
&= -0.01wp + 0.03wp \\
&= 0.02wp = 0.02g
\end{aligned}
$$

So the relative rate of change of gross productivity at time t is 2 percent.

3.7 **Exercises**

1. Work through Example 1 with the following version of the growth model: $P(t) = P_0 a^t$, where a is treated as the (initially unknown) "growth constant." (**Hint:** Since P doubles every hour, $P(1) = 2P(0)$ gives $P_0 \cdot a^1 = 2P_0$.)

2. In an effort to control vegetation overgrowth, 100 rabbits are released in an isolated area that is free of predators. After one year, it is estimated that the rabbit population has increased to 500. Assuming exponential population growth, what will the population be after another 6 months?

3. The population of a certain inner-city area is estimated to be declining according to the model $P(t) = 237{,}000 e^{-0.018t}$, where t is the number of years from the present. What does this model predict the population will be in 10 years?

4. A population of squirrels is growing in a Louisiana forest with a monthly growth constant of 6 percent. If the initial count is 100 squirrels, how many are there in a year? (**Hint:** Let $N(t)$ stand for the number of squirrels after t months, and note that $N'(t) = 0.06N(t)$. Mimic the steps of Example 1 or, alternatively, make use of the fact that $\frac{d}{dt}(a^t) = (\ln a)a^t$.)

5. The process of radioactive decay is akin to population growth in the sense that the rate of decay is proportional to the amount of material present at any given time. Therefore, it shouldn't come as a surprise that this process can be modeled with the same type of function. Suppose that $A(t)$ stands for the amount of a certain radioactive material at time t, and that it is decaying in a way that the rate of decay satisfies $\frac{d}{dt}A(t) = -0.1A(t)$ (note the negative sign), where t is measured in days. If we start with 1000 g of material, how much is left after 10 days?

6. In Exercise 67 of Section 1.2, we defined the *half-life* of a radioactive substance to be the amount of time required for half of the substance to decay. Find the half-life of the material in Exercise 5.

7. Carbon-11 has a radioactive half-life of approximately 20 minutes, and is useful as a diagnostic tool in certain medical applications. Because of the relatively short half-life, time is a crucial factor when conducting experiments with this element.

   a. Determine a so that $A(t) = A_0 a^t$ describes the amount of carbon-11 left after t minutes (as usual, A_0 is the amount at time $t = 0$).

   b. How much of a 2 kg sample of carbon-11 would be left after 30 minutes?

   c. How much of a 2 kg sample of carbon-11 would be left after 6 hours?

8. The half-life of radium-226 is approximately 4 days. Determine what percentage of the initial amount is left after two weeks.

9. According to Newton's Law of Cooling, the rate of change of temperature of a cooling object is proportional to the temperature difference between the object and the surrounding medium, that is, $\frac{dT(t)}{dt} = k[T(t) - T_s]$, where $T(t)$ is the temperature of the object at time t, and T_s is the temperature of its surroundings. Suppose that a cup of 180 °F coffee is left in a 72 °F room and cools to 122 °F in five minutes. How long does it take for the coffee to cool down to 85 °F? (**Hint:** Introduce a new variable for the temperature difference: Let $D(t) = T(t) - 72$, and observe that Newton's law translates into the equation $D'(t) = k \cdot D(t)$. Now mimic the procedure seen in Example 1.)

10.* According to the Stefan-Boltzmann Law, the radiation energy emitted by a hot object of temperature T is $R(T) = kT^4$, where T is measured in kelvins. Use this to find a formula for the rate of change of energy emitted by the coffee of Exercise 9. (**Hint:** Use the following formula to convert degrees Fahrenheit to kelvins: $K = \frac{5}{9}(F - 32) + 273$.)

11. A snowplow is moving at a constant speed of 3 m/s, and the snow it is pushing is accumulating at a rate of 100 kg/s. What extra force is necessary for the engine of snowplow to maintain constant speed despite the increasing mass?

12. When washing his car, Brad is aiming a water hose at the side of the car, with water leaving the hose at a rate of 1 liter per second, with a speed of 15 m/s. If we ignore any "splash backs," what force does the water exert on the side of the car? (**Hint:** Use the equation $F = dP/dt$. See Example 2.)

13. Use Example 3 to find a formula for the force (in newtons) exerted on a scale by a rope dropped on it if the rope is 2 meters long and each centimeter of it weighs 20 grams. (As in the text, let x stand for the length of the segment of the rope that has already landed on the scale.)

14. Referring back to Exercise 12, suppose that after washing the car, with almost half of the contents of his 20-liter bucket still left, Brad pours it with a quick move into a much smaller 5-liter container that is sitting on the ground. Find the force exerted on the bottom of the smaller container by the incoming water at the instant when it starts overflowing. (**Hint:** Modify and use the result of Example 3.)

15. Suppose that the snowplow of Exercise 11 is 3500 kg and starts accelerating from 3 m/s^2 at a rate of 0.1 m/s^2. Assuming there is no snow accumulation this time, find the force exerted by the engine.

16.* Consider the accelerating snowplow of Exercise 15, this time assuming the same accumulation rate for the snow as in Exercise 11. Find the force exerted by the engine at $t = 2$ seconds.

17. Show that the velocity v of an object that has fallen a distance x from rest satisfies the equation $v^2 = 2xg$. (**Hint:** Velocity increases at a constant rate from 0 to v, so the distance x can be calculated as the product of the average velocity and time: $x = v_{ave} \cdot t = \left[(0 + gt)/2\right]t = \frac{1}{2}gt^2$.)

18. Derive the result of Exercise 17 in an alternative way, using the fact that if air resistance is ignored, the potential energy of an object with mass m at altitude x, which is calculated as $E_p = mgx$, is turned into kinetic energy upon impact, which is calculated as $E_{kin} = \frac{1}{2}mv^2$.

19. In an attempt to escape from a predator, a small fish is swimming vertically downward at a rate of 75 cm/s. Find the rate of change of water pressure around the fish. Express your answer in atm/s. (**Hint:** Use the fact that underwater pressure at a depth of d meters is approximately $P(d) = 1 + 0.097d$ standard atmospheres (atm), where 1 atm = 101.325 kPa.)

20. Hydrogen may be obtained from water by a process called electrolysis, according to the process $2H_2O \rightarrow 2H_2 + O_2$. If we measure the production of hydrogen at 2.5 mol/(L·h) (i.e., 2.5 moles of hydrogen per liter per hour), what will be the concentration of the newly obtained oxygen in 3 hours?

21. The combustion of ammonia gas (NH_3) produces nitrogen and water according to the process $4NH_3 + 3O_2 \rightarrow 2N_2 + 6H_2O$. Supposing that the rate of combustion is 0.5 mol/(L·s), what is the rate of the production of water? How many milliliters of water are produced in two seconds? (**Hint:** Use the fact that the approximate molar mass of hydrogen is 1 g, while that of oxygen is 16 g. Also, the mass of 1 mL of water is 1 g.)

22. Magnesium is a flammable metal, and because of its bright light it has traditionally been used in camera flashes, illumination of mine shafts, fireworks, and flares. The reaction itself is described by $2\text{Mg} + \text{O}_2 \rightarrow 2\text{MgO}$. If magnesium burns in a chamber at an initial rate of 1.5 g/s, find the rate (in $\text{mol}/(\text{L} \cdot \text{s})$) at which the concentration of O_2 is decreasing in the chamber. (**Hint:** The approximate molar mass of magnesium is 24 g.)

23–29 Use the technique of linearization to determine the answer.

23. A manufacturer of small remote-controlled cars found its weekly revenue to be $R(x) = 160x - 0.3x^2$ dollars when x units are produced and sold.

   a. Use marginal revenue to estimate the extra revenue when production is increased from 12 to 13 units.

   b. Use the revenue function to calculate the actual revenue increase. Compare your answers.

24. Suppose the monthly cost of producing x units of a particular commodity is $C(x) = 75x^2 + 200x + 5100$ dollars, while the revenue function is $R(x) = 32x(120 - x)$.

   a. Use marginal cost to find the added expense in increasing production from 5 to 6 units.

   b. Use marginal revenue to estimate the revenue generated by raising production from 5 to 6 units.

   c. Find the actual increases in cost and revenue by producing and selling the sixth unit, and compare these numbers to your estimations from parts a. and b.

25. Repeat Exercise 24 for $C(x) = \frac{486}{7}x^2 + 98x + 120$ and $R(x) = -19.5x^2 + 2526x + 442$.

26. A manufacturer models the total cost of producing n hundreds of a particular pocket calculator by the function $C(n) = \frac{1}{900}n^2 + 2500n + 3100$ dollars. Market research shows that all products will be sold at the price of $p(n) = \frac{2}{5}(120 - \frac{1}{4}n)$ dollars per calculator.

   a. Use the marginal cost function to estimate the cost of raising the level of production from 1000 to 1100 calculators.

   b. Use the marginal profit function to estimate the additional profit if the level of production is raised from 1000 to 1100 calculators.

   c. Find the actual increases in cost and profit when production is raised as in parts a. and b., and compare these values with your estimates obtained in the previous parts.

27. A lumber company estimates the cost of producing x units of a product to be $C(x) = 0.1x^2 + 2250x + 1450$ dollars, while the price of each unit has to be $p(x) = 50(128 - 0.2x)$ in order to sell all x units. However, seasonal supply of raw materials makes x dependent on time so that $x(t) = -0.35(t - 6)^2 + 192$, where t is measured in months. Use the value of the marginal profit at $t = 2$ to estimate the change in profit during the third month.

28. Repeat Exercise 27 if $x(t) = 185 + 10\sin^2\left(\frac{\pi}{6}t\right)$.

29. A child playing on the beach is pouring sand from a bucket, forming a sand cone that is growing in such a way that its height is always half of the radius of its circular base. Estimate the change in volume of the sand cone as its height grows from 3 to 4 inches.

30–39 Slightly generalize the technique of linearization to find the answer. For cases where Δx is not 1, estimate the change in a function $f(x)$ by $\Delta f = f'(x)\Delta x$.

30. Suppose the cost of manufacturing x units of a certain commodity was found to be $C(x) = 35x^2 + 20x + 780$ dollars, and that the current production level is 50 units. Use linearization to estimate how the cost changes if production is raised to 50.25 units.

31. Answer the question of Exercise 30 when production is increased from 13 to 13.5 units.

32. Answer the questions of Exercise 24 when production is increased from 6 to 6.75 units.

33. Use marginal analysis to estimate the changes in cost and profit of Exercise 26 when production is decreased from 1000 to 950 calculators. Then find the actual changes and compare them with your estimates.

34. An ice cube with a side length of 1.5 inches starts melting in such a way that its sides are decreasing by 0.1 inches per minute. Use linearization to estimate the change in the cube's volume during the second minute.

35. According to MRI scans, a benign tumor in a patient had a radius of 1.6 cm when it was first discovered, and it is growing by 1 mm each month. Assuming the tumor is spherical, estimate the change in its volume during the first week.

36. The daily output of a small factory is $n(I) = 50\sqrt{I}$ units, where I is the owner's investment measured in dollars. If the current investment is \$100,000, use linearization to estimate how much additional capital is needed to increase the daily output by 5%.

37. Suppose that $w(t)$, the number of workers at a certain factory at time t, has been decreasing at a rate of 2.5 percent, relative to the size of the workforce, due to a recent recession. At the same time, however, the workers' average productivity $p(t)$ has been increasing by 4 percent due to extra training and the inherent fear of a job loss. Find the change in gross productivity. (**Hint:** See Example 5.)

38. The daily output of a factory is $n(I) = 200\sqrt{I(t)}\sqrt[3]{g(t)}$ units, where I is the total investment measured in dollars and g stands for gross productivity. If the total investment is decreasing by 1 percent while the gross productivity is increasing by 2.1 percent (both in the relative sense), find the relative change in the daily output of the factory.

39. Repeat Exercise 38 if $I(t)$ is decreasing by 2 percent, the number of workers is decreasing by 3 percent, but worker productivity is increasing by 5.1 percent. (**Hint:** Recall the equation from Example 5: $g(t) = w(t)p(t)$.)

3.8 Related Rates

In Section 3.7, we saw how the concept of rate of change pervades a wide variety of contexts and disciplines. This section continues that exploration, but with a focus on applications in which two or more rates are present. As we will see, the ability to quantify how the different rates of change relate to one another is the key to solving many problems.

Topic One

Constructing and Solving Related Rates Equations

Commonly, a *related rates* problem centers on two or more physical features that can be measured (such as distance, area, volume, pressure, temperature, etc.) and that are changing over time. In such cases, the rates of change of interest are derivatives with respect to time, and we will demonstrate our solution strategy with such a problem.

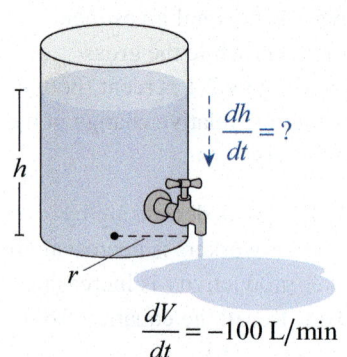

$$\frac{dh}{dt} = ?$$

$$\frac{dV}{dt} = -100 \text{ L/min}$$

Figure 1

Example 1

A cylindrical water tank is being emptied at a rate of 100 liters/minute (L/min). How quickly does the height h of the water level fall if the radius of the tank is 1 meter? How does the answer change if the radius is 10 meters? (Note that 1 liter = 0.001 cubic meters.)

Solution

It is important to identify what it is exactly that we are seeking to determine and what it is that we know. As is typical, the problem as it is given does not contain such mathematical language as "derivative" or even "rate of change," but a careful reading of the problem reveals that we are seeking dh/dt and that we are given information about dV/dt, the rate of change of the volume V of water in the tank.

The next step is to relate the relevant quantities in the problem. In addition to height h and volume V, the radius r of the tank plays a role. These three physical quantities are related by the volume formula $V = \pi r^2 h$. Note that we are *not* trying to relate rates of change at this point—that will come in the following step.

In this problem, two of the three quantities, V and h, change with respect to time t. But we also know how V depends on h. Using the Chain Rule, we now relate dV/dt to dh/dt as follows:

$$\frac{dV}{dt} = \frac{dV}{dh} \cdot \frac{dh}{dt} = \left(\pi r^2\right)\frac{dh}{dt},$$

so

$$\frac{dh}{dt} = \left(\frac{1}{\pi r^2}\right)\frac{dV}{dt}.$$

We are given that $dV/dt = -100\,\text{L/min}$ and $1\,\text{L} = 0.001\,\text{m}^3$, so $dV/dt = -0.1\,\text{m}^3/\text{min}$ (note the negative sign, signifying a loss of volume with respect to time). So if the radius r of the tank is 1 m, then

$$\frac{dh}{dt} = \left(\frac{1}{\pi r^2}\right)\frac{dV}{dt} = \left(\frac{1}{\pi \cdot 1\,\text{m}^2}\right)\left(-0.1\,\text{m}^3/\text{min}\right) \approx -0.03\,\text{m/min},$$

whereas if the radius is 10 m, then

$$\frac{dh}{dt} = \left(\frac{1}{\pi r^2}\right)\frac{dV}{dt} = \left(\frac{1}{\pi \cdot 100\,\text{m}^2}\right)\left(-0.1\,\text{m}^3/\text{min}\right) \approx -0.0003\,\text{m/min}.$$

The first rate of drop in water level is 3 centimeters per minute, which would be perceptible, while the second rate of change would be a very slow 0.3 millimeters per minute.

The steps in solving the problem in Example 1 can be summarized as follows.

Strategy for Solving Related Rates Problems

Step 1: Read the given information carefully and identify what is known and what needs to be found. Typically, the overall goal is to determine the rate of change of one particular variable with respect to time.

Step 2: Identify and appropriately label the relevant quantities and draw a diagram, if possible, as an aid in determining the relationship between them. Express the known and desired quantities in terms of your chosen labels.

Step 3: Write an equation expressing the relationship between the relevant quantities. Be sure you know which of them are functions of time t.

Step 4: Using the Chain Rule, differentiate both sides of the equation with respect to t.

Step 5: Using the given information, solve for the desired rate of change. Check your answer to see if it makes sense as a solution to the problem. Such characteristics as sign, relative magnitude, and the units of your answer are helpful cues.

The remainder of this section will illustrate how this strategy applies to different related rates problems. In each case, note the use of the steps outlined above.

Example 2

A hot air balloon rising straight up is being tracked by a film crew stationed 200 meters away on level ground from its takeoff point. At the moment when the balloon is 150 meters up it is rising at a rate of 50 meters per minute. How fast is the camera angle (measured with respect to the ground) increasing at that moment?

Solution

We know how fast the balloon is rising at a particular moment in time, and we are being asked to determine the rate of change of the camera angle at that moment. Since the balloon is rising straight up from level ground, a right triangle such as the one in Figure 2 captures the essential elements of the problem. The camera crew is stationed at a fixed distance of 200 meters from the takeoff point, but the balloon's height above the ground is changing. In the diagram, y represents the balloon's height, and θ is the camera angle measured from the ground as it tracks the balloon's ascent. We need to determine $d\theta/dt$ at the moment when $y = 150$ m and $dy/dt = 50$ m/min.

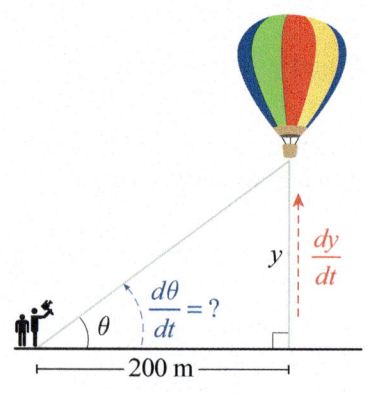

Figure 2

The equation $\tan\theta = y/200$ defines the relationship between y and θ. Now we differentiate both sides with respect to t and solve for $d\theta/dt$.

$$\tan\theta = \frac{y}{200}$$

$$\frac{d}{dt}(\tan\theta) = \frac{d}{dt}\left(\frac{y}{200}\right)$$

$$\sec^2\theta \frac{d\theta}{dt} = \frac{1}{200}\frac{dy}{dt}$$

$$\frac{d\theta}{dt} = \frac{\cos^2\theta}{200}\frac{dy}{dt}$$

We know that $dy/dt = 50$ m/min, but we need to perform one more computation in order to determine $\cos^2\theta$ at the moment when $y = 150$ m. From the Pythagorean Theorem, the hypotenuse at that moment is $\sqrt{(200)^2 + (150)^2} = 250$ m, and so $\cos^2\theta = \left(\frac{200}{250}\right)^2 = \frac{16}{25}$. Hence, we calculate the answer as follows:

$$\frac{d\theta}{dt} = \frac{1}{200}\cos^2\theta\frac{dy}{dt}$$

$$= \left(\frac{1}{200\text{ m}}\right)\left(\frac{16}{25}\right)(50\text{ m/min})$$

$$= 0.16\text{ rad/min}$$

Note that this rate of change of the camera angle is in radians per minute because we implicitly assumed radian measure in applying our differentiation rules; it can also be expressed as approximately 9.2 degrees per minute.

Example 3

Susan is in her car, leading a group of other drivers to a restaurant. She has just completed a right turn at an intersection and is accelerating away. At the moment that she is 40 ft from the intersection and heading north at 25 ft/s, the second car in the convoy is still 40 ft away from the intersection and is approaching it at 15 ft/s. How fast is the "as the crow flies" distance between Susan and the second car increasing at that particular moment?

Solution

We know how quickly Susan is pulling away from the intersection, how quickly the second car is approaching it, and their respective distances from the intersection at a given moment in time. We want to determine the rate of change of the actual distance between the two cars at that moment.

A right triangle is again the appropriate diagram, though this time both legs (as well as the hypotenuse) of the triangle are changing with respect to time. With the labels as shown, we know that $dy/dt = 25$ ft/s when $y = 40$ ft and that $dx/dt = -15$ ft/s when $x = 40$ ft (note that dx/dt is negative since x is decreasing). We want to determine dh/dt at that particular moment.

Differentiating both sides of the relationship $x^2 + y^2 = h^2$ with respect to t yields the following:

$$2x\frac{dx}{dt} + 2y\frac{dy}{dt} = 2h\frac{dh}{dt}$$

$$\frac{x}{h}\frac{dx}{dt} + \frac{y}{h}\frac{dy}{dt} = \frac{dh}{dt}$$

At the moment when $y = x = 40$, $h = 40\sqrt{2}$. Substituting these numbers and the two rates that we are given, we calculate the answer.

$$\frac{dh}{dt} = \frac{40}{40\sqrt{2}}(-15 \text{ ft/s}) + \frac{40}{40\sqrt{2}}(25 \text{ ft/s})$$

$$= \frac{10}{\sqrt{2}} \text{ ft/s} \approx 7.07 \text{ ft/s}$$

The sign and magnitude of this answer are reasonable. The magnitude is comparable to the other two rates of change, and since Susan is pulling away from the intersection (thus increasing the relative distance) at a faster rate than the second car is approaching the intersection (and shortening the distance), we would expect the rate of change to be positive.

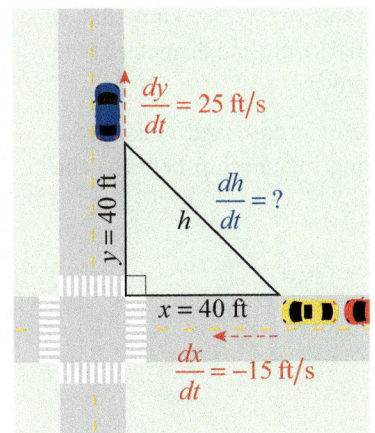

$\frac{dy}{dt} = 25$ ft/s

$y = 40$ ft

$\frac{dh}{dt} = ?$

h

$x = 40$ ft

$\frac{dx}{dt} = -15$ ft/s

Figure 3

Example 4

Corn is falling from the end of a conveyor belt at a rate of $10 \text{ m}^3/\text{min}$ and is forming a conical pile below. As the corn falls, the height of the pile is remaining equal to the diameter of the base. How fast are the height and radius of the pile changing at the moment when the pile is 4 m high?

Solution

If we let V denote the volume of corn in the conical pile, h the height of the pile, and r the radius of the base, then we have been given information about dV/dt and want to determine dh/dt and dr/dt. We know that h and r are related by the equation $h = 2r$ since the height and diameter of the base remain equal to one another as the pile grows.

Beginning with the formula for the volume of a cone and making the substitution $r = h/2$, we obtain the following relationship between the three variables.

$$V = \frac{1}{3}\pi r^2 h = \frac{1}{3}\pi \left(\frac{h}{2}\right)^2 h = \frac{1}{12}\pi h^3$$

Differentiating with respect to t, we have

$$\frac{dV}{dt} = \frac{1}{4}\pi h^2 \frac{dh}{dt},$$

and substituting our given values for the rate of change of volume and the height $h = 4$ m, we find that

$$10 \text{ m}^3/\text{min} = \frac{1}{4}\pi \left(16 \text{ m}^2\right)\frac{dh}{dt} = \left(4\pi \text{ m}^2\right)\frac{dh}{dt},$$

so

$$\frac{dh}{dt} = \frac{5}{2\pi} \text{ m/min} \approx 0.80 \text{ m/min}.$$

Note that the equation $r = h/2$ tells us that the rate of change of the radius is half the rate of change of the height, so $dr/dt \approx 0.40 \text{ m/min}$.

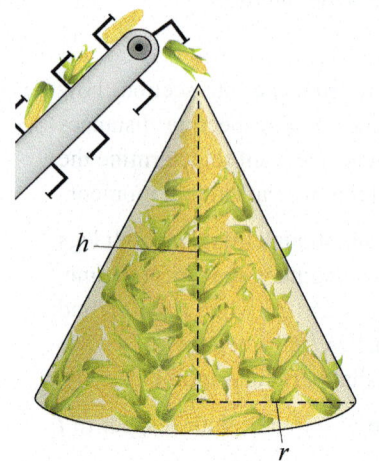

Figure 4

As you might expect, not every related rates problem can be solved with a straightforward and literal application of our solution strategy. But the principles in the strategy are still good guidelines and can be adapted as necessary. Our final example illustrates this point.

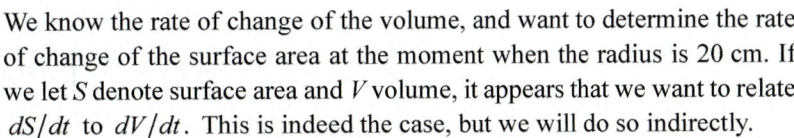

Example 5

A spherical balloon is being filled with helium at a rate of $200 \text{ cm}^3/\text{s}$. At the moment when the radius is 20 cm, how fast is the surface area of the balloon increasing?

Solution

We know the rate of change of the volume, and want to determine the rate of change of the surface area at the moment when the radius is 20 cm. If we let S denote surface area and V volume, it appears that we want to relate dS/dt to dV/dt. This is indeed the case, but we will do so indirectly.

We can begin with the relationships that we know:

$$V = \frac{4}{3}\pi r^3 \text{ and } S = 4\pi r^2$$

V, S, and r are all functions of time and we can relate the following rates easily enough:

$$\frac{dV}{dt} = \frac{4}{3}\left(3\pi r^2 \frac{dr}{dt}\right) = 4\pi r^2 \frac{dr}{dt} \text{ and } \frac{dS}{dt} = 8\pi r \frac{dr}{dt}$$

Since we know dS/dt in terms of dr/dt and can express dr/dt in terms of dV/dt, our path forward is beginning to appear. Note that $dV/dt = 200 \text{ cm}^3/\text{s}$ and that $r = 20$ cm at the moment in time we are interested in. So at that moment,

$$\frac{dr}{dt} = \frac{1}{4\pi r^2}\frac{dV}{dt} = \frac{1}{4\pi\left(400 \text{ cm}^2\right)}\left(200 \text{ cm}^3/\text{s}\right) = \frac{1}{8\pi} \text{ cm/s}.$$

Now we can calculate the rate of change of the surface area.

$$\frac{dS}{dt} = 8\pi r \frac{dr}{dt} = 8\pi\left(20 \text{ cm}\right)\left(\frac{1}{8\pi} \text{ cm/s}\right) = 20 \text{ cm}^2/\text{s}$$

Note that our units of measurement are consistent with a change of area over time—such dimensional verification is a useful way to catch errors.

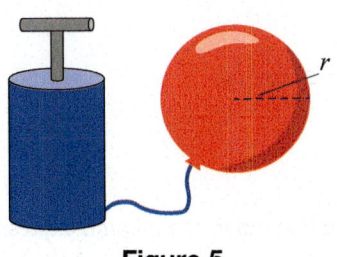

Figure 5

Topic Two
Interlude: A Cautionary Tale

We answered the question in Example 5 by appropriately modifying the steps in our strategy for solving related rates problems. Namely, since we didn't have a single equation in S and V that we could differentiate with respect to t, we worked with two separate equations and substituted information about dr/dt from one into the other. More generally, note that

$$\frac{dr}{dt} = \frac{1}{4\pi r^2}\frac{dV}{dt}$$

and so

$$\frac{dS}{dt} = 8\pi r \frac{dr}{dt} = \frac{8\pi r}{4\pi r^2} \frac{dV}{dt} = \frac{2}{r} \frac{dV}{dt}.$$

If $r = 20$ cm and $dV/dt = 200$ cm^3/s, then we have

$$\frac{dS}{dt} = \left(\frac{2}{20 \text{ cm}} \right)(200 \text{ cm}^3/\text{s}) = 20 \text{ cm}^2/\text{s},$$

the answer we obtained before.

But we might try to attack the problem in a different manner. If we *could* express S in terms of V, we could differentiate both sides with respect to t and again arrive at a relation between dS/dt and dV/dt. In pursuit of this, note that

$$\frac{r}{3}S = \frac{r}{3}\left(4\pi r^2\right) = \frac{4}{3}\pi r^3 = V,$$

so

$$S = \frac{3}{r}V.$$

But then this seems to imply that

$$\frac{dS}{dt} = \frac{dS}{dV} \cdot \frac{dV}{dt} = \frac{3}{r} \frac{dV}{dt},$$

which is not the result of $(2/r)(dV/dt)$ that we found above! What has gone wrong?

The answer is that just as V is a function of r, r can also be said to be a function of V; that is, as V changes, it certainly implies a change in r. In fact,

$$r = cV^{1/3},$$

where $c = \left(\frac{3}{4\pi}\right)^{1/3}$ (this comes from solving $V = \frac{4}{3}\pi r^3$ for r). So,

$$S = \frac{3}{r}V = \frac{3}{cV^{1/3}}V = \frac{3}{c}V^{2/3}.$$

We leave it as an exercise (Exercise 15) to show that if this last equation is differentiated with respect to t and simplified, the result is again $dS/dt = (2/r)(dV/dt)$.

Moral: Watch out for hidden functional relationships between the variables in your equations!

3.8 **Exercises**

1. A theme park ride is descending on a parabolic path that can be approximated by the equation $y = -\frac{1}{90}x^2 + 90$ (distance is measured in feet). If the horizontal component of its velocity is a constant 6 ft/s, find the rate of change of its elevation when $x = 22.5$.

2. Adapt your solution from Exercise 1 to find dx/dt at $x = 30$ feet if the equation of the ride's path is $y = 0.01(x - 95)^2 - 2.25$ and $dy/dt = -20$ ft/s.

3–10 Find the rate of change using the given information.

3. $\dfrac{dy}{dt}$ at $y = 3$, if $y = \sqrt{x+2}$ and $\dfrac{dx}{dt} = 1$

4. $\dfrac{dy}{dt}$ at $x = 2$, if $x^2 + y^2 = 5$, $y > 0$, and $\dfrac{dx}{dt} = 3$

5. $\dfrac{dx}{dt}$ at $y = 0.5$, if $y = \dfrac{1}{x}$ and $\dfrac{dy}{dt} = -2$

6. $\dfrac{dx}{dt}$ at $x = 1$, if $xy^2 = \dfrac{1}{4}$ and $\dfrac{dy}{dt} = -0.25$

7. $\dfrac{dy}{dt}$ at $x = 0$, if $y = \dfrac{x+2}{x^2+1}$ and $\dfrac{dx}{dt} = -5.2$

8. $\dfrac{dy}{dt}$ at $y = \dfrac{1}{2}$, if $y = \dfrac{1}{2}e^{-x}$ and $\dfrac{dx}{dt} = 25$

9. $\dfrac{dy}{dt}$ at $x = -\dfrac{3\pi}{4}$, if $y = 2\sin\left(x + \dfrac{\pi}{4}\right)$ and $\dfrac{dx}{dt} = 7.4$

10. $\dfrac{dx}{dt}$ at $y = \dfrac{\pi}{4}$, if $x = \cot y$ and $\dfrac{dy}{dt} = -3.35$

11. The length of a rectangle is increasing at a rate of 5 in./s, while its width is decreasing at 2 in./s. Find the rate of change of its area when its length is 45 in. and its width is 25 in.

12. Find a formula for the rate of change of the distance from the origin of a point moving on the graph of $f(x) = x^2$ when $x = 2$ and $dx/dt = 3$ units per second.

13. Find the rate of separation between the moving points (x_1, y_1) and (x_2, y_2) on the graph of $y = \sin x$ when $x = \pi/2$ if they start at the origin at the same time, and the horizontal components of their velocities are $dx_1/dt = 1/2$ units per second and $dx_2/dt = -1/2$ units per second, respectively.

14. A Ferris wheel of radius 34 feet needs 3 minutes to complete a full revolution. At what rate is a rider descending when she is 51 feet above ground level?

15. Using the notation of Example 5, use the Chain Rule to differentiate the equation $S = (3/c)V^{2/3}$ with respect to time to obtain $dS/dt = (2/r)(dV/dt)$. (**Hint:** After differentiating, make use of the equation $r = cV^{1/3}$ again.)

16. Rework Example 1 assuming that the tank is a rectangular prism with a 2 m by 2 m square base.

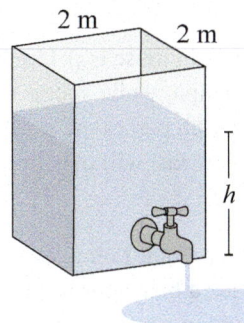

17. Rework Example 1 again, this time assuming that the tank is an inverted, right square pyramid of height 4 m and a 2 m by 2 m base. How fast is the level of water falling when its depth is 2 m?

18. A spectator is tracking a stunt plane at an air show with his video camera. If the plane is on a near-vertical path, rising at a speed of 100 feet per second, and the camera is 400 feet from the point on the ground directly below the plane, how fast is the camera angle changing when the plane's altitude is 400 feet? How fast is the distance between the camera and the plane increasing at that instant?

19. A cistern in the form of an inverted circular cone is being filled with water at the rate of 75 liters per minute. If the cistern is 5 meters deep, and the radius of its opening is 2 meters, find the rate at which the water level is rising in the cistern half an hour after the filling process began. (**Hint:** $1 \text{ m}^3 = 1000 \text{ L.}$)

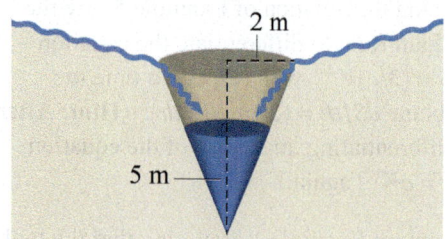

20. Repeat Exercise 19, this time assuming that the cistern is in the form of a pyramid with a 4-by-4-meter square opening.

21. A ship passes a lighthouse at 3:15 p.m., sailing to the east at 10 mph, while another ship sailing due south at 12 mph passes the same point half an hour later. How fast will they be separating at 5:45 p.m.?

22. A tourist at scenic Point Loma, California uses a telescope to track a boat approaching the shore. If the boat moves at a rate of 10 meters per second, and the lens of the telescope is 35 meters above water level, how fast is the angle of depression of the telescope (θ) changing when the boat is 200 meters from shore?

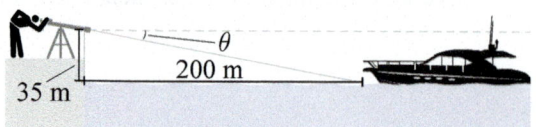

23. When preparing cereal for her child, a mother is pouring milk into a bowl, the shape of which can be approximated by a hemisphere with a radius of 6 in. If milk is being poured at a rate of $4 \text{ in.}^3/\text{s}$, how fast is the level of milk rising in the bowl when it is 1.5 inches deep? (**Hint:** The volume of fluid of height h in a hemispherical bowl of radius r is $V = \pi h^2 \left(r - \frac{1}{3}h\right)$.)

24. Suppose that in Exercise 33 of Section 3.7, the sand is being poured at a rate of 8 cubic inches per second. Find the rate of change of the height of the cone when it is 4 inches tall.

25. When finished playing in the sand, the child of Exercise 24 takes advantage of a nice wind and starts flying his kite on the beach. When the kite reaches an altitude of 60 feet the wind starts blowing it horizontally away from the child at a rate of 15 feet per second while maintaining the altitude of the kite. How fast does the child have to be letting out the string when 100 feet are already out?

26. A passenger airplane, flying at an altitude of 6.5 miles at a ground speed of 585 miles per hour, passes directly over an observer who is on the ground. How fast is the distance between the observer and the plane increasing 3 minutes later?

27. A military plane is flying directly toward an air traffic control tower, maintaining an altitude of 9 miles above the tower. The radar detects that the distance between the plane and the tower is 15 miles and that it is decreasing at a rate of 950 miles per hour. What is the ground speed of the plane?

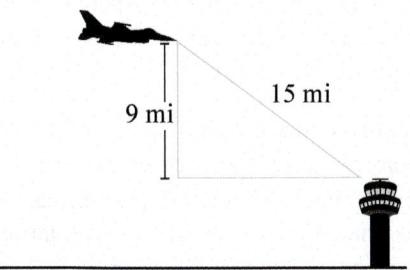

28. A child is retrieving a wheeled toy that is attached to a string by pulling in the string at a rate of 1 foot per second. If the child's hands are 3 feet from the ground, at what rate is the toy approaching when 5 feet of the string are still out?

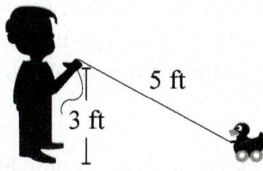

29. A fisherman is reeling in a fish at a rate of 20 centimeters per second. If the tip of his fishing rod is 4.5 meters above the water, and we are assuming that the fish is near the water surface throughout the process, how fast is it approaching when 7.5 meters of fishing line are still out? How fast is the angle θ between the fishing line and the water increasing at that instant?

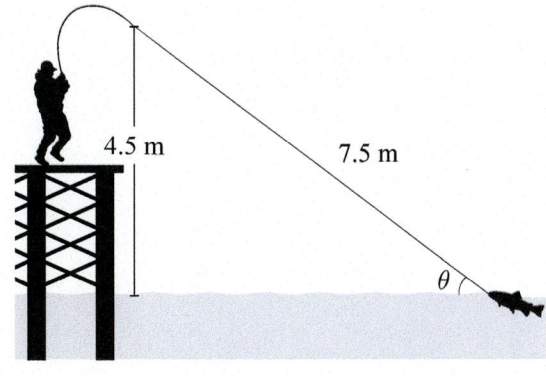

30. A construction worker is using a winch to pull a 9-foot column to a vertical position. If the winch is in the exact position where the top of the installed column is supposed to be, and the rope is being pulled at the rate of 6 inches per second, at what rate is the angle between the column and the ground changing when it is $\pi/6$ radians? At what rate is the top of the column rising vertically at that instant?

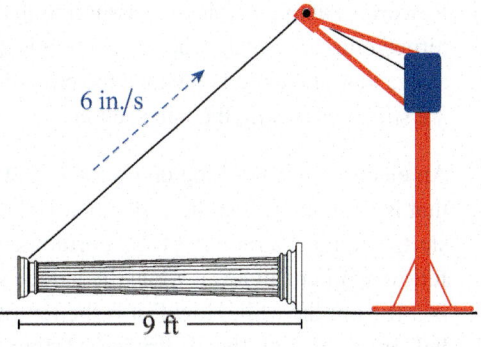

31. The volume of a cube is decreasing at a rate of 150 mm^3/s. What is the rate of change of the cube's surface area when its edges are 30 mm long?

32. The acute angles of a rhombus are increasing at a rate of 0.25 radians per second. If the sides of the rhombus are 20 cm, at what rate is the area of the rhombus increasing when the acute angles are $\pi/3$ radians?

33. A 35-foot-by-18-foot pool, whose depth increases uniformly from 3 feet to 8 feet (along the 35-foot side), is being filled with water at the rate of 4.5 cubic feet per minute. You observe that water appears to be "creeping up" on the angled bottom much faster than it rises along the vertical walls. Find the rate at which the water rises along the angled bottom at the instant when the water level is 2 feet at the deep end of the pool.

34. Considering again the pool of Exercise 33, suppose that it is measured that the water is climbing upward along the angled bottom at a rate of 3.02 in./min when the water level is 1 foot at the deep end. Assuming that the pump is working at the same rate of 4.5 cubic feet per minute, use this information to prove that the pool has a leak, and find the rate at which water is leaking out of the pool.

35. A trough that is 5 meters long and 1 meter across at the top has a cross-section in the form of an isosceles trapezoid and both of its endplates are vertical. The altitude of the trapezoid is 40 centimeters, and the shorter base is 20 centimeters long. If the trough is being filled at the rate of 30 liters per minute, how fast is the water level rising at the instant when the water's depth is 20 centimeters?

36. Rework Example 3, this time assuming that Susan turned onto a highway whose elevation is 20 feet above that of the road on which the other drivers are still approaching the intersection.

37. An electrician is working on top of a 15 ft ladder that is leaning against the wall when its bottom starts sliding at a rate of 1 ft/s. Fortunately, a fellow worker catches it when the ladder's bottom is 5 ft from the wall. How fast is the top of the ladder (along with the electrician) sliding down the wall at that instant?

38. Adam is arriving home one evening in his SUV and is slowly approaching his garage door at a rate of 5 ft/s when the sensor lights come on. If the lights are mounted directly above the door at a height of 15 ft from the ground and Adam's SUV is 6 ft tall, at what rate is the length of the car's shadow shrinking when it is 25 ft from the garage door? What is the speed of the tip of the car's shadow?

39. A baseball player is running from first base to second base at 25 feet per second. At what rate is his distance increasing from home plate when he is 22.5 feet from second base? (**Hint:** The baseball diamond is a 90-foot-by-90-foot square.)

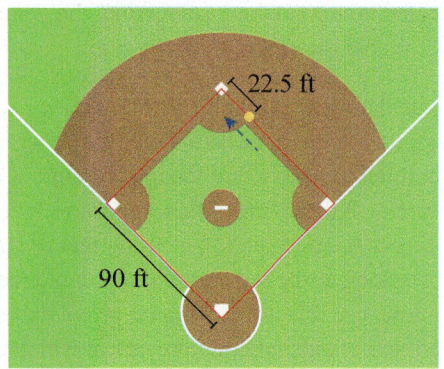

40.* A container in the shape of a cone, standing on its circular base, is being filled with water at the rate of 1.5 cubic feet per minute. If the radius of the base is 2 feet and the height of the cone is $2\sqrt{3}$ feet, how fast is the water level rising when it is 2 feet deep? (**Hint:** The volume of liquid in a partially filled conical tank is $V = \frac{1}{3}\pi d\left(R^2 + Rr + r^2\right)$, where R is the radius of the base, r is the radius of the top of the liquid, and d is its depth.)

41.* Italian police are chasing a criminal down a narrow street at a speed of 90 kilometers per hour. If the blue light on the top of the car is rotating counterclockwise at a rate of 1 rotation per second, and the buildings are only 3 meters from the car on the right, how fast is the beam moving on the wall at the instant when it is already 6 meters ahead of its source?

42.* When studying for a calculus test, Roger accidentally pushes his book over the edge of his 2.5 ft high desk. If his 6 ft tall lamp is standing 3 ft from where the textbook fell down, how fast was the book's shadow moving when the text hit the ground? (Ignore air resistance. Use $g \approx 32$ ft/s^2.)

43.* A wall clock has a 10 in. minute hand and a 6 in. hour hand. At what rate are the tips of the hands approaching each other at 3 o'clock?

44.* The *lens equation*, easily derivable from geometric similarity for a thin converging lens, is

$$\frac{1}{o} + \frac{1}{i} = \frac{1}{f},$$

where *o* (the *object distance*) and *i* (the *image distance*) are the respective distances of the object and the image from the lens, and *f* is the *focal length* of the lens. Suppose a 100 mm high object is being slowly moved away from a lens at a speed of 5 mm/s. The focal length of the lens is 200 mm.

a. Find the rate at which the image changes its location when the object distance is 600 mm.

b. Find the rate at which the image changes its size at the same instant.

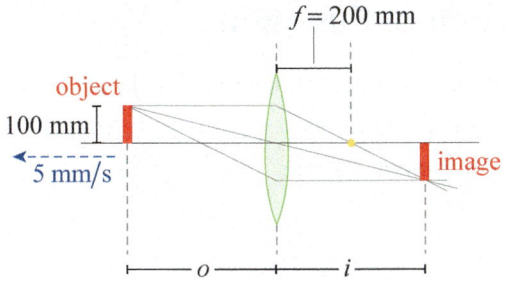

45. Suppose that the torque output of an automobile engine, as a function of engine speed, is approximated by

$$T(s) = \left(-0.001/150^4\right)(s - 3000)^4 + 160 \text{ lb} \cdot \text{ft},$$

where *s* is measured in revolutions per minute (rpm), and that the engine revs up from 0 to 5000 rpm (assume no gear shift takes place).

a. Use a graphing calculator or computer algebra system to graph the torque as a function of *s* on the interval $[0, 5000]$ (this is called the engine's *torque curve*).

b. If the power output of the engine, measured in horsepower (hp), is calculated by $P = \frac{1}{5252} sT(s)$ hp, and the engine is revving up according to the function $s(t) = 1000t$ (*t* is measured in seconds), find the rate of change of the power output at $t = 3$ seconds.

3.9 Linearization and Differentials

As we have seen, the line tangent to a (differentiable) curve at a given point captures the trend of the curve, in the sense that the tangent line's rise or fall matches that of the curve. We say that the tangent line *approximates* the curve in the neighborhood of the point, and in many instances such an approximation is a convenient and sufficient substitution for the curve. This is illustrated in Figure 1 with the curve $y = x^2 + 1$ and its tangent line at $(1, 2)$.

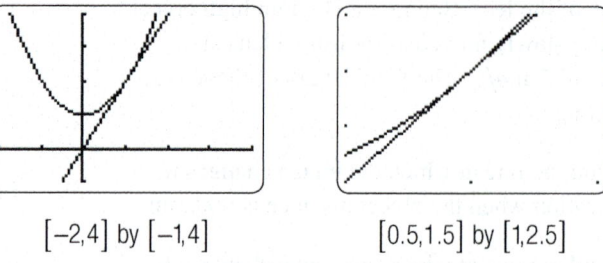

$[-2,4]$ by $[-1,4]$ $[0.5,1.5]$ by $[1,2.5]$

Figure 1

Topic One

Linear Approximation of Functions

Given a function f, which is differentiable at the point $(c, f(c))$, the line tangent to f at that point is described by the equation $y = f(c) + f'(c)(x - c)$. Such a line, which itself is a function of x, merits a name.

Definition

Linear Approximation

Given a function f, differentiable at the point $(c, f(c))$, we call

$$L(x) = f(c) + f'(c)(x - c)$$

the **linear approximation** or **linearization** of f at the point c. Because L is a first-degree polynomial in x, such a formula is also referred to as a **first-order approximation**, a phrase that hints at the higher-order approximations we will discuss when we study Taylor series later in this text.

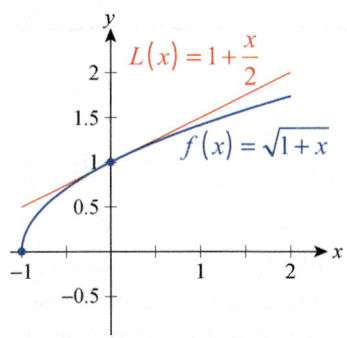

$L(x) = 1 + \dfrac{x}{2}$

$f(x) = \sqrt{1+x}$

Figure 2

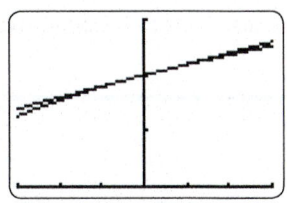

Figure 3

$f(x)$ and $L(x)$

on $[-0.6, 0.6]$ by $[0, 1.5]$

Example 1

Find the linearization of $f(x) = \sqrt{1+x}$ at $x = 0$.

Solution

Since $f(0) = 1$ and

$$f'(x) = \frac{1}{2}(1+x)^{-1/2},$$

$f'(0) = \frac{1}{2}$ and hence

$$L(x) = f(0) + f'(0)(x-0) = 1 + \frac{1}{2}(x-0) = 1 + \frac{x}{2}.$$

The graph in Figure 2 is an illustration of how f and L relate to one another. In this example, note that the linearization lies above the graph of f in the neighborhood of $x = 0$, so using L to approximate f for nearby points would result in an overestimate. The critical observation, however, is that the difference between f and L is vanishingly small as $x \to 0$, as shown in Figure 3.

While linear approximation was once useful as an aid to the actual numerical evaluation of functions, calculators and computers have largely reduced the importance of this particular use. Linearization still has great value, however, in developing mathematical models and understanding the behavior of complicated functions over small intervals. We do this by making use of the fact that

$$f(x) \approx L(x) = f(c) + f'(c)(x - c)$$

for all x close to the point c.

As a simple example, and as a generalization of Example 1, consider the behavior of the family of functions $f(x) = (1+x)^k$ near $x = 0$, where k can take on any real number. Following the same procedure as before,

$$(1+x)^k \approx f(0) + f'(0)(x-0) = 1 + kx$$

for x near 0 and for any k. This approximation holds true even if the argument of the function appears more complex, as long as it approaches 0 as x itself does:

$$\frac{1}{1+x} = (1+x)^{-1} \approx 1 + (-1)x = 1 - x,$$

$$\frac{1}{\sqrt{1+3x}} = (1+3x)^{-1/2} \approx 1 + \left(-\frac{1}{2}\right)(3x) = 1 - \frac{3}{2}x,$$

and $\quad \dfrac{1}{\sqrt[3]{1-2x^4}} = \left[1 + \left(-2x^4\right)\right]^{-1/3} \approx 1 + \left(-\frac{1}{3}\right)\left(-2x^4\right) = 1 + \frac{2}{3}x^4.$

As an example of linearization in the development of models, physicists routinely make use of the fact that $\sin\theta \approx \theta$ for θ close to 0. In fact, the standard model describing the motion of a pendulum, a model that has had great practical utility over the last several centuries, requires this approximation in order for the motion to be expressed in terms of elementary functions. As the next example indicates, this is not unreasonable for small θ.

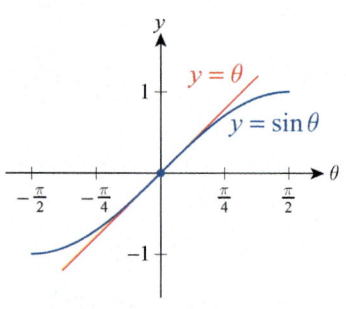

Figure 4

Example 2

Find the linearizations of $\sin\theta$, $\cos\theta$, and $\tan\theta$ at $\theta = 0$.

Solution

We have $\sin 0 = 0$ and $\left.\dfrac{d\left(\sin\theta\right)}{d\theta}\right|_{\theta=0} = \cos 0 = 1$, so $\sin\theta \approx 0 + \left(1\right)\left(\theta - 0\right) = \theta$.

Similarly, we find that $\cos\theta \approx 1$ and $\tan\theta \approx \theta$ at $\theta = 0$.

How close are these approximations? As we will learn later when we study Taylor series, the difference between $\sin\theta$ and θ is no larger than $\left|\theta^3/6\right|$. So for example, over the interval $\left[-0.1, 0.1\right]$, the error in the approximation for $\sin\theta$ is no larger than $0.000\overline{16}$.

Figure 4 indicates how closely $\sin\theta$ and its linearization correspond over small intervals centered at 0.

Topic Two

Differentials

When the Leibniz notation dy/dx was introduced in Section 3.1, we mentioned in passing that such notation does indeed represent, as it appears to, a ratio. To this point, however, we have not dealt with the components of the ratio separately; we are now ready to do so.

Definition

Differentials

Given a differentiable function $y = f\left(x\right)$, the **differential** *dx* is defined to be an independent variable and the **differential** *dy* is a dependent variable defined by

$$dy = f'\left(x\right)dx.$$

That is, *dx* can take on any real number value, while the value of *dy* depends on the values of $f'\left(x\right)$ and *dx*.

Don't be misled by the unusual nature of the symbols used for differentials. Remember that the symbol chosen to represent a given variable really has no purpose other than (at best) to suggest the meaning or role of the variable. In usage, dx typically denotes a small quantity, and often appears in expressions such as $c + dx$. If this reminds you of the usage of Δx, that is no coincidence—the two often play the same role. Similarly, dy is reminiscent of Δy, and like Δy it represents a vertical change. But while for a given function f and a given fixed value of x we may define dx and Δx to be identical, we cannot assume dy and Δy to be equal. The distinction between the two, as well as the connection between differentials and the linearization of a function, is illustrated in Figure 5.

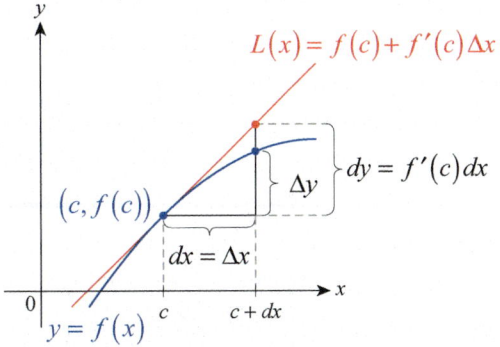

Figure 5 Differentials and Differences

As Figure 5 points out, Δy is the exact change in the value of the function over the interval $(c, c + \Delta x)$; that is, $\Delta y = f(c + \Delta x) - f(c)$. The differential dy, on the other hand, is the rise or fall in the tangent line $L(x)$ over the interval $(c, c + \Delta x)$. The differential dy is defined so that

$$\frac{dy}{dx} = f'(x) = \lim_{\Delta x \to 0} \frac{\Delta y}{\Delta x}$$

and the difference between dy and Δy goes to 0 as $\Delta x \to 0$.

Differential notation also appears in expressions such as df, and the meaning of df is the same as that of dy if $y = f(x)$. All of our differentiation rules can also be stated in terms of differentials. For example, if u and v are functions of x, then

$$\frac{d}{dx}(uv) = \frac{du}{dx} v + u \frac{dv}{dx},$$

so multiplication by the differential dx results in

$$d(uv) = (du)v + u(dv).$$

Example 3

Determine the requested values of the differentials.

a. Find dy if $y = 3x^{4/3} + \ln x$ when $x = 8$ and $dx = \frac{1}{10}$.

b. Find $d(\sin\theta)$ if $\theta = 0$ and $d\theta = 0.05$.

Solution

a. $dy = \left(4x^{1/3} + \frac{1}{x}\right)dx$

Substituting $x = 8$ and $dx = \frac{1}{10}$, we have the following:

$$dy = \left(4 \cdot 8^{1/3} + \frac{1}{8}\right)\left(\frac{1}{10}\right) = \frac{65}{80} = \frac{13}{16}$$

b. Note that $d(\sin\theta) = \cos\theta \, d\theta$, so if $\theta = 0$ and $d\theta = 0.05$,

$$d(\sin\theta) = (\cos 0)(0.05) = 0.05.$$

Finally, differentials play a very useful role in the practical applications of calculus. Since measurement errors and approximate values are inescapable in such fields as physics, chemistry, and economics, it is important to have a solid understanding of how estimates in fundamental variables propagate through later computations.

Example 4

The radius of a ball bearing is measured by a micrometer to be 1.2 mm with a margin of error of 0.05 mm. Given this possible error, estimate the maximum error in the calculated volume of the ball bearing.

Solution

Since $V = \frac{4}{3}\pi r^3$, $dV = 4\pi r^2 dr$. We will use a value of $r = 1.2$ mm for the radius, and note that the actual radius could be as much as $dr = 0.05$ mm larger. So

$$dV = 4\pi(1.2 \text{ mm})^2 (0.05 \text{ mm}) \approx 0.90 \text{ mm}^3.$$

To put this into perspective, it is useful to estimate the *percentage error* for both the radius and the volume.

$$\text{Percentage error in radius: } \frac{dr}{r} = \frac{0.05}{1.2} \approx 0.04 = 4\%$$

$$\text{Percentage error in volume: } \frac{dV}{V} \approx \frac{0.90}{(4/3)\pi(1.2)^3} \approx 0.12 = 12\%$$

So the propagated error in the calculated volume, in percentage terms, is three times larger than the margin of error in the measured radius. This factor of three will always hold true for such volume calculations; in Exercise 50 you will show that

$$\frac{dV}{V} = 3\frac{dr}{r}.$$

3.9 **Exercises**

1–12 Find the linearization of the function at the given value.

1. $f(x) = x^3 - x; \quad x = 1$

2. $g(x) = \sqrt{x-3}; \quad x = 4$

3. $h(x) = (x^4 - 5x^2 + 1)^7; \quad x = 0$

4. $k(x) = (x^2 + 1)^{-2}; \quad x = 2$

5. $C(\theta) = \cos\theta; \quad \theta = 0$

6. $T(\theta) = \tan\theta; \quad \theta = 0$

7. $F(t) = (t^2 + 5t - 6)^{-1/3}; \quad t = 2$

8. $r(x) = \dfrac{1}{x+4}; \quad x = -3$

9. $t(u) = \dfrac{u+2}{u^2 - 15}; \quad u = -4$

10. $v(x) = \sin \pi x; \quad x = \dfrac{1}{6}$

11. $G(z) = e^z; \quad z = 0$

12. $U(s) = \ln(s^4 + 1); \quad s = 1$

13–24 Determine the differential dy for the given values of x and dx.

13. $y = 3x^2 + x; \quad x = 1, \quad dx = 0.2$

14. $y = x\sqrt{x-5}; \quad x = 6, \quad dx = 0.01$

15. $y = \dfrac{4x+1}{x-3}; \quad x = 2, \quad dx = 0.1$

16. $y = \sec x; \quad x = \dfrac{\pi}{4}, \quad dx = \dfrac{1}{8}$

17. $y = x^{3/2} + x^{-3/2}; \quad x = 4, \quad dx = \dfrac{1}{16}$

18. $y = \ln x + \dfrac{1}{\ln x}; \quad x = e, \quad dx = 0.01$

19. $y = x\tan x; \quad x = -\dfrac{\pi}{4}, \quad dx = \dfrac{1}{4}$

20. $y = e^{\sqrt{x^2+3}}; \quad x = 1, \quad dx = 0.001$

21. $y = \sqrt{\ln(x+1)}; \quad x = e-1, \quad dx = \dfrac{-1}{e^2}$

22. $y = \arctan x; \quad x = -1, \quad dx = \dfrac{-1}{2^5}$

23. $y = \dfrac{\tan x}{x^2 + 1}; \quad x = \dfrac{\pi}{3}, \quad dx = -0.1$

24. $y = \cos(\arcsin x); \quad x = 0.6, \quad dx = -0.16$

25–28 Calculate the values of dy and Δy and then use graph paper to draw the curve near the given point, indicating all three of the line segments dx, dy, and Δy.

25. $y = \dfrac{1}{2}x^2; \quad x = 1, \quad dx = \dfrac{1}{2}$

26. $y = \tan x; \quad x = 0, \quad dx = \dfrac{\pi}{6}$

27. $y = 2^x; \quad x = 1, \quad dx = \dfrac{1}{4}$

28. $y = \dfrac{1}{x^2}; \quad x = 1, \quad dx = -\dfrac{1}{4}$

29–40. Find the values of Δy and compare them with dy at the indicated points for the curves given in Exercises 13–24.

41–48 Use linear approximation to approximate the given number. Compare this approximation to the actual value obtained using a calculator or computer algebra system. Round your answer to four decimal places. (**Hint:** First identify $f(x)$ and c; then find and appropriately evaluate $L(x)$.)

41. $\sqrt{9.1}$

42. $(1.01)^3$

43. $(7.9)^{2/3}$

44. $\dfrac{1}{10.1}$

45. $\sqrt[5]{31}$

46. $\cos 1$

47. $\ln 2.7$

48. $e^{1.05}$

49. Prove the power and quotient rules for differentials.

   a. $d(x^n) = nx^{n-1}dx$

   b. $d\left(\dfrac{u}{v}\right) = \dfrac{vdu - udv}{v^2}$

50. Use the equations for V and dV from Example 4 to prove that the propagated error in the calculated volume of a sphere, in percentage terms, is three times larger than the margin of error in the measured radius; that is,

$$\frac{dV}{V} = 3\frac{dr}{r}.$$

51. Prove or disprove that an analogous equation to that obtained in Exercise 50 is true for a cube; that is, if the measured side length of a cube is a units with a margin of error of da, then

$$\frac{dV}{V} = 3\frac{da}{a}.$$

52–71 Use differentials or linearization to provide the requested approximations.

52. The side of a square was measured to be 9.5 cm with a possible error of 0.5 mm. Approximate the propagated error in the calculated area of the square. Express your answer as a percentage error.

53. The radius of a circular disk was measured to be $10\frac{1}{8}$ inches. Estimate the maximum allowable error in the measurement of the radius if the percentage error in the calculated area of the disk cannot exceed 2.5 percent.

54. The base and altitude of a triangle were measured to be 7 and 9 inches, respectively. If the possible error in both cases is $\frac{1}{16}$ inches, approximate the propagated error when computing the area of the triangle.

55. Two sides of a triangle were measured to be 60 and 80 mm, respectively, while the included angle is 60 degrees. If the margin of error of the linear measurements is 0.1 mm, while that of the angle measurement is 0.1 degrees, find the possible propagated error in the calculated area of the triangle.

56. A box in the shape of a rectangular prism has a square base. If the edge of the base is 25 cm and the height is 50 cm, both with a possible measurement error of 0.2 mm, estimate the propagated errors in both the computed volume and surface area of the box. Express both answers as percentage errors.

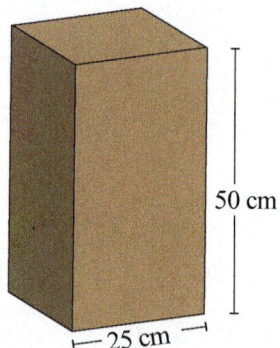

50 cm

25 cm

57. A piston of diameter 84 mm is being manufactured for an automobile engine. If the maximum percentage error in the measurement of the diameter is 0.05%, estimate the greatest possible value of the propagated error in the computed cross-sectional area of the piston. Express your answer as a percentage error.

84 mm

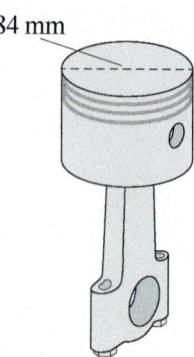

58. If the radius of an inflated balloon is 10 inches and the thickness of its wall is 0.002 inches, estimate the volume of the material it is made of. (Assume the balloon is perfectly spherical.)

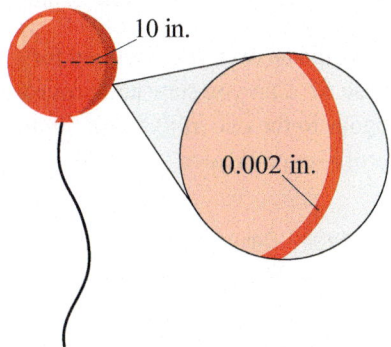

59. A tin can has a circular base of radius 2 inches and a height of 6 inches. If the thickness of its walls is 0.01 inches, estimate the volume of the material it is made of.

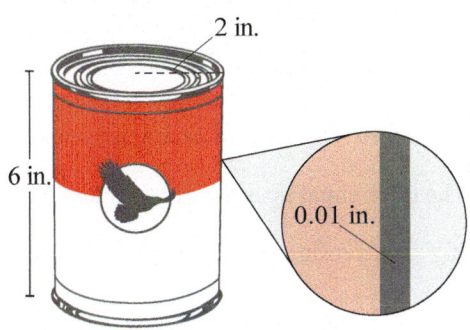

60. The exterior of a small private observatory needs to be painted. The building is approximately a circular cylinder with a hemisphere on top. The radius of the base is 3.5 feet and the height of the entire structure is 10 feet. Express the volume as a function of the radius of the base and use linearization to estimate the amount of paint that will provide a coat that is $\frac{1}{32}$ inches thick.

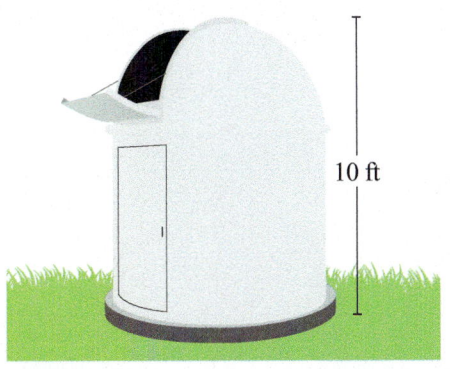

61. A trigonometry student stands 15 meters from a building and measures the angle of elevation to the top of the building as 60°. How accurate does her angle measurement have to be if she wants her propagated percentage error in estimating the height of the building to be no more than 5%?

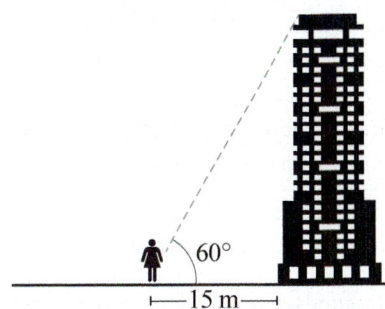

62. Referring to Exercise 44 of Section 3.8, estimate the change in image distance when the object distance increases from 60 cm to 61 cm.

63. The magnetic force experienced by a wire carrying a current I in an external magnetic field of uniform strength B is found from the equation

$$F = BIL\sin\theta,$$

where L is the length of the wire (measured in meters), and θ is the angle between the directions of B and I.

a. Find the magnetic force on a 50 cm wire if $B = 0.03\ \text{N}/(\text{A}\cdot\text{m})$, $I = 25$ amperes (A), and $\theta = 30°$.

b. Estimate the change in force if θ is increased to 33°.

c. Calculate the "true value" of the change and compare it with your approximation.

64. Estimate the change in the force in Exercise 63 if θ is increased to 33°, I is increased to 27 A, and B is decreased to $0.025\ \text{N}/(\text{A}\cdot\text{m})$.

65. The kinetic energy (in J) of a moving object is found from the equation $E_{kin} = \frac{1}{2}mv^2$, where m is the mass (in kg) of the object and v is its velocity (in m/s). Estimate the change in kinetic energy of a 1400 kg car that is accelerating from 100 km/h (approx. 62 mph) to 112 km/h (approx. 70 mph). What is the estimated percentage change?

66. When air resistance is negligible, the speed of impact of an object falling from height h is $v_i = \sqrt{2hg}$. Suppose that a rock is dropped from a height of 5 meters.

   a. Find the speed of impact as the rock hits the ground.

   b. Approximate the height from which the rock has to be dropped in order to increase the speed of impact by 10 percent. Express the height difference in both absolute and relative (percentage) terms.

   c. Find the "true value" of the above height and compare it with your approximation.

67.* The volume of a cube of side length a is being determined by immersing the cube into a container of water and measuring the volume of the displaced water and then the surface area is calculated. Estimate the percentage error we can allow in the measurement of the volume if the calculated surface area cannot differ from the true value by more than 2%. Can you generalize the result?

68. The profit function for a company is found to be $P(x) = -1.2x^2 + 500x - 2600$, where x is the number of units manufactured. If the current production level is 100 units, estimate the percentage change in profit if production is raised to 110 units.

69. For the company in Exercise 68, estimate how much the company has to increase production from 100 units in order to achieve a 10 percent profit increase.

70.* The diameter of the bottom of a 4.5-inch-tall paper cup is 2.5 inches, while the diameter of its opening is 3.5 inches. If the cup is filled with iced soda to a depth of 4.3 inches and an additional 1-cubic-inch ice cube is dropped in, predict whether the cup will overflow. (**Hint:** See Exercise 40 of Section 3.8 for help in finding the volume of soda in the cup.)

71. Suppose the velocity function of a moving object is $v(t) = 1/(1+t^2)$, and that it is moving in the positive direction along the x-axis. If you know that its location at $t = 2$ is $x = 5$, estimate its position half a second later.

72. The actual error in measurement is sometimes called absolute error, while the percentage error is referred to as relative. Write a short paragraph comparing absolute, relative, and propagated errors. Illustrate with a concrete example.

73. Examine the answer you obtained for Exercise 57. Can you state and prove a result, analogous to the one in Exercise 50, for the radius and cross-sectional area? Explain.

74–80 *True or False?* Determine whether the given statement is true or false. In case of a false statement, explain or provide a counterexample.

74. Since the differential dx is an increment, its value is always positive.

75. If $f(x) = k$, then $df = 0$.

76. If f is linear, then $\Delta f / \Delta x = df / dx$.

77. If f is differentiable at c, then $\lim_{\Delta x \to 0} (\Delta f / \Delta x) = df / dx$.

78. Propagated error is also called percentage error.

79. The differential dy is always a bit less than Δy.

80. If f is increasing and $df < 0$, then $dy > \Delta y$.

3.9 **Technology Exercises**

81–92. Use a graphing calculator to graph the functions given in Exercises 13–24 in the same viewing window along with their linear approximations at the specified x-values. Use the "Zoom" and "Trace" features to find the maximum value for dx so that the approximation is accurate to 0.01.

Chapter 3
Review Exercises

1–2 Find the derivative of the given function at the specified point and express your answer using the differential notation due to Leibniz.

1. $f(x) = x^3 + x; \quad x = 1$

2. $g(x) = \dfrac{2}{x}; \quad x = 2$

3–4 Find the derivative of the function and use the differentiation operator D_x to express your answer.

3. $s(x) = \sqrt{x-2}$

4. $t(x) = \dfrac{1}{x^2+1}$

5–6 Find the first, second, and third derivatives of the function.

5. $f(x) = x^2 - 1$

6. $g(x) = x^4$

7–10 Find all the points where the function is not differentiable. For each of those points, find the one-sided derivatives (if they exist).

7. $f(x) = \sqrt[3]{x}$

8. $g(x) = |x+1| + |x-3|$

9. $h(x) = [\![x]\!] + x$

10. $F(t) = \begin{cases} t^2 \sin\dfrac{1}{t} & \text{if } t \ne 0 \\ 0 & \text{if } t = 0 \end{cases}$

11–20 Use differentiation rules to find the derivative of the function.

11. $f(x) = 0.2x^5 - 2x^4 + x^3 + 0.5x^2 + 2^{3/4}$

12. $g(x) = \sqrt{3}x + \sqrt{3x} + \dfrac{1}{\sqrt{3x}}$

13. $h(x) = (2x-1)(x^2+4)$

14. $k(x) = \dfrac{x-2}{x} e^{x-2}$

15. $F(t) = (t-1)\left(t^2 + \dfrac{1}{t}\right)(\sqrt{t} + t)$

16. $G(s) = \dfrac{1}{2s + s^2}$

17. $u(x) = \dfrac{2e^x + 1}{3e^x + 5}$

18. $t(x) = \dfrac{\dfrac{2}{x} + \dfrac{1}{x^2} + 4}{\dfrac{2}{x^2} - \dfrac{1}{x}}$

19. $v(x) = \ln(x^2 + 2)$

20. $w(x) = \sin(\sin(\sin x))$

21–24 Find the first, second, and third derivatives of the function.

21. $f(x) = \dfrac{x}{x+1}$

22. $f(x) = 3\sqrt{x}$

23. $f(x) = \tan x$

24. $f(x) = \arctan x$

25–26 Find a function f that satisfies the given conditions. (**Hint:** A polynomial is the most natural choice. Answers will vary.)

25. $f(0) = 0$, $f'(1) = 1$, and $f''(2) = 4$.

26. $f(0) = 2$ and $y = 2x + 1$ is tangent to the graph at $x = -1$.

27–28 Find the equation of the line tangent to the graph of the function at the given point.

27. $f(x) = \dfrac{1}{\sqrt{3x^2 + 1}}; \quad \left(1, \dfrac{1}{2}\right)$

28. $f(x) = \tan(\sin x); \quad (\pi, 0)$

29. Find the equation(s) of the line(s) tangent to the graph of $f(x) = x^2 + 3x + 1$ through the point $(2, 2)$, which is not on the graph of f.

30. Assuming f is differentiable, find the derivative of $y = \ln\sqrt{[f(x)]^2 + 1}$.

31–32 Find the indicated limit.

31. $\displaystyle\lim_{x \to 0} \dfrac{-\sin 2x}{4x}$

32. $\displaystyle\lim_{x \to 0^+} \dfrac{x\cos x}{1 - \cos x}$

33. The position function of a moving particle is given by $x(t) = \dfrac{50t}{t+1}$ feet at t seconds. Find its velocity and acceleration at $t = 1$ second.

34. An object is moving along a straight line so that its distance from the start at t seconds is given by $d(t) = 12t - t^3$ meters. Find its position and acceleration at the instant when its velocity changes directions.

35. The radius of a spherical balloon being inflated increases according to the function $r(t) = 3 + 4\sqrt[3]{t}$, where r is measured in centimeters and t in seconds. Find the rate of change of the balloon's volume and surface area with respect to time at $t = 1$ second.

36–37 Find dy/dx by implicit differentiation.

36. $x^3 + y^3 = 2$

37. $6(x^2 + y^2) = 15xy$

38–39 Find dx/dy by implicit differentiation.

38. $x\sin(x + y) = y^2 + 6$

39. $6(y^2 - x^2) = y^4$

40–43 Use implicit differentiation to find the equation of the line tangent to the curve at the indicated point.

40. $\dfrac{1}{x^3} + \dfrac{1}{y^3} = 2; \quad (1,1)$

41. $x^3 + y^2 = 2x + 1; \quad (0,1)$

42. $\dfrac{3(x+y)}{xy} = 16\sqrt{x+y}; \quad \left(\dfrac{3}{4}, \dfrac{1}{4}\right)$

43. $3\sqrt{x} + \dfrac{2}{\sqrt{y}} = xy; \quad (1,4)$

44. Find all points on the lemniscate $(x^2 + y^2)^2 = 2(x^2 - y^2)$ where its graph has horizontal tangent lines.

45. Use implicit differentiation to find d^2y/dx^2 for $x^{2/3} + y^{2/3} = 1$.

46–49 Use the Derivative Rule for Inverse Functions to determine $\left(f^{-1}\right)'(a)$ for the indicated value of a. (The domain of f is assumed to have been restricted so that the inverse exists and is differentiable, whenever appropriate.)

46. $f(x) = x^3 + x; \quad a = 10$

47. $f(x) = \sqrt[4]{x+1}; \quad a = 2$

48. $f(x) = \dfrac{2}{x^2}; \quad a = \dfrac{1}{2}$

49. $f(x) = e^x + x; \quad a = 1$

50–53 Determine the derivative of the given function.

50. $f(x) = \log\sqrt{x^2 + 1}$

51. $f(x) = \tan^{-1}\sqrt{x}$

52. $f(x) = e^{\arcsin x}$

53. $f(x) = \sec^{-1}(\ln x)$

54–55 Use logarithmic differentiation to find y'.

54. $y = \dfrac{\sqrt[3]{x^2 + 1}(x+3)}{x^{2/3}\sqrt{2x^2 + 3}}$

55. $y = \left(\sqrt{x}\right)^{\ln x}$

56. A fast-growing population of bacteria doubles every half hour. If the initial count is 1000, how many bacteria are there in 100 minutes?

57. A 350 °F pizza is left on the counter and cools to 250 °F in 4 minutes. If the room temperature is 70 °F, determine the total time it takes for the pizza to cool down from 350 °F to 185 °F. (**Hint:** See Exercise 9 of Section 3.7.)

58. Find the rate of change of the distance from the origin of a point moving on the graph of $f(x) = x^3$ when $x = 1$ and $dx/dt = 2$ units per second.

59. A spherical balloon is being filled with helium at a rate of $20 \text{ in.}^3/\text{s}$. How fast is the radius increasing at the instant when the radius is 4 in.?

60. A small plane, flying at an altitude of 0.1 miles at a ground speed of 85 miles per hour, passes directly over an observer. How fast is the distance between the observer and the plane increasing a minute later?

61. Radar is tracking a rocket that was launched vertically upward. It is found that the rocket's distance from the radar is increasing at a rate of 1200 km/h at the instant when that distance is 5 km. If the radar station is 4 km from the launch site, find the speed of the rocket.

62. A ship sailing west at 9 miles per hour passes a buoy 20 minutes before another ship sailing due north at 12 miles per hour passes the same buoy. How fast will they be separating an hour later?

63. Tiffany walks toward a light source that is 8 feet above ground. If the speed of the tip of her shadow is three times that of her walking speed, how tall is Tiffany?

64–65 Find the linearization of the function at the given value.

64. $f(x) = \dfrac{1}{(x-1)^2}; \quad x = 2$

65. $f(x) = \sin x; \quad x = \dfrac{\pi}{4}$

66–67 Use linear approximation to approximate the given number. Round your answer to four decimal places.

66. $\sqrt[3]{8.2}$

67. $\arctan 0.9$

68. The diameter of a large bouncy ball was measured to be 65 cm with a possible error of 1 mm. Approximate the propagated errors in the calculated volume and surface area of the ball, respectively. Express your answers as percentage errors.

69. The proper dosage d of a certain over-the-counter medicine for children depends on body weight w according to the function $d(w) = \frac{5}{4} w^{3/5}$, where d is measured in milligrams and w in pounds. Use differentials to estimate how accurately (in terms of percentage error) we need to know a 32-pound child's weight if we cannot stray from the proper dosage by more than 6 percent.

70. A manufacturing business found its daily revenue to be $R(x) = 150x - \frac{1}{4}x^2$ dollars when x units are produced and sold.

a. Use linearization and marginal revenue to estimate the extra revenue when production is increased from 100 to 102 units.

b. Use the revenue function to calculate the actual revenue increase. Compare your answers.

71. Use the concept of the derivative function to explain why the graph of $y = x^a$, $a > 1$ curves upward, while the graph of $y = x^b$, $0 < b < 1$ curves downward.

72–82 *True or False?* Determine whether the given statement is true or false. In case of a false statement, explain or provide a counterexample.

72. If both one-sided derivatives of $f(x)$ exist at c, then f is continuous at c.

73. If $p(x)$ is a polynomial of degree n, then all k^{th}-order derivatives of $p(x)$ for $k > n$ are 0.

74. If $y = \pi^n \sin x$, then $y' = n\pi^{n-1} \sin x + \pi^n \cos x$.

75. If $y = \dfrac{1}{x^2 - 3x + 1}$, then $y' = \dfrac{1}{2x - 3}$.

76. If $y = \ln(3x+1)$, then $y' = \dfrac{1}{3x+1}$.

77. If $y = x^x$, then $y' = x \cdot x^{x-1}$.

78. Since $\left(e^x\right)' = e^x$, therefore $\left(e^{e^x}\right)' = e^{e^x}$.

79. If $f(x) = x$, then $df = dx$.

80. If $f(x)$ is linear, then its linearization at any point is itself.

81. If $x \to 0$, then $\Delta x \to dx$ and $\Delta y \to dy$.

82. If $\Delta x \to 0$, then $\Delta y / \Delta x \to dy/dx$.

Chapter 3
Technology Exercises

83–85 Use a graphing calculator or computer algebra system to graph the function and identify all points where the function is not differentiable. Explain.

83. $f(x) = |x^2 - x|$

84. $f(x) = |x|(x+2)$

85. $f(x) = \sqrt[4]{x^2 - 1}$

86. Use the differentiation capabilities of a computer algebra system to find the derivative of $f(x) = 2\cos^2 x - \cos 2x$. Then find the derivative by hand, applying a trigonometric identity before differentiating. Does your answer agree with that of your technology? If not, what do you think is the reason? Can you "force" your CAS to represent its answer in a simpler form?

87. Repeat Exercise 86 for the function $f(x) = 2\sin(x/2)\cos(x/2)$.

88. Use a graphing calculator or computer algebra system to graph the functions $y = \ln x$, $y = a^x$, $a > 1$ and $y = x^b$, $0 < b < 1$ for various values of the parameters a and b. By zooming out appropriately, compare their relative growth rates; that is, conjecture "who wins the race toward infinity" in general among these three types of functions. Use the concept of the derivative to support your conjecture.

89. The displacement of a mass attached to a spring is given by the function $h(t) = e^{-t/6}\cos 2t$.

   a. Use a graphing calculator or computer algebra system to graph the function and explain why it is realistic.

   b. Use a graphing calculator or computer algebra system to graph the velocity and acceleration functions together with $h(t)$ on the same screen. What seems to be the position of the mass when velocity is maximum? When is velocity 0? When is acceleration maximum, and when is it 0?

Chapter 3
Project

The following table shows the atmospheric pressure p at the altitude of k feet above sea level (pressure is measured in mm Hg; note that this unit of pressure is approximately the pressure generated by a column of mercury 1 millimeter high).

k (ft)	0	1000	2000	3000	4000	5000	6000	7000	8000	9000	10,000
p (mm Hg)	760	733	707	681	656	632	609	586	564	543	523

1. Find the average rate of change of air pressure from sea level to 2000 feet of altitude.

2. Find the average rate of change of air pressure between the altitudes of 4000 and 10,000 feet.

3. Use a *symmetric difference quotient*

$$\frac{p(c+h)-p(c-h)}{2h}$$

to estimate the instantaneous rate of change of air pressure at 7000 ft by choosing $h = 1000$ ft.

4. Tell whether you expect the answer to Question 2 or 3 to better approximate the instantaneous rate of change of air pressure at altitude 7000 ft. Explain. (**Hint:** Plotting the data on paper may help.)

5.* Explain why you expect the symmetric difference quotient $\dfrac{f(c+h)-f(c-h)}{2h}$ in general to be a better approximation of the instantaneous rate of change of f at $x = c$ than the "regular" difference quotient $\dfrac{f(c+h)-f(c)}{h}$.

6. Use a graphing calculator or computer algebra system to find an exponential regression curve to the given data and plot the curve along with the data on the same screen.

7. Use the exponential function you found in Question 6 to estimate the instantaneous rate of change of air pressure at 7000 ft, and compare with your estimate given in Question 3.

8. Is the instantaneous rate of change increasing or decreasing with altitude? Explain.

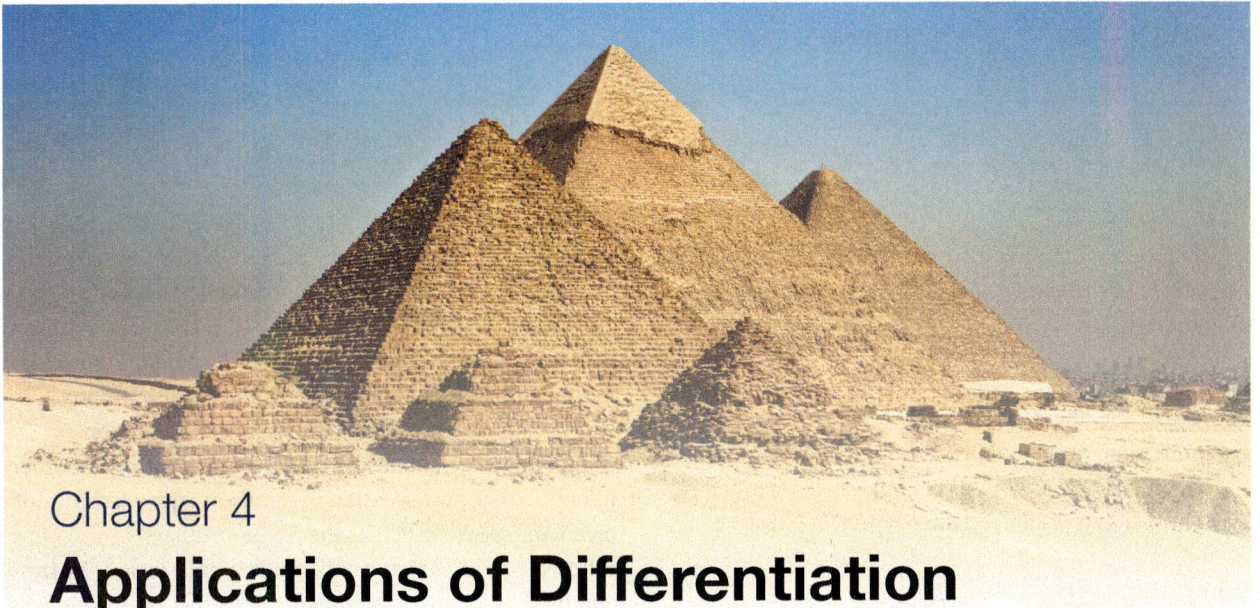

Chapter 4
Applications of Differentiation

Introduction

While Chapter 3 focused primarily on developing the techniques of differentiation, this chapter puts those techniques to use by focusing attention on applications of differentiation.

As we will see, the applications are drawn from many disciplines and at first glance the objectives vary widely. But the underlying theme of the chapter can be summarized as a prolonged answer to this question:

What does the derivative of a function tell us about the function itself?

If we are seeking more accurate and precise graphs of functions, we can use differentiation to identify intervals where a function is consistently increasing or decreasing, and thus also identify the relative "high spots" and "low spots" on a function's graph. Further, we can use a function's second derivative to reveal deeper and more subtle layers of meaning in its graph.

Much of the calculus theory of the 17th, 18th, and 19th centuries was developed by mathematicians working to solve problems arising in such fields as physics, biology, and engineering. Frequently, a given problem's solution hinged upon being able to determine the minimum and/or maximum values of an associated function, establishing the connection with differentiation. In order to apply their new solution methods to entire families of problems, people such as the Swiss brothers Jakob (1655–1705) and Johann Bernoulli (1667–1748) and the French mathematician Augustin-Louis Cauchy (1789–1857) developed new general theorems of calculus. One of these theorems, Cauchy's Mean Value Theorem, will be used in this chapter—it is a deep result that lies at the heart of several other insights. Another theorem, known as l'Hôpital's Rule but first formulated by Johann Bernoulli, uses the power of differentiation to easily evaluate limits that would otherwise be extremely difficult to determine.

Augustin-Louis Cauchy
(1789–1857)

The chapter closes with the most extreme variation of the underlying question: *If we have perfect knowledge of the derivative of a function, can we use that to completely determine the function itself?* We will see that the answer is a qualified "yes," and an exploration of the ramifications of this answer will lead to the Fundamental Theorem of Calculus, the principal subject of study in Chapter 5.

Jakob Bernoulli
(1655–1705)

Johann Bernoulli
(1667–1748)

4.1 **Extreme Values of Functions**

In this section we introduce the techniques by which the derivative leads to a deeper understanding of the graph of a function. Understanding and identifying features of a function's graph enables us to better determine its roots, its minimum and/or maximum values, and its limit behavior.

Topic One

Absolute and Relative Extrema

Some of the most distinctive and noteworthy points on the graph of a function are those where it takes on its largest and smallest values. But anything beyond a cursory discussion of such points (called *extrema*, plural of the Latin *extremum*) leads quickly to the realization that we need to refine our notions of "largest" and "smallest" before we can make much progress in finding them.

Definition

Absolute (Global) Extrema

We say the function f has an **absolute maximum** (or **global maximum**) on the domain D at the point c if $f(c) \geq f(x)$ for all $x \in D$. Consequently $f(c)$ is called the **maximum value** of f on D.

Similarly, if $f(c) \leq f(x)$ for all $x \in D$, we would say f has an **absolute minimum** (or **global minimum**) at c. We call $f(c)$ the **minimum value** of f on D.

These extreme values of f are termed *absolute* or *global* because they are, respectively, the largest and smallest values of the function over D.

Having defined absolute extrema, it is important to realize that a function may not actually possess an absolute maximum or an absolute minimum on a given domain D. Example 1 shows that the definition of both the function and the domain under consideration are critical when looking for extrema.

Example 1

Identify the absolute extrema, if they exist.

a. $f(x) = x^2$ on $D = (-\infty, \infty)$ **b.** $f(x) = x^2$ on $D = (0,1]$

c. $f(x) = x^2$ on $D = [-1,0) \cup (0,1]$ **d.** $f(x) = x^2$ on $D = (0,1)$

e. $f(x) = \begin{cases} \dfrac{1}{x} & \text{if } x \neq 0 \\ 1 & \text{if } x = 0 \end{cases}$ on $D = [0,1]$

f. $f(x) = \sin\dfrac{1}{x}$ on $D = (0,1)$

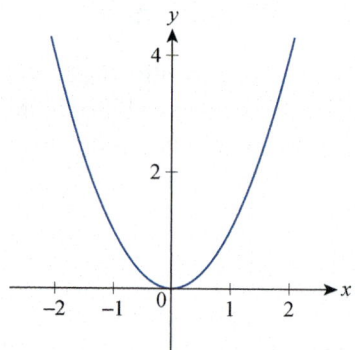

Figure 1

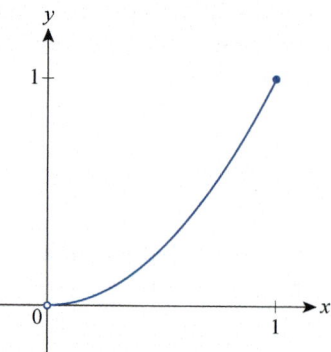

Figure 2

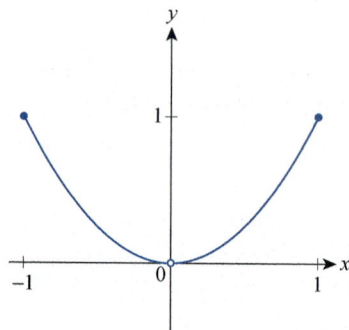

Figure 3

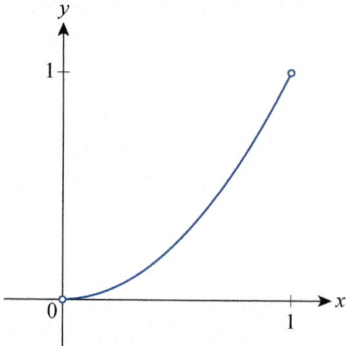

Figure 4

Solution

a. The function $f(x) = x^2$ increases without bound as $x \to \pm\infty$, so it has no absolute maximum over $(-\infty, \infty)$. But clearly $f(0) = 0$ is the absolute minimum over the domain. (See Figure 1.)

b. On the interval $(0, 1]$, $f(x) = x^2$ has an absolute maximum of $f(1) = 1$, but no absolute minimum value. (See Figure 2.)

c. On the domain $[-1, 0) \cup (0, 1]$, $f(x) = x^2$ has an absolute maximum of $f(-1) = f(1) = 1$, but no absolute minimum. (See Figure 3.)

d. On the interval $(0, 1)$, $f(x) = x^2$ has neither an absolute maximum nor an absolute minimum. (See Figure 4.)

e. On the domain $[0, 1]$, $f(x) = \begin{cases} 1/x & \text{if } x \neq 0 \\ 1 & \text{if } x = 0 \end{cases}$ has an absolute minimum of $f(0) = f(1) = 1$, but no absolute maximum. (See Figure 5; note the different scales on the axes.)

f. On the interval $(0, 1)$, $f(x) = \sin(1/x)$ attains both its absolute maximum of 1 and its absolute minimum of -1 infinitely often. (See Figure 6.)

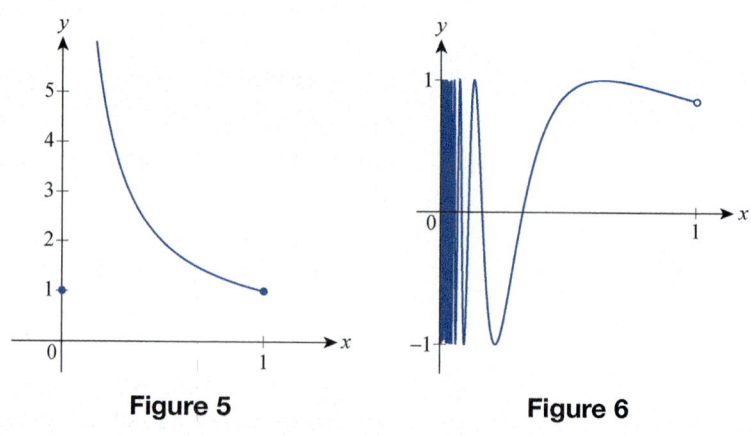

Figure 5   **Figure 6**

As Example 1 illustrates, absolute extrema may or may not exist for a given function and domain. Fortunately, there is a powerful theorem that guarantees the existence of both an absolute maximum and an absolute minimum under very common conditions. This theorem, while intuitively reasonable and easy to understand, actually requires a deep knowledge of the properties of the real line to prove; its proof (which typically is studied in a course like Advanced Calculus or Real Analysis) is tied to similarly deep statements about the real numbers such as the Bolzano-Weierstrass Theorem (see Appendix E for the statement of this theorem).

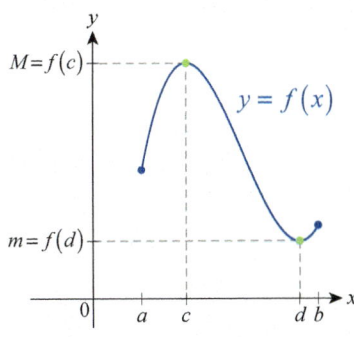

Figure 7 Extreme Values

Theorem

The Extreme Value Theorem

If f is a continuous function defined on the closed interval $[a,b]$, then there exist points c and d in $[a,b]$ where f attains both its absolute maximum value $M = f(c)$ and absolute minimum value $m = f(d)$. (See Figure 7 for an illustrative depiction.)

The Extreme Value Theorem tells us quite a lot: first, the absolute extrema exist for a continuous function defined on a closed and bounded interval, and second, there are points in the interval where those extreme values are actually attained.

Given a continuous function f defined on an interval $[a,b]$, we are still left with the question of *how*, exactly, we find the absolute extrema. Visually, the graph of f over $[a,b]$ is likely to suggest fairly obvious candidates for the absolute extreme values—those points on the graph which are, in comparison to nearby points, either high spots or low spots. Our mathematical language for such points is as follows.

Definition

Relative (Local) Extrema

We say the function f has a **relative maximum** (or **local maximum**) at the point c if $f(c) \geq f(x)$ for all x in some open interval containing c.

Similarly, if $f(c) \leq f(x)$ for all x in some open interval containing c, we say f has a **relative minimum** (or **local minimum**) at c.

We can extend the definition to the endpoints of an interval by saying f has a relative extremum at an endpoint of $[a,b]$ if f attains its maximum or minimum value at that endpoint in a half-open interval containing it.

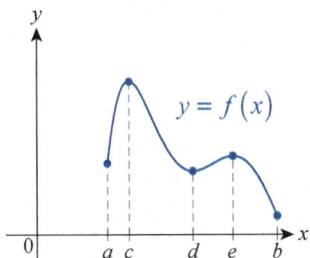

Figure 8

Example 2

Identify the relative and absolute extrema for the function f over the interval $[a,b]$ graphed in Figure 8.

Solution

The function has relative maxima at the points c and e, since $f(c)$ and $f(e)$ are the largest values f attains over small open intervals containing c and e. To illustrate this fact, a small portion of the graph focusing on $(c, f(c))$ is reproduced in Figure 10; note that $f(c) \geq f(x)$ for all x near c.

Similarly, a, d, and b are points where f has relative minima (note that a and b are also the endpoints of the interval).

By comparing the values of f at each of the five points a through e, we see that $f(c)$ is the absolute maximum and $f(b)$ is the absolute minimum.

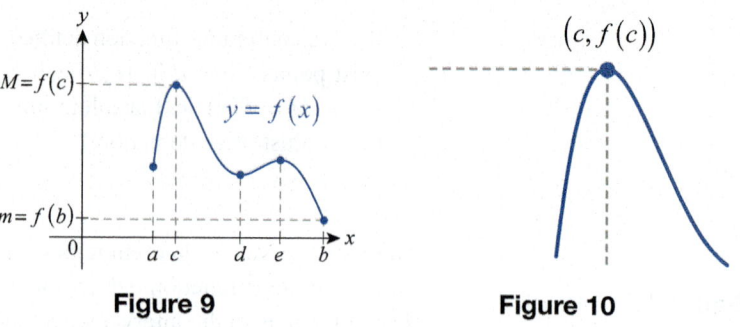

Figure 9   **Figure 10**

Topic Two

Finding Extrema

Up to this point, our definitions and examples in this section have not called upon any ideas unique to calculus—the notions of absolute and relative extrema make sense independently of calculus. The following theorem, the essence of which was conceived by the French mathematician Pierre de Fermat (1601–1665) even before the formalization of calculus, shows how differentiation greatly simplifies the task of finding extrema.

Theorem

Fermat's Theorem

If f has a relative extremum at a point $c \in (a,b)$, and if $f'(c)$ exists, then $f'(c) = 0$.

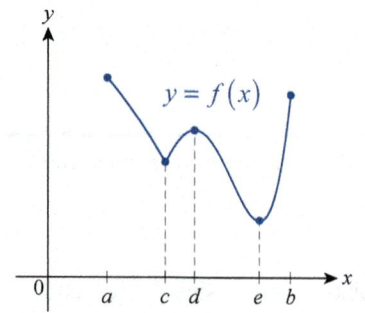

Figure 11
Applying Fermat's Theorem

Before proving Fermat's Theorem, it's useful to consider a graph such as the one sketched in Figure 11. For this function, there are three points in (a,b), referred to as the *interior* of the interval $[a,b]$, where relative extrema occur. The first point, $(c, f(c))$, is a cusp of the graph, and so $f'(c)$ does not exist—Fermat's Theorem does not apply here. At points d and e, the function f makes a transition from rising to falling and then from falling to rising. This means that the sign of f' will pass from positive to negative and then back to positive as we scan the graph from left to right through d and e, and it makes intuitive sense to suppose that f' will equal 0 at the two transition points. The proof is merely a rigorous rendition of this argument.

— **Proof** —

We will first assume that f has a relative maximum at $c \in (a,b)$ and that $f'(c)$ exists; from these two facts, we will prove that $f'(c) = 0$. A similar argument can be used to show that if f has a relative minimum at c and if $f'(c)$ exists, then again $f'(c) = 0$.

Since $f'(c)$ exists, we know that the limit $\lim\limits_{x \to c} \dfrac{f(x) - f(c)}{x - c}$ exists (recall that this is one formulation of the difference quotient), and since f has a relative maximum at c, $f(x) - f(c) \leq 0$ for all x close to c; this is just a different way of expressing the fact that $f(c) \geq f(x)$ for all x in some open interval containing c. The existence of the two-sided limit above means both one-sided limits exist and are equal to $f'(c)$, so

$$f'(c) = \lim_{x \to c^-} \frac{f(x) - f(c)}{x - c} \geq 0 \qquad \text{\small\textcolor{blue}{$x - c < 0$ as $x \to c$ from the left}}$$

and

$$f'(c) = \lim_{x \to c^+} \frac{f(x) - f(c)}{x - c} \leq 0 \qquad \text{\small\textcolor{blue}{$x - c > 0$ as $x \to c$ from the right}}$$

The only number that is both nonnegative and nonpositive is 0, so $f'(c) = 0$.

The only change necessary if we begin with the assumption that f has a relative minimum at c is to note that $f(x) - f(c) \geq 0$ for all x sufficiently close to c and to reverse the two inequalities above.

With Fermat's Theorem as inspiration, we are led to the following definition.

Definition

Critical Point

Assume that f is defined on an open interval containing c. We say c is a **critical point** of the function f if $f'(c) = 0$ or $f'(c)$ does not exist.

Thus if f is defined on a closed interval $[a, b]$, the only candidates for extrema (relative or absolute) are critical points in (a, b) and the two endpoints a and b.

Caution!

Don't read too much into the definition of critical point and Fermat's Theorem. Points where the derivative fails to exist are not guaranteed to be relative extrema, as the first graph in Figure 12 illustrates. Similarly, the converse of the theorem is false: points at which the derivative is 0 are not necessarily relative extrema, as seen in the second graph of Figure 12.

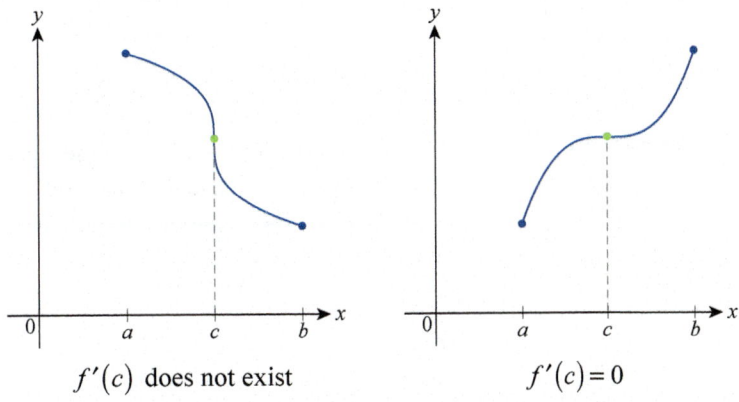

$f'(c)$ does not exist $f'(c) = 0$

Figure 12 Critical Points That Are Not Extreme Points

Nevertheless, the Extreme Value Theorem and Fermat's Theorem together provide a very useful method for locating the absolute extrema of a function defined on a closed and bounded interval.

Finding Absolute Extrema

Assume that f is continuous on the interval $[a,b]$. To find the absolute extrema of f on $[a,b]$, perform the following steps.

Step 1: Find the critical points of f in (a,b).

Step 2: Evaluate f at each of the critical points in (a,b) and at the two endpoints a and b.

Step 3: Compare the values of the function found in Step 2. The largest is the absolute maximum of f on $[a,b]$ and the smallest is the absolute minimum of f on $[a,b]$.

Example 3

Find the absolute extrema of $f(x) = 3x^4 - 4x^3 - 12x^2$ on the interval $[-2,3]$.

Solution

$$f'(x) = 12x^3 - 12x^2 - 24x$$
$$= 12x(x^2 - x - 2) = 12x(x-2)(x+1)$$

The derivative of f is defined for all real numbers. So the only critical points are those where $f'(x) = 0$; the solutions of this equation are 0, 2, and -1, all of which are in the interval $[-2,3]$. We now compute f at the endpoints and the three critical points in the given interval.

$f(x) = 3x^4 - 4x^3 - 12x^2$	Point in $[-2,3]$
$f(-2) = 32$	Left endpoint
$f(-1) = -5$	Critical point
$f(0) = 0$	Critical point
$f(2) = -32$	Critical point
$f(3) = 27$	Right endpoint

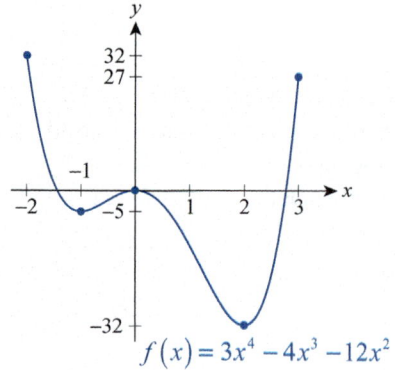

$f(x) = 3x^4 - 4x^3 - 12x^2$

Figure 13

Comparing these values, we see that the absolute maximum of f on $[-2,3]$ is $f(-2) = 32$ and that the absolute minimum is $f(2) = -32$.

Example 4

Find the absolute extrema of $f(x) = x^{2/3}$ on the interval $[-1, 2]$.

Solution

$$f'(x) = \frac{2}{3}x^{-1/3} = \frac{2}{3x^{1/3}}$$

The derivative of f is never 0. Since f' is, however, undefined at $x = 0$, 0 is the one critical point of f. Comparing the values of f at the two endpoints and 0 we have:

$$f(-1) = (-1)^{2/3} = \left[(-1)^2\right]^{1/3} = 1,$$

$$f(0) = 0, \text{ and}$$

$$f(2) = 2^{2/3} = \left(2^2\right)^{1/3} = 4^{1/3} \approx 1.59.$$

So the absolute minimum value of f on $[-1, 2]$ is 0 and the absolute maximum value is $\sqrt[3]{4}$.

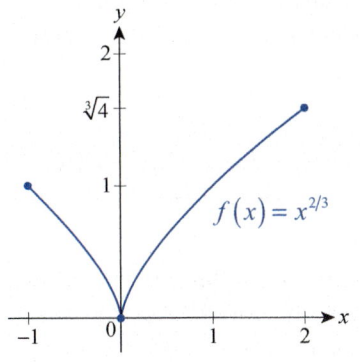

Figure 14

Example 5

Find the absolute extrema of $f(x) = x^{3/5}(2-x)$ on the interval $[-1, 2]$.

Solution

We can use the Product Rule to find f', as follows.

$$f'(x) = \frac{3}{5x^{2/5}}(2-x) + x^{3/5}(-1)$$

$$= \frac{6-3x}{5x^{2/5}} - \frac{5x^{2/5}x^{3/5}}{5x^{2/5}}$$

$$= \frac{6-8x}{5x^{2/5}}$$

From this, we see that f' does not exist at $x = 0$, and that $f'\left(\frac{3}{4}\right) = 0$. We now note that $f(-1) = -3$, $f(0) = 0$, $f\left(\frac{3}{4}\right) = \left(\frac{3}{4}\right)^{3/5}\left(\frac{5}{4}\right) \approx 1.05$, and $f(2) = 0$. So f has its absolute minimum value of -3 at the left endpoint of $[-1, 2]$, and its absolute maximum value of approximately 1.05 at $x = \frac{3}{4}$.

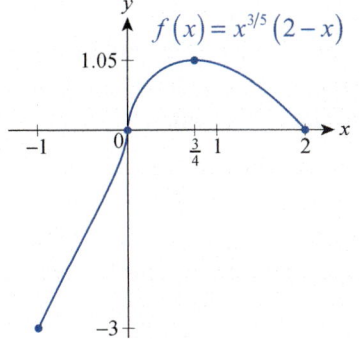

Figure 15

Our final example will be a brief introduction to a class of problems called *optimization problems*, which we will study in depth in Section 4.6. In these sorts of problems, we put the tools of calculus to use in finding an optimal solution to a problem from among many possible solutions.

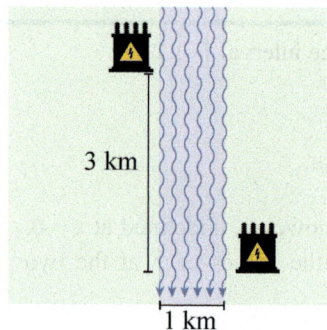

3 km

1 km

Figure 16

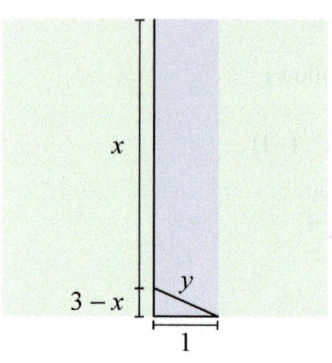

x

$3 - x$

y

1

Figure 17

Example 6

A new electrical power substation is being built on the right bank of a river. Cables to the substation need to be laid underground and underwater from another substation 3 kilometers upstream and on the opposite bank. The river follows a straight route through this stretch and has a fairly constant width of 1 kilometer. Given that the cost of laying cable underground is $30,000/km and the cost of laying cable underwater is $50,000/km, how should the cable be laid in order to minimize cost?

Solution

Even bearing in mind that the shortest distance between two points is a straight line, there are an infinite number of what could be deemed as reasonable solutions to this problem.

One very elementary approach would be to simply lay all the cable diagonally underwater in one straight shot, but given the fact that underwater installation is more expensive than underground installation, this is unlikely to be the most economical solution. Another approach would be to cross the river in the shortest possible path (so as to minimize underwater installation) and then run cable 3 kilometers along the bank of the river. But any combination of the two approaches is also a possibility, and some combination may lead to the absolute minimum cost. Calculus allows us to find that optimal solution.

We will begin with a schematic that defines and relates the variable quantities that we have at our disposal. We can assume that the underground installation follows the left bank of the river, as shown in Figure 17; let x represent the length of the underground cable in kilometers. Let y then represent the length of the underwater installation. (We could have proceeded by first crossing the river and then laying cable down the right bank, but for given lengths x and y the cost would be the same.)

Note that $x = 0$ corresponds to the solution in which all the cable is underwater, and $x = 3$ corresponds to the solution in which the shortest length of underwater installation ($y = 1$) is called for. Any value for x in the interval $[0,3]$ corresponds to an intermediate solution, as discussed above.

From the diagram, $y^2 = 1^2 + (3 - x)^2$, so $y = \sqrt{1 + (3 - x)^2}$ and the total cost of laying the cable for a given value of x is as follows.

$$C = 30,000x + 50,000y = 30,000x + 50,000\sqrt{1 + (3 - x)^2}$$

We are looking for the absolute minimum of the function C on the interval $[0,3]$, so we proceed to find the critical points of C.

$$C(x) = 30,000x + 50,000\left[1 + (3-x)^2\right]^{1/2}$$

$$C'(x) = 30,000 + 50,000\left(\frac{1}{2}\right)\left[1 + (3-x)^2\right]^{-1/2}\left[2(3-x)\right](-1)$$

$$= 30,000 + \frac{50,000(x-3)}{\sqrt{1 + (3-x)^2}}$$

C' exists for all x, but there are two values of x for which $C'(x) = 0$.

$$30,000 + \frac{50,000(x-3)}{\sqrt{1 + (3-x)^2}} = 0$$

$$\frac{(x-3)}{\sqrt{1 + (3-x)^2}} = -\frac{30,000}{50,000}$$

$$\frac{(x-3)^2}{1 + (3-x)^2} = \frac{9}{25} \qquad \text{Square both sides.}$$

$$25(x-3)^2 = 9 + 9(x-3)^2 \qquad \text{Note that } (3-x)^2 = (x-3)^2.$$

$$16(x-3)^2 = 9$$

$$x - 3 = \pm\frac{3}{4} \qquad \text{Divide by 16 and take the square root.}$$

$$x = 3 \pm \frac{3}{4}$$

The point $3 + \frac{3}{4}$ does not actually solve the original equation, so we only have to evaluate C at the two endpoints 0 and 3 and at the critical point $3 - \frac{3}{4} = \frac{9}{4}$.

$$C(0) = 50,000\sqrt{10} \approx \$158,114$$

$$C\left(\frac{9}{4}\right) = \$130,000$$

$$C(3) = \$140,000$$

From this comparison, we see that the minimal cost of installation can be achieved by laying the cable with an underground run of $\frac{9}{4} = 2\frac{1}{4}$ kilometers and a diagonal run underwater of $\sqrt{1 + \left(\frac{3}{4}\right)^2} = \sqrt{\frac{25}{16}} = 1\frac{1}{4}$ kilometers.

4.1 **Exercises**

1–4 Use the graph as an aid to identify the absolute extrema, if they exist, for the given function on the specified domain.

1. $f(x) = -x^2 + 1; \quad D = [-1, 0) \cup (0, 1]$

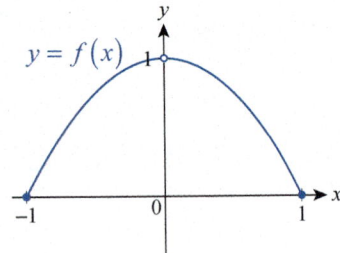

2. $f(x) = -\cos x; \quad D = (0, 2\pi]$

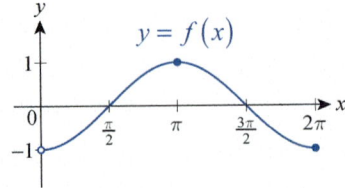

3. $f(x) = 2\sin x; \quad D = [0, 2\pi]$

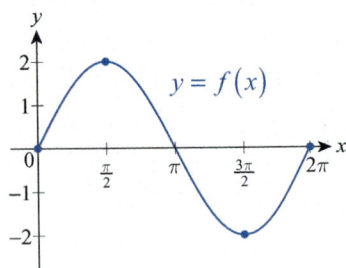

4. $f(x) = |x - 1|; \quad D = (-1, 2]$

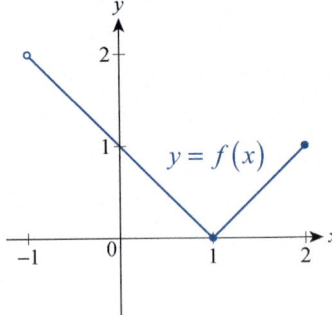

5–8 Use the graph to decide whether each highlighted point is a critical point, and then find and classify all relative and absolute extrema for the function over the given interval.

5.

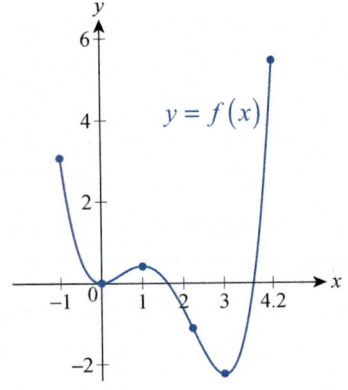

6.

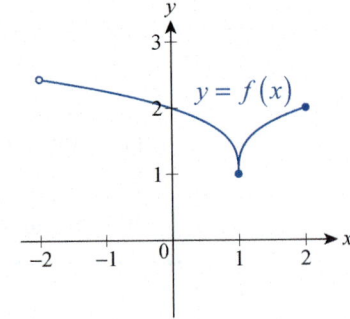

7.

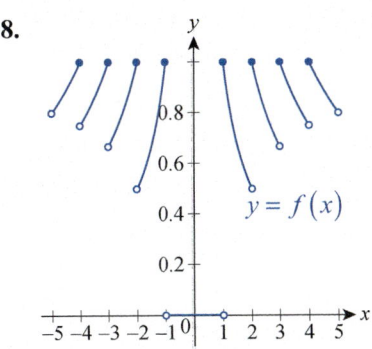

8.

9–20 Use graph paper to sketch the graph of the given function on the specified domain, and then use the graph to visually identify and classify any absolute extrema.

9. $f(x) = 2x + 1; \quad D = [0,3]$

10. $g(x) = -x - 1; \quad D = (-1,2]$

11. $h(x) = \dfrac{1}{2}x - 3; \quad D = \mathbb{R}$

12. $u(x) = -x^2; \quad D = (-1,1)$

13. $v(x) = (x+1)(x-3); \quad D = [-2,4]$

14. $k(x) = (x-4)^4; \quad D = \mathbb{R}$

15. $K(x) = x^7; \quad D = \mathbb{R}$

16. $m(x) = e^{-x+2}; \quad D = [2,\infty)$

17. $n(x) = \cos \pi x; \quad D = \left(0, \dfrac{3}{2}\right]$

18. $F(x) = \dfrac{1}{(x+1)^2}; \quad D = \mathbb{R}$

19. $G(x) = \dfrac{1}{x^2 + 1}; \quad D = \mathbb{R}$

20. $H(t) = \arcsin t; \quad D = [-1,1]$

21–37 Sketch by hand the graph of a function f on the specified domain, with the specified properties. (Answers will vary.)

21. Defined on $[2,4]$, absolute maximum at 2, absolute minimum at 4

22. Defined on $[-1,2]$, absolute maximum at 0, absolute minimum at 1

23. Defined on $[-5,5]$, absolute maximum at 1, absolute minimum at 5

24. Defined on $\mathbb{R}$, absolute minimum at 2, no absolute maximum

25. Defined on $[-3,2]$, absolute maximum at –2, absolute minimum at 0, relative maximum at 1

26. Defined on $[0,6]$, absolute maximum at 2, relative minimum at 4, no absolute minimum

27. Defined on $[-1,1]$, absolute maximum occurs twice, no minimum

28. Defined on $(1.5,7)$, continuous, has relative maximum and minimum, but no absolute maximum or minimum

29. Defined on $(1.5,7)$, continuous, has both absolute maximum and minimum

30. Defined on $[-2,4]$, two relative maxima, but no absolute maximum

31. Defined on $(0,\infty)$, continuous, no relative or absolute extrema

32. Defined on $(-1,3]$, continuous, no absolute minimum, one relative minimum, absolute maximum occurs twice

33. Defined on $\mathbb{R}$, both the absolute maximum and absolute minimum occur infinitely often

34. Defined on $(0,\infty)$, infinitely many relative maxima and minima, no absolute maximum or minimum

35. Differentiable on $\mathbb{R}$, has one critical point, but no extrema

36. Defined on $(0,10)$, nondifferentiable at 5, but absolute maximum occurs at 5

37. Defined on $(0,10)$, discontinuous at 5, but absolute maximum occurs at 5

38–55 Find all critical points, if they exist, for the given function.

38. $f(x) = x^2 - 7x + 1.5$

39. $g(x) = 2x^3 + 3x^2 - 12x + 1.5$

40. $h(x) = x^3 + 1.5x^2 + 3x - 2.5$

41. $u(x) = -\dfrac{3}{2}x + 2$

42. $v(x) = x^4 - \dfrac{16}{3}x^3 + 2x^2 + 24x - 1$

43. $k(x) = |2x - 3|$

44. $K(x) = |3x^2 + 3x - 18|$

45. $m(x) = \dfrac{2-x}{x^2 - x + 2}$

46. $n(x) = \dfrac{|x^2 - 2|}{2x^2 + 4}$ **47.** $F(t) = \sqrt{3 + 3t^2}$

48. $G(x) = x^{3/2} - 3\sqrt{x}$ **49.** $T(s) = 2\sqrt[3]{s}\,(s-2)$

50. $r(v) = \dfrac{v-1}{\sqrt{v}}$

51. $s(\alpha) = \cos\alpha + \cos^2\alpha$

52. $u(z) = \cot z + 2z$ **53.** $t(x) = \sqrt{x}\ln x$

54. $U(t) = e^t \sin t$ **55.** $a(t) = \cos(\arctan t)$

56–77 Find all absolute extrema of the function on the given closed interval.

56. $f(x) = 4x - x^2$ on $[0,6]$

57. $g(x) = 3x^2 - 30x + 7$ on $[0,8]$

58. $h(x) = x^3 + 1.5x^2 - 6x + 3.5$ on $[-4,3]$

59. $u(x) = 3x^4 - 8x^3 + 6x^2 - 24x - 9$ on $[0,3]$

60. $v(x) = \dfrac{x^4}{4} - 2x^2 + 4$ on $[-2,2]$

61. $k(x) = \dfrac{x^4}{2} + 2x^3 - x^2 - 6x + \dfrac{1}{2}$ on $[-3,1]$

62. $f(x) = |x + 3| \cdot |x - 3|$ on $[-4,4]$

63. $m(x) = |x + 3| + |x - 3|$ on $[-4,4]$

64. $n(x) = \dfrac{3x}{2x^2 + 2}$ on $[-4,4]$

65. $g(x) = \dfrac{x^2 + 5}{x + 2}$ on $[-1.5,1.5]$

66. $F(x) = \dfrac{1}{1 + x^2}$ on $[-10,10]$

67. $G(t) = \dfrac{1}{\sqrt{t}} + \sqrt{t}$ on $\left[\dfrac{1}{4}, 4\right]$

68. $k(s) = (s^2 - 1)\sqrt{s}$ on $[0,2]$

69. $r(z) = \sin(\arccos z)$ on $[-1,1]$

70. $G(x) = \arctan x$ on $[-1,1]$

71. $w(x) = x\sqrt{8 - x^2}$ on $\left[-2\sqrt{2}, 2\sqrt{2}\right]$

72. $T(s) = s^2 e^{-s}$ on $[0,10]$

73. $r(x) = (\cos x)e^x$ on $\left[0, \dfrac{3\pi}{2}\right]$

74. $L(x) = x\ln x$ on $\left[\dfrac{1}{e^2}, e\right]$

75. $t(x) = \ln((e-1)\sin\pi x + 1)$ on $[0,1]$

76. $U(x) = \sqrt[3]{x}\,(x-3)$ on $[-1,3]$

77. $V(x) = 5 + (5 + x)x^{5/7}$ on $[-5,1]$

78–89 Find and classify the absolute extrema, if they exist, of the function over the given domain.

78. $f(x) = 3x - 2; \quad D = (0, 2]$

79. $g(x) = x^2 - 4; \quad D = (-2, 2)$

80. $h(x) = 2x^3 - 5x; \quad D = \mathbb{R}$

81. $K(z) = \sqrt{4 - z^2}; \quad D = (-2, 2)$

82. $r(z) = -\dfrac{2}{z}; \quad D = [2, \infty)$

83. $n(x) = \dfrac{1}{(x+3)^2}; \quad D = (-3, \infty)$

84. $t(x) = 10^{x/2}; \quad D = \mathbb{R}$

85. $L(x) = \ln(x + 1); \quad D = [0, \infty)$

86. $F(x) = \sec x; \quad D = \left(-\dfrac{\pi}{2}, \dfrac{\pi}{2}\right)$

87. $t(z) = 2\cos \pi z + 2; \quad D = \mathbb{R}$

88. $u(x) = x - [\![x]\!]; \quad D = [1, 3)$

89. $v(x) = \arctan x, \quad D = \mathbb{R}$

90. Find two numbers whose sum is 50 and whose product is as large as possible. (**Hint:** Denote the numbers by x and $50 - x$, and maximize the product.)

91. A 50-inch piece of wire is cut into two pieces, which are then bent into a square and a circle, respectively. Where should the wire be cut in order to minimize the sum of the areas of these two shapes? (**Hint:** Start with the notation of Exercise 90, and use appropriate formulas from geometry.)

50 in.

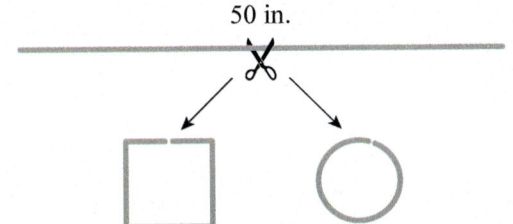

92. A lighthouse is 5 miles off a straight shoreline. Ten miles down the coast is a restaurant where the lighthouse keeper is planning to meet his friends. If he can row at 2.5 mph and walk at 4 mph, where should he land in order to make the fastest possible time to the restaurant?

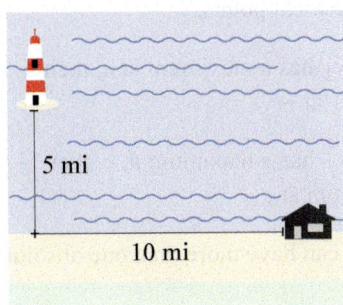

5 mi

10 mi

93. Referring to Exercise 26 of Section 3.7, find the number of calculators that have to be produced in order to maximize profit.

94. The power output of a 12-volt car battery when a resistor is connected to it, is given by the formula $P = 12I - (R + r)I^2$, where I is the current (in amperes), and r stands for the (typically very small) so-called internal resistance of the battery. Suppose we are starting a car with a starter motor of resistance $R = 0.16$ ohms, and that the internal resistance of the battery is $r = 0.016$ ohms. Find the current that corresponds to the battery's maximum power output.

95–105 *True or False?* Determine whether the given statement is true or false. In case of a false statement, explain or provide a counterexample.

95. If f attains both its absolute minimum and absolute maximum values on a closed interval, then f is a continuous function.

96. A continuous function on a closed interval can attain its absolute extrema only at critical points.

97. If $f(x)$ is a differentiable function and k is a constant, then $f(x)$ and $f(x) + k$ have the same critical points.

98. If $f(x)$ is a differentiable function and k is a nonzero constant, then $f(x)$ and $kf(x)$ have the same critical points.

99. If $f(x)$ is a differentiable function and k is a nonzero constant, then $f(x)$ and $f(x+k)$ have the same critical points.

100. If $f(x)$ is a differentiable function and k is a nonzero constant, then $f(x)$ and $f(kx)$ have the same critical points.

101. If $f(x)$ has a maximum at c, then so does $f(-x)$ at $-c$.

102. If $f(x)$ has a maximum at c, then $-f(x)$ has a minimum at c.

103. $f(x)$ can have more than one absolute maximum value.

104. If $f(x)$ is continuous on a closed interval I, then it attains its minimum value on I.

105. If $f(x)$ has no maximum on a closed interval I, then $f(x)$ must be discontinuous on I.

4.1 **Technology Exercises**

106–127. Use a graphing calculator or computer algebra system to verify the answers you obtained for Exercises 56–77.

4.2 **The Mean Value Theorem**

In this section, we introduce and prove the Mean Value Theorem (MVT), a result that relates the behavior of a differentiable function to that of its derivative in a fundamental and useful way. As we will see, one use of the MVT is to further strengthen the connection between average rate and instantaneous rate of change; another is in the proof of an impressively named result that we will encounter in Chapter 5 called the Fundamental Theorem of Calculus.

Topic One

Rolle's Theorem and the Mean Value Theorem

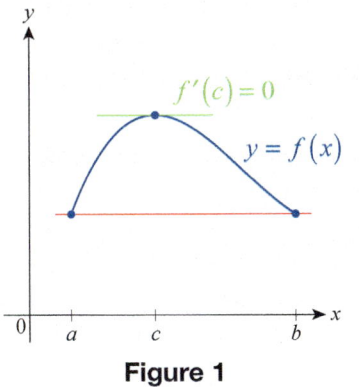

Figure 1

Our development begins with a preliminary observation first proved by the French mathematician Michel Rolle (1652–1719) in 1691. Suppose f is a function that is continuous on the interval $[a,b]$, and suppose further that f has the same value at the two endpoints a and b, as illustrated in Figure 1. Your intuition may suggest that if the graph of f begins and ends with the same value and varies in a sufficiently "smooth" manner over the course of the interval, then there must be at least one point c where it has a horizontal tangent line. This is exactly what Rolle's Theorem guarantees, and the precise property corresponding to a "smoothly" varying graph is differentiability on the open interval (a,b).

> **Theorem**
>
> **Rolle's Theorem**
>
> If f is continuous on the closed interval $[a,b]$ and differentiable on (a,b), and if $f(a) = f(b)$, then there is at least one point $c \in (a,b)$ for which $f'(c) = 0$.

― Proof ―

Because f is continuous on the closed interval $[a,b]$, the Extreme Value Theorem tells us that f attains its absolute maximum and minimum values on $[a,b]$. If f attains both its absolute maximum and absolute minimum at an endpoint of the interval, then f is a constant function on $[a,b]$ (note that in this case $f(x) = f(a) = f(b)$ for all x in the interval) and so we know that $f'(x) = 0$ for all $x \in (a,b)$, proving the claim.

If, on the other hand, f assumes either its absolute maximum or absolute minimum at an interior point $c \in (a,b)$, then $f(c)$ is a relative extremum of the function and by Fermat's Theorem $f'(c) = 0$, since by assumption $f'(x)$ exists for all x in (a,b).

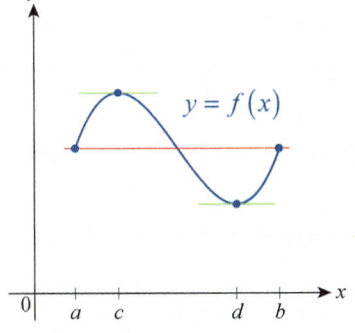

Figure 2

There may be more than one point in (a,b) where f has a horizontal tangent line when the hypotheses of Rolle's Theorem are satisfied, but in practice it suffices to know for certain that there is at least one such point. Theorems of this sort are known as *existence* theorems, and they often provide great insight into the nature of objects under study. Note that in Figure 2, $f'(c) = f'(d) = 0$. In Figure 3,

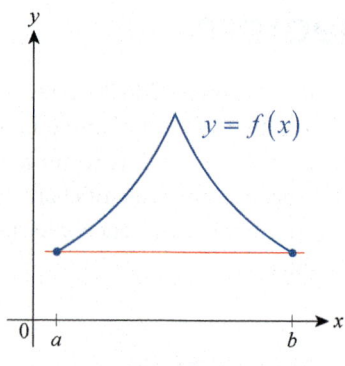

Figure 3

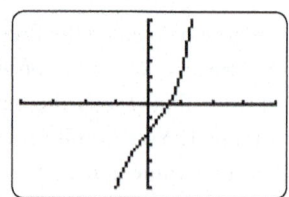

Figure 4

$$f(x) = x^5 + 3x - 2$$

on $[-4,4]$ by $[-6,6]$

there is no point c in $[a,b]$ for which $f'(c) = 0$, but this does not violate Rolle's Theorem as the function is not differentiable everywhere on (a,b).

One of the most notable uses of Rolle's Theorem is as a stepping-stone toward the Mean Value Theorem, but it can also be put to more immediate use.

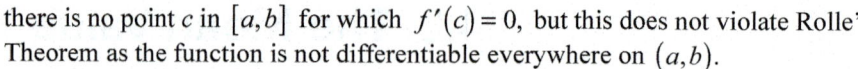

Example 1

Show that the equation $x^5 + 3x = 2$ has exactly one real solution.

Solution

Let $f(x) = x^5 + 3x - 2$. Then $f'(x) = 5x^4 + 3$, which is defined and positive (in fact, greater than or equal to 3) for all real x. If there were two solutions $x = a$ and $x = b$ of the equation $x^5 + 3x = 2$, then we would have $f(a) = 0 = f(b)$, and by Rolle's Theorem there would then be a point c between a and b for which $f'(c) = 0$, contradicting $f'(x) > 0$ for all x. So $x^5 + 3x = 2$ has no more than one solution.

The fact that it does indeed have a solution follows from the observation that $f(0) = -2 < 0$ and $f(1) = 2 > 0$, so by the Intermediate Value Theorem there is a point c between 0 and 1 for which $f(c) = 0$ (see Section 2.5 for a refresher on the use of the IVT in locating zeros of functions).

Example 2

Suppose that $f(t)$ is a differentiable function describing the location at time t of an object moving along a straight line. If $f(t_1) = f(t_2)$ for two points in time t_1 and t_2, then there must be a time t_3 between t_1 and t_2 for which $f'(t_3) = 0$. That is, the velocity of the object at t_3 is 0.

As a particular example, if $h(t)$ represents the vertical height of a thrown object with, say, $h(t_1) = h(t_2) = 0$, then $h'(t_3) = 0$ and $h(t_3)$ represents the maximum height achieved by the object. (Note that in this example, $h(t)$ represents the motion up and down relative to ground level; the object may also travel horizontally, as shown in Figure 5, but such motion happens independently of $h(t)$ and would be measured by another function.)

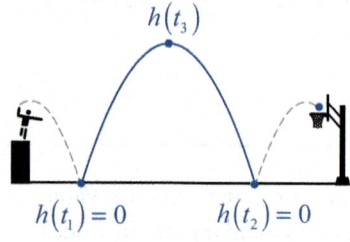

$h(t_3)$

$h(t_1) = 0$ $h(t_2) = 0$

Figure 5

The Mean Value Theorem (MVT), first stated in a form similar to the following by Joseph-Louis Lagrange (1736–1813), is a generalization of Rolle's Theorem—it provides the same sort of insight into the behavior of a function defined on an interval $[a,b]$ while relaxing one of the hypotheses.

Theorem

The Mean Value Theorem

If f is continuous on the closed interval $[a,b]$ and differentiable on (a,b), then there is at least one point $c \in (a,b)$ for which $f'(c) = \dfrac{f(b) - f(a)}{b - a}$.

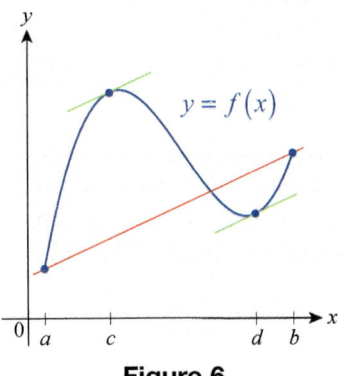

Figure 6

In preparation for proving the MVT, it's important to understand geometrically what it says. Figure 6 shows how the MVT can be viewed as a "slanted" version of Rolle's Theorem, in that it guarantees the existence of at least one point $c \in (a,b)$ at which the line tangent to the graph of f is parallel to the secant line through $(a, f(a))$ and $(b, f(b))$. In Figure 6, this conclusion is actually true for two points c and d in the interval.

Proof

The sort of depiction shown in Figure 6 points the way toward a proof of the MVT: if we can modify the function under consideration by "unslanting" it, we can then apply Rolle's Theorem. To do this, we want to consider a new function defined as the difference between f and the function whose graph is the secant line passing through $(a, f(a))$ and $(b, f(b))$, shown in red in the figure.

The slope of the secant line is

$$m_{sec} = \frac{f(b) - f(a)}{b - a},$$

so $y - f(a) = m_{sec}(x - a)$ is an equation for the secant line in point-slope form; solving for y, the secant line is the graph of the function $L(x) = f(a) + m_{sec}(x - a)$. If we now define a new function g by

$$\begin{aligned} g(x) &= f(x) - L(x) \\ &= f(x) - f(a) - m_{sec}(x - a), \end{aligned}$$

then we have, by design, constructed a function for which

$$g(a) = f(a) - f(a) - m_{sec}(a - a) = 0$$

and

$$\begin{aligned} g(b) &= f(b) - f(a) - m_{sec}(b - a) \\ &= f(b) - f(a) - \left[f(b) - f(a) \right] = 0. \end{aligned}$$

That is, $g(a) = g(b)$, satisfying one of the hypotheses of Rolle's Theorem.

The other two hypotheses of Rolle's Theorem are also satisfied by g. Since L is continuous on $[a,b]$ and differentiable on (a,b), as is f, then $g = f - L$ possesses these two properties as well. Rolle's Theorem thus tells us there is at least one point c in (a,b) for which $g'(c) = 0$. Since $g'(x) = f'(x) - L'(x) = f'(x) - m_{sec}$, we have:

$$\begin{aligned} 0 &= g'(c) \\ &= f'(c) - \frac{f(b) - f(a)}{b - a}, \end{aligned}$$

or

$$f'(c) = \frac{f(b) - f(a)}{b - a}.$$

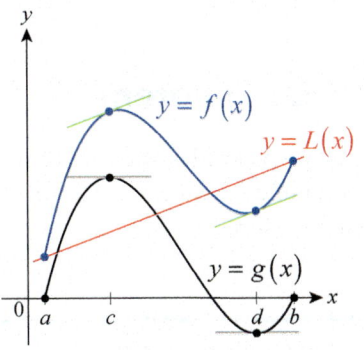

Figure 7

Figure 7 illustrates the process of subtracting the secant line L from the function f of Figure 6 in preparation for applying Rolle's Theorem. Note that c and d are points where g has horizontal tangent lines and f has tangent lines with slope m_{sec}.

Example 3

The function $f(x) = \frac{1}{3}x^3 - 2x^2 + 5x + 1$ is differentiable everywhere, so it satisfies the hypotheses of the Mean Value Theorem over any closed, bounded interval. Find the point(s) satisfying the conclusion of the MVT over the interval $[0, 4]$.

Solution

We begin by calculating the slope of the secant line over the interval.

$$m_{sec} = \frac{f(4) - f(0)}{4 - 0} = \frac{\frac{31}{3} - 1}{4} = \frac{7}{3}$$

The Mean Value Theorem guarantees the existence of at least one point $c \in (0, 4)$ where $f'(c) = 7/3$, but we want to go one step further and actually find the point (or points). Since $f'(x) = x^2 - 4x + 5$, we proceed to solve the equation $x^2 - 4x + 5 = 7/3$.

$$x^2 - 4x + 5 = \frac{7}{3} \qquad \text{Set } f'(x) = m_{sec}.$$

$$x^2 - 4x + \frac{8}{3} = 0$$

$$3x^2 - 12x + 8 = 0$$

$$x = \frac{12 \pm \sqrt{144 - 96}}{6} \qquad \text{Apply the quadratic formula.}$$

$$x = 2 \pm \frac{2\sqrt{3}}{3}$$

These two solutions, approximately equal to 0.85 and 3.15, both satisfy the conclusion of the MVT. (See Figure 8.)

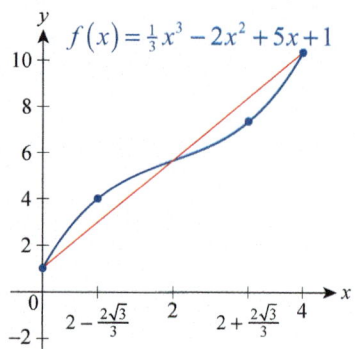

Figure 8

$f(x) = \frac{1}{3}x^3 - 2x^2 + 5x + 1$

Topic Two

Consequences of the Mean Value Theorem

Recall that a discussion of average and instantaneous velocities was one of the motivating examples we used to develop the concept of differentiation. The MVT provides a more quantitative connection between these ideas.

Example 4

An aerial surveillance crew is monitoring traffic on the roads below and timing the passage of cars between painted road marks that are 1 mile apart. The crew spots a distinctive yellow Lotus Elise and records its passage between two marks at 47 seconds. What speed can the crew assert the Elise must have reached during the 47 seconds?

Solution

If we let $f(t)$ denote the position of the Elise along the stretch of road over the time interval $[0,47]$, then $f'(t)$ denotes the car's velocity and the Mean Value Theorem guarantees that there is at least one $c \in (0,47)$ for which

$$f'(c) = \frac{f(47) - f(0)}{47 - 0} = \frac{1 \text{ mile}}{47 \text{ s}} \cdot \frac{3600 \text{ s}}{1 \text{ hr}} \approx 77 \text{ mph.}$$

That is, at least once over the mile-long stretch of road, the Elise was moving at a rate of approximately 77 miles per hour.

We already know that a constant function has a derivative of zero at all points of its domain, but we don't yet know that the converse is true. That is, does the fact that $f'(x) = 0$ for all x in a given interval mean that f is a constant function? The Mean Value Theorem provides an elegant answer to this question.

Theorem

Corollary 1

If $f'(x) = 0$ for each x in an open interval (a,b), then f is constant on (a,b).

Proof

We will show that given any two distinct points x_1 and x_2 in (a,b), $f(x_1) = f(x_2)$. Since $x_1 \neq x_2$, one is less than the other; we can assume $x_1 < x_2$. The fact that f is differentiable on all of (a,b) means that f is continuous on $[x_1, x_2]$ and differentiable on (x_1, x_2), so by the MVT there is a point $c \in (x_1, x_2)$ such that

$$f'(c) = \frac{f(x_2) - f(x_1)}{x_2 - x_1}.$$

But since f' is 0 throughout (a,b), this means that

$$f(x_2) - f(x_1) = f'(c) \cdot (x_2 - x_1) = 0 \cdot (x_2 - x_1) = 0,$$

so $f(x_1) = f(x_2)$.

While much of the text to this point has focused on the theory and mechanics of differentiation (that is, gaining knowledge about a function's derivative), Corollary 1 is a good illustration of just the reverse: what does knowledge of a function's derivative tell us about the function itself? The next corollary is similar in nature, and we will see the theme continue to develop in successive sections.

> **Theorem**
>
> **Corollary 2**
>
> If $f'(x) = g'(x)$ for all x in an open interval (a,b), then there is a constant C such that $f(x) = g(x) + C$ for all $x \in (a,b)$.

Proof

Since $f'(x) = g'(x)$, we know that $(f-g)'(x) = f'(x) - g'(x) = 0$ for all x in (a,b). So by Corollary 1, it must be the case that $f - g$ is a constant function on (a,b). That is, there is a constant C such that $(f-g)(x) = C$ for all $x \in (a,b)$, or $f(x) = g(x) + C$.

Example 5

Suppose that $f(0) = -2$ and that $f'(x) \leq 3$ for all x. What is the largest possible value of $f(4)$?

Solution

The implication is that f' exists everywhere, so f is certainly continuous on $[0,4]$ and differentiable on $(0,4)$. So by the MVT,

$$f'(c) = \frac{f(4) - f(0)}{4 - 0} = \frac{f(4) + 2}{4}$$

for some $c \in (0,4)$. Since $f'(x) \leq 3$ for all x, rearranging the above equation gives us the following.

$$f(4) + 2 = 4f'(c)$$
$$f(4) = 4f'(c) - 2 \leq 4 \cdot 3 - 2$$
$$f(4) \leq 10$$

Hence the largest possible value of $f(4)$ is 10.

Example 6

Find the unique function f whose derivative is $3x^2$ and whose graph passes through $(1,5)$.

Solution

By now, we have differentiated enough polynomials to recognize $3x^2$ as the derivative of x^3. If we let $g(x) = x^3$, then g and f have the same derivative and by Corollary 2 it must be the case that $f(x) = g(x) + C$ for some constant C. We can determine C as follows.

$$f(1) = 5 \qquad \text{The graph of } f \text{ passes through } (1,5).$$
$$g(1) + C = 5$$
$$1^3 + C = 5$$
$$C = 4$$

Now it can be easily verified that $f(x) = x^3 + 4$ satisfies the given criteria, and by Corollary 2 it is the unique function to do so.

4.2 **Exercises**

1–4 Use the graph of the function to visually estimate the value of c in the given interval that satisfies the conclusion of the Mean Value (or Rolle's) Theorem; then check your guess by calculation. If such a c doesn't exist, say why.

1. $f(x) = -x^2 + 6x - 4$ on $[1,4]$

2. $g(x) = -2\sqrt{|x-2|} + 2$ on $[-2,6]$

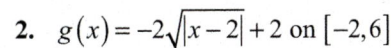

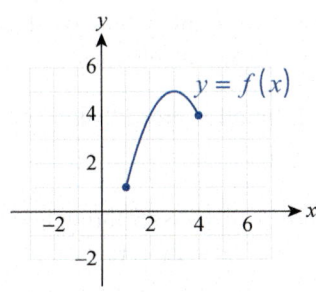

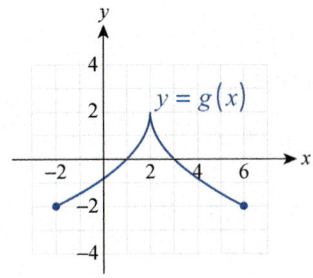

3. $h(x) = \dfrac{x^3}{3} - 2x^2 + \dfrac{11x}{3}$ on $[1,4]$

4. $k(x) = -\dfrac{1}{(x-3)^2} + 3$ on $[0,6]$

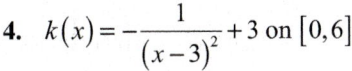

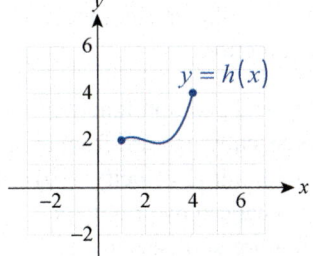

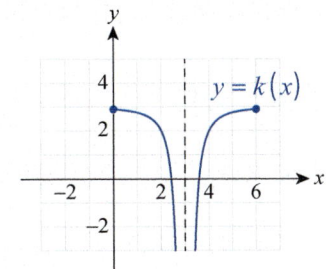

5–8 Prove that the equation has exactly one real solution on the given interval.

5. $x^5 - 3x^2 = 25$ on $[2,3]$

6. $5x^3 + 7x = 9$ on $\mathbb{R}$

7. $\arctan x = 3 - x$ on $[0,3]$

8. $\tan x = \cos x$ on $\left(0, \dfrac{\pi}{2}\right)$

9–20 Determine whether Rolle's Theorem applies to the function on the given interval. If so, find all possible values of c as in the conclusion of the theorem. If the theorem does not apply, state the reason.

9. $f(x) = -x^2 + 4x - 3$ on $[1,3]$

10. $g(x) = x^3 - 5x^2 + 2x + 10$ on $[-1,4]$

11. $h(x) = 2x^4 - x^3 + 6x^2 + x - 8$ on $[1,3]$

12. $F(x) = \dfrac{3}{(x-1)^2}$ on $[0,2]$

13. $G(x) = \dfrac{1}{x^2 + 1}$ on $[-3,3]$

14. $k(x) = \dfrac{x}{x^2 - 1}$ on $[-3,3]$

15. $H(x) = x^{4/5} - 10$ on $[-10,10]$

16. $m(x) = -\cos x$ on $[0, 4\pi]$

17. $T(z) = \cot z$ on $[0, 5\pi]$

18. $F(x) = \sec x$ on $\left[-\dfrac{\pi}{3}, \dfrac{\pi}{3}\right]$

19. $w(t) = \csc t$ on $\left[-\dfrac{\pi}{4}, \dfrac{\pi}{4}\right]$

20. $A(x) = |x+1| - 5$ on $[-6,4]$

21–32 Determine whether the Mean Value Theorem applies to the function on the given interval. If so, find all possible values of c as in the conclusion of the theorem. If the theorem does not apply, state the reason.

21. $f(x) = |x-1| + 3$ on $[-1,2]$

22. $g(x) = -\dfrac{1}{2}x + 4$ on $[0,8]$

23. $h(x) = -x^2 + 4x + 4$ on $[-1,4]$

24. $F(x) = \dfrac{1}{2}x^3 - x^2$ on $[-2,2]$

25. $G(x) = \dfrac{2x}{x+1}$ on $[-2,3]$

26. $k(x) = \dfrac{x+1}{2x-5}$ on $[-1,2]$

27. $H(x) = \dfrac{5}{x^2 + 5}$ on $[-5,5]$

28. $m(x) = (x-4)^{2/3} + 2$ on $[1,5]$

29. $C(x) = \sqrt{1 - x^2}$ on $[-1,0]$

30. $L(t) = \ln|t+1|$ on $[-3,1]$

31. $u(z) = \tan z$ on $[0, \pi]$

32. $v(s) = \arctan s$ on $[0,1]$

33. If $f(-2) = 2$ and $f'(x) \le 2$ for all x, what is the largest possible value of $f(2)$?

34. If $g(1) = 4$ and $g'(x) \ge -3.5$ for all x, what is the smallest possible value of $g(3)$?

35. One of your classmates claims that he found a function f such that $f(-5) = -1$, $f(5) = 1$, and $f'(x) \le 0.1$ for all x. Explain how you know that he made an error in his calculations.

36. If $|F'(x)| \le 2.5$ for all x, prove that $|F(7) - F(3)| \le 10$.

37. Find the function f that passes through $(3,1)$ and whose derivative is $3x^2 - 2x + 1$.

38. Find the function g that passes through $(\pi/2, 0)$ and whose derivative is $2\cos x$.

39. Suppose that the velocity function of a moving object is $v(t) = -2t + 5$ and that its position at $t = 1$ is -8 units from the origin. Find a formula for the position function.

40. Find the velocity and position functions of an object thrown vertically upward from an initial height of 20 ft with initial velocity $v(0) = 42$ ft/s. (The acceleration caused by gravity is $a = -32\,\text{ft/s}^2$. Ignore air resistance.)

41. Generalize Example 1 by proving the following: if $f(x)$ is differentiable on $[a,b]$, $f'(x) \ne 0$ on (a,b), and $f(a)$ and $f(b)$ have opposite signs, then the equation $f(x) = 0$ has exactly one real solution.

42. Two highway patrol cars are stationed 7 miles apart along a straight highway where the speed limit is 65 mph. The first patrol car clocks a red Porsche at 64 mph, and five minutes later, the second police car clocks it at 61.5 mph. Explain why the driver is pulled over and issued a ticket.

43. A plane leaves London Heathrow Airport to arrive at Houston Intercontinental Airport ten and a half hours later. The distance between the two airports is 4830 miles. Explain why the plane must have reached a speed of 450 mph at least twice during the trip.

44. Two MotoGP riders finish a race in a tie. Show that there was at least one moment during the race when the two riders had the exact same speed. (**Hint:** Ignoring the difference in their starting positions, consider the difference of the two position functions and use Rolle's Theorem.)

45. Suppose that $f'(x) = x$ for all $x \in \mathbb{R}$. Prove that there exists a constant C such that $f(x) = \frac{1}{2}x^2 + C$. (**Hint:** Use Corollary 2 from the text.)

46. Use Corollary 2 of this section to prove the well-known trigonometric identity $\cos^2 x + \sin^2 x = 1$. (**Hint:** Use the corollary for the functions $f(x) = \cos^2 x + \sin^2 x$ and $g(x) = 1$, and argue that necessarily $C = 0$.)

47. Follow the hint given in Exercise 46 to prove the identity $\sin^{-1} x + \cos^{-1} x = \pi/2$.

48. Let $f(x) = \cot \pi x$ and $g(x) = \cot \pi x + [\![x]\!]$. Show that $f'(x) = g'(x)$, but $f(x) - g(x)$ is not constant. Why does this not contradict Corollary 2 of this section?

49. Suppose that for a function $f(x)$, the second derivative $f''(x)$ exists for all x on an interval I. Prove that if f has three zeros on I, then its second derivative f'' also has a zero. (**Hint:** Apply Rolle's Theorem on f and then on f'.)

50. Suppose that f is twice differentiable on $\mathbb{R}$ and a and b are two successive zeros of f'. Prove that f can have at most one zero on (a,b). (**Hint:** Start by assuming that f has at least two zeros, and apply Rolle's Theorem.)

51. Suppose that f is continuous on $[a,b]$, differentiable on (a,b), and $c \in (a,b)$ with $f'(c) = 0$. Does it follow that $f(a) = f(b)$? Explain.

52.* Suppose that f is twice differentiable on $\mathbb{R}$ and $f'(c) = 0$ for some $c \in \mathbb{R}$. If $f''(c) \neq 0$, prove that there exist $a, b \in \mathbb{R}$ such that $c \in (a,b)$ and $f(a) = f(b)$.

53. Suppose that the rabbit population at a game preserve is observed monthly and is found to have increased from 550 to 1150 rabbits in a year's time. Explain why there must exist a time during the year when the population is increasing at a rate of 50 rabbits per month.

54. Show that if f is a quadratic function, that is, $f(x) = Ax^2 + Bx + C$, then on any interval $[a,b]$, the point c satisfying the conclusion of the Mean Value Theorem is the midpoint of the interval.

55.* Prove that for all $0 < \alpha < \beta < \pi/2$, $(\beta - \alpha)\cos \beta \leq \sin \beta - \sin \alpha \leq (\beta - \alpha)\cos \alpha$. (**Hint:** Apply the Mean Value Theorem for the function $f(x) = \sin x$ on the interval $[\alpha, \beta]$.)

56. Use the Mean Value Theorem to prove the inequality $\ln(1+x) \leq x$ for all $x \geq 0$. (**Hint:** For $x > 0$, consider $f(t) = \ln(1+t)$ on the interval $[0,x]$ and apply the Mean Value Theorem.)

57. Prove the inequality $\sqrt[3]{1+x} \leq 1 + \frac{1}{3}x$ for $x \geq 0$. (Adapt and follow the hint given in Exercise 56.)

58.* Use the Mean Value Theorem to show that $\lim_{x \to \infty}\left(\sqrt{x} - \sqrt{x-5}\right) = 0$. (**Hint:** Apply the theorem to the function $f(x) = \sqrt{x}$ on the interval $[x-5, x]$ for a fixed x-value. Then let x approach infinity.)

59. Use the Mean Value Theorem to prove the following inequality for all $\alpha,\ \beta \in \mathbb{R}$.

$$\left| \arctan \alpha - \arctan \beta \right| \le \left| \alpha - \beta \right|$$

60.* Generalizing Exercise 59, prove that if f is differentiable and $\left| f'(x) \right| \le M$ for some $M > 0$, then for all $x,\ y$ in the domain of f, $\left| f(x) - f(y) \right| \le M \left| x - y \right|$. (Such a function f is said to have the Lipschitz property with constant M.)

61. Use Exercise 60 to prove that $f(x) = \cos 3x$ has the Lipschitz property with constant 3.

62.* We say that the function $f(x)$ has a *fixed point* $c \in \mathbb{R}$ (or that it leaves c fixed) if $f(c) = c$. Prove that if f is differentiable on $\mathbb{R}$ and $f'(x) < 1$ for all x, then f can have no more than one fixed point.

63.* Prove Cauchy's Mean Value Theorem, which states the following: If f and g are both continuous on the closed interval $[a,b]$ and differentiable on (a,b), and if $g'(x) \ne 0$ on (a,b) and $g(a) \ne g(b)$, then there is a point

$$c \in (a,b) \text{ for which } \frac{f'(c)}{g'(c)} = \frac{f(b) - f(a)}{g(b) - g(a)}.$$

(**Hint:** Apply Rolle's Theorem to the function

$$F(x) = f(x) - f(a) - \frac{f(b) - f(a)}{g(b) - g(a)} \big(g(x) - g(a) \big).)$$

64–68 *True or False?* Determine whether the given statement is true or false. In case of a false statement, explain or provide a counterexample.

64. There are situations when the Mean Value Theorem applies but Rolle's Theorem doesn't.

65. There are situations when Rolle's Theorem applies but the Mean Value Theorem doesn't.

66. If a and b are zeros of the function $f(x)$, then there exists a point c in (a,b) such that $f'(c) = 0$.

67. If a and b are zeros of the polynomial $p(x)$, then there exists a point c in (a,b) such that $p'(c) = 0$.

68. If $f'(x) = 0$ for all x in the domain of f, then $f(x)$ is a constant function.

4.2 **Technology Exercises**

69–72 Use a graphing calculator or computer algebra system to graph the function on the interval $[a,b]$ along with its secant line through the points $(a, f(a))$ and $(b, f(b))$ on the same screen. Then find and graph the line(s) tangent to the graph that are parallel to the secant line.

69. $f(x) = \sqrt{x}$ on $[0,9]$

70. $f(x) = \dfrac{x^3}{2} - x^2 + x + 7.5$ on $[1,3]$

71. $f(x) = \dfrac{3x}{x+2}$ on $[-1,1]$

72. $f(x) = \sin^2 x + 3x$ on $[0,\pi]$

4.3 **The First and Second Derivative Tests**

In this section, we continue to explore what we can learn about a function from its derivatives, specifically concentrating on what a function's first and second derivatives tell us about its graph.

Topic One

First Derivative Tools

In our first example of an application of Rolle's Theorem in Section 4.2, we used the fact that $f'(x)$ was positive for all x to show that the equation $f(x) = 0$ couldn't have more than one solution. The reasoning in that example can be extended to make the following more general observations.

Theorem

Monotonicity Test

Assume f is differentiable at all points of an interval I.

1. If $f'(x) > 0$ for all $x \in I$, then f is strictly increasing on I.

2. If $f'(x) < 0$ for all $x \in I$, then f is strictly decreasing on I.

In other words, if f' is either entirely positive or entirely negative on I, then f is strictly monotonic on I. (Note that I may be an unbounded interval.)

Proof

Following the definition of strict monotonicity (Section 1.1), we start by assuming $x_1, x_2 \in I$ with $x_1 < x_2$. Since f' exists on all of I, we know that f is continuous on $[x_1, x_2]$ and differentiable on (x_1, x_2); by our definition of one-sided differentiation, this conclusion is true even if x_1 or x_2 is an endpoint of I.

Now, by the Mean Value Theorem, there is a point $c \in (x_1, x_2)$ such that

$$f(x_2) - f(x_1) = f'(c)(x_2 - x_1).$$

Since $x_2 - x_1 > 0$, the right-hand side of the above equation is either positive if $f'(c)$ is positive or negative if $f'(c)$ is negative, leading to the respective results of $f(x_2) > f(x_1)$ or $f(x_2) < f(x_1)$. Since this can be done for each such pair of points x_1 and x_2, we are done.

In practice, the Monotonicity Test is used to find intervals where a function is monotonic by first locating the function's critical points. If a and b are successive critical points of f and if f' exists on all of (a, b), then f' must be either positive or negative on (a, b). The easiest way to determine which is to evaluate f' at what is called a *test point* in (a, b).

Example 1

Determine the intervals of monotonicity of the function

$$f(x) = 3x^4 + 4x^3 - 12x^2 + 1.$$

Solution

$$f'(x) = 12x^3 + 12x^2 - 24x = 12x(x+2)(x-1)$$

The critical points of f are -2, 0, and 1. These divide the real line (the domain of f) into the open intervals $(-\infty, -2)$, $(-2, 0)$, $(0, 1)$, and $(1, \infty)$. The following table indicates the sign of f' and the corresponding monotonicity of f in each of these intervals as determined by evaluating f' at a convenient test point.

Interval	$(-\infty, -2)$	$(-2, 0)$	$(0, 1)$	$(1, \infty)$
f' at test point	$f'(-3) = -144$	$f'(-1) = 24$	$f'\left(\dfrac{1}{2}\right) = -\dfrac{15}{2}$	$f'(2) = 96$
Sign of f'	$-$	$+$	$-$	$+$
Monotonicity of f	Decreasing	Increasing	Decreasing	Increasing

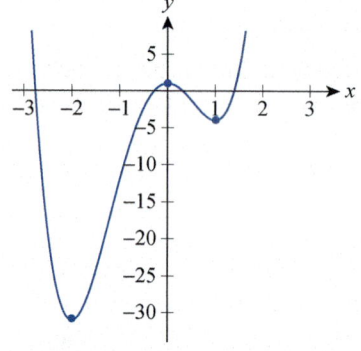

Figure 1

$f(x) = 3x^4 + 4x^3 - 12x^2 + 1$

So, f is decreasing on $(-\infty, -2)$ and $(0, 1)$ and increasing on $(-2, 0)$ and $(1, \infty)$, as shown in Figure 1.

(By our definition of monotonicity, it would also be correct to identify the intervals as $(-\infty, -2]$, $[-2, 0]$, $[0, 1]$, and $[1, \infty)$, but we are typically more interested in making a distinction between the critical points and the intervals between them.)

Note that we are actually only interested in the sign of f' at each test point, and this can typically be determined very quickly, especially if f' has been factored as in this example—there is no need to determine the actual numerical value of f'. In much of our work to come, we will make use of this shortcut.

Given a differentiable function defined on an open interval, Fermat's Theorem tells us that relative extrema occur only at critical points, but we've seen that not every critical point is a relative extremum—we need a test that we can apply to critical points. The conclusions of the previous theorem can be used for just this purpose.

First Derivative Test

Suppose c is a critical point of a function f that is continuous on an open interval containing c and differentiable on the same interval, except possibly at c itself. Moving through c from left to right,

1. if f' changes from positive to negative at c, then f has a relative maximum at c;

2. if f' changes from negative to positive at c, then f has a relative minimum at c;

3. if f' does not change sign at c (that is, if it is either positive on both sides of c or negative on both sides of c), then f does not have a relative extremum at c.

Proof

The conclusions of the First Derivative Test follow from applying the Monotonicity Test and determining whether f is increasing or decreasing to the left and to the right of a given critical point c. If, for instance, f is an increasing function to the left of c and a decreasing function to the right of c, then f has a relative maximum value at c. Figure 2 illustrates some of the possible combinations of intervals of monotonicity and critical points.

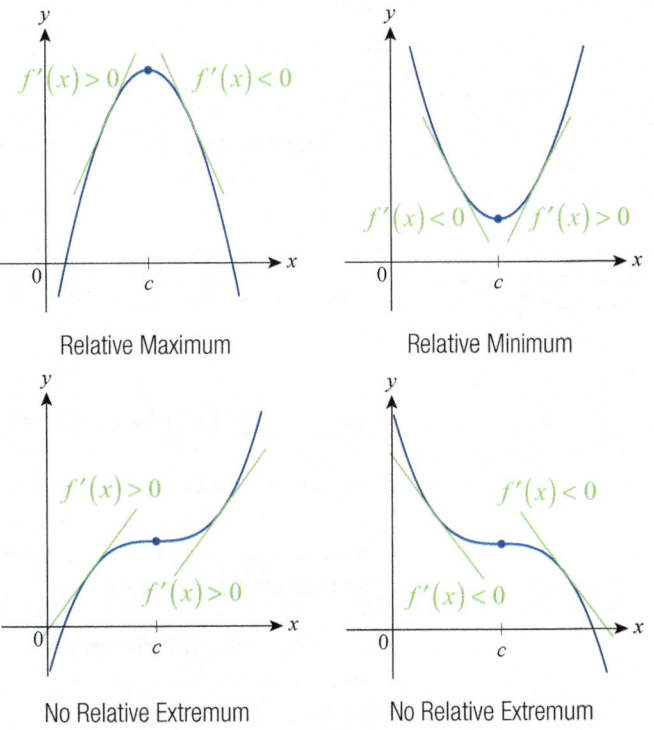

Figure 2

Example 2

Find the critical points of $f(x) = (x^2 - 7)x^{1/3}$, and then use the First Derivative Test to locate the relative extrema of f.

Solution

We note first that f is continuous everywhere, and then proceed to find f'.

$$f'(x) = (2x)x^{1/3} + \frac{1}{3}(x^2 - 7)x^{-2/3} \qquad \text{Product Rule}$$

$$= \frac{1}{3}x^{-2/3}(6x^2 + x^2 - 7) \qquad \text{Factor out } \tfrac{1}{3}x^{-2/3}.$$

$$= \frac{7(x^2 - 1)}{3x^{2/3}} = \frac{7(x-1)(x+1)}{3x^{2/3}}$$

The second step above illustrates that it is often useful to factor out an expression raised to the lowest power that it appears among the terms, in order to write f' in a convenient form (and factoring out the fraction $\frac{1}{3}$ has a similar benefit). Doing so makes it clear that the three critical points of f are 0 (where f' is undefined) and ± 1 (where f' is 0).

We can now quickly determine the sign of f' on each of the intervals between critical points, making use of the factored form above:

Interval	$(-\infty, -1)$	$(-1, 0)$	$(0, 1)$	$(1, \infty)$
f' at test point	$f'(-2) \approx 4.41$	$f'\left(-\frac{1}{2}\right) \approx -2.78$	$f'\left(\frac{1}{2}\right) \approx -2.78$	$f'(2) \approx 4.41$
Sign of f'	$+$	$-$	$-$	$+$
Monotonicity of f	Increasing	Decreasing	Decreasing	Increasing

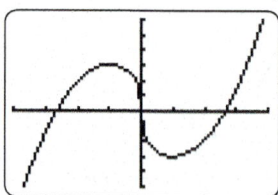

Figure 3

$f(x) = (x^2 - 7)x^{1/3}$ on $[-4,4]$ by $[-10,12]$

The First Derivative Test now tells us that f has a local maximum at -1, a local minimum at 1, and 0 is neither a local maximum nor minimum. The values of these local extrema are, respectively, $f(-1) = 6$ and $f(1) = -6$.

Figure 3 confirms our findings.

Topic Two

Second Derivative Tools

In reading a detailed analysis of, say, national economic trends, you might run across a complicated-sounding phrase like "the rate of growth is expected to decrease" or "the pace of increase picked up in the third quarter." What do these statements actually mean?

A careful study of such phrases shows that they are saying something about the rate of change of a rate of change—in other words, they refer to the second derivative of a quantity. Geometrically, they tell us whether the slopes of the tangent lines of a function are decreasing or increasing as we scan the graph from left to right. Figure 4a illustrates the first sort of situation and Figure 4b the second.

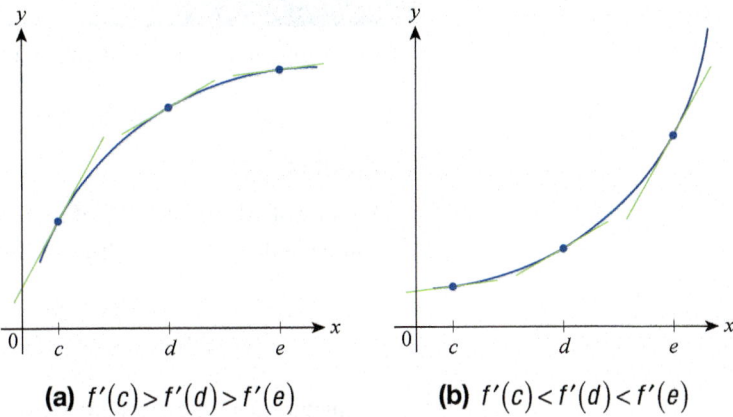

(a) $f'(c) > f'(d) > f'(e)$ **(b)** $f'(c) < f'(d) < f'(e)$

Figure 4 Decreasing and Increasing Slopes of Tangent Lines

This notion of decreasing or increasing slopes of tangent lines lies at the heart of the following definition.

Definition

Concavity

Given a differentiable function f on an interval I, we say f is **concave up** on I if f' is an increasing function on the interval and **concave down** on I if f' is a decreasing function on the interval.

Equivalently, we could define a concave up function as one that lies above all of its tangent lines on an interval and a concave down function as one that lies below all of its tangent lines; note how the graphs in Figure 4 illustrate downward and upward concavity by either definition.

There is yet another way we want to think about concavity, though it has the slight disadvantage of applying only to functions that are twice differentiable on an interval. Assuming that f'' does exist on an interval I, the Mean Value Theorem applied to f' indicates that $f'' > 0$ on I implies that f' is increasing, and hence that f is concave up. Similarly, $f'' < 0$ on I implies that f is concave down. This observation is formally stated as the Concavity Test.

Theorem

Concavity Test

Assume f is twice differentiable at all points of an interval I.

1. If $f''(x) > 0$ for all $x \in I$, then f is concave up on I.

2. If $f''(x) < 0$ for all $x \in I$, then f is concave down on I.

Example 3

Determine the intervals of concavity of the function

$$f(x) = 3x^4 + 4x^3 - 12x^2 + 1.$$

Solution

We identified the intervals of monotonicity for this function in Example 1, and we will resume our study of f by determining its second derivative.

$$f'(x) = 12x^3 + 12x^2 - 24x$$
$$f''(x) = 36x^2 + 24x - 24 = 12(3x^2 + 2x - 2)$$

Using the quadratic formula to solve the equation $f''(x) = 0$ yields the two solutions of $(-1 \pm \sqrt{7})/3$, or approximately -1.22 and 0.55. Since these are the only two points where f'' is 0, we would expect (and Figure 5 would lead us to believe) that f'' will be either consistently positive or consistently negative on each of the three intervals defined by the two points. We proceed then to make a table with these signs of f''.

Interval	$\left(-\infty, \dfrac{-1-\sqrt{7}}{3}\right)$	$\left(\dfrac{-1-\sqrt{7}}{3}, \dfrac{-1+\sqrt{7}}{3}\right)$	$\left(\dfrac{-1+\sqrt{7}}{3}, \infty\right)$
f'' at test point	$f''(-2) = 72$	$f''(0) = -24$	$f''(1) = 36$
Sign of f''	$+$	$-$	$+$
Concavity of f	Concave up	Concave down	Concave up

The three test points of -2, 0, and 1 were chosen merely for convenience—they lie in the interior of each interval, and it is relatively easy to compute f'' at each. Note the behavior of f at each of the two points where $f''(x) = 0$, as shown in the graph in Figure 5.

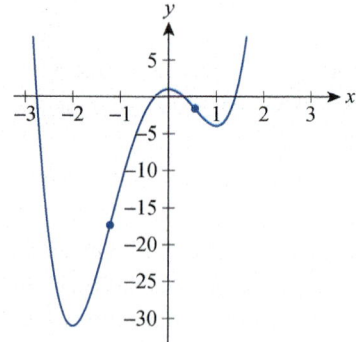

Figure 5

$f(x) = 3x^4 + 4x^3 - 12x^2 + 1$

In Example 3, we determined the intervals of concavity of f by locating those points where f'' changed sign, a process very similar to determining intervals of monotonicity. Informally, Darboux's Theorem (Section 3.1) says that derivatives always possess the Intermediate Value Property, so f'' (being the derivative of f') can only change sign at a point where it is zero or where it is undefined. Keep in mind, however, that points where f'' is either undefined or zero are only candidates for points where the concavity of f may change, and the concavity on either side of such a point must be checked in order to determine whether it actually does change. For example, if $f(x) = x^4$, $f''(x) = 12x^2$ and hence $f''(0) = 0$, but f is concave up everywhere (f'' is always nonnegative).

As you might expect, those points where a function changes concavity are deserving of a name.

Inflection Point

A point on the graph of a function f is called an **inflection point** if f is continuous there and changes concavity from upward to downward or from downward to upward. If the graph has a tangent line at the inflection point, the graph will cross the tangent line at that point.

(An alternate definition found in some settings *requires* the graph of f to actually possess a tangent line at the point of inflection, a stronger condition than just being continuous and one that guarantees the crossing behavior.)

Note that, as shown in Figure 6, f' can have any value or be undefined at a point of inflection. f'', however, will either be zero or undefined at such points.

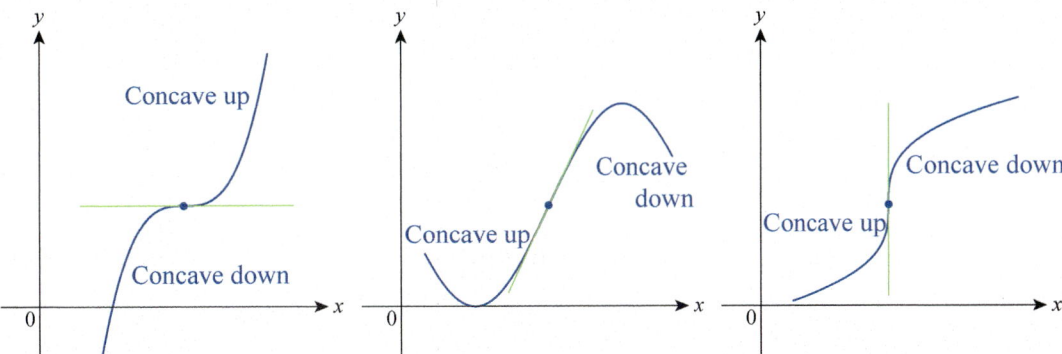

Figure 6

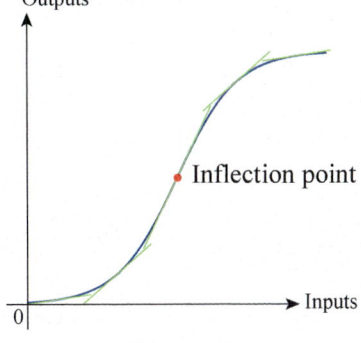

Figure 7

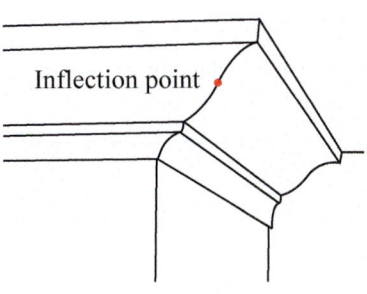

Figure 8

Inflection points occur in many different settings and go by many different names. The commonly heard phrase "point of diminishing returns" actually refers to a point of inflection—although often used loosely, in economics the phrase means that point past which increased inputs (worker hours, raw materials, etc.) yield smaller increases in outputs (finished products). It doesn't mean that production actually starts to decline at that point, but it does mean that the *rate of growth* of productivity becomes negative; that is, the graph of the outputs versus the inputs changes concavity. (See Figure 7.)

In architecture, an *ogee curve* refers to a shape in which a concave curve and a convex curve join in a point—ogee curves are often found, for example, in crown molding and in the designs of building columns. (See Figure 8.)

One last application of the second derivative will be mentioned here. Suppose that $f''(c) < 0$ on an open interval containing c and $f'(c) = 0$. Then f is concave down on the interval and the graph of f has a horizontal tangent line at c. A sketch of how f must roughly appear close to c will likely convince you that f must have a relative maximum at c. This is, in fact, true, and the formal statement of the observation and a parallel one for $f''(c) > 0$ comprise the last theorem of this section.

Theorem

Second Derivative Test

Suppose that $f'(c) = 0$ and that f'' exists at c.

1. If $f''(c) < 0$, then f has a relative maximum at c.

2. If $f''(c) > 0$, then f has a relative minimum at c.

If $f''(c) = 0$, the test does not apply and the graph of f may have a relative maximum, a relative minimum, or neither at c.

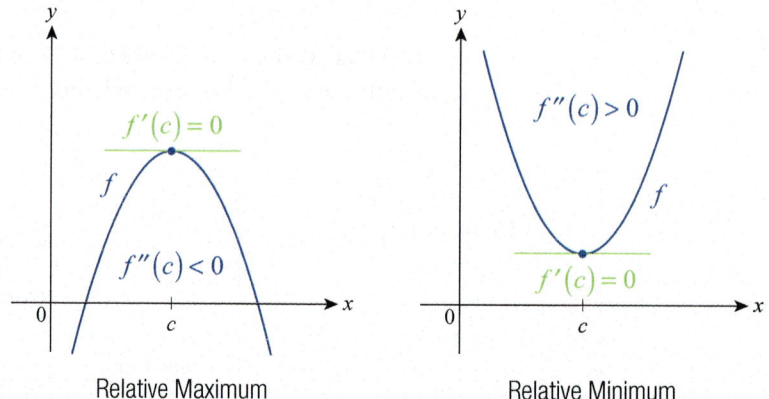

Relative Maximum   Relative Minimum

Figure 9

— **Proof** —

Recall that "Differentiability Implies Continuity"; this is one of our theorems from Section 3.1. So the fact that f'' exists at c means that f' is differentiable at c and hence continuous at c. This, in turn, means that there exists an open interval containing c for which f' is defined. If we know further that $f''(c) < 0$ and that $f'(c) = 0$, then for all h sufficiently close to 0,

$$\frac{f'(c+h) - f'(c)}{h} = \frac{f'(c+h)}{h} < 0$$

and so $f'(c+h)$ must be positive for $h < 0$ and negative for $h > 0$. Restated, this means that f' is positive to the left of c and negative to the right of c, so by the First Derivative Test we can conclude that f has a relative maximum at c. The proof of the second statement is similar in nature.

To see that the test does not apply if $f''(c) = 0$, consider the graphs of $f(x) = x^3$, $f(x) = x^4$, and $f(x) = -x^4$ at $x = 0$.

Example 4

Use the first and second derivatives of $f(x) = x^4 - 2x^3 + 1$ to identify the intervals of monotonicity, extrema, intervals of concavity, and inflection points on its graph.

Solution

$$f'(x) = 4x^3 - 6x^2 = 2x^2(2x - 3) \quad \text{and} \quad f''(x) = 12x^2 - 12x = 12x(x - 1)$$

Solving the equation $f'(x) = 0$ gives us the critical points 0 and $\frac{3}{2}$.

Solving $f''(x) = 0$ gives us the possible inflection points 0 and 1 (note that 0 is both a critical point and a potential inflection point).

We proceed to evaluate the signs of f' and f'' on the intervals of interest.

Interval	$(-\infty, 0)$	$\left(0, \dfrac{3}{2}\right)$	$\left(\dfrac{3}{2}, \infty\right)$
Sign of f'	$f'(-1) = -10 < 0$	$f'(1) = -2 < 0$	$f'(2) = 8 > 0$
Monotonicity of f	Decreasing	Decreasing	Increasing

Interval	$(-\infty, 0)$	$(0, 1)$	$(1, \infty)$
Sign of f''	$f''(-1) = 24 > 0$	$f''\left(\dfrac{1}{2}\right) = -3 < 0$	$f''(2) = 24 > 0$
Concavity of f	Concave up	Concave down	Concave up

We now have a wealth of information about f at our disposal, and can easily answer all of the questions about its graph.

First, the two tables above identify the intervals of constant monotonicity and concavity and whether f is increasing, decreasing, concave up, or concave down on each. Note how those intervals correspond to the graph of f shown in Figure 10.

Second, we can use the First Derivative Test to realize that f does not have a relative extremum at the critical point 0, but does have a relative minimum at $\frac{3}{2}$. And since the concavity changes from up to down at 0 and from down to up at 1, both 0 and 1 are inflection points.

We could also have used the Second Derivative Test to determine that f has a relative minimum at the critical point $\frac{3}{2}$, even though we never actually evaluated f'' at the point. The reason is that we know f'' exists on the open interval $(1, \infty)$ containing $\frac{3}{2}$ (indeed, f'' exists everywhere), and that it is positive on the interval. In particular, it is positive at $\frac{3}{2}$.

The last step is to evaluate f at the two inflection points and the one relative extremum—those values are shown in Figure 10. Note that f has no absolute maximum value, but the relative minimum at $\frac{3}{2}$ is also its absolute minimum.

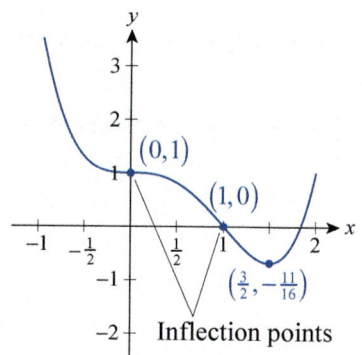

Figure 10

$f(x) = x^4 - 2x^3 + 1$

Example 4 illustrates that the tools of calculus can easily answer questions about particular points of interest on the graph of a function. We will explore this idea further in Section 4.5, but we will acquire one additional tool before doing so.

4.3 **Exercises**

1–22 Determine the intervals of monotonicity of the given function.

1. $f(x) = x^2 - 4x + 1$

2. $g(x) = \dfrac{3}{2}x + 5$

3. $h(x) = \dfrac{2}{3}x^3 + 4x^2 - 10x + \dfrac{5}{3}$

4. $F(x) = 0.75x^4 + x^3 - 15x^2 + 24x + 7$

5. $G(x) = -x^4 - 2x^3 + 8x^2 - 6x + 1$

6. $k(x) = (x^2 - 4)(x^2 - 3)$

7. $m(x) = -\dfrac{x^6}{3} - \dfrac{2x^5}{5} + \dfrac{x^4}{4} + \dfrac{x^3}{3} + \dfrac{x^2}{2} + x$

8. $n(x) = \dfrac{-2}{x+1}$

9. $H(x) = \dfrac{x^2 + 3}{x + 3}$

10. $R(t) = \dfrac{t+2}{t^2 - 9}$

11. $C(x) = \dfrac{3x^2 + 1}{x - 2}$

12. $A(x) = |x + 4| - 1$

13. $f(t) = 2.5 - |t - 3.125|$

14. $t(x) = x^{2/3} + 2$

15. $w(s) = \sqrt{s}\,(s - 1)$

16. $F(x) = 6 - (x - 3)^{3/5}$

17. $G(x) = \sin^2 x + 1$

18. $m(x) = 2^{-x} + 2^{2x}$

19. $g(t) = -2\sqrt{t}\,e^{-2t}$

20. $L(x) = x \ln x$

21. $A(x) = 0.5x^{1/5}(x^2 - 4)$

22. $U(s) = s\sqrt{3 - s^2}$

23–42. Use the First Derivative Test to classify the relative extrema, if any, of the functions given in Exercises 3–22.

43–50 Identify all intervals of monotonicity as well as intervals of concavity for the graphed function. Find all local extrema and inflection points, if any.

43. $f(x) = x^2 + 2x$

44. $g(x) = -\dfrac{1}{3}x^3 + 4x$

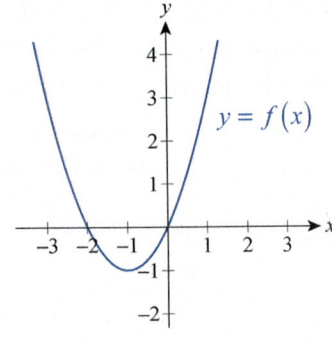

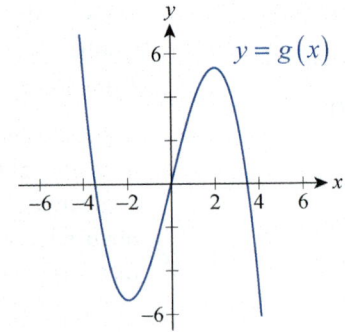

45. $h(x) = \dfrac{12x}{x^2 + 4}$ (Serpentine of Newton)

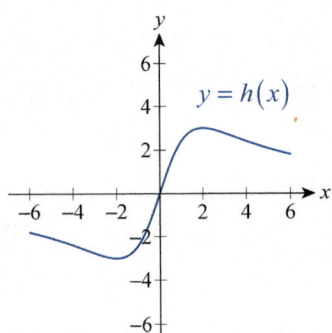

46. $F(x) = \dfrac{1}{16}x^4 - \dfrac{5}{12}x^3 + \dfrac{1}{4}x^2 + 2x$

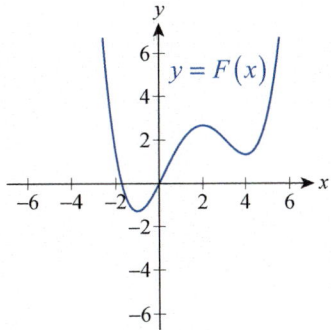

47. $G(x) = 1.5x^5 - 2.5x^3$

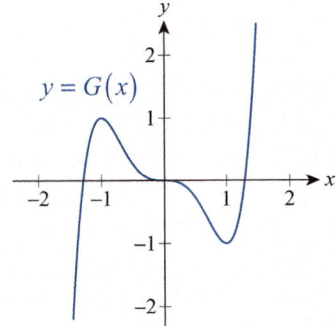

48. $H(x) = \dfrac{|1 - x|}{x}$

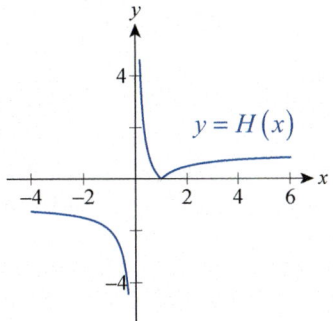

49. $k(x) = 2x - \tan x, \quad |x| < \dfrac{\pi}{2}$

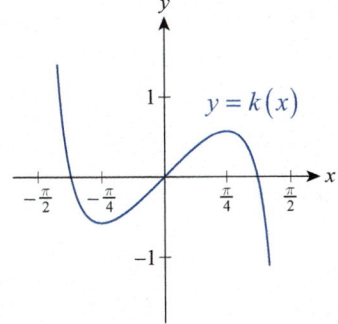

50. $m(x) = \cos x - \dfrac{x}{2}, \quad |x| \le 2\pi$

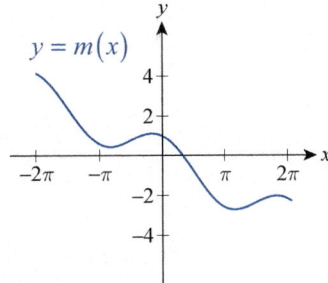

51–62 Determine the intervals of concavity of the given function.

51. $f(x) = 4x - x^2$

52. $g(x) = \dfrac{x^3}{3} - \dfrac{x^2}{2} + x + 1$

53. $h(x) = 4x^3 - 3x^2 - 36x + 5$

54. $k(x) = -x^4 + 5x + 1$

55. $F(x) = x^4 + 4x^3 + 72x$

56. $G(x) = -x^4 + 6x^3 + 24x^2 - 4x + 2$

57. $k(x) = 0.3x^5 + x^4 - 3x^3 + 12x + 4$

58. $v(x) = \dfrac{5}{x-2}$

59. $m(x) = \dfrac{x^2 + 5}{x - 5}$ **60.** $F(x) = \dfrac{2^x}{x}$

61. $H(x) = 9^x + 3^{-x}$ **62.** $u(x) = (x-2)^{5/7}$

63–82 Use the first and second derivatives to identify the intervals of monotonicity, extrema, intervals of concavity, and inflection points of the given function.

63. $f(x) = \dfrac{1}{2}x + 5$ **64.** $g(x) = x^2 - 8x + 3.5$

65. $h(x) = -\dfrac{1}{2}x^2 + 5x + \dfrac{8}{3}$ **66.** $F(x) = 2x^3 + 3x^2 - 7$

67. $G(x) = -4x^3 - 3x^2 + 18x + 10$

68. $K(x) = 0.5x^4 + 2x^3 - 6x^2 - 16x + 19.5$

69. $L(x) = -x^4 + 12x^2 - 20x + 3$

70. $m(x) = -\dfrac{3}{x^2}$

71. $n(x) = \dfrac{x+1}{x-2}$ **72.** $H(x) = \dfrac{2x}{x^2 - 4}$

73. $r(x) = \dfrac{2x^2 + 1}{x - 4}$ **74.** $t(x) = \dfrac{5}{4}\left(x - \dfrac{4}{5}\right)^{4/5}$

75. $F(x) = \dfrac{(x-1)^2}{x^2 - 1}$ **76.** $f(x) = \sqrt[3]{x} - x$

77. $g(x) = x^{2/3}\left(\dfrac{2}{3} - x\right)$ **78.** $h(x) = x\sqrt{9 - x^2}$

79. $u(x) = \sqrt{x}e^{-x}$ **80.** $v(x) = \sin^2 x - 2\cos x$

81. $k(x) = \cos x - \sin x$ **82.** $L(x) = -x^2 \ln|x|$

83–90 Sketch a graph of a function satisfying the given conditions. (Answers will vary.)

83. f is differentiable on $\mathbb{R}$, has both a local maximum and minimum, but no global extrema.

84. f is differentiable on $\mathbb{R}$, f' has a zero, and f has no local extrema.

85. f is differentiable on $\mathbb{R}$, f is an odd function, $f'(x) > 0$ on $(-4, 4)$, $f'(x) \le 0$ elsewhere, and $\lim_{|x| \to \infty} f(x) = 0$.

86. $\lim_{|x| \to \infty} f(x) = 1$ and $f(0) = 0$ is a global minimum.

87. $f(x)$ is everywhere positive on $\mathbb{R}$, but its derivative is everywhere negative.

88. $f'(x) < 0$ and $f''(x) > 0$ for all $x \in \mathbb{R}$.

89. f has vertical asymptotes at $x = \pm 2$, a horizontal asymptote at $y = 1$, f is an even function, $f'(x) < 0$ on $(0, 2)$ and $(2, \infty)$, f has a local maximum at 0, $f''(x) < 0$ on $(0, 2)$, $f''(x) > 0$ on $(2, \infty)$, and f has no absolute extrema.

90. $f(0) = f(2) = 0$, $f'(1) = f'(3) = f'(4) = 0$, $f''(1) < 0$, $f''(2) = 0$, and $\lim_{x \to \infty} f(x) = 0$.

91–96 The function $p(t)$ gives the position, relative to its starting point, of an object moving along a straight line. Identify the time intervals when the object is moving in the positive versus negative direction, as well as those intervals when it is accelerating or slowing down. Find the times when the object changes direction as well as when its acceleration is zero.

91. $p(t) = 2t^2 - 3t + 2.5$

92. $p(t) = 5t - \dfrac{1}{2}t^2$

93. $p(t) = 2t^3 - 15t^2 + 24t$

94. $p(t) = -2t^3 + 22.5t^2 - 66t$

95. $p(t) = e^{-t}\sin t$

96. $p(t) = \dfrac{1-t}{2t+2}$

97–102 The graph of the derivative $f'(x)$ of the function $f(x)$ is given. Use it to sketch an approximate graph of $f''(x)$. Then try to sketch a possible graph of $f(x)$ as well. (**Note:** There are many possible correct answers for the graph of f. Can you see why?)

97.

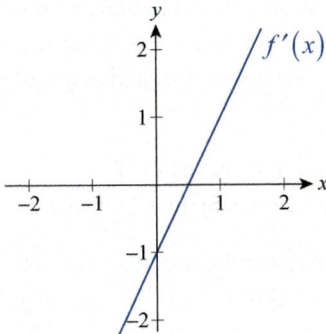

98.

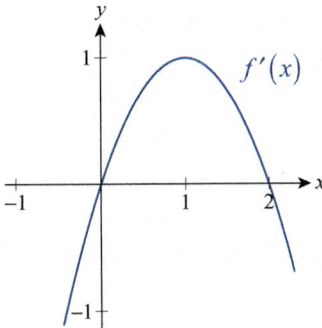

99.

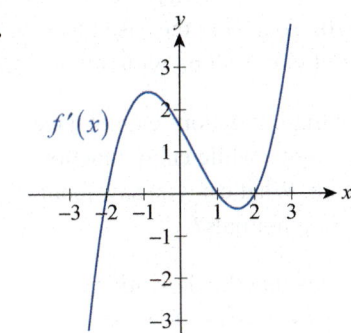

100.

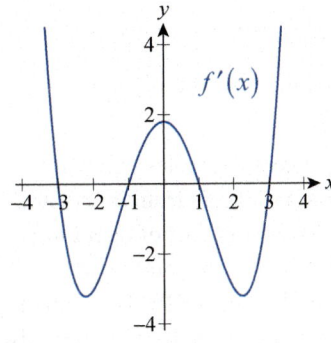

101.

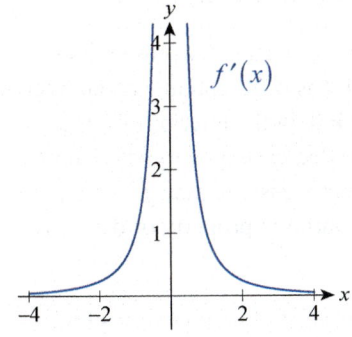

102.

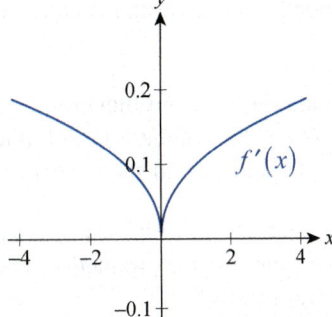

103. Suppose $T(t)$ is the outside temperature, over a 24-hour period on a typical spring day where you live (t is measured in hours, with $t = 0$ corresponding to midnight). Given the following data, what time(s) of day might c_i represent ($i = 1, 2, \ldots, 6$)? Explain your choice(s).

a. $T'(c_1) > 0, \quad T''(c_1) < 0$

b. $T'(c_2) > 0, \quad T''(c_2) > 0$

c. $T'(c_3) = 0, \quad T''(c_3) > 0$

d. $T'(c_4) < 0, \quad T''(c_4) < 0$

e. $T'(c_5) < 0, \quad T''(c_5) > 0$

f. $T'(c_6) = 0, \quad T''(c_6) < 0$

104. The graph below shows the profit (in thousands of dollars) from a product per hundreds of units sold. Use the graph to visually estimate the production level at which the marginal profit starts increasing.

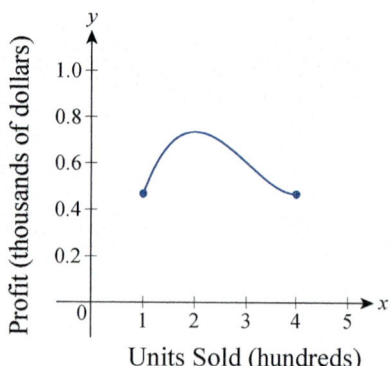

105. An aftermarket auto accessories company manufactures StopTheMess trunk liners and organizers. The overhead cost of operating the plant is $5000 per month and the cost of manufacturing each item is $20. The company estimates that 200 liners can be sold monthly for $50 apiece, and that sales will increase by 10 liners per month for each dollar decrease in price.

a. Find a formula for the profit function $P(n)$, where n is the number of trunk liners manufactured (suppose $200 \le n \le 350$).

b. Identify the intervals on which P is increasing or decreasing, and find the production level that maximizes profits.

106. The strength of an electric field due to a charged ring obeys the equation

$$E = \frac{kqx}{\left(x^2 + R^2\right)^{3/2}}$$

where q is the electric charge measured in coulombs (C), $k \approx 8.99 \cdot 10^9 \ \text{Nm}^2/\text{C}^2$, R is the radius of the ring and x is the distance to the charge in meters. Find a formula for the rate of change of E as x increases. What happens to E and dE/dx as $x \to \infty$?

107. Suppose that $f'(x) = (x+4)(x+1)^2(x-3)^3$. By examining the zeros of $f'(x)$, identify the x-coordinates of the local maxima and minima of $f(x)$. (**Hint:** Recall what you learned about multiplicities of zeros and sign changes of polynomials.)

108. Repeat Exercise 107 for $g(x)$ if $g'(x) = 3x(x-2)(x+5)^4\left(x-\tfrac{1}{3}\right)^2(x+1)^3$.

109. Use derivatives to prove that if $x \in (-\infty, 1)$, then $1/(1-x) \ge 1+x$. (**Hint:** Start by assuming that $x \ge 0$. Rewriting the inequality as $f(x) \ge g(x)$, show that $D(x) = f(x) - g(x)$ is increasing, while $D(0) = 0$. To handle the case of $x < 0$, use the fact that $D'(x) < 0$ along with $D(0) = 0$.)

110. Prove that quadratic functions cannot have any inflection points, while cubic functions have exactly one. What can you say about fourth-degree polynomials?

111. Suppose we know that the derivative of a function f is $f'(x) = a/(x^2 + 1)$ for some nonzero $a \in \mathbb{R}$. Prove that f is increasing or decreasing everywhere on $\mathbb{R}$.

112. Generalize Exercise 111 by proving the following: If f is differentiable on an interval I and $f'(x) \ne 0$ in the interior of I, then f is increasing or decreasing everywhere on I. (**Hint:** Indirectly assume that f' changes signs, and use the Darboux property of derivative functions.)

113. Use Exercise 112 to find conditions under which the general cubic polynomial $p(x) = ax^3 + bx^2 + cx + d$ is decreasing everywhere on $\mathbb{R}$.

114. Prove that a cubic polynomial $p(x) = ax^3 + bx^2 + cx + d$ has exactly one inflection point. Find its first coordinate, assuming that the three real roots of $p(x)$ are x_1, x_2, and x_3.

115. Determine the values of the coefficients so that the third-degree polynomial $p(x) = ax^3 + bx^2 + cx + d$ has a local maximum at $x = 0$ and a local minimum at $x = 4$.

116.* Consider the following function:

$$f(x) = \begin{cases} 2x^2 & \text{if } x \text{ is irrational} \\ 4x^2 & \text{if } x \text{ is rational} \end{cases}$$

Use f to explain why the changing of signs of the derivative is not necessary for a function to have a local extremum.

117. By considering $f(x) = \begin{cases} x & \text{if } x \le 1 \\ 3 - x & \text{if } x > 1 \end{cases}$, explain

why the First Derivative Test can't be used for a discontinuous function.

118. Suppose that $f(x)$ and $g(x)$ are at least twice differentiable and that both their first and second derivatives are positive everywhere on an interval I. Which of the following can you prove from these conditions? Prove those statements that are true, and provide counterexamples for the rest.

a. $f(x) + g(x)$ is increasing on I.

b. $f(x) + g(x)$ is concave up on I.

c. $f(x) \cdot g(x)$ is increasing on I.

d. $f(x) \cdot g(x)$ is concave up on I.

e. $f(g(x))$ is increasing on I.

f. $f(g(x))$ is concave up on I.

119.* Use mathematical induction to prove the following generalization of the Second Derivative Test: Suppose that the derivatives of all orders of the function f exist at c, up to $f^{(2k)}(c)$, and that $f'(c) = f''(c) = \cdots = f^{(2k-1)}(c) = 0$, but $f^{(2k)}(c) \ne 0$. Then if $f^{(2k)}(c) < 0$, f has a relative maximum at c; if $f^{(2k)}(c) > 0$, f has a relative minimum at c.

120.* Use mathematical induction to prove that if f is $(2k+1)$-times differentiable at c, $f'(c) = f''(c) = \cdots = f^{(2k)}(c) = 0$, but $f^{(2k+1)}(c) \ne 0$, then f has a point of inflection at c.

121–128 *True or False?* Determine whether the given statement is true or false. In case of a false statement, explain or provide a counterexample.

121. Not all fourth-degree polynomials have inflection points.

122. A function with no inflection points cannot change concavity.

123. If $f'(x)$ is negative on $(-\infty, c)$ and positive on (c, ∞), then f has a minimum at c.

124. If $f(x)$ and $g(x)$ are decreasing, then so is $(f + g)(x)$.

125. If $f(x)$ and $g(x)$ are decreasing, then so is $(f \cdot g)(x)$.

126. A polynomial of degree n cannot have more than $n - 1$ extrema on $\mathbb{R}$.

127. If $c \in \mathbb{R}$ is a critical point, then the function has a local minimum or a local maximum at c.

128. If $f''(c) = 0$, then c is an inflection point.

4.3 Technology Exercises

129–130 Use a graphing calculator or computer algebra system to graph the given function along with its first and second derivatives on the same screen. Use your graphs to explain the behavior of the function with regard to the signs and values of its derivatives.

129. $f(x) = (x^2 + 1)\sqrt{9 - x^2}$ on $[-3,3]$ **130.** $g(x) = \sqrt{x}\cos x - \sin 2x$ on $[0, 4\pi]$

131. The table below shows the temperature of a pediatric patient over a 24-hour period (measurements were taken every two hours, starting at midnight). Use the regression capabilities of a graphing calculator or computer algebra system to approximate the data by a fourth-degree polynomial. When do you estimate the patient's temperature to have been the highest? The lowest? When did the highest rates of increase and decrease occur?

Time	12 a.m.	2 a.m.	4 a.m.	6 a.m.	8 a.m.	10 a.m.	12 p.m.
Temp (°F)	99.9	99.5	99.1	98.9	98.7	98.8	99.4
Time	2 p.m.	4 p.m.	6 p.m.	8 p.m.	10 p.m.	12 a.m.	
Temp (°F)	100.0	102.1	101.9	101.3	101.0	99.9	

132. In the first few months following the launch of a new product, monthly sales were given by the function

$$S(t) = \frac{300t^2}{t^2 + 2}, \text{ where } t \text{ is measured in months.}$$

a. Use a graphing calculator or computer algebra system to graph the function over the first year, and estimate when the rate of growth in sales was greatest.

b. Use the differentiation capabilities of a computer algebra system to check your estimate in part a.

133–136 Use a graphing calculator or computer algebra system to graph the given function for different values of the parameter(s). Examine how the values of the parameter(s) affect the number of local extrema. How about inflection points?

133. $f(x) = x^4 + cx^3$; $\quad 1 \le c \le 3$ **134.** $g(x) = 0.5x^5 + cx^4 - dx$; $\quad 0 \le c, d \le 3$

135. $h(x) = \cos x - \sin(cx)$; $\quad 0 \le c \le 4$ **136.** $k(x) = \sin^2(cx)\cos(dx)$; $\quad 0 \le c, d \le 5$

4.4 L'Hôpital's Rule

In this section, we develop a calculus-based tool that often allows us to evaluate, fairly easily, limits that would otherwise pose a considerable challenge. The tool is called *l'Hôpital's Rule*, and the limits it applies to are called *limits of indeterminate form*.

Topic One
L'Hôpital's Rule

As motivation, consider a limit of the form

$$\lim_{x \to c} \frac{f(x)}{g(x)},$$

where $f(c) = g(c) = 0$, $f'(c)$ and $g'(c)$ both exist, and $g'(c) \neq 0$. Such a limit is said to be of **indeterminate form 0/0**, and the limit cannot be determined by simply evaluating $f(c)/g(c)$. However, we can rewrite the limit in a form that can be evaluated, as follows.

$$\lim_{x \to c} \frac{f(x)}{g(x)} = \lim_{x \to c} \frac{f(x) - 0}{g(x) - 0}$$

$$= \lim_{x \to c} \frac{f(x) - f(c)}{g(x) - g(c)} \qquad f(c) = g(c) = 0$$

$$= \lim_{x \to c} \frac{\dfrac{f(x) - f(c)}{x - c}}{\dfrac{g(x) - g(c)}{x - c}} \qquad \text{Divide top and bottom by } x - c.$$

$$= \frac{\displaystyle\lim_{x \to c} \frac{f(x) - f(c)}{x - c}}{\displaystyle\lim_{x \to c} \frac{g(x) - g(c)}{x - c}}$$

$$= \frac{f'(c)}{g'(c)}$$

So if the values of these derivatives are known (and $g'(c) \neq 0$), the evaluation of the limit *can* be accomplished with a different substitution. This observation is a simple form of l'Hôpital's Rule, named after the French nobleman Guillaume François Antoine de l'Hôpital (1661–1704) in whose introductory calculus textbook it first appeared—the result is actually due to the Swiss mathematician Johann Bernoulli (1667–1748).

Example 1

Determine $\lim\limits_{x \to 0} \dfrac{5x - \sin 2x}{x}$.

Solution

Although we have learned ways to algebraically manipulate fractions of this sort in order to find the limit, l'Hôpital's Rule makes it strikingly easy.

$$\lim_{x \to 0} \frac{5x - \sin 2x}{x} = \lim_{x \to 0} \frac{5 - 2\cos 2x}{1} \qquad \text{Differentiate top and bottom.}$$
$$= 5 - 2(1) \qquad \text{Substitute } x = 0.$$
$$= 3$$

Before presenting the full form of l'Hôpital's Rule, we will state one other useful theorem—it is a formulation of the Mean Value Theorem due to Augustin-Louis Cauchy (1789–1857). Cauchy's MVT is the basis for one of the more elegant proofs of l'Hôpital's Rule (a hint on proving Cauchy's MVT appears in Exercise 63 of Section 4.2).

Theorem

Cauchy's Mean Value Theorem

Suppose that f and g are continuous on $[a,b]$ and differentiable on (a,b), $g'(x) \neq 0$ on (a,b), and $g(a) \neq g(b)$. Then there is a point $c \in (a,b)$ such that

$$\frac{f'(c)}{g'(c)} = \frac{f(b) - f(a)}{g(b) - g(a)}.$$

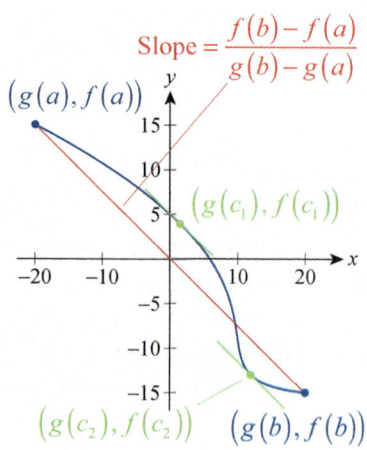

Figure 1

Cauchy's Mean Value Theorem

Note that Cauchy's Mean Value Theorem reduces to the simpler Mean Value Theorem if $g(x) = x$.

The simpler version of the MVT guarantees the existence of a point where the tangent to a function is parallel to a secant line, and Cauchy's MVT does something similar. Given two functions f and g with the properties above, the collection of ordered pairs $\{(g(x), f(x)) \mid x \in [a,b]\}$ defines a curve in $\mathbb{R}^2$ such as the one depicted in Figure 1, and the red line segment connecting $(g(a), f(a))$ and $(g(b), f(b))$ is also called a secant line. Specifically, the blue curve in Figure 1 is defined by the two functions $f(x) = x^2 - 5x - 9$ and $g(x) = x^3 + x + 10$, and the interval $[a,b]$ is $[-3,2]$; the slope of the secant line is $-\frac{3}{4}$. Cauchy's MVT tells us that for at least one number $c \in (a,b)$, the line tangent to the curve at $(g(c), f(c))$ is parallel to the secant line. (Curves defined in this manner are called *parametric curves*, and we will study them in detail in Chapter 9.)

We are now ready for a stronger form of l'Hôpital's Rule.

Theorem

L'Hôpital's Rule

Suppose f and g are differentiable at all points of an open interval I containing c, and that $g'(x) \neq 0$ for all $x \in I$ except possibly at $x = c$. Suppose further that either

$$\lim_{x \to c} f(x) = 0 \text{ and } \lim_{x \to c} g(x) = 0$$

or

$$\lim_{x \to c} f(x) = \pm\infty \text{ and } \lim_{x \to c} g(x) = \pm\infty.$$

Then

$$\lim_{x \to c} \frac{f(x)}{g(x)} = \lim_{x \to c} \frac{f'(x)}{g'(x)},$$

assuming the limit on the right is a real number or ∞ or $-\infty$.

Further, the rule is true for one-sided limits at c and for limits at infinity; that is, $x \to c$ can be replaced with $x \to c^+$, $x \to c^-$, $x \to -\infty$, or $x \to \infty$, assuming always that the limit on the right is a real number or ∞ or $-\infty$.

We have already mentioned limits of indeterminate form $0/0$ in passing; the other type of limit described in l'Hôpital's Rule is of **indeterminate form ∞/∞**, and we will discuss other variations soon.

— Proof

We will prove only the case in which

$$\lim_{x \to c} f(x) = 0 \text{ and } \lim_{x \to c} g(x) = 0,$$

and we will prove that the claim is true as $x \to c^-$; the corresponding result for $x \to c^+$ is nearly identical, and the two one-sided limits together prove the theorem.

Suppose that $x \in I$ is a number lying to the left of c. Then $g'(x) \neq 0$, and we can apply Cauchy's MVT to the interval $[x, c]$. Thus, there is a point $\tilde{c} \in (x, c)$ such that

$$\frac{f'(\tilde{c})}{g'(\tilde{c})} = \frac{f(c) - f(x)}{g(c) - g(x)}$$

$$= \frac{f(x)}{g(x)}. \qquad\qquad {\color{blue} f(c) = g(c) = 0}$$

As we let $x \to c^-$, $\tilde{c} \to c^-$ as well since $\tilde{c}$ always lies between x and c. Hence,

$$\lim_{x \to c^-} \frac{f(x)}{g(x)} = \lim_{\tilde{c} \to c^-} \frac{f'(\tilde{c})}{g'(\tilde{c})} = \lim_{x \to c^-} \frac{f'(x)}{g'(x)}.$$

The value of l'Hôpital's Rule comes from the fact that differentiating the numerator and denominator of a fraction of indeterminate form often results in a fraction that is not, making the limit easier to determine.

Example 2

Determine $\displaystyle\lim_{x \to \infty} \frac{\ln x}{2\sqrt{x}}$.

Solution

First, note that l'Hôpital's Rule is indeed applicable: this limit at infinity is of indeterminate form ∞/∞. So we evaluate the limit as follows.

$$\lim_{x \to \infty} \frac{\ln x}{2\sqrt{x}} = \lim_{x \to \infty} \frac{\dfrac{d}{dx}(\ln x)}{\dfrac{d}{dx}\left(2\sqrt{x}\right)} = \lim_{x \to \infty} \frac{\dfrac{1}{x}}{\dfrac{1}{\sqrt{x}}}$$

$$= \lim_{x \to \infty} \frac{1}{\sqrt{x}} = 0$$

Example 3

Determine $\displaystyle\lim_{x \to 0} \frac{a^x - 1}{x}$, where $a > 0$ is a constant.

Solution

The limit is of indeterminate form $0/0$, so we proceed using l'Hôpital's Rule.

$$\lim_{x \to 0} \frac{a^x - 1}{x} = \lim_{x \to 0} \frac{(\ln a)a^x}{1} = \ln a$$

Caution!

L'Hôpital's Rule says the limit of a quotient of two functions is equal to the limit of the quotient of their derivatives—don't mistakenly apply the Quotient Rule of differentiation! After verifying that the conditions of l'Hôpital's Rule are satisfied, proceed by differentiating the numerator and denominator individually.

L'Hôpital's Rule is obviously useful, but there is even more power in it than might appear at first glance. Suppose the rule is applied to a limit of indeterminate form, and the resulting limit is again of indeterminate form. Does that mean the rule fails to tell us anything? Often, no—if the new limit is also indeterminate, then we can again apply l'Hôpital's Rule. This process can be repeated as often as necessary, *as long as we stop as soon as we reach a limit that is not indeterminate.* The next example illustrates how repeated application of the rule works, and the sort of mistake that can arise if we misuse it.

Example 4

Evaluate the following limits.

a. $\displaystyle\lim_{x\to\infty}\frac{e^x}{5x^2-3x+2}$

b. $\displaystyle\lim_{x\to 0}\frac{1-\cos x}{3x^2+7x}$

Solution

a. The limit is of indeterminate form ∞/∞, so we can apply l'Hôpital's Rule.

$$\lim_{x\to\infty}\frac{e^x}{5x^2-3x+2}=\lim_{x\to\infty}\frac{e^x}{10x-3}\qquad\text{Limit is still of indeterminate form } \infty/\infty.$$
$$\text{Apply l'Hôpital's Rule again.}$$

$$=\lim_{x\to\infty}\frac{e^x}{10}\qquad\text{Limit can now be determined; } e^x\to\infty \text{ as } x\to\infty$$

$$=\infty$$

b. The limit is of indeterminate form $0/0$ and we apply l'Hôpital's Rule.

$$\lim_{x\to 0}\frac{1-\cos x}{3x^2+7x}=\lim_{x\to 0}\frac{\sin x}{6x+7}=\frac{0}{7}=0.$$

If we mistakenly continued to apply the rule, we would have obtained

$$\lim_{x\to 0}\frac{\sin x}{6x+7}=\lim_{x\to 0}\frac{\cos x}{6}=\frac{1}{6},$$

which is incorrect.

Topic Two

Limits of Indeterminate Form

In addition to the two indeterminate forms we have already seen, l'Hôpital's Rule can be used to evaluate other potentially challenging limits. The remaining examples illustrate how indeterminate products, differences, and powers can be rewritten in such a way that l'Hôpital's Rule applies.

Example 5

Determine $\displaystyle\lim_{x\to 0^+}\sqrt{x}\,\ln x$.

Solution

We say that a limit of a product fg is of **indeterminate form $0\cdot\infty$** if one of the functions approaches 0 and the other approaches ∞ or $-\infty$. The limit, if it exists, depends on which function dominates—the product could tend toward 0, could grow unbounded, or could approach some nonzero real number if the two functions balance one another just right.

We can apply l'Hôpital's Rule if we can rewrite the product as a quotient in either the indeterminate form $0/0$ or ∞/∞. That is, we rewrite fg as either

$$\frac{f}{1/g}\quad\text{or}\quad\frac{g}{1/f},$$

whichever is easier to work with. In this example,

$$\lim_{x \to 0^+} \sqrt{x} \ln x = \lim_{x \to 0^+} \frac{\ln x}{\dfrac{1}{\sqrt{x}}} \qquad \text{Rewrite to obtain the indeterminate form } \infty/\infty.$$

$$= \lim_{x \to 0^+} \frac{\dfrac{1}{x}}{-\dfrac{1}{2}x^{-3/2}} = \lim_{x \to 0^+} \left(-2\sqrt{x} \right) = 0.$$

Example 6

Determine $\displaystyle \lim_{x \to 1} \left(\frac{x}{x-1} - \frac{1}{\ln x} \right)$.

Solution

A limit of a difference $f - g$ is of **indeterminate form** $\infty - \infty$ if $f \to \infty$ and $g \to \infty$. Again, such limits are usually not trivial: if f dominates, the difference will tend to ∞; and if g dominates, the difference will tend to $-\infty$; but it is also possible for the two to balance out and result in a finite limit.

To use l'Hôpital's Rule, we need to rewrite the difference as a quotient. In this case, we can do so by combining the two fractions using a common denominator; in other such problems, rationalization or factoring out a common factor may be helpful.

$$\lim_{x \to 1} \left(\frac{x}{x-1} - \frac{1}{\ln x} \right) = \lim_{x \to 1} \frac{x \ln x - x + 1}{(x-1)\ln x}$$

$$= \lim_{x \to 1} \frac{\ln x + 1 - 1}{\ln x + \dfrac{x-1}{x}} \qquad \text{Apply l'Hôpital's Rule once.}$$

$$= \lim_{x \to 1} \frac{x \ln x}{x \ln x + x - 1} \qquad \text{Simplify fraction.}$$

$$= \lim_{x \to 1} \frac{\ln x + 1}{\ln x + 1 + 1} \qquad \text{Apply l'Hôpital's Rule again.}$$

$$= \frac{1}{2}$$

Figure 2 visually confirms the limit we found.

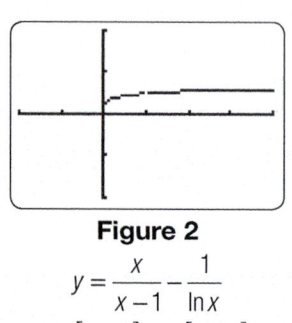

Figure 2

$$y = \frac{x}{x-1} - \frac{1}{\ln x}$$

on $[-2,4]$ by $[-2,2]$

The next three limits are all of the form f^g, and constitute limits of **indeterminate forms 1^∞, 0^0, and ∞^0**. As with the other indeterminate forms, there is competition between the effects of the two functions in the limit. In order to use l'Hôpital's Rule, we set $y = f^g$ and take the natural logarithm of both sides to obtain $\ln y = g \ln f$. If we can determine the limit of $g \ln f$, we can determine the limit of $y = e^{\ln y} = e^{g \ln f}$.

Example 7

Determine $\lim\limits_{x\to 0^+}(1+x)^{1/x}$.

Solution

Note that the base of the expression $(1+x)^{1/x}$ goes to 1 and the exponent goes to ∞ as $x\to 0^+$. We let $y=(1+x)^{1/x}$.

$$\ln y = \left(\frac{1}{x}\right)\ln(1+x)$$

$$= \frac{\ln(1+x)}{x}$$

The limit of this last expression is of the form $0/0$, and we can apply l'Hôpital's Rule.

$$\lim_{x\to 0^+}\frac{\ln(1+x)}{x} = \lim_{x\to 0^+}\frac{\dfrac{1}{1+x}}{1} = 1$$

Since $\lim\limits_{x\to 0^+}\ln y = 1$, $\lim\limits_{x\to 0^+}y = \lim\limits_{x\to 0^+}e^{\ln y} = e^1 = e$. Hence, $\lim\limits_{x\to 0^+}(1+x)^{1/x} = e$.

Example 8

Determine $\lim\limits_{x\to 0^+}x^x$.

Solution

Both the base and the exponent approach 0. Letting $y=x^x$, we arrive at

$$\ln y = x\ln x = \frac{\ln x}{\dfrac{1}{x}},$$

a limit of indeterminate form ∞/∞. So,

$$\lim_{x\to 0^+}\frac{\ln x}{\dfrac{1}{x}} = \lim_{x\to 0^+}\frac{\dfrac{1}{x}}{-\dfrac{1}{x^2}} = \lim_{x\to 0^+}(-x) = 0,$$

and hence $x^x \to e^0 = 1$ as $x\to 0^+$ (don't forget this last step!).

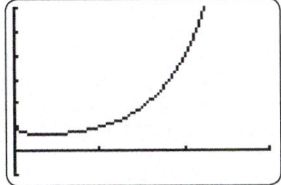

Figure 3

$y = x^x$ on $[0,3]$ by $[-1,6]$

Example 9

Determine $\lim\limits_{x\to\infty} x^{1/x}$.

Solution

The base has a limit of ∞ and the exponent has a limit of 0. We proceed as in the last two examples.

$$y = x^{1/x}$$

$$\ln y = \frac{1}{x}\ln x = \frac{\ln x}{x} \qquad\qquad \text{Indeterminate form } \infty/\infty$$

Applying l'Hôpital's Rule,

$$\lim_{x\to\infty} \ln y = \lim_{x\to\infty}\frac{\ln x}{x} = \lim_{x\to\infty}\frac{\dfrac{1}{x}}{1} = 0$$

and therefore $\lim\limits_{x\to\infty} x^{1/x} = \lim\limits_{x\to\infty} y = e^0 = 1$.

4.4 Exercises

1–12 Evaluate the limit using the theorems of Chapter 2. Then decide whether l'Hôpital's Rule is applicable and, if so, use it to check your answer.

1. $\lim\limits_{x\to 3}\dfrac{2x^2-18}{x-3}$

2. $\lim\limits_{x\to -2}\dfrac{x^3+8}{x+2}$

3. $\lim\limits_{x\to 0}\dfrac{\sin x}{2x}$

4. $\lim\limits_{x\to 0}\dfrac{x^2}{1-\cos x}$

5. $\lim\limits_{x\to 0}\dfrac{\cos x}{x}$

6. $\lim\limits_{x\to\infty}\dfrac{6x^2-x+7}{x-3x^2}$

7. $\lim\limits_{x\to -\infty}\dfrac{5x^2-2x+1}{2.5x^3-3x^2+6}$

8. $\lim\limits_{x\to 0}\dfrac{2x}{\sqrt{x+3}-\sqrt{3}}$

9. $\lim\limits_{x\to 0}\dfrac{\sec x}{x}$

10. $\lim\limits_{x\to 0^+}\left(\sqrt{x}\right)^{1/x}$

11. $\lim\limits_{x\to 0}\left(\dfrac{1}{x}-\dfrac{1}{x\sqrt{x+1}}\right)$

12. $\lim\limits_{x\to 0}\dfrac{x}{3\tan x}$

13–16 Two functions are in competition to determine the indicated limit. Identify the type of the indeterminate form, and fill out the table to decide which function dominates.

13. $\lim\limits_{x\to\infty} f(x)$, where $f(x) = \dfrac{\sqrt{5x^3+7}}{0.2x^2+1}$

x	1	10	100	1000	10,000	100,000
$f(x)$						

14. $\lim\limits_{x\to\infty} g(x)$, where $g(x) = \dfrac{0.5\sqrt{x}}{\ln(x+1)}$

x	1	10	100	1000	10,000	100,000
$g(x)$						

15. $\lim\limits_{x\to\infty} h(x)$, where $h(x) = x^{100}e^{-x}$

x	1	10	100	1000	10,000	100,000
$h(x)$						

16. $\lim\limits_{x\to 0} k(x)$, where $k(x) = (\sin x)^x$

x	1	0.5	0.1	0.01	0.001	0.0001
$k(x)$						

17–48 Check whether l'Hôpital's Rule applies to the given limit. If it does, use it to determine the value of the limit. If it does not, find the limit some other way. (When necessary, apply l'Hôpital's Rule several times.)

17. $\lim\limits_{x\to\infty} \dfrac{2x+5}{x^2-7}$

18. $\lim\limits_{x\to\infty} \dfrac{4-2.5x}{x+3}$

19. $\lim\limits_{x\to-\infty} \dfrac{1.5x^3-2x^2+x+9}{x^2+2.1x-4}$

20. $\lim\limits_{x\to-\infty} \dfrac{4.5x^4+x^3-2}{3-1.5x^4}$

21. $\lim\limits_{x\to 0} \dfrac{\sqrt{x}}{\ln x}$

22. $\lim\limits_{x\to\infty} \dfrac{x\sin x}{e^{-x}}$

23. $\lim\limits_{x\to\infty} \dfrac{\dfrac{1}{x}+2}{2x+1}$

24. $\lim\limits_{t\to 0} \dfrac{t}{\sqrt{2t+9}-3}$

25. $\lim\limits_{x\to\infty} \dfrac{\sin x+2\ln x}{x^2+5}$

26. $\lim\limits_{x\to 0} \dfrac{\sin x-x}{1-\cos x}$

27. $\lim\limits_{t\to 0} \dfrac{1-\cos t}{3t}$

28. $\lim\limits_{x\to-1^+} \dfrac{\sin\sqrt{x+1}}{x+1}$

29. $\lim\limits_{x\to 0} \dfrac{x^{3/2}}{\ln(\cos x)}$

30. $\lim\limits_{x\to 0} \dfrac{\ln(\sec^2 x)}{\sqrt{x}}$

31. $\lim\limits_{x\to\infty} \dfrac{\ln x}{\ln(x^2+3x)}$

32. $\lim\limits_{x\to 0} \dfrac{\log_{10}(x^2+2x+1)}{\log_{10}(x+1)}$

33. $\lim\limits_{x\to 0} \dfrac{x}{3^{x/2}-1}$

34. $\lim\limits_{x\to\infty} \dfrac{2^x}{x^2-3x+4}$

35. $\lim\limits_{x\to 0} \dfrac{\sin x-x}{3x^2}$

36. $\lim\limits_{\phi\to 0^+} \dfrac{1-\cos\phi}{\csc\phi}$

37. $\lim\limits_{\alpha\to 0} \dfrac{\alpha}{e^{\sin\alpha}-1}$

38. $\lim\limits_{\theta\to 0} \dfrac{\theta\tan\theta}{1-\cos\theta}$

39. $\lim\limits_{t\to\infty} \dfrac{\ln(t+1)}{e^{-t}\sin t}$

40. $\lim\limits_{t\to\pi} \dfrac{(\cos(2t)-1)^2}{t-\pi}$

41. $\lim\limits_{\theta\to\pi/2} \dfrac{\left(\theta-\dfrac{\pi}{2}\right)^2}{\ln(\sin\theta)}$

42. $\lim\limits_{x\to\infty} \dfrac{x+2^x}{5^x-x}$

43. $\lim\limits_{x\to\infty} \dfrac{4^x+x^2}{3^x-x}$

44. $\lim\limits_{x\to\infty} \dfrac{\ln(\ln x)}{x\ln x}$

45. $\lim\limits_{x\to 0^+} \dfrac{\log_2(1+x)}{\log_3(\sin x+1)}$

46. $\lim\limits_{x\to\infty} \dfrac{\log_4(2x+1)}{\log_5(x-4)}$

47. $\lim\limits_{x\to 0^+} \dfrac{\log_4(x+1)}{\log_3 x}$

48. $\lim\limits_{x\to 0} \dfrac{3^x-1}{x3^x}$

49–74 Identify the indeterminate product, quotient, difference, or power, and use l'Hôpital's Rule to find the limit. If the limit is not of indeterminate form, say so and find it by other means.

49. $\lim\limits_{x\to 0^+} x\ln x$

50. $\lim\limits_{x\to\infty} \dfrac{\sqrt{2x^2+1}}{x+3}$

51. $\lim\limits_{x\to 0} x\cos\dfrac{\pi}{x}$

52. $\lim\limits_{x\to\infty} (\ln x)^{-1/x}$

53. $\lim\limits_{x\to 0^+} \left(\dfrac{1}{x}\right)^x$

54. $\lim\limits_{x\to 0^+} (-\ln x)^x$

55. $\lim\limits_{x\to 1^+} \left(\dfrac{1}{\ln x}-\dfrac{2}{x-1}\right)$

56. $\lim\limits_{x\to 0^+} \left(\dfrac{1}{x}+\ln x\right)$

57. $\lim\limits_{x\to 4^+} \left(\dfrac{32}{x^2-16}-\dfrac{x}{x-4}\right)$

58. $\lim\limits_{x\to 0^+} x^{(x^2)}$

59. $\lim\limits_{x\to 0^+} (2^x-x)^{1/x}$

60. $\lim\limits_{x\to 0^+} (1-x)^{1/x}$

61. $\lim\limits_{x \to 0^+} \left(\dfrac{1}{x^2}\right)^{\csc x}$

62. $\lim\limits_{x \to \infty} \left(\sqrt{x^2 - 3x} - \dfrac{3}{x^2 + 1}\right)$

63. $\lim\limits_{x \to \infty} (x - 1)^{1/x}$

64. $\lim\limits_{x \to \infty} \dfrac{\ln x}{x^{7/5}}$

65. $\lim\limits_{x \to 0} (\cos x)^{\cot x}$

66. $\lim\limits_{x \to 0^-} (\cot x)^{\cos x}$

67. $\lim\limits_{x \to 0^+} \tan x \sec x$

68. $\lim\limits_{x \to \infty} \dfrac{x^{100}}{3^x}$

69. $\lim\limits_{x \to \infty} \dfrac{\ln\left(100x^2 + e^x\right)}{100x}$

70. $\lim\limits_{x \to 0^+} 2\sqrt{x}\,\csc x$

71. $\lim\limits_{x \to 0} (1 + 2x)^{1/x}$

72. $\lim\limits_{x \to (\pi/2)^-} \left(\dfrac{\pi^2}{4} - x^2\right)\sec x$

73. $\lim\limits_{x \to 0} \dfrac{\sin 2x}{\tan 3x}$

74. $\lim\limits_{x \to 1} x^{1/(1-x)}$

75–85 Find the limit. If applicable, use l'Hôpital's Rule, even several times when appropriate.

75. $\lim\limits_{x \to \infty} \dfrac{2x^5 + x^3 - 4}{e^x}$

76. $\lim\limits_{x \to \infty} \dfrac{\cos x}{2^x}$

77. $\lim\limits_{x \to \infty} x \sin \dfrac{1}{x}$

78. $\lim\limits_{x \to \infty} x^{1/x^3}$

79. $\lim\limits_{x \to 0^+} x^{x^x}$

80. $\lim\limits_{x \to 0^+} \left(x^x\right)^x$

81. $\lim\limits_{x \to \infty} x^{1/x^n}, \quad n \in \mathbb{Z}^+$

82. $\lim\limits_{x \to \infty} \dfrac{(\ln x)^3}{x^2}$

83. $\lim\limits_{x \to 0} \dfrac{\sin x - x}{2x - e^x + e^{-x}}$

84. $\lim\limits_{x \to 0} \left(\dfrac{\sin x}{x}\right)^{x^2}$

85. $\lim\limits_{x \to 0} \dfrac{\sin x - x}{\tan x - x}$

86–91 Find the error(s) in the limit calculation.

86. $\lim\limits_{x \to 0} \dfrac{1 - \sin x}{x} = \lim\limits_{x \to 0} \dfrac{-\cos x}{1} = -1$ (Incorrect!)

87. $\lim\limits_{x \to 2} \dfrac{x^2 - 2}{x - 2} = \lim\limits_{x \to 2} \dfrac{2x}{1} = 4$ (Incorrect!)

88. $\lim\limits_{x \to -\infty} \dfrac{5^x + 1}{5^x} = \lim\limits_{x \to 2} \dfrac{(\ln 5)5^x}{(\ln 5)5^x} = 1$ (Incorrect!)

89. $\lim\limits_{x \to 0^+} x \cot x = \lim\limits_{x \to 0^+} (1)\left(-\csc^2 x\right)$
$$= (1)(-\infty) = -\infty \quad \text{(Incorrect!)}$$

90. $\lim\limits_{x \to 0} x \sin\dfrac{1}{x} = \lim\limits_{x \to 0} \dfrac{\sin\dfrac{1}{x}}{\dfrac{1}{x}}$

$$= \lim\limits_{x \to 0} \dfrac{-\dfrac{1}{x^2}\cos\dfrac{1}{x}}{-\dfrac{1}{x^2}}$$

$$= \lim\limits_{x \to 0} \cos\dfrac{1}{x}$$

$$= \text{does not exist} \quad \text{(Incorrect!)}$$

91. $\lim\limits_{x \to 0^-} \dfrac{\cos x - x^2 - 1}{x^4 - 2x^3} = \lim\limits_{x \to 0^-} \dfrac{-\sin x - 2x}{4x^3 - 6x^2}$

$$= \lim\limits_{x \to 0^-} \dfrac{-\cos x - 2}{12x^2 - 12x}$$

$$= \lim\limits_{x \to 0^-} \dfrac{\sin x}{24x - 12} = 0 \quad \text{(Incorrect!)}$$

92–106 Convince yourself that the initial use of l'Hôpital's Rule is not helpful in finding the limit. If possible, try to find a way to make use of the theorem, or evaluate the limit in some other way.

92. $\lim\limits_{x \to \infty} \dfrac{\sqrt{x + 2}}{\sqrt{x}}$

93. $\lim\limits_{x \to \infty} \dfrac{\sqrt[3]{x + 1} - 2}{\sqrt{x^2 + 2}}$

94. $\lim\limits_{x \to \infty} \dfrac{2^x + 3^x}{5^x}$

95. $\lim\limits_{x \to \infty} \dfrac{5^x - 6^x}{7^x + 8^x}$

96. $\lim\limits_{x \to \infty} \dfrac{2^{-x}}{x^{-1}}$

97. $\lim\limits_{x \to \infty} \left(\dfrac{1}{x + 1}\right)^{-x^3}$

98. $\lim\limits_{x \to \infty} \left(\dfrac{1}{x^2}\right)^{e^{-x}}$

99. $\lim\limits_{x \to \infty} \dfrac{x}{\sqrt{x^2 + 1}}$

100. $\lim\limits_{x \to 0} \dfrac{\csc x}{\cot x}$

101. $\lim\limits_{x \to 0^+} \left(\cot x - \dfrac{5x + 1}{x}\right)$

102. $\lim\limits_{x \to 1^+} \left(\dfrac{1}{x - 1} - \dfrac{1}{\ln x}\right)$

103. $\lim\limits_{x \to \pi^+} (\cot x)^{\sin x}$

104. $\lim\limits_{x \to \infty} 2^{-x} x \ln x$

105. $\lim\limits_{x \to 0} (\sin x)^{\tan x}$

106. $\lim\limits_{x \to (\pi/2)^-} \left(\dfrac{1}{\dfrac{\pi}{2} - x} - \tan x\right)$

107–110 Find the limit of the sequence by considering the function you obtain after replacing n with the real variable x.

107. $\lim\limits_{n\to\infty} \dfrac{n^2+1}{2^n}$

108. $\lim\limits_{n\to\infty}\left(1+\dfrac{1}{n}\right)^n$

109. $\lim\limits_{n\to\infty}\sqrt[n]{n}$

110. $\lim\limits_{n\to\infty}\dfrac{2^n+5^n}{6^n}$

111–114 Use l'Hôpital's Rule to prove the assertion.

111. $\lim\limits_{x\to 0}\dfrac{\sin(kx)}{x^k}=\infty \quad (k>1)$

112. $\lim\limits_{x\to\infty}\dfrac{p(x)}{e^{kx}}=0 \quad (p(x) \text{ is a polynomial, } k>0)$

113. $\lim\limits_{x\to\infty}\dfrac{(\ln x)^n}{x^k}=0 \quad (n\in\mathbb{N}, k>0)$

114. $\lim\limits_{x\to\infty}\dfrac{a^x}{x^n}=\infty \quad (a>1, n\in\mathbb{N})$

115–122 Find the value(s) of c satisfying the conclusion of Cauchy's Mean Value Theorem. If the theorem doesn't apply, explain why.

115. $f(x)=x, \quad g(x)=x^2+1; \quad [0,1]$

116. $f(x)=x^3-1, \quad g(x)=x^2+2x; \quad [-1,1]$

117. $f(x)=x^3-x, \quad g(x)=-x^2+2x+3; \quad [-1,3]$

118. $f(x)=x^3, \quad g(x)=-x^2; \quad [-2,3]$

119. $f(x)=x^2+3x, \quad g(x)=3x^2-5x+3; \quad [-1,3]$

120. $f(x)=\dfrac{1}{x}, \quad g(x)=\ln x; \quad [1,2]$

121. $f(x)=\cos x, \quad g(x)=\sin x; \quad \left[-\dfrac{\pi}{2},0\right]$

122. $f(x)=x^2-5x-9, \quad g(x)=x^3+x+10; \quad [-3,2]$

123–124 Prove that $f(x)$ has a removable discontinuity at $x=0$. Then find the value of c so as to make f continuous.

123. $f(x)=\begin{cases} \dfrac{3\tan x-2x}{5x^2+3x} & \text{if } x\neq 0 \\ c & \text{if } x=0 \end{cases}$

124. $f(x)=\begin{cases} (e^x-\sin 2x)^{2/x} & \text{if } x\neq 0 \\ c & \text{if } x=0 \end{cases}$

125. Recall the compound interest formula for the value of an investment of P dollars after t years, compounded n times a year at an annual interest rate of r:

$$A=P\left(1+\dfrac{r}{n}\right)^{nt}$$

Use l'Hôpital's Rule to prove that if we let $n\to\infty$, we obtain the continuous compounding formula:

$$A=Pe^{rt}$$

126. The strength of an electric field due to a disk charge is obtained from the formula

$$E(x)=\dfrac{\sigma}{2\varepsilon_0}\left(1-\dfrac{x}{\sqrt{x^2+R^2}}\right)$$

where σ is the electric charge per unit area (in C/m^2), $\varepsilon_0=8.85\cdot 10^{-12}\ C^2/Nm^2$, R is the radius of the ring, and x is the distance to the charge in meters. Use l'Hôpital's Rule to confirm that $E(x)\to\infty$ as $x\to\infty$. How is E affected by σ and R at a given distance? What happens to the rate of change of E as x increases? (**Hint:** Apply l'Hôpital's Rule to dE/dx as $x\to\infty$.)

127. Marquis de l'Hôpital first illustrated the rule named after him in his 1696 textbook, *Analyse des Infiniment Petits*. He used an example where the objective was to find

$$\lim\limits_{x\to a}\dfrac{\sqrt{2a^3x-x^4}-a\sqrt[3]{a^2x}}{a-\sqrt[4]{ax^3}}$$

for $a>0$. Determine the above limit.

4.4 **Technology Exercises**

128–131 Check whether the limit is of indeterminate form, and then use a computer algebra system to evaluate the limit.

128. $\displaystyle\lim_{x\to 1^+}(x-1)^{\ln x}$

129. $\displaystyle\lim_{x\to 0^+}\tan x\ln x$

130. $\displaystyle\lim_{x\to 0^+}x^{x^{x^x}}$

131. $\displaystyle\lim_{x\to 0}\left(\frac{1}{\sin^2 x}-\frac{1}{x^2}\right)$

132–133 Use a graphing calculator or computer algebra system to graph the function for different values of the parameter c. Examine how the values of the parameter affect the indicated limit.

132. $\displaystyle\lim_{x\to\infty}\left(1+\frac{1}{cx}\right)^x$

What happens to the limit when $|c|\to\infty$?

133. $\displaystyle\lim_{x\to 0^+}\frac{1-c^x}{cx}$

What happens to the limit when $c\to\infty$?

4.5 **Calculus and Curve Sketching**

In the last several sections, we acquired a number of tools that provide a deeper understanding of the behavior of a function than can be obtained by algebra alone. In this section, we put all those tools to good use in producing detailed graphs of functions.

Topic One
A Curve-Sketching Strategy

As Example 4 of Section 4.3 illustrated, the knowledge we gain from a function's first and second derivatives can be extremely useful in identifying some of the most pertinent features of its graph. Such knowledge is especially valuable when used in conjunction with the graphing capability found in calculators and software—when such technological aids fail to pick out subtle or unexpected behavior, calculus can usually be used to fill the gap.

All of the purely algebraic curve-sketching techniques that you have learned still apply, but those you have used in the past (for instance, in Section 1.2) can now be augmented with a few additional steps that pick out extrema, intervals of monotonicity and concavity, inflection points, and more complicated limits. Although the curve-sketching strategy below should not be taken as a rigid prescription (the steps may not all be relevant and don't necessarily need to be taken in the order presented), it does serve as a convenient summary of the tools we now have.

Curve-Sketching Strategy

To sketch the graph of the function $f(x)$, perform the following steps.

Step 1: Determine the **domain** of f and any **symmetry** it may have (e.g., symmetry with respect to a certain point or line). Check also to see if f may be **periodic**, a trait frequently found, for example, in trigonometric functions.

Step 2: Identify the **intercepts**, if possible. The y-intercept value of $f(0)$ is normally easy to determine, if 0 is in the domain of f. The x-intercepts are found by solving the equation $f(x) = 0$.

Step 3: Identify and plot any **asymptotes** for f. Recall that these can be horizontal, vertical, or oblique and are especially helpful in sketching the graphs of rational functions. Use l'Hôpital's Rule as necessary to help evaluate limits.

Step 4: Find f' and f''. **Critical points** of f are located where f' is undefined or 0, and potential inflection points are located where f'' is undefined or 0.

Step 5: Use the **Monotonicity Test** and the **Concavity Test** to find intervals where f is increasing, decreasing, concave up, and concave down. Use either the **First Derivative Test** or the **Second Derivative Test**, whichever is more convenient, to locate **relative extrema**. Compare the values of the relative extrema to locate the **absolute extrema**. The actual **inflection points** are points where f is continuous and where the concavity of the graph changes.

Step 6: Combine all of the information found to either sketch a graph by hand or use it in conjunction with a graph generated by calculator or computer. If the graph is being constructed as part of solving some larger problem, the extrema, points of inflection, and asymptotic behavior you have found are likely to be of great use.

Example 1

Use the curve-sketching strategy to construct a graph of

$$f(x) = 10x^3 - 6x^2 + x.$$

Solution

The function f is a cubic polynomial, so from this we know that its domain is all of $\mathbb{R}$ and, since its leading coefficient is positive, we know $\lim_{x \to -\infty} f(x) = -\infty$ and $\lim_{x \to \infty} f(x) = \infty$. The y-intercept is easily seen to be 0, and to determine the x-intercept(s) we need to solve the equation $10x^3 - 6x^2 + x = 0$, or $x(10x^2 - 6x + 1) = 0$; since $10x^2 - 6x + 1 = 0$ has no real number solutions (by the quadratic formula), this tells us that the only x-intercept is also 0.

Since f is a polynomial, the function has no asymptotes. The first and second derivatives are as follows.

$$f'(x) = 30x^2 - 12x + 1$$
$$f''(x) = 60x - 12 = 12(5x - 1)$$

Solving $30x^2 - 12x + 1 = 0$ by the quadratic formula gives us the two critical points of $\left(6 \pm \sqrt{6}\right)/30$, or approximately 0.12 and 0.28. Setting f'' equal to 0, we see that the only potential inflection point is at $\frac{1}{5} = 0.20$. We can now use the derivatives and these three points to make the following table, which succinctly captures a great deal of information about the behavior of f.

Domain	$\dfrac{1}{30}\left(6 - \sqrt{6}\right)$		$\dfrac{1}{5}$		$\dfrac{1}{30}\left(6 + \sqrt{6}\right)$	
Sign of f'	$f'(0) > 0$		$f'\left(\dfrac{1}{5}\right) = -\dfrac{1}{5} < 0$			$f'(1) > 0$
Monotonicity	Increasing		Decreasing			Increasing
Sign of f''	$f''(0) < 0$				$f''(1) > 0$	
Concavity	Concave down				Concave up	

The first row of the table depicts the domain of f with the two critical points and the potential inflection point marked. In the second row, the sign of f' is determined by evaluating f' at three test points on the intervals defined by the critical points. Why were those particular test points used? For no better reason than convenience: the sign of f' is easily determined at 0 and 1, and the test point for the middle interval was selected merely because it lies halfway between the two critical points. Note that all we care about is the sign of f', but to determine this for the middle interval we actually computed the value of f'. Given the signs of f' on the three intervals, we are led immediately to the conclusions about monotonicity on the intervals $\left(-\infty, \left(6-\sqrt{6}\right)/30\right)$, $\left(\left(6-\sqrt{6}\right)/30, \left(6+\sqrt{6}\right)/30\right)$, and $\left(\left(6+\sqrt{6}\right)/30, \infty\right)$ listed in the third row.

The sign of f'' on either side of the potential inflection point is easily calculated and appears in the fourth row, and the corresponding conclusions about concavity on the two intervals $(-\infty, 1/5)$ and $(1/5, \infty)$ appear in the last row.

At this point, the First Derivative Test tells us, from the information in row two, that f has a relative maximum at $x = \left(6-\sqrt{6}\right)/30 \approx 0.12$ and a relative minimum at $x = \left(6+\sqrt{6}\right)/30 \approx 0.28$; we could also use the Second Derivative Test to conclude the same thing, as f'' is negative at the first critical point and positive at the second. And since the concavity does indeed change at $x = \frac{1}{5}$, this potential inflection point actually *is* an inflection point.

We are now ready to put all this knowledge together, and use it to refine any graphs we may obtain by calculator or computer. If we were to sketch the graph entirely by hand, we would calculate f at a few convenient points, paying special attention to the values at the extrema and the inflection point. If we are using technology to construct the graph, we would now know to zoom in on the portions around these same points. The importance of this is illustrated by the fact that our graph in Figure 1 (created with a graphing calculator) may lead us to incorrectly assume that f is an increasing function on its entire domain, but a close-up shows more interesting behavior centered around the inflection point as in Figure 2.

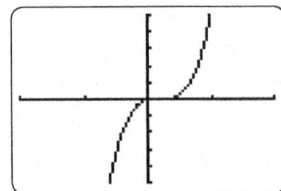

Figure 1

$f(x) = 10x^3 - 6x^2 + x$

on $[-2, 2]$ by $[-5, 5]$

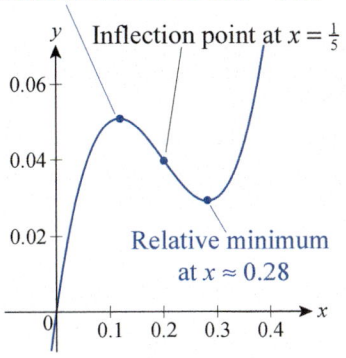

Figure 2

Zoomed-In Graph of

$f(x) = 10x^3 - 6x^2 + x$

In the remaining examples, the steps in the curve-sketching strategy will be referred to more succinctly by number.

Example 2

Use the curve-sketching strategy to construct a graph of $f(x) = \dfrac{\sin x}{2 + \cos x}$.

Solution

Step 1: The denominator $2 + \cos x$ is never 0, so f is defined for all x and the domain is all of $\mathbb{R}$. Since $\sin x$ is an odd function and $1/(2 + \cos x)$ is even, f is an odd function (the product of an odd function and an even function is odd, but you can also verify that $f(-x) = -f(x)$). Moreover, since sine and cosine are both 2π-periodic, f is as well.

Step 2: Note that $f(0) = 0$, and the only solutions of $f(x) = 0$ are the solutions of $\sin x = 0$, namely integer multiples of π. So the origin is the only y-intercept and the x-intercepts are $(n\pi, 0)$, $n \in \mathbb{Z}$.

Step 3: No asymptotes

Step 4:
$$f'(x) = \frac{\cos x(2 + \cos x) - \sin x(-\sin x)}{(2 + \cos x)^2} \qquad \text{\small Quotient Rule}$$

$$= \frac{1 + 2\cos x}{(2 + \cos x)^2} \qquad \text{\small $\cos^2 x + \sin^2 x = 1$}$$

$$f''(x) = \frac{-2\sin x(2 + \cos x)^2 - (1 + 2\cos x) \cdot 2(2 + \cos x)(-\sin x)}{(2 + \cos x)^4}$$

$$= \frac{-4\sin x - 2\sin x \cos x + 2\sin x + 4\sin x \cos x}{(2 + \cos x)^3}$$

$$= \frac{2\sin x(\cos x - 1)}{(2 + \cos x)^3}$$

The first derivative is defined everywhere, and $f'(x) = 0$ only when $1 + 2\cos x = 0$:

$$\cos x = -\frac{1}{2}$$

$$x = \pm\frac{2\pi}{3} + 2n\pi, \quad n \in \mathbb{Z}$$

Similarly, f'' is defined everywhere and equal to 0 only when $\sin x = 0$ or $\cos x = 1$; these points are $x = n\pi$.

Step 5: Since f is 2π-periodic, we will make a chart for just one period of the graph.

Domain	0	$\dfrac{2\pi}{3}$	π	$\dfrac{4\pi}{3}$	2π
Sign of f'	$f'\left(\dfrac{\pi}{3}\right)>0$		$f'(\pi)<0$		$f'\left(\dfrac{5\pi}{3}\right)>0$
Monotonicity	Increasing		Decreasing		Increasing
Sign of f''	$f''\left(\dfrac{\pi}{4}\right)<0$			$f''\left(\dfrac{4\pi}{3}\right)>0$	
Concavity	Concave down			Concave up	

Both derivative tests now tell us that $2\pi/3$ is a relative maximum and that $4\pi/3$ is a relative minimum, and we also now know that f has an inflection point at π. The absolute maximum value (which is attained at each relative maximum point) is $f(2\pi/3)=\sqrt{3}/3 \approx 0.577$, and the absolute minimum value is $f(4\pi/3)=-\sqrt{3}/3$. The concavity changes at each multiple of π, so each is an inflection point.

Step 6: The graph of f over two periods is shown in Figure 3.

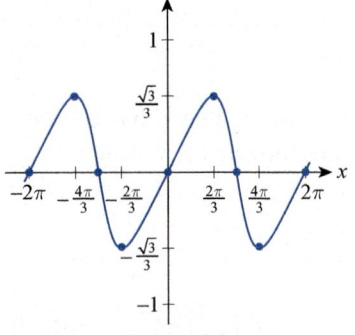

Figure 3

$$f(x)=\frac{\sin x}{2+\cos x} \text{ on } [-2\pi,2\pi]$$

Example 3

Use the curve-sketching strategy to construct a graph of

$$f(x)=x^{2/3}(3-x)^{1/3}.$$

Solution

Step 1: The domain of f is all of $\mathbb{R}$, and since $x^{2/3}$ is nonnegative everywhere and $(3-x)^{1/3}$ is positive to the left of 3 and negative to the right of 3 we know f will behave the same way as this second factor. There is no apparent symmetry or periodicity.

Step 2: Note that $f(0)=0$, and the only solutions of $f(x)=0$ are 0 and 3. So other than the origin, the only x-intercept is $(3,0)$.

Step 3: No asymptotes

Step 4:
$$f'(x)=\frac{2}{3}x^{-1/3}(3-x)^{1/3}+x^{2/3}\frac{1}{3}(3-x)^{-2/3}(-1) \qquad \text{Product Rule}$$

$$=\frac{1}{3}x^{-1/3}(3-x)^{-2/3}\left[2(3-x)-x\right] \qquad \text{Factor out } \tfrac{1}{3}x^{-1/3}(3-x)^{-2/3}.$$

$$=\frac{2-x}{x^{1/3}(3-x)^{2/3}}$$

$$f''(x) = \frac{(-1)x^{1/3}(3-x)^{2/3} - (2-x)\left[\frac{1}{3}x^{-2/3}(3-x)^{2/3} + x^{1/3}\frac{2}{3}(3-x)^{-1/3}(-1)\right]}{x^{2/3}(3-x)^{4/3}}$$ Quotient Rule

$$= \frac{x^{-2/3}(3-x)^{-1/3}\left[-x(3-x) - (2-x)\left(\frac{1}{3}(3-x) - \frac{2}{3}x\right)\right]}{x^{2/3}(3-x)^{4/3}}$$ Factor out $x^{-2/3}(3-x)^{-1/3}$.

$$= \frac{-3x + x^2 - (2-x)(1-x)}{x^{4/3}(3-x)^{5/3}}$$

$$= \frac{-2}{x^{4/3}(3-x)^{5/3}}$$

From the formulas for f' and f'' we see that f has critical points at 0, 2, and 3 and potential inflection points at 0 and 3.

Step 5:

Domain	0		2	3
Sign of f'	$f'(-1) < 0$	$f'(1) > 0$	$f'(2.5) < 0$	$f'(4) < 0$
Monotonicity	Decreasing	Increasing	Decreasing	Decreasing
Sign of f''	$f''(-1) < 0$	$f''(2) < 0$		$f''(4) > 0$
Concavity	Concave down	Concave down		Concave up

The First Derivative Test tells us that f has a relative minimum at 0 and that 3 is not a relative extremum. Both tests indicate that f has a relative maximum at 2. The concavity only changes at 3, so that is the only point of inflection.

Step 6:   The graph of f is shown in Figure 4.

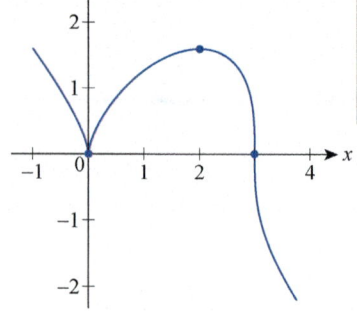

Figure 4

$f(x) = x^{2/3}(3-x)^{1/3}$

Example 4

Use calculus to locate the relative extrema of the rational function $h(x) = (x^2 + 1)/(x-1)$ from Example 3c of Section 1.2.

Solution

We have already determined the asymptotes of h, and our sketch from Section 1.2 contains quite a lot of information. But the relative extrema would be difficult to determine without calculus. We can locate them quickly with h'.

$$h'(x) = \frac{(2x)(x-1) - (x^2 + 1)}{(x-1)^2}$$ Quotient Rule

$$= \frac{x^2 - 2x - 1}{(x-1)^2}$$

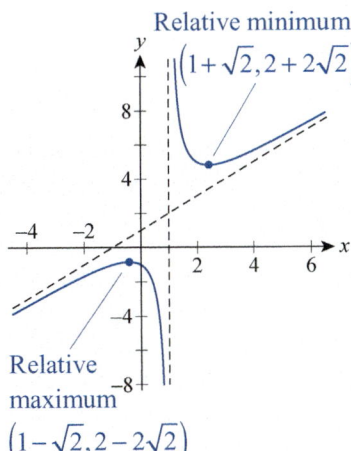

Relative minimum
$\left(1+\sqrt{2}, 2+2\sqrt{2}\right)$

Relative maximum
$\left(1-\sqrt{2}, 2-2\sqrt{2}\right)$

Figure 5 $h(x) = \dfrac{x^2 + 1}{x - 1}$

Since 1 is not in the domain of h, the only critical points are where $x^2 - 2x - 1 = 0$; the solutions are $1 \pm \sqrt{2}$. Evaluating h at these points, we find a relative maximum value of $2 - 2\sqrt{2}$ at $1 - \sqrt{2}$ and a relative minimum value of $2 + 2\sqrt{2}$ at $1 + \sqrt{2}$. (See Figure 5.) If we had not already sketched the graph of h, we could use either the First or Second Derivative Test to show that these were, indeed, relative extrema.

Example 5

Use the curve-sketching strategy to construct a graph of $f(x) = xe^x$.

Solution

Step 1: Note that e^x is defined for all x, so xe^x is as well. There is no apparent symmetry or periodicity.

Step 2: Since e^x is always positive, xe^x will be positive for positive x, negative for negative x, and 0 if, and only if, $x = 0$.

Step 3: It's clear that f grows without bound as $x \to \infty$, but the limit as $x \to -\infty$ is of indeterminate form $0 \cdot \infty$ (or, more precisely, of form $-\infty \cdot 0$). We can determine the limit using l'Hôpital's Rule:

$$\lim_{x \to -\infty} xe^x = \lim_{x \to -\infty} \frac{x}{e^{-x}} = \lim_{x \to -\infty} \frac{1}{-e^{-x}} = 0$$

Step 4: The first and second derivatives are $f'(x) = e^x + xe^x = e^x(x+1)$ and $f''(x) = e^x + e^x(x+1) = e^x(x+2)$, so f has a critical point at -1 and a potential inflection point at -2.

Step 5:

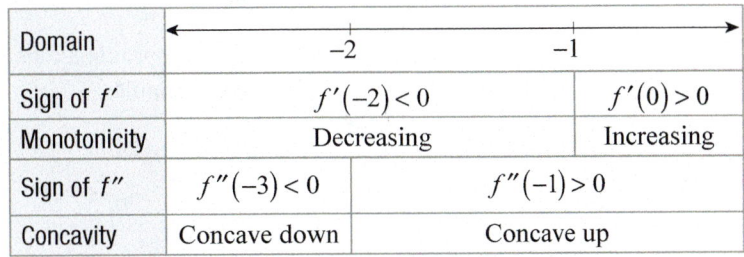

Domain		−2		−1	
Sign of f'		$f'(-2) < 0$		$f'(0) > 0$	
Monotonicity		Decreasing		Increasing	
Sign of f''	$f''(-3) < 0$		$f''(-1) > 0$		
Concavity	Concave down		Concave up		

Both derivative tests tell us that f has a relative minimum at -1. And the concavity changes at -2, so that is an inflection point.

Step 6: The graph of f is shown in Figure 6.

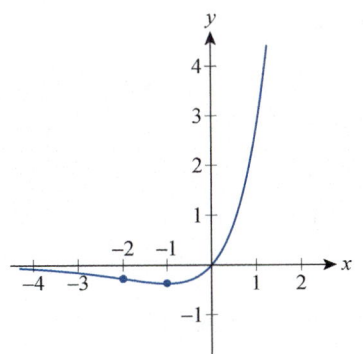

Figure 6 $f(x) = xe^x$

Topic Two

Newton's Method

The second step in the curve-sketching strategy says "Identify the intercepts, if possible." This step is usually fairly important; in fact, depending on the particular application, locating the solutions of the equation $f(x) = 0$ may be the whole motivation for sketching the graph of f in the first place.

Unfortunately, solving the equation $f(x) = 0$ is only a trivial task for a relatively small number of classes of functions. There are formulas for finding the roots of polynomials of degree four or less, but such a formula for degrees five and higher can't exist (the mathematicians Évariste Galois and Niels Henrik Abel, working independently, were the first to show this). And while the roots of the basic trigonometric functions are easy to determine, relatively simple algebraic combinations of them can be quite complicated. In practice, it is very common to wind up working with a function whose roots are elusive. And yet, graphing calculators and computer algebra systems seem to be able to at least give us approximations of roots of functions to seemingly any precision we need. How is that accomplished?

One of the simplest and widely used techniques is called *Newton's method* (or the *Newton-Raphson method*), and questions of this sort belong to an area of mathematics called *numerical analysis*. We have all the knowledge we need at this point to understand how Newton's method works.

Newton's Method

To find an approximation of a root of the function $f(x)$, perform the following steps.

Step 1: Begin with a guess x_1 of a solution of the equation $f(x) = 0$. A sketch of the graph of f can be very helpful in this step.

Step 2: Define new approximations x_2, x_3, and so on by successively applying the following formula.

$$x_{n+1} = x_n - \frac{f(x_n)}{f'(x_n)} \qquad (\text{if } f'(x_n) \neq 0)$$

Step 3: Continue until the desired degree of precision is obtained.

It is important to realize that Newton's method is not guaranteed to work, a characteristic of nearly all numerical analysis techniques. But a certain amount of trial and error, combined with a good understanding of the underlying mathematics, is usually sufficient to overcome any difficulties. The ways in which Newton's method may fail will become clear as we work through its derivation.

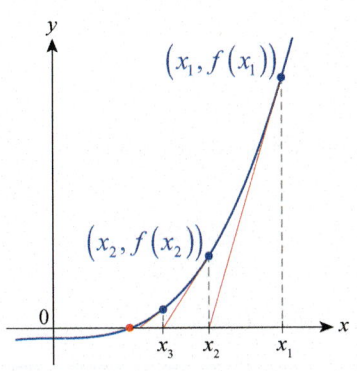

Figure 7 Successive Approximations to the Root

Consider the function f, part of whose graph appears in Figure 7. In this portion of the graph, f has exactly one root (shown as a red dot). If we didn't know the exact value of this root, we might start with the guess depicted as x_1. The linearization of f at $\left(x_1, f\left(x_1\right)\right)$, shown as a red line, is

$$y - f\left(x_1\right) = f'\left(x_1\right)\left(x - x_1\right),$$

and it is easy to determine where this linear approximation crosses the x-axis:

$$0 - f\left(x_1\right) = f'\left(x_1\right)\left(x - x_1\right)$$
$$0 = f\left(x_1\right) + f'\left(x_1\right)\left(x - x_1\right)$$
$$-\frac{f\left(x_1\right)}{f'\left(x_1\right)} = x - x_1$$
$$x = x_1 - \frac{f\left(x_1\right)}{f'\left(x_1\right)}$$

If we label this result x_2 and repeat the process, we will (hopefully) obtain a sequence of points that converges to the actual root of f.

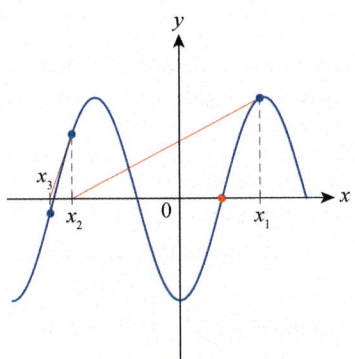

Figure 8
Newton's Method Going Astray

But the process can go awry in several different ways. First, if $f'\left(x_n\right) = 0$ for some x_n, the formula makes no sense. Geometrically, this means the linearization at this point is parallel to the x-axis, so it has no x-intercept. Further, if the initial guess x_1 is not close enough to the desired root, the iterative process can actually result in a sequence that either doesn't converge at all or else converges to a different root. Figure 8 illustrates this second possibility; the desired root is shown in red, but the sequence converges elsewhere.

It is even possible for Newton's method to never converge no matter how close the initial guess is to the actual root; Exercises 83–87 contain examples of this unfortunate outcome. Other approximation techniques exist to handle such unusual cases, however, and are studied in classes like Numerical Analysis.

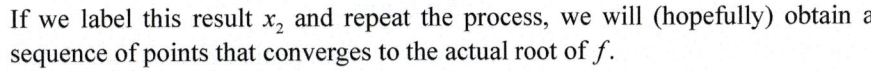

Example 6

Use Newton's method to approximate $\sqrt[3]{2}$ to five decimal places.

Solution

We are looking for the value of $x = \sqrt[3]{2}$; in other words, a solution of the equation $x^3 = 2$. So if we define $f(x) = x^3 - 2$, we see that we are seeking a root of the function f. Since $f'(x) = 3x^2$,

$$x_{n+1} = x_n - \frac{f\left(x_n\right)}{f'\left(x_n\right)} = x_n - \frac{x_n^3 - 2}{3x_n^2} = \frac{2\left(x_n^3 + 1\right)}{3x_n^2}.$$

As a rule of thumb, we can feel comfortable with our approximation if we continue iterating Newton's method until successive results agree to the number of decimal places we desire (more precise statements about convergence are studied in advanced classes). In our case, if we begin with a guess of $x_1 = 2$ we can obtain x_2 as follows.

$$x_2 = \frac{2\left(x_1^3 + 1\right)}{3 \cdot x_1^2} = \frac{2\left(2^3 + 1\right)}{3 \cdot 2^2} = 1.5$$

In a similar way, we obtain the subsequent approximations, truncating our results to six decimal places.

$$x_1 = 2 \qquad x_2 = 1.5 \qquad x_3 \approx 1.296296$$
$$x_4 \approx 1.260932 \qquad x_5 \approx 1.259921 \qquad x_6 \approx 1.259921$$

So to five decimal places, our approximation is $\sqrt[3]{2} \approx 1.25992$.

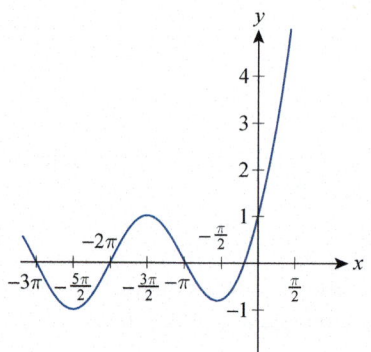

Figure 9

$f(x) = e^x + \sin x$

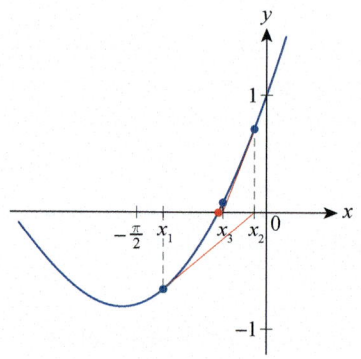

Figure 10

Zoomed-In Graph of

$f(x) = e^x + \sin x$

Example 7

Use Newton's method to approximate the largest negative root of $f(x) = e^x + \sin x$ to five decimal places.

Solution

Solving the equation $e^x + \sin x = 0$ is definitely not trivial, and a sketch of the graph (or at least a mental image of it) is immensely useful. (See Figure 9.)

Since $\sin x$ oscillates between -1 and 1 over all of $\mathbb{R}$, while e^x gets very large for positive x and quickly approaches 0 for negative x, we should expect e^x to dominate to the right of 0 and $\sin x$ to dominate to the left. The graph supports this expectation.

The largest negative root of f lies somewhere to the right of $-\pi/2$, as the close-up in Figure 10 indicates. An initial guess of 0 would work well to begin Newton's method, but for illustrative purposes we use $x_1 = -1.25$. (See Figure 10.) Applying Newton's method, we obtain the following formula for x_{n+1}.

$$x_{n+1} = x_n - \frac{e^{x_n} + \sin x_n}{e^{x_n} + \cos x_n}$$

The first few approximations are as follows.

$$x_1 = -1.25 \qquad x_2 \approx -0.149219$$
$$x_3 \approx -0.534414 \qquad x_4 \approx -0.587419$$
$$x_5 \approx -0.588532 \qquad x_6 \approx -0.588533$$

Since x_5 and x_6 agree to five decimal places, the largest negative root of f is approximate -0.58853.

4.5 **Exercises**

1–4 The graphs of the first and second derivatives of a function f are given. Identify which one is which, and then sketch a possible graph of f. (Answers for the graph of f will vary.)

1.
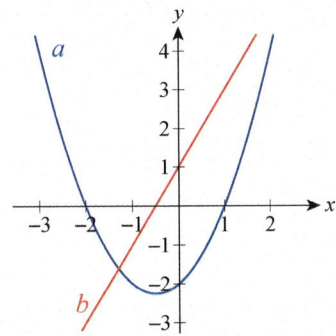

2.
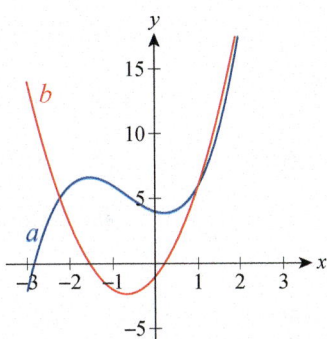

3.
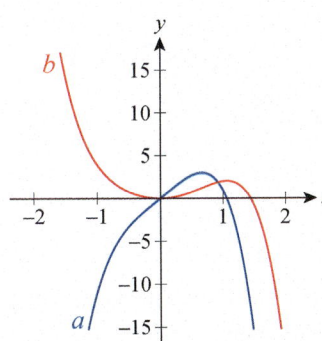

4.
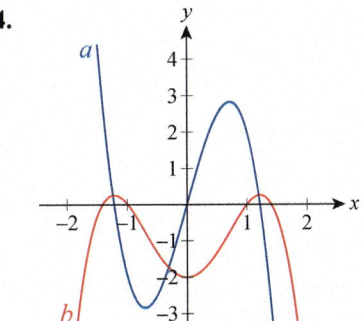

5–48 Use the curve-sketching strategy to construct a graph of the function.

5. $f(x) = x^3 + 3x^2 - 9x$

6. $g(x) = -x^3 + 2x^2 - x + 4$

7. $h(x) = \frac{1}{4}x^4 + \frac{5}{3}x^3 + x^2 - 8x$

8. $F(x) = -\frac{3}{4}x^4 + x^3 + 9x^2 + 2$

9. $G(x) = (x^2 - 1)(x^2 - 2)$

10. $k(x) = x^5 - 2x^3 - 8x + 1$

11. $L(x) = x^5 - 3x^2$

12. $m(x) = 4x^3 - 5x^4$

13. $n(x) = \frac{-3}{x - 2}$

14. $H(x) = \frac{x^2 + 2}{x + 2}$

15. $R(x) = \frac{x}{x^2 - 4}$

16. $r(x) = \frac{2x^2 + 1}{x - 3}$

17. $A(x) = |x - 3| - 2$

18. $f(x) = 1.5 - |x - 2.2|$

19. $w(x) = x^{2/5} + \frac{2}{5}$

20. $u(x) = (x - 2)\sqrt{x}$

21. $F(x) = 2 - (x - 1)^{3/5}$

22. $G(x) = \sin^2 x - 1$

23. $h(x) = e^{-x} + e^{2x}$

24.* $H(x) = -2\sqrt{x} \cdot 2^{-2x}$

25. $P(x) = x \ln x$

26. $H(x) = 0.3\sqrt[3]{x}(x^2 - 1)$

27. $G(x) = x\sqrt{4 - x^2}$

28. $L(x) = \frac{3}{x - 2}$

29. $m(x) = \frac{x^2 + 7}{x - 7}$

30. $K(x) = \frac{e^x}{x}$

31. $F(x) = e^x - e^{-2x}$

32. $v(x) = (x-1)^{3/5}$

33. $m(x) = -\dfrac{5}{(x-2)^2}$

34. $R(x) = \dfrac{x+2}{x-4}$

35. $G(x) = \dfrac{-x}{x^2-1}$

36. $t(x) = \dfrac{2x^2+2}{x-4}$

37. $H(x) = \dfrac{3}{4}\left(x-\dfrac{4}{3}\right)^{4/3}$

38. $w(x) = \dfrac{(x-1)^2}{2x^2-2}$

39. $k(x) = x - \sqrt[3]{x}$

40. $F(x) = x^{4/5}\left(x-\dfrac{4}{5}\right)$

41. $c(x) = x\sqrt[3]{1-x^2}$

42. $G(x) = -\sqrt{x}\,e^{-x}$

43. $Z(x) = 2\sin x - \cos^2 x$

44. $K(x) = \sin x - \cos x$

45. $L(x) = x^{5/3}\ln|x|$

46. $G(x) = \sqrt{4x^2+3}$

47. $u(x) = 7 - \sqrt{9x^2+2x+1}$

48. $z(x) = e^{\cos x}$

49–54 First prove that $\displaystyle\lim_{x\to\pm\infty}(f(x)-g(x)) = 0$. This means that when $x \to \pm\infty$, the graph of $f(x)$ approaches that of $g(x)$. Use this observation as an aid in graphing $f(x)$. (In this case, we say that $f(x)$ is asymptotic to $g(x)$.)

49. $f(x) = \dfrac{x^3+5}{x+2}$, $\quad g(x) = x^2 - 2x + 4$

50. $f(x) = \dfrac{(x+1)^4+2}{3x+3}$, $\quad g(x) = \dfrac{1}{3}(x+1)^3$

51. $f(x) = \sqrt{4x^2+5}$, $\quad g(x) = |2x|$

52. $f(x) = \sqrt{x^2-4x+5}$, $\quad g(x) = |x-2|$

53. $f(x) = \sqrt[3]{x} + \dfrac{1}{x^2}$, $\quad g(x) = \sqrt[3]{x}$

54. $f(x) = \sin x + \dfrac{1}{x}$, $\quad g(x) = \sin x$

55–56 Sketch on paper a few of the tangent lines that are used to approximate the largest root of the indicated function by Newton's method, using the starting values of -1, 0, and 1, respectively. Does the method always work? Explain.

55.

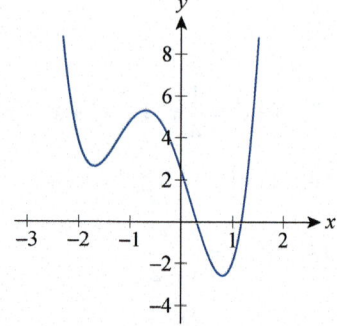

56.

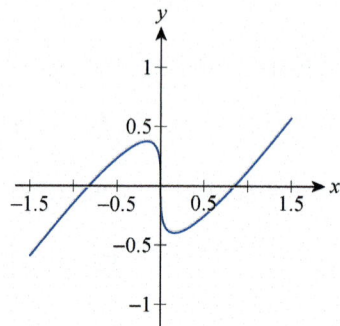

57–60 Use Newton's method to approximate the given number to five decimal places.

57. $\sqrt[4]{50}$

58. $\sqrt[10]{10}$

59. $\ln 5$

60. $\ln 100$

61–70 Use Newton's method to approximate the zero(s) of the given function to five decimal places. Restrict the domain to the given interval where indicated.

61. $f(x) = x^3 - x + 2$

62. $f(x) = 2x^3 + x^2 - 5x + 1$

63. $f(x) = x^4 - 6.1x^3 + 4.7x^2 - 12.2x + 5.4$

64. $f(x) = 0.25x^4 - 2x^2 + x + 0.69$

65. $f(x) = x^5 + x + 1$

66. $f(x) = 2x^5 - 5x^4 + 2x^3 - 4x^2 + 1$

67. $f(x) = 4.2x - \sqrt{x+3}$

68. $f(x) = \sqrt{2+x^2} - 1.1x$

69. $f(x) = 2x^2 - \cos(x-1); \quad \left(0, \dfrac{\pi}{2}\right)$

70. $f(x) = \sin(2x+1) - \dfrac{x}{2}; \quad (0,1)$

71–76 Use Newton's method to solve the equation on the given interval. Approximate the root to six decimal places.

71. $\sin x = x^2$ on $\left(0, \dfrac{\pi}{2}\right)$

72. $2 - x^3 = e^x$ on $\mathbb{R}$

73. $x^4 = \arctan x$ on $(0, \infty)$

74. $\ln x = 2 - \sqrt{x}$ on $(0, \infty)$

75. $\cos x = \tan x$ on $\left(\dfrac{\pi}{2}, \dfrac{3\pi}{2}\right)$

76. $\log_{1/2} x = \sin x$ on $(0, \infty)$

77–80 Recall from Exercise 62 of Section 4.2 that $c \in \mathbb{R}$ is said to be a fixed point of $f(x)$ if $f(c) = c$. Use Newton's method to approximate to four decimal places the fixed point(s) of the function on the given interval.

77. $f(x) = e^{-x}$ on $(0, \infty)$

78. $f(x) = \cos x$ on $\mathbb{R}$

79. $f(x) = 2\cot x$ on $(0, 2\pi)$

80. $f(x) = \log_{1/2} x$ on $(0, \infty)$

81–82 Use Newton's method to find the critical point(s) of the function correct to five decimal places.

81. $f(x) = x^5 - x^3 - 5x$

82. $f(x) = x^2 \sin x, \quad 0 < x < \pi$

83–87 Perform the first few iterations of Newton's method for the given function with the indicated first guess, and explain why the method doesn't work.

83. $f(x) = \sin x - \cos x; \quad x_1 = -\dfrac{\pi}{4}$

84. $f(x) = x^3 - 6x^2 + 12x - 6; \quad x_1 = 3$

85. $f(x) = \begin{cases} -\sqrt{-x} & \text{if } x < 0 \\ \sqrt{x} & \text{if } x \geq 0 \end{cases}; \quad x_1 = a \, (a \neq 0)$

86. $f(x) = \sqrt[3]{x}; \quad x_1 = a \, (a \neq 0)$

87. $f(x) = -x^3 + 9x^2 - 19x + 19; \quad x_1 = 3$

88. The following rule for approximating the square root of a has been known since ancient times.

$$x_{n+1} = \dfrac{1}{2}\left(x_n + \dfrac{a}{x_n}\right)$$

Use Newton's method to derive this rule.
(**Hint:** Start with the equation $x^2 - a = 0$.)

89. Generalizing Exercise 88, use Newton's method to derive a rule for approximating $\sqrt[k]{a}$, $k \geq 3$.

90. Using the approach you have taken in the previous two exercises, derive the following formula approximating $1/a$.

$$x_{n+1} = x_n(2 - ax_n)$$

91–96 Use the formulas you derived in Exercises 88–90 to approximate the given number to five decimal places.

91. $\sqrt{2}$

92. $\sqrt{50}$

93. $\sqrt[3]{10}$

94. $\sqrt[5]{30}$

95. $\dfrac{1}{7}$

96. $\dfrac{1}{19}$

4.5 **Technology Exercises**

97. Use a computer algebra system to approximate π by generating the first 10 iterations of Newton's method for solving the equation $\sin x = 0$ with an appropriate starting value.

98. Repeat Exercise 97 for the equation $(x-5)^{50} = 0$ with the starting value of $x_1 = 6$. What do you find? Graph $f(x) = (x-5)^{50}$, and see if the graph gives insight into why things went wrong.

99–100 Perform the first two iterations of Newton's method with each of the given starting values in an attempt to find the positive root of $f(x)$; then use a computer algebra system to come up with better approximations. What do you find? Graph $f(x)$, and see if the graph gives insight into why things went wrong.

99. $f(x) = x^3 - 2x - 1$; $\quad x_1 = 0.9, \quad x_1 = 0.8, \quad x_1 = -0.4$

100. $f(x) = x^4 - 6x^3 + 9.5x^2 - 1.5x - 4.9375$; $\quad x_1 = 3, \quad x_1 = 2.9, \quad x_1 = 0$

4.6 **Optimization Problems**

We return in this section to the study of optimization problems, a class of problems introduced briefly in Section 4.1; we now have many more calculus techniques that we can bring to bear on the task of finding optimal solutions to problems. Recall that, in this context, an optimal solution corresponds to the point (or points) where a given function attains an absolute minimum or absolute maximum value—both the point and the absolute minimum or maximum value are usually of interest.

To guide us, the general steps we would use in solving any application problem are stated here with a focus on locating absolute extrema.

Optimization Strategy

Step 1: Read the given information carefully and identify what is known and what needs to be found. What quantity must be optimized in order to solve the problem?

Step 2: Identify and appropriately label the relevant quantities and draw a diagram, if possible, as an aid in determining the relationships between them.

Step 3: Assign a label to the quantity that needs to be maximized or minimized, and work toward expressing that quantity (we'll give it the generic name f for now) as a function of one variable. This may require some intermediate steps in which you write equations relating the quantities you have identified and use algebra to write f as a function of one of them. Note the appropriate domain of f for the problem.

Step 4: Using the techniques of this chapter, find the absolute minimum or maximum of f (whichever is desired) on the domain.

Step 5: Check your answer to see if it makes sense as a solution to the problem.

Topic One

Mathematical/Geometrical Problems

Although all of the examples in this section will be rephrased fairly quickly in mathematical terms, the first two that we explore are immediately of a mathematical or geometrical nature.

Example 1

A 15 in. by 24 in. piece of sheet metal is to be formed into an open-top box by cutting out a square from each of the four corners and folding up the sides. How large should each square be in order to achieve a box of maximum volume?

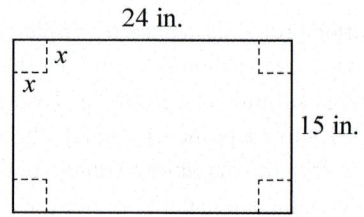

Figure 1

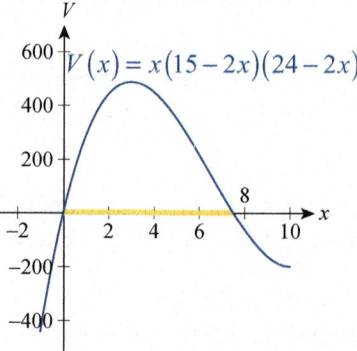

Figure 2

Solution

We begin with a picture, illustrating a rectangular piece of metal with four squares of side length x to be cut out. (See Figure 1.) Once the squares have been removed and the edges folded upward, the volume of the resulting open box will be

$$V(x) = x(15 - 2x)(24 - 2x).$$

Note that in this problem, we have quickly arrived at a function of one variable to be maximized—no intermediate steps were necessary. And in order for V to make sense as a volume, its domain is limited to $[0, 15/2]$.

V is a cubic polynomial, and although a graph of V is not necessary in order to solve this problem, it's never a bad idea to graph your function to be maximized if it can be easily done—such a graph can serve as another self check of your work. In Figure 2, V is graphed over an interval larger than its domain for this problem, with the actual domain $[0, 15/2]$ highlighted in yellow.

The remaining steps are to find the critical point(s) and test them to identify the absolute maximum—from the graph, we expect to find just one critical point in the domain, and that is indeed the case.

$$\begin{aligned} V'(x) &= (15 - 2x)(24 - 2x) + x(-2)(24 - 2x) + x(15 - 2x)(-2) \\ &= 12(x^2 - 13x + 30) \\ &= 12(x - 10)(x - 3) \\ V'(x) &= 0 \Leftrightarrow x = 3 \text{ or } x = 10 \end{aligned}$$

The only critical point in $[0, 15/2]$ is 3. Evaluating V at this critical point and at the endpoints of the domain we see that the maximum possible volume is obtained when $x = 3$: $V(0) = V(15/2) = 0$ and $V(3) = 486$ in.^{3}

Example 2

Find the rectangle of largest possible area that can be inscribed in a semicircle of radius r.

Solution

We begin again with a sketch in order to better understand the problem and to begin labeling relevant quantities and dimensions with names. First, a little experimentation leads to the realization that the rectangle of largest area will have one side on the diameter of the semicircle; any rectangle not drawn like this can be rotated so that it does and then enlarged until the opposite vertices are on the semicircle. It will also, by the same reasoning, be centered in the semicircle. We label the radius of the semicircle r and give the width and height of the inscribed rectangle the labels w and h, respectively. (See Figure 3.)

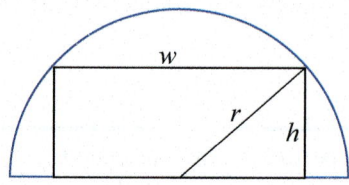

Figure 3

Given the labels we have assigned, the quantity we want to maximize is $A = wh$. At this point, A is a function of the two variables w and h, so we need to use another relationship in order to reduce A to a function of one variable.

From the figure and the Pythagorean Theorem, we see that

$$\left(\frac{w}{2}\right)^2 + h^2 = r^2,$$

or $w^2 + 4h^2 = 4r^2$. We can solve this equation easily for w to obtain $w = 2\sqrt{r^2 - h^2}$ (note that only the positive root makes sense), and then write

$$A(h) = 2h\sqrt{r^2 - h^2}.$$

From the geometry of the problem, as well as the nature of the formula, we know that $h \in [0, r]$; that is, the domain of A is the closed bounded interval $[0, r]$, so absolute extrema are guaranteed to exist.

We now proceed to find the critical point(s):

$$A'(h) = 2\sqrt{r^2 - h^2} - \frac{2h^2}{\sqrt{r^2 - h^2}} = \frac{2r^2 - 4h^2}{\sqrt{r^2 - h^2}}$$

$$A'(h) = 0$$
$$2r^2 - 4h^2 = 0$$
$$h = \pm\frac{r}{\sqrt{2}}$$

A' is also undefined when $h = r$, but that's one of the endpoints of the domain. Because $h = -r/\sqrt{2}$ is not in the domain of A, we evaluate A at the two endpoints and the one critical point $r/\sqrt{2}$ in $(0, r)$:

$$A(0) = 2(0)\sqrt{r^2 - 0} = 0$$

$$A\left(\frac{r}{\sqrt{2}}\right) = 2\frac{r}{\sqrt{2}}\sqrt{r^2 - \frac{r^2}{2}} = r^2$$

$$A(r) = 2r\sqrt{r^2 - r^2} = 0$$

So the maximum possible area for an inscribed rectangle is r^2, and it occurs when $h = r/\sqrt{2}$ and $w = r\sqrt{2}$. One way to quickly check the reasonableness of this answer is to note that r^2 is smaller than $\pi r^2/2$, the area of the semicircle.

Topic Two

Optimization Problems from Other Disciplines

The need to either minimize or maximize some quantity is extremely common, and the next few examples illustrate how optimization problems arise in a variety of contexts.

Example 3

As it prepares to expand into a new market, an American oil company decides to design a half-liter can for its brand of engine oil. The can will have the typical shape of a right circular cylinder. What dimensions for the can will minimize the amount of metal used to form it?

Solution

We will label the radius and height of the cylinder r and h, respectively. And since the can consists of two circles of radius r and a (rolled-up) rectangle of width $2\pi r$ and height h, the area of the metal used to form it is $A = 2\pi r^2 + 2\pi rh$.

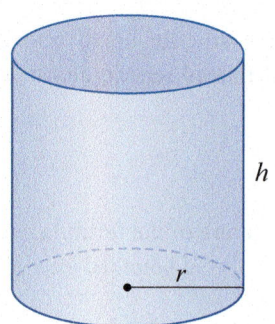

Figure 4

One liter corresponds to a volume of 1000 cm^3, so a half liter of oil occupies 500 cm^3. That is, $V = \pi r^2 h = 500$. We want to minimize A, which currently is a function of two variables. But we can use the volume relationship between r and h to eliminate one of them; it's easier to eliminate h (you should check that this is so).

$$.A = 2\pi r^2 + 2\pi rh = 2\pi r^2 + 2\pi r\left(\frac{500}{\pi r^2}\right) = 2\pi r^2 + \frac{1000}{r}$$

Note that in this problem, the domain of A is $(0, \infty)$. In practice, we know we won't use a radius r close to 0, nor will we use a very large r, but we have no way to narrow down the domain at the moment. However, the calculus techniques we have learned will still resolve the question definitively. We proceed to find A' and then the critical points.

$$A'(r) = 4\pi r - \frac{1000}{r^2}$$
$$A'(r) = 0$$
$$4\pi r = \frac{1000}{r^2}$$
$$r^3 = \frac{1000}{4\pi}$$
$$r = \frac{10}{\sqrt[3]{4\pi}}$$

The only critical point is $r = 10/\sqrt[3]{4\pi} \approx 4.30$ cm.

The First Derivative Test verifies that A does indeed have a relative and absolute minimum at this critical point, as $A'(1) = 4\pi - 1000 < 0$ and $A'(10) = 40\pi - 10 > 0$ (1 and 10 are merely convenient test points). Figure 5 is a graph of A and confirms our findings. The corresponding value of h is as follows.

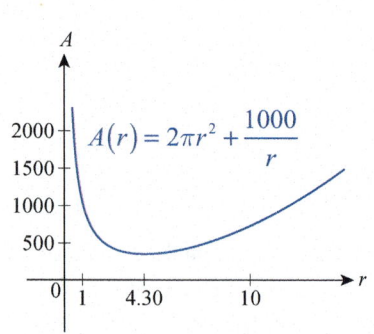

Figure 5

$$h = \frac{500}{\pi r^2} = \frac{500}{\pi\left(\dfrac{10}{\sqrt[3]{4\pi}}\right)^2} = \frac{(5)\left(2^{4/3}\right)}{\pi^{1/3}} = 10\left(\frac{2}{\pi}\right)^{1/3} \approx 8.60 \text{ cm}$$

Thus, to minimize the amount of metal used for the half-liter can, the radius should be approximately 4.30 cm and the height approximately 8.60 cm.

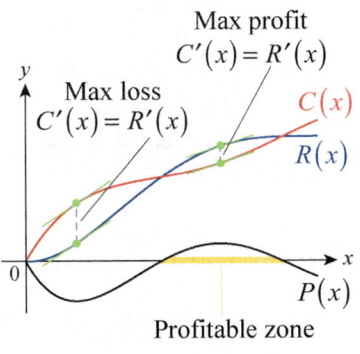

Figure 6

Recall from Sections 2.6 and 3.7 that the profit $P(x)$ realized by selling x units of a given product is related to the cost $C(x)$ to produce x units and the revenue $R(x)$ by the formula $P(x) = R(x) - C(x)$. If we seek to maximize profit, the techniques of this chapter lead us to look at the critical points of P; in particular, if C and R are differentiable functions, we are interested in the solutions of the equation $P'(x) = 0$, which are also the solutions of the equation $R'(x) = C'(x)$. Figure 6 is a graph of typical cost and revenue functions—costs often initially exceed revenue, then fall below revenue as bulk manufacturing and transportation savings are realized, then eventually exceed revenue again as production capacity and market saturation points are reached. Note that the profitable zone is bounded by the two positive break-even points, where $C(x) = R(x)$, and that the points of maximum profit and loss are indeed where $C'(x) = R'(x)$.

Example 4

Suppose that Jan's Custom Calendar Company models the coming year's revenue and cost functions, in thousands of dollars, to be $R(x) = -x^3 + 9x^2$ and $C(x) = 2x^3 - 12x^2 + 30x$, where x represents units of 1000 calendars and where the model is thought to be accurate up to approximately $x = 6$. What is Jan's profitable zone, and what level of production will maximize her profit?

Solution

The profit function in this problem is $P(x) = -3x^3 + 21x^2 - 30x$, which factors as $P(x) = -3x(x-2)(x-5)$. Solving the equation $P(x) = 0$ is equivalent to solving $R(x) = C(x)$, so we know that the positive break-even points are $x = 2$ and $x = 5$. Therefore, Jan's profitable zone is between 2000 and 5000 calendars.

Similarly, we can solve either the equation $P'(x) = 0$ or the equation $R'(x) = C'(x)$ to find the relative extrema of P.

$$P'(x) = -9x^2 + 42x - 30$$
$$0 = -9x^2 + 42x - 30$$
$$x = \frac{7 \pm \sqrt{19}}{3} \qquad \text{Quadratic formula}$$
$$x \approx 0.880 \text{ and } 3.786 \qquad \text{Two critical points}$$

We can use either derivative test to determine that P has a relative minimum at 0.880 and a relative maximum at 3.786, corresponding to production of, respectively, 880 and 3786 calendars. The maximum possible profit is thus $P(3.786) = 24.626$ or $\$24,626$. Figure 7 shows the graphs of R, C, and P. Note how the relative extrema of P correspond to points where $R' = C'$.

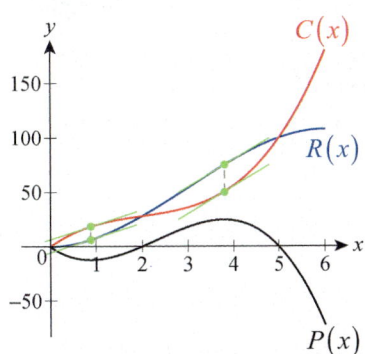

Figure 7

Example 5

Another useful model in business is the average cost function. If $C(x)$ represents the cost of producing x units of a product, then the average cost of producing those x units is $C(x)/x$. Show that if the average cost function is differentiable, then the marginal cost $C'(x)$ is equal to the average cost at the critical points of the average cost function.

Solution

We begin by looking at the critical points of $C(x)/x$.

$$\left[\frac{C(x)}{x}\right]' = \frac{C'(x)x - C(x)}{x^2} \qquad \text{Quotient Rule}$$

We are assuming that $C(x)/x$ is differentiable, so its critical points are where the expression above is 0.

$$\frac{C'(x)x - C(x)}{x^2} = 0$$
$$C'(x)x - C(x) = 0$$
$$C'(x) = \frac{C(x)}{x}$$

Example 6

You are probably familiar with the fact that an object underwater can appear to be at a position slightly offset from its actual position; this is due to refraction. In optics, Fermat's Principle says that the light rays observed in such a situation are those that travel from the object to your eyes in the shortest period of time, and light travels at different speeds through different media (such as water and air).

If v_1 and v_2 represent the speeds of light in, respectively, air and water, and θ_1 and θ_2 are the angles shown in Figure 8, use Fermat's Principle to derive Snell's Law:

$$\frac{\sin\theta_1}{\sin\theta_2} = \frac{v_1}{v_2}$$

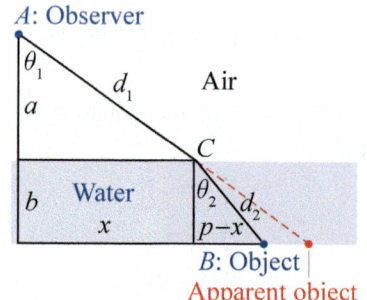

Figure 8

Solution

The given information consists only of the two angles and the two speeds, but in the figure we have already labeled other quantities that might help us relate the angles and speeds and arrive at Snell's Law. Specifically, we have let p denote the horizontal distance between the object B and the point directly beneath the observer at A. And since the point C of refraction (where the rays of light bend) is unknown, we have given its horizontal displacement from A the label x, meaning the horizontal distance between C and B is $p - x$. The vertical distances a and b are fixed, but the lengths of the hypotenuses, d_1 and d_2, will vary as x varies.

At this point, it may very well be unclear how to arrive at Snell's Law from what we have. But we haven't yet applied Fermat's Principle, and we can deduce many relationships between the labeled quantities. To begin with, since distance = rate · time, the time it takes light to travel from B to C is d_2/v_2 and the time it takes to travel from C to A is d_1/v_1. So the total time is expressed as follows.

$$T = \frac{d_1}{v_1} + \frac{d_2}{v_2}$$

We can express the two hypotenuses as functions of x by noting that $a^2 + x^2 = d_1^2$ and $b^2 + (p-x)^2 = d_2^2$, so

$$T(x) = \frac{\sqrt{a^2 + x^2}}{v_1} + \frac{\sqrt{b^2 + (p-x)^2}}{v_2}$$

and the domain of T is $[0, p]$.

The actual distance x must be the value of x that minimizes T, so our next step is to find T'.

$$T'(x) = \frac{x}{v_1\sqrt{a^2 + x^2}} - \frac{p-x}{v_2\sqrt{b^2 + (p-x)^2}}$$

$$= \frac{x}{v_1 d_1} - \frac{p-x}{v_2 d_2} \qquad \text{Substitute } d_1 \text{ and } d_2 \text{ for their formulas.}$$

$$= \frac{\sin\theta_1}{v_1} - \frac{\sin\theta_2}{v_2} \qquad \sin\theta_1 = \frac{x}{d_1} \text{ and } \sin\theta_2 = \frac{p-x}{d_2}$$

Note that $T_+'(0) < 0$ and $T_-'(p) > 0$, so by Darboux's Theorem (Section 3.1), there is a point $x \in [0, p]$ for which $T'(x) = 0$. And by the First Derivative Test, that point must minimize T. Rewriting $T'(x) = 0$, we have developed the formula for Snell's Law.

$$\frac{\sin\theta_1}{v_1} = \frac{\sin\theta_2}{v_2} \quad \text{or} \quad \frac{\sin\theta_1}{\sin\theta_2} = \frac{v_1}{v_2}$$

4.6 **Exercises**

1. Find two integers whose sum is 120 and whose product is as large as possible. (**Hint:** If you denote the first number by x, then the second number is $120 - x$. Now write a formula for the product, and use calculus to find the maximum.)

2–14 Use the strategy suggested in Exercise 1 to find two numbers satisfying the given requirements.

2. The sum is S and the product is a maximum.

3. The difference is 36 and the product is as small as possible.

4. Two positive numbers whose product is 144 and the sum is a minimum.

5. Two positive numbers whose product is n^2 and the sum is a minimum.

6. Two positive numbers whose product is 162 and the sum of twice the first and the second is a minimum.

7. Two positive numbers that are reciprocals of each other and their sum is a minimum.

8. Two positive integers so that the square of the first number plus the second number is 243 and their product is a maximum.

9. The sum of twice the first and three times the second is 480 and their product is a maximum.

10. The product of two positive integers is 32 and the sum of twice the first plus the second is a minimum.

11. Two numbers whose product is 16 and the sum of whose squares is a minimum.

12. Two positive numbers whose sum is 1 and the sum of whose cubes is a minimum.

13. Two nonnegative numbers whose sum is 1 and the sum of whose cubes is a maximum.

14. Repeat Exercises 12 and 13 using fourth powers instead of cubes.

15. Modify Example 2 by inscribing a rectangle in the region bounded by the x-axis and the parabola $y = k - x^2$ $(k > 0)$.

16. A vertex of a rectangle is at the origin; the opposite vertex sits in the first quadrant and on the line $2y + x = 4$. Find the dimensions that maximize the area of such a rectangle.

17. Repeat Exercise 16 with the opposite vertex sitting on the graph of $y = 32 - x^3$.

18. From among all lines through the point $(3, 1)$, find the one forming with the coordinate axes a right triangle of maximum area.

19. Repeat Exercise 18, this time finding the line forming a triangle whose hypotenuse is of minimum length.

20. Suppose that when constructing a trough similar to the one in Exercise 35 of Section 3.8, both the shorter base and the legs of its cross-section are 20 centimeters long. Find the base angle α that maximizes the volume of the trough.

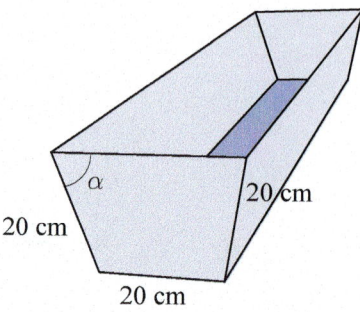

21. Find the coordinates of the point on the graph of $y = \sqrt{x}$ that is closest to the point $(1, 0)$.

22. Find the coordinates of the point on the graph of $y = x^3$ that is closest to the point $(-4, 0)$.

23. Find the dimensions of the rectangle of largest area that can be inscribed in the ellipse $2x^2 + 6y^2 = 12$.

24. Find the equation of the line tangent to the graph of $y = 1 - x^2$ that forms with the coordinate axes the triangle of minimum area in the first quadrant.

25. A farmer has 120 feet of fencing to construct a rectangular pen up against the straight side of a barn, using the barn for one side of the pen. The length of the barn is 100 feet. Determine the dimensions of the rectangle of maximum area that can be enclosed under these conditions. (**Hint:** Be mindful of the domain of the function you are maximizing.)

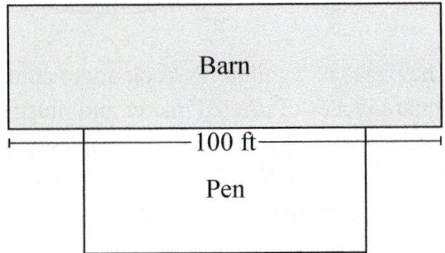

26. A farmer needs to construct two adjoining rectangular pens of identical areas, as shown. If each pen is to have an area of 1200 square feet, what dimensions will minimize the cost of fencing?

1200 ft²	1200 ft²

27. Repeat Exercise 26 if the pens are constructed against a straight wall that serves as a side for each.

28. Repeat Exercise 27 if three identical adjoining pens are to be constructed, as shown.

Wall

1200 ft²	1200 ft²	1200 ft²

29. A supporting beam of rectangular cross-section is to be cut from a log that has an approximately circular cross-section with a radius of r inches. Knowing that the strength of such a beam is directly proportional to the width multiplied by the second power of the height of its cross-section, find the strongest beam that can be cut under these conditions.

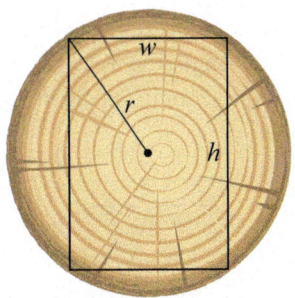

30. An 8-foot fence stands 5 feet from a tall building. A contractor needs to reach the building with a ladder from the outside of the fence. Find the minimum length of the ladder that can do the job.

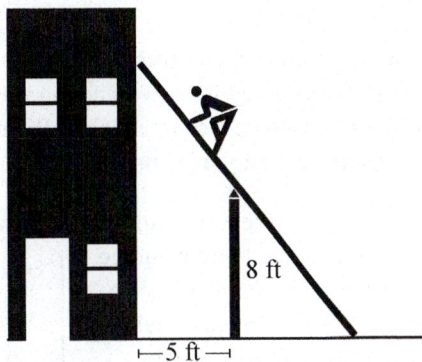

31. A 30 in. piece of wire is cut and the pieces are bent into a circle and a square, respectively. Where should we cut in order to minimize the sum of the areas of these two shapes?

32. Repeat Exercise 31, this time producing an equilateral triangle and a square.

33. Repeat Exercises 31 and 32, this time maximizing the sum of the two areas.

34. Prove that among all rectangles that can be inscribed in a circle, the square has the greatest perimeter.

35. Prove that among all isosceles triangles of a given area, the equilateral triangle has the minimum perimeter.

36. Find the dimensions of the rectangle whose perimeter is P units and area is a maximum.

37. Find the dimensions of the rectangle whose area is A units and perimeter is a minimum.

38. The perimeter of an isosceles triangle is P inches. Find the side lengths so as to minimize the sum of areas of the semicircles drawn onto the sides of the triangle.

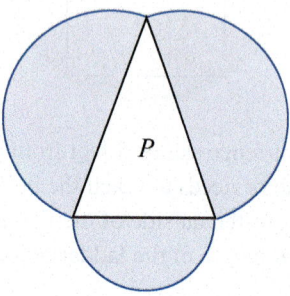

39. Suppose we want to construct a can in the shape of a right circular cylinder with no top whose surface area is to be S square inches. What dimensions will maximize the volume?

40. If we want to make a rectangular box with a square bottom and no top that holds 32 cubic inches, and the construction material costs 3 cents per square inch, what are the dimensions and the cost of the least expensive box that can be made?

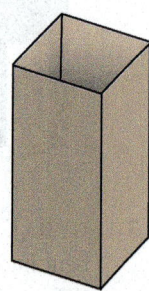

41. If the box to be constructed in Exercise 40 is to hold the same volume, but we need to construct a top from an expensive, heat-resistant material that costs 21 cents per square inch, how does the new requirement change the cost and dimensions of the least expensive box?

42. Determine the dimensions and maximum volume of the rectangular box with no top and a square base if its surface area is A square inches.

43. A cone is to be constructed by cutting out a sector of central angle α of a disk of radius R and gluing the cut lines together to form a cone. Find α that maximizes the volume of the cone.

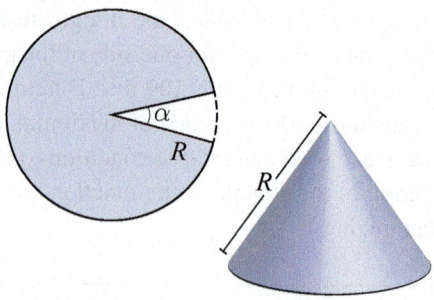

44. The pages of a children's book are to contain 54 square inches of printed matter and illustrations, with margins of 1 inch along the sides and $1\frac{1}{2}$ inches along the top and bottom of each page. Find the dimensions of the page that will require the minimum amount of paper.

45. A poster is to contain 150 square inches of printed matter, surrounded by margins that are 3 inches wide on the top and bottom, and 2 inches on each side. Find the dimensions for the poster that minimize its total area.

46. The sum of squares of lengths of the sides of a right triangle is 64 square inches. Find the side lengths that maximize the area of the triangle.

47. A flower bed is planned in the form of a circular sector. Find the central angle and radius if it is to cover 169 square feet, and its perimeter is to be a minimum.

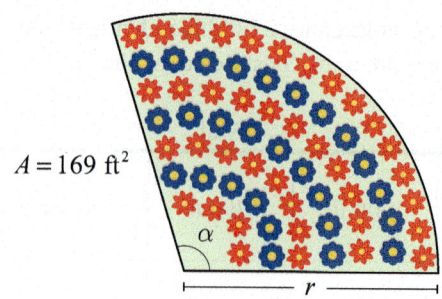

48. The shape of a Norman window can be approximated by a rectangle with a semicircle on top. What dimensions will admit the maximum amount of light if the perimeter of the window is to be P inches?

49. In Exercise 103 of Section 3.6, find the optimum distance s that maximizes the viewing angle.

50. An office building is located right on a riverbank, which is straight. A small power plant is on the opposite bank, 1500 feet downstream from the point directly opposite the office building. The river is 300 feet wide. If we want to connect the power plant and the building by cable, which costs $1700 per foot to lay down underwater and $800 per foot underground, what is the least expensive path for the cable?

51. Two antennas standing 30 feet apart are to be stayed with a single wire. The wire runs from the top of the first antenna, is secured to the ground somewhere between the antennas, and is finally attached to the top of the second antenna. If the height of the first antenna is 12 feet, while that of the second is 8 feet, find the point along the line segment connecting the bases where the wire needs to be staked to the ground if the length of the wire is to be minimal.

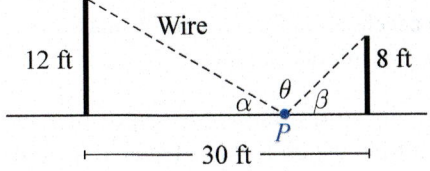

52. If we denote the heights of the antennas in Exercise 51 by h_1 and h_2, respectively, and the distance between them is d, prove that the wire has minimal length if and only if $\alpha = \beta$.

53. In Exercise 51, find the location of P that maximizes the angle θ.

54.* An inverted square pyramid is to be inscribed into a larger square pyramid of volume V, so that the two have a common axis, and the vertex of the inscribed pyramid coincides with the center of the outer pyramid's base. Find the ratio of the pyramids' altitudes so that the volume of the inscribed pyramid is maximal.

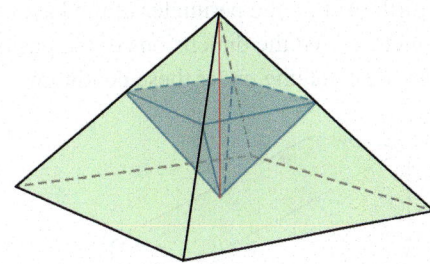

55.* The lower left corner of a letter-sized paper, which is 8.5 in. by 11 in., is folded over to reach the right edge of the paper. Find a way that this can be done so as to produce a crease of minimum length.

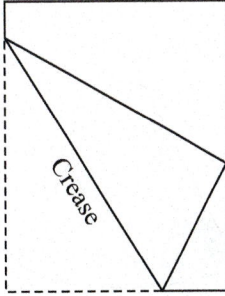

56. Find the radius of the base and the height of the right circular cylinder of largest volume that can be inscribed in a sphere of radius R.

57. Repeat Exercise 56, but inscribe a right circular cone instead of a cylinder in the sphere of radius R.

58. Find the radius of the base and the height of the right circular cylinder of largest volume that can be inscribed in a circular cone if the height of the cone is H and its base has radius R.

59. Repeat Exercise 58, but find the extremum of the surface area of the cylinder instead of its volume.

60. The sum of the height and the radius of the base of a circular cylinder is 12 inches. Find their lengths if the volume of the cylinder is to be a maximum.

61. Suppose that we want to send a parcel in the shape of a square-based rectangular solid, and the Standard Post service limits the sum of the length and girth (girth = the perimeter of the base) to 130 inches. Find the dimensions of the package of the greatest volume under these conditions.

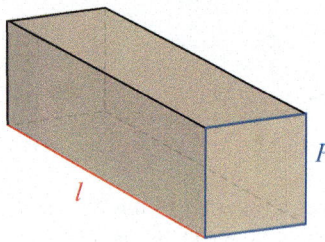

62. Find the maximum volume a right circular cone can have if its slant height is a inches.

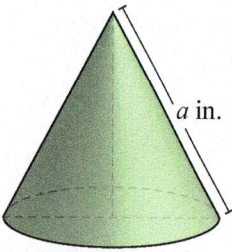

63. An isosceles triangle of perimeter P is rotated around its base. What base length will produce the solid of maximum volume?

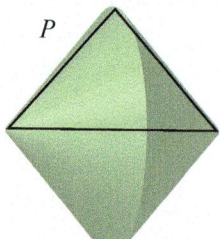

64. A lighthouse is 2 miles off a straight shoreline, and a grocery store is 10 miles down the coast. If the lighthouse keeper can row at 2.4 mph and walk at 4 mph, where should he land in order to make the best time to the store to get supplies? What if he is picked up by a golf cart that can drive at 9.9 mph?

65. Repeat Exercise 64 if the lighthouse keeper uses a motorboat whose top speed is 20.1 mph, and will be picked up by a car that will drive at the posted speed limit of 45 mph.

66. Repeat Exercise 64 if the lighthouse and the store are both on the shore of a circular lake of diameter d at the endpoints of the diameter.

67. At noon on a certain day, a plane is 200 miles south of another airliner and flying north at 550 mph, while the second plane is flying southwest at 600 mph. How much later after this instant is their distance a minimum?

68. A straight two-lane highway intersects a straight interstate at a right angle. A car exits the interstate and starts moving away from it on the two-lane highway at 50 mph. At the same instant, another car, moving at 75 mph on the interstate, is approaching the same intersection, but is still 10 miles from it. When will their distance be a minimum and what will this distance be?

69. The position of an object connected to a spring is given by $d(t) = \sin 3t + \cos 3t$, where d is measured in feet, and t in seconds. Find when the absolute value of its velocity first reaches its maximum and the value of the maximum velocity.

70. In Exercise 69, find when the absolute value of the acceleration first reaches its maximum and the value of this acceleration.

71. Ignoring air resistance, the range r of a projectile fired from the ground in a flat area with an initial velocity of v_0 can be calculated by $r = (v_0^2 / g) \sin 2\theta$, where g is the gravitational acceleration and θ is the launch angle relative to the horizontal. Find the launch angle that maximizes the range if the initial velocity is a given constant.

72. The luminance E_l at distance d from a light source is directly proportional to the light intensity F_l (also called luminous flux) and inversely proportional to the square of distance: $E_l = F_l / (4\pi d^2)$. Suppose two light bulbs are 3 meters apart, with respective light intensities of $F_{l,1} = 1700$ lumens (lm) and $F_{l,2} = 1000$ lm. Where between these light bulbs will the sum of their luminance levels be a minimum?

73. Management and Power, Inc. has found that its seminar on management techniques attracts 800 people when the seminar fee is set to $600. They estimate that for each $15 discount in the charge, an additional 50 people will attend the seminar. Find the amount that Management and Power, Inc. should charge for the seminar to maximize the revenue, and find the maximum revenue.

74. A blueberry farmer owns 1056 plants, each producing p pounds on average during a regular season. He estimates that for each additional dozen of new plants planted on his farm, average production per plant is going to drop by a half percent. What would be the optimum number of plants on the farm in order to maximize production, and what is the optimum production level?

75. The manager of a 115-unit apartment complex finds that all units are rented at a price of $1500 per month. Research shows that for each $20 increase in rent, one additional unit remains vacant. How much should he charge for rent in order to bring in maximum revenue, and how many units are rented then?

76. A moving company sends a truck on a 2000-mile round-trip to move two households. The hourly fuel consumption of the truck is approximated by $2 + \frac{1}{280.1} v^2$ gallons, where v is assumed to be a constant speed somewhere between 35 and 70 miles per hour. If a gallon of diesel fuel costs $2.50 and the driver is paid $22 an hour, what speed will minimize the company's transportation costs?

77. Cool Wheels, a manufacturer of die-cast model cars, has a monthly overhead cost of $6000, material costs of $2 per toy car, and each has associated labor costs of $0.40. When producing and marketing 2500 cars a month, each sells for $30.75. When producing more, it was found that for each additional 100 units, the market conditions cause the price to drop by a dollar. In addition, labor costs go up by 5 cents for each additional 100 units because of expensive overtime pay. Find the production level and selling price that maximize the profit under these conditions.

78. Suppose it costs a candy company $3 to produce and distribute a box of Chi-Can chipotle candy bars, and the number of boxes sold at x dollars a box is approximated by $n = 80/(x-11) + 15(50-x)$. What sale price will bring the maximum profit?

79. Prove that when the company in Exercise 78 maximizes its profit, the marginal cost equals the marginal revenue.

80. Suppose that $R(x) = 2x^3 - 15x^2$ and $C(x) = 3x^3 - 25x^2 + 21x$ are the weekly revenue and cost functions for a particular commodity, where x represents units of 100 individual products and where the model is thought to be accurate up to approximately $x = 10$. What is the profit zone, and what level of production will maximize the profit? (See Example 4.)

81. The cost of manufacturing x units of a commodity is given by $C(x) = x^3 - 15x^2 + 12{,}000x$. Find x that minimizes the average cost of production. (See Example 5.)

4.6 **Technology Exercises**

82–83 Use the graphing and symbolic differentiation capabilities of a computer algebra system to solve the problem.

82. Suppose we have a small supply of craft paint, enough for 1 square foot, and we want to use it to paint a regular tetrahedron and a cube from a children's toy set. What should the dimensions of these solids be if we want to maximize the total volume? How about minimizing the volume?

83. Repeat Exercise 82 for a tetrahedron and a sphere.

4.7 **Antiderivatives**

The underlying goal of this chapter has been to use knowledge about the derivative of a function to gain a better understanding of the function itself. Given that, it's fitting that in the last section of this chapter we ask the following: If we have perfect knowledge of f', can we recover f in its entirety? As we shall see, this question is rooted in more than mere intellectual curiosity. The answer, which we will fully develop in Chapter 5, has a great deal of both theoretical and practical importance.

Topic One
Finding Antiderivatives

Since differentiation is the name we apply to the process of finding the derivative of a function, *antidifferentiation* is a natural name for the reverse process, and a function that results from antidifferentiation is termed an *antiderivative*.

> **Definition**
>
> **Antiderivative**
>
> A function F is called an **antiderivative** of a given function f on an interval I if $F'(x) = f(x)$ for all x in I.

Given all that we have learned about derivatives and differentiation, we already know quite a lot about antiderivatives and antidifferentiation. Example 1 illustrates this.

> **Example 1**
>
> Find an antiderivative of each of the following functions.
>
> **a.** $f(x) = 3x^2$ **b.** $g(x) = \sin x$ **c.** $h(x) = \dfrac{1}{x} + 5$
>
> **Solution**
>
> We have seen functions identical or similar to f, g, and h appear as the derivatives of other functions, so we can use our experience with differentiation to "work backward" and arrive at an antiderivative of each.
>
> **a.** $F(x) = x^3$ qualifies as an antiderivative of f, since $F'(x) = 3x^2$. Note that the interval on which this is valid is $(-\infty, \infty)$.
>
> **b.** To find an antiderivative of $g(x) = \sin x$, we can start with the fact that $\dfrac{d}{dx}(\cos x) = -\sin x$. The result of this differentiation is off by a minus sign, but one of our many differentiation rules allows us to correct for that: if we define $G(x) = -\cos x$, then $G'(x) = \sin x$. Again, this is true on all of $(-\infty, \infty)$.
>
> **c.** We know that $\dfrac{d}{dx}\ln|x| = \dfrac{1}{x}$ and $\dfrac{d}{dx}(5x) = 5$, so by the additive property of differentiation, $H(x) = \ln|x| + 5x$ is an antiderivative of h. This makes sense for all x where H and h are defined, so it's true for the interval $(-\infty, 0)$ and the interval $(0, \infty)$.

With a little thought, though, Example 1 already points out that the answer to the opening question of this section is "No, not uniquely." Another differentiation fact that we know is that two differentiable functions have the same derivative on an interval if and only if they differ by no more than a constant (the "only if" part is the second corollary of the Mean Value Theorem). This has the practical effect that each of the functions in Example 1 has an infinite number of antiderivatives. For instance, $x^3 - 17$, $x^3 + \frac{27}{8}$, and $x^3 + \pi$ all serve equally well as an antiderivative of $3x^2$. But the good news is that once we have found one antiderivative of a function on a given interval, we know the form of all its antiderivatives on that interval. That is the meaning of the following theorem, which is a reformulation of the corollary mentioned above.

Theorem

General Antiderivative Theorem

If F is an antiderivative of the function f on a given interval I, then any other antiderivative of f can be written as $F(x) + C$ for some constant C.

And the preceding theorem is the motivation for the next definition.

Definition

General Antiderivative

If F is an antiderivative of the function f on a given interval I, then the **general antiderivative** of f on I is $F(x) + C$ where C represents an arbitrary constant.

Example 2

Find the general antiderivative of $f(x) = e^x - 2x$. Then find the particular antiderivative F that satisfies $F(0) = -2$.

Solution

Using the same reasoning as in Example 1, we can quickly verify that $e^x - x^2$ is an antiderivative of f, so the general antiderivative is $F(x) = e^x - x^2 + C$. In order to determine the constant C that will satisfy the condition $F(0) = -2$, we can simply solve the equation $-2 = F(0) = e^0 - (0)^2 + C = 1 + C$, so $C = -3$.

Graphically, the family of functions described by the general antiderivative consists of the same curve crossing at different points on the y-axis. The particular antiderivative $e^x - x^2 - 3$ that satisfies the given condition, is depicted in blue in Figure 1, with a few other members of the family in black.

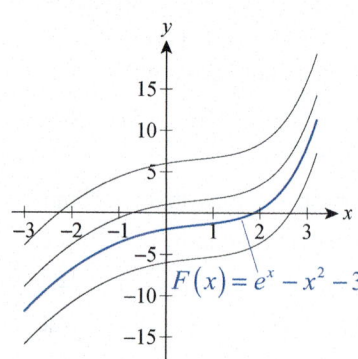

$F(x) = e^x - x^2 - 3$

Figure 1

Over the course of the next several chapters, we will learn some particular techniques for antidifferentiating various classes of functions, but the most basic technique is that which we have used in the first two examples; namely, using rules of differentiation in reverse. Table 1 indicates how a given differentiation rule can be restated as an antidifferentiation rule and, in the interest of brevity, Table 2 simply lists the general antiderivatives relating to basic trigonometric functions. In the tables, F and G represent antiderivatives of f and g, respectively, and k represents a nonzero constant (note the frequent use of the Chain Rule).

Function	General Antiderivative	Justification
$kf(x)$	$kF(x)+C$	$\dfrac{d}{dx}\left[kF(x)+C\right]=kF'(x)=kf(x)$
$f(x)+g(x)$	$F(x)+G(x)+C$	$\dfrac{d}{dx}\left[F(x)+G(x)+C\right]=f(x)+g(x)$
$x^r,\ r\neq -1$	$\dfrac{1}{r+1}x^{r+1}+C$	$\dfrac{d}{dx}\left[\dfrac{1}{r+1}x^{r+1}+C\right]=\dfrac{r+1}{r+1}x^{(r+1)-1}=x^r$
$\dfrac{1}{x}$	$\ln|x|+C$	$\dfrac{d}{dx}\left[\ln|x|+C\right]=\dfrac{1}{x}$
e^{kx}	$\dfrac{1}{k}e^{kx}+C$	$\dfrac{d}{dx}\left[\dfrac{1}{k}e^{kx}+C\right]=\dfrac{k}{k}e^{kx}=e^{kx}$
$a^{kx},\ a>0, a\neq 1$	$\left(\dfrac{1}{k\ln a}\right)a^{kx}+C$	$\dfrac{d}{dx}\left[\left(\dfrac{1}{k\ln a}\right)a^{kx}+C\right]=\dfrac{k\ln a}{k\ln a}a^{kx}=a^{kx}$

Table 1 General Antiderivatives with Justifications

Function	General Antiderivative	Function	General Antiderivative
$\sin(kx)$	$-\dfrac{1}{k}\cos(kx)+C$	$\cos(kx)$	$\dfrac{1}{k}\sin(kx)+C$
$\sec^2(kx)$	$\dfrac{1}{k}\tan(kx)+C$	$\csc^2(kx)$	$-\dfrac{1}{k}\cot(kx)+C$
$\sec(kx)\tan(kx)$	$\dfrac{1}{k}\sec(kx)+C$	$\csc(kx)\cot(kx)$	$-\dfrac{1}{k}\csc(kx)+C$
$\dfrac{1}{\sqrt{1-(kx)^2}},\ |kx|<1$	$\dfrac{1}{k}\sin^{-1}(kx)+C$	$\dfrac{1}{1+(kx)^2}$	$\dfrac{1}{k}\tan^{-1}(kx)+C$
$\dfrac{1}{|kx|\sqrt{(kx)^2-1}},\ |kx|>1$	$\dfrac{1}{k}\sec^{-1}(kx)+C$		

Table 2 General Antiderivatives of Trigonometric Functions

With practice, the general antiderivatives just summarized will be so familiar that you won't need to refer to the tables, but initially it may be convenient to use them as you work through the exercises at the end of this section. The next example shows how this can be done.

Example 3

Find the general antiderivative of each of the following functions.

a. $f(x) = \dfrac{1}{\sqrt{x}} - \csc^2 5x$ **b.** $g(x) = \dfrac{7x^{4/3} + 3x^{1/2}}{2x} - 2^x$

Solution

a. We can rewrite the function as $f(x) = x^{-1/2} - \csc^2 5x$, and we start by noting that the general antiderivative of $x^{-1/2}$ is

$\dfrac{1}{(-1/2)+1}x^{(-1/2)+1} + C_1 = 2x^{1/2} + C_1$ and the general antiderivative of $-\csc^2 5x$ is $\frac{1}{5}\cot 5x + C_2$ where C_1 and C_2 represent arbitrary constants. When we add these two antiderivatives, we get $2x^{1/2} + \frac{1}{5}\cot 5x + C_1 + C_2$. But $C_1 + C_2$ also represents nothing more than an arbitrary constant, and it is customary to combine all arbitrary constants in an antiderivative together and denote their sum with a single symbol. Hence, we write

$$F(x) = 2x^{1/2} + \frac{1}{5}\cot 5x + C$$

as the general antiderivative of $f(x) = \dfrac{1}{\sqrt{x}} - \csc^2 5x$.

(Remember: it is easy to check your work and verify that $F'(x) = f(x)$.)

b. It would be difficult to come up with an antiderivative of g in the form given, but after breaking apart the fraction the way forward becomes clear.

$$g(x) = \frac{7x^{4/3} + 3x^{1/2}}{2x} - 2^x = \frac{7}{2}x^{1/3} + \frac{3}{2}x^{-1/2} - 2^x$$

$$G(x) = \left(\frac{7}{2}\right)\left(\frac{3}{4}\right)x^{4/3} + \left(\frac{3}{2}\right)(2)x^{1/2} - \left(\frac{1}{\ln 2}\right)2^x + C$$

$$= \frac{21}{8}x^{4/3} + 3x^{1/2} - \left(\frac{1}{\ln 2}\right)2^x + C$$

Topic Two

Antiderivatives in Use

Example 2 is a simple example of a type of problem called a *differential equation*, which we will study in greater depth in Chapter 8. Informally, a differential equation is an equation that involves the derivative of an unknown function. Such equations arise naturally in many different settings, as it is often easiest to see how the derivative of a function (as opposed to the function itself) relates to other quantities. And if a differential equation involving $f'(x)$ is accompanied by information about the value of f at a particular point, say $f(x_0) = y_0$, then a particular solution of the equation (as opposed to a family of solutions) can be determined; such equations are called *initial value problems*.

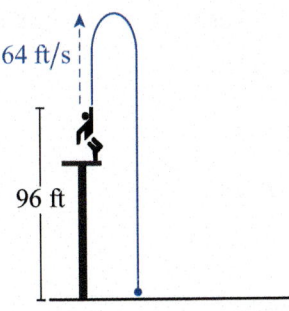

64 ft/s

96 ft

Figure 2

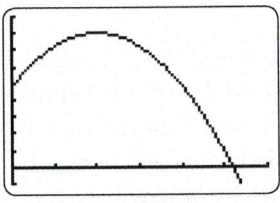

Figure 3

$h(t) = -16t^2 + 64t + 96$

on $[0,6]$ by $[-20,180]$

Example 4

A ball is thrown straight up at 64 ft/s by a person standing on the edge of a platform. At the moment the ball leaves the thrower's hand, it is 96 ft above the ground. On its way back down, the ball misses the platform and hits the ground. When does the ball reach its maximum height? What is that maximum height? When does the ball hit the ground?

Solution

If we let $h(t)$ denote the height (in feet) of the ball at time t (in seconds), then $h'(t)$ represents the ball's vertical velocity and we are given the initial conditions $h'(0) = 64$ and $h(0) = 96$. Recall that the acceleration due to gravity is $-32 \, \text{ft/s}^2$, so the rate of change of h' must be -32; that is, $h''(t) = -32$. Thus, the general antiderivative of h'' is $h'(t) = -32t + C$, and the condition $h'(0) = 64$ tells us that $h'(t) = -32t + 64$.

We know that the ball will transition from a positive (upward) velocity to a negative (downward) velocity as it stops rising and starts to fall, and its maximum height is attained at that time t when $h'(t) = 0$, namely $t = 2$ s. This tells us *when* the ball reaches its maximum height, but not what that maximum height *is*—we still need to determine $h(2)$. To do so, we must perform another antidifferentiation: given $h'(t) = -32t + 64$, we can determine that $h(t) = -16t^2 + 64t + C$, and for the condition $h(0) = 96$ to be true it must be the case that $h(t) = -16t^2 + 64t + 96$. Now, we can see that $h(2) = -16(4) + 64(2) + 96 = 160$ ft.

There is one question remaining, but we have all we need to answer it. The ball hits the ground when $h(t) = 0$, so we solve the equation $-16t^2 + 64t + 96 = 0$ and obtain $t = 2 + \sqrt{10} \approx 5.16$ s (the other solution is negative and doesn't apply).

Figure 3 confirms our findings.

Example 5

Given that $\dfrac{d^2 y}{dx^2} = 9e^{3x} + \sin x + 2$ and that $y'(0) = 3$ and $y(0) = 0$, find y.

Solution

We can determine the general form for $y'(x)$ by antidifferentiating the expression for $y''(x)$ term by term.

$$y'(x) = 3e^{3x} - \cos x + 2x + C$$

And for the condition $y'(0) = 3$ to be satisfied, it must be the case that

$$3 = y'(0) = 3e^0 - \cos 0 + 2(0) + C = 3 - 1 + C,$$

so $C = 1$.

We can now repeat the process with $y'(x) = 3e^{3x} - \cos x + 2x + 1$ to obtain

$$y(x) = e^{3x} - \sin x + x^2 + x + C,$$

and

$$0 = y(0) = e^0 - \sin 0 + (0)^2 + 0 + C = 1 + C$$

implies $C = -1$. So our final answer is

$$y(x) = e^{3x} - \sin x + x^2 + x - 1.$$

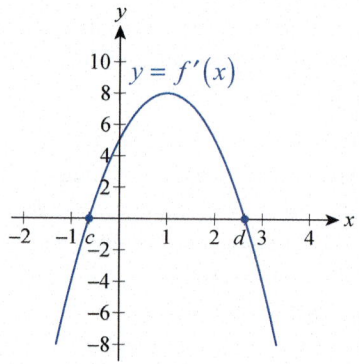

Figure 4

Our last example illustrates that the reasoning we have been using applies even when all we have to work with is a picture.

Example 6

Given the graph of f' in Figure 4 and the knowledge that f passes through $(1, 2)$, sketch the graph of f.

Solution

We begin by observing that f' is positive on the interval (c, d), negative on the intervals $(-\infty, c)$ and (d, ∞), and that $f'(c) = f'(d) = 0$. This tells us that f is increasing on (c, d), decreasing on $(-\infty, c)$ and (d, ∞), and that f has a relative minimum at $x = c$ and a relative maximum at $x = d$. Further, f' has a critical point at 1 and appears to be differentiable there (the graph of f' is nicely smooth), so $f''(1) = 0$; moreover, since f' is increasing to the left of 1 and decreasing to the right of 1, it must be the case that f changes concavity as it passes through the point $(1, 2)$.

Putting all these observations together, the graph of f must be something along the lines of the one in Figure 5, though the actual values of $f(c)$ and $f(d)$ can be nothing more than a rough guess.

The values of f' from the original graph tell us approximately how fast the graph of f rises or falls near a given point—for instance, since $f'(1) = 8$, the "slope" of f at $(1, 2)$ should be 8.

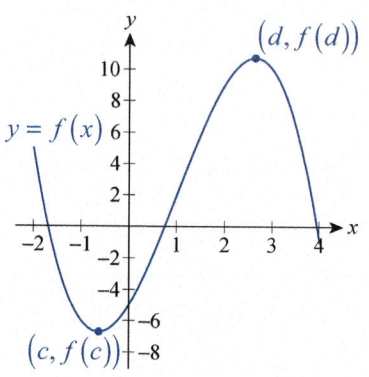

Figure 5

4.7 **Exercises**

1–8 Verify by differentiating that $F(x)$ is an antiderivative of $f(x)$.

1. $f(x) = \dfrac{1}{\sqrt{x}} + \dfrac{1}{x^2}$, $F(x) = 2\sqrt{x} - \dfrac{1}{x}$

2. $f(x) = 2(x-1)(x+5)$, $F(x) = \dfrac{2}{3}x^3 + 4x^2 - 10x$

3. $f(x) = -x(x+2)(x-4)$, $F(x) = -\dfrac{1}{4}x^4 + \dfrac{2}{3}x^3 + 4x^2 + \dfrac{5}{3}$

4. $f(x) = 6\cos 3x$, $F(x) = 2\sin 3x$

5. $f(x) = 5\sec^2(5x+1)$, $F(x) = \tan(5x+1) + 5$ **6.** $f(x) = \dfrac{x^2+1}{\sqrt{x}}$, $F(x) = \dfrac{\sqrt{x}}{5}(10+2x^2)$

7. $f(x) = \dfrac{2x}{x^2+7}$, $F(x) = \ln(x^2+7)$ **8.** $f(x) = \pi^{2x}$, $F(x) = \dfrac{\pi^{2x}}{2\ln\pi}$

9–20 Find an antiderivative of the function.

9. $f(x) = 1$ **10.** $g(x) = 2x+2$ **11.** $h(x) = 4x^3 - x$

12. $u(x) = x^5 + x^3 + \pi$ **13.** $v(x) = \sec^2 x + 3x$ **14.** $k(x) = \dfrac{2}{x}$

15. $f(x) = 5e^x$ **16.** $m(x) = \dfrac{1}{2\sqrt{x}}$ **17.** $u(t) = -\dfrac{4}{t^3}$

18. $v(s) = \dfrac{1}{6s^{2/3}}$ **19.** $w(z) = \dfrac{1}{\sqrt{1-z^2}}$ **20.** $g(s) = \dfrac{1}{1+s^2} + 1 + s^2$

21–32 Find the general antiderivative of $f(x)$; then find the particular antiderivative $F(x)$ that satisfies $F(1) = 1$.

21. $f(x) = 2x - 3$ **22.** $f(x) = 3x^2 + \dfrac{1}{2}$ **23.** $f(x) = \dfrac{1}{\sqrt{x}}$

24. $f(x) = 1$ **25.** $f(x) = 0$ **26.** $f(x) = -\dfrac{1}{x}$

27. $f(x) = x^3 - \dfrac{1}{x^2}$ **28.** $f(x) = \dfrac{-1}{3x^{2/3}}$ **29.** $f(x) = (\ln 10)10^x$

30. $f(x) = \sin x$ **31.** $f(x) = \dfrac{\pi}{4}\sec^2\left(\dfrac{\pi}{4}x\right) + 1$ **32.** $f(x) = \dfrac{2x}{x^2+4}$

33–36 Given the graph of f' and the knowledge that f passes through the point $(3,1)$, sketch a possible graph for f.

33.

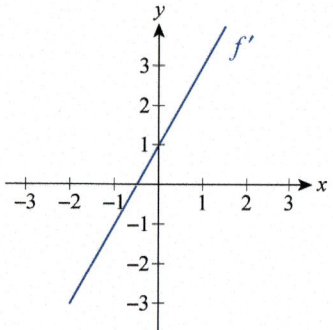

34.

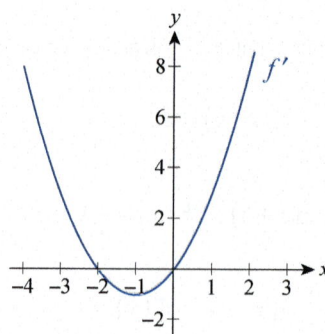

35.

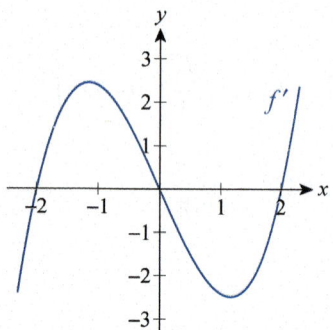

36.

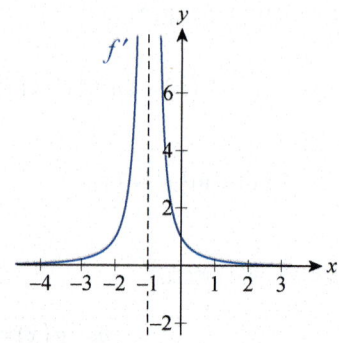

37–60 Find the general antiderivative of the given function, and check your answer by differentiation. (If necessary, rewrite the function before antidifferentiation.)

37. $f(x) = 6x^2 - 4x + 1.5$

38. $g(x) = 5x^3 - \pi x$

39. $h(x) = 3x^5 - 10x^4 + x^2 + 7$

40. $u(x) = -7x^4 + \dfrac{1}{2}x^3 + 6x^2 - 8x + \dfrac{5}{2}$

41. $v(x) = 3(x+6)(2x+1)$

42. $k(x) = -x(x+3)(7x-5)$

43. $h(x) = x^3\sqrt{x}$

44. $m(x) = \dfrac{3}{\sqrt{x}} + 2x\sqrt[3]{x}$

45. $n(x) = \dfrac{x^3 + 7x}{x^2}$

46. $f(t) = \dfrac{t^2 - t}{\sqrt{t} + 1}$

47. $a(y) = \left(\sqrt[3]{y^4} - 1\right)^2$

48. $w(z) = \dfrac{2}{z} + \dfrac{2}{\sqrt{z}}$

49. $g(t) = e^{3t} - 3\sec t \tan t$

50. $s(t) = 2 \cdot 10^{1.5t}$

51. $t(\theta) = \theta + \cos\theta$

52. $c(\theta) = \theta^2 + \csc^2\theta$

53. $v(x) = (\csc x - \cot x)\csc x$

54. $t(x) = -\sec^2 x(\cos^2 x + \sin^2 x)$

55. $w(x) = \dfrac{\cos x}{\cos^2 x - 1}$

56. $u(x) = \dfrac{2\tan 2x}{2\cos^2 x - 1}$

57. $a(x) = \dfrac{5}{1 + 9x^2}$

58. $b(x) = \dfrac{1}{\sqrt{1 - 4x^2}}$

59. $c(x) = \dfrac{4}{|5x|\sqrt{25x^2 - 1}}$

60. $d(x) = \dfrac{-3}{\sqrt{4 - 36x^2}}$

61–76 Find $f(x)$ that satisfies the specified conditions. (When no initial conditions are specified, find the general antiderivative.)

61. $f''(x) = \pi$, $f'(1) = 0$, $f(1) = 0$

62. $f''(x) = 1 - 4x$, $f'(-1) = 1$, $f(-1) = -4$

63. $f'''(x) = 0$, $f''(2) = 2$, $f'(2) = 2$, $f(2) = 2$

64. $f'''(x) = x + 1$, $f''(0) = 1$, $f'(0) = 2$, $f(0) = 3$

65. $f''(x) = \sqrt[3]{x}$, $f'(1) = 0$, $f(1) = \dfrac{1}{7}$

66. $f''(x) = x + \dfrac{1}{\sqrt{x}}$, $f'(4) = 6$, $f(4) = 0$

67. $f'''(x) = \sqrt{x} + 1$, $f''(0) = 1$, $f'(0) = -1$, $f(0) = 7$

68. $f''(x) = \sqrt[3]{x}(x - 3)$, $f'(0) = 0$, $f(0) = 0$

69. $f'(x) = \dfrac{4}{1 + 4x^2}$, $f\left(\dfrac{1}{2}\right) = \pi$

70. $f'(x) = \dfrac{-1}{\sqrt{1 - 3x^2}}$, $f\left(\dfrac{\sqrt{3}}{3}\right) = 0$

71. $f'''(x) = -\cos 2x$, $f''(0) = 1$, $f'(0) = 1$, $f(0) = -1$

72. $f'''(x) = \cos x - \sin x$

73. $f''(x) = 2^{5x}$

74. $f''''(x) = e^x + e$, $f''(0) = 1$, $f'(0) = 2$, $f(0) = 3$

75. $f''(x) = \cos x - e^{2x}$, $f'(0) = -2$, $f(0) = 1$

76. $f'''(x) = \sin 10x + 10x + 10$, $f''(0) = 0$, $f'(0) = 3.5$, $f(0) = -0.5$

77–85 Use -32 ft/s^2 for the acceleration caused by gravity (-9.81 m/s^2 in the metric system). Ignore air resistance. (**Hint:** See Example 4.)

77. A soccer ball is kicked upward from a height of 3 feet with an initial velocity of 48 feet per second. How high will it go?

78. A student drops a pen from a classroom window on the fourth floor of the mathematics building. If the window is 48 ft above ground level, how long is the pen in the air and at what velocity does it hit the ground?

79. A hiker throws a pebble into a canyon that is 350 meters deep, with a downward initial velocity of 10 m/s. For how many seconds is the pebble in the air and what is the velocity of impact?

80. A baseball is thrown upward from a height of 1.5 meters with an initial velocity of 30 meters per second. How high will it go, and for how long is it going to rise?

81.* With what initial velocity do we need to throw a tennis ball vertically upward in order for it to reach the top of a 60 ft campus flagpole?

82. An air rifle shoots a pellet at 1200 feet per second. What is the horizontal range of the rifle, that is, how far from where the pellet is shot will it hit the ground, if we shoot horizontally from a height of 5 feet?

83. A golf ball is hit horizontally at 40 meters per second from the top of a slight hill that is 1.5 meters high. If the terrain around the hill is nearly flat, approximately how far will the golf ball fly?

84. Prove that the position function of an object thrown vertically from an initial height of h_0 feet with an initial velocity of v_0 feet per second is $h(t) = -16t^2 + v_0 t + h_0$.

85. Repeat Exercise 84 using the metric system (meters and seconds) to arrive at the formula $h(t) = -4.905t^2 + v_0 t + h_0$.

86. The acceleration due to gravity on the lunar surface is approximately -5.25 ft/s^2. How high would the soccer ball of Exercise 77 fly on the Moon?

87. Find out what would happen in the situation described in Exercise 83 under lunar conditions. (See Exercise 86 for the acceleration due to gravity on the Moon.)

88. The rate of growth of a rabbit population in a certain state park, where food supply is limited and predators are present, is proportional to $e^{-0.1t}$, where t is time measured in months. If the initial population size is 300 rabbits, which grows to 400 in three months, find the population size in a year. (**Hint:** Let $P(t)$ stand for population size, and use $\dfrac{d}{dt}P(t) = ke^{-0.1t}$.)

89. The rate of growth of a population of a certain virus in a medical experiment is proportional to $\sqrt[3]{t}$, where t is time measured in days. If the initial population size is 1000, which grows to 1500 in a day, find the population size in five days. (See and appropriately modify the hint given in Exercise 88.)

90. * A modern Formula 1 car is able to come to a complete stop from 200 km/h (124.3 mph) using a braking distance of only about 65 meters. Assuming constant deceleration (which is not fully realistic), what multiple of g is this? (**Hint:** 1 m/s = 3.6 km/h.)

91. The Bugatti Veyron, the fastest production grand tourer to date, can go from 0 to 100 km/h in 2.5 seconds. Find its position function when accelerating from a standstill and the distance covered during the first 1.5 seconds. What is the car's acceleration time from 0 to 60 mph? (Use the simplifying assumption that acceleration is constant. Also see the hint provided in Exercise 90.)

92. Jerry the mouse is running towards his hole at a steady speed of 11 ft/s. Still 20 feet from his destination, he is discovered by Tom the cat, who is 2 feet behind Jerry at that moment. If Tom can reach his top speed of 40 ft/s in 3 seconds, will he be able to catch Jerry? (Suppose the locations of Tom, Jerry, and the mousehole remain collinear throughout the pursuit.)

93. * Assume that an airplane needs to reach a liftoff speed of 180 mph and that it can achieve the same on a runway that is 0.8 miles long. Assuming constant acceleration during takeoff, what would this acceleration be?

94. The acceleration function of a particle moving along the x-axis is $a(t) = 3\sqrt{t} - \dfrac{1}{\sqrt{t}}$ units/s^2. If it starts at the origin with an initial velocity of 2 units per second, find the position function of the particle. Where will it be in 5 seconds?

95. Repeat Exercise 94 for the acceleration function $a(t) = (2-t)\sqrt{t}$, if the particle starts from rest at the point $(3, 0)$. Where will it be in 5 seconds, and when will its instantaneous velocity be zero?

96. It follows from our discussions in Section 3.6 as well as the present section that an antiderivative of $-1/\sqrt{1-x^2}$ can be written as $-\sin^{-1}x$. Use the graphs of inverse trigonometric functions provided in Section 3.6 to argue that $\cos^{-1}x$ is also an antiderivative of $-1/\sqrt{1-x^2}$. (It follows that the general antiderivative of $-1/\sqrt{1-(kx)^2}$, $|kx| < 1$ is $(1/k)\cos^{-1}(kx) + C$. See also Exercise 67 of Section 3.6.)

97–103 *True or False?* Determine whether the given statement is true or false. In case of a false statement, explain or provide a counterexample.

97. If $f(x)$ has an antiderivative on an interval I, then it has infinitely many antiderivatives on the same interval.

98. All polynomials have antiderivatives on the entire real line $\mathbb{R}$.

99. It is possible for a function to have an unique antiderivative on an interval I.

100. Whenever F_1 and F_2 are both antiderivatives of f on an open interval, then $F_1 - F_2$ is a constant function.

101. If a function has an antiderivative on the interval $(-a, a)$ for some $a > 0$, then it has exactly one antiderivative whose graph goes through the origin.

102. Every antiderivative of a polynomial function of degree n has degree $n + 1$.

103. If $F(x)$ is an antiderivative of $f(x)$, and $G(x)$ is an antiderivative of $g(x)$ on an interval I, then $F(x) \cdot G(x)$ is an antiderivative of $f(x) \cdot g(x)$ on the same interval.

Chapter 4
Review Exercises

1–4 Sketch by hand the graph of a function f on the specified domain, with the specified properties. (Answers will vary.)

1. Defined on $(0,2)$, absolute minimum at 1, no absolute maximum

2. Defined on $[-2,2]$, absolute maximum occurs twice, no absolute minimum

3. Defined on $(-1,1)$, no absolute or relative extrema

4. Differentiable on $(0,1)$, no critical points, no extrema

5–14 Find all absolute extrema of the function on the indicated domain.

5. $f(x) = x^2 - \dfrac{4}{3}x^3; \quad D = [-1,1]$

6. $f(x) = -x^4 + 8x^3 - 16x^2; \quad D = [-1,5]$

7. $f(x) = \left|x^2 - 2x - 8\right|; \quad D = [-3,5]$

8. $f(x) = (x+2)|x|; \quad D = [-2,2]$

9. $f(x) = \sqrt{x}(1-x); \quad D = [0,2]$

10. $f(x) = \dfrac{x^2+1}{x+1}; \quad D = [0,1]$

11. $f(x) = 3x^5 - 2x^3 + 1; \quad D = \mathbb{R}$

12. $f(x) = \sqrt{1-x^4}; \quad D = (-1,1)$

13. $f(x) = \csc\dfrac{x}{2}; \quad D = (0,2\pi)$

14. $f(x) = x^2 - x[\![x]\!]; \quad D = [1,2]$

15–16 Prove that the equation has exactly one real solution on the given interval.

15. $3x^3 - x^4 = 1$ on $(0,1)$

16. $x \arcsin x = e^{-x}$ on $(0,1)$

17–18 Determine whether Rolle's Theorem applies to the function on the given interval. If so, find all possible values of c as in the conclusion of the theorem. If the theorem does not apply, state the reason.

17. $f(x) = x^3 + x^2 - 8x - 12$ on $[-2,3]$

18. $g(x) = x^4 + 2x^2 - 2$ on $[0,1]$

19–20 Determine whether the Mean Value Theorem applies to the function on the given interval. If so, find all possible values of c as in the conclusion of the theorem. If the theorem does not apply, state the reason.

19. $f(x) = \left|x^4 - 3x\right|$ on $[2,3]$

20. $f(x) = \left|x^4 - 3x\right|$ on $[0,2]$

21. If $\left|f'(x)\right| \le 3$ for all x, prove that $\left|f(10) - f(2)\right| \le 24$.

22. If $g(-5) = -1$ and $g'(x) \le 4$ for all x, what is the largest possible value of $g(1)$?

23. Find the function f that passes through $(0,3)$ and whose derivative is $\cos x + e^x$.

24. An object is moving along the x-axis, starting at $x_0 = -4$ with velocity function $v(t) = 3t^2 - 2t$ ($0 \le t \le 5$). Find the time t when it reaches the origin.

25. A car driving at 70 mph passes a mile marker, and then exactly 48 seconds later, still driving at 70 mph, passes the next mile marker.

   a. Prove that there was at least one instant when the car traveled at 75 mph between the markers.

   b. Prove that there was at least one instant when the car's acceleration was zero.

26–29 Use the first derivative to determine where the function is increasing and decreasing.

26. $f(x) = |2x - 2|$ 27. $g(x) = -x^2 + 6x + 7$

28. $h(x) = x^3 - 6x^2 + 9x + 1$

29. $k(x) = x^4 - 4x^3 - 20x^2 + 96x$

30–33 Determine the intervals of concavity of the given function.

30. $f(x) = \dfrac{x^3}{6} - x^2 - 2$ **31.** $g(x) = \dfrac{e^x}{x}$

32. $h(x) = (x-2)\sqrt[3]{x}$ **33.** $k(x) = \dfrac{x+3}{x^2-1}$

34–35 Use the first and second derivatives to identify the intervals of monotonicity, extrema, intervals of concavity, and inflection points of the given function.

34. $f(x) = 3x^5 - 20x^3$ **35.** $g(x) = \arctan(x^2)$

36–37 The function $p(t)$ gives the position, relative to its starting point, of an object moving along a straight line. Identify the time intervals when the object is moving in the positive versus negative direction, as well as those intervals when it is accelerating or slowing down. Find the times when the object changes direction as well as those when its acceleration is zero.

36. $p(t) = t^3 - 3t^2$, $0 \le t \le 5$

37. $p(t) = 3t^2 - \dfrac{t^4}{2}$, $0 \le t \le 3$

38–49 Check whether l'Hôpital's Rule applies to the given limit. If it does, use it to determine the value of the limit. If it does not, find the limit some other way. (When necessary, apply l'Hôpital's Rule several times.)

38. $\displaystyle\lim_{x\to 0} \dfrac{\sqrt{x+4}-2}{x}$ **39.** $\displaystyle\lim_{x\to 0} \dfrac{x - \tan x}{\sec x - 1}$

40. $\displaystyle\lim_{x\to\infty} \dfrac{e^{-x}}{\ln x}$ **41.** $\displaystyle\lim_{x\to 0^+} \cot x \csc x$

42. $\displaystyle\lim_{x\to 0} (1+4x^2)^{1/x^2}$

43. $\displaystyle\lim_{x\to(1/2)^+} \left(\dfrac{1}{4x-2} - \dfrac{1}{\ln 2x} \right)$

44. $\displaystyle\lim_{x\to 0^+} x^{\sqrt{x}}$ **45.** $\displaystyle\lim_{x\to 0^+} \left(\sqrt{x}\right)^{\ln x}$

46. $\displaystyle\lim_{x\to 0^-} x \cot x$ **47.** $\displaystyle\lim_{x\to 0} \dfrac{\arctan x}{\arctan 2x}$

48. $\displaystyle\lim_{x\to 0^+} \sin x \ln x$ **49.** $\displaystyle\lim_{x\to 0} \left(\dfrac{1}{x} - \csc x \right)$

50. By examining the limits $\displaystyle\lim_{x\to\infty} \dfrac{\ln x}{x^a}$ and $\displaystyle\lim_{x\to\infty} \dfrac{x^a}{b^x}$ using l'Hôpital's Rule, compare the relative growth rates of the functions $y = \ln x$, $y = x^a$, and $y = b^x$ ($a > 0$, $b > 1$). (See Exercise 88 in the Chapter 3 Review.)

51–60 Use the curve-sketching strategy to construct a graph of the function.

51. $f(x) = \dfrac{x^3}{3} - x^2 - 15x$

52. $g(x) = -x^3 + 12x - 16$

53. $h(x) = x^4 - 2x^3$ **54.** $m(x) = 3x^3 - 4x^5$

55. $f(x) = \dfrac{2x}{x^2+1}$ **56.** $f(x) = \dfrac{2x}{x^2-1}$

57. $f(x) = |x|(x-1)$ **58.** $f(x) = x\sqrt{9-x^2}$

59. $f(x) = \dfrac{x^3}{x^2-4}$ **60.** $f(x) = \sin^2 x \cos x$

61–62 Use Newton's method to approximate the given number to five decimal places.

61. $\sqrt[5]{30}$ **62.** $\log 11$

63–64 Use Newton's method to solve the equation on the given interval. Approximate the root to six decimal places.

63. $2x^5 = 1 - x^2$ on $(0,1)$ **64.** $\ln x = \cos x$ on $(0,\infty)$

65. Use Newton's method to approximate to four decimal places the fixed point(s) of $f(x) = 1 - \tan x$ on $(0, \pi/2)$. (See Exercise 62 of Section 4.2.)

66. Find a positive number that is greater than its own cube by the greatest possible amount.

67. Generalize Exercise 66 to the nth power of a number ($n \ge 2$).

68. Find a number a so that for given $a_1, a_2, a_3 \in \mathbb{R}$, the quantity $S_3 = (a-a_1)^2 + (a-a_2)^2 + (a-a_3)^2$ is minimal.

69. Generalize Exercise 68 for n given numbers to minimize the quantity:
$$S_n = (a - a_1)^2 + (a - a_2)^2 + \cdots + (a - a_n)^2.$$

70. A wire of length l is bent into an L shape. Where should we bend in order to minimize the distance between the two endpoints?

71. Find the length l and width w of the rectangle inscribed in the unit circle for which $l^2 w$ is maximal.

72. Find the dimensions of the rectangle whose diagonal is d units and area is maximum.

73. A book page of area 500 cm² is required to have 1 cm margins on the side, while the margins on the top and bottom are to be 2 cm. Find the dimensions of the page that maximize the printed area.

74. Among all isosceles triangles whose legs are l units long, find the base angle that maximizes the area.

75. A vertex of a rectangle is at the origin and the opposite vertex sits in the first quadrant and on the graph of $y = \dfrac{2 - x}{x + 1}$. Find the maximum possible area for such a rectangle.

76. Find the point on the graph of $y = 1 - x^2$ that is closest to the point $(-3, 1)$.

77. Among all isosceles triangles that can be inscribed in the circle of radius R, find the one with maximum area.

78. A vending machine sells 500 bars of a certain type of candy when the price is $1.50. It was discovered that 10 fewer customers will buy the candy bar for each 5¢ increase in price. What is the price that will bring maximum revenue from the sales of this type of candy bar?

79. Maximize the surface area of the can in Example 3 of Section 4.6. Explain your findings.

80. Minimize the cost of producing the can in Example 3 of Section 4.6 if the top and bottom are produced using a material that is 50% more expensive than the material used for the side.

81. Nate needs to reach a restaurant that is 600 ft upstream on the other side of a 150 ft wide river. Find the point where he has to reach the other side in order to make the best time if he can swim at 5 ft/s and walk at 9 ft/s. (Ignore the flow of the river.)

82–89 Find the general antiderivative of the given function and check your answer by differentiation. (If necessary, rewrite the function before antidifferentiation.)

82. $f(x) = 2x^3 - 6x^2 + 3x$

83. $f(x) = 5x^4 - 4.8x^3 + e^2$

84. $f(x) = x(x + 2)(2x - 3)$

85. $f(x) = 0.4x\sqrt{x} - \dfrac{2}{\sqrt{x}}$

86. $f(x) = \dfrac{x^4 - 4x}{x^2}$

87. $f(x) = 2(x + \sec^2 2x)$

88. $f(x) = 6e^{3x}$ **89.** $f(x) = \dfrac{3}{4x^2 + 1}$

90–91 Find $f(x)$ that satisfies the specified conditions.

90. $f''(x) = x, \quad f'(1) = 1, \quad f(1) = 0$

91. $f'''(x) = 2, \quad f''(2) = -1, \quad f'(2) = 2, \quad f(2) = 3$

92. A tennis ball is thrown upward from an initial height of 4 feet with an initial velocity of 56 feet per second. How high will it go and for how long is it rising? (Ignore air resistance.)

93. With what initial velocity do we need to throw a golf ball vertically upward in order for it to rise 100 feet high? (Ignore the initial height and air resistance.)

94. A pebble is shot horizontally using a slingshot at 10 meters per second from the top of a building that is 20 meters high. If the terrain around the building is nearly flat, approximately how far from the building will the pebble hit the ground? (Use the approximation $g \approx 10 \text{ m/s}^2$ and ignore air resistance.)

95–101 *True or False?* Determine whether the given statement is true or false. In case of a false statement, explain or provide a counterexample.

95. A continuous function on a finite interval always attains its maximum and minimum.

96. If $f(x)$ has a relative maximum or minimum at $x = c$, then $f'(c) = 0$.

97. If $f(x)$ has a relative maximum or minimum at $x = c$, then c is a critical point of f.

98. A cubic polynomial has exactly one inflection point.

99. If $f(x)$ is a polynomial, then between two consecutive local extrema there must be an $x = c$ so that $f''(c) = 0$.

100. If $f(x)$ is a polynomial and c is a critical point, then there is a relative maximum or minimum at $x = c$.

101. If $f'''(c) = 0$, then $f'(x)$ has a point of inflection at $x = c$.

Chapter 4
Technology Exercises

102–111. Use a graphing calculator or computer algebra system to verify the answers you obtained for Exercises 51–60.

112–113. Use a graphing calculator or computer algebra system to verify the conclusions of Exercises 15 and 16.

Chapter 4
Project

Consider a function $f(x)$ that is at least twice differentiable. In this project, you will show that the second derivative of $f(x)$ at $x = c$ can be found as the limit of so-called **second-order differences**:

$$f''(c) = \lim_{h \to 0} \frac{f(c+h) - 2f(c) + f(c-h)}{h^2}$$

1. Instead of working with a secant line through the points $(c, f(c))$ and $(c+h, f(c+h))$ like we did when approximating the first derivative, suppose that

$$y = a_1 x^2 + a_2 x + a_3$$

   is the parabola through the following three points on the graph of f: $(c-h, f(c-h))$, $(c, f(c))$, and $(c+h, f(c+h))$. Do you expect to always be able to find coefficients $a_1, a_2, a_3 \in \mathbb{R}$ such that the resulting parabola satisfies the desired conditions? Why or why not? Why would you expect $2a_1$ to be "close" to $f''(c)$ if h is "small"? What will happen to $2a_1$ as $h \to 0$? Write a short paragraph answering the above questions.

2. By substituting the points $(c-h, f(c-h))$, $(c, f(c))$, and $(c+h, f(c+h))$ into $y = a_1 x^2 + a_2 x + a_3$, obtain a system of linear equations in unknowns a_1, a_2, and a_3. Solve the system for the unknown a_1.

3. Use Questions 1 and 2 to argue that $f''(c)$ is the limit of the second-order differences:

$$f''(c) = \lim_{h \to 0} \frac{f(c+h) - 2f(c) + f(c-h)}{h^2}$$

4. Use l'Hôpital's Rule to verify the result you found in Question 3.

Chapter 5
Integration

Introduction

Up to this point, we have focused on the concept of the derivative of a function and on developing methods to find and use derivatives. We have seen that a differentiable function f gives rise to a potentially endless list of successive derivatives f', f'', and so on. In particular, we closed Chapter 4 with the question: *If we have perfect knowledge of the derivative of a function, can we use that to completely determine the function itself?*

One way to approach the subject of this chapter is to literally reverse the differentiation process. That is, given a function f, is it possible to find a function F such that $F' = f$? As we learned in Section 4.7, if such a function F exists, we call it an antiderivative of f. But another approach begins with questions that are seemingly unrelated to antidifferentiation, such as *How do we calculate the area bounded between an axis and the graph of a function?* and *If we know an object's instantaneous velocity and its starting point, can we determine its location at any later moment in time?* We call the process of answering such questions *integration*, and the genius of Sir Isaac Newton (1643–1727) and Gottfried Wilhelm Leibniz (1646–1716) lay in their ability to fully grasp the relationship between integration and antidifferentiation.

The *Fundamental Theorem of Calculus* is the name we now give to this relationship. As is often the case with deep fundamental truths, the theorem required a lengthy period of fermentation and the contributions of many others before Newton and Leibniz articulated it in the form we use today. The mathematicians James Gregory (1638–1675) and Newton's teacher Isaac Barrow (1630–1677) were two such forerunners, but the philosophical basis of the Fundamental Theorem of Calculus appeared as early as the 3rd century BC, when Archimedes used the "method of exhaustion" to find the area under a parabola and the volumes of certain objects.

Informally, the underlying philosophy might be stated in this manner:

The sum of all the infinitesimal changes of a function over a region is equal to the net change in the function.

As we will see by the time we finish Chapter 15 of this text, the underlying philosophy takes many specialized forms and is used to solve problems of astonishing diversity, ranging from the areas under parabolas and volumes of objects to problems in electrostatics, gravitational mechanics, and fluid flow. The insights of early Greek mathematicians from millennia ago, expanded and unified by mathematicians of the 17th century, continue to be principal means of discovery and innovation today.

Isaac Newton
(1643–1727)

Gottfried Leibniz
(1646–1716)

5.1 **Area, Distance, and Riemann Sums**

We began our development of the derivative by discussing two broad problem types: the velocity problem and the tangent problem. In a similar fashion, we will develop the central topic of this chapter, the *integral*, by analyzing area and distance problems. By the end of the chapter, we will have sketched the outlines of an area of math called *integral calculus*, complementing the *differential calculus* we have studied to this point. As we will see in Section 5.3, the Fundamental Theorem of Calculus serves as the linchpin connecting these two areas.

Topic One
The Area Problem

As with so many concepts in mathematics, that of *area* possesses both elementary features and features with surprising depth—it is the discovery of such hidden depth that often lures people into further study of math. You are familiar with formulas for the areas of simple plane figures such as rectangles and triangles, and by extension it is easy to see how, in principle, the area of any region bounded by a polygon could be determined. For instance, the polygonal region in Figure 1 can be decomposed into a finite number of triangular regions in any number of ways, so we can (again, in principle) extend our knowledge of triangular area and *define* the area of the polygon to be the sum of the areas of the triangles. Note the subtle shift in perspective: for a complicated enough polygonal region, we are more likely to specify that its area is the sum of areas that we know how to calculate than we are to look for a general formula such as, for example, $(\text{base})(\text{height})/2$. If we make such definitions wisely, our extended definition of area leads to no contradictions—that is, the area in Figure 1 will be the same no matter how we chop it up into triangles.

We can take a similar approach to develop reasonable definitions of area for many other figures, but for figures not bounded by straight lines we are likely to have to include a limiting step. Just as the limit process allowed us to make the transition from average velocity to instantaneous velocity and to extend the notion of tangency, limits will allow us to refine our definition of *area* and further understand the relationship between position and velocity. To see how, we begin with a simple problem: find the area of the region bounded by the x-axis and the graph of $f(x) = x^2$ between $x = 0$ and $x = 1$.

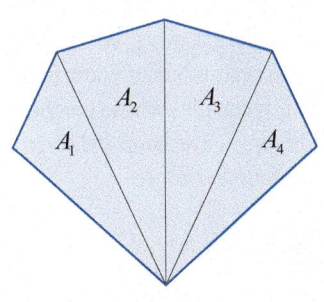

Figure 1

$$A_{\text{Polygon}} = A_1 + A_2 + A_3 + A_4$$

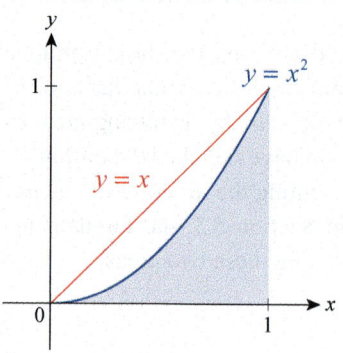

Figure 2

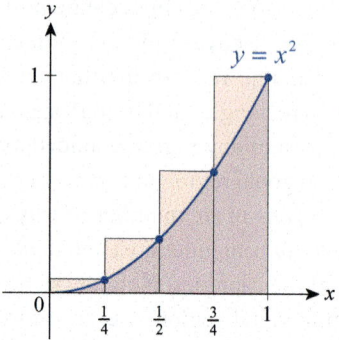

Figure 3

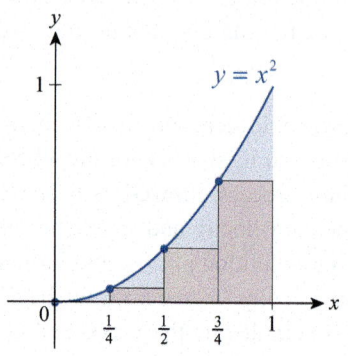

Figure 4

Example 1

Find the area under the graph of $f(x) = x^2$ and above the x-axis on the interval $[0,1]$.

Solution

At this point, we have nothing to work with other than our existing knowledge of area. We can certainly make some rough statements concerning the requested area. For instance, the area under the parabola on the interval $[0,1]$ is less than half a square unit, which is the area of the triangular region bounded by the x-axis and the line $y = x$. But we can also estimate the area under the curve by approximating it with regions whose areas we *do* know.

Rectangles are easy to work with, and we can approximate the region under the parabola with a collection of rectangles in many different ways. We could, for example, divide the interval $[0,1]$ into four subintervals of equal width and overestimate the desired area by covering it with four rectangles whose heights are $f\left(\frac{1}{4}\right)$, $f\left(\frac{1}{2}\right)$, $f\left(\frac{3}{4}\right)$, and $f(1)$, as shown in Figure 3. The combined area of these four rectangles, denoted O_4 for "Over," is as follows.

$$O_4 = \frac{1}{4} \cdot f\left(\frac{1}{4}\right) + \frac{1}{4} \cdot f\left(\frac{1}{2}\right) + \frac{1}{4} \cdot f\left(\frac{3}{4}\right) + \frac{1}{4} \cdot f(1)$$

$$= \frac{1}{4}\left(\frac{1}{16} + \frac{1}{4} + \frac{9}{16} + 1\right) = \frac{15}{32} = 0.46875$$

We could also underestimate the area by adding up the areas of the four rectangles whose heights are $f(0)$, $f\left(\frac{1}{4}\right)$, $f\left(\frac{1}{2}\right)$, and $f\left(\frac{3}{4}\right)$ (see Figure 4). This would give us the approximation U_4 (U for "Under").

$$U_4 = \frac{1}{4} \cdot f(0) + \frac{1}{4} \cdot f\left(\frac{1}{4}\right) + \frac{1}{4} \cdot f\left(\frac{1}{2}\right) + \frac{1}{4} \cdot f\left(\frac{3}{4}\right)$$

$$= \frac{1}{4}\left(0 + \frac{1}{16} + \frac{1}{4} + \frac{9}{16}\right) = \frac{7}{32} = 0.21875$$

Knowing that the area under the parabola is actually something between these two approximations, we might estimate the area with their average: $(O_4 + U_4)/2 = 0.34375$.

To obtain a better approximation of the desired area, we can increase the number of rectangles. Graphs of the results, such as those from Figure 5 for $n = 10$, certainly seem to indicate that the errors in the approximations get smaller as n increases, but we can prove that this is indeed the case by noting the differences between O_n and U_n. The following table contains O_n, U_n, their difference, and their average for a few values of n.

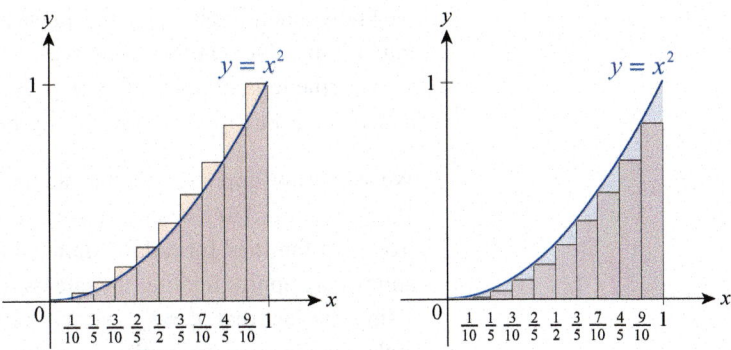

Figure 5 Approximating the Area with 10 Rectangles

n	O_n	U_n	$O_n - U_n$	$(O_n + U_n)/2$
10	0.385	0.285	0.1	0.335
100	0.33835	0.32835	0.01	0.33335
1000	0.33383	0.33283	0.001	0.33333

Given the results in the table, we would be justified in guessing that the area under $f(x) = x^2$ over the interval $[0,1]$ is $\frac{1}{3}$; we already know this guess is correct to several decimal places. We will soon return to this problem and prove that this guess is indeed correct.

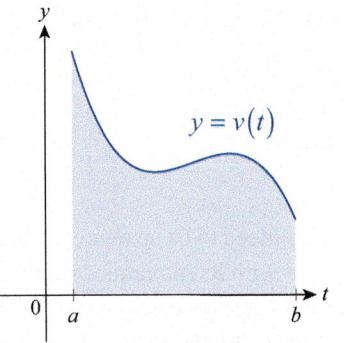

Figure 6 Graph of v

Topic Two
The Distance Problem

Our second motivating example is, in a sense that will soon be made precise, the opposite of the velocity problem of Chapter 2. Suppose that an object is moving along a straight line and that $v(t)$ tells us the velocity of the object at time t. Can we use this knowledge of its velocity to tell us the position of the object?

We will answer this question with a generic velocity function $v(t)$ defined on a time interval $[a,b]$ so that our solution can be applied to, say, a thrown ball whose vertical velocity is known, a car traveling along a straight road, or an atomic particle whose velocity is under the influence of an electromagnetic field. Suppose the graph in Figure 6 is a depiction of our generic $v(t)$. Since the graph is not constant, we cannot simply use the fact that distance is the product of rate (velocity) and time to determine how far the object has traveled over the interval $[a,b]$. But as with the area problem, we can develop better and better *approximations* of this distance by dividing $[a,b]$ into subintervals and assuming v to be constant on each one; as the subintervals decrease in width and the number of subintervals increases, our approximations will approach the true distance.

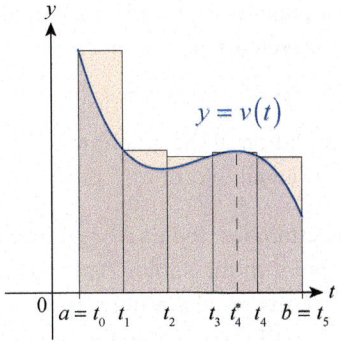

Figure 7

In Figure 7, the interval $[a,b]$ has been divided into five subintervals of equal width, labeled $[t_0, t_1]$ through $[t_4, t_5]$, and the height of each rectangle has been set to the maximum value of v on each one. If we know that v is continuous on $[a,b]$ (as it certainly appears to be), then by the Extreme Value Theorem we know that there is a point t_i^* in each subinterval $[t_{i-1}, t_i]$ for which $v(t_i^*)$ is the maximum value on that subinterval. For the five subintervals in Figure 7, note that $t_1^* = t_0$ (the left endpoint of $[t_0, t_1]$), $t_2^* = t_1$, $t_3^* = t_3$ (the right endpoint of $[t_2, t_3]$), and $t_5^* = t_4$, but t_4^* is a point strictly between t_3 and t_4.

We can now approximate the actual distance traveled over the time interval $[t_0, t_1]$ by $v(t_0)\Delta t$, where $\Delta t = t_1 - t_0 = (b-a)/5$. The product $v(t_0)\Delta t$ comes from the familiar formula *distance = rate × time*, since the length of time is Δt and we are approximating the rate over the interval with the constant value $v(t_0)$. Using the fact that $t_1^* = t_0$, we will rewrite $v(t_0)\Delta t$ as $v(t_1^*)\Delta t$, and proceed to make similar approximations for the distance traveled over the remaining four subintervals. Thus, the sum of the five expressions and our approximation for the distance traveled over $[a, b]$ is:

$$v(t_1^*)\Delta t + v(t_2^*)\Delta t + v(t_3^*)\Delta t + v(t_4^*)\Delta t + v(t_5^*)\Delta t$$

We encounter sums of this form so frequently that it is convenient to make use of **sigma notation** to denote them. Using this notation, the above sum is written as follows.

$$\sum_{i=1}^{5} v(t_i^*)\Delta t$$

The symbol Σ is the capital Greek letter sigma (standing for "sum"), and the letter i is the **index of summation**. In general, an expression such as $\sum_{i=1}^{n} a_i$ stands for the sum of all the terms a_1 through a_n; that is,

End with $i = n$

Begin with $i = 1$

$$\sum_{i=1}^{n} a_i = a_1 + a_2 + \cdots + a_n.$$

Our approximation of the distance at this point is guaranteed to be an overestimate, since we have approximated $v(t)$ on each subinterval with its maximum value on that subinterval. As in Example 1, however, we could just as easily arrive at an underestimate by approximating v with its minimum value on each subinterval. With five equal subdivisions, the heights of the corresponding rectangles using minimum values are as shown in Figure 8. In keeping with our previous usage, if we define t_i^* to be a point in $[t_{i-1}, t_i]$ where v attains its *minimum* value, then our underestimate approximation to the distance traveled is again given by the expression $\sum_{i=1}^{5} v(t_i^*)\Delta t$. Note that by this definition, it is clear that $t_1^* = t_1$, $t_3^* = t_2$, and $t_5^* = t_5$, but we may be uncertain (given the resolution of the graph) if t_2^* is t_2 or a little bit less than t_2, and it's not obvious if t_4^* is equal to t_3 or t_4.

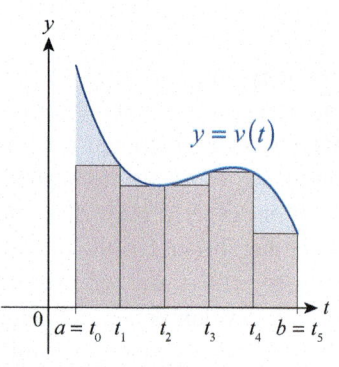

Figure 8

Fortunately, we don't need to worry about such uncertainties. In this context, each t_i^* is called a **sample point** in $[t_{i-1}, t_i]$, and *any* point in the subinterval can be chosen as the sample point. Remember, our last step is to take a limit as n, the number of subintervals, increases to infinity; as we will soon see, the limit does not depend on the exact choice of sample points.

Example 2

A stone is released from a high cliff and falls for many seconds before hitting the canyon floor. Assuming air resistance is negligible, how far does the stone fall in the interval $t = 1$ second to $t = 3$ seconds?

Solution

Recall that the acceleration due to gravity is $-32 \, \text{ft/s}^2$, meaning that with every passing second a falling object changes in velocity by another $-32 \, \text{ft/s}$ (recall also that the negative signs reflect the fact that objects fall downward). So if we let t denote the time in seconds after release, the stone falls with velocity $v(t) = -32t$. Since we are interested only in determining the difference between the stone's position at time $t = 1$ and time $t = 3$, we will work with the speed of the stone $s(t) = |v(t)| = 32t$.

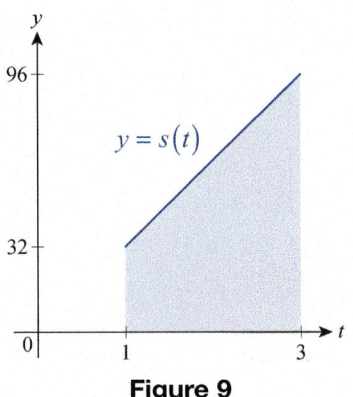

Figure 9

The graph of s is linear, and the distance traveled by the stone between $t = 1$ and $t = 3$ corresponds to the shaded area between the graph and the t-axis. (See Figure 9.) You will show in Exercise 75 that the area of this polygonal region is 128 square units, so the distance traveled between these two points in time is 128 feet.

Although we now know the answer to our question, it is worthwhile to obtain it again through the approximation process we have developed—this will give us a chance to practice the techniques and notation that we will use for the next several sections. To this end, let O_n denote the overestimate that results by dividing the interval $[1,3]$ into n subintervals and defining t_i^* to be the right endpoint of the subinterval $[t_{i-1}, t_i]$; note that by this definition, $s(t_i^*)$ is the maximum value of s on the subinterval $[t_{i-1}, t_i]$, since s is an increasing function.

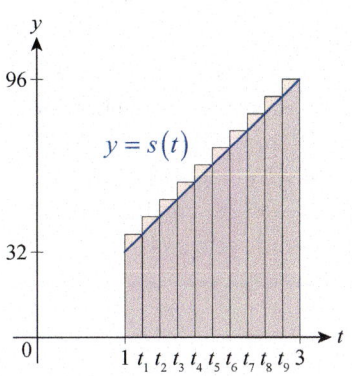

Figure 10 Depiction of O_{10}

Using sigma notation, $O_n = \sum_{i=1}^{n} s(t_i^*) \Delta t$. While we could pick a specific value for n, calculate the n terms $s(t_i^*) \Delta t$, and add them up to arrive at an approximation, it is possible to achieve much more. We will do so by keeping n unspecified (other than knowing it represents a positive integer) and replacing each of the factors in the above generic sum with a formula based on our particular function s and interval $[1,3]$. Note that

$$\Delta t = \frac{b-a}{n} = \frac{3-1}{n} = \frac{2}{n},$$ The width of each subinterval is $b - a$ divided by n.

$$t_i^* = a + (\Delta t)i = 1 + \left(\frac{2}{n}\right)i = 1 + \frac{2i}{n},$$ Add Δt to the left endpoint a a total of i times.

and

$$s(t_i^*) = 32t_i^* = 32\left(1 + \frac{2i}{n}\right).$$ Evaluate $s(t)$ at t_i^*.

Hence,

$$O_n = \sum_{i=1}^{n} s(t_i^*)\Delta t = \sum_{i=1}^{n} 32\left(1+\frac{2i}{n}\right)\left(\frac{2}{n}\right) = \sum_{i=1}^{n} 32\left(\frac{2}{n}+\frac{4i}{n^2}\right).$$

Remember that this notation signifies nothing more than a sum and that all the usual properties of arithmetic hold. So we can reorder the terms as we like, factor out constants, and so forth:

$$\sum_{i=1}^{n} 32\left(\frac{2}{n}+\frac{4i}{n^2}\right) = 32\left[\frac{2}{n}+\frac{4(1)}{n^2}\right]+32\left[\frac{2}{n}+\frac{4(2)}{n^2}\right]+\cdots+32\left[\frac{2}{n}+\frac{4(n)}{n^2}\right]$$

$$= 32\left[\frac{2}{n}(n)+\frac{4}{n^2}(1+2+\cdots+n)\right]$$

$$= 32\left(2+\frac{4}{n^2}\sum_{i=1}^{n} i\right) = 64+\frac{128}{n^2}\sum_{i=1}^{n} i$$

Sums such as $\sum_{i=1}^{n} i$ arise frequently, and it can be shown that $\sum_{i=1}^{n} i = n(n+1)/2$ (see Exercise 79). A catalog of other similar formulas will be given shortly. Using this fact in the expression above, we have the following.

$$O_n = 64+\frac{128}{n^2}\sum_{i=1}^{n} i = 64+\frac{128}{n^2}\left[\frac{n(n+1)}{2}\right] = 64+64\left(\frac{n+1}{n}\right)$$

In other words, we now have a simple expression giving us the overestimate resulting from approximating with n rectangles—there is no need to calculate this number repeatedly for different values of n. More importantly, we can study this expression to see what happens as $n\to\infty$. Since $[(n+1)/n]\to 1$ as $n\to\infty$, we see that $O_n\to 128$, confirming that the distance the stone falls between the first and third seconds is 128 feet.

Topic Three
Riemann Sums and Summation Formulas

We have now seen sums of the form $\sum_{i=1}^{n} f(x_i^*)\Delta x$ arise several times, and we will study them further as we develop the notion of the *integral*. Such an expression is an example of a **Riemann sum** of the function f over an interval $[a,b]$, named in honor of the German mathematician Bernhard Riemann (1826–1866). It is very helpful to keep in mind the geometric meaning of the individual pieces of a Riemann sum. Specifically, remember that each x_i^* represents a sample point in the ith subinterval, that $\Delta x = (b-a)/n$ is the width of each subinterval, and that $f(x_i^*)$ is nothing more than the value of f at the ith sample point.

In working with Riemann sums, the following facts about finite sums will be very useful. The first three are nothing more than restatements, using sigma notation, of elementary arithmetic facts, while the last three are formulas that you will prove in Exercises 79–83.

Summation Facts and Formulas

Constant Rule for Summations	$\sum\limits_{i=1}^{n} c = nc,$ for any constant c
Constant Multiple Rule for Summations	$\sum\limits_{i=1}^{n} ca_i = c\sum\limits_{i=1}^{n} a_i,$ for any constant c
Sum/Difference Rule for Summations	$\sum\limits_{i=1}^{n} (a_i \pm b_i) = \sum\limits_{i=1}^{n} a_i \pm \sum\limits_{i=1}^{n} b_i$
Sum of the First n Positive Integers	$\sum\limits_{i=1}^{n} i = \dfrac{n(n+1)}{2}$
Sum of the First n Squares	$\sum\limits_{i=1}^{n} i^2 = \dfrac{n(n+1)(2n+1)}{6}$
Sum of the First n Cubes	$\sum\limits_{i=1}^{n} i^3 = \left[\dfrac{n(n+1)}{2}\right]^2$

Example 3

Simplify the following sums.

a. $\displaystyle\sum_{i=1}^{9} (3i+4)$ **b.** $\displaystyle\sum_{j=1}^{n} \frac{-2}{n}$ **c.** $\displaystyle\sum_{i=1}^{n} (i-1)^2$

Solution

a. $\displaystyle\sum_{i=1}^{9} (3i+4) = 3\sum_{i=1}^{9} i + \sum_{i=1}^{9} 4 = 3\left(\frac{9 \cdot 10}{2}\right) + 9 \cdot 4 = 135 + 36 = 171$

b. $\displaystyle\sum_{j=1}^{n} \frac{-2}{n} = \left(\frac{-2}{n}\right)n = -2$

c. $\displaystyle\sum_{i=1}^{n} (i-1)^2 = \sum_{i=1}^{n} (i^2 - 2i + 1) = \sum_{i=1}^{n} i^2 - 2\sum_{i=1}^{n} i + \sum_{i=1}^{n} 1$

$\displaystyle = \frac{n(n+1)(2n+1)}{6} - 2 \cdot \frac{n(n+1)}{2} + n$

$\displaystyle = \frac{2n^3 + 3n^2 + n - 6n^2 - 6n + 6n}{6}$

$\displaystyle = \frac{2n^3 - 3n^2 + n}{6}$

We will conclude this section by returning to our question in Example 1, this time proving that the area under the graph of $f(x) = x^2$ over the interval $[0,1]$ is indeed exactly $\frac{1}{3}$.

Example 4

Find the area under the graph of $f(x) = x^2$ and above the x-axis on the interval $[0,1]$.

Solution

For variety, we will construct an expression for U_n (the underestimate based on n subintervals) and evaluate $\lim_{n \to \infty} U_n$. In Exercise 77 you will show that the same answer is obtained by evaluating $\lim_{n \to \infty} O_n$.

If we divide $[0,1]$ into n subintervals of equal width, each one has width $\Delta x = 1/n$. The minimum value of f on $[x_{i-1}, x_i]$ occurs at $x_i^* = x_{i-1}$, that is, at the left endpoint of each subinterval. So $x_1^* = 0$, $x_2^* = 1/n$, $x_3^* = 2/n$, and in general $x_i^* = (i-1)/n$. So we have the following:

$$U_n = \sum_{i=1}^{n} f(x_i^*)\Delta x = \sum_{i=1}^{n} \left(\frac{i-1}{n}\right)^2 \left(\frac{1}{n}\right) = \frac{1}{n^3}\sum_{i=1}^{n}(i-1)^2$$

We have already simplified the sum $\displaystyle\sum_{i=1}^{n}(i-1)^2$ in part c. of Example 3.

$$U_n = \frac{1}{n^3}\left(\frac{2n^3 - 3n^2 + n}{6}\right) = \frac{2n^3 - 3n^2 + n}{6n^3}$$

Hence, $\displaystyle\lim_{n \to \infty} U_n = \lim_{n \to \infty} \frac{2n^3 - 3n^2 + n}{6n^3} = \frac{2}{6} = \frac{1}{3}$.

5.1 Exercises

1–2 Use $(O_4 + U_4)/2$ to estimate the area under the graph and above the x-axis on the interval $[0,8]$.

1. $f(x) = \left(\dfrac{x}{4}\right)^3$

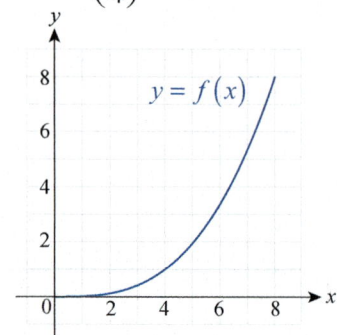

2. $g(x) = 3\sqrt[3]{x}$

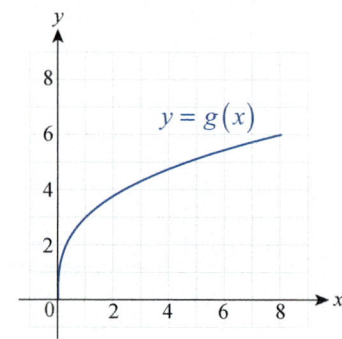

3–4. Repeat Exercises 1–2 using eight rectangles.

5. The figure below shows the upward velocity (in feet per second) of a model rocket during its rise. Use the method of Example 1 to estimate how high the rocket rose by calculating O_6. Is your estimate an over- or underestimate?

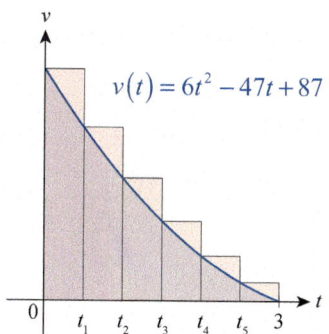

$$v(t) = 6t^2 - 47t + 87$$

6. The velocity of an object undergoing simple harmonic motion is given by the graph below (time is measured in seconds, distance in feet). Using subintervals of width $\frac{1}{4}$,

   a. give an overestimate for the total distance covered from $t = 1$ s to $t = 6$ s;

   b. estimate the total displacement from $t = 1$ s to $t = 6$ s.

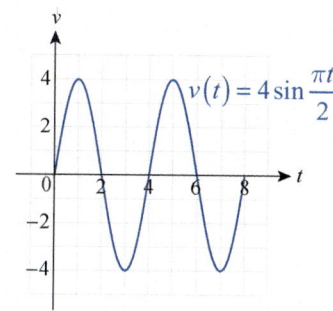

$$v(t) = 4\sin\frac{\pi t}{2}$$

7. The given table contains the velocity data recorded by an automotive testing device during an acceleration test.

   a. Use 12 subintervals to give over- and underestimates of the distance covered by the car during the acceleration run (i.e., find O_{12} and U_{12}).

   b. Approximate the above distance using 6 subintervals of equal width and choosing the midpoint of each as the sample point (we shall call the resulting quantity M_6).

   c. Compare M_6 with $(O_{12} + U_{12})/2$. Which one is greater? Explain why this is the case.

Time (s)	0.5	1	1.5	2	2.5	3	3.5	4	4.5	5	5.5	6
v (m/s)	3	6.6	9.8	13	16.1	19.1	21.6	23.8	25.8	27.6	28.5	29.1

8. In order to estimate the length of the runway, a passenger on an airplane jotted down some velocity data during takeoff from the on-board entertainment screen. From the resulting table given below, calculate $(O_8 + U_8)/2$ to find his estimate.

Time (s)	6	12	18	24	30	36	42	48
v (mph)	30	79	115	150	180	204	223	230

9–14 Use four rectangles to estimate the area between the graph of the given function and the x-axis on the given interval. Construct three estimates for each function: the first using the left endpoints of the subintervals as the sample points, the second using the right endpoints of the subintervals, and the third using the midpoints of the subintervals. Can you tell which are guaranteed to be underestimates or overestimates? (**Hint:** Consider the increasing/decreasing and concavity features of the graph. It is helpful to make a sketch.)

9. $f(x) = \sqrt{x}$ on $[0,4]$

10. $f(x) = \dfrac{x^3}{16}$ on $[0,4]$

11. $f(x) = \dfrac{1}{x}$ on $[1,5]$

12. $f(x) = \sqrt{4 - x^2}$ on $[-2,2]$

13. $f(x) = \cos\dfrac{x}{2}$ on $[0,\pi]$

14. $f(x) = e^{2-x}$ on $[0,2]$

15–24 Write the given sum using sigma notation.

15. $3+6+9+\cdots+99$

16. $1+2+9+28+\cdots+\left(25^3+1\right)$

17. $1+\dfrac{1}{4}+\dfrac{1}{9}+\dfrac{1}{16}+\cdots+\dfrac{1}{10,000}$

18. $1-\dfrac{1}{2}+\dfrac{1}{3}-\dfrac{1}{4}+\cdots-\dfrac{1}{50}$

19. $a_{-3}+a_{-1}+a_{1}+a_{3}+a_{5}+\cdots+a_{77}$

20. $b_0+b_3+b_6+b_9+b_{12}+\cdots+b_{297}$

21. $f\left(\dfrac{3}{n}\right)+f\left(\dfrac{6}{n}\right)+f\left(\dfrac{9}{n}\right)+\cdots+f\left(\dfrac{(n-1)3}{n}\right)+f(3)$

22. $g\left(c_0\right)+g\left(c_5\right)+g\left(c_{10}\right)+g\left(c_{15}\right)+\cdots+g\left(c_{650}\right)$

23. $f\left(x_0^*\right)\Delta x+f\left(x_1^*\right)\Delta x+f\left(x_2^*\right)\Delta x+\cdots+f\left(x_n^*\right)\Delta x$

24. $s\left(t_1^*\right)\Delta t+s\left(t_2^*\right)\Delta t+s\left(t_3^*\right)\Delta t+\cdots+s\left(t_{n-1}^*\right)\Delta t$

25–30 Assuming that $\displaystyle\sum_{i=0}^{n}a_i=36$ and $\displaystyle\sum_{i=0}^{n}b_i=100,$ find the given sum.

25. $\displaystyle\sum_{i=0}^{n}\left(b_i-a_i\right)$

26. $\displaystyle\sum_{i=0}^{n}\left(2a_i+3b_i\right)$

27. $\displaystyle\sum_{i=0}^{n}\left(5b_i+1\right)$

28. $\displaystyle\sum_{i=0}^{n}\left(\dfrac{a_i}{6}-\dfrac{b_i}{2}\right)$

29. $\displaystyle\sum_{j=0}^{n}\left(\dfrac{4a_j}{3}-\dfrac{b_j}{4}+2\right)$

30. $\displaystyle\sum_{k=1}^{n}\left(\dfrac{4}{3}-2a_k+\dfrac{b_k}{2}\right)$

31–42 Find the value of each sum. Use the summation formulas when possible.

31. $\displaystyle\sum_{i=2}^{4}\dfrac{1}{i-1}$

32. $\displaystyle\sum_{j=-2}^{2}\sqrt{j+2}$

33. $\displaystyle\sum_{i=1}^{10}\left(5i-2\right)$

34. $\displaystyle\sum_{i=1}^{n}\left(1-3i\right)$

35. $\displaystyle\sum_{j=1}^{n}\dfrac{4j+5n}{2}$

36. $\displaystyle\sum_{k=1}^{n}\dfrac{6k^2+2k}{3}$

37. $\displaystyle\sum_{j=1}^{30}\left(2j^2-4j+1\right)$

38. $\displaystyle\sum_{i=1}^{100}\left(2i-1\right)\left(3-i\right)$

39. $\displaystyle\sum_{j=1}^{n}\left(3j+1\right)^2$

40. $\displaystyle\sum_{i=1}^{n}\left(i^3-2i^2+\dfrac{1}{n}\right)$

41. $\displaystyle\sum_{j=1}^{n}2j^2\left(j-2\right)$

42. $\displaystyle\sum_{k=0}^{n}k\left(k+1\right)\left(k+2\right)$

43–48 Write out the first few terms as well as the last few terms of the sum. Find a way to simplify and use your observation to evaluate the sum. (Sums of this type are called *collapsing sums*.)

43. $\displaystyle\sum_{i=1}^{10}\left(\dfrac{1}{i}-\dfrac{1}{i+1}\right)$

44. $\displaystyle\sum_{k=3}^{n}\left[\dfrac{1}{k^3}-\dfrac{1}{(k+1)^3}\right]$

45. $\displaystyle\sum_{j=1}^{n}\left(\sqrt{j}-\sqrt{j+1}\right)$

46. $\displaystyle\sum_{k=1}^{n+1}\ln\dfrac{k}{k+1}$

47. $\displaystyle\sum_{j=2}^{n+3}\left(e^j-e^{j+1}\right)$

48. $\displaystyle\sum_{k=1}^{2n+1}\left[\sin k\pi-\sin(k+1)\pi\right]$

49–52 A *geometric sum* (or *geometric progression*) is a sum of the following form.

$$a + ar + ar^2 + \cdots + ar^n = \sum_{i=0}^{n} ar^i, \quad r \ne 1$$

(Notice that each term is a constant multiple of the preceding term; this constant is called the *common ratio* and is denoted by *r*.)

Use the formula $\displaystyle\sum_{i=0}^{n} ar^i = a\frac{1-r^{n+1}}{1-r}$ to evaluate the sum.

49. $\displaystyle\sum_{i=0}^{10} 3^i$

50. $\displaystyle\sum_{j=0}^{8} 5\left(\frac{1}{2}\right)^j$

51. $\displaystyle\sum_{k=0}^{99} (-1)^k \left(\frac{2}{3}\right)^k$

52. $\displaystyle\sum_{n=0}^{1000} 4.9(-3.9)^n$

53. Prove the formula for the sum of the first $n + 1$ terms of a geometric progression given in the directions preceding Exercises 49–52. (**Hint:** Let S denote the sum, recognize $S - rs$ as a collapsing sum, evaluate, and solve for S.)

54–57 Sometimes, sums become easier to manage (the general term becomes simpler) after an appropriate shift in the index. For example, $\displaystyle\sum_{i=1}^{n+1}(i-1)^2$ can be rewritten as $\displaystyle\sum_{i=0}^{n} i^2$. Perform an appropriate shift in the indexing of the given sum to simplify its general term.

54. $\displaystyle\sum_{i=5}^{25} 2(i-4)^3$

55. $\displaystyle\sum_{j=3}^{n} \frac{1}{j-2}$

56. $\displaystyle\sum_{k=0}^{n} n(k-3)^2$

57. $\displaystyle\sum_{l=4}^{20} \cos\big((2l+2)\pi\big)$

58–63 Follow the lead of Examples 2 and 4 in using the limit process to find the area under the graph of $f(x)$ and above the x-axis on the given interval. (In Exercises 58 and 59, use a formula from geometry to check your answer.)

58. $f(x) = \dfrac{1}{2}x + 1$ on $[0,4]$

59. $f(x) = 5 - x$ on $[1,3]$

60. $f(x) = x^2$ on $[1,2]$

61. $f(x) = x - x^3$ on $[0,1]$

62. $f(x) = x^2 + 3x$ on $[-2,2]$

63. $f(x) = (1-x^2)(1+x)$ on $[0,1]$

64–67 Identify the region whose area is the given limit. Do not evaluate the limit. (**Hint:** For guidance, see Example 4 and Exercises 58–63.)

64. $\displaystyle\lim_{n\to\infty} \frac{1}{n}\sum_{i=1}^{n}\left(1-\frac{i}{n}\right)^2$

65. $\displaystyle\lim_{n\to\infty} \frac{3}{n}\sum_{i=1}^{n}\left[\left(\frac{3i}{n}\right)^3 + \frac{6i}{n}\right]$

66. $\displaystyle\lim_{n\to\infty} \frac{2}{n}\sum_{i=1}^{n-1}\sqrt{1+\frac{2i}{n}}$

67. $\displaystyle\lim_{n\to\infty} \frac{\pi}{2n}\sum_{i=1}^{n-1}\sin\frac{\pi i}{2n}$

68. A fruit vendor stacks apples in a rectangular, pyramid-like pile. If the foundational layer consists of 8 rows of 10 apples, and the top layer is a single row of apples, find how many apples are in the stack. Generalize to the case of an $m \times n$ bottom layer of fruit.

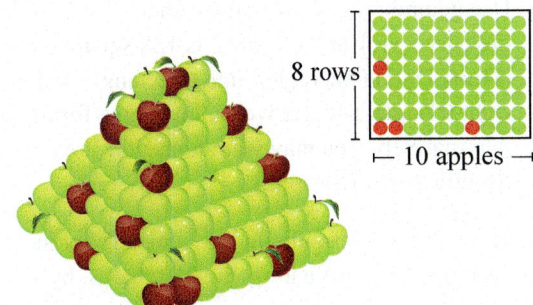

8 rows · ⊢ 10 apples ⊣

69. In statistics, the standard deviation of a data set $x_1, x_2, ..., x_n$ is defined to be the square root of the average of the squares of deviations of the data from their mean $\bar{x}$.

$$s = \sqrt{\frac{\left(x_1 - \bar{x}\right)^2 + \left(x_2 - \bar{x}\right)^2 + \cdots + \left(x_n - \bar{x}\right)^2}{n}}$$

Rewrite the definition of s using sigma notation, and use summation facts to show the following:

$$s^2 = \left(\frac{1}{n}\sum_{i=1}^{n} x_i^2\right) - \bar{x}^2$$

70. Find the distance covered by the pebble in Exercise 79 of Section 4.7 from $t = 2$ seconds to $t = 5$ seconds. (**Hint:** Find the velocity function first.)

71. Assuming constant acceleration, use the method of Exercise 70 to find the distance covered by the Bugatti in Exercise 91 of Section 4.7 from $t = 1$ second to $t = 3$ seconds.

72. Assuming constant deceleration, use the method of Example 2 to find the distance covered by the braking race car in Exercise 90 of Section 4.7 from $t = 1$ second to $t = 2$ seconds. (See the hint given in Exercise 70.)

73. The velocity function of a moving object is given by $v(t) = 9 - 0.5t^2$ m/s from $t = 0$ s to $t = 3$ s. Find the distance covered by the object during this time.

74. Repeat Exercise 73 for $v(t) = 4 - 0.5t^3$ on the interval $[0, 2]$.

75. Use geometry to show that the shaded area under the curve $s(t)$ in Example 2 is 128 square units. (**Hint:** Divide the region into a rectangle and a right triangle, and use well-known area formulas. Alternatively, you may want to use the area formula for a trapezoid.)

76. Show that you can obtain the same answer in Example 2 by evaluating $\lim_{n \to \infty} U_n$, that is, by choosing $t_i^* = t_{i-1}$ for every index value i.

77. Show that you can obtain the same answer in Example 4 by evaluating $\lim_{n \to \infty} O_n$, that is, by choosing $x_i^* = x_i$ for every index value i.

78. Show that you can obtain the same answer in Example 2 by choosing t_i^* to be the midpoint of the ith interval for every index value i.

79. Use an elementary argument to prove the following summation formula:

$$\sum_{i=1}^{n} i = \frac{n(n+1)}{2}$$

(**Hint:** Letting $S = \sum_{i=1}^{n} i$, add to S its terms in "reverse order"; that is, calculate $2S$ as

$$2S = \sum_{i=1}^{n} i + \sum_{j=0}^{n-1}(n - j), \text{ and notice that, after}$$

rearranging terms, this latter sum equals

$$(1 + n) + (2 + (n - 1)) + (3 + (n - 2)) + \cdots =$$
$$(n + 1) + (n + 1) + (n + 1) + \cdots.$$

Use this observation to complete the argument. Note that this argument is attributed to C. F. Gauss, who discovered it as a barely nine-year-old elementary school student.)

80.* Use mathematical induction to establish the summation formula of Exercise 79.

81.* Use mathematical induction to establish the following summation formula:

$$\sum_{i=1}^{n} i^2 = \frac{n(n+1)(2n+1)}{6}$$

82.* Use mathematical induction to establish the following summation formula:

$$\sum_{i=1}^{n} i^3 = \left[\frac{n(n+1)}{2}\right]^2$$

83.* Prove the summation formula of Exercise 82 by making use of the following identity:

$$(i + 1)^4 - i^4 = 4i^3 + 6i^2 + 4i + 1$$

84.* Inscribe a regular n-gon in a circle of radius r. Use radii to divide the n-gon into n isosceles triangles, and add the areas of the triangles to obtain the area of the inscribed n-gon. Finally, let $n \to \infty$ to obtain the area formula for the circle.

85–87 *Double summations* are important in many areas of mathematics, statistics, computer science, and the sciences in general. They have the form $\displaystyle\sum_{i=1}^{n}\sum_{j=1}^{m}a_{ij}$.

Evaluate the given double sum.

85. $\displaystyle\sum_{i=1}^{4}\sum_{j=1}^{5}(i+j)$

86. $\displaystyle\sum_{i=1}^{5}\sum_{j=1}^{6}ij$

87. $\displaystyle\sum_{i=1}^{n}\sum_{j=1}^{m}ij$

5.1 **Technology Exercises**

88–91 Use a computer algebra system to express the area under the graph of $f(x)$ and above the x-axis on the indicated interval as a limit. Then use technology to evaluate the limit to find the area.

88. $f(x)=x^{6}$ on $[0,1]$

89. $f(x)=\sin x$ on $[0,\pi]$

90. $f(x)=e^{x}$ on $[1,2]$

91. $f(x)=x+\cos^{2}(\pi x)$ on $\left[0,\dfrac{\pi}{2}\right]$

5.2 The Definite Integral

We continue our exploration of Riemann sums in this section, with the goal of developing the *definite integral* of a function. As we will see, the properties possessed by the definite integral arise directly from its definition as the limit of Riemann sums.

Topic One

Riemann Sums and the Definite Integral

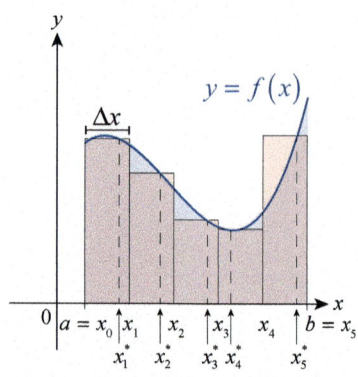

Figure 1 Riemann Sum with $n = 5$

Recall that for a function f defined on an interval $[a,b]$, any expression of the form $\sum_{i=1}^{n} f(x_i^*)\Delta x$ is called a Riemann sum of f over $[a,b]$. Implicit in this notation are the assumptions that $\Delta x = (b-a)/n$, that $[a,b]$ has been divided into n subintervals of equal width, $[x_0, x_1]$, $[x_1, x_2]$, ..., $[x_{n-1}, x_n]$, and that the ith sample point x_i^* lies in $[x_{i-1}, x_i]$ (note that $x_0 = a$ and $x_n = b$). If $f(x) \geq 0$ on $[a,b]$, as depicted in Figure 1, then $\sum_{i=1}^{n} f(x_i^*)\Delta x$ is an approximation of the area bounded by the graph of f and the x-axis between $x = a$ and $x = b$, and it makes sense to *define* this area as $\lim_{n \to \infty} \sum_{i=1}^{n} f(x_i^*)\Delta x$, if the limit exists. This is the inspiration for what follows.

Definition

Definite Integral

Given a function f defined on an interval $[a,b]$, let $\Delta x = (b-a)/n$, let $[a,b]$ be divided into n subintervals of equal width, $[x_0, x_1]$, $[x_1, x_2]$, ..., $[x_{n-1}, x_n]$, where $x_i = a + i\Delta x$, and let the ith sample point satisfy $x_i^* \in [x_{i-1}, x_i]$. Then the **(Riemann) definite integral of f from a to b** is denoted $\int_a^b f(x)\,dx$ and is defined to be

$$\int_a^b f(x)\,dx = \lim_{n \to \infty} \sum_{i=1}^{n} f(x_i^*)\Delta x,$$

provided the limit exists and is the same for every choice of the sample points. If so, we say f is **integrable** on $[a,b]$, and call f the **integrand** of the integral. In this context, a and b are called, respectively, the **lower** and **upper limits of integration**.

This definition is our first in what is sometimes called the *theory of integration*, which is an expansive area of mathematics. It is worth devoting a few paragraphs to a brief discussion of different interpretations of the definite integral and an introduction to some of the more advanced branches of the theory.

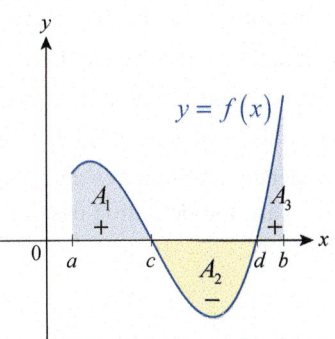

Figure 2 Signed Areas

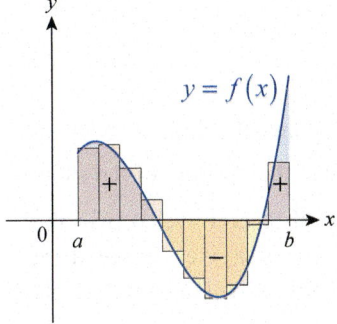

Figure 3 Signed Riemann Sum

First, note that there is nothing in the definition that requires f to be nonnegative on $[a,b]$. While it has been convenient up to this point to assume this is so, we now need a way to visualize the meaning of $\int_a^b f(x)\,dx$ when f takes on both positive and negative values. It is still the case that each term $f(x_i^*)\Delta x$ represents the product of a width Δx by the value of f at a sample point $x_i^* \in [x_{i-1}, x_i]$, but if $f(x_i^*) < 0$ then that particular term will be negative. A natural way to interpret such a situation is to use the notion of *signed area*. In Figure 2, for example, if we let A_1, A_2, and A_3 denote the (positive) areas of the regions bounded by the graph of f and the x-axis, then $\int_a^b f(x)\,dx = A_1 - A_2 + A_3$. That is, the integral of f between c and d is a negative number, and we would indicate this by writing $\int_c^d f(x)\,dx = -A_2 < 0$. Again, this interpretation follows directly from the definition of the definite integral as a limit of Riemann sums; in Figure 3, note that the signed area of the first four rectangles is positive, that of the next five rectangles is negative, and the signed area of the last rectangle is again positive.

The symbol $\int$ was introduced by Leibniz and represents a stylized "S" as a reminder that the basis of the definite integral is summation. Similarly, the process of calculating the value of a definite integral is termed **integration**—this reflects the fact that we arrive at a definite integral by adding up (integrating) a finite number of summands, and taking the limit as the number of summands approaches infinity.

For the moment, the "dx" part of the notation simply tells us that the variable in use is x, but its presence is not superfluous. It does (as the notation suggests) represent the differential dx, and it corresponds (as its location suggests) to the Δx in the Riemann sum; we will pay special attention to it later in this text when we make changes in the variable of integration. It plays a more active role in alternate definitions of the integral (studied in later math courses) and, importantly for our purposes, the "dx" tells us to integrate with respect to x, an important instruction when more than one variable is present. Note, though, that the particular symbol used to denote a given variable is immaterial. For a given function f defined on $[a,b]$, $\int_a^b f(x)\,dx$, $\int_a^b f(t)\,dt$, and $\int_a^b f(\square)\,d\square$ all mean the same thing—the symbol used for the variable is just a placeholder, and for that reason is sometimes referred to (somewhat dismissively) as a "dummy variable."

Finally, the definition given for the definite integral is actually a simplified version of a more general one. Remember that the choice of sample point x_i^* in each subinterval is irrelevant—in the limit, any one choice is as good as another. More generally, even the subintervals can be defined liberally. Given an interval $[a,b]$, a **partition** P of $[a,b]$ is any finite set $P = \{x_0, x_1, \ldots, x_n\}$ where $a = x_0 < x_1 < \cdots < x_{n-1} < x_n = b$ (note that the x_i's don't have to be evenly spaced). The **norm** of the partition P is defined to be $\|P\| = \max_{1 \le i \le n}\{\Delta x_i\}$ where $\Delta x_i = x_i - x_{i-1}$; that is, $\|P\|$ is the maximum of the subinterval widths. Using the epsilon-delta formulation of limit, we say that f is integrable on $[a,b]$ and that its definite integral is the number we denote as $\int_a^b f(x)\,dx$ if for every $\varepsilon > 0$ there is a $\delta > 0$ so that $\|P\| < \delta \Rightarrow \left| \int_a^b f(x)\,dx - \sum_{i=1}^n f(x_i^*)\Delta x_i \right| < \varepsilon$.

In words, the definite integral of f on the interval $[a,b]$ is the number $\int_a^b f(x)\,dx$ if whenever the partition P is "fine enough" (meaning the norm is smaller than delta), then the Riemann sum over that partition is within epsilon of $\int_a^b f(x)\,dx$. In proving theorems and in some applications, such as when only an approximation of an integral is possible, this more general definition is desirable (see Exercises 4–13, 28, and 29). But for the most part, the definition based on equally spaced partitions is sufficient and easier to apply.

Example 1

George uses a slingshot to shoot a small stone straight up into the sky with a vertical velocity of $v(t) = 64 - 32t$ ft/s. What is the height of the stone, relative to its initial height, when $t = 4$ s?

Solution

Following our discussion in Section 5.1, we will take the limit of Riemann sums of the form $\sum_{i=1}^n v(t_i^*)\Delta t$ in order to deduce the position of the stone after 4 seconds—it is again the case that the distance traveled over the interval $[t_{i-1}, t_i]$ is approximately $v(t_i^*)\Delta t$. However, as we will see, there is one important difference between this example and Example 2 of the previous section.

To set the stage, note that $\Delta t = (4-0)/n = 4/n$ and, if we decide to let t_i^* be the right endpoint of each subinterval, then $t_1^* = 0 + (4/n) = 4/n$, $t_2^* = 0 + 2(4/n) = 8/n$, and, in general, $t_i^* = 4i/n$. Using the integral notation we write the following:

$$\int_0^4 v(t)\,dt = \lim_{n \to \infty} \sum_{i=1}^n v(t_i^*)\Delta t$$

$$= \lim_{n \to \infty} \sum_{i=1}^n v\left(\frac{4i}{n}\right)\left(\frac{4}{n}\right)$$

$$= \lim_{n \to \infty} \sum_{i=1}^n \left[64 - 32\left(\frac{4i}{n}\right)\right]\left(\frac{4}{n}\right)$$

$$= \lim_{n \to \infty} \frac{4}{n}\sum_{i=1}^n \left(64 - \frac{128i}{n}\right) \qquad \text{Pull } \frac{4}{n} \text{ outside the sum.}$$

Continuing to use the summation formulas we listed in the previous section,

$$\int_0^4 v(t)\,dt = \lim_{n \to \infty} \frac{4}{n}\left(\sum_{i=1}^n 64 - \frac{128}{n}\sum_{i=1}^n i\right) \qquad \sum_{i=1}^n (a_i \pm b_i) = \sum_{i=1}^n a_i \pm \sum_{i=1}^n b_i \text{ and } \sum_{i=1}^n ca_i = c\sum_{i=1}^n a_i$$

$$= \lim_{n \to \infty} \frac{4}{n}\left[64n - \frac{128}{n}\cdot\frac{n(n+1)}{2}\right] \qquad \sum_{i=1}^n c = cn \text{ and } \sum_{i=1}^n i = \frac{n(n+1)}{2}$$

$$= \lim_{n \to \infty}\left[256 - 256\left(\frac{n+1}{n}\right)\right]$$

$$= 256 - 256\lim_{n \to \infty}\left(\frac{n+1}{n}\right)$$

$$= 0.$$

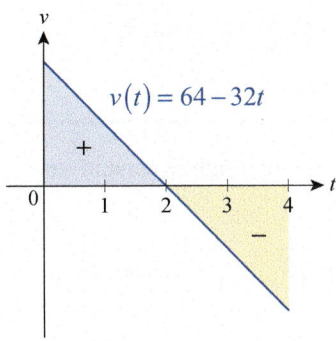

$v(t) = 64 - 32t$

Figure 4

What does this answer mean? Surely the stone has traveled, so does an answer of 0 feet make sense? It does, but only if we understand exactly what the definite integral of velocity tells us. If we take a look at the graph of v, we see that it is positive for the first 2 seconds and negative for the last 2 seconds—that is, the stone first rises and then falls. At the end of 4 seconds, the stone has the same height that it had at time 0. This is the physical meaning of the result and a consequence of the fact that the signed areas in Figure 4 cancel one another out exactly. If instead of the position after 4 seconds we want to know how far the stone travels over the interval $[0,4]$, we need to determine $\int_0^4 |v(t)|\, dt$, a task you will undertake in Exercise 2.

Now that the definite integral of a function has been defined, the question of which functions are integrable naturally arises; this is the counterpart to the question of differentiability we faced in Chapter 3. The next theorem, the proof of which is usually seen in a course such as Advanced Calculus or Real Analysis, tells us that those in a large and important class are integrable.

Theorem

Piecewise Continuous Functions Are Integrable

If f is a continuous function defined on $[a,b]$, or if f is a piecewise continuous function on $[a,b]$ (meaning it is continuous except for a finite number of jump discontinuities), then f is integrable on $[a,b]$.

Functions that are not integrable under our definition can be found without too much trouble (see Exercises 49–51), and the goal of enlarging the class of integrable functions is one motivation for some of the alternate definitions of integration that exist. Some advanced applications in physics also call for alternative formulations of the integral. But throughout this text the Riemann integral will serve our needs admirably; given that, and in preparation for our work to follow, we now define some shorthand notation for two Riemann sums we use frequently.

Definition

Left and Right Riemann Sums

Given a function f defined on the interval $[a,b]$, let L_n and R_n denote the Riemann sums obtained by dividing $[a,b]$ into n subintervals of equal width, $[x_0, x_1]$, $[x_1, x_2]$, ..., $[x_{n-1}, x_n]$, and defining the ith sample point to be, respectively, the left and right endpoint of the ith subinterval. That is,

$$L_n = \sum_{i=1}^{n} f(x_{i-1}) \Delta x \text{ and } R_n = \sum_{i=1}^{n} f(x_i) \Delta x,$$

where $\Delta x = (b-a)/n$ and $x_i = a + i\Delta x$.

Note in particular that, if f is integrable, $\int_a^b f(x)\,dx = \lim_{n\to\infty} L_n = \lim_{n\to\infty} R_n$.

Example 2

Evaluate $\int_{-1}^{3} x^3\,dx$ by taking the limit of the associated right Riemann sum.

Solution

From the expression $\int_{-1}^{3} x^3\,dx$, we see that $f(x) = x^3$, $a = -1$, and $b = 3$.

So

$$\Delta x = \frac{3-(-1)}{n} = \frac{4}{n}, \quad x_i = -1 + \frac{4i}{n} = \frac{-n+4i}{n},$$

and

$$f(x_i) = f\left(\frac{-n+4i}{n}\right) = \frac{(-n+4i)^3}{n^3} = \frac{-n^3 + 12n^2 i - 48ni^2 + 64i^3}{n^3}.$$

Hence,

$$\int_{-1}^{3} x^3\,dx = \lim_{n\to\infty} \sum_{i=1}^{n} \left(\frac{-n^3 + 12n^2 i - 48ni^2 + 64i^3}{n^3}\right)\left(\frac{4}{n}\right)$$

$$= \lim_{n\to\infty} \frac{4}{n^4} \sum_{i=1}^{n} \left(-n^3 + 12n^2 i - 48ni^2 + 64i^3\right) \qquad \text{Factor out } \frac{4}{n^4}.$$

$$= \lim_{n\to\infty} \frac{4}{n^4} \left(\sum_{i=1}^{n} -n^3 + \sum_{i=1}^{n} 12n^2 i + \sum_{i=1}^{n} -48ni^2 + \sum_{i=1}^{n} 64i^3\right) \; \scriptstyle \sum_{i=1}^{n}(a_i \pm b_i) = \sum_{i=1}^{n} a_i \pm \sum_{i=1}^{n} b_i$$

$$= \lim_{n\to\infty} \frac{4}{n^4} \left(-n^4 + 12n^2 \sum_{i=1}^{n} i - 48n \sum_{i=1}^{n} i^2 + 64 \sum_{i=1}^{n} i^3\right). \quad \scriptstyle \sum_{i=1}^{n} c = cn \text{ and } \sum_{i=1}^{n} ca_i = c\sum_{i=1}^{n} a_i$$

We can now apply the last three summation formulas from Section 5.1 to obtain

$$\int_{-1}^{3} x^3\,dx = \lim_{n\to\infty} \frac{4}{n^4} \left(-n^4 + 12n^2 \cdot \frac{n(n+1)}{2} - 48n \cdot \frac{n(n+1)(2n+1)}{6} + 64 \cdot \frac{n^2(n+1)^2}{4}\right),$$

which, after significant simplification, yields

$$\int_{-1}^{3} x^3\,dx = \lim_{n\to\infty} \frac{4}{n^4}\left(5n^4 + 14n^3 + 8n^2\right)$$

$$= \lim_{n\to\infty}\left(20 + \frac{56}{n} + \frac{32}{n^2}\right) = 20.$$

Topic Two

Properties of the Definite Integral

Clearly, evaluation of definite integrals by the process demonstrated in Example 2 requires a significant amount of work. Even worse, most integrals we are likely to run across simply can't be determined by this process—the limit that results is often one we can't evaluate (Exercises 30–33 ask you to construct, but not evaluate, Riemann sums corresponding to specific definite integrals). However, we faced a similar quandary as we developed the notion of the derivative, and resolved it by uncovering certain general properties that greatly simplified the task of differentiation. We will do the same thing for the integral, beginning here with some simple but useful properties.

First, always bear in mind the geometric meaning of the expression $\int_{a}^{b} f(x)\,dx$, namely, that it can be interpreted as the signed area of a region bounded by the graph of f and the x-axis between $x = a$ and $x = b$. In some cases, we can use this meaning to evaluate an integral on the basis of knowledge we already possess.

Example 3

Evaluate $\int_{-2}^{2} -\sqrt{4 - x^2}\ dx$.

Solution

If we construct a Riemann sum based on the function $f(x) = -\sqrt{4 - x^2}$, we arrive at a limit that is quite difficult to evaluate. But if we think about the graph of f and what $\int_{-2}^{2} -\sqrt{4 - x^2}\ dx$ represents, we see that elementary geometry suffices to evaluate the integral.

The graph of the function f corresponds to the graph of the equation $y = -\sqrt{4 - x^2}$, which upon squaring both sides yields $y^2 = 4 - x^2$. Rewritten as $x^2 + y^2 = 4$, this equation is familiar to us—its graph is the circle of radius 2 centered at the origin. If we think about what $y = -\sqrt{4 - x^2}$ then represents, we realize its graph is the lower half of the circle (and $y = \sqrt{4 - x^2}$ is the upper half); the graph of f is shown in Figure 5. Hence, $\int_{-2}^{2} -\sqrt{4 - x^2}\ dx$ represents the negative of half of the area of a circle of radius 2. That is, $\int_{-2}^{2} -\sqrt{4 - x^2}\ dx = -\dfrac{\pi(2)^2}{2} = -2\pi$.

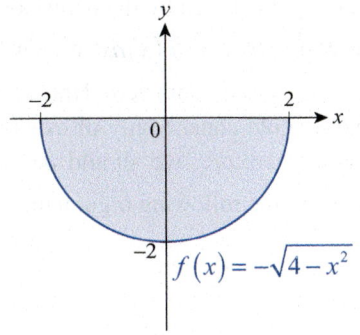

$f(x) = -\sqrt{4 - x^2}$

Figure 5

Other properties of the definite integral are simply a consequence of its definition as a limit of Riemann sums. The following list of many such properties begins with two logical extensions of the definition. The proofs of the properties have much in common; some will be supplied here, and others will be left as exercises.

Properties of the Definite Integral

Given the integrable functions f and g on the interval $[a,b]$ and any constant k, the following properties hold.

1. $\displaystyle\int_a^a f(x)\,dx = 0$

2. $\displaystyle\int_b^a f(x)\,dx = -\int_a^b f(x)\,dx$

3. $\displaystyle\int_a^b k\,dx = k(b-a)$

4. $\displaystyle\int_a^b kf(x)\,dx = k\int_a^b f(x)\,dx$

5. $\displaystyle\int_a^b \left[f(x)\pm g(x)\right]dx = \int_a^b f(x)\,dx \pm \int_a^b g(x)\,dx$

6. $\displaystyle\int_a^c f(x)\,dx + \int_c^b f(x)\,dx = \int_a^b f(x)\,dx,$ assuming each integral exists

7. If $f(x)\le g(x)$ on $[a,b]$, then $\displaystyle\int_a^b f(x)\,dx \le \int_a^b g(x)\,dx.$

8. If $m = \displaystyle\min_{a\le x\le b} f(x)$ and $M = \displaystyle\max_{a\le x\le b} f(x)$, then
$$m(b-a) \le \int_a^b f(x)\,dx \le M(b-a).$$

Selected Proofs and Comments

The first two properties result from thinking about our definition of the Riemann integral and making logical extensions to cases not yet covered by the definition. Specifically, a reasonable interpretation of the expression $\int_a^a f(x)\,dx$ is that it represents the signed area of a rectangle of height $f(a)$ and width zero. Note also that any Riemann sum approximating the integral could consist only of exactly one term, namely $f(a)\Delta x$, with $\Delta x = a - a = 0$. Therefore, we should define $\int_a^a f(x)\,dx$ to be 0. For the second property, consider the following argument.

$$\int_a^b f(x)\,dx = \lim_{n\to\infty} \sum_{i=1}^n f\left(x_i^*\right)\Delta x$$

$$= \lim_{n\to\infty} \sum_{i=1}^n f\left(x_i^*\right)\left(\frac{b-a}{n}\right)$$

$$= -\lim_{n\to\infty} \sum_{i=1}^n f\left(x_i^*\right)\left(\frac{a-b}{n}\right) \qquad \text{Factor out } -1.$$

$$= -\int_b^a f(x)\,dx \qquad\qquad \text{Note the reversal of the limits of integration.}$$

This is our justification for stating that reversing the order of the limits of integration of a given function changes the sign of the result.

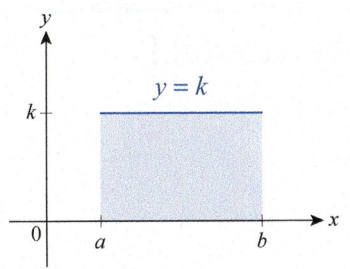

Figure 6 $\int_a^b k\,dx = k(b-a)$

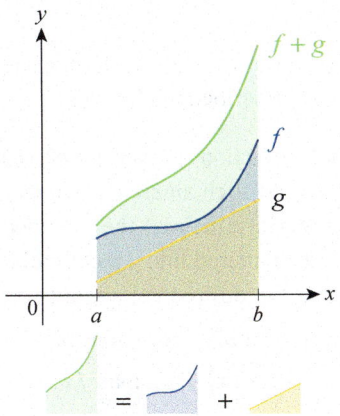

Figure 7

$\int_a^b [f(x)+g(x)]dx$
$= \int_a^b f(x)dx + \int_a^b g(x)dx$

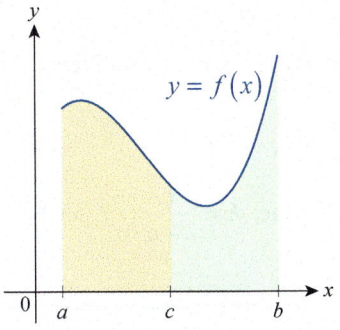

Figure 8

$\int_a^c f(x)dx + \int_c^b f(x)dx = \int_a^b f(x)dx$

To prove Property 3, it suffices to note that if $f(x) = k$, then

$$\sum_{i=1}^n f(x_i^*)\Delta x = \sum_{i=1}^n k\left(\frac{b-a}{n}\right) = k\left(\frac{b-a}{n}\right)n = k(b-a)$$

for every n and for any choice of sample points. It is also useful, though, to visualize the meaning of $\int_a^b k\,dx$. This definite integral corresponds to the signed area of the shaded rectangle in Figure 6, which is $k(b-a)$.

Similarly, Property 5 can be proved algebraically and interpreted graphically.

$$\int_a^b [f(x)\pm g(x)]dx = \lim_{n\to\infty}\sum_{i=1}^n [f(x_i^*)\pm g(x_i^*)]\Delta x$$

$$= \lim_{n\to\infty}\sum_{i=1}^n f(x_i^*)\Delta x \pm \lim_{n\to\infty}\sum_{i=1}^n g(x_i^*)\Delta x$$

$$= \int_a^b f(x)\,dx \pm \int_a^b g(x)\,dx$$

Figure 7 shows the graphs of three functions f, g, and $f+g$ and illustrates the fact that the sum of the areas under the graphs of f and g equals the area under the graph of $f+g$.

The proof of Property 6 is only slightly more technical in nature and reflects the idea that appears in Figure 8. In this graph, $a < c < b$, and geometrically it is easy to believe that the signed area under f on the interval $[a,b]$ can be expressed as the sum of the signed areas over the intervals $[a,c]$ and $[c,b]$. More interestingly, Property 6 remains true even if c lies outside of $[a,b]$, a fact you will prove in Exercise 99.

Example 4

Given that $\int_0^{3\pi/2} \sin^2 x\,dx = 3\pi/4$ and $\int_\pi^{3\pi/2} \sin^2 x\,dx = \pi/4$, determine $\int_0^\pi \sin^2 x\,dx$.

Solution

By Property 6, we know that $\int_0^\pi \sin^2 x\,dx + \int_\pi^{3\pi/2} \sin^2 x\,dx = \int_0^{3\pi/2} \sin^2 x\,dx$, and so $\int_0^\pi \sin^2 x\,dx + (\pi/4) = 3\pi/4$. Hence,

$$\int_0^\pi \sin^2 x\,dx = \frac{3\pi}{4} - \frac{\pi}{4} = \frac{\pi}{2}.$$

Example 5

Use Property 8 to find lower and upper bounds on the integral $\int_0^1 e^x \, dx$.

Solution

The function $f(x) = e^x$ is an increasing function, so

$$m = \min_{0 \le x \le 1} f(x) = f(0) = 1 \text{ and } M = \max_{0 \le x \le 1} f(x) = f(1) = e.$$

Hence, $1(1-0) \le \int_0^1 e^x \, dx \le e(1-0)$, or $1 \le \int_0^1 e^x \, dx \le e \approx 2.7183$.

We will discover a broad array of uses for the definite integral in the coming sections, but we will close this section with a quickly described application.

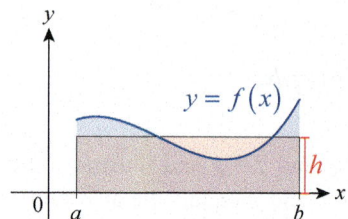

$y = f(x)$

Figure 9 Average Value of f

Consider the signed area bounded by an integrable function f over $[a, b]$, as shown in Figure 9. We now know how to make sense of such an area, no matter how "curvy" the graph of f is. Given that we know (in principle) the bounded area, it's fairly easy to imagine how the region could be altered into a rectangular region of the same area—simply define the width of the desired rectangle to be $b - a$, and specify the height h so that $h(b-a) = \int_a^b f(x) \, dx$. We even already know some bounds on h, as it must be the case that $\min_{a \le x \le b} f(x) \le h \le \max_{a \le x \le b} f(x)$.

An alternative approach to this same relation is to consider how we would define the *average* of an infinite number of numbers. Specifically, if we wanted to determine the average value of $f(x)$ for all $a \le x \le b$, we might start by considering the approximation

$$\frac{f(x_1) + f(x_2) + \cdots + f(x_n)}{n},$$

where again $x_i = a + i\Delta x$ and $\Delta x = (b-a)/n$. This average can be rewritten as

$$\frac{1}{n}\sum_{i=1}^n f(x_i) = \frac{\Delta x}{b-a}\sum_{i=1}^n f(x_i) = \frac{1}{b-a}\sum_{i=1}^n f(x_i)\Delta x,$$

and as $n \to \infty$ we obtain a formula for the average value: $h = \dfrac{1}{b-a}\int_a^b f(x) \, dx$.

Example 6

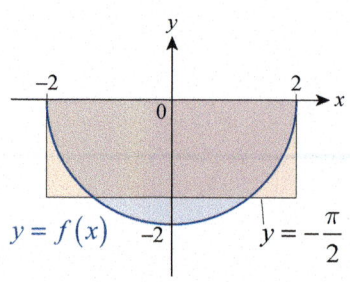

$y = f(x)$ -2 $y = -\dfrac{\pi}{2}$

Figure 10

Determine the average value of $f(x) = -\sqrt{4 - x^2}$ on the interval $[-2, 2]$.

Solution

We have already determined in Example 3 that

$$\int_{-2}^2 -\sqrt{4 - x^2}\, dx = -2\pi, \text{ so } h = \frac{1}{2 - (-2)}(-2\pi) = -\frac{\pi}{2}.$$

Note that, geometrically, this corresponds to transforming a half circle into a rectangle, as shown in Figure 10.

5.2 **Exercises**

1. Use the given graph along with appropriate formulas from geometry to evaluate each of the indicated definite integrals. (Note that the graph of f consists of linear pieces and a semicircle.)

   a. $\int_{-3}^{1} f(x)\,dx$ **b.** $\int_{-3}^{9} f(x)\,dx$ **c.** $\int_{0}^{6} |f(x)|\,dx$ **d.** $\int_{0}^{9} [f(x)-2]\,dx$

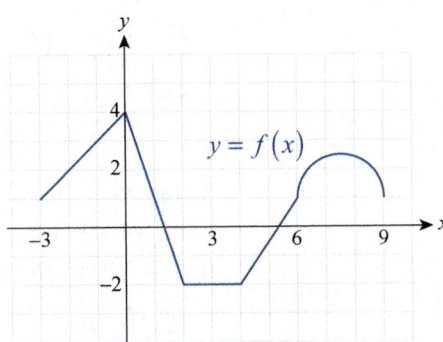

2. Calculate the total distance traveled by the stone of Example 1 by evaluating $\int_{0}^{4} |v(t)|\,dt$.

3. Suppose that in Example 1, George shoots the stone upward while standing near the edge of a deep canyon, and this time pulls the slingshot a bit harder, achieving a velocity function of $v(t) = 80 - 32t$ ft/s. What is the height of the stone, relative to its initial height, at $t = 4$ seconds? How about at $t = 6$ seconds? 10 seconds?

4–13 Use the given partition and sample points to approximate the definite integral of $f(x)$ on the indicated interval. (Note that the subintervals do not always have to be of equal width, and the sample points may be unevenly spaced.)

4. $f(x) = \dfrac{1}{3}x + 1$, $x_0 = 0 < 1 < 2 < 3 < 4 < 5 < 6 = x_6$, $x_i^* = x_i$

5. $f(x) = x^2 + x + 2$, $x_0 = -1 < 0 < 1 < 2 < 3 = x_4$, $x_i^* = x_{i-1}$

6. $f(x) = -x - \dfrac{3}{2}$, $x_0 = -2 < -1.5 < -0.9 < 0 < 1 = x_4$, $x_1^* = -1.8, x_2^* = -1, x_3^* = -0.4, x_4^* = 0.5$

7. $f(x) = \dfrac{1}{x^2}$, $x_0 = 1 < 2 < 3 < 4 = x_3$, $x_i^* = \dfrac{x_i - x_{i-1}}{2}$

8. $f(x) = \dfrac{1}{1+x^2}$, $x_0 = -3 < -2 < -1 < 0 < 1 < 2 < 3 = x_6$, $x_i^* = \dfrac{x_i - x_{i-1}}{2}$

9. $f(x) = x^3 - x$, $x_0 = 0 < 0.3 < 0.5 < 1 < 1.5 = x_4$, $x_1^* = 0.25, x_2^* = 0.5, x_3^* = 1, x_4^* = 1.2$

10. $f(x) = \sin x$, $x_0 = 0 < \dfrac{\pi}{6} < \dfrac{\pi}{4} < \dfrac{\pi}{3} < \dfrac{\pi}{2} < \dfrac{2\pi}{3} < \dfrac{3\pi}{4} < \dfrac{5\pi}{6} < \pi = x_8$, $x_i^* = x_{i-1}$

11. $f(x) = \ln(x+1)$, $x_0 = -0.5 < 1 < 2 < 2.5 = x_3$, $x_1^* = 0, x_2^* = e - 1, x_3^* = 2$

12. $f(x) = 10^{-x}$, $x_0 = 0 < 0.05 < 0.15 < 1 = x_3$, $x_1^* = 0.01, x_2^* = 0.1, x_3^* = 1$

13. $f(x) = \sqrt{x}$, $x_0 = 0 < \dfrac{1}{25} < \dfrac{4}{25} < \dfrac{9}{25} < \dfrac{16}{25} < 1 = x_5$, $x_i^* = x_i$

14–27 Use the concept of the definite integral to find the total area between the graph of $f(x)$ and the x-axis, by taking limits of the associated Riemann sums. When setting up the Riemann sums, make your choice between the left-endpoint, right-endpoint, and midpoint strategies. (**Hint:** Extra care is needed on those intervals where $f(x) < 0$. Remember that the definite integral represents a signed area.)

14. $f(x) = 2x + 4$ on $[0, 2]$

15. $f(x) = x - 1$ on $[0, 5]$

16. $f(x) = \dfrac{3 - x}{2}$ on $[0, 5]$

17. $f(x) = x^2$ on $[1, 3]$

18. $f(x) = x^2 - 1$ on $[-1, 1]$

19. $f(x) = x^2 - 4x$ on $[0, 5]$

20. $f(x) = \dfrac{x^2}{2} + 2$ on $[-2, 2]$

21. $f(x) = 3x^2 - 3$ on $[-1, 1]$

22. $f(x) = x^2 - 2x - 3$ on $[-1, 4]$

23. $f(x) = x^3$ on $[0, 1]$

24. $f(x) = 4x^3 - 32$ on $[0, 2]$

25. $f(x) = x^3 + 3x^2 + 1$ on $[0, 3]$

26. $f(x) = \begin{cases} 1 - (x - 1)^2 & \text{if } 0 \le x \le 3 \\ x - 6 & \text{if } 3 \le x \le 4 \end{cases}$

27. $f(x) = \begin{cases} x^3 & \text{if } 0 \le x \le 2 \\ 8x - 2x^2 & \text{if } 2 \le x \le 4 \end{cases}$

28. Generalize Exercise 13 to n subintervals and find the definite integral $\int_0^1 \sqrt{x}\, dx$ by letting $n \to \infty$. (**Hint:** Let $x_i^* = i^2/n^2$.)

29. Use the same approach as in Exercise 28 to find $\int_0^2 \sqrt[3]{x}\, dx$. (**Hint:** Let $x_i^* = 2i^3/n^3$.)

30–33 Express the integral as a limit of Riemann sums. (Do not attempt to evaluate the limit.)

30. $\displaystyle\int_1^3 \frac{1}{x}\, dx$

31. $\displaystyle\int_0^4 (x^2 - \log_2 x)\, dx$

32. $\displaystyle\int_{-a}^a \frac{1}{x^2 + 1}\, dx$

33. $\displaystyle\int_2^b \sqrt[4]{x}\, dx$

34–45 Sketch the region whose (signed) area is represented by the definite integral, and then use appropriate formulas from geometry to evaluate the integral.

34. $\displaystyle\int_{-1}^5 3\, dx$

35. $\displaystyle\int_{2.5}^{12} (-2)\, dx$

36. $\displaystyle\int_4^2 (1 - x)\, dx$

37. $\displaystyle\int_8^3 \left(4 - \frac{1}{2}x\right) dx$

38. $\displaystyle\int_0^4 |2x - 3|\, dx$

39. $\displaystyle\int_{-1}^5 (5 - |2x|)\, dx$

40. $\displaystyle\int_0^{10} (|x - 2| - |7 - x|)\, dx$

41. $\displaystyle\int_{-5}^0 \sqrt{25 - x^2}\, dx$

42. $\displaystyle\int_{-a/2}^a \sqrt{a^2 - x^2}\, dx, \quad a > 0$

43. $\displaystyle\int_{-2}^5 (2 - [\![x]\!])\, dx$

44. $\displaystyle\int_{-3}^8 [\![3x - 1]\!]\, dx$

45. $\displaystyle\int_{-4}^6 (x - [\![x]\!])\, dx$

46. Use Riemann sums resulting from midpoint estimates to prove $\int_a^b x\,dx = \left(b^2 - a^2\right)/2$.
(**Hint:** Notice that after using $(b+a)(b-a) = b^2 - a^2$, each Riemann sum becomes a collapsing sum.)

47. Provide an alternate proof for Exercise 46 by making a sketch and using areas of triangles.

48.* Mimic the argument used in Exercise 46, but using $x_i^* = \sqrt{\left(x_{i-1}^2 + x_{i-1}x_i + x_i^2\right)/3}$, to prove the formula $\int_a^b x^2\,dx = \left(b^3 - a^3\right)/3$.

49. The Dirichlet function is defined as follows.

$$\xi(x) = \begin{cases} 0 & \text{if } x \text{ is rational} \\ 1 & \text{if } x \text{ is irrational} \end{cases}$$

Prove that $\xi(x)$ is not integrable. (**Hint:** For a given n, form a Riemann sum by choosing each sample point x_i^* to be rational, then see what happens if each x_i^* is irrational. Use your observation to argue that $\lim\limits_{n\to\infty}\sum\limits_{i=1}^n f\left(x_i^*\right)\Delta x$ does not exist.)

50. Prove that the function $f(x) = \begin{cases} 1/x^2 & \text{if } x \neq 0 \\ 0 & \text{if } x = 0 \end{cases}$ is not integrable on $[0,1]$.

(**Hint:** By examining the first term of each R_n, show that $\lim\limits_{n\to\infty} R_n$ does not exist.)

51. Repeat Exercise 50 for $g(x) = \begin{cases} 1/x & \text{if } x \neq 0 \\ 0 & \text{if } x = 0 \end{cases}$ on $[0,1]$.

(**Hint:** Show that arbitrarily large Riemann sums can be constructed by choosing appropriate x_i^*'s.)

52–59 Decide whether the function is integrable on the indicated interval. If not, say why. (Do not evaluate the integral.)

52. $f(x) = \dfrac{1}{\sqrt{x+2}}$ on $[-1,1]$

53. $g(x) = \dfrac{2}{x}$ on $[-2,2]$

54. $h(x) = \dfrac{-3}{x-1}$ on $[0,5]$

55. $F(x) = \dfrac{x}{|x|}$ on $[-3,4]$

56. $G(x) = x \cdot [\![x]\!]$ on $[-2,2]$

57. $H(x) = \begin{cases} \dfrac{\sin x}{x} & \text{if } x \neq 0 \\ 0 & \text{if } x = 0 \end{cases}$ on $[-1,1]$

58. $u(x) = \begin{cases} \cos\dfrac{1}{x} & \text{if } x \neq 0 \\ 0 & \text{if } x = 0 \end{cases}$ on $[-1,1]$

59. $v(x) = \begin{cases} 3.14 & \text{if } x \text{ is rational} \\ \pi & \text{if } x \text{ is irrational} \end{cases}$ on $[0,2]$

60–65 Match the given property of the definite integral to the relevant illustration (labeled A–F).

60. $\int_a^b k\,dx = k(b-a)$ (Property 3)

61. $\int_a^b kf(x)\,dx = k\int_a^b f(x)\,dx$ (Property 4)

62. $\int_a^b \left[f(x) \pm g(x) \right] dx = \int_a^b f(x)\,dx \pm \int_a^b g(x)\,dx$ (Property 5)

63. $\int_a^c f(x)\,dx + \int_c^b f(x)\,dx = \int_a^b f(x)\,dx$ (Property 6)

64. If $f(x) \le g(x)$ on $[a,b]$, then $\int_a^b f(x)\,dx \le \int_a^b g(x)\,dx$.   (Property 7)

65. If $m = \min_{a \le x \le b} f(x)$ and $M = \max_{a \le x \le b} f(x)$, then $m(b-a) \le \int_a^b f(x)\,dx \le M(b-a)$.   (Property 8)

A.

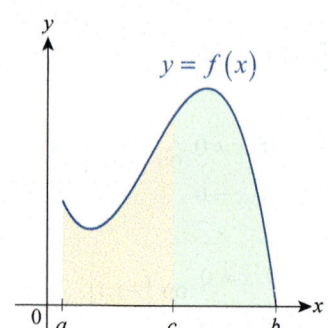

B.

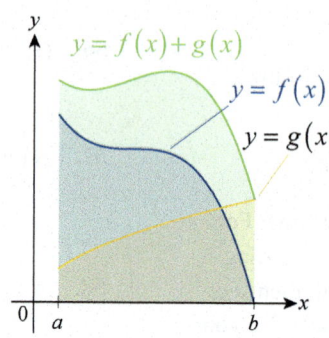

C.

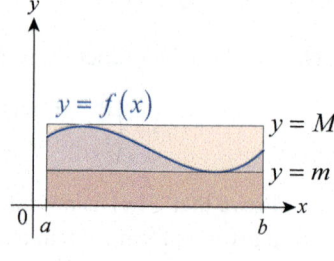

D.

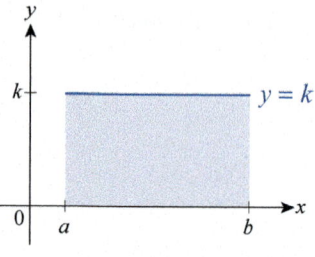

E.

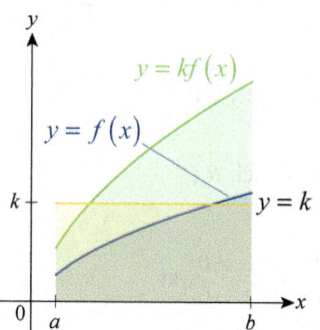

F.

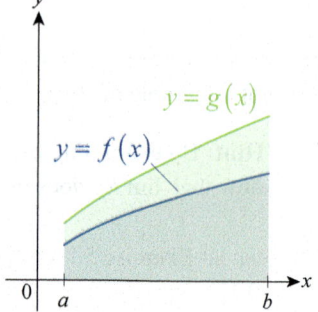

66–75 Use the properties of the definite integral to find the given integral, if possible, given that $\int_a^b f(x)\,dx = 3$, $\int_c^b f(x)\,dx = -1$, and $\int_a^b g(x)\,dx = -5$.

66. $\int_a^b \left[f(x) - g(x) \right] dx$

67. $\int_a^c \left[2f(x) + 1 \right] dx$

68. $\int_c^a 10f(x)\,dx$

69. $\int_a^a f(x)g(x)\,dx$

70. $\int_a^b \left[4f(x) + \dfrac{g(x)}{10} \right] dx$

71. $\int_b^a \dfrac{\sqrt{2}}{2} g(x)\,dx$

72. $\int_a^b \left[f(x) + 2g(x) - 2 \right] dx$

73. $\int_a^b \left[f(x) \right]^2 dx$

74. $\int_a^b \dfrac{5}{g(x)}\,dx$

75. $\int_a^b \left[\dfrac{f(x)}{3} - \pi g(x) \right] dx$

76–83 Use the results from Exercises 46 and 48, along with the properties of the definite integral and formulas from geometry, to evaluate the given integral.

76. $\int_0^2 (3x-1)\,dx$

77. $\int_{\sqrt{2}}^{-1}\left(1-\dfrac{\sqrt{2}}{2}x\right)dx$

78. $\int_{-1}^4\left(x^2+5\right)dx$

79. $\int_1^4\left(2x^2-x\right)dx$

80. $\int_0^3\left(t^2+\dfrac{t}{4}+4\right)dt$

81. $\int_0^1\left(2\sqrt{x}+x\right)dx$

82. $\int_2^0\left(\dfrac{\sqrt[3]{x}}{4}-x^2\right)dx$

83. $\int_{-2}^2\left(u-3\sqrt[3]{u}\right)du$

84. Suppose that f is an even function, g is odd, and both are integrable on $[-a,a]$. Use the properties of the definite integral to prove the following:
$$\int_{-a}^a f(x)\,dx = 2\int_0^a f(x)\,dx,\text{ and }\int_{-a}^a g(x)\,dx = 0.$$

85–90 Suppose that f is an even function, g is odd, both are integrable on $[-2,2]$, and we know that $\int_0^2 f(x)\,dx = 1$, while $\int_0^2 g(x)\,dx = 2.5$. If possible, find the integral.

85. $\int_{-2}^2\left[f(x)+g(x)\right]dx$

86. $\int_{-2}^2\left[2f(x)-3g(x)\right]dx$

87. $\int_{-2}^2\left|g(x)\right|dx$

88. $\int_{-2}^2 f(x)g(x)\,dx$

89. $\int_{-1}^1\left[f(x)\right]^2 dx$

90. $\int_0^2\left|g(x)\right|dx$

91. Use Property 8 of the definite integral to prove the validity of the following upper and lower estimates: $12 \le \int_0^4 \sqrt{x^2+9}\,dx \le 20$.

92–96 Use an argument similar to the one you gave in Exercise 91 to give upper and lower estimates for the given definite integral.

92. $\int_{-1}^4 \sqrt{5+x}\,dx$

93. $\int_2^3 \sqrt{3-x}\,dx$

94. $\int_4^5 \dfrac{1}{x-2}\,dx$

95. $\int_0^6\left(\dfrac{x^2}{32}-\dfrac{x}{4}+\dfrac{3}{2}\right)dx$

96. $\int_1^{\sqrt{3}} \arctan x\,dx$

97. Use Property 7 of the definite integral to prove the following inequalities.

a. $\int_0^1 \sqrt{1-x}\,dx \le \int_0^1 \sqrt{1-x^2}\,dx$

b. $\int_0^{\pi/2} \cos x\,dx \le \int_0^{\pi/2}\dfrac{\sin x}{x}\,dx$

98. Prove Property 4 of the definite integral. (**Hint:** Write a typical Riemann sum for f on $[a,b]$; use the Constant Multiple Rule for Summations, followed by properties of limits.)

99. Prove Property 6 of the definite integral in general; that is, prove that the property $\int_a^c f(x)\,dx+\int_c^b f(x)\,dx=\int_a^b f(x)\,dx$ holds irrespective of the order of the points a, b, and c. (**Hint:** The standard case of $a<c<b$ is discussed in the text. To start you off with the remaining cases assume, for example, that $a<b<c$. By an argument analogous to the one given in the text, we see that $\int_a^b f(x)\,dx+\int_b^c f(x)\,dx=\int_a^c f(x)\,dx$. Observe by Property 2 that $\int_b^c f(x)\,dx=-\int_c^b f(x)\,dx$, and rearrange the terms. Handle the remaining cases in a similar fashion.)

100. Prove Property 7 of the definite integral. (**Hint:** For a particular partition of $[a,b]$ and choice of sample points, argue that $\displaystyle\sum_{i=1}^n f\left(x_i^*\right)\Delta x \le \sum_{i=1}^n g\left(x_i^*\right)\Delta x$, and take the limits as $n\to\infty$.)

101. Use Property 7 to prove that the definite integral of a nonnegative function is nonnegative: If $f(x)\ge 0$ on $[a,b]$, then $\int_a^b f(x)\,dx \ge 0$. Then state and prove the analogous statement for nonpositive functions.

102. Prove Property 8 of the definite integral. (**Hint:** Use Property 7 with the constant function $g(x)=M$. The other inequality can be handled in a similar manner.)

103. Use Properties 4 and 7 to prove the following: If f is integrable on $[a,b]$, then $\left| \int_a^b f(x)\,dx \right| \le \int_a^b |f(x)|\,dx$. (**Hint:** Let $k = 1$ or $k = -1$ so that $\left| \int_a^b f(x)\,dx \right| = k \cdot \int_a^b f(x)\,dx$. Use the fact that $k \cdot f(x) \le |f(x)|$, along with Properties 4 and 7.)

104–115 Find the average value of the function over the given interval. (**Hint:** Instead of using Riemann sums, try using the results from Exercises 46 and 48 along with formulas from geometry and the properties of the definite integral.)

104. $f(x) = 3x - 1$ on $[0,4]$

105. $g(x) = -1 - \dfrac{1}{2}x$ on $[-2,2]$

106. $h(x) = x^2 - 2$ on $[-1,5]$

107. $F(x) = -3x^2 + 7x + 12$ on $[-2,3]$

108. $G(x) = 9x - x^3$ on $[-4,4]$

109. $H(x) = x^3 - 2x^2 - 1$ on $[0,2]$

110. $k(x) = |x - 4| - 2$ on $[0,7]$

111. $m(x) = |x| + |x + 1|$ on $[-3,2]$

112. $u(x) = \sqrt{1 - (x-1)^2}$ on $[0,2]$

113. $v(x) = [\![x]\!]$ on $\left[\dfrac{1}{2}, 3 \right]$

114. $t(x) = \sqrt{x} - 1$ on $[0,5]$

115. $w(x) = \sqrt[3]{x} - x$ on $[-1,8]$

116–119 Recognize the given limit as a Riemann sum of a function over an interval and then use geometry to evaluate it.

116. $\displaystyle\lim_{n\to\infty} \sum_{i=1}^{n} \dfrac{1}{n}\left(2 - \dfrac{i}{n} \right)$

117. $\displaystyle\lim_{n\to\infty} \sum_{i=1}^{n} \dfrac{3}{n}\left(\dfrac{2i}{3n} + 4 \right)$

118. $\displaystyle\lim_{n\to\infty} \sum_{i=1}^{n-1} \left(\dfrac{2}{n} + \dfrac{4i}{n^2} \right)$

119. $\displaystyle\lim_{n\to\infty} \sum_{i=1}^{n} \dfrac{2}{n}\sqrt{4 - \left(\dfrac{2i}{n} \right)^2}$

120. Prove that if $f(x)$ is an increasing nonnegative function on $[a,b]$, then for every n, $L_n \le \int_a^b f(x)\,dx \le R_n$. Then state and prove the analogous statement for a decreasing function $g(x)$ on the same interval.

121. Prove that L_n corresponding to $f(x)$ of Exercise 120 is increasing, while R_n is decreasing. Then state and prove the analogous statement for $g(x)$.

122. Use geometry and a fundamental trigonometric identity to find $\int_0^\pi \sin^2 x\,dx$. (**Hint:** Start out by comparing the given integral with $\int_0^\pi \cos^2 x\,dx$.)

123. Use the result of Exercise 122 to evaluate $\int_0^\pi (2\sin^2 x + x^2 - 3x)\,dx$.

124.* Suppose that the nonnegative function $R(x)$ has the property that $R(x) = 0$ whenever x is rational. If R is integrable on the interval $[a,b]$, prove that $\int_a^b R(x)\,dx = 0$.

125–130 *True or False?* Determine whether the given statement is true or false. In case of a false statement, explain or provide a counterexample.

125. If f and g are both integrable on $[a,b]$, then $\int_a^b f(x) \cdot g(x)\,dx = \int_a^b f(x)\,dx \cdot \int_a^b g(x)\,dx$.

126. The integral $\int_a^b f(x)\,dx$ is numerically equal to the area between the graph of $f(x)$ and the x-axis.

127. A Riemann sum for $f(x)$ on $[a,b]$ can be based upon a division of $[a,b]$ into subintervals of unequal width.

128. If $|f(x)|$ is integrable on $[a,b]$, then so is $f(x)$.

129. If $f(x)$ is positive and increasing on $[a,b]$, then $\int_a^b f(x)\,dx \ge f(a)(b-a)$.

130. If $\int_a^b f(x)\,dx < 0$, then $f(x) \le 0$ on $[a,b]$.

5.3 The Fundamental Theorem of Calculus

The appropriately named Fundamental Theorem of Calculus (FTC) ties together the two branches of differential and integral calculus and allows us to express the definite integral of a function in terms of an antiderivative. It thus serves as both an important theoretical achievement and as an eminently practical computational tool. Newton and Leibniz, drawing upon the prior work of a few mathematical predecessors, are justly renowned for their development of the FTC and their insight into its power and broad applicability.

Topic One
The Fundamental Theorem, Part I

The Fundamental Theorem is traditionally presented in two parts; the first shows that differentiation and integration are inverse operations of each other, and the second indicates how to use antiderivatives to calculate definite integrals. The two parts can be thought of as the theoretical and computational facets of the FTC, and our development begins with the following observation.

Theorem

The Mean Value Theorem for Definite Integrals

If f is a continuous function on the interval $[a,b]$, then there exists a point c in $[a,b]$ for which

$$f(c) = \frac{1}{b-a}\int_a^b f(x)\,dx.$$

Proof

The proof of this statement follows from the application of the Intermediate Value Theorem to a slight rephrasing of one of the definite-integral properties seen in the previous section. Recall that for a continuous function f on $[a,b]$,

$$m(b-a) \le \int_a^b f(x)\,dx \le M(b-a),$$

where $m = \min\limits_{a \le x \le b} f(x)$ and $M = \max\limits_{a \le x \le b} f(x)$.

Dividing each side of the inequality by $b - a$, we see that

$$m \le \frac{1}{b-a}\int_a^b f(x)\,dx \le M.$$

Since f is continuous, the Intermediate Value Theorem tells us that f takes on every value between m and M at least once on the interval $[a,b]$. In particular, there must exist a point $c \in [a,b]$ such that $f(c)$ is the value

$$\frac{1}{b-a}\int_a^b f(x)\,dx.$$

Example 1

Find every point c in the interval $[-2,2]$ at which the function $f(x) = -\sqrt{4-x^2}$ takes on its average value.

Solution

In Example 6 of Section 5.2, we graphed this function and determined that its average value over the interval is $-\pi/2$. So to find each point c at which f assumes its average value, we simply solve the equation $-\sqrt{4-x^2} = -\pi/2$.

$$-\sqrt{4-x^2} = -\frac{\pi}{2}$$

$$4-x^2 = \frac{\pi^2}{4}$$

$$x^2 = \frac{16-\pi^2}{4}$$

$$x = \pm\frac{\sqrt{16-\pi^2}}{2} \approx \pm 1.24$$

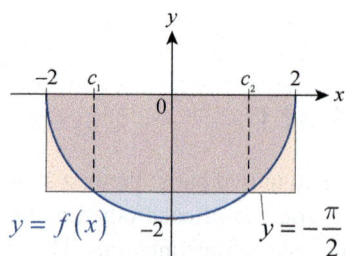

Figure 1

These two solutions are labeled c_1 and c_2 in Figure 1.

Suppose now that f is continuous on an interval I and that a is a fixed point in I. Then for every $x \in I$, we can define the value $F(x)$ by

$$F(x) = \int_a^x f(t)\, dt.$$

This definition is valid even if $x < a$ (so long as $x \in I$), since we defined right-to-left integration in Section 5.2. Visually, $F(x)$ represents the signed area shown in Figure 2; be sure to note that x is really the variable of interest in this definition, and that we have chosen to use t as the so-called "dummy variable" (the variable of integration).

Figure 2 Defining $F(x)$

The crux of the FTC is the realization that the function F, as defined, is an antiderivative of f. We have the necessary machinery now to prove this fact with relative ease.

The Fundamental Theorem of Calculus, Part I

Given a continuous function f on an interval I and a fixed point $a \in I$, define the function F on I by $F(x) = \int_a^x f(t)\, dt$. Then $F'(x) = f(x)$ for all $x \in I$.

── **Proof** ──────────────────────

We will prove that $F'(x) = f(x)$ by showing that

$$\lim_{h \to 0} \frac{F(x+h) - F(x)}{h} = f(x)$$

for all $x \in I$. In what follows, we will assume that each h is sufficiently small so that both $x - h \in I$ and $x + h \in I$. Further, if x happens to be an endpoint of I, then $F'(x)$ should be interpreted as an one-sided derivative and the limit above restricted to $h > 0$ (if x is the left endpoint of I) or $h < 0$ (if x is the right endpoint).

Note that one of the properties of the definite integral tells us that

$$F(x+h) - F(x) = \int_a^{x+h} f(t)\,dt - \int_a^x f(t)\,dt = \int_x^{x+h} f(t)\,dt,$$

and so, for each h sufficiently small,

$$\frac{F(x+h) - F(x)}{h} = \frac{1}{h} \int_x^{x+h} f(t)\,dt.$$

By the Mean Value Theorem for Definite Integrals, there is a point c between x and $x + h$ for which $(1/h)\int_x^{x+h} f(t)\,dt = f(c)$; note that $c \in [x, x+h]$ if h is positive and $c \in [x+h, x]$ if h is negative. As Figures 3a and 3b illustrate (for a positive h), $F(x+h) - F(x)$ represents the signed area bounded by the graph of f between x and $x + h$, and this signed area is equal to the area of the rectangle with width $(x+h) - x = h$ and height $f(c)$. As $h \to 0$, it must be the case that $c \to x$, as c is squeezed between x and $x + h$. Hence,

$$F'(x) = \lim_{h \to 0} \frac{F(x+h) - F(x)}{h} = \lim_{h \to 0} \frac{h \cdot f(c)}{h} = \lim_{h \to 0} f(c) = \lim_{c \to x} f(c) = f(x).$$

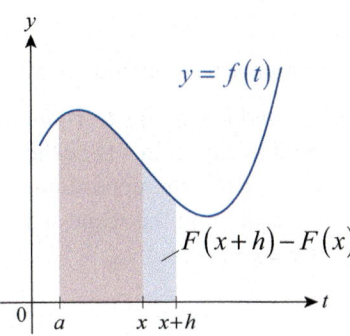

Figure 3a

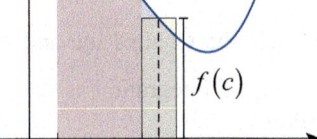

Figure 3b

──────────────────────

Using our alternate notation for differentiation makes it even more apparent that differentiation "undoes" integration, as we have just shown that

$$\frac{d}{dx} \int_a^x f(t)\,dt = f(x).$$

Example 2

Given a function F defined by $F(x) = \int_4^x \left(\sqrt{t} + 1 \right) dt$, determine the interval on which the Fundamental Theorem of Calculus can be applied and find $F'(x)$ on that interval.

Solution

First, note that $f(t) = \sqrt{t} + 1$ is continuous on $I = [0, \infty)$ and that 4 (the lower limit of integration) is an element of I. So the FTC applies for each nonnegative x, and $F'(x) = f(x) = \sqrt{x} + 1$. In the particular case of $x = 0$, this means $F'_+(0) = \sqrt{0} + 1 = 1$.

Example 3

Use the FTC to evaluate the following:

a. $\dfrac{d}{dx}\displaystyle\int_a^{x^2}\dfrac{1}{2+t}\,dt$ **b.** $\dfrac{d}{dx}\displaystyle\int_{3-5x}^{7}e^{t-3}\,dt$ **c.** $\dfrac{d}{dx}\displaystyle\int_{x^2+3}^{1+5x}\cos t\,dt$

Solution

a. The expression $y=\displaystyle\int_a^{x^2}\dfrac{1}{2+t}\,dt$ does indeed represent a function in x, but because the upper limit of integration is x^2 and not x, dy/dx is not simply $1/(2+x)$. To understand why we shouldn't expect this to be the case, note that for a small change Δx from x to $x+h$, the upper limit changes from x^2 to $x^2+2xh+h^2$. But we can easily find the requested derivative by using the Chain Rule: if we let $u=x^2$, then

$$\frac{dy}{dx}=\frac{dy}{du}\cdot\frac{du}{dx}$$

$$=\left(\frac{d}{du}\int_a^u\frac{1}{2+t}\,dt\right)\left(\frac{du}{dx}\right)$$

$$=\left(\frac{1}{2+u}\right)(2x)$$

$$=\left(\frac{1}{2+x^2}\right)(2x) \qquad \text{Replace } u \text{ with } x^2.$$

$$=\frac{2x}{2+x^2}.$$

b. To determine $\dfrac{d}{dx}\displaystyle\int_{3-5x}^{7}e^{t-3}\,dt$, we must again use the Chain Rule and also account for the fact that the variable limit of integration is the lower one, not the upper. We can do so as follows.

$$\frac{d}{dx}\int_{3-5x}^{7}e^{t-3}\,dt=\frac{d}{dx}\left(-\int_7^{3-5x}e^{t-3}\,dt\right) \qquad \int_a^b f(x)\,dx=-\int_b^a f(x)\,dx$$

$$=\frac{d}{du}\left(-\int_7^u e^{t-3}\,dt\right)\left(\frac{du}{dx}\right) \qquad \text{Set } u=3-5x.$$

$$=\left(-e^{u-3}\right)(-5)$$

$$=5e^{3-5x-3} \qquad \text{Replace } u \text{ with } 3-5x.$$

$$=5e^{-5x}$$

c. The limits in the integral $\displaystyle\int_{x^2+3}^{1+5x}\cos t\,dt$ are both functions of x, but the two examples above guide us on how to proceed. Note that since the integrand is continuous everywhere, we can break the integral apart into a sum of two integrals, and the choice of a is immaterial—any fixed $a\in\mathbb{R}$ will do.

$$\frac{d}{dx}\int_{x^2+3}^{1+5x}\cos t\,dt=\frac{d}{dx}\left[\int_{x^2+3}^{a}\cos t\,dt+\int_a^{1+5x}\cos t\,dt\right] \qquad \int_a^b f(x)\,dx=\int_a^c f(x)\,dx+\int_c^b f(x)\,dx$$

$$=\frac{d}{dx}\left(-\int_a^{x^2+3}\cos t\,dt\right)+\frac{d}{dx}\int_a^{1+5x}\cos t\,dt \qquad \begin{array}{l}\text{Reverse the limits}\\\text{in the first integral.}\end{array}$$

$$=\left[-\cos\left(x^2+3\right)\right](2x)+\left[\cos\left(1+5x\right)\right](5) \quad \text{Use the Chain Rule twice.}$$

$$=-2x\cos\left(x^2+3\right)+5\cos\left(1+5x\right)$$

Example 4

Given the function $f(x) = (x-2)/(x+1)$, define functions F, G, and H by

$$F(x) = \int_5^x f(t)\,dt, \quad G(x) = \int_{10}^x f(t)\,dt, \quad \text{and } H(x) = \int_{-10}^x f(t)\,dt.$$

Determine the intervals on which the FTC applies for each integral, and find $F'(x)$, $G'(x)$, and $H'(x)$ on those intervals. How are F, G, and H similar and how are they different?

Solution

The function f has a single point of discontinuity at $x = -1$, and so the FTC is applicable for F and G on the interval $(-1, \infty)$ and for the function H on the interval $(-\infty, -1)$ (note that the lower limit of integration is the key to determining the interval on which the FTC can be used). Next, the FTC tells us that $F'(x) = G'(x) = f(x)$ for all $x \in (-1, \infty)$ and that $H'(x) = f(x)$ for all $x \in (-\infty, -1)$.

The last question is the most interesting. By Corollary 2 of the Mean Value Theorem from Section 4.2, we know that two functions with the same derivative on an open interval differ only by a constant. So it must be the case that $F(x) = G(x) + C$ on $(-1, \infty)$ for some constant C. And while it's also true that $H' = f$, the domain of H does not overlap with the domain of F or G at all.

In Figures 4 and 5, the graphs of f, F, G, and H are shown. We will soon develop the tools needed to find explicit formulas for F, G, and H, but for the moment we will take the graphs as given. What we *can* do now is confirm that F, G, and H possess the properties that such functions should— remember that each was designed to return the signed area bounded by the graph of f over a (varying) interval.

For instance, it should be the case that $H(-10) = 0$, since $H(-10) = \int_{-10}^{-10} f(t)\,dt$, and that is indeed the case. Further, $H(x)$ should be positive for $x \in (-10, -1)$, since f is positive on the interval and we are integrating from left to right for such x's. Similarly, $H(x)$ should be negative for $x \in (-\infty, -10)$, since f is again positive on the interval but for these values of x we are integrating from right to left. Note that for such an x, $H(x) = \int_{-10}^x f(t)\,dt = -\int_x^{-10} f(t)\,dt < 0$.

Turning to F and G, we can say a bit more. First, both F and G should have a horizontal tangent line at $x = 2$, since $f(2) = 0$ and the FTC tells us that $F'(2) = f(2)$ and $G'(2) = f(2)$. And as with H, we know that $F(5) = \int_5^5 f(t)\,dt = 0$ and that $G(10) = \int_{10}^{10} f(t)\,dt = 0$. All of these facts are reflected in the graphs of F and G.

One last observation is in order. We have already determined that $F(x) = G(x) + C$ for some constant C on $(-1, \infty)$, and the graphs seem to confirm that fact. But what is the actual value of C? Based on Figure 5, it appears to be approximately 3—note the difference, for example, between $F(2)$ and $G(2)$ or between $F(10)$ and $G(10)$. More precisely, we

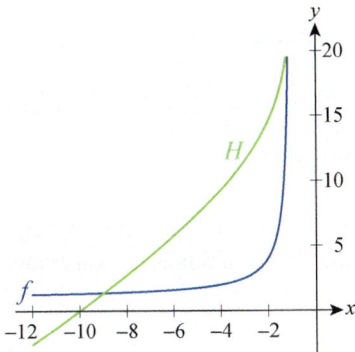

Figure 4 Graphs of f and H

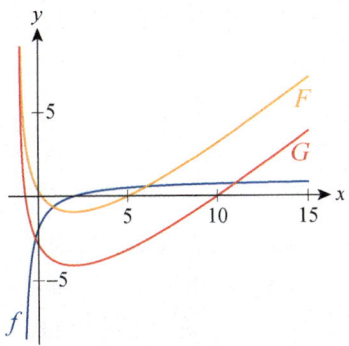

Figure 5 Graphs of f, F, and G

can use two of the properties of the definite integral to note that for any $x \in (-1, \infty)$,

$$C = F(x) - G(x)$$

$$= \int_5^x f(t)\, dt - \int_{10}^x f(t)\, dt$$

$$= \int_5^x f(t)\, dt + \int_x^{10} f(t)\, dt \qquad \int_a^b f(x)\, dx = -\int_b^a f(x)\, dx$$

$$= \int_5^{10} f(t)\, dt. \qquad \int_a^c f(x)\, dx + \int_c^b f(x)\, dx = \int_a^b f(x)\, dx$$

With the knowledge that we shall soon acquire, we will be able to determine that $\int_5^{10} f(t)\, dt = 5 + 3\ln(6/11) \approx 3.18$.

Topic Two

The Fundamental Theorem, Part II

Part I of the FTC shows us how to construct an antiderivative of a function through the use of definite integration. The true power of that construction is demonstrated by the second part of the FTC.

Theorem

The Fundamental Theorem of Calculus, Part II

If f is a continuous function on the interval $[a,b]$ and if F is any antiderivative of f on $[a,b]$, then $\int_a^b f(x)\, dx = F(b) - F(a)$.

Proof

First, we know the definite integral $\int_a^b f(x)\, dx$ exists, since f is continuous on the interval. Further, by Part I of the FTC, we know that the function $G(x) = \int_a^x f(t)\, dt$ exists and is an antiderivative of f on $[a,b]$. This means $G'(x) = f(x)$ for all $x \in [a,b]$ and we know $F'(x) = f(x)$ for all $x \in [a,b]$ (since F is an antiderivative of f), so $F'(x) = G'(x)$ for all $x \in [a,b]$. Hence, by Corollary 2 of the Mean Value Theorem for derivatives (Section 4.2) it must be the case that $F(x) = G(x) + C$ for all $x \in (a,b)$ and for some constant C—we have already seen this consequence in Example 4. Since differentiability implies continuity,

$$F(a) = \lim_{x \to a^+} F(x) = \lim_{x \to a^+} G(x) + C = G(a) + C$$

and

$$F(b) = \lim_{x \to b^-} F(x) = \lim_{x \to b^-} G(x) + C = G(b) + C.$$

So we have

$$F(b)-F(a) = G(b)+C-\left[G(a)+C\right]$$
$$= G(b)-G(a)$$
$$= \int_a^b f(t)\,dt - \int_a^a f(t)\,dt$$
$$= \int_a^b f(t)\,dt - 0$$
$$= \int_a^b f(x)\,dx.$$

The relationship between differentiation and integration is now complete, and we can exploit the above result to evaluate definite integrals without resorting to a limit of Riemann sums whenever we know an antiderivative of the integrand. And since expressions of the form $F(b)-F(a)$ arise so frequently in this context, the following notation is useful:

$$F(x)\Big]_a^b = F(b)-F(a)$$

This may also appear as $\left[F(x)\right]_a^b = F(b)-F(a)$.

Example 5

Evaluate the following integrals.

a. $\displaystyle\int_0^9 e^x\,dx$ b. $\displaystyle\int_1^3 \frac{1}{x}\,dx$ c. $\displaystyle\int_{-2}^1 x^5\,dx$

Solution

In each case, our knowledge of derivatives will be sufficient to "reverse differentiate" and find a suitable antiderivative—you may want to refer to Section 4.7 for a refresher. Remember, any antiderivative of the integrand will do.

a. Since $\dfrac{d}{dx}\left(e^x\right)=e^x$, we will use $F(x)=e^x$. Hence,

$$\int_0^9 e^x\,dx = e^x\Big]_0^9 = e^9 - e^0 = e^9 - 1.$$

b. Recall that $\dfrac{d}{dx}\ln|x| = \dfrac{1}{x}$, so $\displaystyle\int_1^3 \frac{1}{x}\,dx = \ln|x|\Big]_1^3 = \ln 3 - \ln 1 = \ln 3.$

c. Note that $\dfrac{d}{dx}\left(\dfrac{x^6}{6}\right)=x^5$, so $\displaystyle\int_{-2}^1 x^5\,dx = \dfrac{x^6}{6}\Big]_{-2}^1 = \dfrac{1^6}{6} - \dfrac{(-2)^6}{6} = -\dfrac{21}{2}.$

Example 6

Construct a function $f(x)$ whose derivative is $\sin x$ and which has the value -3 at 0.

Solution

We can begin by noting that the function $\int_0^x \sin t\, dt$ has the desired derivative, by Part I of the FTC. We can modify this function to have the correct value at 0 by defining $f(x) = \int_0^x \sin t\, dt - 3$; now,

$$f(0) = \int_0^0 \sin t\, dt - 3 = 0 - 3 = -3.$$

While f in this form defines a valid function, we can use the fact that $\dfrac{d}{dt}(-\cos t) = \sin t$ to rewrite our answer as follows:

$$f(x) = \int_0^x \sin t\, dt - 3$$
$$= \left[-\cos t\right]_0^x - 3$$
$$= -\cos x - (-\cos 0) - 3$$
$$= -\cos x + 1 - 3 = -\cos x - 2$$

Example 7

Recall that the difference between the functions F and G in Example 4 was determined to be $\int_5^{10} \dfrac{x-2}{x+1}\, dx$. Evaluate this integral.

Solution

As always, an antiderivative of the integrand will make the evaluation of the integral an easy task. In this case, though, it will take a bit more thought to arrive at an antiderivative—it's not immediately clear what sort of function has a derivative of $(x-2)/(x+1)$.

We will learn many techniques for systematically developing antiderivatives in coming sections, but in this case rewriting the integrand as follows will suffice.

$$\frac{x-2}{x+1} = \frac{(x+1)-3}{x+1} = 1 - \frac{3}{x+1}$$

The same result can be obtained by dividing $x - 2$ by $x + 1$.

$$
\begin{array}{r}
1 \\
x+1{\overline{\smash{\big)}\,x-2}} \\
\underline{-(x+1)} \\
-3
\end{array}
$$

$$\frac{x-2}{x+1} = 1 - \frac{3}{x+1}$$

Also, since $\dfrac{d}{dx}\ln(x+1)=\dfrac{1}{x+1}$ (you should verify this),

$$\int_5^{10}\frac{x-2}{x+1}\,dx=\int_5^{10}\left(1-\frac{3}{x+1}\right)dx$$

$$=\left[x-3\ln(x+1)\right]_5^{10}$$

$$=(10-3\ln 11)-(5-3\ln 6)$$

$$=5-3\ln 11+3\ln 6=5+3\ln\frac{6}{11}.$$

5.3 **Exercises**

1–8 Find every point c in the given interval at which $f(x)$ takes on its average value.

1. $f(x)=x^3;\quad [0,2]$

2. $f(x)=\dfrac{x(6-x)}{2};\quad [0,6]$

3. $f(x)=\dfrac{-x^4}{4}+4;\quad [-2,2]$

4. $f(x)=e^x;\quad [0,1]$

5. $f(x)=\sin x;\quad [0,\pi]$

6. $f(x)=x-\sqrt{x+1};\quad [0,8]$

7. $f(x)=\csc^2 x;\quad \left[\dfrac{\pi}{4},\dfrac{3\pi}{4}\right]$

8. $f(x)=\dfrac{x^2+2}{x^2};\quad [1,3]$

9–10 Let $F(x)=\displaystyle\int_0^x f(t)\,dt$. Use the graph of f to answer the questions. (Note that the graph in Exercise 10 consists of linear and parabolic pieces.)

9. a. Evaluate $F(2)$, $F(4)$, $F(6)$, $F(8)$, and $F(10)$.

b. Give a formula for $F(x)$. (**Hint:** It will be a piecewise defined function.)

c. Sketch the graph of $F(x)$.

10. a. Evaluate $F(0)$, $F(2)$, $F(4)$, and $F(7)$.

b. Give a formula for $F(x)$. (**Hint:** It will be a piecewise defined function.)

c. Sketch the graph of $F(x)$.

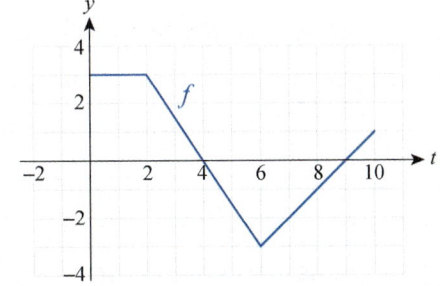

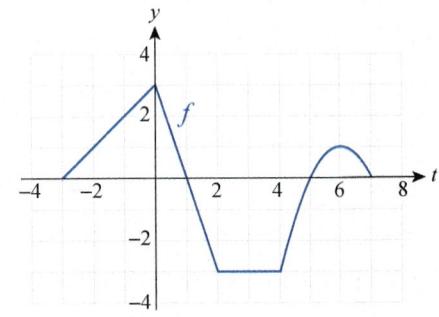

11–16 Find the area between the graph of $f(x)$ and the x-axis on the indicated interval.

11. $f(x) = \sqrt{x}$ on $[1,4]$

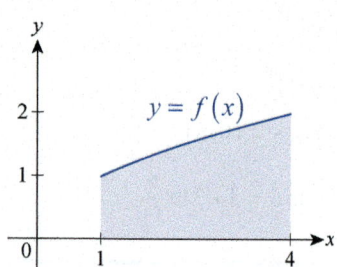

12. $f(x) = 6x - x^2$ on $[0,6]$

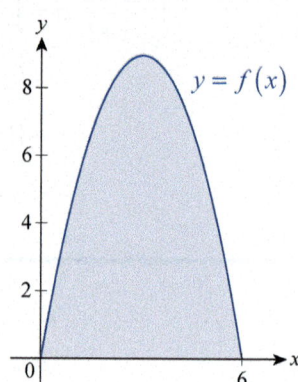

13. $f(x) = \dfrac{1}{\sqrt{x}}$ on $[0.5, 2]$

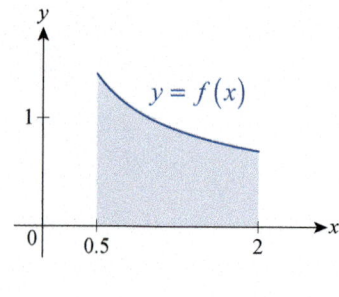

14. $f(x) = e^{-x} + 0.6x$ on $[0,2]$

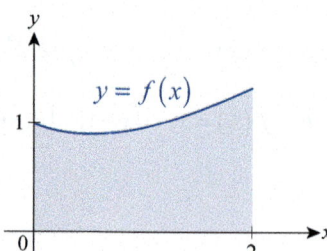

15. $f(x) = -2.5x^{4/3} + 5x$ on $[0,8]$

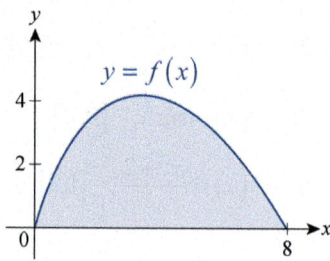

16. $f(x) = 0.5\sec^2 x$ on $\left[0, \dfrac{\pi}{3}\right]$

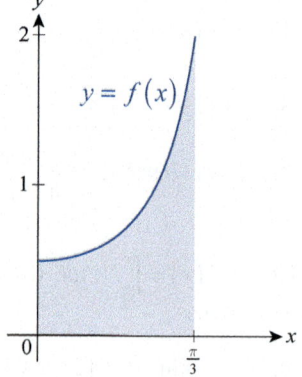

17–32 Use Part I of the Fundamental Theorem of Calculus to find the derivative of the given function.

17. $F(x) = \displaystyle\int_0^x \frac{1}{3}\left(t^2 + \sqrt{t}\right)dt$

18. $F(x) = \displaystyle\int_{1/2}^x \ln s\, ds$

19. $G(x) = \displaystyle\int_{-4}^x \frac{t^4}{t^4 + 4}\, dt$

20. $G(x) = \displaystyle\int_2^x \sqrt[3]{u^2 - u}\, du$

21. $y = \displaystyle\int_x^1 \sin\sqrt{t+1}\, dt$

22. $y = \displaystyle\int_x^0 t\arccos t\, dt$

23. $y = \displaystyle\int_{-5}^{3x}\left(t^2 + 3\right)e^{t-2}\, dt$

24. $y = \displaystyle\int_0^{x^2} \sec^{2/3}\sqrt{t}\, dt$

25. $y = \displaystyle\int_0^{\sin x}\left(t^2 + e^t\right)dt$

26. $y = \displaystyle\int_{\sqrt{x}}^1 \log t\, dt$

27. $y = \displaystyle\int_x^{\pi x} \sin t\, dt$

28. $y = \displaystyle\int_{\sqrt{x}}^{x^2} \cos\left(z^2\right)dz$

29. $F(x) = \displaystyle\int_0^{\cos^{-1}x} \sqrt{1 + \sqrt{1 + \sec^2 t}}\, dt$

30. $G(x) = \displaystyle\int_{x-c}^{x+c} \sin t\, dt$

31. $H(x) = \displaystyle\int_{\ln x}^x \ln t\, dt$

32. $K(x) = \displaystyle\int_x^{x^2} \sqrt{1 + t^4}\, dt$

33–38 Find a formula for $F(x)$ that is free of the integral symbol. Then differentiate it to verify Part I of the Fundamental Theorem of Calculus.

33. $F(x) = \int_1^x 2\,dt$

34. $F(x) = \int_{-3}^x (5-t)\,dt$

35. $F(x) = \int_x^1 (t^2 + t)\,dt$

36. $F(x) = \int_x^8 \dfrac{w+2}{\sqrt[3]{w}}\,dw$

37. $F(x) = \int_1^{\sqrt{x}} \dfrac{1}{s^2}\,ds$

38. $F(x) = \int_0^{\tan x} (1+u^2)\,du$

39–65 Use Part II of the Fundamental Theorem of Calculus to evaluate the definite integral.

39. $\int_{-2}^4 (-5)\,dx$

40. $\int_0^{1/\pi} 3\pi^2\,dx$

41. $\int_2^9 (4x+3)\,dx$

42. $\int_{-2.5}^6 (1-5u)\,du$

43. $\int_{-2}^4 (1.5x^2 - x + 3)\,dx$

44. $\int_0^3 (5s-1)(2+s)\,ds$

45. $\int_1^7 (2.4x^3 - 4x^2 + 1)\,dx$

46. $\int_{-1}^1 (2x^2 + 1)^2\,dx$

47. $\int_1^2 \left(1 - \dfrac{2}{x}\right)dx$

48. $\int_1^3 \left(\dfrac{1}{x^2} + \dfrac{2}{x} + 3\right)dx$

49. $\int_{-2}^{-1} \dfrac{2x^5 - 4x^2}{x^3}\,dx$

50. $\int_0^3 \dfrac{2x^2 - \sqrt{x}}{4}\,dx$

51. $\int_1^2 \left(x\sqrt{x} - \dfrac{1}{\sqrt{x}}\right)dx$

52. $\int_2^4 \dfrac{5x^2 - 3x + 2}{\sqrt{x}}\,dx$

53. $\int_{1/8}^1 \left(2\sqrt[3]{t} - \sqrt[3]{\dfrac{2}{t}}\right)dt$

54. $\int_0^1 \dfrac{x + 3\sqrt{x}}{\sqrt[5]{x}}\,dx$

55. $\int_0^{\pi/2} \left(\dfrac{\sin x}{2} - \sqrt{x}\right)dx$

56. $\int_0^{\pi/3} \dfrac{2}{\cos^2 \theta}\,d\theta$

57. $\int_{\sqrt{2}/2}^1 \dfrac{3}{\sqrt{1-t^2}}\,dt$

58. $\int_{\sqrt{3}/3}^1 \dfrac{-5}{1+x^2}\,dx$

59. $\int_0^{\pi/3} (e^x + \sec x \tan x)\,dx$

60. $\int_{-3}^3 2^x\,dx$

61. $\int_{-2}^4 |x(x-2)|\,dx$

62. $\int_1^3 f(x)\,dx$, where $f(x) = \begin{cases} \sin\dfrac{\pi x}{4} & \text{if } 0 < x \le 2 \\ (x-2)^2 + 1 & \text{if } 2 < x \le 3 \end{cases}$

63. $\int_{\pi/4}^{3\pi/4} (1 - \csc\theta\cot\theta)\,d\theta$

64. $\int_{\pi/4}^{\pi/2} \dfrac{2}{1 - \cos^2 x}\,dx$

65. $\int_{-1}^1 g(x)\,dx$, where $g(x) = \begin{cases} \sqrt{x+1} & \text{if } -1 < x \le 0 \\ e^x & \text{if } 0 < x \le 1 \end{cases}$

66–69 Recognize the given limit as a Riemann sum of a function over an interval and then use the Fundamental Theorem of Calculus to evaluate it.

66. $\lim\limits_{n\to\infty} \sum\limits_{i=1}^n \dfrac{\sqrt{i}}{n^{3/2}}$

67. $\lim\limits_{n\to\infty} \sum\limits_{i=1}^n \dfrac{2}{n}\left(\dfrac{2i}{n}\right)^4$

68. $\lim\limits_{n\to\infty} \sum\limits_{i=1}^n \dfrac{\pi}{2n}\cos\dfrac{\pi i}{2n}$

69. $\lim\limits_{n\to\infty} \sum\limits_{i=1}^{n-1} \dfrac{e-1}{n+i(e-1)}$

70–78 Find the area of the region between the graph of the given function and the x-axis on the indicated interval.

70. $y = \cos x$ on $\left[-\dfrac{\pi}{2}, \dfrac{3\pi}{2}\right]$

71. $y = -x^3$ on $[-2, 2]$

72. $y = -x^2 + 1$ on $[-1, 2]$

73. $y = \dfrac{1}{x}$ on $\left[\dfrac{1}{e}, e^2\right]$

74. $y = -2|x-3| + 6$ on $[0, 10]$

75. $y = -x^3 + 7x^2 - 10x$ on $[0, 6]$

76. $y = 2\sqrt{x} - x$ on $[0, 9]$

77. $y = \dfrac{1-2x}{2x+1}$ on $[0, 1]$

78. $y = x^4 - x^2$ on $[-1, 1]$

79–87 Use the method of Example 7 to evaluate the definite integral.

79. $\int_3^4 \frac{x}{x-2}\,dx$

80. $\int_5^7 \frac{x+5}{x-4}\,dx$

81. $\int_0^{e-1} \frac{2x-5}{x+1}\,dx$

82. $\int_{-3/2}^0 \frac{3x-1}{2x+4}\,dx$

83. $\int_0^1 \frac{3x^2+4}{x^2+1}\,dx$

84. $\int_0^2 \frac{5x^2-1}{2x^2+4}\,dx$

85. $\int_4^6 \frac{3x^2+2x-9}{x^2-3}\,dx$

86. $\int_0^2 \frac{2x^2+4x+11}{x^2+x+5}\,dx$

87. $\int_{-1}^1 \frac{x^3+5x^2+4x+1}{x^3+2x^2+1}\,dx$

88–90 The function $v(t)$ gives the velocity, in units per second, of a particle moving along the x-axis, having started from the origin. Find **a.** the position of the particle at $t=t_0$ and **b.** the total distance traveled by the particle in the time interval $[0,t_0]$.

88. $v(t)=1-(t-1)^2$; $t_0=4$

89. $v(t)=\dfrac{t-1}{2(t+1)}$; $t_0=3$

90. $v(t)=t(t-3)(t-5)$; $t_0=6$

91. Find a formula for $f(x)$ if $\int_0^x f(t)\,dt=\sin 2x+x$.

92.* Repeat Exercise 91 if $\int_0^{x^2} f(t)\,dt=x^3$.

93.* Let $f(x)=x-[\![x]\!]$ and $F(x)=\int_0^x f(t)\,dt$. Prove that F is continuous, briefly discuss its graph, and sketch it on paper.

94. Show that the piecewise defined function
$$F(x)=\begin{cases}-\frac{1}{2}x^2 & \text{if } x\le 0\\ \frac{1}{2}x^2 & \text{if } x>0\end{cases}$$
is an antiderivative of $f(x)=|x|$. Then find an easier formula for $F(x)$ and use the Fundamental Theorem of Calculus to evaluate $\int_a^b |x|\,dx$.

95. Write a paragraph entitled "Differentiation and Integration as Inverse Operations." Quote the Fundamental Theorem of Calculus and include concrete examples.

96.* Use the properties of the definite integral to show directly that if $f(x)$ is integrable on $[a,b]$, then $F(x)=\int_a^x f(t)\,dt$ is continuous on the same interval. (**Hint:** Argue that there is an M so that $|f(x)|\le M$ on $[a,b]$ and use the result of Exercise 103 of Section 5.2.)

97. Let $l(x)$ be defined as the integral function of $1/x$, that is, $l(x)=\int_1^x (1/t)\,dt$. Show that $l(x)=\ln x$. (**Hint:** See the discussion in Example 4.)

98.* Use the definition from Exercise 97 to show the following well-known property of logarithms: For positive $a,b\in\mathbb{R}$, $l(a\cdot b)=l(a)+l(b)$. (**Hint:** Use the definition to show that $l'(ax)=l'(x)$, which implies $l(ax)=l(x)+C$ for some constant C. Argue that $l(a)=C$. Finally, let $x=b$.)

99.* Use the definition from Exercise 97 to show that $l(1/x)=-l(x)$.

100. Taking a cue from Exercises 97–99, let
$$\ln x=\int_1^x \frac{1}{t}\,dt, x>0$$
be the definition of the natural logarithm function—that is, let all our knowledge of the natural logarithm function be determined by this particular definite integral. Note that, by the Fundamental Theorem of Calculus (which applies, since $1/t$ is continuous on the interval $0<t<\infty$), the natural logarithm is a differentiable function and
$$\frac{d}{dx}(\ln x)=\frac{1}{x}.$$

a. Prove that $\ln 1 = 0$ and that $\lim\limits_{x \to \infty} \ln x = \infty$.

(**Hint:** Construct a Riemann sum based on the given figure to show that

$$\int_1^x \frac{1}{t}\,dt > \frac{1}{2} + \left(\frac{1}{3} + \frac{1}{4}\right) + \left(\frac{1}{5} + \frac{1}{6} + \frac{1}{7} + \frac{1}{8}\right) + \cdots$$

$$> \frac{1}{2} + 2\left(\frac{1}{4}\right) + 4\left(\frac{1}{8}\right) + \cdots$$

$$= \frac{1}{2} + \frac{1}{2} + \frac{1}{2} + \cdots$$

for sufficiently larger x.)

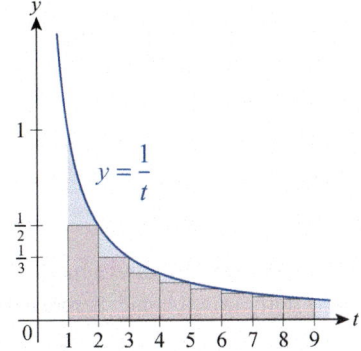

Note that these two facts, along with the fact that $\ln x$ is a continuous function, implies that $\ln x$ takes on every positive real value over the interval $1 < x < \infty$ (and also, given the result of Exercise 99, every negative real value over the interval $0 < x < 1$).

b. Prove that $\ln x$ is one-to-one and hence has an inverse function. (**Hint:** Prove that $\ln x$ is strictly increasing). Given this fact, define e^x to be the inverse of $\ln x$; that is, define e^x by $e^x = \ln^{-1} x$. In particular, define $e = \ln^{-1}(1)$.

c. Use L'Hôpital's Rule to prove

$$\lim_{u \to 0} \frac{u}{\ln(1+u)} = 1.$$

d. Use the result from part c. to prove

$$\lim_{h \to 0} \frac{e^h - 1}{h} = 1.$$ (**Hint:** Let $u = e^h - 1$ and note that $u \to 0$ as $h \to 0$.)

101. Archimedes (287–212 BC) discovered that the area under a parabolic arch is two-thirds the length of the base times its height. Sketch the graph of $y = h - ax^2$, the general parabolic arch with vertex at $(0, h)$ and use the FTC to verify Archimedes' formula. (Note the interesting parallel between Archimedes' formula and that of the area of an isosceles triangle of the same base and height.)

102. The marginal cost of production of baby toys at a small company has been determined to be $C'(x) = 200 / \left(3\sqrt[3]{x}\right)$ dollars. How much will it cost to increase production from 400 to 500 toys?

103–108 *True or False?* Determine whether the given statement is true or false. In case of a false statement, explain or provide a counterexample.

103. If $f(x)$ is continuous on $[a,b]$ and $c \in [a,b]$ is the point guaranteed by the Mean Value Theorem for Definite Integrals, then $y = f(x)$ and the constant function $y = f(c)$ both have the same definite integral on $[a,b]$.

104. When evaluating a definite integral using the Fundamental Theorem of Calculus, we can use *any* of the antiderivatives of the integrand.

105. If $f(x)$ is a continuous, odd function on $\mathbb{R}$ and $F(x) = \int_{-a}^{x} f(t)\,dt$ for some $a > 0$, then $F(x)$ has a zero at $x = a$.

106. If $f(x)$ is integrable on $\mathbb{R}$, then $\int_a^x f(t)\,dt$ and $\int_b^x f(t)\,dt$ have the same derivative for all $a, b \in \mathbb{R}$.

107. $\dfrac{d}{dx}\displaystyle\int_a^{x^3} (t+1)^3\,dt = \left(x^3 + 1\right)^3$

108. If f is continuous on $[a,b]$ and F is any antiderivative of f, then the area of the region bounded by the graph of f and the x-axis is $F(b) - F(a)$.

5.3 Technology Exercises

109. Joy bought a new compact car for $15,000. She estimates that the rate of depreciation will be $f(t) = 15,000(\ln 1.25)0.8^t$ dollars, where t is measured in years. At the same time, all additional expenses (except for fuel costs) are expected to accumulate at the rate of $g(t) = 350t + 200$ dollars. This gives rise to the following formula for the average cost of ownership.

$$C(t) = \frac{1}{t}\int_0^t \left[f(s) + g(s) \right] ds$$

   a. Explain why the above formula makes sense. What is the average ownership cost during the first year? The first two years?

   b. Joy plans to replace her car at the time when her average cost starts to rise. When is that expected to happen? (**Hint:** Use a computer algebra system to find the minimum of $C(t)$.)

110. During a nighttime drive on an unfamiliar two-lane road in the California desert, Adam notices that his odometer is broken. He checks to see that his current speed is 52 mph and asks his passenger to jot down their speed every five minutes for the next hour. The table below shows the results.

t (min)	5	10	15	20	25	30	35	40	45	50	55	60
v (mph)	56	48	52	57	63	58	53	49	55	59	62	60

   a. Use the regression capabilities of a graphing calculator or computer algebra system to find a model for the velocity function, and denote it by $v(t)$. (**Hint:** It is advisable to convert minutes to hours before calculations.)

   b. Use the Fundamental Theorem of Calculus to estimate the distance they covered during the experiment by integrating $v(t)$ over the one-hour interval.

   c. Calculate the right-endpoint and midpoint estimates R_{12} and M_6, respectively, and compare them to the answer you gave in part b.

111. The owner of a large gas station wants to estimate the number of customers on a typical day between 6 a.m. and 6 p.m. He records the number of cars pulling in during a one-minute period at the top of each hour, from 6 a.m. to 5 p.m. The table below shows the results.

6 a.m.	7 a.m.	8 a.m.	9 a.m.	10 a.m.	11 a.m.	12 p.m.	1 p.m.	2 p.m.	3 p.m.	4 p.m.	5 p.m.
1	3	4	1	2	1	5	4	2	1	3	5

   a. Use the regression features of a graphing calculator or computer algebra system to find a model for the above data. (Make your own choice, but, for example, a sine regression works well here. Answers will vary.)

   b. Use your model and the integration capabilities of a computer algebra system to estimate the number of cars from 6 a.m. to 6 p.m.

   c. Find the average number of cars pulling in each minute during the above 12-hour time period.

112. Use the integrating and graphing capabilities of a computer algebra system to complete Exercise 93.

113–115 Find a formula for $F(x) = \int_0^x f(t)\,dt$, and then use a graphing calculator or computer algebra system to graph $f(x)$ and $F(x)$ together on the same screen over the given interval. Discuss how the main features of f (such as signs, intervals of monotonicity, zeros, critical points) are reflected in the graph of F.

113. $f(x) = x\sin\sqrt{x};\;\; [0,17]$ **114.** $f(x) = x\ln x;\;\; [0, 3.5]$ **115.** $f(x) = \sin 3x \cos 5x;\;\; [-1,1]$

5.4 **Indefinite Integrals and the Substitution Rule**

The Fundamental Theorem of Calculus solidly establishes the connection between integration and antidifferentiation and provides the single most powerful tool for calculating definite integrals. Given this connection, it isn't surprising that the notation typically used to denote antiderivatives also incorporates the integral sign. In this section, we introduce that notation and learn the first of many techniques for finding antiderivatives of a given function.

Topic One
The Meaning of Indefinite Integration

Remember that the definite integral of a function f over an interval $[a,b]$, if it exists, is a number defined by the limit of Riemann sums—even if, in practice, we avoid actually calculating a definite integral by this manner whenever possible. The FTC gives us the ability to determine the definite integral $\int_a^b f(x)\,dx$ by instead evaluating $F(b)-F(a)$, where F is any antiderivative of f. This connection is the basis of our next definition.

Definition

Indefinite Integral

Given a function f, the **indefinite integral of f** is denoted by $\int f(x)\,dx$. This notation stands for the set of all antiderivatives of f—that is, $\int f(x)\,dx = F(x)+C$, where C represents an arbitrary constant and F is any particular antiderivative of f.

At this point, since we are integrating real-valued functions of a single variable, C can be taken to represent an arbitrary real number. More generally, it is important to remember that C represents an arbitrary term, which is constant with respect to the variable of integration, so that its derivative with respect to that variable is 0. (This will be relevant when we later consider integrals involving more than one variable or, in other settings, when the underlying set is not the set of real numbers.)

Example 1

Evaluate the following indefinite integrals.

a. $\int \left(3x^2 - 4x + 7\right) dx$ **b.** $\int \sec^2 t\, dt$ **c.** $\int \dfrac{1}{x} dx$

Solution

The instruction "evaluate" in this context means to describe the family of antiderivatives without the use of the integral sign, if possible.

a. Since $\dfrac{d}{dx}\left(x^3 - 2x^2 + 7x\right) = 3x^2 - 4x + 7$, we know that

$$\int \left(3x^2 - 4x + 7\right) dx = x^3 - 2x^2 + 7x + C.$$

While this is, in some sense, the most natural way to describe the family of antiderivatives of the integrand, the number of equivalent answers is infinite. Note that

$$\int \left(3x^2 - 4x + 7\right) dx = x^3 - 2x^2 + 7x - 13 + C$$

and

$$\int \left(3x^2 - 4x + 7\right) dx = x^3 - 2x^2 + 7x + 15\pi + C$$

are also correct (though admittedly strange), because as C takes on all possible real values, each formulation of the answer describes the same family of functions.

b. $\dfrac{d}{dt}(\tan t) = \sec^2 t$, so we write $\int \sec^2 t\, dt = \tan t + C.$

c. Since $\dfrac{d}{dx}(\ln|x|) = \dfrac{1}{x}$, we write $\int \dfrac{1}{x}\, dx = \ln|x| + C.$

Caution!

Always remember that definite integrals are numbers, while indefinite integrals correspond to families of functions.

The relationship between definite and indefinite integration is sometimes seen as follows:

$$\int_a^b f(x)\, dx = \left[\int f(x)\, dx\right]_a^b$$

This makes sense, as long as it is recognized that the expression on the right refers to the evaluation of *any one* antiderivative of f at the endpoints a and b; since $F(b) - F(a)$ results in the same number for every antiderivative F of f, the expression $\left[\int f(x)\, dx\right]_a^b$ is the same no matter which antiderivative is used. Remember, though, the conditions under which the Fundamental Theorem of Calculus applies.

Example 2

Use the FTC to evaluate the following, if possible. If the FTC does not apply, indicate why not.

a. $\displaystyle\int_{-3}^{2}\left(3x^2 - 4x + 7\right)dx$ **b.** $\displaystyle\int_{-5}^{-1}\frac{1}{x}\,dx$ **c.** $\displaystyle\int_{-1}^{1}\frac{1}{x^2}\,dx$

Solution

a. We have already determined that $\displaystyle\int\left(3x^2 - 4x + 7\right)dx = x^3 - 2x^2 + 7x + C$, so we can select any one of this family of functions and proceed. If we set $C = 0$, we find the following:

$$\int_{-3}^{2}\left(3x^2 - 4x + 7\right)dx = \left[x^3 - 2x^2 + 7x\right]_{-3}^{2}$$
$$= \left[2^3 - 2\left(2^2\right) + 7\left(2\right)\right] - \left[\left(-3\right)^3 - 2\left(-3\right)^2 + 7\left(-3\right)\right]$$
$$= 14 - \left(-66\right) = 80$$

The result is the same for every choice of C, since C is both added and subtracted in the expression $F(2) - F(-3)$. Moreover, since $3x^2 - 4x + 7$ is continuous over any interval $[a,b]$, we could evaluate the integral for any choice of a and b.

b. We know the definite integral exists because the integrand is continuous over the interval $[-5,-1]$. Further, since $1/x$ is negative over the interval, we should expect the integral (which represents signed area) to be negative as well. This is in fact what we see.

$$\int_{-5}^{-1}\frac{1}{x}\,dx = \left[\ln|x|\right]_{-5}^{-1} = \ln 1 - \ln 5 = -\ln 5 \approx -1.61$$

c. Our first observation has to be that the FTC does not apply, since $1/x^2$ is not a continuous function over the interval $[-1,1]$. If we failed to notice that fact and tried to evaluate the integral, we would obtain a very misleading result. We might (in error) say that $F(x) = -1/x$ is an antiderivative, since $F'(x) = 1/x^2$. However, F only qualifies as an antiderivative on $(-\infty,0)$ or $(0,\infty)$, not on an interval that contains 0. If we misapplied the FTC, we would obtain

$$\int_{-1}^{1}\frac{1}{x^2}\,dx = \left[-\frac{1}{x}\right]_{-1}^{1} = -1 - \left(-1\right) = 0,$$

which can't be true since $1/x^2$ is positive everywhere it's defined.

Using indefinite integral notation, the set of antiderivative facts that we have accumulated can be summarized as follows. Compare this to the tables of antiderivatives in Section 4.7.

$\int kf(x)\,dx = k\int f(x)\,dx$	$\int [f(x) \pm g(x)]\,dx = \int f(x)\,dx \pm \int g(x)\,dx$
$\int x^r\,dx = \dfrac{1}{r+1}x^{r+1} + C, \quad r \neq -1$	$\int \dfrac{1}{x}\,dx = \ln|x| + C$
$\int e^{kx}\,dx = \dfrac{1}{k}e^{kx} + C$	$\int a^{kx}\,dx = \left(\dfrac{1}{k\ln a}\right)a^{kx} + C, \quad a > 0, a \neq 1$
$\int \sin(kx)\,dx = -\dfrac{1}{k}\cos(kx) + C$	$\int \cos(kx)\,dx = \dfrac{1}{k}\sin(kx) + C$
$\int \sec^2(kx)\,dx = \dfrac{1}{k}\tan(kx) + C$	$\int \csc^2(kx)\,dx = -\dfrac{1}{k}\cot(kx) + C$
$\int \sec(kx)\tan(kx)\,dx = \dfrac{1}{k}\sec(kx) + C$	$\int \csc(kx)\cot(kx)\,dx = -\dfrac{1}{k}\csc(kx) + C$
$\int \dfrac{1}{\sqrt{1-(kx)^2}}\,dx = \dfrac{1}{k}\sin^{-1}(kx) + C, \quad |kx| < 1$	$\int \dfrac{1}{1+(kx)^2}\,dx = \dfrac{1}{k}\tan^{-1}(kx) + C$
$\int \dfrac{1}{|kx|\sqrt{(kx)^2-1}}\,dx = \dfrac{1}{k}\sec^{-1}(kx) + C, \quad |kx| > 1$	

Table 1 Indefinite Integrals

Topic Two

The Substitution Rule

Up to this point, we have found antiderivatives using nothing more than our knowledge of derivatives; since we know a substantial amount about differentiation, this approach shouldn't be discounted. Even in some cases where the integrand appears at first intractable (as far as antidifferentiation is concerned), a little algebraic manipulation can go a long way—for instance, we discovered in Section 5.3 that

$$\int \frac{x-2}{x+1}\,dx = x - 3\ln(x+1) + C$$

by rewriting $\dfrac{x-2}{x+1}$ as $1 - \dfrac{3}{x+1}$.

But as we encounter a wider variety of integrands, a larger toolbox of integration techniques will be very welcome. At heart, all of these techniques have properties of derivatives as their basis, so in a sense we are simply continuing to use knowledge of differentiation to work backward—the only difference is that the upcoming techniques have been refined and systematized over the years. Our first such technique is called the *Substitution Rule* or, more informally, *u*-substitution.

Theorem

The Substitution Rule

If $u = g(x)$ is a differentiable function whose range is the interval I, and if f is continuous on I, then

$$\int f(g(x))g'(x)\,dx = \int f(u)\,du.$$

Hence, if F is an antiderivative of f on I, $\int f(g(x))g'(x)\,dx = F(g(x)) + C$.

— Proof

The fact that $u = g(x)$ is differentiable tells us that $du/dx = g'(x)$, so the differentials du and dx are related by the equation $du = g'(x)\,dx$ (see Section 3.9). In this respect, the Substitution Rule is an example of a change of variables—if we restate the integral in terms of the variable u instead of x, then

$$\int f(g(x))g'(x)\,dx = \int f(u)\,du. \qquad g(x) = u \text{ and } g'(x)\,dx = du$$

But beyond this observation, the Substitution Rule is actually just a rephrasing of the Chain Rule of differentiation. Since f is continuous on the interval I, Part I of the FTC tells us that it has an antiderivative F. Applying the Chain Rule to $F \circ g$,

$$\frac{d}{dx}\Big[F(g(x))\Big] = F'(g(x))g'(x) = f(g(x))g'(x). \qquad F' = f$$

So $F(g(x))$ is an antiderivative of $f(g(x))g'(x)$ or, using integral notation,

$$\int f(g(x))g'(x)\,dx = F(g(x)) + C.$$

Example 3

Evaluate the following indefinite integrals.

a. $\int x\cos(x^2 + 1)\,dx$

b. $\int \dfrac{4x}{\sqrt{3 - x^2}}\,dx$

Solution

a. The key to applying u-substitution is examining the integrand and identifying a part of it that can be labeled u so that the rest (including the differential dx) corresponds to du. It is vital to remember that *all* of the integral needs to be restated in terms of the new variable u—an integral with a mixture of x's and u's is of no use at all. Our familiarity with differentiation guides us in defining u—in this particular integral, note that if we define $u = x^2 + 1$, then

$$\frac{du}{dx} = 2x, \text{ so } du = 2x\,dx.$$

This is almost exactly what we are after, and it is easy to fix it and make it perfect. Note that $x\,dx = \frac{1}{2}\,du$, so we evaluate the integral as follows:

$$\int x\cos(x^2 + 1)\,dx = \int \cos(x^2 + 1)x\,dx$$

$$= \int \cos(u)\left(\frac{1}{2}\right)du \qquad\qquad x\,dx = \frac{1}{2}\,du$$

$$= \frac{1}{2}\int \cos u\,du$$

$$= \frac{1}{2}\sin u + C$$

$$= \frac{1}{2}\sin(x^2 + 1) + C \qquad\qquad \text{Replace } u \text{ with } x^2 + 1.$$

And, as always, you can easily check your answer by differentiating it and verifying that you obtain the original integrand.

b. Part of this integrand is a second-degree polynomial and the rest is a first-degree polynomial. Specifically, if we set $u = 3 - x^2$, then

$$\frac{du}{dx} = -2x, \text{ so } du = -2x\,dx.$$

Proceeding as in part a., we evaluate the integral.

$$\int \frac{4x}{\sqrt{3 - x^2}}\,dx = \int \frac{4}{\sqrt{u}}\left(-\frac{1}{2}\right)du \qquad\qquad x\,dx = -\frac{1}{2}\,du$$

$$= -2\int u^{-1/2}\,du$$

$$= -2\left(2u^{1/2}\right) + C \qquad\qquad \frac{d}{du}\left(2u^{1/2}\right) = u^{-1/2}$$

$$= -4\sqrt{3 - x^2} + C \qquad\qquad u = 3 - x^2$$

Our choice of u in each integral of Example 3 was guided by the fact that part of the integrand was a second-degree polynomial while the rest of the integrand was, after adjusting by a constant factor, the derivative of that polynomial. The way forward may not always be so apparent in applying the Substitution Rule, but there are some general guidelines to keep in mind. One good way to begin is to define u to be the inner expression in a composition of functions; this is another way of describing what we did in Example 3. Another approach to take is to rewrite the integrand in an algebraically equivalent form, as illustrated in parts b. and c. of Example 4. And, as always, don't be discouraged if your first attempts don't result in an easier integral—u-substitution, like many other techniques in mathematics, is something of an art and it will become second nature with practice.

Example 4

Evaluate the following indefinite integrals.

a. $\displaystyle\int x^3 \sqrt{x^2 + 1}\, dx$ **b.** $\displaystyle\int \tan x\, dx$ **c.** $\displaystyle\int \frac{dx}{3^x + 3^{-x}}$

Solution

a. It is worth seeing what happens if we set $u = x^2 + 1$, even though in this case the rest of the integrand is a polynomial one degree higher than u, not lower. If $u = x^2 + 1$, then $du = 2x\, dx$ or $\frac{1}{2} du = x\, dx$.

$$\int x^3 \sqrt{x^2 + 1}\, dx = \int x^2 \cdot x\sqrt{x^2 + 1}\, dx$$

$$= \frac{1}{2}\int x^2 \sqrt{u}\, du \qquad\qquad x\, dx = \tfrac{1}{2} du$$

We don't yet have an integral we can do anything with (both x's and u's appear in the integrand), but the x^2 that remains can be expressed in terms of u. Since $u = x^2 + 1$, it's also true that $x^2 = u - 1$.

$$\frac{1}{2}\int x^2 \sqrt{u}\, du = \frac{1}{2}\int (u-1)u^{1/2}\, du$$

$$= \frac{1}{2}\int \left(u^{3/2} - u^{1/2}\right) du$$

$$= \frac{1}{2}\left(\frac{2}{5}u^{5/2} - \frac{2}{3}u^{3/2}\right) + C$$

$$= \frac{1}{5}\left(x^2 + 1\right)^{5/2} - \frac{1}{3}\left(x^2 + 1\right)^{3/2} + C$$

b. At first glance, it's hard to see what we can do with $\int \tan x\, dx$, since the integrand consists of a single factor and we haven't yet encountered a function whose derivative is tangent. Some algebraic rewriting is necessary.

$$\int \tan x\, dx = \int \frac{\sin x}{\cos x}\, dx \qquad\qquad \tan x = \frac{\sin x}{\cos x}$$

$$= \int \left(\frac{1}{\cos x}\right) \sin x\, dx \qquad \text{Set } u = \cos x; \text{ then } du = -\sin x\, dx.$$

$$= -\int \frac{1}{u}\, du$$

$$= -\ln|u| + C$$

$$= -\ln|\cos x| + C \qquad \text{Alternatively, } -\ln|\cos x| = \ln\left(|\cos x|^{-1}\right) = \ln|\sec x|.$$

c. This integral is difficult to get a handle on in its original form, but it is more tractable if we multiply the numerator and denominator by 3^x.

$$\int \frac{dx}{3^x + 3^{-x}} = \int \frac{1}{3^x + 3^{-x}} \left(\frac{3^x}{3^x} \right) dx \qquad \text{Multiply the integrand by 1.}$$

$$= \int \frac{3^x}{3^{2x} + 1} dx \qquad \text{Set } u = 3^x; \text{ so } du = 3^x (\ln 3) dx.$$

$$= \frac{1}{\ln 3} \int \frac{1}{u^2 + 1} du \qquad 3^{2x} = (3^x)^2 = u^2$$

$$= \frac{1}{\ln 3} \tan^{-1} u + C$$

$$= \frac{1}{\ln 3} \tan^{-1} (3^x) + C$$

5.4 Exercises

1–15 Evaluate the integral, definite or indefinite, as indicated. (**Hint:** See Examples 1 and 2 and the subsequent table of integrals.)

1. $\int (12x^5 + 7.5x^4 - x^3 + 2) dx$

2. $\int (-3x^4 + 0.8x^3 - 6x^2 + 4x - \pi) dx$

3. $\int 2(x+1)(5x-2) dx$

4. $\int_{-1}^{1} -x(x+4)(2x-1) dx$

5. $\int_0^1 x\sqrt[3]{x} \, dx$

6. $\int \frac{x^4 - 3\sqrt{x}}{x^2} dx$

7. $\int \frac{x^2 - 2x}{\sqrt{x} + \sqrt{2}} dx$

8. $\int \left(\sqrt[3]{x^2} + \frac{1}{\sqrt{x}} \right)^2 dx$

9. $\int (\pi \sec x - \tan x) \sec x \, dx$

10. $\int (e^{3x} + 2^{2x/3}) dx$

11. $\int \frac{\cot 2x}{2 \sin x \cos x} dx$

12. $\int_0^{\sqrt{3}/2} \frac{2}{1 + 4x^2} dx$

13. $\int_0^{1/2} \frac{2}{\sqrt{1 - 4x^2}} dx$

14. $\int \frac{3}{|4x| \sqrt{16x^2 - 1}} dx$

15. $\int \frac{-7}{\sqrt{1 - 25x^2}} dx$

16–36 Perform the suggested substitution to evaluate the given indefinite integral.

16. $\int 6x(3x^2 + 5)^7 dx; \quad u = 3x^2 + 5$

17. $\int x^3 \sqrt{x^4 + 2} \, dx; \quad u = x^4 + 2$

18. $\int \frac{2x}{4x^2 + 1} dx; \quad u = 4x^2 + 1$

19. $\int \frac{2}{4x^2 + 1} dx; \quad u = 2x$

20. $\int 4e^{2t+3} dt; \quad u = 2t + 3$

21. $\int 4e^{2t+3} dt; \quad u = e^{2t+3}$

22. $\int \cos 5\theta \, d\theta; \quad u = 5\theta$

23. $\int \frac{5^{\arctan x}}{1 + x^2} dx; \quad u = \arctan x$

24. $\int \frac{\sin 2x}{\sqrt{\sin^2 x + 1}} dx; \quad u = \sin^2 x + 1$

25. $\int \dfrac{\sec^2\left(1+\sqrt{s}\right)}{\sqrt{s}}\,ds; \quad u = 1+\sqrt{s}$

26. $\int \dfrac{1}{3s^{2/3}\sqrt{1-s^{2/3}}}\,ds; \quad u = \sqrt[3]{s}$

27. $\int \dfrac{1}{x^2}\sec^2\dfrac{1}{x}\,dx; \quad u = \dfrac{1}{x}$

28. $\int \sqrt{x}\cos^3\left(x^{3/2}\right)dx; \quad u = x^{3/2}$ (**Hint:** Use $u = x^{3/2}$ and $\cos^3 u = \cos^2 u\cos u = \left(1-\sin^2 u\right)\cos u$, and then perform another substitution $w = \sin u$.)

29. $\int \sqrt{1+\cot^2 x}\,\cot x\csc^2 x\,dx; \quad u = 1+\cot^2 x$

30. $\int (x+1)(x-7)^8\,dx; \quad u = x-7$

31. $\int \dfrac{z}{2z-1}\,dz; \quad u = 2z-1$

32. $\int z\sqrt{z^2-5}\,dz; \quad u = z^2-5$

33. $\int x\sqrt{x-5}\,dx; \quad u = x-5$

34. $\int \dfrac{\left(\sqrt{x}-2\right)\sqrt{1+\sqrt{x}}}{\sqrt{x}}\,dx; \quad u = 1+\sqrt{x}$

35.* $\int \sqrt{1-x^2}\,dx; \quad x = \sin u$

36. $\int \sin^4 x\cos^3 x\,dx; \quad u = \sin x$

37–84 Use an appropriate substitution to evaluate the indefinite integral.

37. $\int 3(3x-2)^7\,dx$

38. $\int -2x\sqrt{9-x^2}\,dx$

39. $\int (4x+3)\left(2x^2+3x\right)^{20}\,dx$

40. $\int -2x^3\sqrt{9-x^2}\,dx$

41. $\int \cot z\,dz$

42. $\int (z-2)\left(3z^2-12z\right)^{99}\,dz$

43. $\int x^2\left(x^3-5\right)^{19}\,dx$

44. $\int 6x^3\left(x^4+2\right)^{14}\,dx$

45. $\int e^{\sin x}\cos x\,dx$

46. $\int e^x\csc^2\left(e^x\right)dx$

47. $\int (5-s)^{37}\,ds$

48. $\int s\sqrt{s^2+1}\,ds$

49. $\int x^3\sqrt[3]{x^4+11}\,dx$

50. $\int \sqrt[3]{5x+9}\,dx$

51. $\int \dfrac{2x+1}{\left(x^2+x-7\right)^2}\,dx$

52. $\int \dfrac{x^2}{x^3-1}\,dx$

53. $\int \dfrac{-x^3}{\sqrt{2+x^4}}\,dx$

54. $\int \dfrac{4x^3+10x}{x^4+5x^2+6}\,dx$

55. $\int \left(1+\dfrac{1}{t}\right)^3\dfrac{1}{t^2}\,dt$

56. $\int \dfrac{3t+9}{\sqrt{2t^2+12t}}\,dt$

57. $\int \dfrac{2}{t\ln 2t}\,dt$

58. $\int \dfrac{\ln x}{x}\,dx$

59. $\int xe^{x^2-3}\,dx$

60. $\int x\sin\left(x^2\right)dx$

61. $\int \cos \pi x\,dx$

62. $\int \dfrac{x}{\cos^2\left(x^2\right)}\,dx$

63. $\int \sin^3 2t\cos 2t\,dt$

64. $\int 3x\sec^2\left(x^2+1\right)dx$

65. $\int \dfrac{\csc^2 v}{e^{\cot v-1}}\,dv$

66. $\int \dfrac{5^{\sqrt{3v}}}{\sqrt{3v}}\,dv$

67. $\int \dfrac{\sqrt{x}}{\left(30-x^{3/2}\right)^2}\,dx$

68. $\int \dfrac{1}{\sqrt{x}\left(1+\sqrt{x}\right)^2}\,dx$

69. $\int \dfrac{\sin 2x}{\sin^2 x+2}\,dx$

70. $\int \dfrac{\tan \sqrt{x}\sec \sqrt{x}}{\sqrt{x}}\,dx$

71. $\int \cos x\cos(\sin x)\,dx$

72. $\int \dfrac{\ln\left(x^3\right)}{x}\,dx$

73. $\int \dfrac{\sin 2x}{1-\cos 2x}\,dx$

74. $\int \dfrac{e^{2t}}{e^t-2}\,dt$

75. $\int \tan^3 2x\sec 2x\,dx$

76. $\int \dfrac{z^3}{1-z^2}\,dz$

77. $\int \dfrac{\sqrt{1+\sqrt{w}}}{\sqrt{w}}\,dw$

78. $\int \sqrt{1+\sqrt{w}}\,dw$

79. $\int (x-5)(x+1)^{11}\,dx$

80. $\int 2x\sqrt{x+2}\,dx$

81. $\int \dfrac{x+2}{4x+1}\,dx$

82. $\int \dfrac{x^2+x+1}{x-2}\,dx$

83. $\int \dfrac{(2\ln x+5)(1-\ln x)^3}{2x}\,dx$

84. $\int \sqrt{2+\sqrt{x}}\,dx$

85–90 Find the function that satisfies the given conditions.

85. $\dfrac{df}{dx} = 4x\sqrt{4x^2 + 4}; \quad f(0) = 1$

86. $\dfrac{dg}{ds} = \dfrac{2\ln s}{s}; \quad g(1) = 5$

87. $\dfrac{dy}{dt} = \dfrac{\sin 2t}{\sin^2 t + e}; \quad y(0) = 0$

88. $y'(x) = \dfrac{3\sqrt{x}}{\left(1 - x^{3/2}\right)^2}; \quad y(0) = 0$

89. $y''(x) = \cos 4x; \quad y'(0) = 1; \quad y(0) = 0$

90. $\dfrac{d^2 y}{dt^2} = 2\cot t \csc^2 t; \quad y'\left(\dfrac{\pi}{2}\right) = 0; \quad y\left(\dfrac{\pi}{2}\right) = 0$

91. A particle that started at the origin and is moving along the *x*-axis has a velocity function given by

$$v(t) = \dfrac{1 + \sqrt{t+1}}{\sqrt{t+1}} \text{ units/s}.$$

What is the particle's position at $t = 3$ seconds?

92. A particle is undergoing simple harmonic motion along the *y*-axis around the equilibrium $y = 0$, while its acceleration is given by

$$a(t) = 4\pi^2 \sin \dfrac{\pi(1 + 8t)}{4} \text{ units/s}^2.$$

Find the particle's position at $t = 1.5$ seconds and the total distance covered by the particle from $t = 0$ seconds to $t = 1.5$ seconds.

93–98 *True or False?* Determine whether the given statement is true or false. In case of a false statement, explain or provide a counterexample.

93. If f is defined on the interval $[a,b]$ and has an antiderivative, then the indefinite integral of f on $[a,b]$ is a number.

94. If f is defined on the interval $[a,b]$ and has an antiderivative, then the indefinite integral of f on $[a,b]$ is a function.

95. Two different elements of $\int f(x)\,dx$ can only differ by a constant; that is, if both $F(x)$ and $G(x)$ are elements of the set, then there is a constant C such that $F(x) = G(x) + C$.

96. The Substitution Rule can be interpreted as the Chain Rule in reverse.

97. $\displaystyle \int (\cos x + 1)^2\, dx = \dfrac{(\cos x + 1)^3}{3} + C$

98. $\displaystyle \int \dfrac{1}{x^2 + x + 1}\, dx = \ln\left| x^2 + x + 1 \right| + C$

5.5 **The Substitution Rule and Definite Integration**

In this last section of the chapter, we introduce a refinement of the Substitution Rule as it applies to definite integrals and then apply our developing integration skills to variations of the area problem with which we began the chapter.

Topic One
Substitution and Definite Integration

If we use the Substitution Rule to find an antiderivative, we can use it to evaluate a definite integral as long as we ensure that it is an antiderivative *of the original integrand*. Remember that a change of variables occurs in applying *u*-substitution (hence the name), and it is all too easy to forget to restate the integrand in terms of the original variable before evaluating it at the limits of integration. To avoid this problem, and to shorten the process of determining definite integrals slightly, the following variation of the Substitution Rule is frequently used.

Theorem

The Substitution Rule for Definite Integrals

If g' is a continuous function on the interval $[a,b]$, and if f is continuous on the range of $u = g(x)$, then

$$\int_a^b f(g(x))g'(x)\,dx = \int_{g(a)}^{g(b)} f(u)\,du.$$

— Proof

First, we know that both integrals exist since both integrands are continuous on their respective intervals: $(f \circ g)g'$ is a product of two continuous functions on $[a,b]$, and f is continuous on the interval $[g(a),g(b)]$. Our task is simply to prove that the integrals are equal.

As in the first version of the Substitution Rule, let F be an antiderivative of f.

$$\int_a^b f(g(x))g'(x)\,dx = F(g(x))\Big]_{x=a}^{x=b} \qquad \frac{d}{dx}F(g(x)) = f(g(x))g'(x)$$

$$= F(g(b)) - F(g(a))$$

$$= F(u)\Big]_{u=g(a)}^{u=g(b)}$$

$$= \int_{g(a)}^{g(b)} f(u)\,du$$

Example 1

Evaluate $\displaystyle\int_0^2 x^3 \sqrt{x^2+1}\, dx$.

Solution

Using the substitution $u = x^2 + 1$, we determined in Example 4a of Section 5.4 that

$$\int x^3 \sqrt{x^2+1}\, dx = \frac{1}{5}u^{5/2} - \frac{1}{3}u^{3/2} + C$$

$$= \frac{1}{5}\left(x^2+1\right)^{5/2} - \frac{1}{3}\left(x^2+1\right)^{3/2} + C,$$

so

$$\int_0^2 x^3 \sqrt{x^2+1}\, dx = \left[\frac{1}{5}\left(x^2+1\right)^{5/2} - \frac{1}{3}\left(x^2+1\right)^{3/2} + C\right]_{x=0}^{x=2}$$

$$= \left(\frac{1}{5}(5)^{5/2} - \frac{1}{3}(5)^{3/2}\right) - \left(\frac{1}{5}(1)^{5/2} - \frac{1}{3}(1)^{3/2}\right)$$

$$= \left(5\sqrt{5} - \frac{5\sqrt{5}}{3}\right) - \left(\frac{1}{5} - \frac{1}{3}\right) = \frac{2}{15}\left(25\sqrt{5}+1\right).$$

Alternatively, we can change every part of the definite integral, including the limits of integration: when $x = 0$, $u = 0^2 + 1 = 1$, and when $x = 2$, $u = 2^2 + 1 = 5$.

$$\int_{x=0}^{x=2} x^3 \sqrt{x^2+1}\, dx = \frac{1}{2}\int_{u=1}^{u=5}(u-1)\sqrt{u}\, du \qquad \tfrac{1}{2}\, du = x\, dx$$

$$= \left[\frac{1}{5}u^{5/2} - \frac{1}{3}u^{3/2}\right]_{u=1}^{u=5}$$

$$= \left(\frac{1}{5}(5)^{5/2} - \frac{1}{3}(5)^{3/2}\right) - \left(\frac{1}{5} - \frac{1}{3}\right)$$

$$= \frac{2}{15}\left(25\sqrt{5}+1\right)$$

Caution!

The two methods illustrated in Example 1 result in the same answer; the choice usually depends only on which seems easier to apply. Just be careful not to mix and match—don't evaluate an antiderivative expressed in the variable u with limits that belong to the original variable! To keep track of which variable the limits correspond to, it can be useful to explicitly note the variable in the integral symbol as we did above.

It's always worthwhile to try to keep the big picture in mind in mathematics, and doing so while evaluating definite integrals can save a significant amount of work. In many practical applications, for instance, integrals over an interval centered at 0 appear. If the integrand in such a case is an even or odd function, the following theorem applies.

Theorem

Definite Integrals of Even and Odd Functions

Assume f is continuous on the interval $[-a, a]$. Then

1. if f is even, $\displaystyle\int_{-a}^{a} f(x)\,dx = 2\int_{0}^{a} f(x)\,dx$;

2. if f is odd, $\displaystyle\int_{-a}^{a} f(x)\,dx = 0$.

Proof

We will prove the first statement here. Note the use of the Substitution Rule for Definite Integrals in the second-to-last step and, as an aid in making the change of variable, the explicit mention of the variable corresponding to the limits in two of the integrals. The symmetry with respect to the y-axis in Figure 1 illustrates why the signed area of an even function over the interval $[-a, a]$ is twice its signed area over the interval $[0, a]$.

$$\int_{-a}^{a} f(x)\,dx = \int_{-a}^{0} f(x)\,dx + \int_{0}^{a} f(x)\,dx \qquad \text{\footnotesize $\int_{a}^{b} f(x)\,dx = \int_{a}^{c} f(x)\,dx + \int_{c}^{b} f(x)\,dx$}$$

$$= -\int_{0}^{-a} f(x)\,dx + \int_{0}^{a} f(x)\,dx \qquad \text{\footnotesize $\int_{a}^{b} f(x)\,dx = -\int_{b}^{a} f(x)\,dx$}$$

$$= -\int_{x=0}^{x=-a} f(-x)\,dx + \int_{0}^{a} f(x)\,dx \qquad \text{\footnotesize Since f is even, $f(x) = f(-x)$.}$$

$$= \int_{u=0}^{u=a} f(u)\,du + \int_{0}^{a} f(x)\,dx \qquad \text{\footnotesize $u = -x$, $du = -dx$, and a change in limits}$$

$$= 2\int_{0}^{a} f(x)\,dx. \qquad \text{\footnotesize $\int_{0}^{a} f(u)\,du = \int_{0}^{a} f(x)\,dx$}$$

The proof of the second statement is similar, with the difference that in the third step above we will use $-f(x) = f(-x)$ since f is odd.

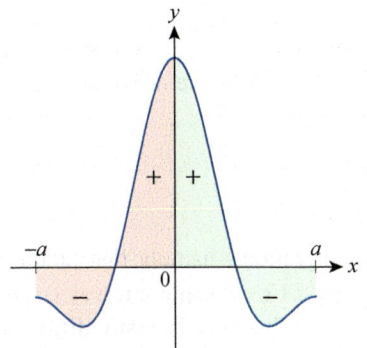

Figure 1
Signed Area of an Even Function

Example 2

Evaluate $\displaystyle\int_{-\pi}^{\pi} \frac{\sin x}{\sqrt{1 + x^2 + x^4}}\,dx.$

Solution

It would be very difficult to find an antiderivative of this integrand, but fortunately we don't have to. The integrand

$$f(x) = \frac{\sin x}{\sqrt{1 + x^2 + x^4}}$$

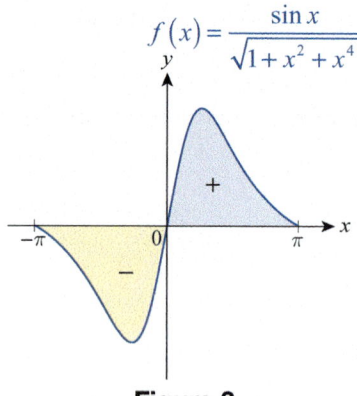

$$f(x) = \frac{\sin x}{\sqrt{1 + x^2 + x^4}}$$

Figure 2

is continuous and odd since it is a product of an odd continuous function ($\sin x$) and an even continuous function $1/\sqrt{1 + x^2 + x^4}$ (remember that the product of an odd function and an even function is odd). Since we are integrating over an interval centered at 0, we can say immediately that

$$\int_{-\pi}^{\pi} \frac{\sin x}{\sqrt{1 + x^2 + x^4}} \, dx = 0.$$

Geometrically, this reflects the fact that the signed areas in Figure 2 cancel each other out.

Topic Two
Area between Curves

Returning to the issue of finding areas bounded by curves, we are now equipped not only with powerful computational techniques, but also with an idea for how to attack such problems. In general, if we know the rate of change of an area A with respect to some variable, say x, over an interval $[a, b]$, then we know that

$$A = \int_a^b A'(x) \, dx.$$

In hindsight, we now know that this is simply one way of expressing the Fundamental Theorem of Calculus. If we rewrite it slightly, we arrive at a more philosophical statement:

$$A = \int_a^b A'(x) \, dx = \int_a^b \frac{dA}{dx} \, dx = \int_a^b dA$$

This version reflects the key idea at the heart of integration—namely, that "adding up" (that is, integrating) all of the small elements dA over some interval $[a, b]$ results in the total quantity A. We will apply this philosophy in many different contexts through the remainder of this text.

To see how this approach looks when applied to the specific problem of finding the area bounded by two curves $y = f(x)$ and $y = g(x)$ over the interval $[a, b]$, consider the assertion expressed in Figure 3.

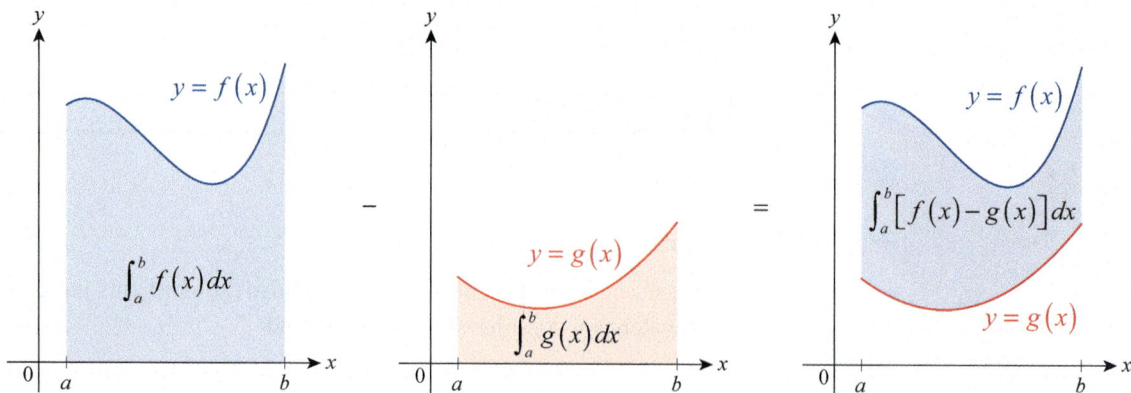

Figure 3 Area Bounded between f and g

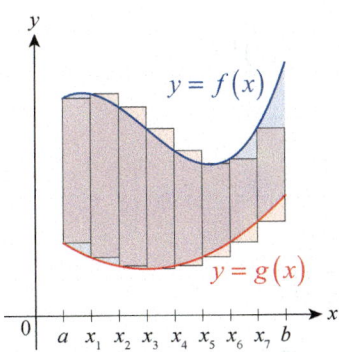

Figure 4
A Riemann Sum Approximation of A

Algebraically, we know that $\int_a^b f(x)\,dx - \int_a^b g(x)\,dx = \int_a^b \left[f(x) - g(x)\right]dx$, so this is not surprising. But we could also have determined A by integrating the area differential dA, with a formula for dA suggested by Figure 4.

Using this approach, each small area element ΔA is approximated by a rectangle of height $f(x) - g(x)$ and width Δx. In terms of differentials, $dA = \left[f(x) - g(x)\right]dx$, so we arrive at the same formula.

$$A = \int_a^b dA = \int_a^b \left[f(x) - g(x)\right]dx$$

Definition

Area of a Region between Two Curves

Given two continuous functions f and g defined on an interval $[a,b]$, with $f(x) \geq g(x)$ for all $x \in [a,b]$, the **area of the region bounded between the graphs of f and g over $[a,b]$** is $A = \int_a^b \left[f(x) - g(x)\right]dx$.

The next two examples further illustrate the use of the principle $A = \int_a^b dA$.

Example 3

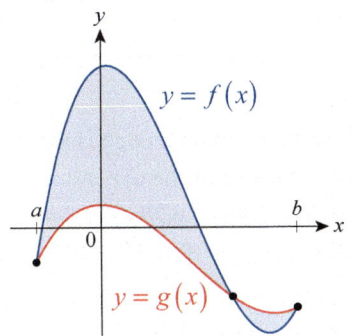

Figure 5

The graphs of the functions

$$f(x) = \frac{3x^3}{2} - 6x^2 + x + 7 \quad \text{and} \quad g(x) = \frac{x^3}{2} - 2x^2 + 1$$

are shown in Figure 5. Determine the total area of the shaded regions.

Solution

Using the graphs depicted in Figure 5, it appears that f is the upper function on the first part of the interval $[a,b]$ and that g is the upper function on the remainder of the interval. So in order to find the area of the shaded regions, we have to determine not only a and b but also the point inside the interval $[a,b]$ where f and g intersect.

All three are points where $f(x) = g(x)$, so we begin by solving this equation.

$$\frac{3x^3}{2} - 6x^2 + x + 7 = \frac{x^3}{2} - 2x^2 + 1$$
$$x^3 - 4x^2 + x + 6 = 0$$

Remember that if a polynomial equation has a rational number solution, it must be of the form p/q, where p is a factor of the constant term and q is a factor of the leading coefficient. Since the leading coefficient is 1 in this case, a rational root of the polynomial $x^3 - 4x^2 + x + 6$ must be one of the numbers $\pm\{1, 2, 3, 6\}$. As you can verify by synthetic or long division, $-1, 2$, and 3 are roots, so we know that $a = -1$, $b = 3$, and $x = 2$ must be the third point where $f(x) = g(x)$.

So the area of the shaded regions can be found as follows:

$$\text{Area} = \int_{-1}^{2}\left[f(x)-g(x)\right]dx + \int_{2}^{3}\left[g(x)-f(x)\right]dx$$

$$= \int_{-1}^{2}\left(x^3 - 4x^2 + x + 6\right)dx + \int_{2}^{3}\left(-x^3 + 4x^2 - x - 6\right)dx$$

$$= \left[\frac{x^4}{4} - \frac{4x^3}{3} + \frac{x^2}{2} + 6x\right]_{-1}^{2} + \left[-\frac{x^4}{4} + \frac{4x^3}{3} - \frac{x^2}{2} - 6x\right]_{2}^{3}$$

$$= \left[\frac{22}{3} - \left(-\frac{47}{12}\right)\right] + \left[-\frac{27}{4} - \left(-\frac{22}{3}\right)\right] = \frac{71}{6}$$

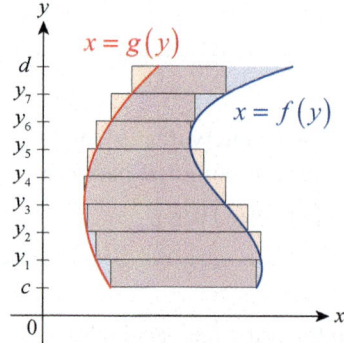

Figure 6 Integration in y

There is nothing sacred about vertical strips—if a region can be more naturally "deconstructed" in some other way, we should take advantage of that fact. If we are aiming to express the area of a region as an integral (or a sum of integrals), it is likely that we will think of the region as being composed of a collection of thin strips, but a horizontal arrangement, as depicted in Figure 6, may make more sense. In such a case, we can express each differential element of area as $dA = \left[f(y) - g(y)\right]dy$, and proceed to integrate with respect to y over the interval $[c,d]$. This can save us considerable effort, as shown in our next example.

Example 4

Find the area of the region bounded by the graphs of the equations $y = 0$, $3x - 5y = 12$, and $y = \sqrt{x}$.

Solution

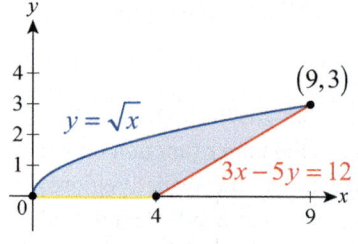

Figure 7

Two edges of the described region are lines, and the remaining edge is the graph of the function $y = \sqrt{x}$, which we can also think of as the upper half of the parabola $x = y^2$. As always, if it's possible to sketch a picture of what we are doing, such a sketch is bound to be helpful—in this case, it's easy to graph the region we're discussing. (See Figure 7.)

Note that we need two integrals if we want to express the area of the region as an integral in x, since the lower function changes at $x = 4$. On the interval $[0,4]$, the lower function is $g(x) = 0$ (corresponding to the equation $y = 0$), and on the interval $[4,9]$ the lower function is $g(x) = \frac{3}{5}(x - 4)$, which we obtain by solving $3x - 5y = 12$ for y. The upper function is $f(x) = \sqrt{x}$ for the entire interval $[0,9]$. So the total area of the described region is

$$A = \int_{0}^{4} \sqrt{x}\,dx + \int_{4}^{9}\left[\sqrt{x} - \frac{3}{5}(x - 4)\right]dx.$$

This is certainly doable, and you are asked to evaluate the above integrals in Exercise 52. But if we think of the region as being composed of horizontal strips, we see that the left edge can be described as a single function of y, and the same is true for the right edge—we don't have to divide the interval of integration into subintervals. Specifically, the left edge of the region is the function $g(y) = y^2$ (corresponding to the equation $x = y^2$) and the right edge is the function $f(y) = \frac{5}{3}y + 4$ (obtained by solving $3x - 5y = 12$ for x).

So each horizontal differential element of area can be written as

$$dA = \left[f(y) - g(y) \right] dy = \left(\frac{5}{3} y + 4 - y^2 \right) dy$$

thus the area is calculated as follows:

$$A = \int_0^3 dA = \int_0^3 \left(\frac{5}{3} y + 4 - y^2 \right) dy$$

$$= \left[\frac{5}{6} y^2 + 4y - \frac{1}{3} y^3 \right]_0^3$$

$$= \frac{15}{2} + 12 - 9 = \frac{21}{2}$$

5.5 Exercises

1–51 Evaluate the definite integral. Whenever possible, take advantage of symmetry.

1. $\displaystyle\int_0^1 2(2x+1)^5 \, dx$

2. $\displaystyle\int_0^2 2x\sqrt{4-x^2} \, dx$

3. $\displaystyle\int_0^2 (x^2-1)(x^3-3x)^8 \, dx$

4. $\displaystyle\int_0^4 (x-2)(2x^2-8x)^{49} \, dx$

5. $\displaystyle\int_1^2 w^2 (w^3+4)^{99} \, dw$

6. $\displaystyle\int_1^{10} 2x^3 (x^4-1)^{49} \, dx$

7. $\displaystyle\int_1^3 (2-x)^6 \, dx$

8. $\displaystyle\int_0^2 x\sqrt{x^2+1} \, dx$

9. $\displaystyle\int_0^1 x^3 \sqrt[3]{x^4+1} \, dx$

10. $\displaystyle\int_0^3 \sqrt{2x+1} \, dx$

11. $\displaystyle\int_1^4 \frac{2x+1}{(x^2+x+1)^2} \, dx$

12. $\displaystyle\int_0^1 \frac{z^2}{z^3+2} \, dz$

13. $\displaystyle\int_0^2 \frac{x^3}{\sqrt{x^4+9}} \, dx$

14. $\displaystyle\int_1^3 \frac{8x^3+20x}{x^4+5x^2+6} \, dx$

15. $\displaystyle\int_1^2 \left(1+\frac{1}{t^2}\right)^2 \frac{1}{t^3} \, dt$

16. $\displaystyle\int_1^2 \frac{6x+10.5}{\sqrt{2x^2+7x}} \, dx$

17. $\displaystyle\int_e^{e^2} \frac{1}{s\ln(s^3)} \, ds$

18. $\displaystyle\int_{10}^{100} \frac{\log x}{x\ln 10} \, dx$

19. $\displaystyle\int_1^{\sqrt{2}} (\ln 2)x \cdot 2^{x^2-1} \, dx$

20. $\displaystyle\int_0^{\sqrt{\pi}} 4x\cos\frac{x^2}{2} \, dx$

21. $\displaystyle\int_0^1 \sin \pi x \, dx$

22. $\displaystyle\int_{\sqrt{\pi/4}}^{\sqrt{3\pi/4}} \frac{-x}{\sin^2(x^2)} \, dx$

23. $\displaystyle\int_0^{\pi/4} \sin^2 2x \cos 2x \, dx$

24. $\displaystyle\int_0^1 x\sec^2(2x^2-1) \, dx$

25. $\displaystyle\int_{\pi^2/16}^{9\pi^2/16} \frac{\cot\sqrt{x}\csc\sqrt{x}}{\sqrt{x}} \, dx$

26. $\displaystyle\int_{-\pi}^{\pi} \sin x \sin(\cos x) \, dx$

27. $\displaystyle\int_1^e \frac{\ln(2x^2)}{x} \, dx$

28. $\displaystyle\int_0^1 \frac{e^{2t}}{e^t+1} \, dt$

29. $\displaystyle\int_{-2}^2 \frac{t^3}{t^2+1} \, dt$

30. $\displaystyle\int_0^{\pi/4} \frac{\csc^2\theta}{e^{\cot\theta}} \, d\theta$

31. $\displaystyle\int_{\pi^2/16}^{\pi^2/4} \frac{\csc^2\left(\frac{\pi}{4}+\sqrt{t}\right)}{\sqrt{t}} \, dt$

32. $\displaystyle\int_0^2 \frac{e^{\sqrt{2x}}}{\sqrt{2x}} \, dx$

33. $\displaystyle\int_0^1 \frac{\sqrt{v}}{\sqrt{v^{3/2}+1}} \, dv$

34. $\int_0^2 \dfrac{1}{\sqrt{x}\left(2+\sqrt{x}\right)^2}\,dx$

35. $\int_0^{\sqrt{3}} \dfrac{e^{\arctan x}}{1+x^2}\,dx$

36. $\int_0^1 \dfrac{1}{x^{2/3}\left(1+x^{2/3}\right)}\,dx$

37. $\int_0^{\pi/2} \dfrac{\sin 2x}{\cos^2 x+1}\,dx$

38. $\int_0^1 -e^{-x}\sec\left(e^{-x}-1\right)\tan\left(e^{-x}-1\right)dx$

39. $\int_4^9 \dfrac{\sqrt{2+\sqrt{t}}}{\sqrt{t}}\,dt$

40. $\int_0^4 \sqrt{2+\sqrt{t}}\,dt$

41. $\int_{-1}^1 \dfrac{x^2+x+1}{x-2}\,dx$

42. $\int_1^e \dfrac{(\ln x+1)(2\ln x+3)^2}{x}\,dx$

43. $\int_2^3 (x+3)(x-2)^7\,dx$

44. $\int_{-1}^0 x\sqrt[3]{x-1}\,dx$

45. $\int_{-1}^1 x\sqrt{4-x^2}\,dx$

46. $\int_4^{e+3} \dfrac{2x+5}{x-3}\,dx$

47. $\int_{-4}^0 \dfrac{x^2-x+3}{x+5}\,dx$

48. $\int_1^{e^2} \dfrac{(2\ln x+3)(\ln x-1)^2}{x}\,dx$

49. $\int_1^9 \dfrac{\left(2\sqrt{x}+1\right)\sqrt{1+\sqrt{x}}}{\sqrt{x}}\,dx$

50. $\int_0^{\pi^{2/3}} \sqrt{x}\sin^2\left(x^{3/2}\right)\cos^3\left(x^{3/2}\right)dx$

51. $\int_0^{\pi/4} \tan x\sec^3 x\,dx$

52. Evaluate $A = \int_0^4 \sqrt{x}\,dx + \int_4^9 \left[\sqrt{x}-\dfrac{3}{5}(x-4)\right]dx$ from Example 4.

53–56 Find the area of the region bounded by the graphs of the given equations, as shown.

53.

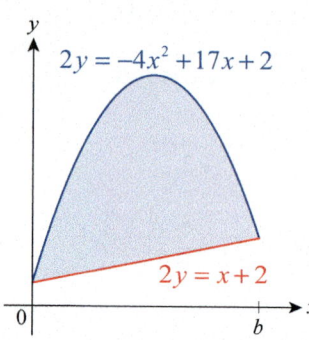

$2y = -4x^2 + 17x + 2$

$2y = x + 2$

54.

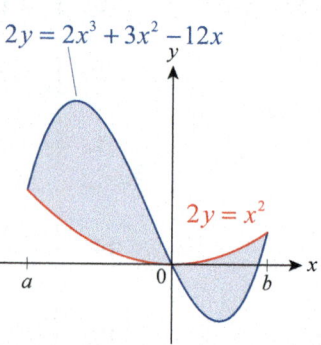

$2y = 2x^3 + 3x^2 - 12x$

$2y = x^2$

55.

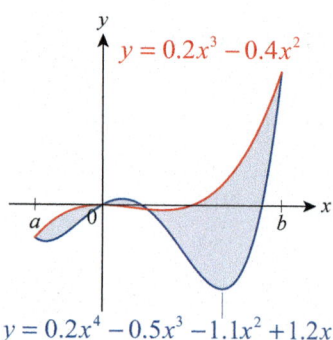

$y = 0.2x^3 - 0.4x^2$

$y = 0.2x^4 - 0.5x^3 - 1.1x^2 + 1.2x$

56.

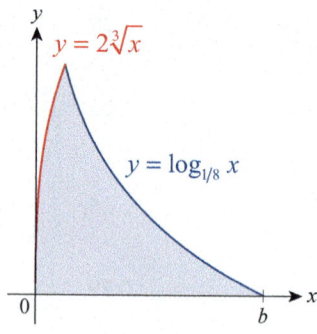

$y = 2\sqrt[3]{x}$

$y = \log_{1/8} x$

57–96 Find the area of the region bounded by the graphs of the given equations. Be careful to find intersection points, if applicable, and to identify the upper and lower functions on each interval. If convenient or necessary, divide the region into horizontal rather than vertical strips and integrate with respect to y. Whenever possible, take advantage of symmetry.

57. $y = x^2, \quad y = 2x$

58. $y = x^2, \quad y = 2$

59. $y = 4x - x^2, \quad y = x$

60. $y = 1 - x^4, \quad y = |x| - 1$

61. $y = |x^3 - 1|, \quad 3y = 5x + 11$

62. $y = x^2, \quad y = x^4$

63. $y = 2x - x^2, \quad y = x^3$

64. $y = \sqrt{x}, \quad y = 2 - x$

65. $x = y^2, \quad y = x^3, \quad x \geq 0$

66. $y = \sqrt[3]{x}, \quad y = \sqrt[7]{x}$

67. $2xy - y = 3 - 2x, \quad y = x, \quad y = 0$

68. $x + 30y = 2y^3 + 5, \quad 9y - 31 = x + y^3$

69. $y = \sqrt{x+1}, \quad y = x^2 - 1, \quad y = 0$

70. $y = x^3, \quad y = x$

71. $y = \dfrac{1}{x}, \quad y = \dfrac{1}{x^2}, \quad x = e$

72. $y = x^3, \quad y = \dfrac{3}{2} - \dfrac{x}{2}, \quad y = 0$

73. $y = 4x^2 - x^4, \quad y = -5x^2$

74. $y = 3x^2 - x - 4, \quad y = x^2 + 3x + 2$

75. $y^2 - x = 2, \quad y = x$

76. $1 - y^2 = x, \quad (1 - y)^2 = x$

77. $3y - x = 3y^2, \quad 6y^3 = x + 6y^2$

78. $y = x^3 - 3x^2 + 2x, \quad y = x^2 - x$

79. $y = x^3 - 6x^2 + 5, \quad y = -x^3 + 12x - 11$

80. $y = x^4 - \dfrac{x^3}{2} - 4x^2, \quad y = -\dfrac{x^3}{2}$

81. $y = 2\sqrt[3]{x+1}, \quad y = 2 - x, \quad y = 0$

82. $y = \sqrt{x}, \quad y = \dfrac{1}{x^2}, \quad x = 4$

83. $x = y^3 - 10y^2 + 20y, \quad x = 4y^2 - 33y + 40$

84. $x = y^4 - 3y^3 - y^2, \quad x = 5y^2 - 8y$

85. $y = \sqrt[4]{x}, \quad y = -\dfrac{1}{16}x^2 + 18, \quad y = 0$

86. $(y - 1)^2 = \dfrac{1}{x}, \quad x - 2 = 4y, \quad x = 0$

87. $y = \cos x, \quad y = \sin 2x, \quad x = 0, \quad x = \pi$

88. $y = \dfrac{2}{x^2 + 1}, \quad y = x^2$

89. $y = \sin x, \quad y = \sqrt{2} - \sin x, \quad x = \dfrac{\pi}{4}, \quad x = \dfrac{3\pi}{4}$

90. $y = \cot x, \quad y = 2\cos x, \quad 0 < x < \pi$

91. $y = x + 1, \quad y = \sqrt{18 - (x+1)^2}$

92. $y = \arctan x, \quad y + x = 1 + \dfrac{\pi}{4}, \quad x = 0$

93. $y = \cos\left(\dfrac{\pi}{2}x\right), \quad y = x^4 - 1$

94. $y = 2\sqrt{2}\sin x, \quad y = \csc^2 x, \quad 0 < x < \pi$

95. $y = \sin^2 x, \quad y = \cos^2 x, \quad x = 0, \quad x = \pi$

96. $y = \tan^2 x, \quad x = 0, \quad x = \dfrac{\pi}{4}$

97. Use the Substitution Rule to prove the following property of the definite integral.

$$\int_a^b f(x)\,dx = \int_{a+c}^{b+c} f(x-c)\,dx$$

Note that the above is often referred to as the *translation invariant property* of the definite integral. Using a generic $f(x)$, make a sketch of both integrals and explain the reason for the name of this property.

98. Use Exercise 97 to explain why the following definite integrals are equal.

$$\int_{-1}^{2} 4x\sqrt[3]{x+2}\,dx = \int_{1}^{4} (4x-8)\sqrt[3]{x}\,dx$$

99. Italian mathematician Bonaventura Cavalieri (1598–1647), who can be considered as one of the early forerunners of modern calculus, discovered what we today call *Cavalieri's Principle*. Use our discussion preceding Example 3 to prove the following version of Cavalieri's Principle: Suppose two plane regions are included between the lines $x = a$ and $x = b$, and are bounded by graphs of integrable functions. If they have the property that any vertical line intersects both regions in line segments of the same length, then the regions have equal areas.

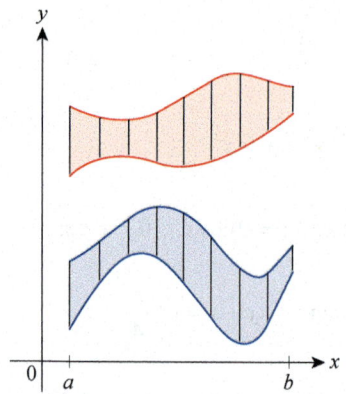

100. Consider the region bounded by the graphs of the equations $y = \sqrt{x}$, $x = 9$, and the x-axis. Find the vertical line $x = a$ that bisects the region in two subregions of equal area.

101. The graphs below show the velocities of two bikes at a motorcycle race right after the start (velocity is measured in km/h). Use the figure to answer the following questions.

a. Which bike is ahead initially?

b. What happens at the instant when the curves intersect?

c. Do the curves suggest that a pass happened, and if so, approximately when?

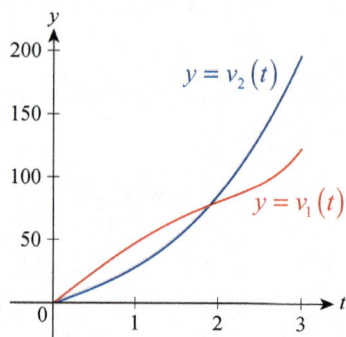

102. Suppose that the function $B(t) = 85 \cdot (1.1163)^t$ approximates the birth rate of a rabbit population on an isolated island, while the death rate is $D(t) = 21 \cdot (1.0811)^t$ (t is measured in months). Find the area between the graphs of these two functions on the interval $[0,12]$. Use your own words to give a real-life interpretation to this number.

5.5 Technology Exercises

103–107 Use a computer algebra system to plot the graphs of $f(x)$ and $g(x)$ on the same screen. After choosing the appropriate viewing window, identifying intersection points, and finding the region bounded by the curves, use the integration features of your technology to find the area of the region. (**Hint:** As in Examples 3 and 4, be sure to identify the upper and lower functions on each subinterval and integrate accordingly. As a final step, you may want to check your answer by evaluating $\int_a^b |f(x) - g(x)| dx$, where a and b are the first and last of the intersection points. Do you obtain the same answer?)

103. $f(x) = 35x - 9, \quad g(x) = 6x^3 - 4.95x^2 - 3.04x - 22.2525$

104. $f(x) = 3 \sin x, \quad g(x) = 0.3x$

105. $f(x) = e^x, \quad g(x) = \dfrac{1}{2}x + 2$

106. $f(x) = 2x^4 - 8x^3 - 6.5x^2 + 29x - 12, \quad g(x) = 2x^3 - 4x^2 - 3.5x + 2.5$

107. $f(x) = \dfrac{0.8^x \sin 2x}{2}, \quad g(x) = \dfrac{1}{2}\sqrt{x}$

Chapter 5
Review Exercises

1–2 Use $(O_4 + U_4)/2$ to estimate the area under the graph of the function and above the x-axis on the given interval.

1. $f(x) = \dfrac{x^2}{2}$ on $[0,2]$

2. $f(x) = \sin x$ on $\left[0, \dfrac{\pi}{2}\right]$

3–6 Write the given sum using sigma notation.

3. $\dfrac{1}{2} - \dfrac{1}{9} + \dfrac{1}{28} - \dfrac{1}{65} + \cdots - \dfrac{1}{1,000,001}$

4. $a_1 + a_5 + a_9 + a_{13} + \cdots + a_{97}$

5. $f\left(\dfrac{2}{n^2}\right) + f\left(\dfrac{4}{n^2}\right) + f\left(\dfrac{6}{n^2}\right) + \cdots + f\left(\dfrac{100}{n^2}\right)$

6. $g\left(t_{-2}^*\right)\Delta t + g\left(t_{-1}^*\right)\Delta t + g\left(t_0^*\right)\Delta t + \cdots + g\left(t_{2n}^*\right)\Delta t$

7–8 Assuming that $\displaystyle\sum_{i=0}^{n} a_i = 50$ and $\displaystyle\sum_{i=0}^{n} b_i = 80$, find the given sum.

7. $\displaystyle\sum_{i=0}^{n}(a_i + 2b_i + 2)$

8. $\displaystyle\sum_{i=0}^{n}\left(\dfrac{a_i}{5} - \dfrac{b_i}{4}\right)$

9–10 Use summation formulas to find the value of the sum.

9. $\displaystyle\sum_{i=1}^{10}(3i^3 - 1)$

10. $\displaystyle\sum_{j=1}^{n}\dfrac{(2j+1)(j-2)}{2}$

11. Find the value of the sum $\displaystyle\sum_{i=1}^{n}\left[\dfrac{1}{i^2} - \dfrac{1}{(i+1)^2}\right]$.

(**Hint:** Write out the first few terms as well as the last few terms.)

12–13 Evaluate the geometric sum using the formula proven in Exercise 53 of Section 5.1.

12. $\displaystyle\sum_{i=0}^{10}\dfrac{3}{2^i}$

13. $\displaystyle\sum_{j=0}^{6}(-1)^j(0.3)^j$

14–15 Evaluate the given double sum.

14. $\displaystyle\sum_{i=3}^{11}\sum_{j=1}^{4}(2i - j)$

15. $\displaystyle\sum_{i=1}^{n}\sum_{j=1}^{m}i^2 j$

16–18 Find the area under the graph of $f(x)$ and above the x-axis on the given interval, by taking the limit of the associated Riemann sums.

16. $f(x) = x^2 + 1$ on $[0,2]$

17. $f(x) = x^3$ on $[0,1]$

18. $f(x) = \sqrt{x}$ on $[0,4]$ (**Hint:** Choose $x_i^* = \dfrac{4(i-1)^2}{n^2}$.)

19. Identify the region whose area is the limit given by $\displaystyle\lim_{n\to\infty}\dfrac{2}{n}\sum_{i=1}^{n}\left[\dfrac{4i}{n} - \left(\dfrac{2i}{n}\right)^2\right]$. Then use summation formulas to evaluate the limit.

20–21 Prove that the function is not integrable on the interval $[0,1]$.

20. $f(x) = \dfrac{1}{x^2}$

21. $g(x) = \begin{cases} 1 & \text{if } x \text{ is rational} \\ -1 & \text{if } x \text{ is irrational} \end{cases}$

22–25 Suppose that f is an even function, g is odd, both are integrable on $[-a,a]$, and we know that $\int_0^a f(x)\,dx = 2$, while $\int_0^a g(x)\,dx = 0.5$ $(a>0)$. If possible, find the integral.

22. $\displaystyle\int_{-a}^{a}[5f(x) + 4g(x)]\,dx$

23. $\displaystyle\int_{-a}^{a}[f(x)]^2 g(x)\,dx$

24. $\displaystyle\int_{-a}^{a}f(x)[g(x)]^2\,dx$

25. $\displaystyle\int_{-a}^{0}[f(x) + g(x)]\,dx$

26–27 Find the average value of $f(x)$ over the given interval and identify all points in the domain where $f(x)$ assumes its average value.

26. $f(x) = 4x - x^2$ on $[0,4]$

27. $f(x) = |x-2| - 1$ on $[1,5]$

28. Use the Fundamental Theorem of Calculus to evaluate the limit $\displaystyle\lim_{n\to\infty}\dfrac{1}{n}\sum_{i=1}^{n}\left(\sqrt{\dfrac{i}{n}} - \dfrac{i}{n}\right)$ by recognizing it as a Riemann sum of a function over an interval.

29–30 Use Part I of the Fundamental Theorem of Calculus to find the derivative of the given function.

29. $F(x) = \int_0^x \sqrt{1+t^2}\, dt$ **30.** $G(x) = \int_0^{x^2} e^{t^2}\, dt$

31–38 Use Part II of the Fundamental Theorem of Calculus to evaluate the definite integral.

31. $\int_1^2 (2x^4 + 3x^2 - 2)\, dx$ **32.** $\int_0^2 (3x+2)(5-x)\, dx$

33. $\int_1^4 \left(\frac{1}{t} - \frac{2}{t^2} + 1\right) dt$ **34.** $\int_1^9 \frac{x^2 - 2\sqrt{x} + 2}{x}\, dx$

35. $\int_0^1 \frac{2}{\sqrt{1-x^2}}\, dx$

36. $\int_{\pi/4}^{3\pi/4} (2\csc^2 x - \cos x)\, dx$

37. $\int_2^3 \frac{x+2}{x-1}\, dx$ **38.** $\int_0^1 \frac{2x^2 - 1}{x^2 + 1}\, dx$

39–40 Find the area of the region between the graph of the given function and the x-axis on the indicated interval.

39. $y = \frac{1}{2x^2}$ on $[1,10]$

40. $y = 2\sqrt{x} - x^2$ on $[0,2]$

41. Find a formula for $f(x)$ if $\int_0^{x^3} f(t)\, dt = \sin(x^3)$.

42. The velocity function of a particle moving along the x-axis is $v(t) = 3t - t^2$ units per second. If it started at the origin, find **a.** the position of the particle at $t = 5$ seconds and **b.** the total distance traveled by the particle in the time interval $[0,5]$.

43–54 Use an appropriate substitution (when necessary) to evaluate the indefinite integral.

43. $\int \frac{-2}{\sqrt{1-x^2}}\, dx$ **44.** $\int \frac{-2x}{\sqrt{1-x^2}}\, dx$

45. $\int \sec x(\sec x + \tan x)\, dx$

46. $\int \sec^2 x \tan x\, dx$

47. $\int 6x^2 (2x^3 - 7)^9\, dx$ **48.** $\int x^4 \sqrt{x^5 - 3}\, dx$

49. $\int \frac{4x}{(x^2+1)^2}\, dx$ **50.** $\int \frac{e^{\arcsin x}}{\sqrt{1-x^2}}\, dx$

51. $\int \frac{\sec^2 \sqrt{x}}{\sqrt{x}}\, dx$ **52.** $\int \frac{1}{x\ln(x^2)}\, dx$

53. $\int \frac{e^{2x}}{e^x + 1}\, dx$ **54.** $\int \frac{1}{x^2}\sin\left(\frac{x+1}{x}\right) dx$

55–56 Find $y(x)$ that satisfies the given conditions.

55. $\frac{dy}{dx} = \frac{1}{\sqrt{x}(\sqrt{x}-1)^2}$; $y(9) = 1$

56. $y''(x) = 1 - \sin x$; $y'(0) = 2$; $y(0) = 0$

57. A particle is moving along the x-axis in the positive direction with a velocity function of $v(t) = \frac{t}{t^2 + 1}$ units per second. If it started at the point $(1,0)$, what is the particle's position at $t = 4$ seconds?

58–63 Evaluate the definite integral.

58. $\int_0^1 3x^5 (x^6 - 1)^{12}\, dx$

59. $\int_0^4 (x+1)\sqrt{x^2 + 2x}\, dx$

60. $\int_1^2 \frac{x^3}{x^4 + 1}\, dx$ **61.** $\int_{-1/2}^0 \frac{2^t}{\sqrt{1-4^t}}\, dt$

62. $\int_e^{e^2} \frac{1}{x(\ln x)^2}\, dx$ **63.** $\int_1^4 \frac{dx}{\sqrt{x}(\sqrt{x}+1)^2}$

64–69 Find the area of the region bounded by the graphs of the given equations. (If convenient or necessary, integrate with respect to y rather than x.)

64. $y = x^3 - 4x$, $3y = 15x$, $x \ge 0$

65. $y = 1 - x^2$, $y = 1 - x^6$, $x \ge 0$

66. $y = 2\sqrt[4]{x}$, $y = 4 - 2x$, $y = 0$

67. $y = \ln x$, $(e-1)y = x - 1$

68. $y = \frac{1}{1+x^2}$, $2y = 1$

69. $y = \sin x$, $y = \sin x \cos x$, $0 \le x \le \pi$

70. Consider the region bounded by the graph of $y = 1/x$ and the x-axis over the interval $[1, a]$ $(a > 1)$. Find the vertical line $x = c$ that bisects the region in two subregions of equal area.

71. Consider the function $f(x) = 1/x^2$ defined on some interval $[a, b]$. Partition $[a, b]$ and in each subinterval $[x_{i-1}, x_i]$ choose the sample point $x_i^* = \sqrt{x_{i-1}x_i}$ (the geometric mean of the endpoints). Show that

$$\frac{1}{\left(x_i^*\right)^2}\Delta x_i = \frac{1}{x_{i-1}} - \frac{1}{x_i}$$

and use this observation to prove the following formula.

$$\int_a^b \frac{1}{x^2}\,dx = \frac{1}{a} - \frac{1}{b}$$

72. Prove that if the conditions of Part I of the Fundamental Theorem of Calculus are satisfied and $F(x) = \int_{g(x)}^{h(x)} f(t)\,dt$, where $g(x)$ and $h(x)$ are differentiable, then $F'(x) = f\big(h(x)\big)h'(x) - f\big(g(x)\big)g'(x)$. (**Hint:** See Example 3 of Section 5.3.)

73. Prove that if f is a linear function, then its definite integral on an interval $[a, b]$ is the average of its left and right Riemann sums, that is,

$$\int_a^b f(x)\,dx = \frac{L_n + R_n}{2}.$$

What is your expectation regarding the integral and the average above if f is concave up? Concave down?

74–81 *True or False?* Determine whether the given statement is true or false. In case of a false statement, explain or provide a counterexample.

74. If $n_1 < n_2$, then the Riemann sum R_{n_2} is always a better approximation of the integral than R_{n_1}.

75. If f is piecewise continuous on a closed interval, then the limit of its Riemann sums always exists.

76. When applying the Fundamental Theorem of Calculus, we must choose the antiderivative with $C = 0$.

77. $\int \dfrac{1}{e^x}\,dx = \ln\left(e^x\right) + C = x + C$

78. The definite integral of the velocity function of a moving object on $[t_1, t_2]$ is equal to the total distance traveled by the object from time $t = t_1$ to $t = t_2$.

79. $\int_a^b f(x)\,dx > 0$ if and only if $f(x) > 0$ on $[a, b]$.

80. $\int \sec x\,dx = \sec x \tan x + C$

81. $\int_{-1}^1 \dfrac{1}{x^3}\,dx = \dfrac{-1}{2x^2}\Bigg]_{-1}^1 = -\dfrac{1}{2} - \left(-\dfrac{1}{2}\right) = 0$

Chapter 5
Technology Exercises

82. Use the summation feature of a computer algebra system to verify your answers for Exercises 9–15.

83. Write a program for a graphing calculator or computer algebra system that calculates the nth Riemann sum for a given function on a given interval, using subintervals of equal width and sample points of your choice. Use your program to verify your answers for Exercises 16–18.

84. Use a computer algebra system to evaluate the limit of Exercise 28. What do you find? (Answer will vary depending on the capabilities of the particular software used.)

Chapter 5
Project

The topic of this project is the so-called *sine integral function*, which is important for its applications, most notably in electrical engineering and signal processing.

1. Consider the following piecewise defined function:

$$f(t) = \begin{cases} \dfrac{\sin t}{t} & \text{if } t > 0 \\ 1 & \text{if } t = 0 \end{cases}$$

Prove that for any $x \geq 0$, $f(t)$ is integrable on $[0, x]$.

2. The **sine integral function** is defined as follows:

$$\text{Si}(x) = \int_0^x f(t)\,dt, \text{ for } x \geq 0$$

Prove that $\text{Si}(x)$ is continuous.

3. Find the derivative $\dfrac{d}{dx}\text{Si}(x)$.

4. Without graphing first, write a short paragraph on why you would expect the graph of $\text{Si}(x)$ to be oscillating. Explain why its amplitude is expected to decrease as $x \to \infty$.

5. Find the x-values where the relative maxima and minima of $\text{Si}(x)$ occur.

6. Extend the definition of $\text{Si}(x)$ to negative x-values and prove that for any $a > 0$,

$$\int_{-a}^{a} \text{Si}(x)\,dx = 0.$$

7. Use a computer algebra system to plot the graph of $\text{Si}(x)$ on the interval $[-8\pi, 8\pi]$.

8. Use a computer algebra system to approximate the range of $y = \text{Si}(x)$ to four decimal places.

Appendices

Appendix A **Fundamentals of *Mathematica***

Mathematica is a powerful and flexible software package with a wide variety of uses. To begin with, *Mathematica* (along with similar products such as Maple, MATLAB, and Derive) can be viewed as a sort of supercalculator. It also understands the rules of algebra, has a huge number of built-in functions ranging from the trivial to the exotic, and is very good at generating high-quality graphs in one, two, and three dimensions. Beyond that, a package such as *Mathematica* is also a programming environment; it is this aspect of *Mathematica* that allows the user to extend its capabilities to suit specialized needs.

The optional use of *Mathematica* and similar technology in this text requires only a basic familiarity; this appendix will serve as a quick guide to the use of *Mathematica*. It should also be noted that a *complete* guide to *Mathematica* can be found within the program itself. Once it is installed and running on your computer, clicking on the "Help" button located in the top toolbar (see Figure 1) gives you access to an electronic version of a very large *Mathematica* user's manual. After clicking on "Help" a drop-down menu appears. By clicking on "Documentation Center" the full selection of "Help" categories appears; a good place to start is with "Get Started." This leads to a set of videos that highlight many useful examples of how *Mathematica* can be used to solve different sorts of problems.

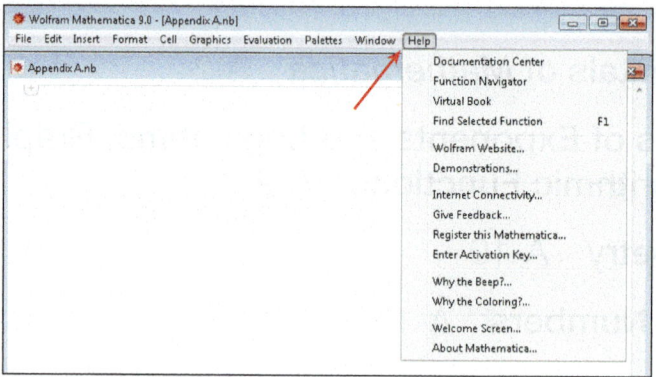

Figure 1 Getting On-Screen Help

At first, you will probably be making use of built-in *Mathematica* commands such as **Plot**, **Fit**, and **Solve** (as opposed to using your own user-defined commands). It is important to realize that *Mathematica* is case sensitive and that all built-in commands begin with a capital letter. Once a command has been typed in, you'll need to tell *Mathematica* to execute it. This can be done in one of two ways—either by pressing [Shift] and [Enter] together (known as [Shift] + [Enter]) or, if you are using an extended keyboard, by using the [Enter] that appears in the numeric keypad area. Pressing [Enter] alone will simply move the cursor to the next line and allow you to continue typing but will not execute any commands.

Each time you press [Shift] + [Enter] *Mathematica* will execute all the commands contained in a single cell. Different *Mathematica* cells are demarcated by brackets along the right-hand edge of the work area, and you can always start a new cell by positioning the mouse cursor over a blank part of the area (you will notice that the cursor symbol becomes horizontal rather than vertical) and clicking the left mouse button once.

The remainder of this appendix contains examples of a few of the basic *Mathematica* commands used in this text, arranged roughly in the order in which they appear. For instant on-screen help on any command, type the command into *Mathematica* and then press [F1]. Doing so will bring up the relevant help pages and, more often than not, provide examples of how the command is used.

Basic *Mathematica* Commands

Defining Functions

A few rules of syntax must be observed in order to define your own functions in *Mathematica*. The first is that each variable serving as a placeholder in the definition must be followed by the underscore symbol "_" when it appears on the left side of the definition and without the underscore when it appears on the right. The second rule is that ":=" (a colon followed by an equal sign) is used in the definition, as opposed to "=" (see the on-screen *Mathematica* help for detailed explanations of these rules). Figure 2 illustrates the definition of the two functions $f(x) = x^2 + 5$ and $g(x,y) = 3x - 7y$, followed by an evaluation of each.

In[1]:= **f[x_] := x^2 + 5**

In[2]:= **g[x_, y_] := 3 x - 7 y**

In[3]:= **f[-2]**

Out[3]= 9

In[4]:= **g[5, 2]**

Out[4]= 1

Figure 2 Defining Functions

Plot

The basic usage of the **Plot** command is **Plot[f, {x, x_{min}, x_{max}}]**, where f is an expression in x representing a function to be plotted and x_{min} and x_{max} define the endpoints of the interval on the x-axis over which f is to be graphed. However, the **Plot** command also recognizes many options that modify the details of the resulting picture; these options are best explored via the on-screen help. Figure 3 illustrates the use of **Plot** in graphing the function $f(x) = x^3 - x^2 - 3x + 5$ over the interval $[-3, 4]$.

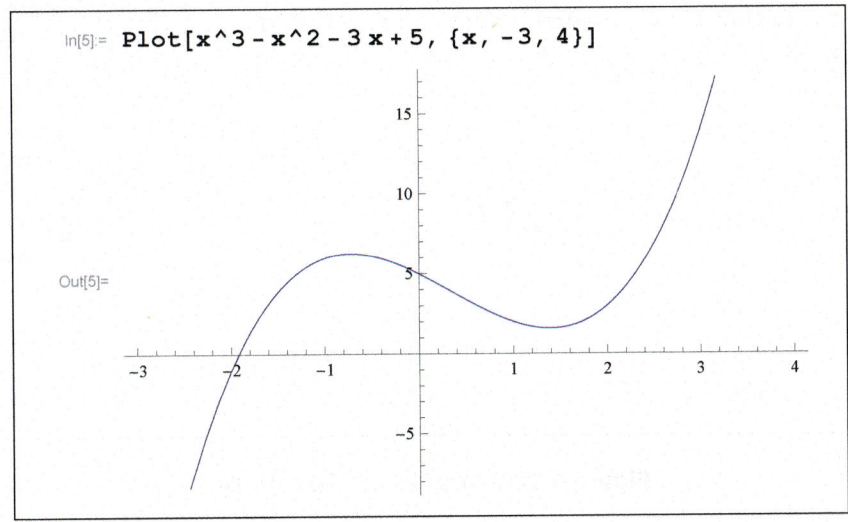

In[5]:= **Plot[x^3 - x^2 - 3 x + 5, {x, -3, 4}]**

Out[5]=

Figure 3 Basic Use of the Plot Command

Curve Fitting (Fit Command)

The *Mathematica* command **Fit** can be used to construct a function of specified form (such as linear, quadratic, exponential, etc.) to a given set of data (i.e., ordered pairs) using the least-squares method. Figure 4 illustrates the use of **Fit** to construct both a linear and a quadratic function that best fits the given set of four data points. Note also the use of the **ListPlot**, **Plot**, and **Show** commands to create graphs of the data and the two best-fitting functions. Two options are shown in the **ListPlot** usage, one of which (**PlotStyle**) specifies the color and size of the points to be plotted, and the other of which (**AxesOrigin**) positions the axes in a certain manner. (For an exponential fit, try out the command **Fit[data, {1, Exp[x]}, x]**.)

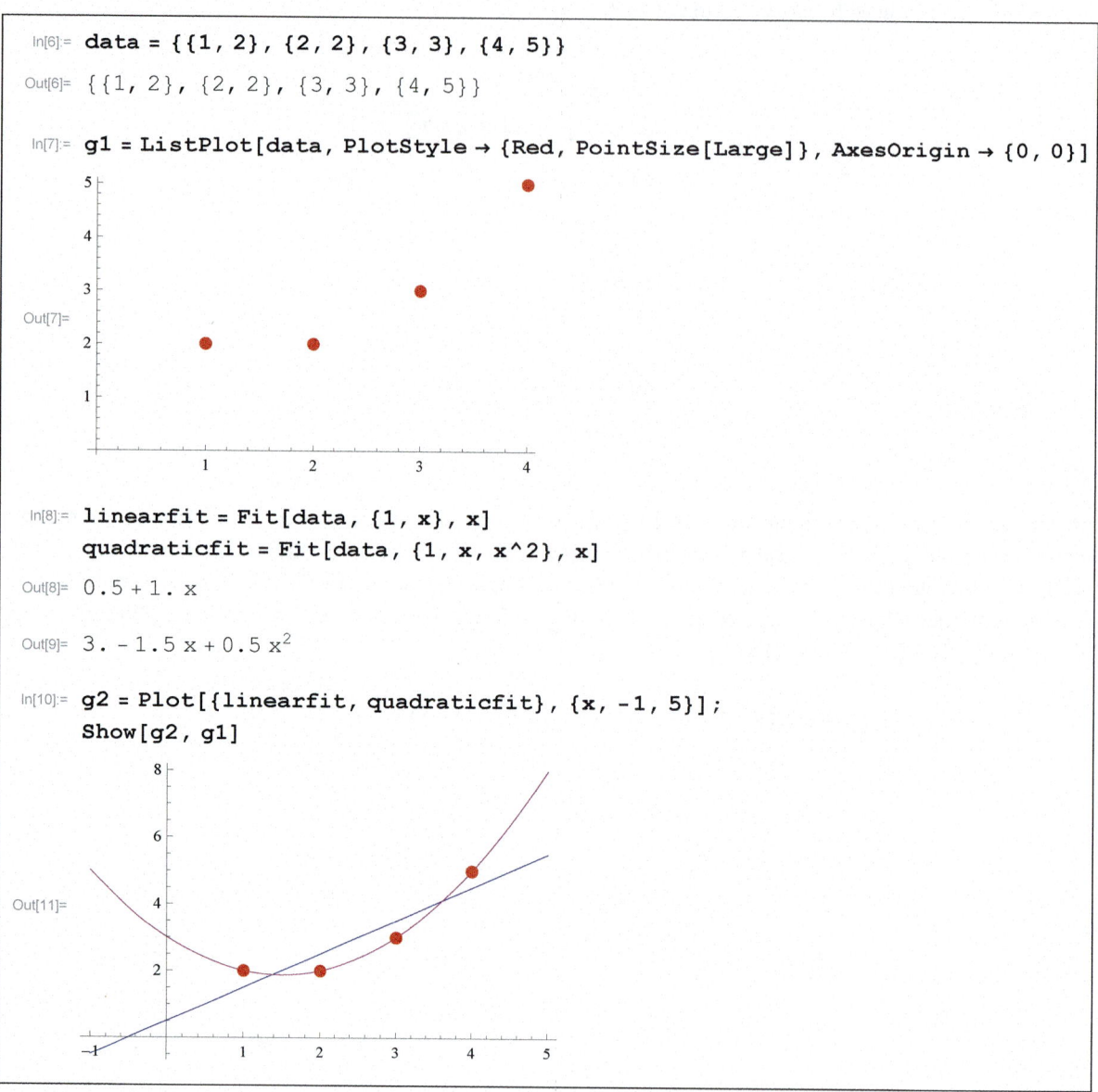

Figure 4 Linear and Quadratic Curve Fitting

Piecewise

The **Piecewise** command allows us to easily create and use functions in *Mathematica* that correspond to the piecewise defined functions referred to in this and many other math texts. See Section 1.2 for an example of the use of the **Piecewise** command.

Manipulate

The **Manipulate** command is a powerful tool that is useful in making dynamic models in *Mathematica*. Such models are especially useful in exploring the effect of changing the value(s) of parameter(s); see Section 1.5 for an example of such usage.

Limit

The built-in command **Limit** is used to direct *Mathematica* to try to determine the limit of a function at a specified point, with the option of asking for one-sided limits from either direction. See Section 2.2 and Figure 5 for examples of the command's use.

```
In[12]:= Limit[ (2 x - 1) / (x - 1), x → 1, Direction → 1]
         Limit[ (2 x - 1) / (x - 1), x → 1, Direction → -1]
         Limit[ (2 x - 1) / (x - 1), x → Infinity]

Out[12]= -∞

Out[13]= ∞

Out[14]= 2
```

Figure 5a Use of the Limit Command

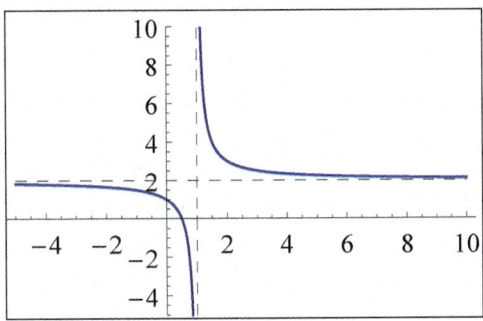

Figure 5b $y = \dfrac{2x - 1}{x - 1}$

Solve

The **Solve** command is very powerful, and can be used in several different ways. Its basic usage is **Solve[** *expr* **,** *vars* **]**, where *expr* represents one or more equations and *vars* represents one or more variables. If more than one equation is to be solved, the collection of equations must be enclosed in a set of braces, separated by commas. Similarly, if more than one variable is to be solved for, the variables must be enclosed in a set of braces. Figure 6 shows the use of **Solve** to first solve one equation for one variable, and then to solve a collection of three equations for all three variables. Note how *Mathematica* expresses the solution in each case.

In[15]:= **Solve[3 x - x * y == 9 y, y]**

Out[15]= $\left\{\left\{ y \to \dfrac{3\,x}{9+x} \right\}\right\}$

In[16]:= **Solve[{3 x + 2 y - 4 z == 8, 4 x - 5 z == -3, 7 y + z == 12}, {x, y, z}]**

Out[16]= $\left\{\left\{ x \to -\dfrac{50}{3},\; y \to \dfrac{53}{15},\; z \to -\dfrac{191}{15} \right\}\right\}$

Figure 6 Two Uses of the Solve Command

It is important to note that equations in *Mathematica* are expressed with two "=" symbols, as seen in Figure 6. The use of just one "=" is reserved for assigning a permanent value to something. For instance, the expression **x=3** assigns the value of 3 to the symbol *x*, while the expression **x==3** represents the equation $x = 3$ in *Mathematica*.

NSolve

The **NSolve** command is used in a manner similar to **Solve**, but typically in situations where an exact solution is either not desired or not feasible. See Section 2.5 for an example of the use of the command in finding a numerical approximation of a solution.

Differentiation (D Command)

The basic usage of the built-in differentiation command **D** is **D[** *f* **,** *x* **]**, where *f* is a function of the variable *x*. Figure 7 illustrates such use in finding the derivative of a given rational function; note the optional use of the **Together** command (discussed later in this appendix) to express the derivative as a single fraction.

In[17]:= **f[x_] := (x^2 - 3 x + 1) / (x + 5)**

In[18]:= **D[f[x], x]**

Out[18]= $\dfrac{-3+2\,x}{5+x} - \dfrac{1-3\,x+x^2}{(5+x)^2}$

In[19]:= **Together[D[f[x], x]]**

Out[19]= $\dfrac{-16+10\,x+x^2}{(5+x)^2}$

Figure 7 Differentiation

If f is a function of more than one variable, the **D** command can be used to find partial derivatives.

FindRoot

The **FindRoot** command uses numerical methods (such as Newton's method, Section 4.5) to find approximate roots of functions, and is especially useful when neither **Solve** nor **NSolve** is able to provide a satisfactory result. Its basic usage is **FindRoot[** f **,** $\{x, x_0\}$ **]** when the goal is to find a root of the function f near a given point x_0, but it can also be used to find a numerical solution of the equation $lhs = rhs$ near x_0 if used in the form **FindRoot[** lhs **==** rhs **,** $\{x, x_0\}$ **]** (note the "double equal sign" used by *Mathematica* to denote an equation).

FindMaximum and FindMinimum

The usage of the commands **FindMaximum** and **FindMinimum** is similar to that of **FindRoot**, and both also rely on numerical methods to obtain results. To approximate the location and value of a local maximum of the function f near a given point x_0, the syntax is **FindMaximum[** f **,** $\{x, x_0\}$ **]**; the use of **FindMinimum** is identical. Figure 8 illustrates the use of **FindMinimum** to identify the radius r that minimizes the surface area of the cylinder of Example 3 in Section 4.6.

In[20]:= **FindMinimum[2 * Pi * r^2 + 1000 / r, {r, 5}]**

Out[20]= $\{348.734, \{r \to 4.30127\}\}$

Figure 8 Use of FindMinimum

Integrate

The **Integrate** command can be used for both indefinite and definite integration, with the goal determined by the options used with the command. Figure 9 illustrates how *Mathematica* provides both the indefinite integral of the rational function $1/(x^2 + 1)$ and the definite integral of the same function over the interval $[-1.5, 1.5]$. (Note that *Mathematica* does not provide an arbitrary constant when evaluating indefinite integrals.)

In[21]:= **Integrate[1 / (x^2 + 1), x]**

Out[21]= ArcTan[x]

In[22]:= **Integrate[1 / (x^2 + 1), {x, -1.5, 1.5}]**

Out[22]= 1.96559

Figure 9 Integration

Other Useful Commands

Simplify

The **Simplify** command is used to simplify mathematical expressions according to the usual rules of algebra. The basic syntax is **Simplify[*expr*]**, where *expr* is the expression to be simplified. Note the examples shown in Figure 10.

In[23]:= **Simplify[x * (4 x - 2 x * y) / (6 x^2)]**

Out[23]= $\dfrac{2 - y}{3}$

In[24]:= **Simplify[(a^2 - b^2) / (a - b)]**

Out[24]= $a + b$

Figure 10 Use of Simplify

Expand

This command is used to multiply out factors in an expression. The syntax for the command is **Expand[*expr*]**. Figure 11 shows the use of the command in multiplying out the expression $(x - y)^5$.

In[25]:= **Expand[(x - y) ^5]**

Out[25]= $x^5 - 5 x^4 y + 10 x^3 y^2 - 10 x^2 y^3 + 5 x y^4 - y^5$

Figure 11 Use of Expand

Factor

The **Factor** command is the reverse of the **Expand** command when applied to polynomials. Its basic usage is **Factor[*poly*]**, where *poly* is a polynomial expression to be factored.

Together

The **Together** command is used primarily to express a sum (or difference) of two or more rational expressions as one with a common denominator, automatically canceling any common factors that may appear. The basic syntax for the command is **Together[*expr*]**.

Appendix B **Properties of Exponents and Logarithms, Graphs of Exponential and Logarithmic Functions**

For ease of reference, the basic algebraic properties of exponents and logarithms and the general forms of exponential and logarithmic graphs appear below. Interestingly, the Scottish mathematician John Napier (1550–1617) introduced logarithms as an aid to computation, and their use led to the development of various types of slide rules and logarithm tables. It was only later that mathematicians made the connection between logarithmic and exponential functions, namely that they are inverses of each other (more precisely, an exponential function of a given base is the inverse function of the logarithmic function with the same base, and vice versa). This fact appears explicitly as the first property of logarithms below, with the other properties reflecting, directly or indirectly, the same fact.

Properties of Exponents

Given real numbers x and y and positive real numbers a and b, the following properties hold.

1. $a^{x+y} = a^x a^y$

2. $a^{x-y} = \dfrac{a^x}{a^y}$

3. $\left(a^x\right)^y = a^{xy}$

4. $\left(ab\right)^x = a^x b^x$

Properties of Logarithms

Given positive real numbers x, y, a, and b, with $a \neq 1$ and $b \neq 1$, and real number r, the following properties hold.

1. $y = a^x \Leftrightarrow \log_a y = x$

2. $\log_a \left(a^x\right) = x$

3. $a^{\log_a x} = x$

4. $\log_a \left(xy\right) = \log_a x + \log_a y$

5. $\log_a \dfrac{x}{y} = \log_a x - \log_a y$

6. $\log_a \left(x^r\right) = r \log_a x$

Change of logarithmic base: $\log_b x = \dfrac{\log_a x}{\log_a b}$

Change of exponential base: $a^x = b^{\log_b\left(a^x\right)} = b^{x \log_b a}$

(in particular, $a^x = e^{\ln\left(a^x\right)} = e^{x \ln a}$)

Graphs of Exponential and Logarithmic Functions

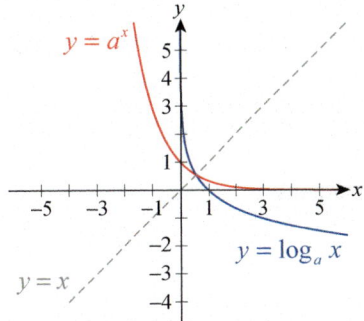

Figure 1 Case 1: $0 < a < 1$

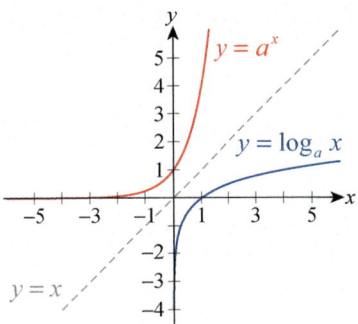

Figure 2 Case 2: $a > 1$

Appendix C **Trigonometry**

The historical records of trigonometry date back to the second millennium BC, and we know of a number of different cultures (Egyptian, Babylonian, Indian, and Greek among them) that studied and used the properties of triangles. Our word "trigonometry" comes from an ancient Greek word meaning "triangle measuring," and the names of the individual trigonometric functions have similarly ancient roots. The study of how different cultures independently discovered the basic tenets of trigonometry, how trigonometric knowledge was further developed and disseminated, and how early civilizations used trigonometry for scientific and commercial purposes is fascinating in its own right and well worth exploring. Many excellent resources for such exploration are available online, in books, and in scholarly articles.

For the purpose of quick reference, this appendix contains the basic definitions used in trigonometry, graphs of the six basic trigonometric functions and their inverses, and the most commonly used trigonometric identities.

Basic Definitions and Graphs

Radian and Degree Measure

$180° = \pi$ radians

$1° = \dfrac{\pi}{180}$ radians $1\text{ radian} = \dfrac{180°}{\pi}$

$x° = x\left(\dfrac{\pi}{180}\right)$ radians $x\text{ radians} = x\left(\dfrac{180°}{\pi}\right)$

Angular Speed

$\omega = \dfrac{\theta}{t}$

Linear Speed

$v = \dfrac{s}{t} = \dfrac{r\theta}{t} = r\omega$

Trigonometric Functions

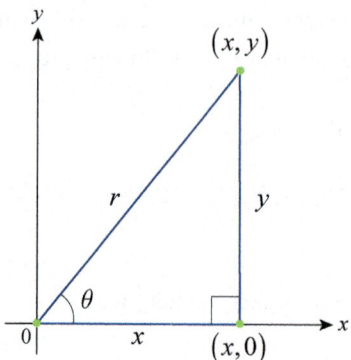

Arc Length

$s = \left(\dfrac{\theta}{2\pi}\right)(2\pi r) = r\theta$

Area of a Sector

$A = \left(\dfrac{\theta}{2\pi}\right)(\pi r^2) = \dfrac{r^2\theta}{2}$

$\sin\theta = \dfrac{y}{r}$ $\csc\theta = \dfrac{r}{y}$ (for $y \neq 0$)

$\cos\theta = \dfrac{x}{r}$ $\sec\theta = \dfrac{r}{x}$ (for $x \neq 0$)

$\tan\theta = \dfrac{y}{x}$ (for $x \neq 0$) $\cot\theta = \dfrac{x}{y}$ (for $y \neq 0$)

Commonly Encountered Angles

θ	Radians	$\sin\theta$	$\cos\theta$	$\tan\theta$
0°	0	0	1	0
30°	$\dfrac{\pi}{6}$	$\dfrac{1}{2}$	$\dfrac{\sqrt{3}}{2}$	$\dfrac{1}{\sqrt{3}}$
45°	$\dfrac{\pi}{4}$	$\dfrac{1}{\sqrt{2}}$	$\dfrac{1}{\sqrt{2}}$	1
60°	$\dfrac{\pi}{3}$	$\dfrac{\sqrt{3}}{2}$	$\dfrac{1}{2}$	$\sqrt{3}$
90°	$\dfrac{\pi}{2}$	1	0	—
180°	π	0	−1	0
270°	$\dfrac{3\pi}{2}$	−1	0	—

Trigonometric Graphs

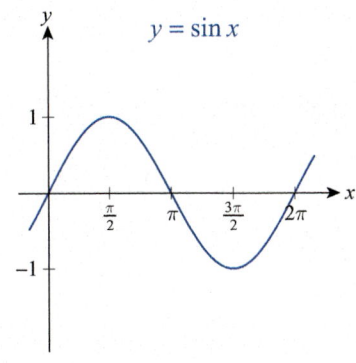

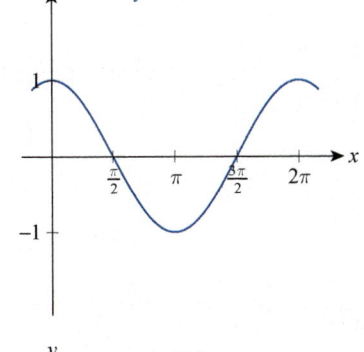

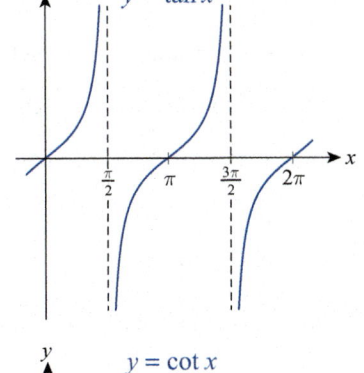

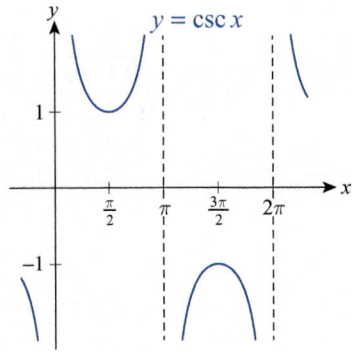

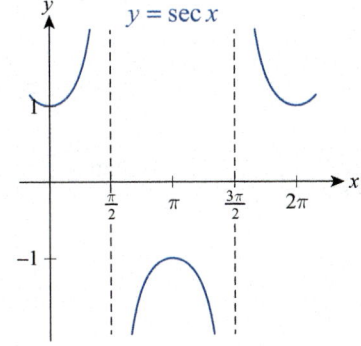

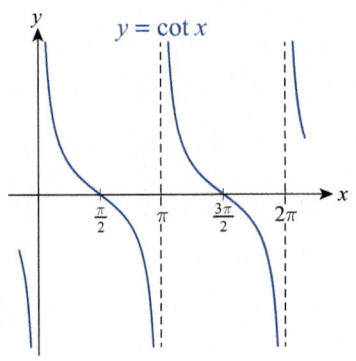

Trigonometric Identities

Reciprocal Identities

$$\csc x = \frac{1}{\sin x} \qquad \sec x = \frac{1}{\cos x} \qquad \cot x = \frac{1}{\tan x}$$

$$\sin x = \frac{1}{\csc x} \qquad \cos x = \frac{1}{\sec x} \qquad \tan x = \frac{1}{\cot x}$$

Cofunction Identities

$$\cos x = \sin\left(\frac{\pi}{2} - x\right) \qquad \sin x = \cos\left(\frac{\pi}{2} - x\right)$$

$$\csc x = \sec\left(\frac{\pi}{2} - x\right) \qquad \sec x = \csc\left(\frac{\pi}{2} - x\right)$$

$$\cot x = \tan\left(\frac{\pi}{2} - x\right) \qquad \tan x = \cot\left(\frac{\pi}{2} - x\right)$$

Quotient Identities

$$\tan x = \frac{\sin x}{\cos x} \qquad\qquad \cot x = \frac{\cos x}{\sin x}$$

Period Identities

$$\sin(x + 2\pi) = \sin x \qquad \csc(x + 2\pi) = \csc x$$
$$\cos(x + 2\pi) = \cos x \qquad \sec(x + 2\pi) = \sec x$$
$$\tan(x + \pi) = \tan x \qquad \cot(x + \pi) = \cot x$$

Even/Odd Identities

$$\sin(-x) = -\sin x \quad \cos(-x) = \cos x \quad \tan(-x) = -\tan x$$
$$\csc(-x) = -\csc x \quad \sec(-x) = \sec x \quad \cot(-x) = -\cot x$$

Pythagorean Identities

$$\sin^2 x + \cos^2 x = 1 \qquad \tan^2 x + 1 = \sec^2 x$$
$$1 + \cot^2 x = \csc^2 x$$

Sum and Difference Identities

$$\sin(u + v) = \sin u \cos v + \cos u \sin v$$

$$\sin(u - v) = \sin u \cos v - \cos u \sin v$$

$$\cos(u + v) = \cos u \cos v - \sin u \sin v$$

$$\cos(u - v) = \cos u \cos v + \sin u \sin v$$

$$\tan(u + v) = \frac{\tan u + \tan v}{1 - \tan u \tan v}$$

$$\tan(u - v) = \frac{\tan u - \tan v}{1 + \tan u \tan v}$$

Double-Angle Identities

$$\sin 2u = 2 \sin u \cos u$$

$$\cos 2u = \cos^2 u - \sin^2 u = 2\cos^2 u - 1 = 1 - 2\sin^2 u$$

$$\tan 2u = \frac{2 \tan u}{1 - \tan^2 u}$$

Power-Reducing Identities

$$\sin^2 x = \frac{1 - \cos 2x}{2} \qquad \cos^2 x = \frac{1 + \cos 2x}{2}$$

$$\tan^2 x = \frac{1 - \cos 2x}{1 + \cos 2x}$$

Half-Angle Identities

$$\sin\frac{x}{2} = \pm\sqrt{\frac{1 - \cos x}{2}} \qquad \cos\frac{x}{2} = \pm\sqrt{\frac{1 + \cos x}{2}}$$

$$\tan\frac{x}{2} = \frac{1 - \cos x}{\sin x} = \frac{\sin x}{1 + \cos x}$$

Product-to-Sum Identities

$$\sin x \cos y = \frac{1}{2}\left[\sin(x+y)+\sin(x-y)\right]$$

$$\cos x \sin y = \frac{1}{2}\left[\sin(x+y)-\sin(x-y)\right]$$

$$\sin x \sin y = \frac{1}{2}\left[\cos(x-y)-\cos(x+y)\right]$$

$$\cos x \cos y = \frac{1}{2}\left[\cos(x+y)+\cos(x-y)\right]$$

Sum-to-Product Identities

$$\sin x + \sin y = 2\sin\left(\frac{x+y}{2}\right)\cos\left(\frac{x-y}{2}\right)$$

$$\sin x - \sin y = 2\cos\left(\frac{x+y}{2}\right)\sin\left(\frac{x-y}{2}\right)$$

$$\cos x + \cos y = 2\cos\left(\frac{x+y}{2}\right)\cos\left(\frac{x-y}{2}\right)$$

$$\cos x - \cos y = -2\sin\left(\frac{x+y}{2}\right)\sin\left(\frac{x-y}{2}\right)$$

The Laws of Sines and Cosines

The Law of Sines

$$\frac{\sin A}{a} = \frac{\sin B}{b} = \frac{\sin C}{c}$$

The Law of Cosines

$$a^2 = b^2 + c^2 - 2bc\cos A$$
$$b^2 = a^2 + c^2 - 2ac\cos B$$
$$c^2 = a^2 + b^2 - 2ab\cos C$$

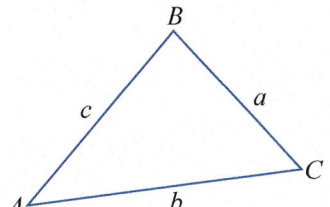

Inverse Trigonometric Functions

Arcsine, Arccosine, and Arctangent

Function	Notation	Domain	Range
Inverse Sine	$\arcsin y = \sin^{-1} y = x \Leftrightarrow y = \sin x$	$[-1,1]$	$\left[-\dfrac{\pi}{2},\dfrac{\pi}{2}\right]$
Inverse Cosine	$\arccos y = \cos^{-1} y = x \Leftrightarrow y = \cos x$	$[-1,1]$	$[0,\pi]$
Inverse Tangent	$\arctan y = \tan^{-1} y = x \Leftrightarrow y = \tan x$	$(-\infty,\infty)$	$\left(-\dfrac{\pi}{2},\dfrac{\pi}{2}\right)$

Inverse Trigonometric Graphs

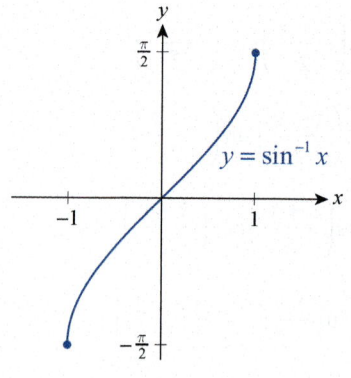

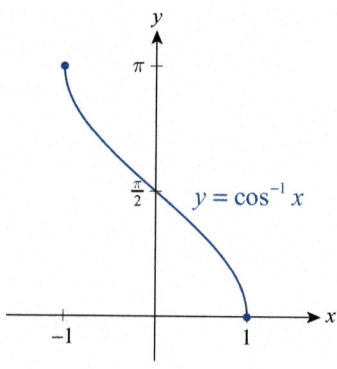

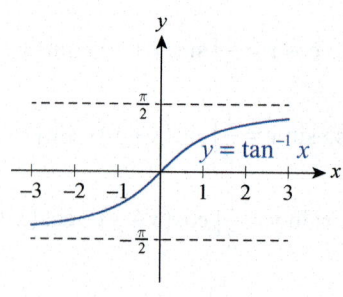

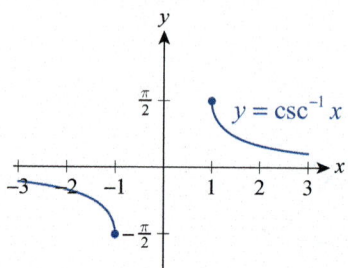

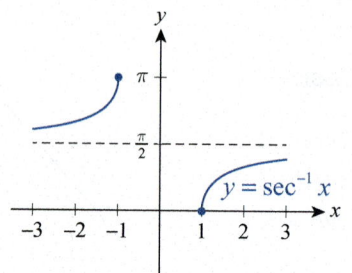

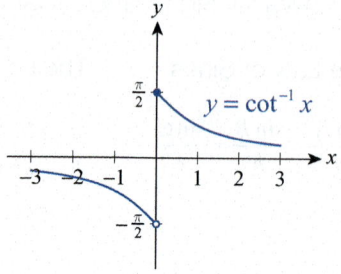

Inverse Trigonometric Identities

$$\csc^{-1} x = \sin^{-1}\left(\frac{1}{x}\right)$$

$$\sec^{-1} x = \cos^{-1}\left(\frac{1}{x}\right)$$

$$\cot^{-1} x = \tan^{-1}\left(\frac{1}{x}\right), \text{ with } \cot^{-1} 0 = \frac{\pi}{2}$$

Appendix D **Complex Numbers**

The **complex numbers**, an extension of the real numbers, consist of all numbers that can be expressed in the form $a + bi$, where a and b are real numbers and i, representing the **imaginary unit**, satisfies the equation $i^2 = -1$. Complex numbers expand the real numbers to a set that is *algebraically closed*, a concept belonging to the branch of mathematics called abstract algebra. Girolamo Cardano (1501–1576) and other Italian Renaissance mathematicians were among the first to recognize the benefits of defining what we now call complex numbers; by allowing such "imaginary" numbers as i, which is a solution of the equation $x^2 + 1 = 0$, mathematicians were able to devise and make sense of formulas solving polynomial equations up to degree four. Later mathematicians conjectured that every nonconstant polynomial function, even those with complex coefficients, has at least one root (a number at which the polynomial has the value of 0), assuming complex roots are allowed. Repeated application of this assertion then implies, counting multiplicities of roots, that a polynomial of degree n has n roots; stated another way, an nth-degree polynomial equation has n solutions (some of which may be repeated solutions). The first reasonably complete proof of this conjecture, now known as the Fundamental Theorem of Algebra, was provided by Carl Friedrich Gauss (1777–1855) in 1799 in his doctoral dissertation.

Unlike real numbers, often identified with points on a line, complex numbers are typically depicted as points in the **complex plane**, also known as the **Argand plane**, which is named after the French Swiss mathematician Jean-Robert Argand (1768–1822). The complex plane has the appearance of the Cartesian plane, with the horizontal axis referred to as the **real axis** and the vertical axis as the **imaginary axis**. A given complex number $a + bi$ is associated with the ordered pair (a, b) in the plane, where a represents the displacement along the real axis and b the displacement along the imaginary axis (see Figure 1 for examples). In this context, a is called the **real part** of $a + bi$ and b the **imaginary part**. Real numbers are thus complex numbers for which the imaginary part is 0 (they can be written in the form $a + 0 \cdot i$), and **pure imaginary numbers** are complex numbers of the form $0 + bi$; the origin of the plane represents the number $0 + 0 \cdot i$ and is usually simply written as 0. Two complex numbers $a + bi$ and $c + di$ are **equal** if and only if $a = b$ and $c = d$ (that is, their real parts are equal and their imaginary parts are equal).

Sums, differences, and products of complex numbers are easily simplified and written in the form $a + bi$ by treating complex numbers as polynomial expressions in the variable i, remembering that $i^2 = -1$. (Keep in mind, though, that i is not, in fact, a variable—this treatment is simply a convenience.) Example 1 illustrates the process.

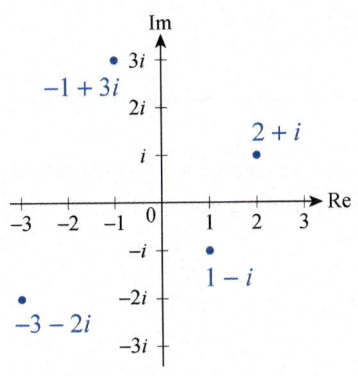

Figure 1

Example 1

Express each of the following in the form $a + bi$.

a. $(4 + 3i) + (-5 + 7i)$

b. $(-2 + 3i) - (-3 + 3i)$

c. $(3 + 2i)(-2 + 3i)$

d. $(2 - 3i)^2$

Solution

a. $(4 + 3i) + (-5 + 7i) = (4 - 5) + (3 + 7)i$
$$= -1 + 10i$$

b. $(-2 + 3i) - (-3 + 3i) = (-2 + 3) + (3 - 3)i$
$$= 1$$

c. $(3 + 2i)(-2 + 3i) = -6 + 9i - 4i + 6i^2$
$$= -6 + (9 - 4)i - 6 \qquad \text{Replace } i^2 \text{ with } -1.$$
$$= -12 + 5i$$

d. $(2 - 3i)^2 = (2 - 3i)(2 - 3i)$
$$= 4 - 6i - 6i + 9i^2$$
$$= 4 - 12i - 9 \qquad \text{Replace } i^2 \text{ with } -1.$$
$$= -5 - 12i$$

Division of complex numbers is slightly more complicated, but a quotient can also be simplified and written in the form $a + bi$ by making use of the following observation:

$$(a + bi)(a - bi) = a^2 - abi + abi - b^2i^2 = a^2 + b^2$$

Given a complex number $z = a + bi$, the complex number $\bar{z} = a - bi$ is called its **complex conjugate**. We simplify a quotient of complex numbers by multiplying the numerator and denominator by the complex conjugate of the denominator, as illustrated in Example 2.

Example 2

Express each of the following in the form $a + bi$.

a. $\dfrac{2+3i}{3-i}$ **b.** $(4-3i)^{-1}$ **c.** $\dfrac{1}{i}$

Solution

a. $\dfrac{2+3i}{3-i} = \dfrac{2+3i}{3-i}$

$\qquad = \dfrac{(2+3i)(3+i)}{(3-i)(3+i)}$ Multiply the numerator and denominator by the conjugate.

$\qquad = \dfrac{6+2i+9i+3i^2}{9+3i-3i-i^2}$

$\qquad = \dfrac{6+11i-3}{9+1}$ Replace i^2 with -1.

$\qquad = \dfrac{3+11i}{10} = \dfrac{3}{10} + \dfrac{11}{10}i$

b. $(4-3i)^{-1} = \dfrac{1}{4-3i}$

$\qquad = \dfrac{1(4+3i)}{(4-3i)(4+3i)}$ Multiply the numerator and denominator by the conjugate.

$\qquad = \dfrac{4+3i}{16+12i-12i-9i^2}$

$\qquad = \dfrac{4+3i}{16+9} = \dfrac{4}{25} + \dfrac{3}{25}i$

c. $\dfrac{1}{i} = \dfrac{1(-i)}{i(-i)} = \dfrac{-i}{-i^2} = \dfrac{-i}{1} = -i$

Endowed with the operations of addition and multiplication, the set of complex numbers, like the set of real numbers and the set of rational numbers, form what is known as a **field**, another concept from the realm of abstract algebra. The following table summarizes the properties possessed by a field; note that each of the three sets of numbers mentioned above possesses all the properties. Also note, by way of contrast, that the set of natural numbers, the set of integers, and the set of irrational numbers are not fields, as each set fails to possess one or more of the field properties.

Field Properties

In this table, a, b, and c represent arbitrary elements of a given field. The first five properties apply individually to the two operations of addition and multiplication, while the last combines the two.

Name of Property	Additive Version	Multiplicative Version
Closure	$a + b$ is an element of the field	ab is an element of the field
Commutative	$a + b = b + a$	$ab = ba$
Associative	$a + (b + c) = (a + b) + c$	$a(bc) = (ab)c$
Identity	$a + 0 = 0 + a = a$	$a \cdot 1 = 1 \cdot a = a$
Inverse	$a + (-a) = 0$	$a \cdot \dfrac{1}{a} = 1,\ \text{assuming } a \neq 0$
Distributive	$a(b + c) = ab + ac$	

The introduction of the imaginary unit i allows us to now define the **principal square root** $\sqrt{a}$ of any real number a, as follows: Given a positive real number a, $\sqrt{a}$ denotes the positive real number whose square is a, and $\sqrt{-a} = i\sqrt{a}$. An application of this definition explains the restriction in one of the properties of exponents (specifically, the exponent $1/2$). Recall that if a and b are both positive, then

$$\sqrt{ab} = (ab)^{1/2} = a^{1/2}b^{1/2} = \sqrt{a}\sqrt{b}.$$

To see why a and b are required to be positive, note that

$$\sqrt{(-9)(-4)} = \sqrt{36} = 6,$$

but

$$\sqrt{-9}\sqrt{-4} = \left(i\sqrt{9}\right)\left(i\sqrt{4}\right) = (3i)(2i) = 6i^2 = -6.$$

Complex numbers can also be expressed in *polar form*, based on the polar coordinates of a given complex number in the plane. We say the **magnitude** $|z|$ of a complex number $z = a + bi$, also known as its **modulus**, **norm**, or **absolute value**, is its distance from 0 in the complex plane—that is, the nonnegative real number

$$|z| = \sqrt{a^2 + b^2}.$$

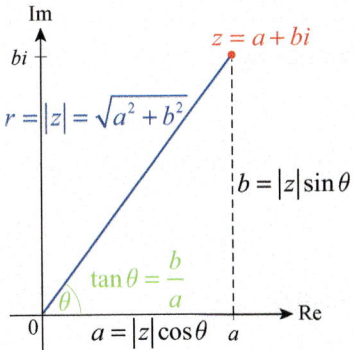

Figure 2

The **argument** of z, denoted $\arg(z)$, is the radian angle θ between the positive real axis and the line joining 0 and z. The quantities $|z|$ and $\arg(z)$ thus play the same roles, respectively, as the polar coordinates r and θ of a point in the plane. The argument of the complex number 0 is undefined, while the argument of every other complex number is not unique (any multiple of 2π added to the argument of a given complex number describes the same number, since 2π corresponds to a complete rotation around the origin). Given these definitions, and letting $\theta = \arg(z)$, the **polar form** of $z = a + bi$ is then

$$z = r(\cos\theta + i\sin\theta), \text{ where } r = |z|$$

(see Figure 2 for a depiction of the relationship between a, b, r, $|z|$, and θ).

Example 3

Write each of the following complex numbers in polar form.

a. $1 + i\sqrt{3}$ **b.** $-1 + i$

Solution

a. The magnitude of $1 + i\sqrt{3}$ is $\sqrt{1^2 + \left(\sqrt{3}\right)^2} = 2$, and its argument is $\tan^{-1}\sqrt{3} = \pi/3$ (see Figure 3). Hence,

$$1 + i\sqrt{3} = 2\left(\cos\frac{\pi}{3} + i\sin\frac{\pi}{3}\right).$$

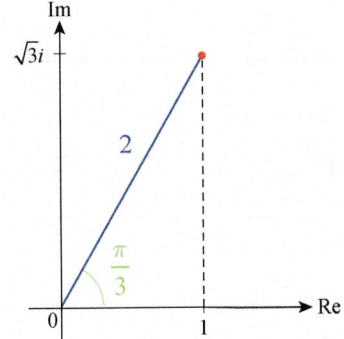

Figure 3

b. The magnitude of $-1 + i$ is $\sqrt{(-1)^2 + 1^2} = \sqrt{2}$, and its argument is $3\pi/4$ (note that this complex number lies in the second quadrant of the plane, as shown in Figure 4). Hence,

$$-1 + i = \sqrt{2}\left(\cos\frac{3\pi}{4} + i\sin\frac{3\pi}{4}\right).$$

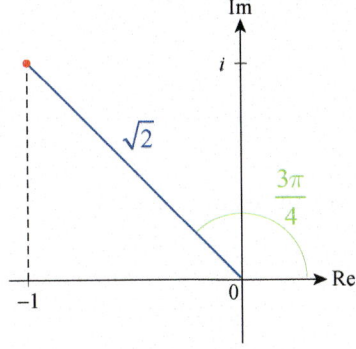

Figure 4

Euler's Formula $e^{i\theta} = \cos\theta + i\sin\theta$, derived in Section 10.9, allows us to express the polar form of a complex number as a complex exponential:

$$z = re^{i\theta}, \text{ where } r = |z| \text{ and } \theta = \arg(z).$$

With this observation, the following formulas regarding products and quotients of complex numbers are easily proved (they can also be proved by using the trigonometric sum and difference identities).

Products and Quotients of Complex Numbers

Given the complex numbers

$$z_1 = r_1\left(\cos\theta_1 + i\sin\theta_1\right) \text{ and } z_2 = r_2\left(\cos\theta_2 + i\sin\theta_2\right),$$

the following formulas hold:

Product Formula	$z_1 z_2 = r_1 r_2 \left[\cos\left(\theta_1 + \theta_2\right) + i\sin\left(\theta_1 + \theta_2\right)\right]$
Quotient Formula	$\dfrac{z_1}{z_2} = \dfrac{r_1}{r_2}\left[\cos\left(\theta_1 - \theta_2\right) + i\sin\left(\theta_1 - \theta_2\right)\right]$, assuming $z_2 \neq 0$.

Proof

Writing each complex number as a complex exponential,

$$z_1 z_2 = \left(r_1 e^{i\theta_1}\right)\left(r_2 e^{i\theta_2}\right) = r_1 r_2 e^{i(\theta_1 + \theta_2)} = r_1 r_2 \left[\cos\left(\theta_1 + \theta_2\right) + i\sin\left(\theta_1 + \theta_2\right)\right]$$

and

$$\frac{z_1}{z_2} = \frac{r_1 e^{i\theta_1}}{r_2 e^{i\theta_2}} = \frac{r_1}{r_2} e^{i(\theta_1 - \theta_2)} = \frac{r_1}{r_2}\left[\cos\left(\theta_1 - \theta_2\right) + i\sin\left(\theta_2 - \theta_2\right)\right]$$

The following statement regarding positive integer powers of complex numbers can be similarly proved.

De Moivre's Theorem

Given a complex number $z = r\left(\cos\theta + i\sin\theta\right)$ and positive integer n,

$$z^n = r^n\left(\cos n\theta + i\sin n\theta\right).$$

Proof

Again writing z as a complex exponential,

$$z^n = \left(r e^{i\theta}\right)^n = r^n e^{in\theta} = r^n\left(\cos n\theta + i\sin n\theta\right).$$

De Moivre's Theorem can be used to determine roots of complex numbers. The first step is to note that if $w = re^{i\theta}$ is a nonzero complex number, and if n is a positive integer, then w has n n^{th} roots. This follows from the Fundamental Theorem of Algebra, which tells us that the equation $z^n = w$ has n solutions (here, z represents a complex variable). One n^{th} root is easily determined: if we let

$$z_0 = r^{1/n} e^{i(\theta/n)},$$

then

$$z_0^n = \left[r^{1/n} e^{i(\theta/n)} \right]^n = re^{i\theta} = w.$$

But as we know, replacing θ with $\theta + 2k\pi$ results in an equivalent complex number for any integer k, leading to the following formula for the nth roots of w.

Theorem

Roots of a Complex Number

Let $w = r(\cos\theta + i\sin\theta)$ and let n be a positive integer. The nth roots of w are given by

$$z_k = r^{1/n} e^{i\left(\frac{\theta + 2k\pi}{n}\right)}, \, k = 0, 1, \ldots, n-1.$$

Alternatively,

$$z_k = r^{1/n} \left[\cos\left(\frac{\theta + 2k\pi}{n}\right) + i\sin\left(\frac{\theta + 2k\pi}{n}\right) \right], \, k = 0, 1, \ldots, n-1.$$

The n n^{th} roots of a given complex number all have the same magnitude and are equally distributed around a circle in the complex plane with radius equal to that common magnitude.

Example 4

Determine the specified roots of the given complex numbers, and graph the roots and the original complex numbers in the plane.

a. 5th roots of 1

b. 4th roots of $-1 - i\sqrt{3}$

Solution

a. The easiest way to determine the 5th roots of 1 is to write 1 as a complex exponential and then apply the above formula with $n = 5$:

$$1 = e^{i \cdot 0},$$

so the 5th roots of 1 are

$$\left\{1,\, e^{i(2\pi/5)},\, e^{i(4\pi/5)},\, e^{i(6\pi/5)},\, e^{i(8\pi/5)}\right\}.$$

Since the complex number 1 (shown as a blue point in Figure 5) has a magnitude of 1, all of the five 5th roots (shown as red points in the figure) lie on a circle of radius 1. Note that 1 is, itself, one of the 5th roots of 1.

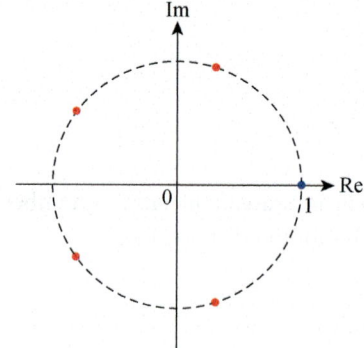

Figure 5

b. The first step is to again express the complex number as a complex exponential. Note that

$$\left|-1 - i\sqrt{3}\right| = \sqrt{(-1)^2 + \left(-\sqrt{3}\right)^2} = \sqrt{4} = 2$$

and

$$\tan \theta = \frac{-\sqrt{3}}{-1} \Rightarrow \theta = \frac{4\pi}{3},$$

so

$$-1 - i\sqrt{3} = 2e^{i(4\pi/3)}.$$

Hence, the 4th roots of $-1 - i\sqrt{3}$ are

$$\left\{2^{1/4} e^{(i/4)\left[(4\pi/3) + 2k\pi\right]}\right\}_{k=0,1,2,3} = \left\{2^{1/4} e^{i(\pi/3)},\, 2^{1/4} e^{i(5\pi/6)},\, 2^{1/4} e^{i(4\pi/3)},\, 2^{1/4} e^{i(11\pi/6)}\right\}.$$

Figure 6 shows the original point $-1 - i\sqrt{3}$ in blue and its four 4th roots in red.

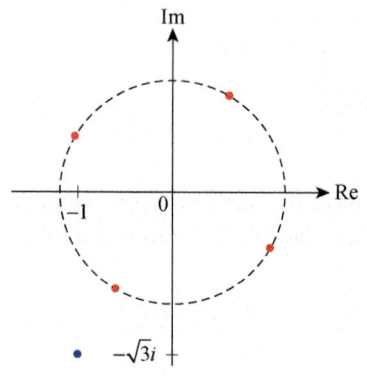

Figure 6

Appendix E **Proofs of Selected Theorems**

In this appendix we provide proofs (or in one case just a statement) of theorems used in the main body of the text. While some of the proofs here are more technical in nature than those presented elsewhere, they are worth studying in order to (1) gain additional insight into the rigorous nature of mathematical thinking and (2) develop a sense of the deeper mathematics to come in later courses.

Section 2.4

Theorem

Basic Limit Laws

Let f and g be two functions such that both $\lim_{x \to c} f(x)$ and $\lim_{x \to c} g(x)$ exist, and let k be a fixed real number. Then the following laws hold.

Sum Law	$\lim_{x \to c}\left[f(x) + g(x)\right] = \lim_{x \to c} f(x) + \lim_{x \to c} g(x)$
Difference Law	$\lim_{x \to c}\left[f(x) - g(x)\right] = \lim_{x \to c} f(x) - \lim_{x \to c} g(x)$
Constant Multiple Law	$\lim_{x \to c}\left[kf(x)\right] = k \lim_{x \to c} f(x)$
Product Law	$\lim_{x \to c}\left[f(x) g(x)\right] = \lim_{x \to c} f(x) \cdot \lim_{x \to c} g(x)$
Quotient Law	$\lim_{x \to c} \dfrac{f(x)}{g(x)} = \dfrac{\lim_{x \to c} f(x)}{\lim_{x \to c} g(x)}$, provided $\lim_{x \to c} g(x) \neq 0$

── **Proof** ──────────────────────────

We already proved the Sum Law in Section 2.4. We proceed to prove the Product Law and Quotient Law, from which the remaining laws will follow quickly.

As in the proof of the Sum Law, let $L = \lim_{x \to c} f(x)$ and $M = \lim_{x \to c} g(x)$, and assume $\varepsilon > 0$ is given. Our goal is to show there exists $\delta > 0$ such that $|f(x) g(x) - LM| < \varepsilon$ for all $0 < |x - c| < \delta$. One way to determine δ is to employ a strategy of adding and subtracting the same quantity, in this case $f(x)M$:

$$
\begin{aligned}
|f(x)g(x) - LM| &= |f(x)g(x) - f(x)M + f(x)M - LM| \\
&\leq |f(x)\left[g(x) - M\right]| + |M\left[f(x) - L\right]| \qquad \text{\color{blue}Triangle Inequality} \\
&= |f(x)||g(x) - M| + |M||f(x) - L|
\end{aligned}
$$

Since $L = \lim_{x \to c} f(x)$, we know there exists $\delta_1 > 0$ for which

$$0 < |x - c| < \delta_1 \Rightarrow |f(x) - L| < 1,$$

so $|f(x)| = |f(x) - L + L| \le |f(x) - L| + |L| < 1 + |L|$.

Similarly, there exists $\delta_2 > 0$ for which

$$0 < |x - c| < \delta_2 \Rightarrow |f(x) - L| < \frac{\varepsilon}{2(1 + |M|)}.$$

And finally, since $M = \lim_{x \to c} g(x)$, there exists $\delta_3 > 0$ for which

$$0 < |x - c| < \delta_3 \Rightarrow |g(x) - M| < \frac{\varepsilon}{2(1 + |L|)}.$$

So if we let $\delta = \min\{\delta_1, \delta_2, \delta_3\}$, $0 < |x - c| < \delta$ will guarantee each of the above three outcomes, meaning

$$|f(x)g(x) - LM| \le |f(x)||g(x) - M| + |M||f(x) - L|$$

$$< (1 + |L|)\frac{\varepsilon}{2(1 + |L|)} + |M|\frac{\varepsilon}{2(1 + |M|)}$$

$$< \frac{\varepsilon}{2} + \frac{\varepsilon}{2} = \varepsilon,$$

thereby proving the Product Law.

To prove the Quotient Law, we will see that it suffices to prove

$$\lim_{x \to c} \frac{1}{g(x)} = \frac{1}{M},$$

under the assumption that $M \ne 0$. In doing so, we will use the fact that $\big||a| - |b|\big| \le |a - b|$ for arbitrary real numbers a and b. This follows from the observation that

$$|a| = |a - b + b| \le |a - b| + |b| \qquad \text{Triangle Inequality}$$

so $|a| - |b| \le |a - b|$. Similarly, interchanging a and b in the same argument shows that $|b| - |a| \le |b - a| = |a - b|$, and the two facts together prove that $\big||a| - |b|\big| \le |a - b|$.

Now, given $\varepsilon > 0$, we need to show there exists $\delta > 0$ for which $0 < |x - c| < \delta$ implies

$$\left| \frac{1}{g(x)} - \frac{1}{M} \right| < \varepsilon.$$

Since

$$\left| \frac{1}{g(x)} - \frac{1}{M} \right| = \left| \frac{M - g(x)}{g(x)M} \right| = \frac{1}{|g(x)|} \cdot \frac{1}{|M|} \cdot |g(x) - M|,$$

we want to choose δ in such a manner that $|g(x) - M|$ is sufficiently small and so that $|g(x)|$ is far enough away from 0 to make

$$\frac{1}{|g(x)|} \cdot \frac{1}{|M|}$$

also sufficiently small. The fact that $M = \lim_{x \to c} g(x)$ tells us there is a $\delta_1 > 0$ for which $0 < |x - c| < \delta_1$ implies $|g(x) - M| < |M|/2$, meaning

$$0 < |x - c| < \delta_1 \Rightarrow \left||g(x)| - |M|\right| \le |g(x) - M| < \frac{|M|}{2} \qquad \text{Using } \left||a| - |b|\right| \le |a - b|$$

and so

$$-\frac{|M|}{2} < |g(x)| - |M| < \frac{|M|}{2}.$$

Adding $|M|$ throughout results in the equivalent compound inequality

$$\frac{|M|}{2} < |g(x)| < \frac{3|M|}{2}$$

whenever $0 < |x - c| < \delta_1$. For our present purposes, we actually only care about the fact that $|g(x)| > |M|/2$, which means that $1/|g(x)| < 2/|M|$. There is also a $\delta_2 > 0$ such that

$$0 < |x - c| < \delta_2 \Rightarrow |g(x) - M| < \frac{|M|^2 \varepsilon}{2},$$

so if we let $\delta = \min\{\delta_1, \delta_2\}$, we have

$$0 < |x - c| < \delta \Rightarrow \left|\frac{1}{g(x)} - \frac{1}{M}\right| = \frac{1}{|g(x)|} \cdot \frac{1}{|M|} \cdot |g(x) - M| < \frac{2}{|M|} \cdot \frac{1}{|M|} \cdot \frac{|M|^2 \varepsilon}{2} = \varepsilon,$$

thus proving

$$\lim_{x \to c} \frac{1}{g(x)} = \frac{1}{M}.$$

We can now apply the Product Law, already proved, to obtain the Quotient Law.

$$\lim_{x \to c} \frac{f(x)}{g(x)} = \lim_{x \to c} \left[f(x) \cdot \frac{1}{g(x)} \right]$$

$$= \lim_{x \to c} f(x) \cdot \lim_{x \to c} \frac{1}{g(x)} \qquad \text{Product Law}$$

$$= L \cdot \frac{1}{M} = \frac{\lim_{x \to c} f(x)}{\lim_{x \to c} g(x)}$$

The Constant Multiple Law is a consequence of the Product Law, using $g(x) = k$ as one of the two functions, and the Difference Law follows from applying first the Sum Law and then the Constant Multiple Law (with $k = -1$).

Section 2.4   Theorem

Positive Integer Power Law

Let f be a function for which $\lim_{x \to c} f(x)$ exists, and let m be a fixed positive integer. Then

$$\lim_{x \to c}\left[f(x)\right]^m = \left[\lim_{x \to c} f(x)\right]^m.$$

─ **Proof** ──────────────────────────────────

The statement is trivially true for $m = 1$, so we prove the theorem for $m \geq 2$. We use mathematical induction to do so.

Basis Step: By the Product Law, we have

$$\lim_{x \to c}\left[f(x)\right]^2 = \lim_{x \to c}\left[f(x)f(x)\right] = \left[\lim_{x \to c} f(x)\right]^2.$$

Inductive Step: Assume $\lim_{x \to c}\left[f(x)\right]^k = \left[\lim_{x \to c} f(x)\right]^k$ for some $k \geq 2$. Then again applying the Product Law, we complete the proof as follows.

$$\lim_{x \to c}\left[f(x)\right]^{k+1} = \lim_{x \to c}\left(\left[f(x)\right]^k f(x)\right)$$

$$= \lim_{x \to c}\left[f(x)\right]^k \cdot \lim_{x \to c} f(x) \qquad \text{Product Law}$$

$$= \left[\lim_{x \to c} f(x)\right]^k \cdot \lim_{x \to c} f(x) \qquad \text{Induction hypothesis}$$

$$= \left[\lim_{x \to c} f(x)\right]^{k+1}$$

Section 2.4   Theorem

The Squeeze Theorem

If $g(x) \leq f(x) \leq h(x)$ for all x in some open interval containing c, except possibly at c itself, and if $\lim_{x \to c} g(x) = \lim_{x \to c} h(x) = L$, then $\lim_{x \to c} f(x) = L$ as well.

─ **Proof** ─────────────────────────────────────

Since $g(x) \le f(x) \le h(x)$ for all x in some open interval containing c, there exists $\delta_1 > 0$ such that $0 < |x - c| < \delta_1 \Rightarrow g(x) \le f(x) \le h(x)$. And by the limit definition, given $\varepsilon > 0$ there exist $\delta_2 > 0$ and $\delta_3 > 0$ such that

$$0 < |x - c| < \delta_2 \Rightarrow |g(x) - L| < \varepsilon \Rightarrow L - \varepsilon < g(x) < L + \varepsilon$$

and

$$0 < |x - c| < \delta_3 \Rightarrow |h(x) - L| < \varepsilon \Rightarrow L - \varepsilon < h(x) < L + \varepsilon.$$

By letting $\delta = \min\{\delta_1, \delta_2, \delta_3\}$, all three conclusions are true for x within δ of c. That is,

$$0 < |x - c| < \delta \Rightarrow L - \varepsilon < g(x) \le f(x) \le h(x) < L + \varepsilon \Rightarrow |f(x) - L| < \varepsilon.$$

──

Section 2.4

Theorem

Upper Bound Theorem

If $f(x) \le g(x)$ for all x in some open interval containing c, except possibly at c itself, and if the limits of f and g both exist at c, then

$$\lim_{x \to c} f(x) \le \lim_{x \to c} g(x).$$

─ **Proof** ─────────────────────────────────────

Let $L = \lim_{x \to c} f(x)$ and $M = \lim_{x \to c} g(x)$. Note that, by the Difference Law, the limit of $g(x) - f(x)$ at c exists, and

$$\lim_{x \to c}[g(x) - f(x)] = \lim_{x \to c} g(x) - \lim_{x \to c} f(x) = M - L.$$

Suppose, in contradiction to the claim, that $L > M$. Then $L - M > 0$, and if we let $\varepsilon = L - M$, there exists $\delta > 0$ such that

$$0 < |x - c| < \delta \Rightarrow |g(x) - f(x) - (M - L)| < \varepsilon$$
$$\Rightarrow -\varepsilon < g(x) - f(x) - M + L < \varepsilon$$
$$\Rightarrow M - L < g(x) - f(x) - M + L < L - M.$$

In particular, $g(x) - f(x) - M + L < L - M$, so $g(x) - f(x) < 0$ for all x such that $0 < |x - c| < \delta$, contradicting the fact that $f(x) \le g(x)$ for all x in some open interval containing c. Thus, it must be the case that $L \le M$; that is, $\lim_{x \to c} f(x) \le \lim_{x \to c} g(x)$.

──

Section 2.5   **Theorem**

"Limits Pass Through a Continuous Function"

Suppose $\lim\limits_{x \to c} g(x) = a$ and f is continuous at the point a. Then

$$\lim_{x \to c} f\big(g(x)\big) = f\left(\lim_{x \to c} g(x)\right) = f(a).$$

In words, we say the limit operation passes inside the continuous function f.

Proof

Assume $\varepsilon > 0$ is given. Since f is continuous at a, there exists $\delta_1 > 0$ for which

$$|x - a| < \delta_1 \Rightarrow |f(x) - f(a)| < \varepsilon.$$

And since $\lim\limits_{x \to c} g(x) = a$, there exists $\delta > 0$ such that

$$0 < |x - c| < \delta \Rightarrow |g(x) - a| < \delta_1.$$

Putting these facts together, we see that

$$0 < |x - c| < \delta \Rightarrow |g(x) - a| < \delta_1 \Rightarrow |f(g(x)) - f(a)| < \varepsilon,$$

and hence $\lim\limits_{x \to c} f\big(g(x)\big) = f(a)$.

Section 2.5   **Theorem**

"The Inverse of a Continuous Function Is Continuous"

If f is one-to-one and continuous on the interval (a,b), then f^{-1} is also a continuous function.

Proof

We first show that f is strictly monotonic on (a,b), and we do so by applying the Intermediate Value Property to a number of cases, all of which are similar. If f is neither strictly increasing nor strictly decreasing, then there must be points $x_1 < x_2 < x_3$ in (a,b) for which $f(x_2)$ does not lie between $f(x_1)$ and $f(x_3)$. We will show that cannot happen, using a proof by contradiction.

To that end, suppose x_1, x_2, and x_3 *are* three points in (a,b) for which $x_1 < x_2 < x_3$ and for which $f(x_2)$ does not lie between $f(x_1)$ and $f(x_3)$. Since f is one-to-one, either $f(x_1) < f(x_3)$ or $f(x_1) > f(x_3)$; we will assume that $f(x_1) < f(x_3)$ and leave consideration of the other case to the reader. The assumption that $f(x_2)$ is not between $f(x_1)$ and $f(x_3)$ again leads to two cases, one of which is that $f(x_1) < f(x_3) < f(x_2)$. Let y be a value such that $f(x_3) < y < f(x_2)$. Then by the continuity of f and the Intermediate Value Property (see Section 2.5), there is a point p such that $x_2 < p < x_3$ and $f(p) = y$ (in words, there is a point between x_2 and x_3 at which f takes on the value y, since y lies between the values of f at x_2 and x_3). But since $f(x_1) < f(x_3)$, y also satisfies $f(x_1) < y < f(x_2)$, so there is a point q such that $x_1 < q < x_2$ and $f(q) = y$. But then $p \neq q$ (since x_2 lies strictly between them) and $f(p) = f(q)$, contradicting the fact that f is one-to-one. By the same reasoning, the possibility that $f(x_2) < f(x_1) < f(x_3)$ is also ruled out, as are the two cases for which $f(x_1) > f(x_3)$. Thus, f must be either strictly increasing or strictly decreasing.

To now show that f^{-1} is continuous we will assume f is strictly increasing—the argument that f^{-1} is continuous when f is strictly decreasing is similar in nature. Let y_0 be a point in the image of (a,b) under f, and let $\varepsilon > 0$ be given. Since y_0 is in the image set, there is a (unique) point $x_0 \in (a,b)$ for which $f(x_0) = y_0$. Define

$$\varepsilon_1 = \min\{\varepsilon, x_0 - a, b - x_0\}.$$

Then, since we are assuming f is increasing, the image of the interval $(x_0 - \varepsilon_1, x_0 + \varepsilon_1)$ is the interval $(f(x_0 - \varepsilon_1), f(x_0 + \varepsilon_1))$, and $y_0 \in (f(x_0 - \varepsilon_1), f(x_0 + \varepsilon_1))$. Choose $\delta > 0$ small enough so that $(y_0 - \delta, y_0 + \delta) \subset (f(x_0 - \varepsilon_1), f(x_0 + \varepsilon_1))$. Then for any y such that $|y - y_0| < \delta$, $y \in (f(x_0 - \varepsilon_1), f(x_0 + \varepsilon_1))$ and hence $f^{-1}(y) \in (x_0 - \varepsilon_1, x_0 + \varepsilon_1)$. That is,

$$|y - y_0| < \delta \Rightarrow |f^{-1}(y) - f^{-1}(y_0)| < \varepsilon_1 \leq \varepsilon$$

and hence f^{-1} is continuous at y_0. Since y_0 was arbitrary, we have shown that f^{-1} is continuous on the image of (a,b) under f.

Section 4.1   Theorem

Bolzano-Weierstrass Theorem (Statement Only)

Every bounded sequence of real numbers has a convergent subsequence.

The Bolzano-Weierstrass theorem has many uses, one of which is to help prove the Extreme Value Theorem of Chapter 4. Specifically, its use assures the existence of points in a closed bounded interval at which a continuous function attains its extreme values.

Answer Key

Chapter 1

Section 1.1

1. Dom $= \{-3, 01\}$, Ran $= \{-1, 0, 1, 2, 5\}$

3. Dom $= \{-2, 3, 4\}$, Ran $= \{-8, 0, \cos 3, 5.98\}$

5. Dom $= \{$Tanisha, Don, Peter, David$\}$,

   Ran $= \begin{Bmatrix} \text{swimming, biking,} \\ \text{skating, skateboarding} \end{Bmatrix}$

7. Dom $= \mathbb{Z}$, Ran $= \{\ldots, -1, 1, 3, 5, \ldots\}$

9. Dom $= \mathbb{Z}$, Ran $= \{\ldots, 3, 5, 7, 9, \ldots\}$

11. Dom $= \mathbb{R}$, Ran $= \mathbb{R}$

13. Dom $= \{5\}$, Ran $= \mathbb{R}$

15. Dom $= [-1, \infty)$, Ran $= \mathbb{R}$

17. Dom $= [-2, 4]$, Ran $= [-3, 3]$

19. Dom $= \mathbb{R}$, Ran $= [0, 1]$

21. Dom $= \mathbb{R}$, Ran $= \mathbb{R}$

23. Dom $= \{$Registered students$\}$,

    Ran $= \{$Courses offered$\}$

25. Dom $= \{$People$\}$, Ran $= \{$Fathers$\}$

27. $\{(1,1), (2,2), (3,3), (4,4), (5,5)\}$

29. $\begin{Bmatrix} (1,1), (1,2), (1,3), (1,4), (1,5), \\ (2,2), (2,4), (3,3), (4,4), (5,5) \end{Bmatrix}$

31. Function

33. Not a function; -1 appears twice as first coordinate.

35. Not a function; it doesn't pass the vertical line test.

37. Function   39. Function   41. Function

43. Not a function; it doesn't pass the vertical line test.

45. Function   47. Function

49. Not a function; it doesn't pass the vertical line test. (However, x is a function of y.)

51. Function (F is a function of r.)

53. $y = -\dfrac{1}{3}x + \dfrac{10}{3}$ 55. $y = x^2 + \dfrac{1}{3}x - \dfrac{5}{3}$

57. $y = \dfrac{3x+1}{x^2-1}$

59. a. $\dfrac{4}{3}$ b. $\dfrac{1}{3}x + \dfrac{7}{3}$ c. $\dfrac{1}{3}(x+h) + 2$ d. $\dfrac{1}{3}$

61. a. 1 b. $x^2 + 2x - 2$ c. $(x+h)^2 - 3$ d. $2x + h$

63. a. Undefined   b. $\sqrt{x+1}$
    c. $\sqrt{x+h}$ d. $\dfrac{1}{\sqrt{x+h}+\sqrt{x}}$

65. a. -1 b. $\dfrac{1}{x+2}$ c. $\dfrac{1}{x+h+1}$
    d. $-\dfrac{1}{(x+1)(x+h+1)}$

67. Dom $= \mathbb{N}$, Cod $= \mathbb{N}$, Ran $= \{2, 3, 4, 5, \ldots\}$

69. Dom $= \mathbb{Z}$, Cod $= \mathbb{Z}$, Ran $= \{0, 1, 4, 9, 16, \ldots\}$

71. Dom $= [0, \infty)$, Cod $= \mathbb{R}$, Ran $= [0, \infty)$

73. $(-\infty, -2) \cup (-2, 3) \cup (3, \infty)$

75. $(-\infty, 1) \cup (3, \infty)$ 77. $[0, 2]$

79. $(-\infty, -3/2) \cup (-3/2, \infty)$ 81. $\mathbb{R} - \{2k\pi \mid k \in \mathbb{Z}\}$

83. $C(r) = 2\pi r$ 85. $C(F) = \dfrac{5}{9}(F - 32)$

87. $V(b, h) = \dfrac{1}{3}b^2 h$; a function of two variables

89. Function

91. Not a function   93. Not a function

95. Decreasing on $(-\infty, 1)$; increasing on $(1, \infty)$

97. Increasing on $(-\infty, -2), (2, \infty)$; decreasing on $(-2, 2)$

99. Decreasing on $(-\infty, 1)$; increasing on $(1, \infty)$

101. Decreasing on $(-\infty, -1)$; constant on $(-1, 2)$; increasing on $(2, \infty)$

103. Symmetric with respect to x-axis

105. Symmetric with respect to both axes and the origin

107. Symmetric with respect to origin

109. Symmetric with respect to x-axis

111. $P(A) = 4\sqrt{A}$

113. $V(x) = x(30 - 2x)(20 - 2x)$

115. $V(r) = \dfrac{2\pi r^3}{3}$

117. $C(F) = \dfrac{5}{9}(F - 32)$

Section 1.2

1. Deg.: 1; lead. coeff.: 1/2;
x-int.: 3; y-int.: $-3/2$;
Ran $= \mathbb{R}$

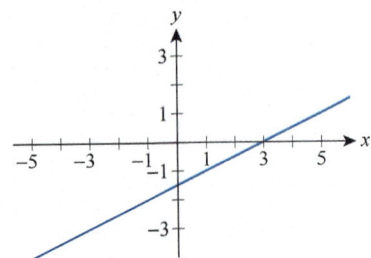

3. Deg.: 2; lead. coeff.: 2;
x-int.: 2, $-1/2$; y-int.: -2;
Ran $= (-25/8, \infty)$

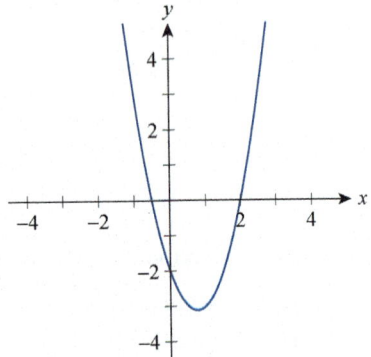

5. Deg.: 3; lead. coeff.: 1;
x-int.: -3, 1, 2; y-int.: 6;
Ran $= \mathbb{R}$

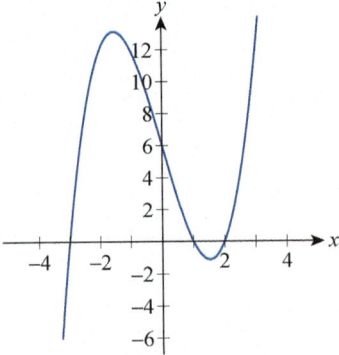

7. Deg.: 4; lead. coeff.: 1/4;
x-int.: $\pm 2\sqrt{2}$, 0; y-int.: 0;
Ran $= [-4, \infty)$

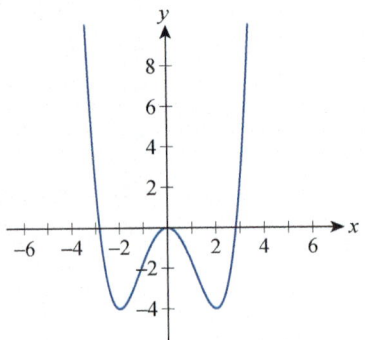

9. Vert. asym.: $x = 1$;
horiz. asym.: $y = 0$;
x-int.: none; y-int.: -5

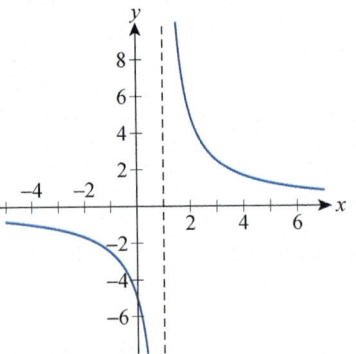

11. Vert. asym.: $x = -3$;
slant asym.: $y = x - 3$;
x-int.: none; y-int.: 1

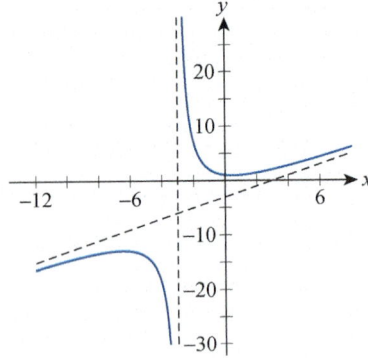

13. Vert. asym.: $x = -1/2$;
horiz. asym.: $y = 1/2$;
x-int.: -1; y-int.: 1

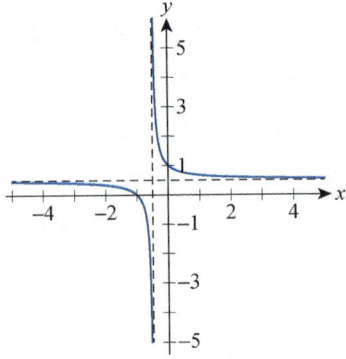

15. Vert. asym.: $x = -1$;
slant asym.: $y = x + 1$;
x-int.: 0, -2; y-int.: 0

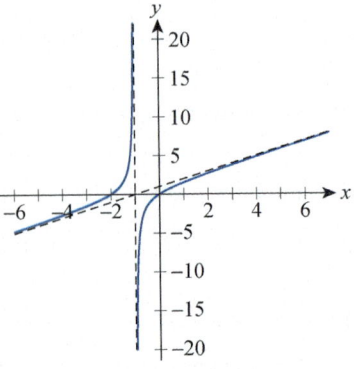

17. Answers will vary.

19. Answers will vary.

21. Answers will vary.

23. Answers will vary.

25. $\cos x$ **27.** $\cos^2 \alpha$ **29.** $\cos^2 \theta$ **31.** $\csc \beta$

33. $\tan^2 x$

35.

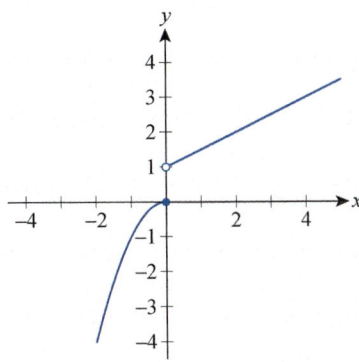

37.

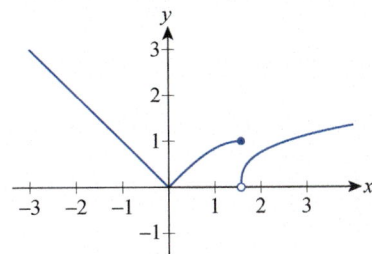

39. $f(x) = \begin{cases} -x+1 & \text{if } x < 1 \\ x-1 & \text{if } x \geq 1 \end{cases}$

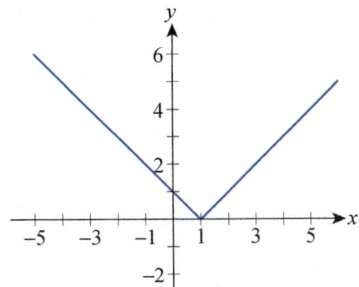

41. $h(x) = \begin{cases} -\sin x & \text{if } 2\pi n - \pi < x < 2\pi n \\ \sin x & \text{if } 2\pi n \leq x \leq 2\pi n + \pi \end{cases}$, $n \in \mathbb{Z}$

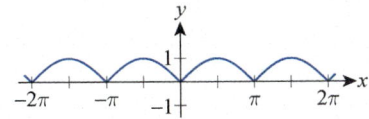

43.

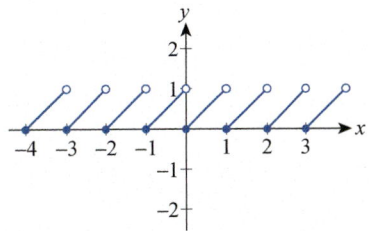

45.

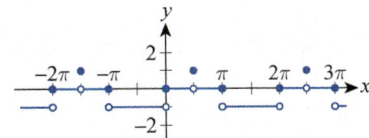

47. 3.5 hours

49. Width = 50 yd; length = 100 yd

51. 50 ft by 100 ft   **53.** $(2,1)$

55. 12 and 24   **57.** 25 sets   **59.** No; yes

61. $y = -4\cos 6\pi t$; it takes a mere $1/3$ s to complete a period.

63. a. 3 years   **b.** 9 years   **65.** $V \approx 178$ people

67. a. 0.999567   **b.** 0.958 g   **c.** 0.648 g

69. Approx. $134,392

71. a. 0.965936   **b.** 0.707 kg   **c.** 7.628 mg

73. True

75. False; the line perpendicular to $y = Ax + B$ has a slope of $-1/A$.

77. True

79. False; consider $\dfrac{x^4}{1+x^2}$, which has no asymptotes.

81. True

Section 1.3

1.

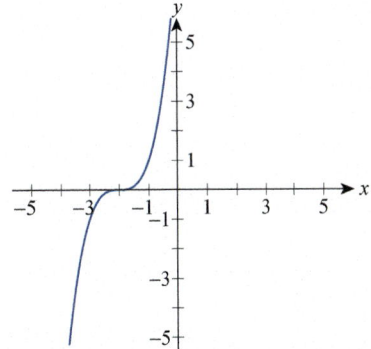

Dom = Ran = $\mathbb{R}$

3.

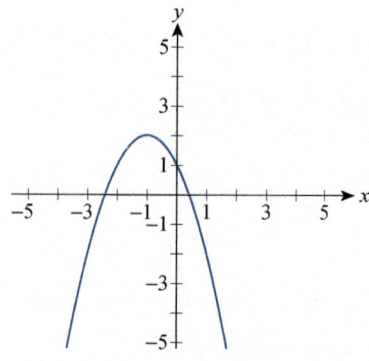

$\text{Dom} = \mathbb{R}, \ \text{Ran} = (-\infty, 2]$

5.

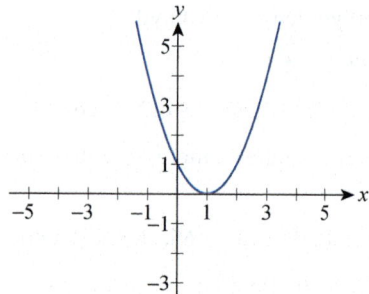

$\text{Dom} = \mathbb{R}, \ \text{Ran} = [0, \infty)$

7.

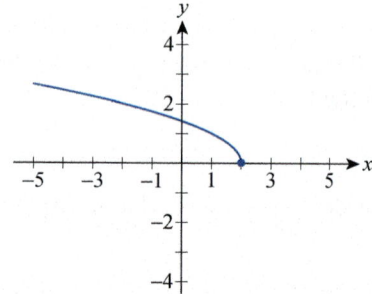

$\text{Dom} = (-\infty, 2], \ \text{Ran} = [0, \infty)$

9.

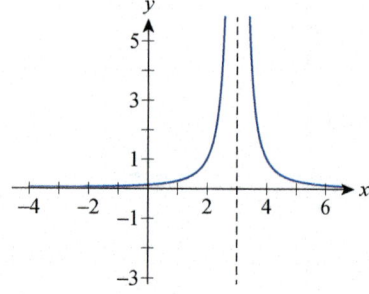

$\text{Dom} = (-\infty, 3) \cup (3, \infty), \ \text{Ran} = (0, \infty)$

11.

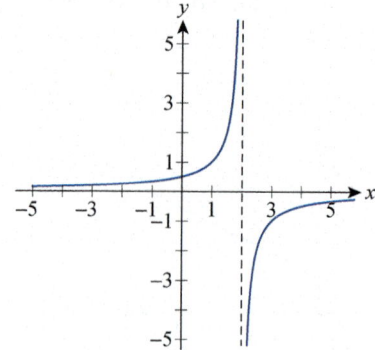

$\text{Dom} = (-\infty, 2) \cup (2, \infty), \ \text{Ran} = (-\infty, 0) \cup (0, \infty)$

13.

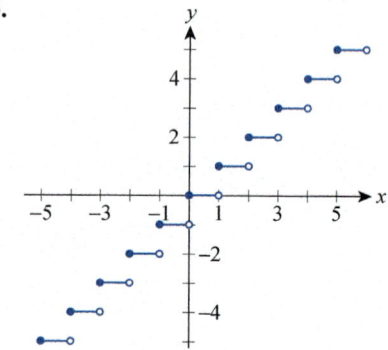

$\text{Dom} = \mathbb{R}, \ \text{Ran} = \mathbb{Z}$

15.

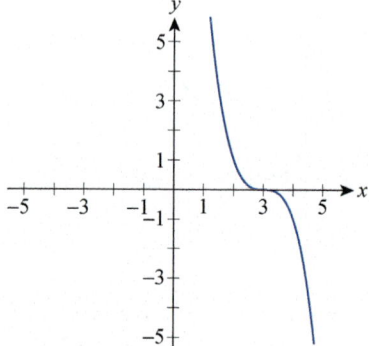

$\text{Dom} = \text{Ran} = \mathbb{R}$

17.

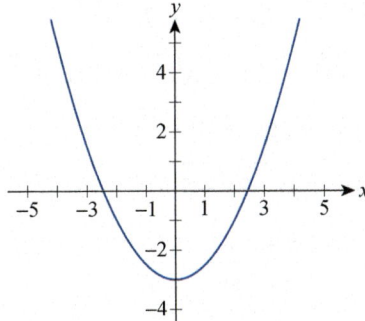

$\text{Dom} = \mathbb{R}, \ \text{Ran} = [-3, \infty)$

19.

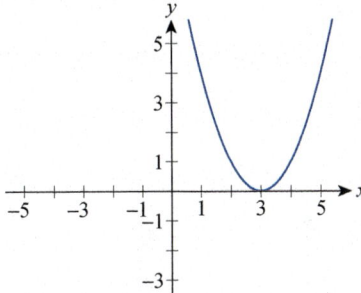

$\text{Dom} = \mathbb{R}, \ \text{Ran} = [0, \infty)$

21.

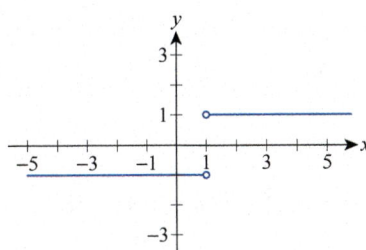

$\text{Dom} = (-\infty, 1) \cup (1, \infty), \ \text{Ran} = \{-1, 1\}$

23.

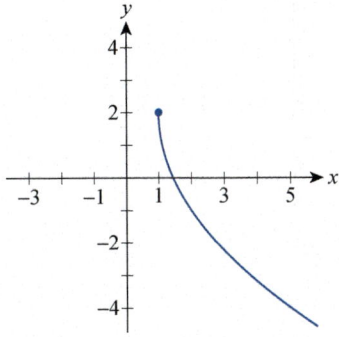

$\text{Dom} = [1, \infty), \ \text{Ran} = (-\infty, 2]$

25. $y = -x^2 + 6$ **27.** $y = -\sqrt{x+5}$

29. $y = -|-x+7|$ **31.** $f(x) = -\sqrt{x+4}$

33. $f(x) = 1 - (x-3)^3$

35. a. -1 **b.** 1 **c.** 0 **d.** 0

37. a. -3 **b.** 1 **c.** 2 **d.** $1/2$

39. a. -4 **b.** 8 **c.** -12 **d.** $-1/3$

41. a. 4 **b.** 3 **c.** $7/4$ **d.** 7

43. a. -1 **b.** 3 **c.** -2 **d.** $-1/2$

45. a. -1 **b.** -3 **c.** -2 **d.** -2

47. a. $x^2 + \sqrt[3]{x} - 1; \ \text{Dom} = \mathbb{R}$

b. $\dfrac{x^2 - 1}{\sqrt[3]{x}}; \ \text{Dom} = (-\infty, 0) \cup (0, \infty)$

49. a. $x^{3/2} + x - 3; \ \text{Dom} = [0, \infty)$

b. $\dfrac{x^{3/2}}{x - 3}; \ \text{Dom} = [0, 3) \cup (3, \infty)$

51. a. $x^3 + \sqrt{x-2} + 4; \ \text{Dom} = [2, \infty)$

b. $\dfrac{x^3 + 4}{\sqrt{x-2}}; \ \text{Dom} = (2, \infty)$

53. a. $x^{2/3} + 6x - 1; \ \text{Dom} = \mathbb{R}$

b. $\dfrac{6x - 1}{x^{2/3}}; \ \text{Dom} = (-\infty, 0) \cup (0, \infty)$

55. $-1699/100$ **57.** Undefined **59.** $-1/3$

61. Undefined **63.** -324

65. π^2 **67.** 4 **69.** 5 **71.** 3 **73.** 2

75. a. $\dfrac{1}{x-1}; \ \text{Dom} = (-\infty, 1) \cup (1, \infty)$

b. $\dfrac{1}{x} - 1; \ \text{Dom} = (-\infty, 0) \cup (0, \infty)$

77. a. $1 - \sqrt{x}; \ \text{Dom} = [0, \infty)$

b. $\sqrt{1 - x}; \ \text{Dom} = (-\infty, 1]$

79. a. $9x^4 + 36x^2 + 35; \ \text{Dom} = \mathbb{R}$

b. $3x^4 + 12x^3 + 12x^2 + 5; \ \text{Dom} = \mathbb{R}$

81. a. $\sqrt{2x}; \ \text{Dom} = [0, \infty)$

b. $2\sqrt{x}; \ \text{Dom} = [0, \infty)$

83. a. $x; \ \text{Dom} = (-\infty, 0) \cup (0, \infty)$

b. $x; \ \text{Dom} = (-\infty, 0) \cup (0, \infty)$

85. a. $x; \ \text{Dom} = \mathbb{R}$

b. $x; \ \text{Dom} = \mathbb{R}$

87. a. $\dfrac{3}{1-3x^2}$;

Dom $= \left(-\infty, -\sqrt{3}/3\right) \cup \left(-\sqrt{3}/3, \sqrt{3}/3\right) \cup \left(\sqrt{3}/3, \infty\right)$

b. $\dfrac{27}{x^2 - 2x + 1}$; Dom $= (-\infty, 1) \cup (1, \infty)$

89. $g(x) = \dfrac{2}{x}$, $h(x) = 5x - 1$, $f(x) = g(h(x))$

91. $g(x) = x + \sqrt{x} - 7$, $h(x) = x + 2$, $f(x) = g(h(x))$

93. $g(x) = \dfrac{\sqrt{x}}{x^2}$, $h(x) = x - 3$, $f(x) = g(h(x))$

95. $S(t) = \pi r \sqrt{r^2 + \dfrac{t^4}{16}}$

97. $c(t) = 2040 + 17,400t - 43.5t^2$

99. $(f \circ g)(x) = \dfrac{-x}{\sqrt[3]{3x^2 - 9}} = -(f \circ g)(-x)$

101. Yes   **103.** Yes   **105.** No   **107.** No

109. True   **111.** True

113.

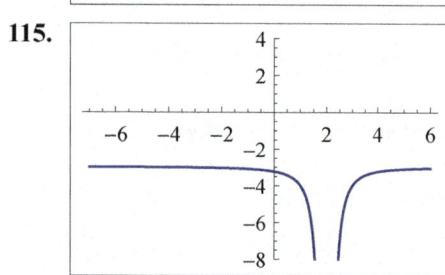

115.

117.

119. $y = -(x+3)^2 + 5$

121. $y = -x^3 + 7$ **123.** $y = (x-2)^3 - 4$

125. $(f+g)(x) = 4 + 12x + 9x^2 + \sqrt{x^2 + 5}$;

$(fg)(x) = 4\sqrt{x^2 + 5} + 12x\sqrt{x^2 + 5} + 9x^2\sqrt{x^2 + 5}$;

$(f \circ g)(x) = 49 + 9x^2 + 12\sqrt{x^2 + 5}$;

$(g \circ f)(x) = \sqrt{(3x+2)^2 + 5}$

127. $(f+g)(x) = \dfrac{2x^2 - x + 1}{x(x-1)}$;

$(fg)(x) = \dfrac{x+1}{x}$;

$(f \circ g)(x) = 1 - 2x$;

$(g \circ f)(x) = \dfrac{2}{x+1}$

Section 1.4

1.

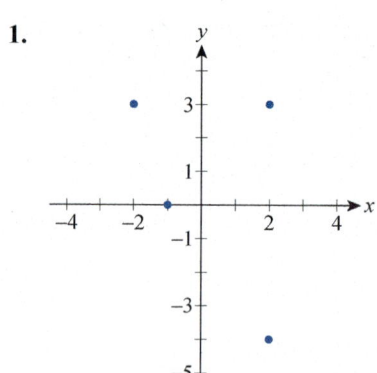

Dom $= \{-2, -1, 2\}$,   Ran $= \{-4, 0, 3\}$

3.

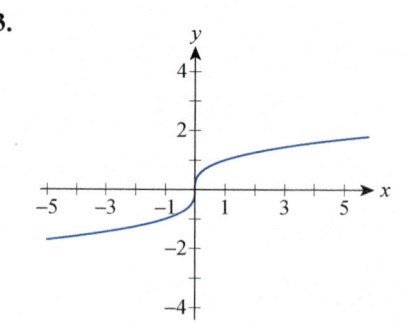

Dom $=$ Ran $= \mathbb{R}$

5.

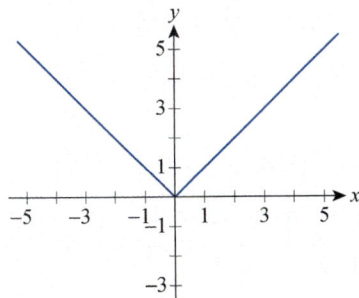

$\text{Dom} = \mathbb{R}, \ \text{Ran} = [0, \infty)$

7.

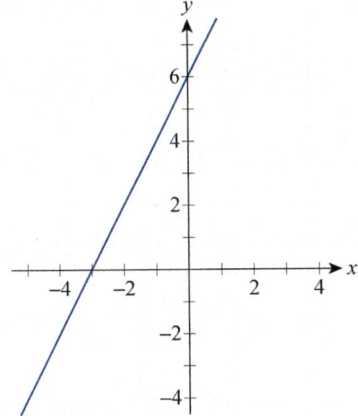

$\text{Dom} = \text{Ran} = \mathbb{R}$

9.

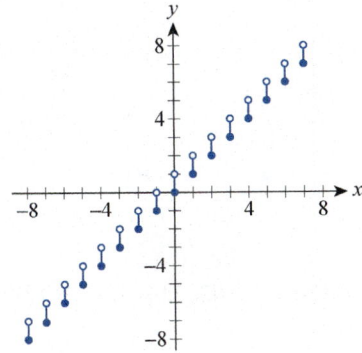

$\text{Dom} = \mathbb{Z}, \ \text{Ran} = \mathbb{R}$

11.

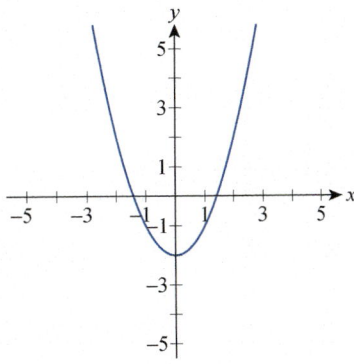

$\text{Dom} = \mathbb{R}, \ \text{Ran} = [-2, \infty)$

13. Restrict to $[0, \infty)$

15. Inverse exists   **17.** Inverse exists

19. Inverse exists   **21.** Inverse exists

23. $f^{-1}(x) = (x+2)^3$ **25.** $r^{-1}(x) = \dfrac{-2x-1}{3x-1}$

27. $F^{-1}(x) = (x-2)^{1/3} + 5$ **29.** $V^{-1}(x) = 2x - 5$

31. $h^{-1}(x) = (x+2)^{5/3}$ **33.** $J^{-1}(x) = \dfrac{x-2}{3x}$

35. $h^{-1}(x) = (x-6)^{1/7}$ **37.** $r^{-1}(x) = x^5/2$

47. F   **49.** C   **51.** D   **53.** G

55. F   **57.** C   **59.** A

61.

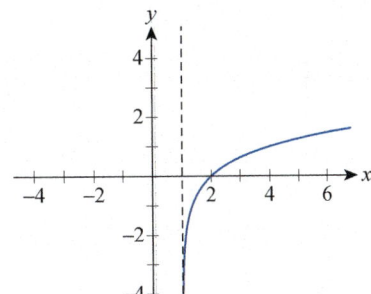

63.

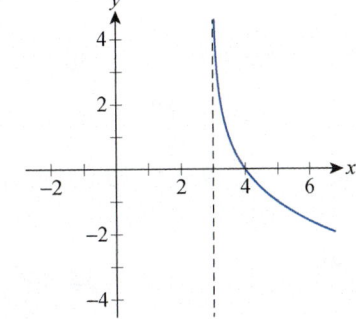

65.

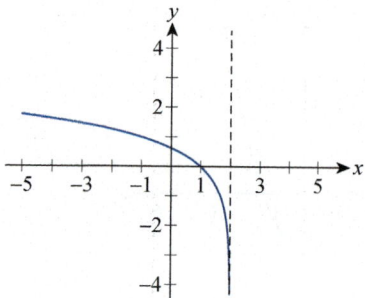

67.

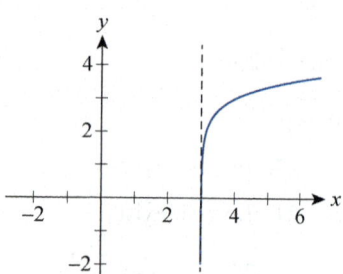

69.

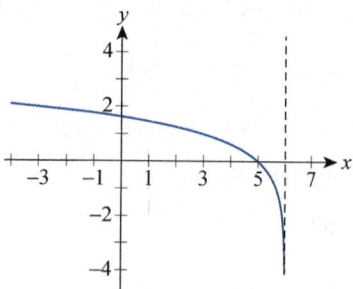

71.

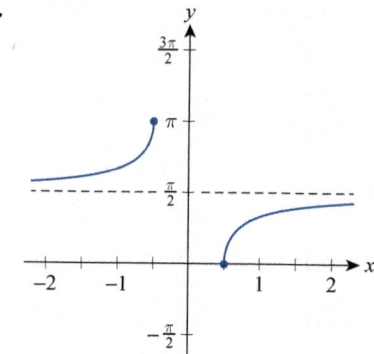

73. 2   **75.** 7   **77.** 1   **79.** 2.45   **81.** 3.30

83. 0.74   **85.** x^2 **87.** $e^2 p/x$ **89.** x^3/y^4

91. $\dfrac{3}{2}\ln x + \ln p + 5\ln q - 7$ **93.** $\log\left(2 + 3\log x\right)$

95. $1 - \dfrac{1}{2}\log\left(x + y\right)$

97. $\log_2\left(y^2 + z\right) - 4 - 4\log_2 x$

99. $2\log_b x + \dfrac{1}{2}\log_b y - \log_b z$ **101.** $\ln\dfrac{3p}{q^2}$

103. $\log\dfrac{x - 10}{x}$ **105.** $\ln\dfrac{e^2}{x^3 y^3}$

107. $-\pi/2$ **109.** $-\pi/6$ **111.** $\pi/4$

113. Does not exist   **115.** -1.3734 **117.** 1.0472

119. $-1/2$ **121.** 2   **123.** 2

125. $\dfrac{\sqrt{x^2 - 4}}{2}$ **127.** $\dfrac{\sqrt{3}x}{3}$ **129.** $\dfrac{4}{\sqrt{x^2 + 16}}$

131.

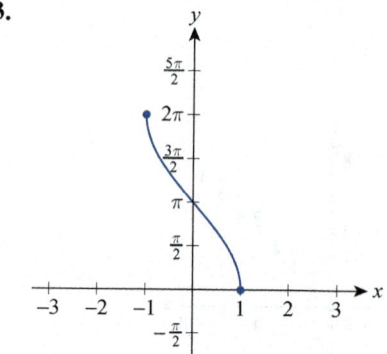

133.

135. 521 73 136 136 73 392 316 −7 73 188 392
217 8 −7 617

137. REMEMBER YOUR SUNBLOCK

139. 6.3   **141.** 13.4 decibels

143. **a.** 0.2898   **b.** 0.4429   **c.** 0.7956

145. True   **147.** True

149. False; the domain of $\arcsin x$ is $\left[-1, 1\right]$.

151. True

Section 1.5

13. a. $y = 3.559x + 22.038$ **b.** 36,274 units

15. a. $y = 10.743x - 25.829$ **b.** \$81,601

17. a. $y = -1.345x + 79.771$
   b. 69.01 cents per pound

19. $x = 0$ **21.** $x = \dfrac{(2k+1)\pi}{6} + \dfrac{2}{3}, \; k \in \mathbb{Z}$

23. $x = -1$ **25.** $x = \pm\sqrt[4]{1/2}$ **27.** $x \le 0$

29. Answers will vary.   **31.** Answers will vary.

33. Answers will vary.   **35.** Answers will vary.

37. Answers will vary.   **39.** Answers will vary.

41. 20   **43.** −21.9744   **45.** −90.8502

47. 49.9990   **49.** 130,783   **51.** 1

53. Answers will vary.   **55.** Answers will vary.

57. Answers will vary.

59. −4.4213, −0.1001, 4.5213

61. 1.1974, 5.6240, 6.8003

63. 0.1586, 3.1462

65. $f(x)$ **67.** $g(x)$ **69.** $g(x)$

71. Answers will vary.   **73.** Answers will vary.

Chapter 1 Review

1. Dom $= \{-2, -3\}$, Ran $= \{-9, -3, 2, 9\}$;
   not a function

3. Dom $= [-6, \infty)$, Ran $= \mathbb{R}$; not a function

5. Dom $= \mathbb{N}$, Cod $= \mathbb{R}$, Ran $= \left\{ \dfrac{3x}{4} \,\middle|\, x \in \mathbb{N} \right\}$

7. Dom $= \mathbb{R}$, Cod $= \mathbb{R}$, Ran $= (0, 1]$

9. a. $2x^2 + 6x - 8$ **b.** $2x^4 + 10x^2$
   c. $4x + 2h + 10$

11. a. $\dfrac{3}{x+1}$ **b.** $\dfrac{3}{x^2+2}$ **c.** $\dfrac{-3}{(x+h+2)(x+2)}$

13. Decreasing on $(-\infty, 2)$; increasing on $(2, \infty)$

15. Odd

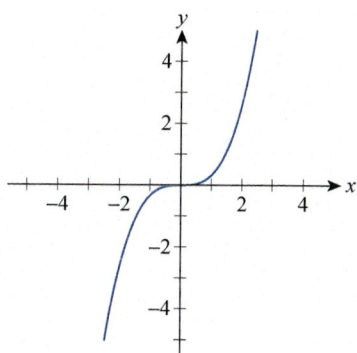

17. Odd

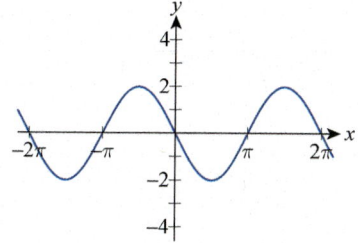

19. Symmetric with respect to y-axis

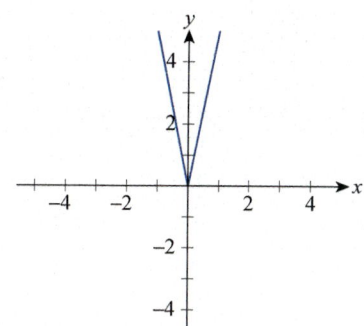

21.

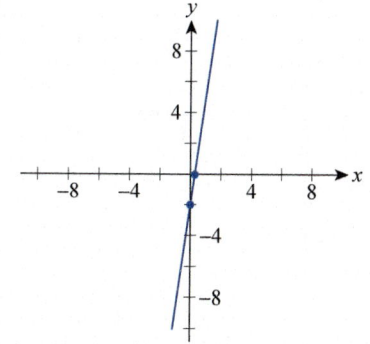

23.

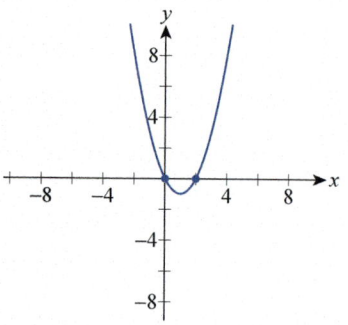

25.

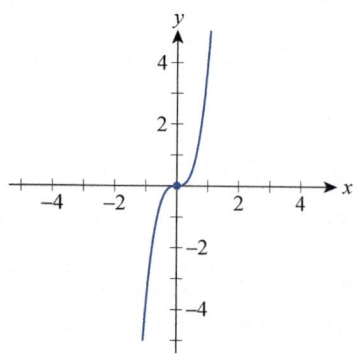

27.

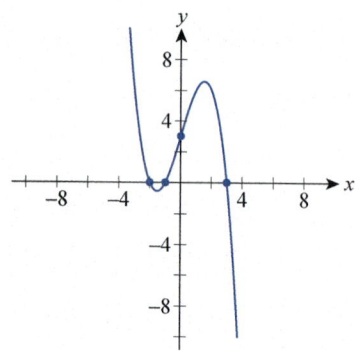

29.

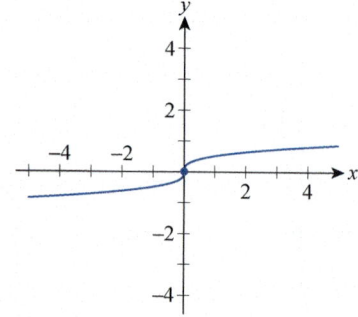

31.

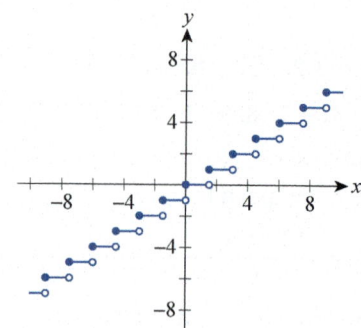

33.

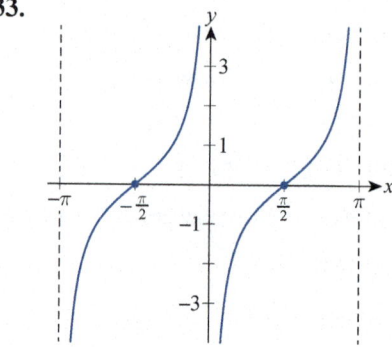

35.

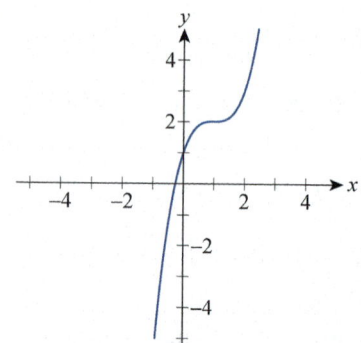

Dom $= \mathbb{R}$, Ran $= \mathbb{R}$

37. $y = \dfrac{1}{x-2} + 3$

39. a. $x^2 + \sqrt{x}$; Dom $= [0, \infty)$

   b. $x^{3/2}$; Dom $= (0, \infty)$

41. a. $x + 2$ **b.** $x - 2$ **c.** 5

43. $g(x) = \dfrac{\sqrt{x}+2}{x^2}$, $h(x) = x+3$; $f(x) = g\big(h(x)\big)$

45.

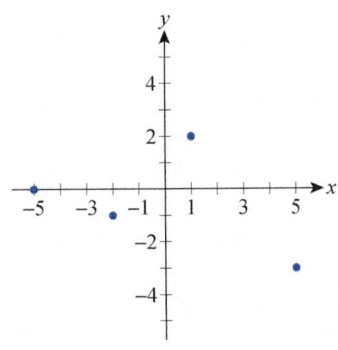

$$\text{Dom} = \{-5, -2, 1, 5\}, \quad \text{Ran} = \{-3, -1, 0, 2\}$$

47. $f^{-1}(x) = \dfrac{x+2}{7x}$ **49.** $f^{-1}(x) = (x+6)^5$

53. $\dfrac{\sqrt{x^2-4}}{2}$ **55.** $\dfrac{1}{3}\left(\log y + 2\log z - 4\log x\right)$

57. $A(t) = A_0 e^{-(\ln 2/10)t} = A_0 (1/2)^{t/10}$; approx. 17.7%

59. 15/2 and 15/2   **61.** Yes

63. False; a function may not be defined on (any subset of) $\mathbb{R}$ and thus may not have a graphical representation in the xy-system. Even graphs of numerical functions are not always curves.

65. False; consider $f(x) = x^2$ with $a = 1$ and $b = -1$.

67. False; consider $f(x) = (x-1)^2$.

69. False; see Example 10 of Section 1.3.

71. False; consider $c = 2$ and $f(x) = x^2$:
$f(cx) = (2x)^2 = 4x^2$, while $cf(x) = 2x^2$.

73. False; consider $y = x^2$. (The statement is only true for invertible functions.)

75. True

77.

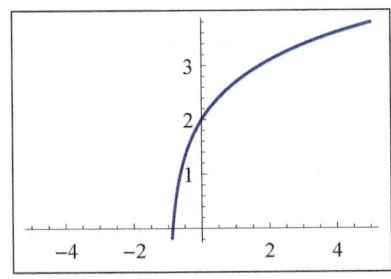

79.

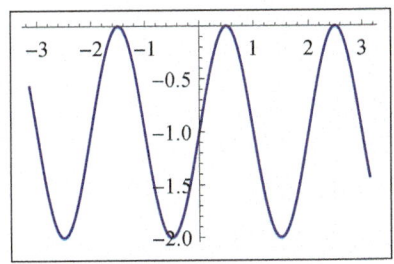

81. 1.4262   **83.** Answers will vary.

Chapter 1 Project

1. $r(t) = 2.6t$

3. Doubling time is approximately 24 minutes and 51 seconds; the time to triple is approximately an additional 19 minutes and 4 seconds. (The exact values are $\sqrt{2} - 1$ hours and $\sqrt{3} - \sqrt{2}$ hours, respectively.)

5. After 3 hours: radius 7.8 km and area approx. 191.13 km^2
After 5.5 hours: radius 14.3 km and area approx. 642.42 km^2

7. Approx. $286.70 \text{ km}^2/\text{h}$

9. Answers will vary, but a major flaw is that the model predicts a hole whose area will grow indefinitely, and is not a periodic function of time. A better approach is to use a periodic function.

Chapter 2

Section 2.1

1. 3   **3.** 0   **5.** $-1/4$ **7.** The exact answer is -2.

9. The exact answer is 2.

11. The exact answer is 2.

13. The exact answer is $1/e$ or approx. 0.3679.

15. The exact answer is $1/(5\ln 10)$ or approx. 0.0869.

17. The exact answer is -4.

19. **a.** 96 ft   **b.** 16 ft/s   **c.** 16   **d.** 2.5 s

21. **a.** 176 ft   **b.** 176 ft/s   **c.** 192 ft/s
d. 64 ft/s   **e.** 6 s

23. **a.** 3 m/s   **b.** 15 m/s   **c.** $6t_0 + 3$

25. **a.** 5 s; 45 m/s   **b.** 10 s   **c.** $-1.83g$

27. a. 87.35 ft **b.** 34.94 ft/s
c. −28 ft/s **d.** −77.47 ft/s

29. a. 60 **b.** 61.44

31. a. 12.7234 **b.** 2.3129

33. a. 1.0536 **b.** 1.0563

35. a. 1.2217 **b.** 1.2218

37.

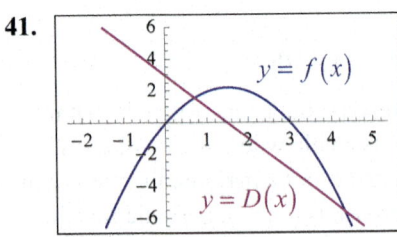

39.

41.

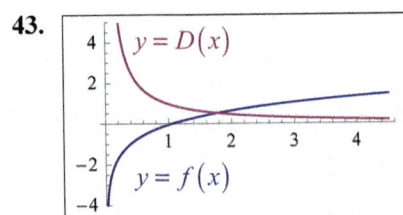

The function value $D(x_0)$ at any given x_0 is approximately equal to the slope of the graph of f at x_0.

43.

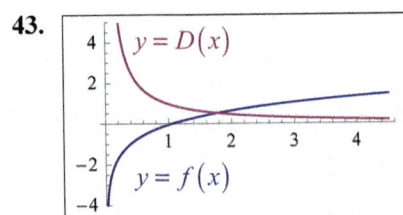

The function value $D(x_0)$ at any given x_0 is approximately equal to the slope of the graph of f at x_0, for any $x_0 > 0$.

45. $x = \pm 2$ **47.** $x = 1$

49. a. 63.9936 **b.** 63.9999

51. a. 11.0508 **b.** 11.0050

53. a. 1.0612 **b.** 1.0612

55. a. 1.222 **b.** 1.222

Section 2.2

1. 4 **3.** 0

5.

X	y
1.0	2.414
1.4	2.814
1.41	2.824
1.414	2.828

The table points to a limit of $2\sqrt{2} \approx 2.828$.

7.

X	y
1.5	113.33
1.1	15.94
1.01	10.46
1.001	10.045

The table points to a limit of 10.

9.

X	y
3	−3.4696
3.14	−3.0048
3.141	−3.0018
3.1415	−3.0003

The table points to a limit of −3.

11.

X	y
7.5	210.5
7.1	994.1
7.01	9814.0
7.001	98014.0

The table points to an undefined limit.

13. $\lim\limits_{x \to 1^-} f(x) = -\infty$; $\lim\limits_{x \to 1^+} f(x) = \infty$

15. $\lim\limits_{x \to 3^-} h(x) = \infty$; $\lim\limits_{x \to 3^+} h(x) = \infty$

17. $\lim\limits_{x \to -1^-} q(x) = \infty$; $\lim\limits_{x \to -1^+} q(x) = -\infty$

19. $\lim\limits_{x \to 1.5^-} v(x) = -\infty$; $\lim\limits_{x \to 1.5^+} v(x) = \infty$

21. $\lim\limits_{x \to \left(\frac{\pi}{2}+k\pi\right)^-} \tan x = \infty$; $\lim\limits_{x \to \left(\frac{\pi}{2}+k\pi\right)^+} \tan x = -\infty$ ($k \in \mathbb{Z}$)

23. $\lim\limits_{x \to 6^-} s(x) = -\infty$; $\lim\limits_{x \to 6^+} s(x) = \infty$

25. $\lim\limits_{x \to \infty} f(x) = 0$; $\lim\limits_{x \to -\infty} f(x) = 0$

27. $\lim\limits_{x \to \infty} h(x) = 0$; $\lim\limits_{x \to -\infty} h(x) = 0$

29. $\lim\limits_{x\to\infty} q(x) = -1;\ \lim\limits_{x\to-\infty} q(x) = -1$

31. $\lim\limits_{x\to\infty} v(x) = \infty;\ \lim\limits_{x\to-\infty} v(x) = -\infty$

33. $\lim\limits_{x\to\infty} \tan x$ does not exist; $\lim\limits_{x\to-\infty} \tan x$ does not exist

35. $\lim\limits_{x\to\infty} s(x) = 1;\ \lim\limits_{x\to-\infty} s(x) = -1$

37. a. -2 **b.** 1

39. a. 1 **b.** 4

41. a. 1 **b.** $-\infty$

43. a. 1 **b.** 1

45. a. Does not exist **b.** 0

47. $\lim\limits_{x\to\infty} f(x) = \infty;\ \lim\limits_{x\to-\infty} f(x) = -\infty$

49. $\lim\limits_{x\to\infty} h(x) = -\infty;\ \lim\limits_{x\to-\infty} h(x) = -\infty$

51. $\lim\limits_{x\to\infty} F(x) = \infty$

53. $\lim\limits_{x\to\infty} H(x) = \infty;\ \lim\limits_{x\to-\infty} H(x) = \infty$

55. $\lim\limits_{x\to\infty} u(x) = \infty;\ \lim\limits_{x\to-\infty} u(x) = \infty$

57. $\lim\limits_{x\to-\infty} s(x) = -\infty$

59. False; see $h(x)$ of Example 5 at $x = 2$.

61. True **63.** True **65.** Does not exist

67. 2 **69.** $3/2$ **71.** 0

75. See the answer given for Exercise 65.

77. See the answer given for Exercise 67.

79. See the answer given for Exercise 69.

81. See the answer given for Exercise 71.

Section 2.3

1. $\delta \approx 0.23$ or smaller

3. $\delta \approx 0.2$ or smaller

5. L is incorrectly quantified and switched with c.

7. ε is incorrectly quantified.

9. The inequality $0 \le |x - c|$ is incorrect.

11. $\delta = 0.02$ or smaller

13. $\delta = 0.1$ or smaller

15. $\delta = \sqrt{0.1}$ or smaller

17. $\delta = 0.0\overline{9}$ or smaller

19. $\delta = e^{0.1} - 1 \approx 0.1052$ or smaller

21. $N = 10$ or larger

23. $N = -10$ or smaller

25. $N = \ln 0.1$ or smaller

27. $\delta = 0.1$ or smaller

29. $\delta = 0.1$ or smaller

31. $\delta = \dfrac{\pi}{2} - \arctan 100 \approx 0.0099997$ or smaller

63. The limit does not exist.

65. The limit does not exist.

67. The limit is 0.

69. 0.02256 mm

71. False; the function value and limit at c need not be equal. (See Example 2.)

73. False; consider $f(x) = -x^2$ and $g(x) = x^2$ at $c = 0$.

75. Answers will vary.

77. Answers will vary.

79. Answers will vary.

81. Answers will vary.

83. Answers will vary.

85. Vert. asym.: $x \approx -1.3340$, $x \approx 1.1759$

87. Vert. asym.: $x = (2n+1)\pi - 6$, $n \in \mathbb{Z}$

89. Vert. asym.: $x = n\pi/2$, $n \in \mathbb{Z}$, $n \ne 0$

Section 2.4

1. a. -4 **b.** 21 **3.** 5 **5.** 7 **7.** 9

9. 22 **11.** -4 **13.** -2 **15.** $\sqrt[3]{2}$

17. $\sqrt[3]{100}$ **19.** 16 **21.** 12 **23.** 11

25. $33/5$ **27.** $1/6$ **29.** $1/(2\sqrt{5})$ **31.** $1/9$

33. 6 **35.** 4 **37.** $1/(2\sqrt{x})$ **39.** $32/7$

41. $-1/6$ **43.** -1 **45.** 8 **47.** 11

49. $9 + 2(2 - c)$ **51.** 1

53. 0 **55.** 0 **57.** $2e^2$

71. The limit is 0. **79.** $r/2$

Section 2.5

1. Points of continuity: $(-\infty,0)\cup(0,3)\cup(3,\infty)$

    Points of discontinuity: $c = 0$, $\lim\limits_{x\to 0} f(x)$ does not

    exist; $c = 3$, $\lim\limits_{x\to 3} f(x) \neq f(3)$

3. Answers will vary.

5. $c = 0$, nonremovable

7. $c = 3$, removable

9. $c = 2$, nonremovable

11. None   13. $c = \pi/2$, nonremovable

15. $c = 4$, removable

17. None   19. None

21. $c = \pm 1$, nonremovable

23. None   25. All integers, nonremovable

27. $c = \pm\sqrt{n}$, n a positive integer, nonremovable

29. $c = 1/n$, n a nonzero integer, nonremovable

35. Continuous on $[-3,-1]\cup(0,2]\cup[3,\infty)$

37. Continuous on $(-\sqrt{3},\sqrt{3})$

39. Continuous except on
    $\{n - 1/\pi \mid n \in \mathbb{Z}\}\cup\{\ln k - 2 \mid k \in \mathbb{N}\}$

45. $f(3) = 7$ will make f continuous.

47. $h(1) = 2$ will make h continuous.

49. $G(0) = 0$ will make G continuous.

51. Continuous on $[-4,4]$

53. Not continuous at the endpoints

55. $a = 5$

57. $a = 3/2$, $b = -5/2$

63. Yes; $c = 2$ 65. Yes; $c = 2$

67. No; G is not continuous on $[-2,2]$

83. **a.** Because speeds of everyday objects are smaller than c by orders of magnitude, the denominator of ΔT is approximately 1.

    b. No moving object can reach the speed of light.

85. False; consider the Dirichlet function.

87. True   89. -0.2895 91. 0.8241 93. -1.7321

95.

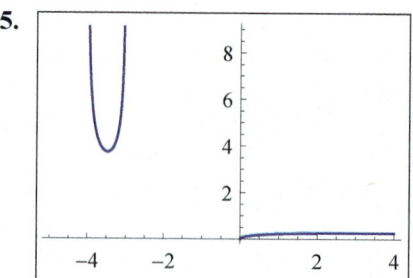

97.

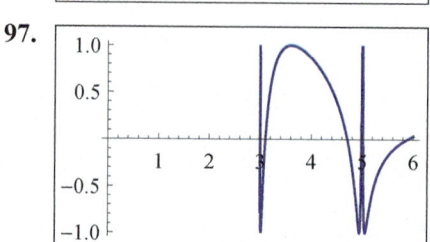

99.

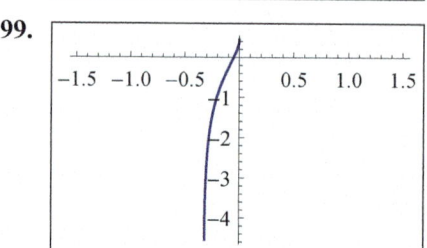

Section 2.6

1. **a.** $f'(-1) = 2$ **b.** $f'(1) = -2$

3. $y = 4x - 6$ 5. $y = \dfrac{1}{2}x + 4$

7. $y = 12x - 16$ 9. $y = \dfrac{1}{2}x + 1$

11. $y = -4x + 4$ 13. $y = -\dfrac{1}{16}x + \dfrac{3}{4}$

15. $f'(x) = 0$; tangent is horizontal for all x

17. $f'(x) = 4$; tangent is never horizontal

19. $f'(x) = 6x$; tangent is horizontal at $x = 0$

21. $f'(x) = x + 5$; tangent is horizontal at $x = -5$

23. $f'(x) = 3x^2 + 1$; tangent is never horizontal

25. $f'(x) = 4x^3$; tangent is horizontal at $x = 0$

27. $f'(x) = -\dfrac{10}{(2x-4)^2}$; tangent is never horizontal

29. $f'(x) = -\dfrac{7}{(x-3)^2}$; tangent is never horizontal

31. $f'(x) = -\dfrac{2x}{\left(x^2+1\right)^2}$; tangent is horizontal at $x = 0$

33. $f'(x) = \dfrac{5}{2\sqrt{5x}}$; tangent is never horizontal

35. $f'(x) = \dfrac{1}{\sqrt{2x+1}}$; tangent is never horizontal

37. $f'(x) = \dfrac{x}{\sqrt{x^2+1}}$; tangent is horizontal at $x = 0$

39. $y = 6x - 6$

41. $y = -\dfrac{1}{2}x + 1$ or $y = -\dfrac{1}{2}x - 1$

43. $54y + x = 27$ **45.** $f'(3.6) = -1/4$

47. $h'(3) = 10$ **49.** $G'(7) = 1/2$

51. $u'(-3) = -1/2$ **53.** $w'(5) = -1/54$

55. $G'(-2) = -32$ **57.** C   **59.** D

61. Answers will vary.   **63.** Answers will vary.

65. Answers will vary.

73. $t = 1$, $t = 5$; the particle stops at 2 s and at 4 s.

75. **a.** The respective velocities are 16 ft/s and -16 ft/s; the speed is 16 ft/s in both cases.

   b. It rises 36 ft above the cliff and reaches the bottom at approx. 4.91 s.

   c. Approx. -109.12 ft/s

77. She seems to have stopped for an extended time period (probably a red light); soon after, she turned around and sped back home (perhaps forgot something at home). Then she sped back to class, this time managing to do so without a red light, and also driving faster.

79. **a.** $P(x) = 0.1x^2 + 40x - 190$

   b. Approx. 5 suitcases

   c. $P(25) = 872.50$, $P(30) = 1100$, $P(40) = 1570$

   d. $P'(x) = 0.2x + 40$

   e. $P'(25) = 45$, $P'(30) = 46$, $P'(40) = 48$

81. $C(x) = 20 + 3x$; $C'(x) = 3$

83.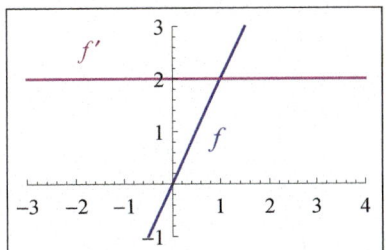

85.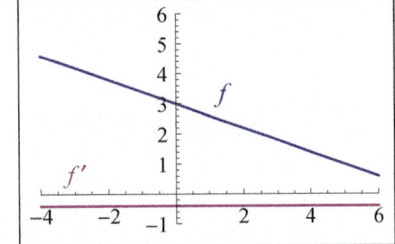

87.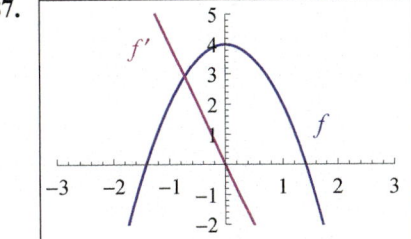

89.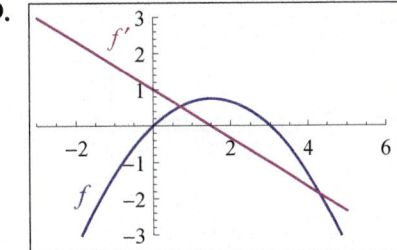

91.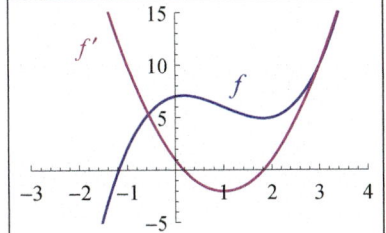

93.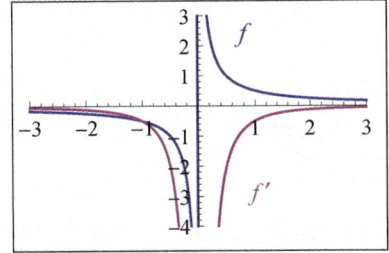

95.

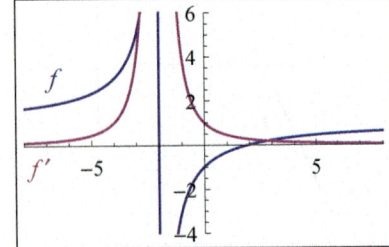

97.

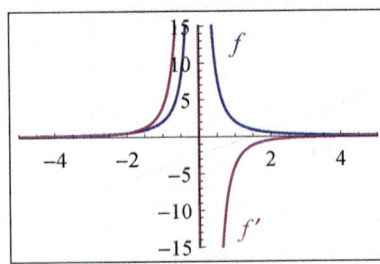

99.

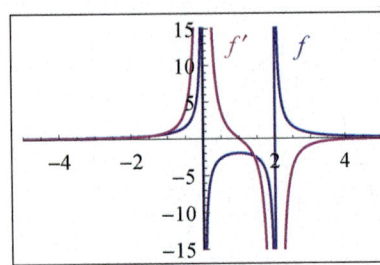

101.

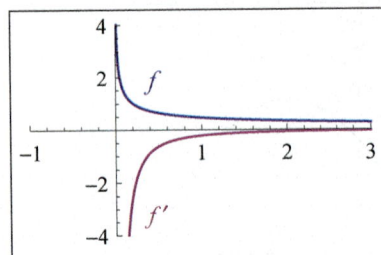

103.

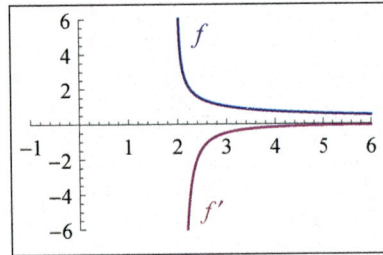

105.

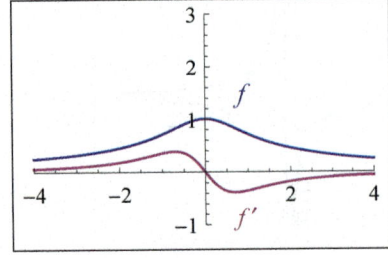

Chapter 2 Review

1. The limit is 3; the function is continuous.

3. The limit is 2; there is a nonremovable (jump) discontinuity at 1.

5. The exact answer is 3.

7. The exact answer is $1/2$.

9. **a.** 598 ft   **b.** 576 ft/s   **c.** 608 ft/s

   d. 544 ft/s   **e.** At 19 seconds

11. 1.5568   **13.** 21.7102

15.

x	y
0.5	0.7071
0.1	0.7943
0.01	0.9550
0.001	0.9931

The table points to a limit of 1.

17.

x	y
0.5	0.47973
0.1	0.11969
0.01	−0.00264
0.001	−0.00194

The table points to a limit of 0.

19. $\delta = 0.005$ or smaller

25. 14   **27.** 7   **29.** 8   **31.** $-1/10$

33. ∞ **35.** $1/4$ **37.** -2 **39.** 5   **41.** $2x$

45. Answers will vary.   **47.** $c = 4$, nonremovable

49. All integers, nonremovable

51. No discontinuities   **57.** $y = 3x - 1$

59. $f'(x) = 2 - 2x$; tangent is horizontal at $x = 1$

61. Answers will vary.

65. **a.** $P(x) = 0.1x^2 + 15x - 247.5$

   b. 15 toys   **c.** $P'(x) = 0.2x + 15$

67. False; there could be, for example, a jump or oscillating discontinuity at $x = c$.

69. False; the condition $B \neq 0$ is needed.

71. True

73. False; while $\lim_{x \to c} f(x) = L$ does follow, $f(c)$ doesn't even have to be defined.

75. 5.3586 and 21.7491, respectively

77. −0.6891 and 0.7729, respectively

79. 1   **81.** 3

Chapter 2 Project

1.

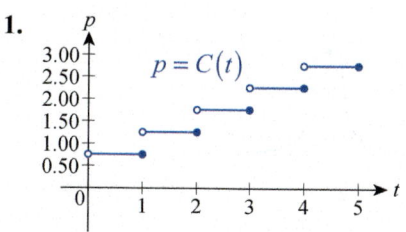

3. No, the left- and right-hand limits are unequal.

5. A call lasting for about two and a half minutes costs $1.75.

7. A call lasting just over three minutes costs $2.25.

9. Does not exist

Chapter 3

Section 3.1

1. $\dfrac{df}{dx} = 0; \dfrac{df}{dx}\Big|_{x=1} = 0$ **3.** $\dfrac{dh}{dx} = x; \dfrac{dh}{dx}\Big|_{x=0} = 0$

5. $\dfrac{dG}{ds} = s^2 - 1; \dfrac{dG}{ds}\Big|_{s=-3} = 8$

7. $\dfrac{dK}{dz} = \dfrac{-15}{(3z+1)^2}; \dfrac{dK}{dz}\Big|_{z=0} = -15$

9. $\dfrac{dw}{dz} = \dfrac{-2z}{(z^2+2)^2}; \dfrac{dw}{dz}\Big|_{z=\sqrt{2}} = \dfrac{-\sqrt{2}}{8}$

11. $\dfrac{dQ}{dy} = \dfrac{-1}{2y\sqrt{3y}}; \dfrac{dQ}{dy}\Big|_{y=1} = \dfrac{-\sqrt{3}}{6}$

13. $D_x f(x) = 0$ **15.** $D_t h(t) = 3t + 10$

17. $D_y G(y) = -1/(3y^2)$ **19.** $D_z S(z) = 6/z^3$

21. $D_v R(v) = 1/\sqrt{2v-3}$ **23.** $D_y X(y) = 8y^3$

25. $f'(x) = 5/2; f''(x) = 0; f'''(x) = 0$

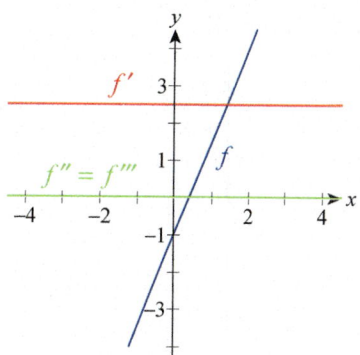

27. $h'(x) = -x+1; h''(x) = -1; h'''(x) = 0$

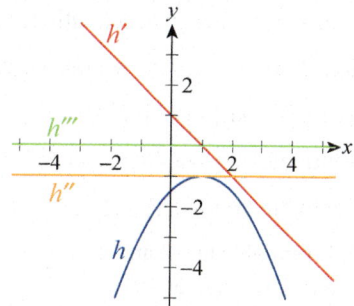

29. $V'(x) = (x-2)^2; V''(x) = 2x-4; V'''(x) = 2$

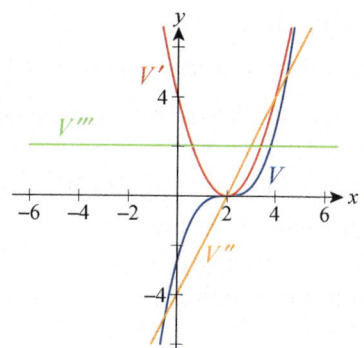

31. $G'(x) = 8(x-1)^3; G''(x) = 24(x-1)^2;$
$G'''(x) = 48x - 48$

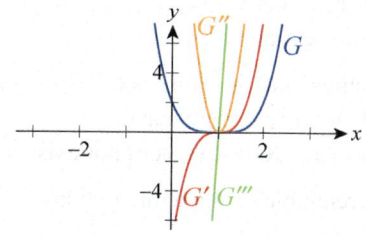

33. $K'(s) = 2/(s-1)^2$; $K''(s) = -4/(s-1)^3$;
$K'''(s) = 12/(s-1)^4$

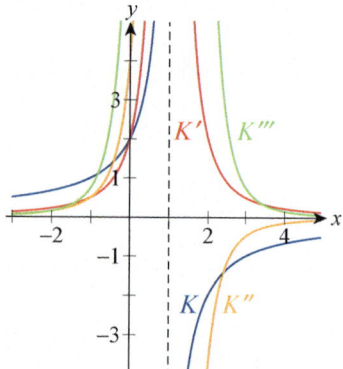

35. Position: k; velocity: g; acceleration: h; jerk: f

37. Position: k; velocity: h; acceleration: g; jerk: f

39. Differentiable on $(-\infty, 3) \cup (3, \infty)$

41. Differentiable on $(-\infty, -1) \cup (-1, 3) \cup (3, \infty)$

43. Differentiable on $(-\infty, 1) \cup (1, \infty)$

45. Not differentiable at -2 and 4
Left-hand derivative at -2: 0
Right-hand derivative at -2: 2
Left-hand derivative at 4: 2
Right-hand derivative at 4: 0

47. Not differentiable at 1.5
One-sided derivatives do not exist at 1.5

49. Undefined and therefore not differentiable on
$(-\infty, -\sqrt{3}) \cup (\sqrt{3}, \infty)$
One-sided derivatives do not exist at $\pm\sqrt{3}$

51. Not differentiable at 1 and 5
Left-hand derivative at 1: -4
Right-hand derivative at 1: 4
Left-hand derivative at 5: -4
Right-hand derivative at 5: 4

53. Not differentiable at $k \in \mathbb{Z}$
Left-hand derivative at each point of
nondifferentiability: 1
Right-hand derivative at each point of
nondifferentiability: 1

55. Undefined and therefore not differentiable for
$z < 0$, not differentiable at 0
Right-hand derivative does not exist at 0

57. Differentiable on the entire real line

61. a. $a(t) = 2.2 \text{ m/s}^2$ **b.** $t = 15 \text{ s}$
c. The runway must be at least 247.5 m long.

63. b. $g(x) = |x|$; $c = 0$

65. True

67. False; consider $f(x) = |x|$ at $c = 0$.

69.

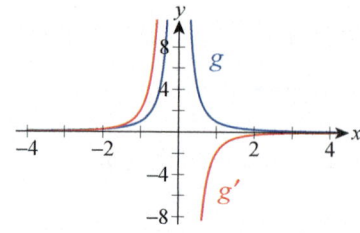

$g' > 0$ and g is increasing for $x < 0$;
$g' < 0$ and g is decreasing for $x > 0$

71.

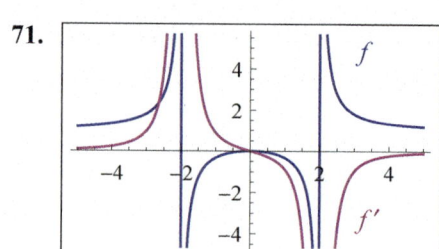

$f' > 0$ and f is increasing for $x < -2$ and
$-2 < x < 2$;

$f' < 0$ and f is decreasing for $0 < x < 2$ and $x > 2$

73.

$f' > 0$ and f is increasing for
$2k\pi < x < (2k+1)\pi$;

$f' < 0$ and f is decreasing for
$(2k-1)\pi < x < 2k\pi$

75.

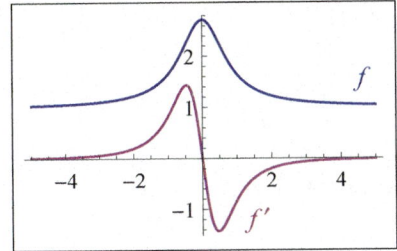

$f' > 0$ and f is increasing for $x < 0$;

$f' < 0$ and f is decreasing for $x > 0$

77. $f'(x) = \dfrac{1}{1+x^2}$; $f''(x) = -\dfrac{2x}{\left(1+x^2\right)^2}$;

$f'''(x) = \dfrac{-2+6x^2}{\left(1+x^2\right)^3}$; $f^{(4)}(x) = -\dfrac{24x\left(-1+x^2\right)}{\left(1+x^2\right)^4}$

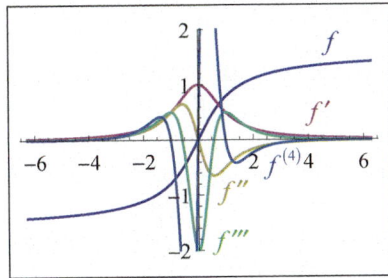

79. $f'(x) = \dfrac{1}{10}x\cos\left(1-\dfrac{x}{10}\right) - \sin\left(1-\dfrac{x}{10}\right)$;

$f''(x) = \dfrac{1}{5}\cos\left(1-\dfrac{x}{10}\right) + \dfrac{1}{100}x\sin\left(1-\dfrac{x}{10}\right)$;

$f'''(x) = -\dfrac{1}{1000}x\cos\left(1-\dfrac{x}{10}\right) + \dfrac{3}{100}\sin\left(1-\dfrac{x}{10}\right)$;

$f^{(4)}(x) = -\dfrac{1}{250}\cos\left(1-\dfrac{x}{10}\right) - \dfrac{1}{10000}x\sin\left(1-\dfrac{x}{10}\right)$

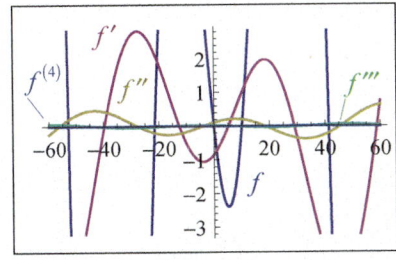

81.

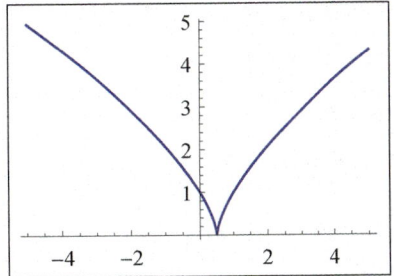

Not differentiable at $t = 1/2$ (cusp)

83.

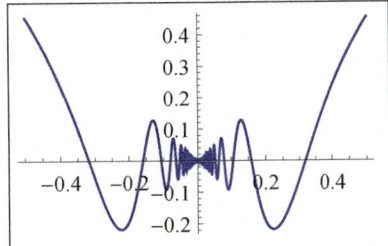

Not differentiable at $t = 0$ (oscillating discontinuity)

Section 3.2

1. $f'(x) = -2$　**3.** $h'(x) = -6x+2$

5. $G'(x) = 2x^3 + 6x^2 - 2x + 3.2$

7. $H'(x) = 8x^7 + 5\sqrt{2}x^4 - 8x^3$

9. $S'(z) = 12z^2 - \dfrac{3}{2\sqrt{z}}$　**11.** $T'(r) = 2\pi r + 2e^r$

13. $6x + 11$　**15.** $8x^3 + 15x^2 - 24x - 30$

17. $-\dfrac{196}{5}x^6 - \dfrac{20}{3}x^4 + \dfrac{56}{5}x^3 + \dfrac{4}{3}x$　**19.** $\dfrac{1}{\sqrt{t}} + 8$

21. $f'(x) = \dfrac{2}{\left(1-2x\right)^2}$

23. $h'(x) = \dfrac{-12x^2 + 20x - 6}{\left(2x^3 - 5x^2 + 3x + 1\right)^2}$

25. $G'(x) = \dfrac{-11}{\left(x-5\right)^2}$

27. $H'(x) = \dfrac{11x^4 + 3x^2 - 6x}{\left(2x^3 + 5x^2 + 1\right)^2}$

29. $B'(u) = \dfrac{3u^{3/2} + 4u}{2\left(\sqrt{u} + 1\right)^2}$

31. $g'(t) = \dfrac{5t - 8\sqrt{t} + 15}{2\sqrt{t}\left(4 - 5\sqrt{t}\right)^2}$

33. $f'(x) = 6 - 10x^4$

35. $h'(x) = -27x^2 + 14x - 5$

37. $G'(x) = 1$ **39.** $f'(t) = \dfrac{3}{2}\left(\dfrac{4t+1}{\sqrt{t}}\right)$

41. $h'(s) = -\dfrac{1}{5s^2} + 10s + 1$

43. $G'(x) = -15x^{-6} - 6x^{-4}$

45. $H'(t) = \dfrac{9}{2\sqrt{t}} - \dfrac{5}{2}t\sqrt{t}$

47. $w'(z) = 2 + \dfrac{4}{4 - \sqrt{z}} + \dfrac{2\sqrt{z}}{\left(4 - \sqrt{z}\right)^2}$

49. $r'(x) = \dfrac{2a^2}{\left(x + a^2\right)^2}$

51. $F'(x) = e^x\left(2 + \sqrt{x} + \dfrac{1}{2\sqrt{x}}\right)$

53. $C'(x) = \dfrac{1}{\left(x+1\right)^2}$

55. $G'(s) = \dfrac{6s}{2e^s + s} - \dfrac{3s^2\left(1 + 2e^s\right)}{\left(2e^s + s\right)^2}$

57. $L'(y) = -18y^8 + 48y^7 + 6y^5 - 15y^4$

59. $H'(s) = -\dfrac{ace^s}{\left(b + ce^s\right)^2}$

61. $L'(t) = \dfrac{1}{2\sqrt{t}} - \dfrac{3\sqrt{t}}{2} - 2te^t$

63. $g'(x) = \dfrac{1}{\left(x+1\right)^2}$; $g''(x) = \dfrac{-2}{\left(x+1\right)^3}$;

  $g'''(x) = \dfrac{6}{\left(x+1\right)^4}$

65. $F'(x) = -3\pi x^2 + 10x - 1$; $F''(x) = -6\pi x + 10$;

  $F'''(x) = -6\pi$

67. $W'(t) = 6t + 3e^t$; $W''(t) = 6 + 3e^t$; $W'''(t) = 3e^t$

69. $f(x) = \dfrac{1}{3}x^3 - 4x$ (Answers will vary.)

71. $f(x) = x^3 + 2x^2 + x + 1$ (Answers will vary.)

73. $g^{(k)}(x) = \dfrac{(-1)^k k!}{x^{k+1}}$

75. $q^{(k)}(x) = e^x\left[\displaystyle\sum_{k=0}^{n}\binom{n}{k}D^k x^n\right]$

77. a. 0 **b.** 18 **c.** 1

79. $y = -\dfrac{1}{2}x + 2$

81. $y = -2ex + 4e$ **83.** $y = -x$

85. $x = 1$ **87.** No horizontal tangent

89. No horizontal tangent **91.** $x = 0$

93. $x = 0$ **95.** No horizontal tangent

97. $a = 2, b = -8$

99. The slopes are $1/2$ and -2, respectively.

101. $y = ex$ **103.** $y = -4x - 4$

105. $\dfrac{f(x) + xf'(x)}{g(x)} - \dfrac{xf(x)g'(x)}{\left[g(x)\right]^2}$

107. $e^x\left[\dfrac{f(x) + f'(x)}{g(x) + 2} - \dfrac{f(x)g'(x)}{\left(g(x) + 2\right)^2}\right]$

111. $2f(x)f'(x);\ 3\left[f(x)\right]^2 f'(x)$

115. At the point $(12, 4)$

117. a. $a(2) = 125/36\ \text{ft/s}^2$ **b.** $a(10) = 5/4\ \text{ft/s}^2$

119. $a = -1.62\ \text{m/s}^2$; $t \approx 1.361\ \text{s}$; $v_{imp} \approx 2.205\ \text{m/s}$

121. True

123. False; y' is found by applying the Reciprocal Rule or the Quotient Rule.

125. True

127. False; $\dfrac{d}{dx}F(x)$ is found by applying the Quotient Rule.

129. True

131.

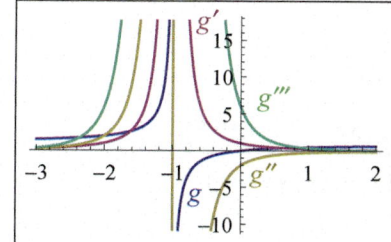

133.

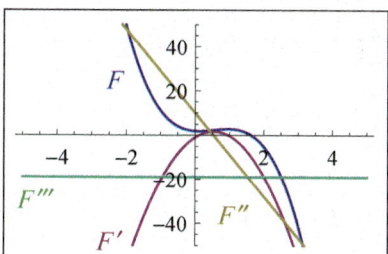

135.

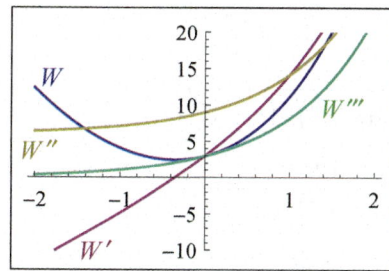

137.

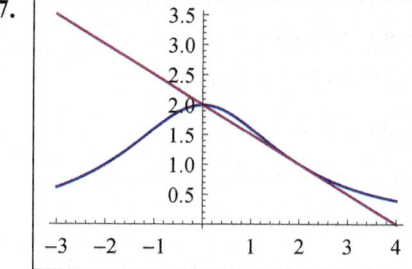

139.

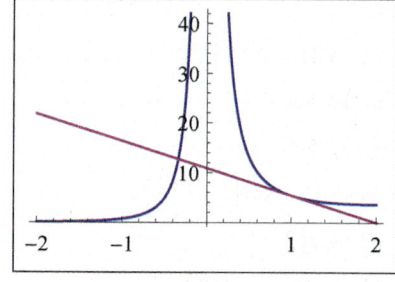

Section 3.3

1. $3/2$ **3.** π **5.** 0 **7.** 2 **9.** 1 **11.** 5

13. $f'(x) = 2\cos x + 5\sin x$

15. $h'(x) = \cos x - x \sin x$

17. $G'(x) = \sec x\left(\dfrac{1}{\sqrt{x}} + 2\sqrt{x}\tan x\right)$

19. $L'(x) = 3e^x(\csc x - 1)(\csc x + \cot x)$

21. $g'(x) = -2\cot x \csc^2 x$

23. $F'(t) = \dfrac{t\sin t + 2\cos t - 2}{t^3}$

25. $R'(z) = \dfrac{e^z(z-1) + z\cos z - \sin z}{z^2}$

27. $B'(x) = -\csc^2 x$

29. $T'(s) = \sec s\left(2\cot s + s\cot s - s\csc^2 s\right)$

31. $x = \dfrac{2\pi}{3} + 2k\pi,\; x = \dfrac{4\pi}{3} + 2k\pi,\; k \in \mathbb{Z}$

33. $x = k\pi,\; k \in \mathbb{Z}$ **35.** $u = (2k+1)\dfrac{\pi}{4},\; k \in \mathbb{Z}$

37. $x = 2k\pi,\; k \in \mathbb{Z}$ **39.** $x = \pm\dfrac{\pi}{6} + k\pi,\; k \in \mathbb{Z}$

41. $y = 2x$ **43.** $y = 1$

45. $F'(0) = 1,\; G'(0) = 2,\; H'(0) = 1$

49. $f^{(4k)}(x) = \sin x,\; f^{(4k+1)}(x) = \cos x,$
$f^{(4k+2)}(x) = -\sin x,\; f^{(4k+3)}(x) = -\cos x$

51. $f^{(4k)}(x) = (-4)^k e^x \sin x,$
$f^{(4k+1)}(x) = (-4)^k e^x(\sin x + \cos x),$
$f^{(4k+2)}(x) = (-4)^k 2e^x \cos x,$
$f^{(4k+3)}(x) = (-4)^k 2e^x(\cos x - \sin x)$

61. a. $F'(x) = mg(\cos x + \mu \sin x)$

b. $x = \operatorname{arccot}(-\mu)$

63. $\left.\dfrac{dx}{d\beta}\right|_{\beta = \pi/6} = -15\sin\dfrac{\pi}{6} = -7.5.$ This rate is negative

since β is decreasing while x is increasing.

65.

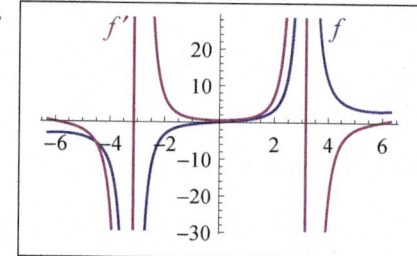

67.

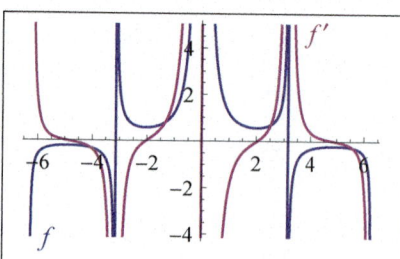

69.

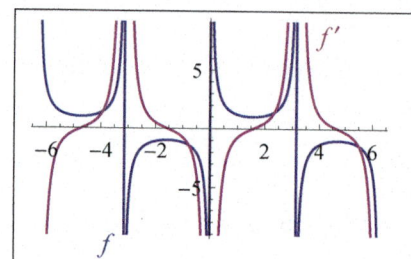

71.

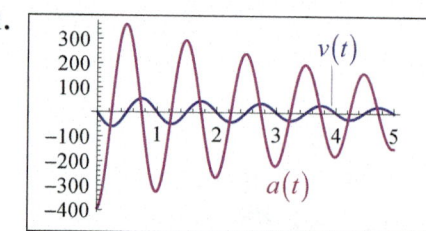

Max. velocity ≈ 54.19 cm/s;

max. acceleration ≈ 358.49 cm/s^2

Section 3.4

1. $f(x) = x^6$, $g(x) = u = 3x - 2.5$

3. $f(x) = 2\sqrt[3]{x}$, $g(x) = u = x^2 - 9$

5. $f(x) = \sin x$, $g(x) = u = \dfrac{1}{x^2 + 1}$

7. $f(x) = \csc x$, $g(x) = u = 3e^x$

9. $f(x) = \dfrac{3}{\sqrt{x}}$, $g(x) = u = \ln x$, $h(x) = x^2 + 1$

11. $g'(x) = 33\left(x^5 - \pi x^2 + 7.5\right)^{10}\left(5x^4 - 2\pi x\right)$

13. $F'(x) = \dfrac{15}{\sqrt{x}\left(5 + 2\sqrt{x}\right)^6}$

15. $k'(x) = \dfrac{2\left(5x^4 - 6x^2 + 10.5\right)}{\left(x^5 - 2x^3 + 10.5x\right)^{7/5}}$

17. $g'(x) = \dfrac{2x - 5}{2\sqrt{x^2 - 5x + 2}}$

19. $q'(x) = \dfrac{4}{3}\left(x^3 - 5x\right)^{-1/3}\left(3x^2 - 5\right)(x+3)^{5/4}$
$+\dfrac{5}{2}\left(x^3 - 5x\right)^{2/3}(x+3)^{1/4}$

21. $k'(z) = \dfrac{4z - 5}{\left(1 + 5z - 2z^2\right)^2}$

23. $S'(v) = \dfrac{\left(v^2 - 5\right)^2\left(6v^2 + 6v + 30\right)}{\left(2v + 1\right)^4}$

25. $T'(s) = \dfrac{-8s}{3\left(s^2 + 1\right)^{1/3}\left(s^2 - 1\right)^{5/3}}$

27. $H'(x) = \dfrac{10x - 3x^3}{\left(x^2 + 2\right)^3\sqrt{x^2 - 2}}$

29. $B'(t) = \dfrac{1 - 2t^2}{3t^{2/3}\left(1 + 2t^2\right)^{4/3}}$

31. $t'(x) = -\sin x \cos(\cos x)$

33. $P'(x) = \tan^2 x + 2x\tan x\sec^2 x$

35. $U'(z) = 10\sec^2 z\tan z$

37. $C'(x) = \sec^2 x\sin(2\tan x)$

39. $V'(x) = (-\sin x)e^{\cos x}$

41. $w'(x) = \dfrac{1}{\sqrt{2x+1}}\left(\begin{array}{l}\cos\sqrt{2x+1}\\ +e^{\tan\sqrt{2x+1}}\sec^2\sqrt{2x+1}\end{array}\right)$

43. $f'(x) = \pi^2(\ln 2)\cos(\pi x)2^{\sin\pi x}$

45. $t'(s) = (\ln 2)2^s\sec^2\left(2^s\right)$

47. $E'(x) = (\ln 5)^2\,5^x5^{5^x}$

49. $N'(x) = 2x\sin\left(x^2\right)\sin\left(2e^{\cos\left(x^2\right)}\right)e^{\cos\left(x^2\right)}$

51. $C'(x) = -2x\sin\left(2x^2\right)$

53. $t'(s) = -\dfrac{(\ln 10)10^s\sin\left(10^s\right)}{2\sqrt{\cos\left(10^s\right)}}$

55. $H'(s) = 2^s\ln 2\left(\cos\left(2^s\right)\tan\left(2^s\right) + \sin\left(2^s\right)\sec^2\left(2^s\right)\right)$

57. $T'(z) = e^z\cos\left(e^z\right) + e^{\sin z}\cos z$

59. $U'(\theta) = \left(1 + \sec^2\left(\theta + \tan(\theta + \tan\theta)\right)\right)$
$\left(1 + \sec^2(\theta + \tan\theta)\right)\left(1 + \sec^2\theta\right)$

61. $y = \dfrac{4}{3}x + \dfrac{1}{3}$ **63.** $y = 1$ **65.** $y = -\dfrac{1}{e\pi^2}x + \dfrac{2}{e\pi}$

67. $y = x + 1$ **69.** $x = 3, x = 4, x = 5$ **71.** $x = 0$

73. $x = 0, x = \pm\sqrt{k\pi - 2}, k \in \mathbb{Z}$ **75.** $x = \pm 1$

77. $p''(x) = 1520x^2\left(x^2 + 5\right)^{18} + 40\left(x^2 + 5\right)^{19}$

79. $g''(x) = 10\left(\sin^2 x - \cos^2 x\right)$

81. $F''(t) = 6t\cos\left(t^2\right) - 4t^3 \sin\left(t^2\right)$

83. $G''(x) = 0$ **85.** $F'(1) + G'(1) = -20$

87. If $n = 4m$, $f^{(n)}(x) = k^n \cos(kx)$;
if $n = 4m + 1$, $f^{(n)}(x) = -k^n \sin(kx)$;
if $n = 4m + 2$, $f^{(n)}(x) = -k^n \cos(kx)$;
if $n = 4m + 3$, $f^{(n)}(x) = k^n \sin(kx)$.

93. $\dfrac{dV}{dt} = 6.4\pi \text{ in.}^3/\text{s}$

95. a. $v_{max} = 1.5 \text{ m/s}$ **b.** $a_{max} = 2250 \text{ m/s}^2$

97. $\dfrac{dP}{dt} \approx 243.63 \text{ Pa/s}$

99. $P_2(x) = 1 - x^2$

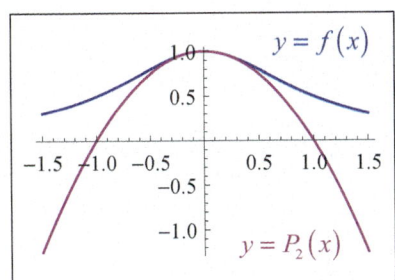

Section 3.5

1. $\dfrac{dy}{dx} = -\dfrac{1}{2y}$ **3.** $\dfrac{dy}{dx} = \dfrac{x}{y}$ **5.** $\dfrac{dy}{dx} = \dfrac{1 - 3y^2}{6xy}$

7. $\dfrac{dy}{dx} = \dfrac{y\left(2\sqrt{x+2} - 1\right)}{2x - 2x\sqrt{x+2} + 4}$

9. $\dfrac{dy}{dx} = \dfrac{1 + y(2\sin x + 1)}{2\cos x - x}$

11. $\dfrac{dy}{dx} = \dfrac{2y^2}{3x^2}$ **13.** $\dfrac{dx}{dy} = 2y$ **15.** $\dfrac{dx}{dy} = -\dfrac{y^2}{x^2}$

17. $\dfrac{dx}{dy} = \dfrac{e^y - 3\cos y - x}{y}$ **19.** $\dfrac{dx}{dy} = \dfrac{5y^3}{6x^2}$

21. Tangent line: $y = x - 1$; normal line: $y = -x + 1$

23. Tangent line: $y = -\dfrac{1}{2}x + 2$;
normal line: $y = 2x - 3$

25. Tangent line: $y = -\dfrac{3}{2}x + \dfrac{9}{2}$;
normal line: $y = \dfrac{2}{3}x - 2$

27. Tangent line: $y = \dfrac{7}{4}x - \dfrac{3}{2}$;
normal line: $y = -\dfrac{4}{7}x + \dfrac{22}{7}$

29. $\dfrac{dy}{dx} = -\dfrac{x^3}{y^3}$ **31.** $\dfrac{dy}{dx} = \dfrac{4xy - 3y^2}{4xy - 3x^2}$ **33.** $\dfrac{dy}{dx} = -1$

35. $\dfrac{dy}{dx} = \dfrac{\sin 2x - 2x\sec^2\left(x^2 + y^2\right)}{\sin 2y + 2y\sec^2\left(x^2 + y^2\right)}$

37. $\dfrac{dy}{dx} = \dfrac{2x^4 - 4x^2y - 6xy^2 - x^2 + 2y^2 - y}{3y^2 - 6x^2y - x}$

39. $\dfrac{dy}{dx} = \dfrac{5\sqrt{2xy} + y}{3\sqrt{2xy} - x}$ **41.** $\dfrac{dy}{dx} = \dfrac{\sec^2 x + 2y}{\cos y - 2x}$

43. $\dfrac{dy}{dx} = \dfrac{\cos x - \sin x}{2\sqrt{\sin x + \cos x}\sec(x+y)\tan(x+y)} - 1$

45. $\dfrac{ds}{dt} = \dfrac{4s - 6s^3t^2}{4s^2t^3 - \sqrt{s}}$ **47.** $\dfrac{d^2y}{dx^2} = \dfrac{4y^2 - x^2}{16y^3}$

49. $\dfrac{d^2y}{dx^2} = \dfrac{\left(y^2 - 1\right)\left(3y^2 + 1\right)}{4x^2y^3}$

51. $\dfrac{d^2y}{dx^2} = \dfrac{-2x\left(x^3 + y^3\right)}{y^5}$

55. Horizontal tangent: $(1/2, 1)$;
vertical tangent: $(-1, -1/2)$

57. Horizontal tangent: $(1, 5/4)$;
vertical tangent: none

59. $y' = x/y$ and $y' = -y/x$, respectively

61. $y' = -1/b$ and $y' = b$, respectively

65. $y = \sqrt{2}x + 2$, $y = -\sqrt{2}x + 2$

67.

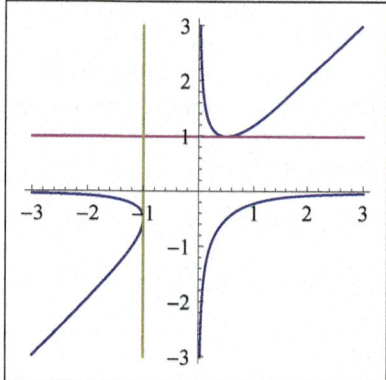

69.

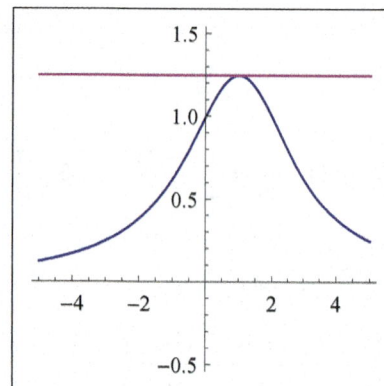

71.

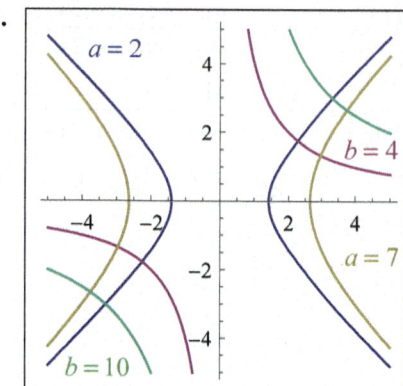

73.

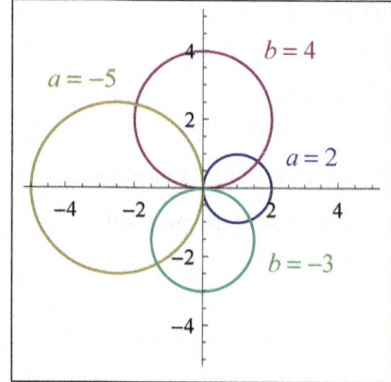

75. Answers will vary.

Section 3.6

1. $1/12$ **3.** 6 **5.** $1/4$ **7.** $-1/2$ **9.** $-3/2$

11. $1/(10\ln 10)$ **13.** 2 **15.** $5/\cos 0.01$

17. $1/39$ **19.** $-\sqrt{5}/9$ **21.** $-5/33$ **23.** $1/37$

25. 7 **27.** 1 **29.** $2/\sec^2 1$ **31.** $f'(x) = 3/x$

33. $h'(x) = \dfrac{2x}{x^2+3}$ **35.** $G'(x) = \ln\sqrt{x^2+4} + \dfrac{x^2}{x^2+4}$

37. $L'(x) = \dfrac{x^2+1-2x^2\ln 2x}{x(x^2+1)^2}$ **39.** $g'(x) = \dfrac{2}{9-x^2}$

41. $F'(t) = \dfrac{t+2\sqrt{t^2+4}}{t^2+4+2t\sqrt{t^2+4}}$ **43.** $T'(x) = -\tan x$

45. $v'(x) = -2\sin 2x\left(\ln(\cos 2x)+1\right)$

47. $w'(x) = \log x + \dfrac{1}{\ln 10}$

49. $y' = (x+2)(x+3)(x+4) + (x+1)(x+3)(x+4)$
$\quad + (x+1)(x+2)(x+4) + (x+1)(x+2)(x+3)$

51. $y' = \sqrt[3]{(2x-1)(x-5)(3x+1)}$
$\quad\left(\dfrac{2}{6x-3} + \dfrac{1}{3x-15} + \dfrac{1}{3x+1}\right)$

53. $y' = \dfrac{\sqrt[3]{x^3-5x^2+7}(x+2)}{x^{2/3}\sqrt{3x^2+4}}$
$\quad\left(\dfrac{3x^2-10x}{3x^3-15x^2+21} + \dfrac{1}{x+2} - \dfrac{2}{3x} - \dfrac{3x}{3x^2+4}\right)$

55. $y' = \dfrac{x^2\sqrt[5]{x^3+3}}{\sqrt[4]{x^4+4}}\left(\dfrac{2}{x} + \dfrac{3x^2}{5(x^3+3)} - \dfrac{x^3}{x^4+4}\right)$

57. $y' = (\sin x)^{1/x}\left(\dfrac{x\cot x - \ln(\sin x)}{x^2}\right)$

59. $y' = (\cos x)^{\sqrt{x}}\left(\dfrac{\ln(\cos x)}{2\sqrt{x}} - \sqrt{x}\tan x\right)$

61. $y' = (\ln x)^x\left(\ln(\ln x) + \dfrac{1}{\ln x}\right)$

63. $y' = x^{x^x}\left(x^x(\ln x+1)\ln x + x^{x-1}\right)$

65. $y' = (\ln x)^{\sin x}\left(\cos x\ln(\ln x) + \dfrac{\sin x}{x\ln x}\right)$

67. $y' = -\dfrac{1}{\sqrt{1-x^2}}$ **69.** $y' = -\dfrac{1}{1+x^2}$

71. $y' = \dfrac{1}{2x^2+2x+1}$ **73.** $y' = \dfrac{1}{(1+x^2)\arctan x}$

75. $y' = \dfrac{-1}{2\sqrt{x-x^2}}$ **77.** $y' = -2\sqrt{1-x^2}$

79. $y' = \dfrac{e^x}{1+e^{2x}}$ **81.** $y' = \dfrac{-2}{(1+x^2)(1+\arctan x)^2}$

83. $y' = \dfrac{6x^2 \arcsin(x^3)}{\sqrt{1-x^6}}$ **85.** $y' = \dfrac{2}{(x^2+1)\sqrt{x^2+2}}$

87. $y' = \dfrac{1}{x^2+1}$

89. $y' = (\arctan x)^x \left[\ln(\arctan x) + \dfrac{x}{(1+x^2)\arctan x} \right]$

91. $y' = \dfrac{2}{(x^2+1)^2 \sqrt{x^2+2}}$ **93.** $y' = 1/x^2$

95. $y = -2.178x + 3.282$ **97.** $y = 1.270x + 1.164$

99. $y = \ln(\pi/4)x + 1 - \ln(\pi/4)$

101. $f'(x) = 0$; as remarked in the text,
arcsin$(1/x) = \operatorname{arccsc} x$ for $|x| \geq 1$.

103. $\theta = \tan^{-1}(28/s) - \tan^{-1}(8/s)$;

$\dfrac{d\theta}{ds} = \dfrac{8}{s^2+64} - \dfrac{28}{s^2+784}$

105. Restrict the domain of $\cot x$ to $(0, \pi)$.

107. False; $y = \log \pi$ is a constant function and therefore $y' = 0$.

109. True

Section 3.7

3. 197,959 people   **5.** Approx. 367.88 g

7. a. Approx. 0.9659   **b.** Approx. 0.707 kg
   c. Approx. 7.628 mg

9. Approx. 8.75 more minutes (i.e., 13.75 total minutes)

11. 300 N   **13.** Approx. 58.86x N

15. 350 N   **19.** 0.073 atm/s

21. 0.75 mol/$(L\cdot s)$; 27 mL of water

23. a. \$152.80   **b.** \$152.50

25. a. \$792.29   **b.** \$2331   **c.** \$861.71; \$2311.50

27. \$1077.22   **29.** Approx. 4.5π in.³

31. The cost increases by \$465.

33. a. Cost decreases by approx. \$1250.

b. Profit decreases by approx. \$1050.

c. The actual decreases are \$1250.01 and \$1052.49, respectively.

35. Approx. 0.8042 cm³   **37.** 1.5 percent increase

39. 0.3 percent decrease

Section 3.8

1. -3 ft/s   **3.** 1/6   **5.** 8   **7.** -5.2 **9.** 0

11. Increasing at 35 in.²/s

13. Approx. 0.8436 units/s

17. -10 cm/min   **19.** 2.641 cm/min

21. 15.524 mph   **23.** 0.0808 in./s   **25.** 12 ft/s

27. 1187.5 mph   **29.** -25 cm/s; 0.02 rad/s

31. -20 mm²/s   **33.** 1.52 in./min   **35.** 1 cm/min

37. $\sqrt{2}/4$ ft/s   **39.** 15 ft/s   **41.** 429.29 km/h

43. Approx. 0.494 in./min

45. a.

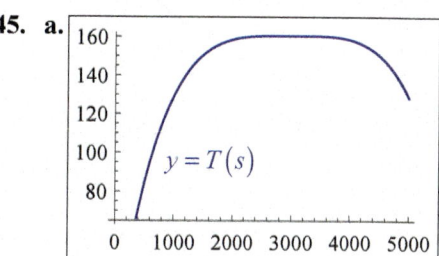

$y = T(s)$

b. Approx. 30.465 hp/s

Section 3.9

1. $L(x) = 2x - 2$ **3.** $L(x) = 1$ **5.** $L(\theta) = 1$

7. $L(t) = -\dfrac{3}{16}t + \dfrac{7}{8}$ **9.** $L(u) = -15u - 62$

11. $L(z) = z + 1$ **13.** $dy = 1.4$ **15.** $dy = -1.3$

17. $dy = 189/1024$ **19.** $dy = -\dfrac{\pi+2}{8}$

21. $dy = -1/(2e^3)$ **23.** $dy \approx -0.1083$

25. $dy = 0.5$; $\Delta y = 0.625$

27. $dy = (\ln 2)/2 \approx 0.3466$; $\Delta y = 2\sqrt[4]{2} - 2 \approx 0.3784$

29. $\Delta y = 1.52$ **31.** $\Delta y = -1.\overline{4}$ **33.** $\Delta y \approx 0.1854$

35. $\Delta y = -0.4678$ **37.** $\Delta y \approx -0.02587$

39. $\Delta y \approx -0.0934$

41. Approx. 3.0167; calculator value: 3.0166

43. Approx. 3.9667; calculator value: 3.9666

45. Approx. 1.9875; calculator value: 1.9873

47. Approx. 0.9933; calculator value: 0.9933

53. 1/8 inches   **55.** 8.1566 mm^2 **57.** 0.1%

59. Approx. 1 in.^{3} **61.** Approx. $\pm 1.24°$

63. **a.** $F = 0.1875$ N   **b.** $dF \approx 0.017$ N

    c. $dF \approx 0.0167$ N

65. $dE_{kin} \approx 129.63$ kJ; about a 24% increase

67. 3%; the generalization is $\dfrac{dA}{A} = \dfrac{2}{3}\dfrac{dV}{V}$

69. Up to 113.615 units   **71.** $x \approx 5.1$

73. Yes, we can prove $\dfrac{dA}{A} = 2\dfrac{dr}{r}$.

75. True   **77.** True

79. False; consider $f(x) = \sqrt{x}$, $x = 1$, $dx = 0.1$.

81.

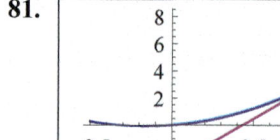
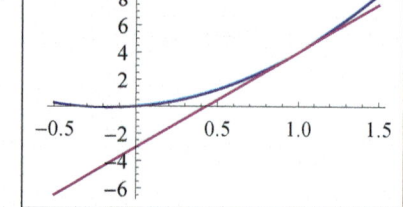

83.

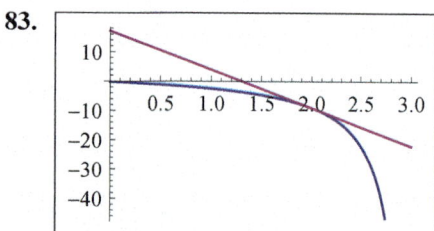

85.

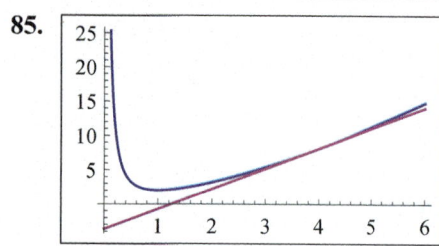

87.

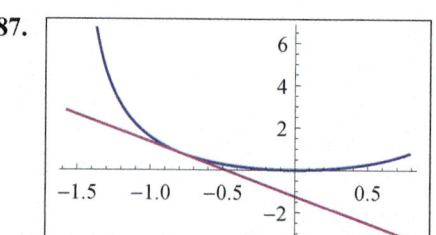

89.

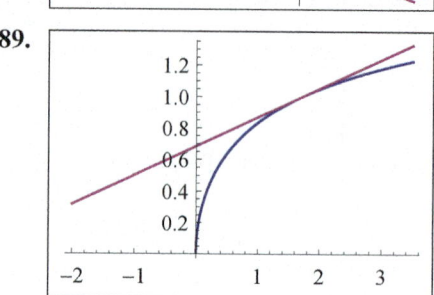

91.

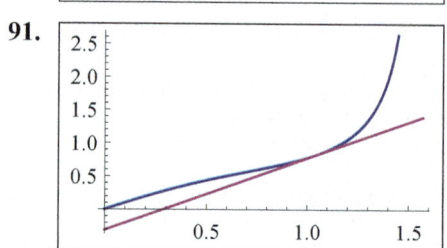

Chapter 3 Review

1. $\dfrac{df}{dx} = 3x^2 + 1$; $\left.\dfrac{df}{dx}\right|_{x=1} = 4$ **3.** $D_x s(x) = \dfrac{1}{2\sqrt{x-2}}$

5. $f'(x) = 2x$; $f''(x) = 2$; $f'''(x) = 0$

7. Not differentiable at 0
One-sided derivatives do not exist at 0

9. Not differentiable at $k \in \mathbb{Z}$
Left-hand derivative at each point of
nondifferentiability: 1
Right-hand derivative at each point of
nondifferentiability: 1

11. $f'(x) = x^4 - 8x^3 + 3x^2 + x$

13. $h'(x) = 6x^2 - 2x + 8$

15. $F'(t) = \dfrac{7}{2}t^{5/2} + 4t^3 + \dfrac{1}{2}t^{-1/2} - \dfrac{5}{2}t^{3/2} - 3t^2 + \dfrac{1}{2}t^{-3/2} + 1$

17. $u'(x) = \dfrac{7e^x}{\left(3e^x + 5\right)^2}$ **19.** $v'(x) = \dfrac{2x}{x^2 + 2}$

21. $f'(x) = \dfrac{1}{(x+1)^2}$; $f''(x) = \dfrac{-2}{(x+1)^3}$;

    $f'''(x) = \dfrac{6}{(x+1)^4}$

23. $f'(x) = \sec^2 x;$ $f''(x) = 2\sec^2 x \tan x;$
$f'''(x) = 2\sec^2 x\left(\sec^2 x + 2\tan^2 x\right)$

25. $f(x) = x^3 - 4x^2 + 6x$ (Answers will vary.)

27. $y = -\dfrac{3}{8}x + \dfrac{7}{8}$

29. $y = x, y = 13x - 24$ **31.** $-1/2$

33. $v(1) = 12.5 \text{ ft/s};$ $a(1) = -12.5 \text{ ft/s}^2$

35. $V'(1) = 784\pi/3 \text{ cm}^3/\text{s};$ $A'(1) = 224\pi/3 \text{ cm}^2/\text{s}$

37. $\dfrac{dy}{dx} = \dfrac{4x - 5y}{5x - 4y}$ **39.** $\dfrac{dx}{dy} = \dfrac{y\left(3 - y^2\right)}{3x}$

41. $y = x + 1$ **43.** $y = -\dfrac{20}{9}x + \dfrac{56}{9}$

45. $\dfrac{d^2 y}{dx^2} = \dfrac{x^{2/3} + y^{2/3}}{3x^{4/3} y^{1/3}}$ **47.** 32   **49.** $1/2$

51. $f'(x) = \dfrac{1}{2\sqrt{x}\left(x + 1\right)}$

53. $f'(x) = \dfrac{1}{x|\ln x|\sqrt{\ln^2 x - 1}}$ **55.** $y' = \dfrac{\left(\sqrt{x}\right)^{\ln x} \ln x}{x}$

57. Approx. 8 minutes (8 minutes 3 seconds)

59. Approx. 0.1 in./s   **61.** 2000 km/h

63. 5 ft 4 in.   **65.** $L(x) = \dfrac{\sqrt{2}}{2}x + \dfrac{\sqrt{2}\left(4 - \pi\right)}{8}$

67. Approx. 0.7354   **69.** Approx. 10%

71. If $a > 1$, the slope of y increases as we scan the graph from left to right, while in the other case it decreases.

73. True

75. False; the Reciprocal Rule along with the Chain Rule (or Quotient Rule) should be used, which yields $y' = \dfrac{3 - 2x}{\left(x^2 - 3x + 1\right)^2}$.

77. False; logarithmic differentiation should be used, which yields $y' = x^x \left(\ln x + 1\right)$.

79. True

81. False; $\Delta x = dx$ and even though $\Delta y \to dy$ might be true under certain circumstances, $x \to 0$ is not a necessary condition.

83.

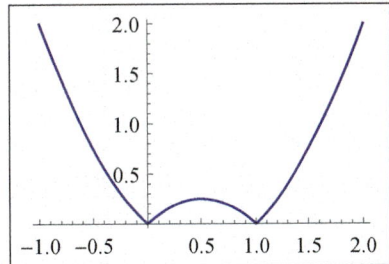

Not differentiable at $x = 0, 1$ (cusps)

85.

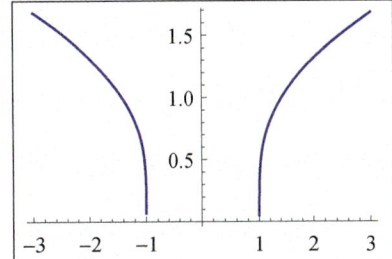

Not differentiable at $x = \pm 1$ (vertical tangents, undefined on $(-1, 1)$)

87. The algorithm employed by the CAS may not simplify before differentiating. Yes, we can simplify with an appropriate command. The simplified answer is $f'(x) = \cos x$.

89. a.

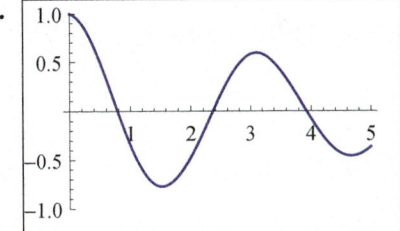

It is an oscillating motion with the amplitude "dying down" as a result of retarding forces.

b.

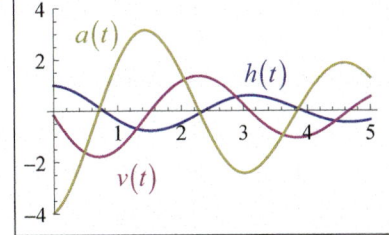

The object is near its equilibrium position when velocity is maximum and near its extreme positions when velocity is zero. Acceleration, on the other hand, is maximum near the extreme positions and zero near equilibrium.

Chapter 3 Project

1. -0.0265 mm Hg/ft   **3.** -0.0225 mm Hg/ft

5. The symmetric difference quotient is the average of the difference quotients $\dfrac{f(c+h)-f(c)}{h}$ and $\dfrac{f(c-h)-f(c)}{-h}$.

7. -0.02302 mm Hg/ft

Chapter 4

Section 4.1

1. Abs. min.: $f(-1)=f(1)=0$; no maximum

3. Abs. min.: $f(3\pi/2)=-2$; abs. max.: $f(\pi/2)=2$

5. Critical points at $x=0,\,1,\,3$; rel. min. at $x=0,\,3$; abs. min. at $x=3$; rel. max. at $x=-1,\,1,\,4.2$; abs. max. at $x=4.2$

7. Critical points at $x=0,\,2$; rel. min. at $x=-2,\,2$; abs. min. at $x=-2$; rel. max. at $x=0$; abs. max. at $x=0$

9. Abs. min.: $f(0)=1$; abs. max.: $f(3)=7$

11. No extrema

13. Abs. min.: $v(1)=-4$; abs. max.: $v(-2)=5$ and $v(4)=5$

15. No extrema

17. Abs. min.: $n(1)=-1$; no maximum

19. No minimum; abs. max.: $G(0)=1$

21. Answers will vary.   **23.** Answers will vary.

25. Answers will vary.   **27.** Answers will vary.

29. Answers will vary.   **31.** Answers will vary.

33. Answers will vary.   **35.** Answers will vary.

37. Answers will vary.   **39.** $x=1,-2$ **41.** None

43. $x=1.5$ **45.** $x=0,\,4$ **47.** $t=0$

49. $s=0,\,0.5$ **51.** $\alpha=k\pi,\,\pm\dfrac{2\pi}{3}+2k\pi,\,k\in\mathbb{Z}$

53. $x=e-2$ **55.** $t=0$

57. Abs. min.: $g(5)=-68$; abs. max.: $g(0)=7$

59. Abs. min.: $u(2)=-49$; abs. max.: $u(3)=0$

61. Abs. min.: $k(-3)=k(1)=-4$; abs. max.: $k(-1)=4$

63. Abs. min.: $m(x)=6,\,x\in[-3,3]$; abs. max.: $m(-4)=m(4)=8$

65. Abs. min.: $g(1)=2$; abs. max.: $g(-1.5)=14.5$

67. Abs. min.: $G(1)=2$; abs. max.: $G(1/4)=G(4)=5/2$

69. Abs. min.: $r(-1)=r(1)=0$; abs. max.: $r(0)=1$

71. Abs. min.: $w(-2)=-4$; abs. max.: $w(2)=4$

73. Abs. min.: $r(5\pi/4)=\left(-\sqrt{2}/2\right)e^{5\pi/4}$; abs. max.: $r(\pi/4)=\left(\sqrt{2}/2\right)e^{\pi/4}$

75. Abs. min.: $t(0)=t(1)=0$; abs. max.: $t(0.5)=1$

77. Abs. min: $V(-25/12)=5-(35/12)(-25/12)^{5/7}$; abs. max.: $V(1)=11$

79. Abs. min.: $g(0)=-4$; no maximum

81. No minimum; abs. max.: $K(0)=2$

83. No extrema

85. Abs. min.: $L(0)=0$; no maximum

87. Abs. min.: $t(2n+1)=0,\,n\in\mathbb{Z}$; abs. max.: $t(2n)=4,\,n\in\mathbb{Z}$

89. No extrema

91. The wire needs to be cut at $21.995\approx22$ inches, and this first piece is to be bent into a circle.

93. Approx. 11,500 calculators

95. False; consider $f(x)=[\![x]\!]$ on $[0,1]$.

97. True

99. False; the critical points shift by k units to the left.

101. True

103. False; f can only have one largest value (however, it can be attained at multiple x-values).

105. True

Section 4.2

1. $c=5/2$ **3.** $c=3$ **9.** $c=2$

11. Rolle's Theorem does not apply; $h(1)\ne h(3)$.

13. $c=0$

15. Rolle's Theorem does not apply; $H(x)$ is not differentiable at $x=0$.

17. Rolle's Theorem does not apply; $T(z)$ is not continuous on $[0,5\pi]$.

19. Rolle's Theorem does not apply; $w(t)$ is not continuous on $[-\pi/4,\pi/4]$.

21. The MVT does not apply; $f(x)$ is not differentiable at $x=1$.

23. $c=3/2$

25. The MVT does not apply; $G(x)$ is not continuous on $[-2,3]$.

27. $c=0$ **29.** $c=-\sqrt{2}/2$

31. The MVT does not apply; $u(z)$ is not continuous on $[0,\pi]$.

33. 10

35. If $f(-5)=-1$ and $f'(x)\le 0.1$, then $f(5)\le 0$.

37. $f(x)=x^3-x^2+x-20$

39. $p(t)=-t^2+5t-12$

43. The average velocity is 460 mph, and thus, by the Mean Value Theorem, there must be a time t_0 when the velocity is $v(t_0)=460$ mph. Use the Intermediate Value Theorem to argue that the equation $v(t)=450$ has solutions on both intervals $[0,t_0]$ and $[t_0,10.5]$.

51. No; consider for example $f(x)=\sin x$ on $[0,2\pi/3]$ with $c=\pi/2$.

53. The average growth rate is 50 rabbits per month, and thus, by the Mean Value Theorem, there must exist a time when the instantaneous growth rate equals the above number.

65. False; Rolle's Theorem is a special case of the Mean Value Theorem.

67. True

69. Tangent line: $y=\dfrac{1}{3}x+\dfrac{3}{4}$

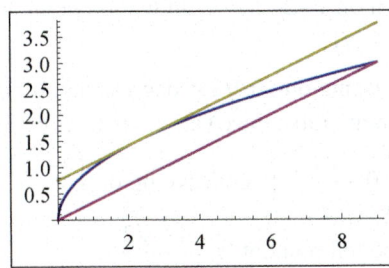

71. Tangent line: $y=2x+7-4\sqrt{3}$

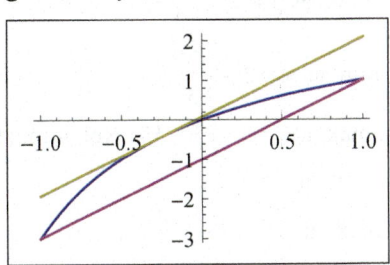

Section 4.3

1. Decreasing on $(-\infty,2)$; increasing on $(2,\infty)$

3. Increasing on $(-\infty,-5),(1,\infty)$; decreasing on $(-5,1)$

5. Increasing on $(-\infty,-3),(1/2,1)$; decreasing on $(-3,1/2),(1,\infty)$

7. Increasing on $(-\infty,1)$; decreasing on $(1,\infty)$

9. Increasing on $\left(-\infty,-3-2\sqrt{3}\right),\left(-3+2\sqrt{3},\infty\right)$; decreasing on $\left(-3-2\sqrt{3},-3\right),\left(-3,-3+2\sqrt{3}\right)$

11. Increasing on $\left(-\infty,2-\dfrac{\sqrt{39}}{3}\right),\left(2+\dfrac{\sqrt{39}}{3},\infty\right)$; decreasing on $\left(2-\dfrac{\sqrt{39}}{3},2\right),\left(2,2+\dfrac{\sqrt{39}}{3}\right)$

13. Increasing on $(-\infty,3.125)$; decreasing on $(3.125,\infty)$

15. Decreasing on $(0,1/3)$; increasing on $(1/3,\infty)$

17. Increasing on $\left(k\pi,k\pi+\dfrac{\pi}{2}\right)$, $k\in\mathbb{Z}$; decreasing on $\left(k\pi+\dfrac{\pi}{2},k\pi\right)$, $k\in\mathbb{Z}$

19. Decreasing on $(0,1/4)$; increasing on $(1/4,\infty)$

21. Increasing on $\left(-\infty,-2/\sqrt{11}\right),\left(2/\sqrt{11},\infty\right)$; decreasing on $\left(-2/\sqrt{11},2/\sqrt{11}\right)$

23. Rel. max. at $x=-5$; rel. min. at $x=1$

25. Rel. max. at $x=-3,1$; rel. min. at $x=0.5$

27. Rel. max. at $x=1$

29. Rel. max. at $x=-3-2\sqrt{3}$; rel. min. at $x=-3+2\sqrt{3}$

31. Rel. max. at $x = 2 - \dfrac{\sqrt{39}}{3}$;

rel. min. at $x = 2 + \dfrac{\sqrt{39}}{3}$

33. Rel. max. at $t = 3.125$ **35.** Rel. min. at $s = 1/3$

37. Rel. min. at $x = k\pi,\ k \in \mathbb{Z}$;

rel. max. at $x = (2k+1)\dfrac{\pi}{2},\ k \in \mathbb{Z}$

39. Rel. min. at $t = 1/4$

41. Rel. max. at $x = -2/\sqrt{11}$; rel. min. at $x = 2/\sqrt{11}$

43. Decreasing on $(-\infty, -1)$; increasing on $(-1, \infty)$; local min. at $x = -1$; concave up on $(-\infty, \infty)$

45. Decreasing on $(-\infty, -2), (2, \infty)$; increasing on $(-2, 2)$; local min. at $x = -2$; local max. at $x = 2$; concave down on $\left(-\infty, -2\sqrt{3}\right), \left(0, 2\sqrt{3}\right)$; concave up on $\left(-2\sqrt{3}, 0\right), \left(2\sqrt{3}, \infty\right)$; inflection points at $x = \pm 2\sqrt{3}, 0$

47. Increasing on $(-\infty, -1), (1, \infty)$; decreasing on $(-1, 1)$; local max. at $x = -1$; local min. at $x = 1$; concave down on $\left(-\infty, -\sqrt{2}/2\right), \left(0, \sqrt{2}/2\right)$; concave up on $\left(-\sqrt{2}/2, 0\right), \left(\sqrt{2}/2, \infty\right)$; inflection points at $x = \pm\sqrt{2}/2, 0$

49. Decreasing on $(-\pi/2, -\pi/4), (\pi/4, \pi/2)$; increasing on $(-\pi/4, \pi/4)$; local min. at $x = -\pi/4$; local max. at $x = \pi/4$; concave up on $(-\pi/2, 0)$; concave down on $(0, \pi/2)$; inflection point at $x = 0$

51. Concave down on $(-\infty, \infty)$

53. Concave down on $(-\infty, 1/4)$; concave up on $(1/4, \infty)$

55. Concave up on $(-\infty, -2), (0, \infty)$; concave down on $(-2, 0)$

57. Concave down on $(-\infty, -3), (0, 1)$; concave up on $(-3, 0), (1, \infty)$

59. Concave down on $(-\infty, 5)$; concave up on $(5, \infty)$

61. Concave up on $(-\infty, \infty)$

63. Increasing on $(-\infty, \infty)$; no extrema; no inflection points

65. Increasing on $(-\infty, 5)$; decreasing on $(5, \infty)$; rel. max. (abs. max.) at $x = 5$; concave down on $(-\infty, \infty)$; no inflection points

67. Decreasing on $(-\infty, -3/2), (1, \infty)$; increasing on $(-3/2, 1)$; rel. min. at $x = -3/2$; rel. max. at $x = 1$; concave up on $(-\infty, -1/4)$; concave down on $(-1/4, \infty)$; inflection point at $x = -1/4$

69. Increasing on $\left(-\infty, \dfrac{-1-\sqrt{21}}{2}\right), \left(1, \dfrac{-1+\sqrt{21}}{2}\right)$;

decreasing on $\left(\dfrac{-1-\sqrt{21}}{2}, 1\right), \left(\dfrac{-1+\sqrt{21}}{2}, \infty\right)$;

rel. max. (abs. max.) at $x = \dfrac{-1-\sqrt{21}}{2}$; rel. max.

at $x = \dfrac{-1+\sqrt{21}}{2}$; rel. min. at $x = 1$; concave

down on $\left(-\infty, -\sqrt{2}\right), \left(\sqrt{2}, \infty\right)$; concave up on $\left(-\sqrt{2}, \sqrt{2}\right)$; inflection points at $x = \pm\sqrt{2}$

71. Decreasing on $(-\infty, 2), (2, \infty)$; no extrema; concave down on $(-\infty, 2)$; concave up on $(2, \infty)$; no inflection points

73. Increasing on $\left(-\infty, 4 - \dfrac{\sqrt{66}}{2}\right), \left(4 + \dfrac{\sqrt{66}}{2}, \infty\right)$;

decreasing on $\left(4 - \dfrac{\sqrt{66}}{2}, 4\right), \left(4, 4 + \dfrac{\sqrt{66}}{2}\right)$;

rel. max. at $x = 4 - \dfrac{\sqrt{66}}{2}$;

rel. min. at $x = 4 + \dfrac{\sqrt{66}}{2}$;

concave down on $(-\infty, 4)$; concave up on $(4, \infty)$; no inflection points

75. Increasing on $(-\infty, -1), (-1, \infty)$; no extrema; concave up on $(-\infty, -1)$; concave down on $(-1, \infty)$; no inflection points

77. Decreasing on $(-\infty, 0), (4/15, \infty)$; increasing on $(0, 4/15)$; rel. min. at $x = 0$; rel. max. at $x = 4/15$; concave up on $(-\infty, -2/15)$; concave down on $(-2/15, \infty)$; inflection point at $x = -2/15$

79. Increasing on $(0, 1/2)$; decreasing on $(1/2, \infty)$; rel. max. (abs. max.) at $x = 1/2$; concave down

on $\left(0, \dfrac{1+\sqrt{2}}{2}\right)$; concave up on $\left(\dfrac{1+\sqrt{2}}{2}, \infty\right)$;

inflection point at $x = \dfrac{1+\sqrt{2}}{2}$

81. For $k \in \mathbb{Z}$: increasing on $\left(\dfrac{3\pi}{4}+2k\pi, \dfrac{7\pi}{4}+2k\pi\right)$;

decreasing on $\left(-\dfrac{\pi}{4}+2k\pi, \dfrac{3\pi}{4}+2k\pi\right)$;

rel. max. (abs. max.) at $x = -\dfrac{\pi}{4}+2k\pi$;

rel. min. (abs. min.) at $x = \dfrac{3\pi}{4}+2k\pi$;

concave down on $\left(-\dfrac{3\pi}{4}+2k\pi, \dfrac{\pi}{4}+2k\pi\right)$;

concave up on $\left(\dfrac{\pi}{4}+2k\pi, \dfrac{5\pi}{4}+2k\pi\right)$;

inflection points at $x = \dfrac{\pi}{4}+2k\pi$

83. Answers will vary.   **85.** Answers will vary.

87. Answers will vary.   **89.** Answers will vary.

91. Direction is negative on $(0, 3/4)$, positive on $(3/4, \infty)$; changes direction at $t = 3/4$; acceleration is constant $a = 4$, never zero

93. Direction is positive on $(0,1), (4,\infty)$, negative on $(1, 4)$; changes direction at $t = 1$ and $t = 4$; acceleration is negative on $(0, 2.5)$, positive on $(2.5, \infty)$, zero at $t = 2.5$

95. Direction is negative on $\left(\dfrac{\pi}{4}+2k\pi, \dfrac{5\pi}{4}+2k\pi\right)$, $k \in \mathbb{Z}$, positive elsewhere; changes direction at $t = \dfrac{\pi}{4}+k\pi$; acceleration is positive on $\left(\dfrac{\pi}{2}+2k\pi, \dfrac{3\pi}{2}+2k\pi\right)$, $k \in \mathbb{N} \cup \{0\}$, negative elsewhere, zero at $t = \dfrac{\pi}{2}+k\pi$

97.

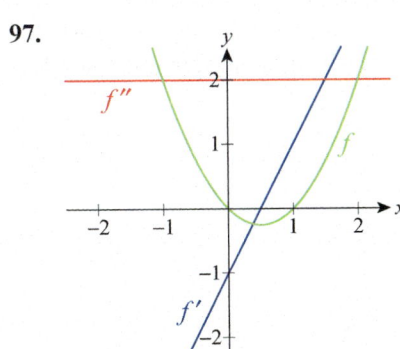

99.

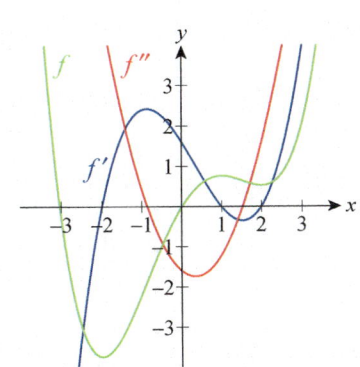

101.

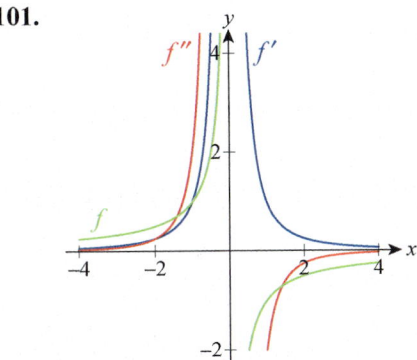

103. a. Midday (temp. is rising, rate of increase is slowing)

b. Morning (temp. is rising, rate of increase is accelerating)

c. Early morning (temp. has local minimum)

d. Evening or nighttime (temp. is falling, rate of decrease is accelerating)

e. Early morning, just before minimum temperature (temp. is falling, rate of increase is slowing)

f. Early afternoon (temp. has reached maximum)

105. a. $P(n) = -\dfrac{n^2}{10} + 50n - 500$

b. P is increasing on $(200, 250)$, decreasing on $(250, 350)$; P is maximal when producing 250 liners

107. Local max. at $x = -4$; local min. at $x = 3$

113. $a < 0, \ 3ac > b^2$ **115.** $p(x) = \dfrac{x^3}{3} - 2x^2 + 5$

121. True

123. False; consider $f(x) = -1/x^2$ with $c = 0$.

125. False; consider $f(x) = g(x) = -x$.

127. False; consider $f(x) = x^3$ with $c = 0$.

129.

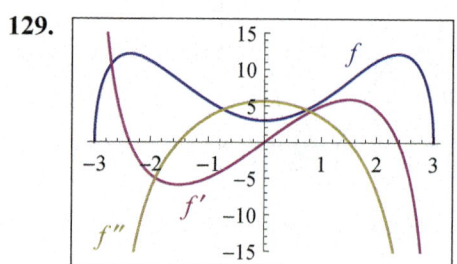

131. Answers will vary.   **133.** Answers will vary.

135. Answers will vary.

Section 4.4

1. 12   **3.** 1/2

5. Limit does not exist; l'Hôpital's Rule does not apply.

7. 0

9. Limit does not exist; l'Hôpital's Rule does not apply.

11. 1/2   **13.** ∞/∞; $0.2x^2 + 1$ dominates

15. $0 \cdot \infty$; e^{-x} dominates   **17.** 0   **19.** $-\infty$

21. L'Hôpital's Rule does not apply; the limit is 0.

23. L'Hôpital's Rule does not apply; the limit is 0.

25. 0   **27.** 0   **29.** $-\infty$ **31.** 1/2

33. $2/\ln 3$ **35.** 0   **37.** 1

39. L'Hôpital's Rule does not apply; the limit does not exist.

41. -2 **43.** ∞ **45.** $\dfrac{\ln 3}{\ln 2}$

47. L'Hôpital's Rule does not apply; the limit is 0.

49. $0 \cdot \infty$; the limit is 0.

51. Not indeterminate form; the limit is 0.

53. ∞^0; the limit is 1.   **55.** $\infty - \infty$; the limit is $-\infty$.

57. $\infty - \infty$; the limit is $-3/2$.

59. 1^∞; the limit is $2/e$.

61. Not indeterminate form; the limit is ∞.

63. ∞^0; the limit is 1.   **65.** 1^∞; the limit is 1.

67. Not indeterminate form; the limit is 0.

69. ∞/∞; the limit is 1/100.   **71.** 1^∞; the limit is e^2.

73. $0/0$; the limit is 2/3.   **75.** 0   **77.** 1

79. L'Hôpital's Rule does not apply; the limit is 0.

81. 1   **83.** 1/2   **85.** $-1/2$

87. $2/0$ is not indeterminate; l'Hôpital's Rule does not apply.

89. L'Hôpital's Rule is applied incorrectly; the factors of a product are differentiated.

91. The third use of l'Hôpital's Rule is unjustified; the limit is $-\infty$.

93. The limit is 0.   **95.** The limit is 0.

97. The limit is ∞.   **99.** The limit is 1.

101. The limit is -5.   **103.** The limit is 1.

105. The limit is 1.   **107.** The limit is 0.

109. The limit is 1.   **115.** $c = 1/2$

117. The theorem does not apply since $g(a) = g(b)$.

119. $c = 1$ **121.** $c = -\pi/4$ **123.** $c = 1/3$

127. The limit is $16a/9$.   **129.** $0 \cdot \infty$; the limit is 0.

131. $\infty - \infty$; the limit is 1/3.

133. The limit of the limits is 0.

Section 4.5

1. First derivative: a; second derivative: b

3. First derivative: b; second derivative: a

5. Dom $= \mathbb{R}$; x-int.: $\dfrac{-3 \pm 3\sqrt{5}}{2}$; y-int.: 0;
no asymptotes; increasing on $(-\infty, -3), (1, \infty)$;
decreasing on $(-3, 1)$; rel. max. at $x = -3$;
rel. min. at $x = 1$; concave down on $(-\infty, -1)$;
concave up on $(-1, \infty)$; inflection point at $x = -1$

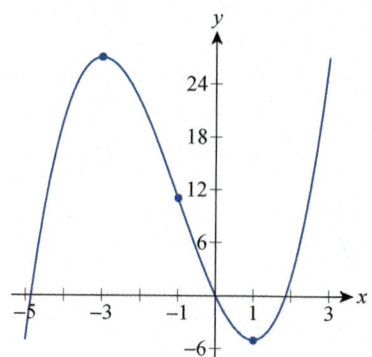

7. Dom = $\mathbb{R}$; y-int.: 0; no asymptotes; decreasing on $(-\infty, -4), (-2, 1)$; increasing on $(-4, -2), (1, \infty)$; rel. min. at $x = -4$; rel. max. at $x = -2$; abs. min. at $x = 1$; concave up on $\left(-\infty, \dfrac{-5 - \sqrt{19}}{3}\right), \left(\dfrac{-5 + \sqrt{19}}{3}, \infty\right)$; concave down on $\left(\dfrac{-5 - \sqrt{19}}{3}, \dfrac{-5 + \sqrt{19}}{3}\right)$; inflection points at $x = \dfrac{-5 \pm \sqrt{19}}{3}$

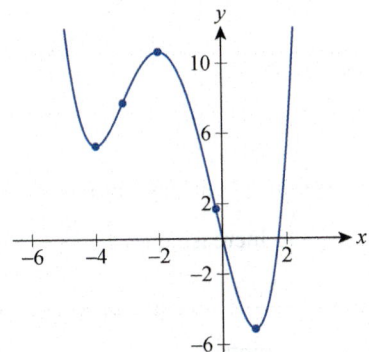

9. Dom = $\mathbb{R}$; x-int.: $\pm 1, \pm\sqrt{2}$; y-int.: 2; no asymptotes; even function; decreasing on $\left(-\infty, -\sqrt{3/2}\right), \left(0, \sqrt{3/2}\right)$; increasing on $\left(-\sqrt{3/2}, 0\right), \left(\sqrt{3/2}, \infty\right)$; rel. max. at $x = 0$; abs. min. at $x = \pm\sqrt{3/2}$; concave up on $\left(-\infty, -\sqrt{2}/2\right), \left(\sqrt{2}/2, \infty\right)$; concave down on $\left(-\sqrt{2}/2, \sqrt{2}/2\right)$; inflection points at $x = \pm\sqrt{2}/2$

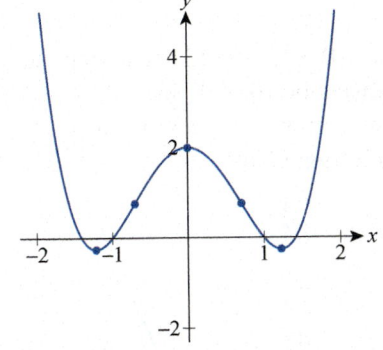

11. Dom = $\mathbb{R}$; x-int.: $0, \sqrt[3]{3}$; y-int.: 0; no asymptotes; increasing on $(-\infty, 0), \left(\sqrt[3]{6/5}, \infty\right)$; decreasing on $\left(0, \sqrt[3]{6/5}\right)$; rel. max. at $x = 0$; rel. min. at $x = \sqrt[3]{6/5}$; concave down on $\left(-\infty, \sqrt[3]{3/10}\right)$; concave up on $\left(\sqrt[3]{3/10}, \infty\right)$; inflection point at $x = \sqrt[3]{3/10}$

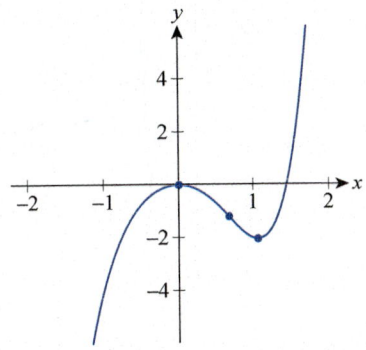

13. Dom $= (-\infty, 2) \cup (2, \infty)$; y-int.: 3/2; vert. asym.: $x = 2$; horiz. asym.: $y = 0$; increasing on $(-\infty, 2), (2, \infty)$; no extrema; concave up on $(-\infty, 2)$; concave down on $(2, \infty)$; no inflection points

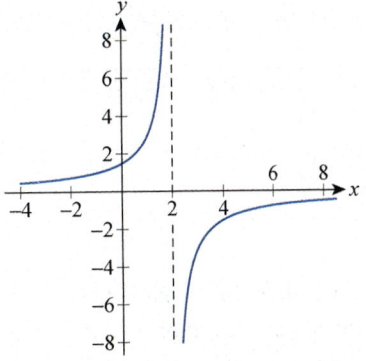

15. Dom $= (-\infty, -2) \cup (-2, 2) \cup (2, \infty)$;
odd function; x-int.: 0; y-int.: 0; vert. asym.:
$x = \pm 2$; horiz. asym.: $y = 0$; decreasing on
$(-\infty, -2), (-2, 2), (2, \infty)$; no extrema; concave
down on $(-\infty, -2), (0, 2)$; concave up on
$(-2, 0), (2, \infty)$; inflection point at $x = 0$

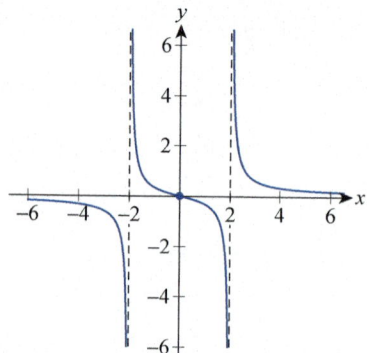

17. Dom $= \mathbb{R}$; x-int.: 1, 5; y-int.: 1; no asymptotes;
decreasing on $(-\infty, 3)$; increasing on $(3, \infty)$;
abs. min. at $x = 3$

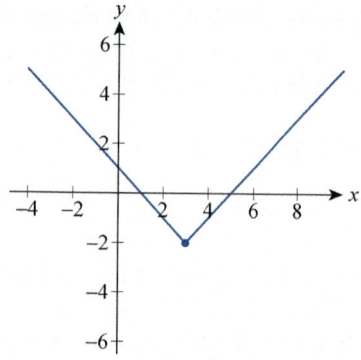

19. Dom $= \mathbb{R}$; even function; y-int.: $2/5$;
no asymptotes; decreasing on $(-\infty, 0)$; increasing
on $(0, \infty)$; abs. min. at $x = 0$; concave down
everywhere on $\mathbb{R}$; no inflection points

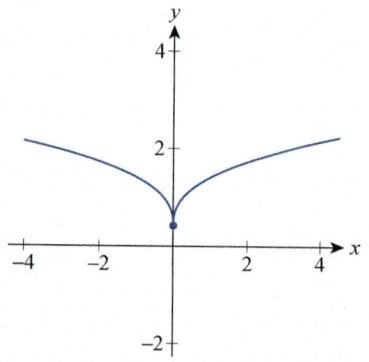

21. Dom $= \mathbb{R}$; y-int.: 3; no asymptotes;
decreasing everywhere on $\mathbb{R}$; no extrema;
concave down on $(-\infty, 1)$; concave up on $(1, \infty)$;
inflection point at $x = 1$

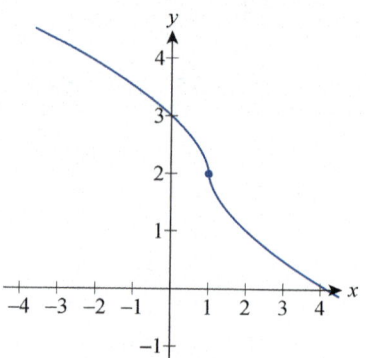

23. Dom $= \mathbb{R}$; y-int.: 2; no asymptotes; decreasing on
$\left(-\infty, -\dfrac{\ln 2}{3}\right)$; increasing on $\left(-\dfrac{\ln 2}{3}, \infty\right)$; abs.
min. at $x = -\dfrac{\ln 2}{3}$; concave up everywhere on $\mathbb{R}$;
no inflection points

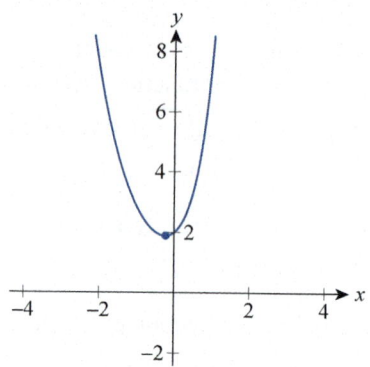

25. Dom $= (0, \infty)$; x-int.: 1; no asymptotes;
decreasing on $(0, 1/e)$; increasing on $(1/e, \infty)$;
abs. min. at $x = 1/e$; concave up on $(0, \infty)$;
no inflection points

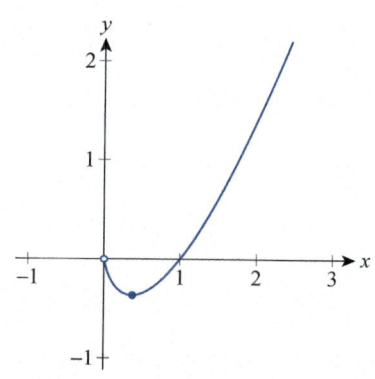

27. Dom $= [-2, 2]$; odd function; x-int.: $0, \pm 2$;
y-int.: 0; no asymptotes; decreasing on
$(-2, -\sqrt{2}), (\sqrt{2}, 2)$; increasing on $(-\sqrt{2}, \sqrt{2})$;
abs. min. at $x = -\sqrt{2}$; abs. max. at $x = \sqrt{2}$;
concave up on $(-2, 0)$; concave down on $(0, 2)$;
inflection point at $x = 0$

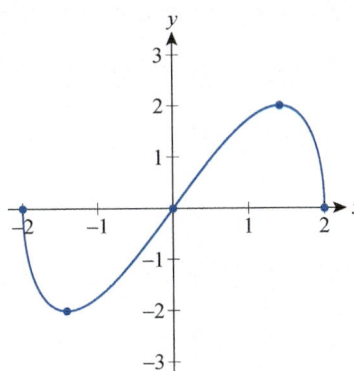

29. Dom $= (-\infty, 7) \cup (7, \infty)$; y-int.: -1;
vert. asym.: $x = 7$; slant asym.: $y = x + 7$;
increasing on $(-\infty, 7 - 2\sqrt{14}), (7 + 2\sqrt{14}, \infty)$;
decreasing on $(7 - 2\sqrt{14}, 7), (7, 7 + 2\sqrt{14})$; rel.
max. at $x = 7 - 2\sqrt{14}$; rel. min. at $x = 7 + 2\sqrt{14}$;
concave down on $(-\infty, 7)$; concave up on $(7, \infty)$;
no inflection points

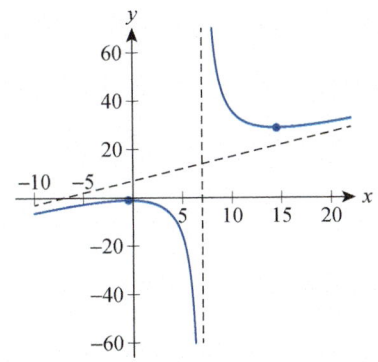

31. Dom $= \mathbb{R}$; x-int.: 0; y-int.: 0; no asymptotes;
increasing everywhere on $\mathbb{R}$; no extrema; concave
down on $\left(-\infty, \dfrac{\ln 4}{3}\right)$; concave up on $\left(\dfrac{\ln 4}{3}, \infty\right)$;
inflection point at $x = \dfrac{\ln 4}{3}$

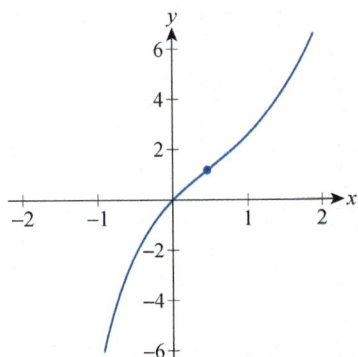

33. Dom $= (-\infty, 2) \cup (2, \infty)$; y-int.: $-5/4$;
vert. asym.: $x = 2$; horiz. asym.: $y = 0$;
decreasing on $(-\infty, 2)$; increasing on $(2, \infty)$;
no extrema; concave down on $(-\infty, 2), (2, \infty)$;
no inflection points

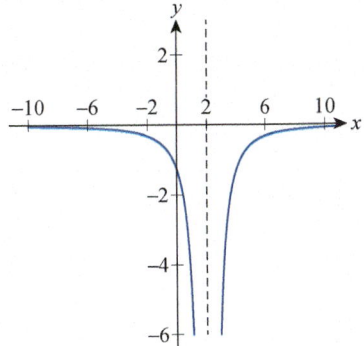

35. Dom $= (-\infty, -1) \cup (-1, 1) \cup (1, \infty)$;
odd function; x-int.: 0, y-int.: 0;
vert. asym.: $x = \pm 1$; horiz. asym.: $y = 0$; increasing
on $(-\infty, -1), (-1, 1), (1, \infty)$; no extrema; concave
up on $(-\infty, -1), (0, 1)$; concave down on
$(-1, 0), (1, \infty)$; inflection point at $x = 0$

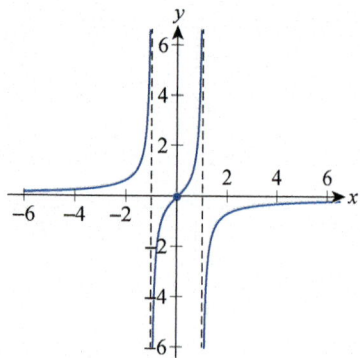

37. Dom $= \mathbb{R}$; x-int.: $4/3$; y-int.: $(4/3)^{1/3}$;
no asymptotes; decreasing on $(-\infty, 4/3)$;
increasing on $(4/3, \infty)$; abs. min. at $x = 4/3$;
concave up everywhere on $\mathbb{R}$; no inflection points

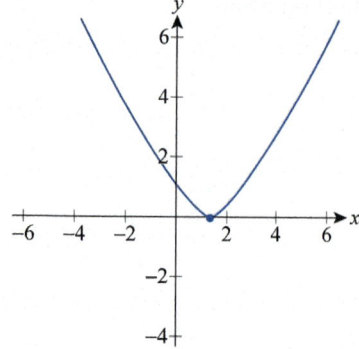

39. Dom $= \mathbb{R}$; odd function; x-int.: 0, ± 1;
y-int.: 0; no asymptotes; increasing on
$\left(-\infty, -\dfrac{1}{3\sqrt{3}}\right), \left(\dfrac{1}{3\sqrt{3}}, \infty\right)$; decreasing
on $\left(-\dfrac{1}{3\sqrt{3}}, \dfrac{1}{3\sqrt{3}}\right)$; rel. max. at $x = -\dfrac{1}{3\sqrt{3}}$;
rel. min. at $x = \dfrac{1}{3\sqrt{3}}$; concave down on $(-\infty, 0)$;
concave up on $(0, \infty)$; inflection point at $x = 0$

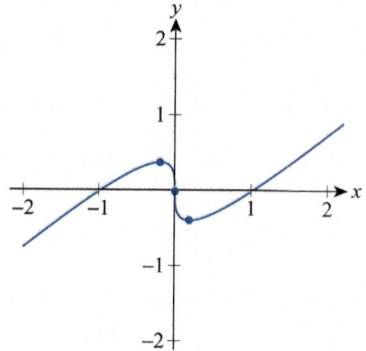

41. Dom $= \mathbb{R}$; odd function;
x-int.: 0, ± 1; y-int.: 0; no asymptotes;
decreasing on $\left(-\infty, -\sqrt{3/5}\right), \left(\sqrt{3/5}, \infty\right)$;
increasing on $\left(-\sqrt{3/5}, \sqrt{3/5}\right)$;
rel. min. at $x = -\sqrt{3/5}$; rel. max. at $x = \sqrt{3/5}$;
concave up on $\left(-\infty, -3/\sqrt{5}\right), (-1, 0), \left(1, 3/\sqrt{5}\right)$;
concave down on $\left(-3/\sqrt{5}, -1\right), (0, 1), \left(3/\sqrt{5}, \infty\right)$;
inflection points at $x = 0, \pm \dfrac{3}{\sqrt{5}}, \pm 1$

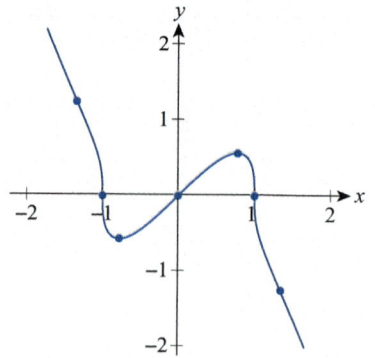

43. Dom = $\mathbb{R}$; 2π-periodic; for the period
$-\pi/2 \le x \le 3\pi/2$ we have the following:
y-int.: -1; no asymptotes; increasing on
$(-\pi/2, \pi/2)$; decreasing on $(\pi/2, 3\pi/2)$;
abs. min. at $x = -\pi/2, 3\pi/2$;
abs. max. at $x = \pi/2$;
concave up on $(-\pi/2, \pi/6), (5\pi/6, 3\pi/2)$;
concave down on $(\pi/6, 5\pi/6)$;
inflection points at $x = \pi/6, 5\pi/6$

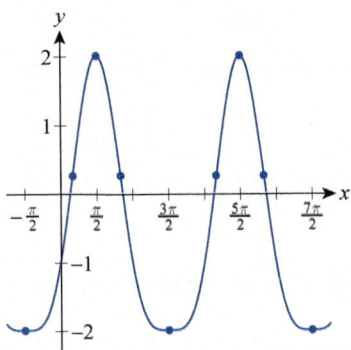

45. Dom $= (-\infty, 0) \cup (0, \infty)$; odd function;
x-int.: ± 1; no asymptotes;
decreasing on $\left(-e^{-3/5}, 0\right), \left(0, e^{-3/5}\right)$;
increasing on $\left(-\infty, -e^{-3/5}\right), \left(e^{-3/5}, \infty\right)$;
rel. max. at $x = -e^{-3/5}$; rel. min. at $x = e^{-3/5}$;
concave down on $\left(-\infty, -e^{-21/10}\right), \left(0, e^{-21/10}\right)$;
concave up on $\left(-e^{-21/10}, 0\right), \left(e^{-21/10}, \infty\right)$;
inflection points at $x = \pm e^{-21/10}$

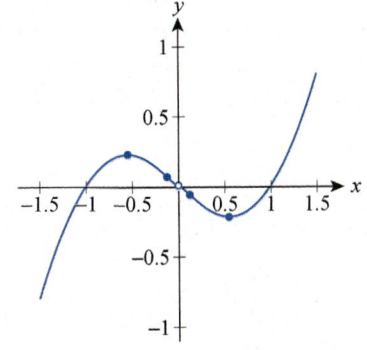

47. Dom = $\mathbb{R}$; x-int.: $\dfrac{-1 \pm \sqrt{433}}{9}$; y-int.: 6; no
asymptotes; increasing on $(-\infty, -1/9)$; decreasing
on $(-1/9, \infty)$; abs. max. at $x = -1/9$; concave
down everywhere on $\mathbb{R}$; no inflection points

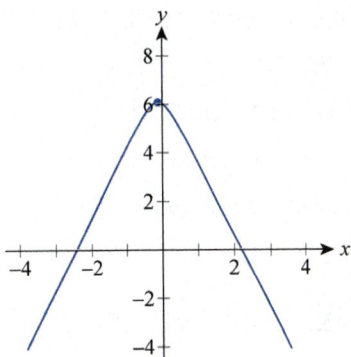

49.

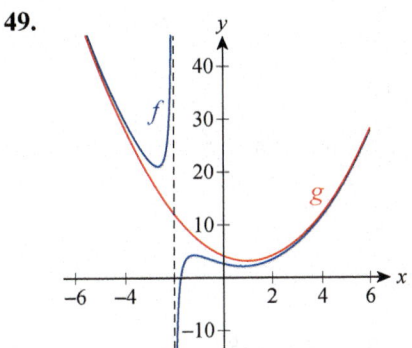

51.

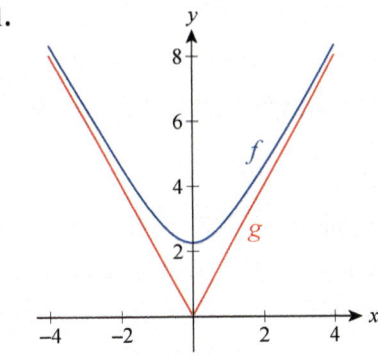

53.

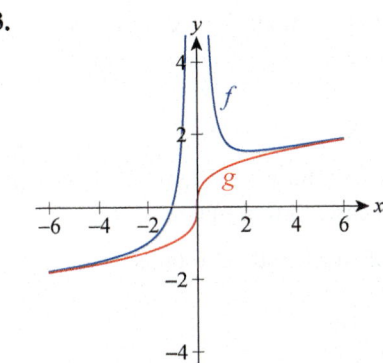

55. No; depending on the starting point, the method may converge to a different root, or not converge at all.

57. 2.65914 **59.** 1.60943 **61.** −1.52137

63. 0.48047, 5.61953 **65.** −0.75488

67. 0.44171 **69.** 0.69005 **71.** 0.876726

73. 0.930407 **75.** 2.475353 **77.** 0.5671

79. 1.0768, 3.6435 **81.** ±1.67071

83. $f'(x_1) = 0$ **85.** $x_{n+1} = -x_n$ for all n.

87. x_n alternates between the values of 3 and 1.

89. $x_{n+1} = \dfrac{k-1}{k} x_n + \dfrac{a}{k(x_n)^{k-1}}$

91. 1.41421 **93.** 2.15443 **95.** 0.14286

99.

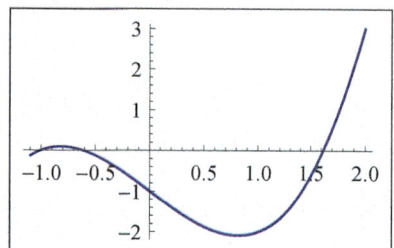

Section 4.6

1. 60 and 60 **3.** −18 and 18

5. n and n **7.** 1 and 1

9. 120 and 80 **11.** 4 and 4 **13.** 1 and 0

15. $w = 2\sqrt{k/3}$; $h = 2k/3$ **17.** $w = 2, h = 24$

19. $y = -\dfrac{1}{3}x + 2$ **21.** $\left(1/2, \sqrt{2}/2\right)$

23. Width: $2\sqrt{3}$, height: 2 **25.** 60 ft by 30 ft

27. Each pen is to be $30\sqrt{2} \times 20\sqrt{2}$ ft, joining along the $20\sqrt{2}$ ft side.

29. Width of cross-section: $w = 2r/\sqrt{3}$ in., depth of cross-section: $h = 2\sqrt{2}r/\sqrt{3}$ in.

31. Cut at $\dfrac{30\pi}{4+\pi}$ in., and use the first piece to form the circle.

33. In the first case, form a circle without cutting. In the second case, form a square.

37. A square of side length $\sqrt{A}$ units

39. $r = h = \sqrt{\dfrac{S}{3\pi}}$ in.

41. A $2 \times 2 \times 8$ in. box is cheapest, costing $2.88.

43. $\alpha = 2\sqrt{2/3}\pi$ **45.** 14×21 in.

47. $\alpha = 2$ rad, $r = 13$ ft **49.** $s = 4\sqrt{14}$ ft

51. The wire needs to be staked 18 ft from the taller antenna.

53. The wire needs to be staked approx. 15.865 ft from the taller antenna.

55. We need to fold $(1/4)(8.5) = 2.125$ in. from the right lower corner.

57. Radius of base: $r = 2\sqrt{2}R/3$, height: $h = 4R/3$

59. Radius of base: $r = \dfrac{HR}{2(H-R)}$, height: $h = \dfrac{H(H-2R)}{2(H-R)}$

61. $\dfrac{65}{3} \times \dfrac{65}{3} \times \dfrac{130}{3}$ in. **63.** $b = P/4$

65. He should land approx. 9 miles from the store.

67. Approx. 10 minutes and 21.2 seconds later

69. $t = \pi/4$ s; $v(\pi/4) = -3\sqrt{2}$ ft/s **71.** $\theta = \pi/2$

73. $420; maximum revenue: $588,000

75. $1900; apartments rented: 95

77. 2600 cars; $29.75 each **81.** $x = 7.5$ units

83. The total volume is maximum when we paint a sphere of radius $\dfrac{1}{2\sqrt{\pi}}$ ft (approx. 3.39 in.) and no tetrahedron. The total volume is a minimum when we paint a sphere of radius approx. 1.63 in., and a tetrahedron of side length approx. 7.99 in.

Section 4.7

9. $F(x) = x$ **11.** $H(x) = x^4 - \dfrac{1}{2}x^2$

13. $V(x) = \tan x + \dfrac{3}{2}x^2$ **15.** $F(x) = 5e^x$

17. $U(t) = 2/t^2$ **19.** $W(z) = \arcsin z$

21. $F(x) = x^2 - 3x + C$, $C = 3$

23. $F(x) = 2\sqrt{x} + C$, $C = -1$

25. $F(x) = C$, $C = 1$

27. $F(x) = \dfrac{x^4}{4} + \dfrac{1}{x} + C$, $C = -1/4$

29. $F(x) = 10^x + C$, $C = -9$

31. $F(x) = \tan\left(\dfrac{\pi}{4}x\right) + x + C,\ C = -1$

33. Answers will vary. Sample graph:

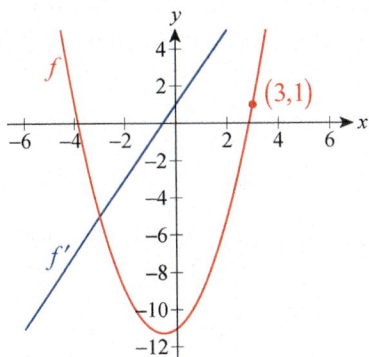

35. Answers will vary. Sample graph:

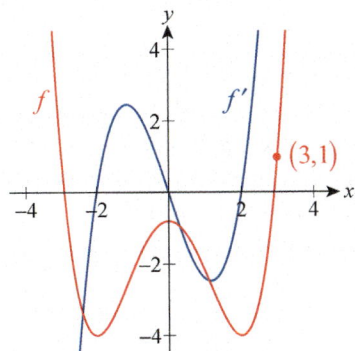

37. $F(x) = 2x^3 - 2x^2 + 1.5x + C$

39. $H(x) = \dfrac{x^6}{2} - 2x^5 + \dfrac{x^3}{3} + 7x + C$

41. $V(x) = 2x^3 + \dfrac{39}{2}x^2 + 18x + C$

43. $H(x) = \dfrac{2x^{9/2}}{9} + C$ **45.** $N(x) = \dfrac{x^2}{2} + 7\ln|x| + C$

47. $A(y) = \dfrac{3}{11}y^{11/3} - \dfrac{6}{7}y^{7/3} + y + C$

49. $G(t) = \dfrac{e^{3t}}{3} - 3\sec t + C$

51. $T(\theta) = \dfrac{\theta^2}{2} + \sin\theta + C$

53. $V(x) = \csc x - \cot x + C = \tan(x/2) + C$

55. $W(x) = \csc x + C$ **57.** $A(x) = \dfrac{5}{3}\arctan 3x + C$

59. $C(x) = \dfrac{4}{5}\operatorname{arcsec} 5x + C$

61. $f(x) = \dfrac{\pi}{2}x^2 - \pi x + \dfrac{\pi}{2}$ **63.** $f(x) = x^2 - 2x + 2$

65. $f(x) = \dfrac{9}{28}x^{7/3} - \dfrac{3}{4}x + \dfrac{4}{7}$

67. $f(x) = \dfrac{8}{105}x^{7/2} + \dfrac{x^3}{6} + \dfrac{x^2}{2} - x + 7$

69. $f(x) = 2\arctan 2x + \dfrac{\pi}{2}$

71. $f(x) = \dfrac{\sin 2x}{8} + \dfrac{x^2}{2} + \dfrac{3x}{4} - 1$

73. $f(x) = \dfrac{2^{5x}}{(\ln 32)^2} + Cx + D$

75. $f(x) = -\cos x - \dfrac{e^{2x}}{4} - \dfrac{3}{2}x + \dfrac{9}{4}$

77. 39 ft

79. It falls for approx. 7.49 s, and the velocity of impact is approx. -83.47 m/s.

81. $16\sqrt{15} \approx 61.97$ ft/s   **83.** Approx. 22.12 m

87. Approx. 54.77 m   **89.** 5275

91. The position function (assuming the initial position is 0) is $p(t) = 5.\overline{5}t^2$. It covers 12.5 m in the first 1.5 s. Its acceleration time from 0 to 60 mph is 2.4 s.

93. 8.25 ft/s²

95. $p(t) = -\dfrac{4}{105}t^{5/2}(3t - 14) + 3$; in 5 seconds it will be at $x = 3 - \dfrac{20\sqrt{5}}{21} \approx 0.87$. Its instantaneous velocity will be zero at $t = 0$ and $t = 10/3$.

97. True

99. False; if $F(x)$ is an antiderivative, then $F(x) + C$ is also an antiderivative for any $C \in \mathbb{R}$.

101. True

103. False; consider $f(x) = g(x) = 1$ with $F(x) = G(x) = x$. Note that $x \cdot x = x^2$ is not an antiderivative of $1 \cdot 1 = 1$.

Chapter 4 Review

1. Answers will vary. 3. Answers will vary.

5. Abs. min.: $f(1) = -1/3$; abs. max.: $f(-1) = 7/3$

7. Abs. min.: $f(-2) = f(4) = 0$;
   abs. max.: $f(1) = 9$

9. Abs. min.: $f(2) = -\sqrt{2}$;
   abs. max.: $f(1/3) = 2/(3\sqrt{3})$

11. No extrema

13. Abs. min.: $f(\pi) = 1$; no maximum

17. $c = 4/3$ 19. $c = \sqrt[3]{65/4}$

23. $f(x) = \sin x + e^x + 2$

27. Increasing on $(-\infty, 3)$; decreasing on $(3, \infty)$

29. Decreasing on $(-\infty, -3), (2, 4)$;
    increasing on $(-3, 2), (4, \infty)$

31. Concave down on $(-\infty, 0)$; concave up on $(0, \infty)$

33. Concave up on $(-\infty, -1), (1, \infty)$;
    concave down on $(-1, 1)$

35. Decreasing on $(-\infty, 0)$; increasing on $(0, \infty)$;
    rel. min. (abs. min.) at $x = 0$;
    concave down on $\left(-\infty, -1/\sqrt[4]{3}\right), \left(1/\sqrt[4]{3}, \infty\right)$;
    concave up on $\left(-1/\sqrt[4]{3}, 1/\sqrt[4]{3}\right)$;
    inflection points at $x = \pm 1/\sqrt[4]{3}$

37. Direction is positive on $\left(0, \sqrt{3}\right)$,
    negative on $\left(\sqrt{3}, 3\right)$; changes direction at $t = \sqrt{3}$;
    acceleration is positive on $(0, 1)$,
    negative on $(1, 3)$, zero at $t = 1$

39. 0

41. L'Hôpital's Rule does not apply; the limit is ∞.

43. $-\infty$ 45. ∞ 47. $1/2$ 49. 0

51. Dom $= \mathbb{R}$; x-int.: $0, \dfrac{3 \pm 3\sqrt{21}}{2}$; y-int.: 0;
    no asymptotes; increasing on $(-\infty, -3), (5, \infty)$;
    decreasing on $(-3, 5)$; rel. max. at $x = -3$;
    rel. min. at $x = 5$; concave down on $(-\infty, 1)$;
    concave up on $(1, \infty)$; inflection point at $x = 1$

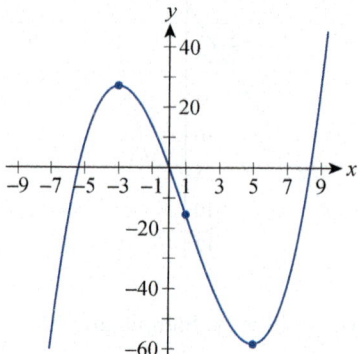

53. Dom $= \mathbb{R}$; x-int.: 0, 2; y-int.: 0; no asymptotes;
    decreasing on $(-\infty, 3/2)$; increasing on $(3/2, \infty)$;
    abs. min. at $x = 3/2$; concave up on
    $(-\infty, 0), (1, \infty)$; concave down on $(0, 1)$;
    inflection points at $x = 0, 1$

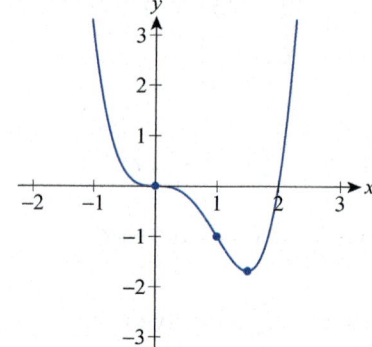

55. Dom = $\mathbb{R}$; odd function; x-int.: 0; y-int.: 0; horiz. asym.: $y = 0$; decreasing on $(-\infty,-1), (1,\infty)$; increasing on $(-1,1)$; abs. min. at $x = -1$; abs. max. at $x = 1$; concave down on $\left(-\infty,-\sqrt{3}\right), \left(0,\sqrt{3}\right)$; concave up on $\left(-\sqrt{3},0\right), \left(\sqrt{3},\infty\right)$; inflection points at $x = 0, \pm\sqrt{3}$

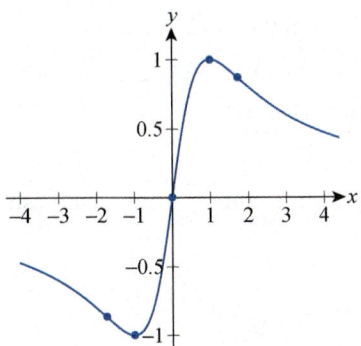

57. Dom = $\mathbb{R}$; x-int.: 0, 1; y-int.: 0; no asymptotes; increasing on $(-\infty,0), (1/2,\infty)$; decreasing on $(0,1/2)$; rel. max. at $x = 0$; rel. min. at $x = 1/2$; concave down on $(-\infty,0)$; concave up on $(0,\infty)$; inflection point at $x = 0$

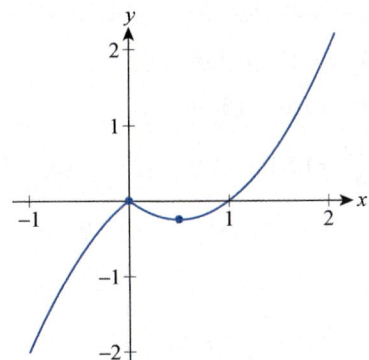

59. Dom = $(-\infty,-2) \cup (-2,2) \cup (2,\infty)$; odd function; x-int.: 0; y-int.: 0; vert. asym.: $x = \pm2$; slant asym.: $y = x$; increasing on $\left(-\infty,-2\sqrt{3}\right), \left(2\sqrt{3},\infty\right)$; decreasing on $\left(-2\sqrt{3},-2\right), (-2,2), \left(2,2\sqrt{3}\right)$; rel. max. at $x = -2\sqrt{3}$; rel. min. at $x = 2\sqrt{3}$; concave down on $(-\infty,-2), (0,2)$; concave up on $(-2,0), (2,\infty)$; inflection point at $x = 0$

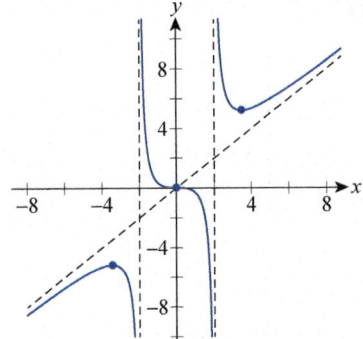

61. 1.97435   **63.** 0.741912   **65.** 0.4797

67. $\dfrac{1}{\sqrt[n]{n}}$ **69.** $a = \left(a_1 + \cdots + a_n\right)/n$

71. $l = 2\sqrt{2/3}$; $w = 2/\sqrt{3}$ **73.** $5\sqrt{10} \times 10\sqrt{10}$

75. $4 - 2\sqrt{3}$ **77.** The equilateral triangle

79. The task is impossible, since $A(r)$ has no maximum. (The surface area approaches infinity as $r \to 0$ or $r \to \infty$.)

81. Nate needs to reach the other side approximately 500 ft downstream from the restaurant.

83. $F(x) = x^5 - 1.2x^4 + e^2 x + C$

85. $F(x) = 0.16x^{5/2} - 4x^{1/2} + C$

87. $F(x) = x^2 + \tan 2x + C$

89. $F(x) = \dfrac{3}{2}\arctan 2x + C$

91. $f(x) = \dfrac{x^3}{3} - \dfrac{5x^2}{2} + 8x - \dfrac{17}{3}$ **93.** 80 ft/s

95. False; consider $y = 1/x$ on $(0,1)$.

97. True   **99.** True

101. False; consider $y = x^5$ at $x = 0$.

113.

Chapter 5

Section 5.1

1. $(O_4 + U_4)/2 = 17$ **3.** $(O_8 + U_8)/2 = 16.25$

5. Approx. 126 ft; an overestimate

7. a. $O_{12} = 112$ m; $U_{12} = 97.45$ m

b. $M_6 = 104.8$ m

c. $(O_{12} + U_{12})/2 = 104.725$ m;
M_6 is greater, because the curve is likely concave down, as indicated by the data.

9. Left-endpoint est. ≈ 4.146 (underestimate);
right-endpoint est. ≈ 6.146 (overestimate);
midpoint est. ≈ 5.384 (overestimate)

11. Left-endpoint est. $= 25/12$ (overestimate);
right-endpoint est. $= 77/60$ (underestimate);
midpoint est. $= 496/315$ (underestimate)

13. Left-endpoint est. ≈ 2.367 (overestimate);
right-endpoint est. ≈ 1.582 (underestimate);
midpoint est. ≈ 2.013 (overestimate)

15. $\displaystyle\sum_{i=1}^{33} 3i$ **17.** $\displaystyle\sum_{i=1}^{100} \frac{1}{i^2}$ **19.** $\displaystyle\sum_{i=-2}^{38} a_{2i+1}$

21. $\displaystyle\sum_{i=1}^{n} f\left(\frac{3i}{n}\right)$ **23.** $\displaystyle\sum_{i=0}^{n} f(x_i^*)\Delta x$ **25.** 64

27. $501 + n$ **29.** $25 + 2n$ **31.** $11/6$ **33.** 255

35. $\dfrac{7n^2 + 2n}{2}$ **37.** $17,080$ **39.** $\dfrac{6n^3 + 15n^2 + 11n}{2}$

41. $\dfrac{n(n+1)(3n^2 - 5n - 4)}{6}$

43. $\dfrac{1}{1} - \dfrac{1}{2} + \dfrac{1}{2} - \cdots + \dfrac{1}{10} - \dfrac{1}{11} = \dfrac{10}{11}$

45. $\sqrt{1} - \sqrt{2} + \sqrt{2} - \cdots + \sqrt{n} - \sqrt{n+1} = 1 - \sqrt{n+1}$

47. $e^2 - e^3 + e^3 - \cdots + e^{n+3} - e^{n+4} = e^2 - e^{n+4}$

49. $88,573$ **51.** $\dfrac{3\left(1 - (2/3)^{100}\right)}{5}$ **55.** $\displaystyle\sum_{j=1}^{n-2} \frac{1}{j}$

57. $\displaystyle\sum_{l=5}^{21} \cos(2l\pi)$ **59.** 6 **61.** $1/4$ **63.** $11/12$

65. The region between the graph of $x^3 + 2x$ and the x-axis from 0 to 3

67. The region between the graph of $\sin x$ and the x-axis from 0 to $\pi/2$

71. Approx. 44.44 m **73.** 22.5 m **85.** 110

87. $\dfrac{mn(m+1)(n+1)}{4}$

89. $\displaystyle\lim_{n\to\infty} \dfrac{-\pi\csc\dfrac{\pi}{2n}\sin\dfrac{\pi - n\pi}{2n} + \pi\csc\dfrac{\pi}{2n}\sin\dfrac{\pi + n\pi}{2n}}{2n} = 2$

91. $\displaystyle\lim_{n\to\infty} \dfrac{3n + \csc\dfrac{\pi}{2n}\sin\dfrac{(2n+1)\pi}{2n}}{8n} = \dfrac{3}{8}$

Section 5.2

1. a. 10 **b.** $(60 + 9\pi)/8 \approx 11.0343$

c. 9 **d.** $(9\pi/8) - 18 \approx -14.4657$

3. 64 ft; -96 ft; -800 ft **5.** 16

7. $7564/11,025 \approx 0.686$

9. $1899/16,000 \approx 0.119$

11. $1 + \dfrac{\ln 3}{2} \approx 1.549$ **13.** $19/25$ **15.** $17/2$

17. $26/3$ **19.** 13 **21.** 4 **23.** $1/4$

25. $201/4$ **27.** $44/3$ **29.** $3/\sqrt[3]{4}$

31. $\displaystyle\lim_{n\to\infty} \sum_{i=1}^{n} \frac{4}{n}\cdot\left[\left(\frac{4i}{n}\right)^2 - \log_2\frac{4i}{n}\right]$

33. $\displaystyle\lim_{n\to\infty} \sum_{i=1}^{n} \frac{b-2}{n}\sqrt[4]{2 + \frac{(b-2)i}{n}}$

35.

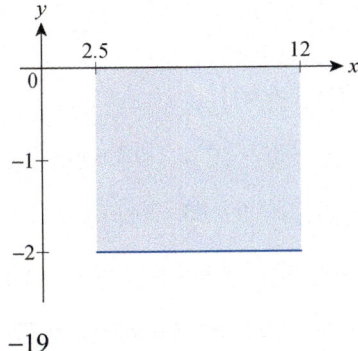

−19

37.

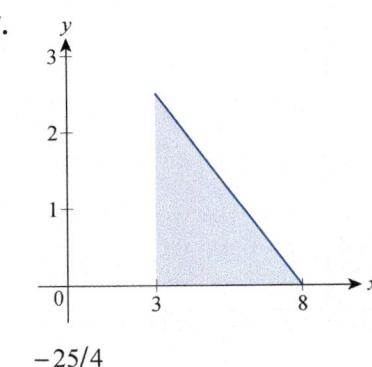

−25/4

39.

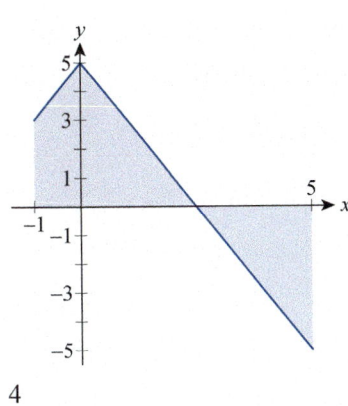

4

41.

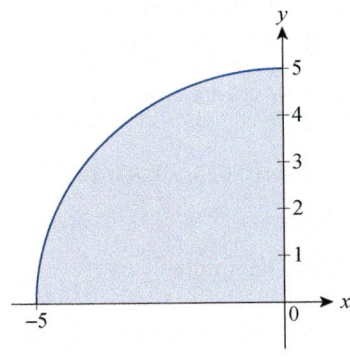

25π/4

43.

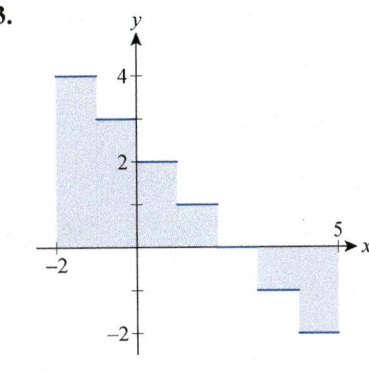

7

45.

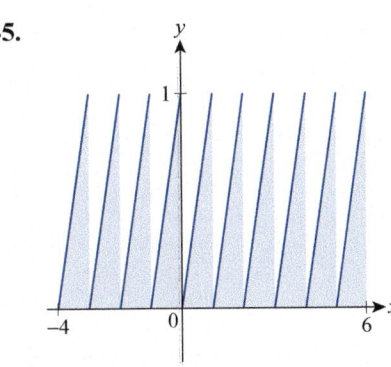

5

53. No; see Exercise 51.   **55.** Yes   **57.** Yes

59. No; see Exercise 49.   **61.** E   **63.** A   **65.** C

67. $8+(c-a)$ **69.** 0   **71.** $5\sqrt{2}/2$

73. Not possible   **75.** $1+5\pi$ **77.** $-1-\dfrac{3\sqrt{2}}{4}$

79. 69/2   **81.** 11/6   **83.** 0   **85.** 2   **87.** 5

89. Not possible   **93.** $0\le\int_{2}^{3}\sqrt{3-x}\,dx\le 1$

95. $6\le\int_{0}^{6}\left(\dfrac{x^2}{32}-\dfrac{x}{4}+\dfrac{3}{2}\right)dx\le 9$

105. −1   **107.** 17/2   **109.** −5/3   **111.** 13/5

113. 6/5   **115.** −9/4   **117.** $\int_{0}^{3}\left(\dfrac{2}{9}x+4\right)dx=13$

119. $\int_{0}^{2}\sqrt{4-x^2}\,dx=\pi$ **123.** $\dfrac{\pi^3}{3}-\dfrac{3\pi^2}{2}+\pi$

125. False; consider $f(x)=g(x)=x$ on $[0,1]$.

Note: $\int_{0}^{1}x^2\,dx\ne\int_{0}^{1}x\,dx\cdot\int_{0}^{1}x\,dx$.

127. True   **129.** True

Section 5.3

1. $c = \sqrt[3]{2}$ **3.** $c = \pm 2/\sqrt[4]{5}$

5. $c = \sin^{-1}(2/\pi), \pi - \sin^{-1}(2/\pi)$

7. $c = \sin^{-1}\left(\sqrt{\pi}/2\right), \pi - \sin^{-1}\left(\sqrt{\pi}/2\right)$

9. a. $F(2) = 6$, $F(4) = 9$, $F(6) = 6$, $F(8) = 2$,
$F(10) = 2$

b. $F(x) = \begin{cases} 3x & \text{if } 0 \le x \le 2 \\ -\dfrac{3}{4}x^2 + 6x - 3 & \text{if } 2 < x \le 6 \\ \dfrac{x^2}{2} - 9x + 42 & \text{if } 6 < x \le 10 \end{cases}$

c.

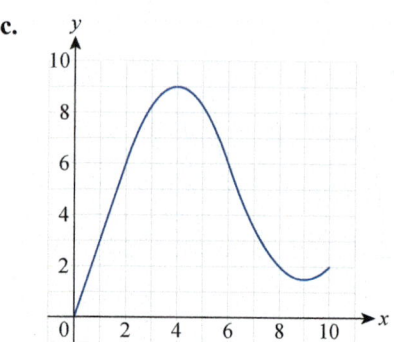

11. $14/3$ **13.** $\sqrt{2}$ **15.** $160/7$

17. $F'(x) = \dfrac{1}{3}\left(x^2 + \sqrt{x}\right)$ **19.** $G'(x) = \dfrac{x^4}{x^4 + 4}$

21. $y' = -\sin\sqrt{x+1}$ **23.** $y' = \left(27x^2 + 9\right)e^{3x-2}$

25. $y' = \left(\sin^2 x + e^{\sin x}\right)\cos x$

27. $y' = \pi\sin(\pi x) - \sin x$

29. $F'(x) = \dfrac{-\sqrt{1 + \sqrt{1 + (1/x^2)}}}{\sqrt{1 - x^2}}$

31. $H'(x) = \ln x - \dfrac{\ln(\ln x)}{x}$

33. $F(x) = 2x - 2$ **35.** $F(x) = -\dfrac{x^3}{3} - \dfrac{x^2}{2} + \dfrac{5}{6}$

37. $F(x) = 1 - \dfrac{1}{\sqrt{x}}$ **39.** -30 **41.** 175

43. 48 **45.** 990 **47.** $1 - \ln 4$

49. $\dfrac{14}{3} + 4\ln 2$ **51.** $\dfrac{8 - 2\sqrt{2}}{5}$ **53.** $\dfrac{45 - 36\sqrt[3]{2}}{32}$

55. $\dfrac{1}{2} - \dfrac{\sqrt{2}}{6}\pi^{3/2}$ **57.** $3\pi/4$ **59.** $e^{\pi/3}$

61. $44/3$ **63.** $\pi/2$ **65.** $e - \dfrac{1}{3}$

67. $\displaystyle\int_0^2 x^4\, dx = 32/5$ **69.** $\displaystyle\int_1^e 1/x\, dx = 1$

71. 8 **73.** 3 **75.** $63/2$ **77.** $2\ln 2 - \ln 3$

79. $1 + 2\ln 2$ **81.** $2e - 9$ **83.** $3 + \dfrac{\pi}{4}$

85. $6 + \ln(33/13)$ **87.** $2 + \ln 2$

89. a. $\dfrac{3}{2} - \ln 4$ **b.** $1/2$ **91.** $f(x) = 2\cos 2x + 1$

93.

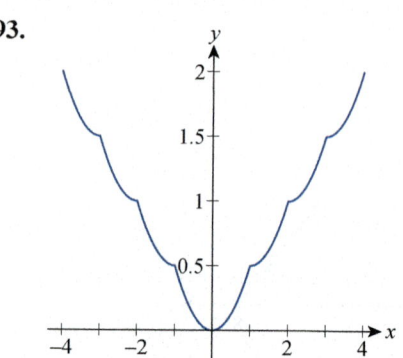

101. Sample graph:

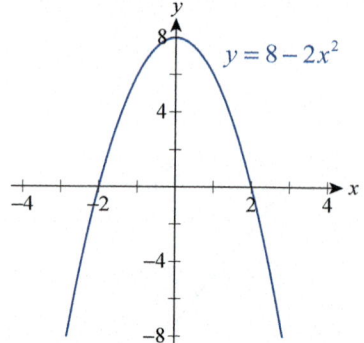

103. True   **105.** True

107. False; see part a. of Example 3.

109. a. $\$3375$; $\$3250$ **b.** 5.377 years

111. The following answers are based on a sine
regression.

a. $N(t) = 1.637\sin(0.023t - 0.936) + 2.583$

b. Approx. 1972 cars

c. Approx. 3 cars per minute

113. $F(x) = (12 - 2x)\sqrt{x}\cos\sqrt{x} + 6(x-2)\sin\sqrt{x}$

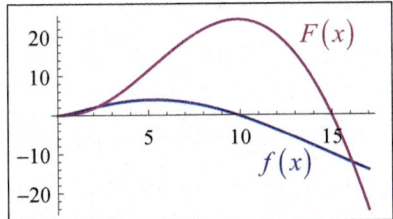

F is increasing where f is positive and decreasing where f is negative. F is concave up where f is increasing and concave down where f is decreasing. The zeros of f are critical points of F, and the critical points of f are potential inflection points for F.

115. $F(x) = \dfrac{\cos^2 x}{2} - \dfrac{\cos 8x + 7}{16}$

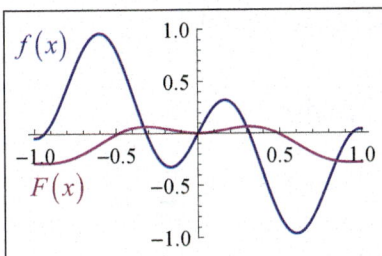

F is increasing where f is positive and decreasing where f is negative. F is concave up where f is increasing and concave down where f is decreasing. The zeros of f are critical points of F, and the critical points of f are potential inflection points for F.

Section 5.4

1. $2x^6 + 1.5x^5 - 0.25x^4 + 2x + C$

3. $\dfrac{10}{3}x^3 + 3x^2 - 4x + C$ **5.** $3/7$

7. $\dfrac{2x^{5/2}}{5} - \dfrac{x^2}{\sqrt{2}} + C$ **9.** $\pi\tan x - \sec x + C$

11. $\dfrac{-\csc 2x}{2} + C$ **13.** $\pi/2$ **15.** $-\dfrac{7}{5}\sin^{-1}(5x) + C$

17. $\dfrac{(x^4 + 2)^{3/2}}{6} + C$ **19.** $\tan^{-1}(2x) + C$

21. $2e^{2t+3} + C$ **23.** $\dfrac{5^{\arctan x}}{\ln 5} + C$

25. $2\tan\left(1 + \sqrt{s}\right) + C$ **27.** $-\tan(1/x) + C$

29. $\dfrac{-\csc^3 x}{3} + C$ **31.** $\dfrac{2z - 1 + \ln|2z - 1|}{4} + C$

33. $\dfrac{(6x + 20)(x - 5)^{3/2}}{15} + C$

35. $\dfrac{x\sqrt{1 - x^2} + \sin^{-1} x}{2} + C$

37. $\dfrac{(3x - 2)^8}{8} + C$ **39.** $\dfrac{(2x^2 + 3x)^{21}}{21} + C$

41. $\ln|\sin z| + C$ **43.** $\dfrac{(x^3 - 5)^{20}}{60} + C$

45. $e^{\sin x} + C$ **47.** $\dfrac{-(5 - s)^{38}}{38} + C$

49. $\dfrac{3(x^4 + 11)^{4/3}}{16} + C$ **51.** $\dfrac{-1}{x^2 + x - 7} + C$

53. $\dfrac{-\sqrt{2 + x^4}}{2} + C$ **55.** $\dfrac{-(t + 1)^4}{4t^4} + C$

57. $2\ln(\ln 2t) + C$ **59.** $\dfrac{e^{x^2 - 3}}{2} + C$

61. $\dfrac{\sin \pi x}{\pi} + C$ **63.** $\dfrac{\sin^4 2t}{8} + C$

65. $e^{1 - \cot v} + C$ **67.** $\dfrac{2}{3(30 - x^{3/2})} + C$

69. $\ln(\sin^2 x + 2) + C$ **71.** $\sin(\sin x) + C$

73. $\ln|\sin x| + C$ **75.** $\dfrac{\sec^3 2x - 3\sec 2x}{6} + C$

77. $\dfrac{4(1 + \sqrt{w})^{3/2}}{3} + C$ **79.** $\dfrac{(x + 1)^{13}}{13} - \dfrac{(x + 1)^{12}}{2} + C$

81. $\dfrac{1 + 4x + 7\ln|1 + 4x|}{16} + C$

83. $\dfrac{(1 - \ln x)^5}{5} - \dfrac{7(1 - \ln x)^4}{8} + C$

85. $f(x) = \dfrac{8}{3}(1 + x^2)^{3/2} - \dfrac{5}{3}$

87. $y(t) = \ln(\sin^2 t + e) - 1$

89. $y(x) = \dfrac{16x + 1 - \cos 4x}{16}$ **91.** $x = 7$

93. False; the indefinite integral is the set of all antiderivatives.

95. True

97. False; the integral doesn't fit the Substitution Rule with $F(g(x)) = \dfrac{(\cos x + 1)^3}{3}$. (You can check the falsity of the answer by differentiation.)

Section 5.5

1. 364/3 **3.** 512/27 **5.** $\dfrac{12^{100} - 5^{100}}{300}$ **7.** 2/7

9. $\dfrac{3(2^{4/3} - 1)}{16}$ **11.** 2/7 **13.** 1 **15.** 129/128

17. $\dfrac{\ln 2}{3}$ **19.** 1/2 **21.** $2/\pi$ **23.** 1/6 **25.** 0

27. $1 + \ln 2$ **29.** 0 **31.** 2 **33.** $\dfrac{4(\sqrt{2} - 1)}{3}$

35. $e^{\pi/3} - 1$ **37.** $\ln 2$ **39.** $\dfrac{20\sqrt{5} - 32}{3}$

41. $6 - 7\ln 3$ **43.** 53/72 **45.** 0 **47.** $33\ln 5 - 32$

49. $\dfrac{608 - 56\sqrt{2}}{15}$ **51.** $\dfrac{2\sqrt{2} - 1}{3}$ **53.** 64/3

55. 10.5845 **57.** 4/3 **59.** 9/2 **61.** 35/4

63. 37/12 **65.** 5/12 **67.** $\ln 2$ **69.** $\dfrac{11 + 5\sqrt{5}}{6}$

71. $1/e$ **73.** 324/5 **75.** 9/2 **77.** 37/32

79. 81 **81.** 7/2 **83.** 937/12

85. $144\sqrt{2} - \dfrac{2656}{15}$ **87.** 1 **89.** $\dfrac{4 - \pi}{\sqrt{2}}$

91. $9\pi/4$ **93.** $\dfrac{8}{5} + \dfrac{4}{\pi}$ **95.** 2

101. a. Bike #1
b. Bike #2 starts gaining ground on #1.
c. Yes, it happens when $\displaystyle\int_0^T v_1(t)\,dt = \int_0^T v_2(t)\,dt$, at about $T = 2.5$ seconds.

103.

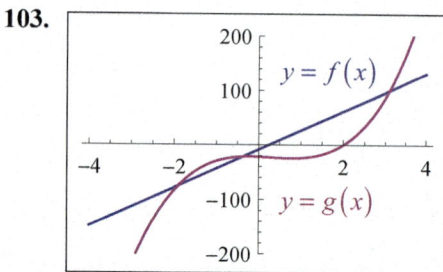

The area is approx. 151.932.

105.

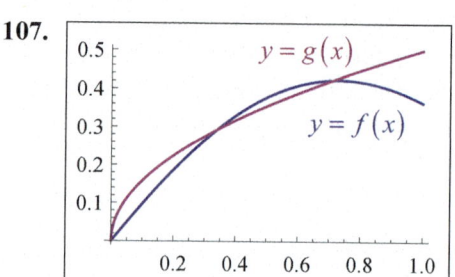

The area is approx. 3.562.

107.

The area is approx. 0.019.

Chapter 5 Review

1. $(O_4 + U_4)/2 = 1.375$ **3.** $\displaystyle\sum_{i=1}^{100} \dfrac{(-1)^{i+1}}{i^3 + 1}$

5. $\displaystyle\sum_{i=1}^{50} f\left(\dfrac{2i}{n^2}\right)$ **7.** $212 + 2n$ **9.** 9065

11. $\dfrac{n^2 + 2n}{(n+1)^2}$ **13.** 0.769399

15. $\dfrac{mn(m+1)(n+1)(2n+1)}{12}$ **17.** 1/4

19. The region between the graph of $2x - x^2$ and the x-axis from 0 to 2; the limit is 4/3.

23. 0 **25.** 1.5 **27.** 1/2; $x = 1/2, 7/2$

29. $F'(x) = \sqrt{1 + x^2}$ **31.** 87/5 **33.** $\dfrac{3}{2} + \ln 4$

35. π **37.** $1 + 3\ln 2$ **39.** 9/20 **41.** $f(x) = \cos x$

43. $-2\sin^{-1} x + C$ **45.** $\sec x + \tan x + C$

47. $\dfrac{(2x^3 - 7)^{10}}{10} + C$ **49.** $\dfrac{-2}{x^2 + 1} + C$

51. $2\tan\sqrt{x} + C$ **53.** $e^x - \ln(e^x + 1) + C$

55. $y(x) = \dfrac{2}{1 - \sqrt{x}} + 2$ **57.** $x = 1 + \dfrac{\ln 17}{2}$

59. $16\sqrt{6}$ **61.** $\dfrac{\pi}{4\ln 2}$ **63.** 1/3 **65.** 4/21

67. $\dfrac{3-e}{2}$ **69.** 2

73. If f is concave up, the integral is expected to be less than the average of L_n and R_n, while if f is concave down, the integral is expected to be greater than the average of L_n and R_n.

75. True

77. False; $\displaystyle\int \frac{1}{e^x}\,dx = \int e^{-x}\,dx = -e^{-x}+C.$

79. False; consider $f(x)=x-1$ on $[0,3]$.

81. False; $1/x^3$ is not continuous on $[-1,1]$.

Chapter 5 Project

3. $\dfrac{d}{dx}\mathrm{Si}(x)=f(x)$

5. Rel. max. at $x=(2k-1)\pi$; rel. min. at $x=2k\pi$, $k\in\mathbb{N}$

7.

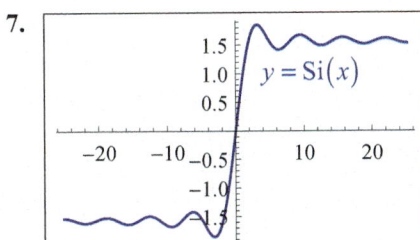

Index